T0301850

FOREST HYDROLOGY

AN INTRODUCTION TO WATER AND FORESTS

THIRD EDITION

FOREST HYDROLOGY

AN INTRODUCTION
TO WATER AND FORESTS

THIRD EDITION

MINGTEH CHANG

CRC Press
Taylor & Francis Group
Boca Raton London New York

CRC Press is an imprint of the
Taylor & Francis Group, an **informa** business

CRC Press
Taylor & Francis Group
6000 Broken Sound Parkway NW, Suite 300
Boca Raton, FL 33487-2742

ISBN-13: 978-1-4398-7994-8 (Hbk)

Library of Congress Cataloging-in-Publication Data

Chang, Mingteh.
    Forest hydrology : an introduction to water and forests / Mingteh Chang. -- 3rd ed.
      p. cm.
    Includes bibliographical references and index.
    ISBN 978-1-4398-7994-8 (hardcover : alk. paper)

1. Forest hydrology. I. Title.
GB842.C5298 2013
2012030330

551.480915'2--dc23

Visit the Taylor & Francis Web site at
http://www.taylorandfrancis.com

and the CRC Press Web site at
http://www.crcpress.com

A nation that fails to plan intelligently for the development and protection of its precious

waters will be condemned to wither because of its shortsightedness. The hard lessons of

history are clear, written on the deserted sands and ruins of once proud civilizations.

It is important that we have a composite, national view of water problems

and needs if we are to attack them intelligently and comprehensively.

Lyndon B. Johnson
XXXVI President of the United States (1963–1969)
"Letter to the President of the Senate and to the Speaker of the House:
Transmitting an Assessment of the Nation's Water Resources," November 18, 1968

Contents

Conversions of Basic Units Used in the Book

Length	1 meter (m) = 10^6 micrometers (μm) = 10^3 millimeters (mm) = 10^2 centimeters (cm) = 10^{-3} kilometers (km) = 39.37 inches (in.) = 3.28 feet (ft) = 6.215 (10^{-4}) miles (mi)
Area	1 hectare (ha) = 10^4 m^2 = 10^{-2} km^2 = 2.47 acres (ac) = 3.86(10^{-3}) mi^2
Volume	1 m^3 = 10^6 cm^3 = 10^3 liters (L) = 35.3 ft^3 = 264.2 gallons (gal) = 40.857(10^{-3}) ft^3 s-day^{-1} = 8.11(10^{-4}) ac-ft = 10^{-4} ha-m
Mass	1 kilogram (kg) = 10^3 grams (g) = 10^6 milligrams (mg) = 2.204 pounds (lb)
Velocity	1 m second^{-1} = 3.28 ft s^{-1} = 3.6 km hour^{-1} = 2.2374 mi h^{-1} = 1.944 knots
Discharge	1 m^3 s^{-1} = 35.3 ft^3 s^{-1} = 2.26(10^7) gal day^{-1} = 15.8(10^3) gal min^{-1}
Density	1 kg m^{-3} = 10^{-3} g cm^{-3} = 10^3 mg L^{-1} = 6.245(10^{-2}) lb ft^{-3}
Force	1 newton (N) = 1 kg-m s^{-2} = 10^5 g-cm s^{-2} (dynes) = 0.225 lb
Work	1 N-m = (1 kg-m s^{-2})(m) = 1 kg-m^2 s^{-2} = 10^7 dyne-cm (or erg) = 1 joule (J)
Power	1 watt (W) = 10^{-3} kW = [(kg-m s^{-2})(m)]/s = 1 J s^{-1} = 1.3406(10^{-3}) horsepower (hp) = 0.7373 ft-lb s^{-1} = 3.413 Btu h^{-1}
Energy	1 calorie (cal) = 4.186 J = 3.968(10^{-3}) British thermal units (Btu) = 3.088 ft-lb
Energy flux	1 cal cm^{-2} minute^{-1} = 1 langley min^{-1} = 697.80 W m^{-2} = 3.6864 Btu ft^{-2} min^{-1}
Pressure	1 millibar (mb) = 10^{-3} bar = 9.872(10^{-4}) atmosphere = 0.75 mm Hg = 0.0145 lb in.$^{-2}$ = 10.20 kg m^{-2} = 1.019 cm of water = 100 pascal (Pa) = 0.1 kPa = 10^3 dynes cm^{-2} = 100 N m^{-2}
Dynamic viscosity	1 poise = 100 centipoise = 1 dyne-s cm^{-2} = 0.1 N-s m^{-2}
Shear stress	1 N m^{-2} = 1 Pa
Soil erosion	1 metric ton ha^{-1} year^{-1} = 0.445 ton ac^{-1} yr^{-1}
Rainfall factor	$R = EI_{30}/100$, which is in: [(m-metric ton/ha-cm)(cm)](cm/h) = m-metric ton-cm/ha-h = 0.5764 ft-ton-in./ac-h
Thermal conductivity	1 cal/[s cm^2 °C cm^{-1}] = 1 cal s^{-1} cm^{-1} °C^{-1} = 418.4 W m^{-1} K^{-1} = 418.4 J cal s^{-1} m^{-1} °C^{-1} = 241.747 Btu cal h^{-1} ft^{-1} °F^{-1} = 1506.24 kJ cal h^{-1} m^{-1} K^{-1}
Electrical conductivity	1 siemen m^{-1} (S m^{-1}) = 10 mmho cm^{-1} = 10^4 μmho cm^{-1}

Preface

This third edition of *Forest Hydrology: An Introduction to Water and Forests* arrives 11 years after the 2002 publication of the first edition. During the past 11-year period, global warming was more apparent, weather-related natural disaster more severe, water shortage more critical, and concern on the forest–water relation more widespread than any of its preceding decades. It becomes necessary to publish a new edition for the text of *Forest Hydrology* to (1) update resources and environmental data, (2) add new research findings and technology, (3) address the concerns and issues regarding forests and climate change, (4) incorporate comments and suggestions, and (5) improve comprehensiveness. The book was 12 chapters and 3 appendices in the original edition, expanded to 14 chapters and 4 appendices in the second edition, and is now 16 chapters and 4 appendices in the third edition. There are also 14 new tables, 24 new figures, and many new sections that have been added to the current edition.

As stated previously, this book is based on lecture material in "forest hydrology," offered to undergraduate and postgraduate students in the Arthur Temple College of Forestry and Agriculture at Stephen F. Austin State University, Texas. As is the case in many forestry programs in the United States, forest hydrology (or watershed management) is the only required course in water sciences in the curriculum. Because students are new to the subject, it is necessary to cover some basic topics in water and water resources before discussing topics in forest hydrology.

Although a few texts on forest hydrology are available for college students, they cover very little or none of the background on water resources. On the other hand, books dealing with water resources do not cover topics on forest–water relations. This book intends to fill that gap and provides an introduction to forest hydrology by bringing water resources and forest–water relations into a single volume and broadly discusses issues that are common to both. It focuses on concepts, processes, and general principles; hydrologic analyses are not emphasized here.

Subjects in the 16 chapters are arranged in two general groups with one linkage in between. The first six chapters (Chapters 1 through 6) deal with the introduction and basic background in water and water resources, while the last nine chapters (Chapters 8 through 16) address the impact and study of forests on water (Chapters 8 through 14), watershed management planning (Chapter 15), and forest hydrology research (Chapter 16). Between these two groups is a chapter (Chapter 7) that describes forests and forest characteristics important to water circulation, sediment movement, stream habitat, and climate change. It serves as an entrance to the study of forest impacts on water resource—as a bridge connecting water and forests.

The impacts of forests on water are separately discussed in terms of precipitation, vaporization, streamflow quantity, streamflow quality, stream sediment, stream habitat, and flood in Chapters 8 through 14. Topics in streamflow quantity and quality jointly discussed in a single chapter in the second edition are now separately discussed, as suggested by Dr. Y. Jun Xu of Louisiana State University at Baton Rouge, LA, United States, in two individual chapters. This is partly due to water quality as an important issue in forest hydrology and partly due to the size of the streamflow chapter in its previous version.

The creation of Chapter 15, Watershed management planning and implementation, is in response to the comments of Dr. D. E. Leaman of Leaman Geophysics in Australia and others. It outlines the transfer of the science of forest hydrology into practices at the watershed scale—the applications of forest hydrology in watershed management. A chapter dealing with watershed management in a forest hydrology text makes the book "more appealing to both forest hydrology and forest watershed management classes," as commented by Dr. Richard Schultz of Iowa State University at Ames, Iowa.

Research is fundamental in all sciences. A chapter designated for research in forest hydrology provides a foundation for students who might need to conduct investigations and resolve watershed problems in carriers. Accordingly, Chapter 16 is designed to deal with research issues, objectives, principles, and methodology in forest hydrology, along with a step-by-step numerical example on watershed calibration and assessment of treatment effects. Such information is basic in watershed research and helpful to those who might pursue graduate studies.

Most books incorporate hydrologic measurements in the main text. This book presents measurements in four appendices. They include precipitation, streamflow, stream sediments, and forest interception; topics on each type of measurement cover general background, available instruments, and sampling procedures. Since there are relevant models presented in the text, the appendices do not cover the measurement of evapotranspiration. Also, hydrologic modeling is beyond the scope of this book.

The use of mathematical expressions is inevitable in subjects such as hydrology and other earth sciences. This book uses mathematical equations in forms for which knowledge of college algebra and trigonometry are sufficient for understanding. Readers with less mathematical background can skip the difficult equations without hindering their comprehension. In such cases, professors may wish to place more emphasis on the forest impacts, as discussed in the latter part of each chapter, and discuss only the basic hydrologic processes important to the understanding of the impact. The book can be used as a text for students in agriculture, forestry, and land resources management, and as a reference for foresters, rangers, geographers, watershed managers, biologists, agriculturists, environmentalists, policy makers, engineers, and others who may need such background in their professions.

Mingteh Chang

Acknowledgment

I am much obliged to many people—colleagues, friends, and students—who provided critical reviews, editing, suggestions, literature information, computer works, and secretarial assistance. They have greatly improved the readability, clarity, and contents of the final manuscript and have made the completion of the book smooth and orderly. They include, but are not limited to, the following people:

Critical reviews: Younes Alila, Scott Beasley, Wayne C. Boring, Douglas G. Boyer, Thomas O. Callaway, Shih-Chieh Chang, Tien-Po Chang, Andrzej Ciepielowski, Richardo Clemente, Dean Coble, Matthew J. Cohen, Theodore A. Endreny, Curt D. Holder, George G. Ice, Ernest Ledger, Kye-Han Lee, D. E. Leaman, James A. Lynch, Darrel L. McDonald, Walt F. Megahan, Ahmad A. Nuruddin, Fred L. Rainwater, David A. Rutherford, Richard Schultz, Donald Richter, Ted Sheng, Peter Siska, William E. Sharpe, Donald J. Turton, Kim L. Wong, Lizhu Wang, Jimmy Williams, and Y. Jun Xu.

Editing: Vijay Bose

Literature and other assistance: Sampurno Bruijnzeel, Mason D. Bryant, Steven H. Bullard, Ian R. Calder, Shih-Chieh Chang, Kenneth Farrish, A. J. Horowitz, George G. Ice, John M. Laflen, Patrick Lane, Medhin Estifanos, Curt. D. Holder, David Kulhavy, Shiyou Li, Matthew McBroom, Brain Oswald, John Stednick, J. P. Summerville, and Donald J. Turton.

Computer works: Mian Ahmad, Micah-John Beierle, Brent Bishop, Lewis Bodden, Misti Compton, Medhin Estifanos, Richard Ford, Rebecca Graves, James Hoard, Crystal Linebarger, Danny McMahon, Adam G. Mouton, Peter Siska, Melissa Watson, Rhonda Barnwell, Jeff Williams, and Yangli Zhang.

Taylor & Francis Group staff: Randy Brehm (senior editor), Jessica Vakili (senior projector coordinator), and Robert Sims (project editor).

Copyright permissions: American Water Resources Association, The American Society of Agricultural Engineers, CSIRO and eWater Limited, John Wiley & Sons, Ltd., Springer-Verlag GmbH & Co. KG, Studentsfriend.com, and UNSECO Press.

Finally, my children Benjamin, Rebecca, and Solomon also helped with some computer programs, proofreading, and pictures. Most importantly, their love is an indispensable strength in my life.

This book could not have been completed without the assistance and encouragement of the aforementioned and other unmentioned persons. I am very grateful and indebted to them all.

Author

Mingteh Chang, with a MS from Pennsylvania State University and a Ph. D. from West Virginia University, is professor emeritus of forest hydrology at the Arthur Temple College of Forestry and Agriculture, Stephen F. Austin State University in Nacogdoches, Texas. He has taught forest hydrology, watershed management, environmental hydrology, and environmental measurements at the undergraduate and graduate levels for 31 years at the college and was named regent professor in 1998.

1

Introduction

Water and forests are two of the most important resources on Earth. They both provide food, energy, habitat, and many other biological, chemical, physical, and socioeconomic functions and services to living things and the environment. Without water, there would be no forests. With forests, the occurrence, distribution, and circulation of water are modified, the quality of water is enhanced, and the timing of flow is altered. Indeed, water and forests impact each other greatly.

1.1 Water Spectrum

Water is essential to life, the environment, and human development. Flora and fauna depend on water for growth, development, and survival, and water sustains human societies; environments without water are hot, dry, uncomfortable, and unsuitable for living. Fortunately, the Earth is blessed with water. More than 70% of Earth's surface is covered by water, and a layer of water vapor up to about 90 km thick embraces the entire planet. Water makes Earth flourish with life in various forms and with places for cultures to develop.

Although water on Earth is abundant in quantity and vast in distribution, it still imposes problems in many regions. Around 1.2 billion people live in areas of physical water scarcity, 0.5 billion are approaching this situation, and another 1.6 billion face economic water shortage where countries lack necessary infrastructure to take water from rivers and aquifers (FAO, 2007). In other words, as much as 3.3 billion people, accounting for 50% of the world population, have water shortage problems. By 2020, water use in the world is expected to increase by 40%, and 17% more water for food production (Palaniappan and Gleick, 2008). Water pollution has worsened the difficulties experienced in regions with water shortages. Water shortages not only create many problems and inconveniences in daily activities but also affect lifestyles and cultural development. At Chungungo, Chile, each villager lives on 13 L of water per day, delivered by trucks once a week. In the northwest plateau of China, it is said that people take only three baths in their lifetimes: when they are born, when they are married, and when they die. Indeed, water is a luxury for many people in many regions.

On the other hand, many people in other parts of the world suffer damage, casualties, and life disruptions from disastrous floods. The flood and storm surge induced by Hurricane Katrina struck a portion of the U.S. coastline along the northern Gulf of Mexico in August 2005, leading to 3 million people without electricity for weeks, 1.2 million under evacuation order, and 1833 fatalities. The estimate of the total damage cost of Katrina is at $40–$80 billion, the costliest flood in U.S. history (Knabb et al., 2006). Probably, the most vulnerable river on Earth is the Yellow River (Huang Ho) in China, a river widely considered the "Sorrow of China." Floods caused the river to change its main course six times between 602 BC and 1855, with a span as much as 800 km in distance (Kuo, 1981). A flood that occurred along the Yellow River in 1887 killed more than 900,000 people (Clark, 1982); its death toll outnumbers that of any natural disasters known worldwide today.

Since water is so crucial to life, people graze livestock in areas with plenty of water and grass, grow crops along floodplains, build cities along major rivers and near seaports, and develop civilizations in regions with mild climates and an abundance of water resources. The Nile River has thus been referred to as the "Spring of Life" and the Yellow River as the "Cradle of Chinese Civilization." However, the uneven distribution of water on Earth has caused major disputes over water, both throughout history and in modern times, between nations, within countries, and among people.

Wars over water between nations were fought in the Mideast throughout ancient history and continue even today. Babylonia, for example, was a center of dispersal of agricultural knowledge in early historical times, about 5000 years ago. Babylonia occupied the lower delta of the Tigris and Euphrates rivers, which was exceptionally fertile because of alluvial deposits from Assyrian highlands and the retreat of the great ice sheet. The constant wars for water supply between Babylonia and Assyria partially contributed to the fall of Babylonian civilization.

In the United States, the nineteenth-century California gold rush produced diverse competition for water between farmers and miners, between cattle grazers and crop growers, and between advocates of public and private irrigation (Pisani, 1992). The availability and distribution of water is a determining factor for the prosperity of cities, industries, agriculture, and tourism in the Western region of the United States, where water scarcity is a common phenomenon. A dam on the Nueces River basin in south Texas was planned in the 1960s but was delayed for almost 20 years due to disputes over water rights and allocations between upstream and downstream water users, along with a conceptual struggle between conservationists and developers. Parties involved in the disputes on water allocations included the Fish and Wildlife Service, the Bureau of Land Reclamation, the Texas Water Commission, and the City of Corpus Christi, Texas, as well as between environmental groups and individuals (Ting, 1989).

Indeed, water is a major concern in our society. People are concerned with too much water, too little water, water that is unsafe to use, a lack of rights to use water, and water allocation. Most universities provide training on subjects relating to water. For example, all 16 state-controlled 4 year universities in Texas offer courses related to water. Knowledge of water properties, problems, environmental significance, and management is essential for effective utilization of this vital natural resource.

1.2 Forest Spectrum

A forest is a biotic community dominated by trees and woody vegetation that covers a large area. It supports a complex array of flora and fauna. The forest forms a microclimate distinct from that produced by other land uses. Of the many different types of forests, each has different characteristics in terms of species composition, size, diversity, and density, with variation depending mainly on temperature and precipitation. No matter what type the forest is, the plant sizes, canopy density, litter floor, and root systems are significantly taller, greater, thicker, and deeper, respectively, than is the case for other vegetation types. These characteristics make forests able not only to provide a number of natural resources but also to perform a variety of environmental functions.

Resources associated with forests may include timber, water, soil, wildlife, aquatic life, vegetation, minerals, and recreation. Except for minerals, all these resources are greatly affected by forestry activities. Some resources can be completely destroyed, depending on the intensity and extent of the forestry activity. Environmental functions performed by forests may include control of water and wind erosion, protection of headwater and reservoir watershed and riparian zone, sand-dune and stream-bank stabilization, landslide and avalanche prevention, preservation of wildlife habitats and gene pools, mitigation of flood damage and wind speed, and sinks for atmospheric carbon dioxide. Many established forests have managed to achieve one or more of these environmental functions, while others are preserved to prevent reduction in biodiversity and degradation of the ecosystem.

The approximately 4.0×10^9 ha of forests and woodlands in 2010 on Earth cover about one-third of the total land surface (FAO, 2010). This is about 80% of the pre-agriculture forested areas (Mather, 1990). Forested areas in the temperate zone (23.5°–66.5° latitudes) have not changed much in recent decades, but deforestation of about 4.1×10^6 ha/year in South America and 3.8×10^6 ha/year in Africa has occurred between 1990 and 2010 (FAO, 2010), mostly in tropical regions. About 93% of the total deforestation in the world occurred on these two continents. Tropical forests make up about 42% of the world's forest cover (Gorte and Sheikh, 2010) and are the major genetic and pharmaceutical pools of the world. In light of the forest environmental functions mentioned earlier, the impacts of such large-scale forest destruction on soil, water, air, and biology have been of great concern among scientists, environmentalists, policy makers, and the general public.

History and modern studies have shown that the misuse of forest resources has caused adverse watershed conditions, depletion of land productivity, disruption of people's routine activities, conversion of arable lands into semiarid or desert lands, and destruction of villages, towns, or even civilizations. Through these experiences, the use of forests has been shifted from single to multiple purposes—from exploitation into preservation and then conservation uses, and from productive into environmental and then ecological functions. The recent growing interest in biological sinks of atmospheric carbon dioxide, global warming, and the balance between production and protection is an example of the awareness of ecological functions. It is our task to seek a balance in forest management between productive and protective functions.

1.3 Issues and Perspectives

Water and forests both cover large portions of Earth and both are crucial to lives and the environment. As the world's population increases with time, so do the pressures of population and the extent of utilization of all natural resources. As a consequence, loss of forests and increasing human use of water are two important trends in natural resources (Cohen, 2008). This makes water and forests two of the most important issues in the twenty-first century. We are concerned about water and forests not only as raw materials required by cultures and industries but also as key factors in the environment. Water and forests are not two independent natural resources; a close linkage exists between the two. Consequently, the study of the interface between these two resources, called forest hydrology, has become an important field. It provides basic knowledge and foundations for watershed management—a discipline and skill for maintaining land productivity and protecting water resources.

Forest distributions enhance the forest–water relation significantly. Forests usually grow and develop in areas with annual precipitation of 500 mm or higher. These are also the areas suitable for certain agricultural activities. Forests cover about 30% of the land, yet this 30% forested land generates 60% of total runoff. In other words, most of our drinking water supplies originate from forested areas. Any activities in, as well as the development and utilization of, forested areas will inevitably destroy forest canopies and disturb forest floors to a certain degree. These actions may affect water quantity through their impacts on transpiration and canopy interception losses, infiltration rate, water-holding capacity, and overland flow velocity. Water quality is also affected because of the exposure of mineral soils to direct raindrop impacts, the loss of soil-binding effect from root systems, the increase in overland flow, and the accelerated decomposition of organic matter. All these may result in an increase in soil erosion and nutrient losses. All forests should be managed with their impacts on water resources in mind so that water resources can be properly protected and utilized.

Many issues concern water and forests, but this book will emphasize the linkages between the two, beginning with a basic explanation of water and water resources and then moving forward to examine the impacts of forests and forest activities. We all recognize that proper management of these two resources is of utmost importance to the well-being of nations. Without fundamental knowledge, adequate studies, and sufficient information, proper management cannot be achieved.

References

Clark, C., 1982, *Flood*, Time-Life, Chicago, IL.

Cohen, M.J., 2008, Book reviews: Forest hydrology, an introduction to water and forests (2nd edition), *For. Sci.*, 54(6), 658–659.

FAO, 2007, UN Water, World Water Day 2007: Coping with water scarcity, challenge of the twenty first century, http://www.worldwaterday07.org/

FAO, 2010, Global Forest Resources Assessment 2010—Main Report, FAO Forestry Paper 163, Rome, Italy.

Gorte, R.W. and Sheikh, P.A., 2010, Deforestation and climate change, CRS report for congress, Congressional Res. Service, 7-5700, http://www.crs.gov/, R41144

Knabb, R.D., Rhome, J.R., and Brown, D.P., 2006, Tropical cyclone report, Hurricane Katrina, August 23–30, 2005, National Hurricane Center, NOAA, http://www.nhc.noaa.gov/pdf/TRC-AL122005_katrina.pdf/

Kuo, S.F., 1981, *Huang Ho, the Great*, Chin-Show Publishers, Taipei, Taiwan (in Chinese).

Mather, A.S., 1990, *Global Forest Resources*, Timber Press, Portland, OR.

Palaniappan, M. and Gleick, P.H., 2008, Peak water, *The World's Water, 2008–2009*, Gleick, P.H., Ed., Island Press, Washington, DC, pp. 1–16.

Pisani, D.J., 1992, *To Reclaim a Divided West: Water, Law, and Public Policy, 1848–1902*, University of New Mexico Press, Albuquerque, NM.

Ting, J.C., 1989, Conflict analysis of allocating freshwater inflows to Bay and Estuaries—A case study on Choke Canyon Dam and Reservoir, Texas, Doctoral dissertation, Stephen F. Austin State University, Nacogdoches, TX.

2

Functions of Water

The significance of water to life and the environment can be discussed from the perspective of its various biological, chemical, physical, socioeconomic, mechanical, political, and military functions. These functions stem from its unique properties, abundant quantity, and vast distribution on Earth, which subsequent chapters will discuss.

2.1 Biological Functions

2.1.1 Necessity of Life

Life cannot exist without water; it began in water and depends on water for survival, growth, and development. Water is the primary constituent of protoplasm, the substance that performs basic life functions. Without water, a plant cannot absorb required nutrients, cannot perform photosynthesis and hydrolytic processes, and cannot maintain its state of vigor. In animals, water removes impurities and the byproducts of metabolism, transports oxygen and carbon dioxide, enhances digestion, and regulates body temperature. It acts as a solvent and as a raw material in the chemical reactions necessary to sustain life and maintains a balance between bases and acids in the body. Tissues can die with minute changes in pH.

A forest can transpire $1000 \, \text{kg/m}^2/\text{year}$ of water while producing $1 \, \text{kg/m}^2/\text{year}$ of dry matter. The body of a newborn calf is 75%–80% water, and 75% of the weight of a living tree is either water or made from water. Water makes up 60%–90% of human body weight. The total daily requirement of water for a person is about 20 L for minimum comfort, while 1–5 L is essential for survival. The human body obtains 60% of its water from drinking, 30% from solid food, and 10% as a byproduct of the chemical process of cellular respiration (Marieb and Hoehn, 2007). We die when 15% of the water in our body is lost through dehydration. A person can survive without food for about 1–2 weeks but only a few days without water.

2.1.2 Habitat of Life

Water is not only the substance of life; it also provides a living environment for about 90% of Earth's organisms. The blue whale (measured up to 30 m long and 200 t) and the hippopotamus, two of the largest mammals on the earth, use oceans and rivers, respectively, as their homes. Oceans, seas, streams, rivers, lakes, ponds, estuaries, wetlands, and even bodies of water in drainage ditches and abandoned containers are habitats to a variety of animals and plants.

2.1.2.1 Wetlands

Wetlands—areas where water is near, at, or above ground level—are considered to be transitional zones between terrestrial and aquatic ecosystems. The five recognized wetland systems are marine, estuarine, lacustrine (associated with lakes), riverine (along rivers and streams), and palustrine (marshes, swamps, and bogs). Biologically, wetlands are one of the richest and most interesting ecosystems on Earth. Hydrologically, they provide mechanisms and resources for aquifer recharge, flood control, sediment control, wastewater treatment, biogeochemical cycling and storage, and water supplies. Richardson (1994) described 5 wetland functions and 19 wetland values only realized in recent decades. Unfortunately, wetlands in the conterminous United States have been reduced by more than 50% that existed at the time of European settlement (U.S. Fish and Wildlife Service, 2002).

In the 1780s, there were 89×10^6 ha of wetlands in the conterminous United States, and wetlands occupied 5% or more of the land area in 34 states. By the 1980s, wetlands had decreased 53% to 42×10^6 ha, and the number of states that had wetlands on 5% or more of the land area had fallen to 19 (Dahl, 1990). The losses of wetlands were due to the cumulative impact of agricultural development, urban development, conversion of wetlands to deepwater habitats, and other types of conversion activities (Dahl and Johnson, 1991).

2.1.2.2 Estuaries

An estuary is the narrow zone along a coastline where fresh water from rivers mixes with a salty ocean. Estuaries receive heavy loads of sediments, suspended and dissolved organic or inorganic materials, and fresh water from rivers, which provide nutrients and maintain salinity levels necessary to sustain marine organisms. These distinctive properties, along with relatively shallow depths that allow deep penetration of sunlight, make estuaries primary spawning areas for a wide variety of finfish and shellfish, nurseries for juvenile marine species, and homes for wildlife and waterfowl (Ting, 1989).

2.1.2.3 Ponds and Lakes

Regardless of their size, ponds and lakes contain three life zones: littoral, limnetic, and profundal. The littoral, at the edge of the lake, is the most richly inhabited. Here, the conspicuous angiosperms, such as cattails and rushes, grow. In this zone, sunlight can penetrate to the bottom and root vegetation can grow. Farther from the shore are water lilies, and often the pond floor is covered with many varieties of weeds. The submerged plants provide shelters for many small organisms. Snails, arthropods, and mosquito larvae feed upon the plants and algae. Large animals such as ducks and geese will in turn feed upon them (Figure 2.1).

In the limnetic zone, the zone of open water, and extending to the limits of light penetration (compensation depth), phytoplankton is usually the dominant photosynthetic organism. This zone provides the habitat for bass, bluegill, and other fish species. The compensation depth is reached when the light intensity in the water reaches 1% of full sunlight (can be detected by a Secchi disk), and photosynthesis balances with respiration. At depths below the limnetic zone to the floor is the profundal zone of deep water where light does not penetrate and no plant life exists. The principal inhabitants are scavenging fish, aquatic worms, bacteria, and other organisms that consume the organic debris filtering down from above.

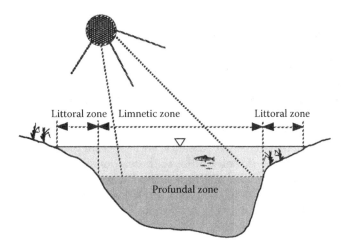

FIGURE 2.1
Life zones in lakes.

2.1.2.4 Streams and Rivers

Species inhabiting streams and rivers are largely affected by flow velocity. In swift streams, most organisms live in the shallows, where small photosynthetic organisms, algae, and mosses cling to the rock surfaces. Insects in both mature and larval form live in and among the rocks and gravel. In slow-moving streams, variations in water temperature, dissolved oxygen levels, nutrients, and light penetration provide diversified habitats for various aquatic species.

2.1.2.5 Oceans

Marine life forms, including plants and animals, fall into three major groups:

1. *Benthos.* These are plants (such as kelp) and animals (such as brittle stars) that live on or depend on the bottom of the ocean. Habitats extend from the shore to the edge of the continental shelf, the continental slopes, and beyond, including the deepest part of the ocean floor.
2. *Nekton.* These are swimming animals, such as fishes and whales, that move independently of water currents and are found in greatest abundance in the relatively shallow and well-lit strata of water above the continental shelf.
3. *Plankton.* These are small to microscopic organisms with limited powers of locomotion carried along with the currents. They are the dominant life form and food source of the ocean. These life forms depend upon organic matter suspended in the water.

2.1.3 Therapy for Illness (Hydrotherapy)

Hot springs, because of their temperature and mineral contents, have been believed in ancient as well as contemporary times to have special healing powers for certain skin diseases, arthritis, and other illnesses. The U.S. National Oceanic and Atmospheric Administration (Berry et al., 1980) lists 3506 thermal springs along with their temperatures and exact locations in the United States. Many of these springs, such as Hot Springs, Arkansas; Palm Springs and Soda Springs in California; Saratoga Springs in New York; Mineral Wells in Texas; Sulfur Springs in West Virginia; and Thermopolis in Wyoming, have been developed as resorts and earned fame as tourist attractions and health treatments.

Warm water will relax spasms; muscular strains or sprains, muscular fatigue, and backache are treated with water in conjunction with heat and pressure (water massage, whirlpool bath). Sitz baths (sitting in hot water) are effective in treatment of swollen, painful hemorrhoids. Muryn (1995) prescribed numerous bath recipes that combine water and herbs as measures for healing, pleasure, beauty, and spiritual growth. In physical therapy, patients can move weak parts of their bodies without contending with the strong force of gravity if they exercise in a buoyant medium such as water.

Mineral Wells is located about 100 km west of Dallas. It attracted some 150,000 health seekers annually in the 1920s and 1930s. Advertising by some of its sanitariums claiming curative powers included "rheumatism, indigestion, insomnia, diabetes, kidney, and liver troubles," "very efficacious in treatment of all female complaints," and "guaranteed cures for cocaine, whiskey, and morphine habits" (Fowler, 1991). Many physicians endorsed the therapy of Mineral Wells water in those early days. Although many of the wilder claims began fading out due to the invention of antibiotics and advancement of medical science in the 1950s, the physical effects of hot springs on blood circulation and tension release that everybody can enjoy conveniently and cheaply at home will never meet with substitutes in the future.

2.2 Chemical Functions

2.2.1 Solvent of Substance

Water molecules are covalently bonded together, comprising one oxygen atom and two much smaller hydrogen atoms. The two hydrogen atoms are separated from each other on one side at an angle of 105°. As a result, the side with the two hydrogens is electropositive and the other is electronegative.

The positive (hydrogen) end causes water molecules to attract (cohere) the negative (oxygen) end of another molecule, causing water molecules to join together in chains and sheets that give water a higher viscosity and greater surface tension. Both positive and negative charges make water molecules attract (adhere) to other materials, which causes water to dissolve a larger variety of substances than any other liquid does. Thus, water is the best cleansing agent for humans, animals, and the environment.

In time, water will dissolve almost any inorganic substance, and about half of the known elements are found dissolved in water. At 20°C, 1 L of water can dissolve 744 g of the fertilizer ammonium sulfate ($(NH_4)_2SO_4$ (Haynes, 2011).

2.2.2 Medium in Chemical Reactions

The majority of chemical reactions take place in aqueous solutions, which affect the chemical properties of many materials. For example, pure water contains dissociated H^+ and OH^- ions in very low concentrations. This makes the pH of materials change when they go into solution in water. Water affects chemicals through the processes of hydrolysis, hydration, and dissolution. Hydrolysis is the process that occurs when water enters into reactions with compounds and ions, while hydration is the attachment of water to a compound. Dissolution refers to the decomposition of ionic compounds in water. These processes result in degradation, alteration, and resynthesis of minerals and materials.

2.3 Physical Functions

2.3.1 Moderator of Climate

Atmospheric water vapor absorbs and reflects portions of incoming solar radiation during the daytime and provides heat energy to Earth in the form of longwave radiation at night. Moreover, due to the high thermal capacity and latent heat of water, water is a good medium for heat storage when air temperature is high, and the stored heat in the water is released when the air temperature is low. Consequently, temperature fluctuations are always smaller over water than land, and climates in areas close to oceans or lakes are always moderate compared to climates of inland areas. If water vapor were not present in the atmosphere, Earth's temperature would be much warmer in the daytime and much colder at night.

The Great Lakes region of the United States illustrates the effects of water on climate. The amount of precipitation and the frequency of thunderstorm and hailstorm activity over the lakes and their downwind areas tend to decrease in the summer and increase in the fall and winter (Changnon and Jones, 1972). Apparently, this is because the waters of these lakes in the fall and winter are warmer than the overlying air. Precipitation over the lakes and their downwind areas is enhanced when moisture and heat are added from the lakes to increase atmospheric instability. On the other hand, the lakes are cooler than the overlying air in summer, tending to stabilize the atmosphere, and precipitation is reduced.

Table 2.1 gives normal monthly air temperatures from 1941 to 1970 for three stations in the United States. They are located at about the same elevation and latitude but differ in distance from the Pacific Ocean. The station in California (Happy Camp) is only about 64 km from the Pacific, and its climate is under direct influence of the ocean. The normal air temperature in January, the coldest month of the year, was 3.5°C at Happy Camp, warmer than Fremont, Nebraska, and Atlantic, Iowa, by as much as 8.8°C and 10.1°C, respectively.

TABLE 2.1

Normal (1941–1970) Monthly Air Temperatures for Three U.S. Stations Located at About the Same Elevation and Latitude but at Different Distances from the Pacific Ocean

Elevation (m)	Latitude (°)	Jan	Feb	Mar	Apr	May	Jun	Jul	Aug	Sep	Oct	Nov	Dec
Happy Camp Ranger Station, California													
369	41.48	3.5	6.6	8.6	11.8	15.7	19.3	23.3	22.3	19.3	13.6	7.9	4.6
Fremont, Nebraska													
366	41.26	−5.3	−2.3	2.7	11.1	17.1	22.1	24.9	24.1	18.8	13.1	4.2	−2.4
Atlantic 1 NE, Iowa													
364	41.25	−6.6	−3.7	1.5	9.9	16.0	20.9	23.5	22.6	17.6	12.0	3.1	−3.6

Source: U.S. National Weather Service, *Climatological Data Annual Summary*, California, Nebraska, and Iowa, National Climatic Data Center, Asheville, NC, 1975.

In July, the warmest month of the year, the normal temperature was 23.3°C at the California station, colder than the Nebraska and Iowa stations by 1.6°C and 0.2°C, respectively.

2.3.2 Agent of Destruction

Although water is so fluid that it fits any shape of container, it is also very mobile, persistent, forceful, and destructive. Moving water over the land surface is responsible for soil erosion, nutrient losses, landslides, stream sedimentation, and various forms of topographical movement and formation. Alternate freezing and thawing, scouring, undercutting, abrasion, hydrolysis, and hydration often break down large rocks into small pieces that finally become small enough to be soil particles. In time, a big rock can be pierced by water drips.

All landscape features, such as valleys, canyons, floodplains, deltas, alluvial fans, caves, gullies, and slopes, are results of the continuous actions of water. The Grand Canyon, created by the Colorado River about 20 million years ago, is a unique example of the cutting power of water. It has been estimated that the average erosion rate of the canyon was about 0.3 mm/year (Bloom, 1978).

The destruction caused by water can be further manifested through damage to agricultural lands, animals, properties, human life, and environment as a result of floods, severe storms, hail, snow, ice rain, tsunamis, and avalanches. On May 30, 1899, an earthen dam at South Fork Lake in Pennsylvania collapsed, spilling 20 million tons of water into the valley and Johnstown below. In a matter of hours, 2209 people lost their lives. The 1927 flood of the Mississippi River measured 160 km wide at some points and caused more than 6.4 million ha of land in seven states to be covered by the flood waters. Total losses in crops were about $102 million, and more than 600,000 people were evacuated from the area (Floyd, 1990).

More recently, the great 1993 flood of the upper Mississippi and lower Missouri rivers covered land for 960 km in length and 320 km in width, submerging 75 small towns in 10 states completely under water for months. It caused 48 flood-related deaths and damaged 50,000 homes, with estimated losses exceeding $20 billion (Dvorchak, 1993). The 2005 flood triggered by Hurricane Katrina along the northern Gulf of Mexico caused $81 billion in property damage (Pielke et al., 2008).

2.3.3 Potential Source of Energy and Power

At 30°C, it takes about 590 cal (2470 J) of energy to change 1 g of water from the liquid state to the vapor state. Conversely, the same amount of energy will be released when the water is condensed from the vapor state to the liquid state. The condensation of water vapor in the atmosphere is the major source of energy to generate violent storms.

Humans have used water as a prime mechanical power source for thousands of years. One such use was the water mill. At a water mill, a dam was built across a river, which raised the water level. Water spilling over the dam provided force to turn a paddle wheel; its shaft drove a millstone for grinding grain.

Electricity is the key to success in industrial societies. It provides power to heat and cool buildings, drive trains, melt metals, and run machines. Water helps generate power through thermoelectric and hydroelectric processes. Thermoelectric plants convert water into steam by heating it with fossil or nuclear fuels. Hydroelectric power can be generated using hydraulic drop in rivers (*hydropower*), ocean waves (*tidal power*), or the

FIGURE 2.2
Three Gorges Dam, measuring 185 m high, 2335 m long, and 40–115 m (crest-base) wide, on Yangtze River, China, is the largest hydropower project of the world. (Image created by N. Lewis Bodden.)

temperature gradient between surface and deep tropical water (*sea thermal power*). In the United States, only about 28% of the maximum potential capacity of hydroelectric power has been developed, which accounts for about 15% of the current total electricity produced.

In hydroelectric power, the force used to drive turbines is the "head" of water held back behind the dam along a river. Total head is the difference in elevation between the turbine and the water level above the turbine. A greater head will generate more power in a given amount of water. The Three Gorges Dam, located in Xilingxia Gorge of the Yangtze River in Hubei Province, China, is the largest hydropower project of the world (Figure 2.2). The dam, completed in 2008, stands 185 m high, 2335 m long, and 40–115 m wide (crest-base), with a reservoir 600 km long and a drainage area of over 1 million km². It generates 84 GkWh/year of electricity—equivalent to 11%–15% of China's energy and to the power used by four cities the size of Los Angeles.

The energy created by the head of water in the river can be applied to generate energy in the ocean. Oceanic tides occur as a result of the gravitational attraction in water bodies of Earth by the sun and moon. Twice a day, a tremendous volume of water flows into and out of bays, producing high or low tides. The kinetic energy in the tidal flows, if differences in water elevation between high and low tides are sufficiently high, can be used to spin turbines to generate electricity. For such purposes, a dam is built, similar to that in a river, with operational gates across the bay. About two dozen sites around the world are suitable for tidal power generation and dam construction (Miller, 1999).

Another potential source of energy is "flammable ice," or "methane hydrate." It is a frozen form of methane (CH_4) surrounded by water molecules and is held together by freezing temperature and extremely high pressure (26 atmospheres or more). The global estimates of methane hydrate are enormous, from 2,000 to 10,000 billion metric tons in oceanic hydrate deposits along the world's continental slopes and from 7.5 to 400 billion metric tons in the vast areas of permafrost in northern latitudes. If warmed or depressurized, it will revert back to water and natural gas. One cubic meter of gas hydrate releases 164 cubic meters of natural gas when brought to the earth's surface. The energy content of methane hydrate form possibly far exceeds the energy content of the earth's coal, oil, and gas resources combined (U.S. Department of Energy, 2007).

A hydrogen fuel cell uses hydrogen as fuel and oxygen as oxidant to produce electricity. The hydrogen and oxygen are generated from water with an input of energy and a catalyst. Such cells have been used to power space crafts and automobiles.

2.3.4 Scientific Standard for Properties

Because of its abundance and distribution on the earth, along with its unique physical and chemical properties, water is used as a scientific standard for mass, specific gravity, heat, specific heat, temperature, and viscosity of other substances.

One kilogram or 1000 g is the mass of water that is free of air at 3.98°C in a volume of 1 L. The ratio between the weight of a given volume of substance and the weight of an equal volume of water is defined as specific gravity. A standard heat unit is expressed in calories. One calorie is defined as the heat required to raise the temperature of 1 g of liquid water 1°C. The number of calories required to raise 1 g of a substance is defined as the thermal capacity of that substance, and the ratio between the thermal capacity of a substance and the thermal capacity of water is called specific heat. A typical soil has a specific heat of 0.25 cal/g/°C, which is about one-fourth that of water.

Temperature scales are based on melting ice and boiling water under standard atmospheric pressure. Four temperature scales are in general use today: Fahrenheit (°F), Rankine (°R), Celsius (°C), and Kelvin or Absolute (K). The temperature of melting ice is set at 32°F, 492°R, 0°C, and 273.16 K, and the temperature of boiling water is set at 212°F, 672°R, 100°C, and 373.16 K. In the United States, the Fahrenheit scale is commonly used, with the Rankine scale used mainly by engineers. The Celsius and Kelvin scales are used internationally for scientific measurements.

2.3.5 Medium of Transport

Nutrients, sediments, seeds, pollen, bacteria, viruses, pathogenic agents, parasitic organisms, insect eggs, pollutants, and many other materials are transported by water. Such transports are beneficial in some areas, but more frequently they are detrimental to human beings and to the environment. Microorganisms that cause typhoid fever, cholera, bacillary dysentery, leptospirosis, gastroenteritis, intestinal worms, hepatitis, giardiasis, shigellosis, and other diseases are transmitted by water. These waterborne and many other water-related diseases are threats to human health and life (Table 2.2).

Waterway transportation is crucial to the development of many countries and to the cultural and technological exchanges among regions and countries. In ancient China, water was spread over the land surface to form ice in freezing weather so that big rocks and stones could be transported from farther north for construction purposes. The Grand Canal of China, built about 2500 years ago, is a 1782 km (1107 miles) waterway between Tientsin in the north and Hangchow in the south. It was the main life artery of commercial and social activities then and is still in use today. The Suez Canal, built in 1914, links the Mediterranean and Red Seas, and the Panama Canal joins the Atlantic and Pacific Oceans and shortens navigation around South America by 20,000 km (12,500 miles).

2.4 Socioeconomic Functions

2.4.1 Source of Comfort

Surface waters are often used in conjunction with recreational activities such as relaxation, fishing, sports, and aesthetic appreciation. Lakes are beautiful to our eyes, thundering

TABLE 2.2

Diseases Associated with Water

Problem of Water	Disease Agent	Disease
Water quality (waterborne)	Bacterial	*Salmonella* (typhoid)
		Enterobacteria (*E. coli*, campylobacter)
		Cholera, leptospirosis, etc.
	Viral	Hepatitis A, poliomyelitis
		Rotaviruses, enteroviruses
	Parasitic	Amoebiasis, giardiasis
		Intestinal protozoa
		Balantidium coli
	Enteric	E.g., a proportion of diarrheas and gastroenteritis
Water quantity (water-washed)	Skin	Scabies, ringworm, ulcers, pyodermitis
	Louse-borne	Typhus and related fevers
	Trepanematoses	Yaws, bejel, pinta
	Eye and ear	Otitis, conjunctivitus, trachoma
Water consuming	Crustacea	Guinea worm, paragonimiasis
	Fish	Diphyllobothriasis
	Shellfish	Flukes, schistosomiasis
Proximity to water	Mosquitoes	Malaria, filariasis, yellow fever
		Dengue, hemorrhagic fever
	Tsetse flies	Trypanosomiasis (sleeping sickness)
	Blackflies	Onchocerciasis

Source: UNESCO, Water and health, IHP Humid Tropics Program Series No. 3, Division of Water Sciences, Paris, France, 1992.

waterfalls are joyous to our ears, and the feel of water is sensational to our body. The soaring ocean surf and pounding tides serve many as a source of peace and tranquility. Most people enjoy water as a source of relaxation in one way or another. Of the recreation areas administered by six U.S. federal agencies from 1981 to 1990, the average million-visitor days per year were highest for Corps of Engineers reservoirs at 556.89. Visitors to national parks, national forests, Bureau of Land Reclamation properties, and Bureau of Land Management properties were 62.0, 42.5, 4.5, and 9.0 million-visitor days per year—only 8.7% that of Corps of Engineers sites. This love for water often gives land with access to water higher value than land without access to water.

2.4.2 Inspiration for Creativity

Water is a subject that inspires writers, poets, artists, and musicians to create many great works. The feeling of being beautiful, peaceful, soothing, intriguing, mystifying, angry, violent, and encompassing triggers streams of inspiration for creativity. Authors such as Shakespeare, Byron, Thoreau, Twain, and Hemingway, among many others, have given water a major role in some of their masterpieces.

Christianity has used water as a symbol of holiness and cleanliness. When a person is baptized in a river or other body of water, it symbolizes the burying of his or her sinful body. The person is a reborn Christian when he or she is pulled out of the water.

Since Jesus was baptized in the Jordan River, thousands of people go there every year to fill bottles with this holy water to baptize their own children. Water in another form—snow—is an indispensable element of Christmas charm and spirit. Songs of winter wonderlands, snowmen, and white Christmas have been around for years and are enjoyed by people of all ages.

Hindus consider the Ganges River holy. Millions of people make pilgrimages to the river and bathe in it to wash away their sins, even though it serves as an open sewer for urban areas. Funeral ashes of deceased loved ones are cast into the Ganges with the belief that their souls will ascend to Heaven.

As long as there is water, there will be activities to enjoy, stories to write, songs to sing, art to create, and rites to follow.

2.4.3 Role Model for Self-Expectation

Water possesses many unique properties that can be personified. It serves as a role model for self-discipline and self-expectation. For example, water accepts all objects that fall into it; water will gradually mix with many of these objects through the dissociation process. Similarly, a person should have the capacity, manner, and patience to accept different people, endure criticisms, and treat all people equally regardless of race, sex, religion, education, and wealth. Water becomes hard as ice when temperature is below freezing. By analogy, people need to be strong and firm under pressure and stress and in harsh environments. That water flows down slopes inspires us to look after those who are less fortunate than us, understand them, be concerned for their needs, and help solve their problems. Water adapts itself to fill all shapes and sizes of containers; we too need to be flexible and adaptable.

Water behaves in many other ways that can be personified and learned by people. If emulated, it is a role model for us to improve our personality and disposition. No wonder Lau Tze, a great ancient Chinese philosopher 3000 years ago, said "Goodness to its utmost stage resembles water."

2.4.4 Medium for Agricultural and Industrial Production

Water is a vital element in agricultural and industrial production. In 2005, agricultural irrigation consumed more than 31% of total off-stream water withdrawals in the United States (Kenny et al., 2009). Grazing lands, barrens, and deserts with sufficient supplies of water can be converted into agricultural production. On the other hand, a deficit of water, due to either climatic conditions or lack of artificial irrigation, can affect plant vigor and growth, the quantity and quality of production, seedling survival, litter production, and fertilizer uptake. Worldwide, about 18% of cropland is irrigated, making crop production two to three times greater than that from rain-watered land (Miller, 1999).

Although irrigation is beneficial to agricultural production, inappropriate water management can create salinization and waterlogging problems on irrigated land. Irrigation water contains salts. Evapotranspiration of irrigation water leaves salts behind in the soil, and the salts accumulate to a harmful level through prolonged irrigation practices. Salinization retards crop growth and yields, and eventually ruins the land. Farmers often reclaim saline soils by applying a large amount of irrigation water to leach salts through the soil profile. However, inadequate underground drainage makes water accumulate under the ground. The water table then gradually rises to the root zone and soaks and eventually kills the plants.

Water is used for the cooling processes in steam–electric generation plants and as a coolant, solvent, lubricant, cleansing agent, conveying agent, screening agent, and reagent in most industrial manufacturing processes. For example, it takes about 55–390 t of water to fabricate 1 t of paper, and 9–190 t of water to make 1 t of textiles. Total industrial use of water accounted for about 49% of total off-stream water withdrawals in the United States in 2005.

2.4.4.1 Hydroponics

Water, by making a nutrient solution containing all the essential elements required by plants for normal growth and development, can be used to culture vegetables, fruits, and flowers in a variety of environments. The technique is referred to as water culture or more professionally as hydroponics, meaning "water working." Today, the term "hydroponics" is used inclusively to refer to the science of growing plants with media other than soils, such as water, gravel, sand, peat, sawdust, vermiculite, or pumice.

The growing of plants in water dates back to several hundred years BC in Babylon, China, Greece, and Egypt, but modern techniques were not developed until plants' macro- and micronutrients were discovered in the late 1880s and early 1890s. During World War II, the U.S. military applied hydroponics to grow fresh vegetables and food for troops stationed on nonarable islands in the Pacific (Resh, 1989). Today, the entire hydroponic system, due to the development of plastics, vinyl, suitable pumps, time clocks, plastic plumbing, solenoid valves, and other equipment, can be automatically operated at costs a fraction of those in the past. Large commercial installations exist throughout the world.

Hydroponics is particularly valuable in countries with little land or limited arable land and large populations. City dwellers and hotel managers may find it attractive to grow plants in living rooms, along hallways, and on porches or windowsills for decoration. Many citizens in Taipei, Taiwan, grow vegetables as supplemental crops on their rooftops. The technique can be used in atomic submarines, spaceships, or space stations.

The crop yield per unit area by hydroponics is about four to 30 times higher than in soil culture under field conditions (Resh, 1989; FAO, 1993). The reasons for this may include high planting density, more efficient use of water and fertilizers, less environmental damage, and low operational costs. Soil culture requires regular control of weeds, and may have problems with soilborne diseases, insects, and animal attacks. It needs crop rotation to overcome buildup of infestation and nutrient deficiency problems. Hydroponics does not encounter these problems.

In traditional soil culture, plants must develop large root systems to absorb elements required for development and growth. However, the availability of nutrients in the soil depends upon soil bacteria to break down organic matter into elements and soil water to dissolve elements into solutions. With hydroponics, nutrients are immediately available to plants. Hydroponics takes only 1/20 to 1/30 the amount of water required by conventional soil gardening.

All these advantages make hydroponics the quickest and simplest method for producing the maximum amount of vegetables from a minimum area (DeKorne, 1992).

2.4.4.2 Fish Culture

Fish culture or farming, in which fish and shellfish are raised for food, began in China around 2000 BC (Lee, 1981). Fish farming supplies 60% of fish consumption in Israel, 40% in China, and 22% in Indonesia (Miller, 1999). In 1995, about 24.4 million Mt or 21% of the total

TABLE 2.3

Production and Utilization of Global
Aquaculture and Fisheries in 2005, 2007, and 2009

Item	2005	2007	2009
Aquaculture, 10^6 t			
Inland	26.8	30.7	35.0
Marine	17.5	19.2	20.1
Total	44.3	49.9	55.1
Capture, 10^6 t			
Inland	9.4	10.0	10.1
Marine	82.7	79.9	79.9
Total	92.1	89.9	90.0
Total production, 10^6 t	136.4	139.9	145.1
Human consumption, 10^6 t	107.3	112.7	117.8
Food fish supply/person, kg	16.5	16.9	17.2

Source: Data from FAO, *The State of World Fisheries and Aquaculture 2010*, FAO Fisheries and Aquaculture Department, Rome, Italy, 2010a.

global fish harvest were raised artificially through fish farming, 12% from inland farming, and 9% from marine farming (NOAA, 2007). In 2009, total global aquaculture was 55.1 million Mt, or about 38% of the total global fish harvest (Table 2.3). This was a steady increase of 2.0 million Mt/year during the 15-year period. Global capture fisheries production has been relatively stable in the recent decade (FAO, 2010a).

Fish farming usually involves stocking fish in ponds or containers until they reach the desired size (inland farming) or holding captured species in fenced-in areas or floating cages in lagoons or estuaries until maturity (marine farming). Surveys conducted by FAO show 1028 species that have been farmed around the world and summarize them into six categories (World Resources Institute et al., 1996):

Freshwater fish: e.g., carp, barbel, and tilapia

Diadromous fish: e.g., sturgeon, river eel, salmon, trout, and smelt

Marine fish: e.g., flounders, cod, redfish, herring, tuna, mackerel, and sharks

Crustaceans: e.g., crabs, lobsters, shrimps, and prawns

Mollusks: e.g., oysters, mussels, scallops, clams, and squid

Others: includes frogs, turtles, and aquatic plants

Although fish farming produces high yields per unit area, the rapid growth of fish farming is subject to some environmental risks. Without proper pollution-control measures, the wastes generated from fish farms can contaminate surface streams, lakes, groundwater, and bay estuaries. Some of the ecologically important mangrove forests in Ecuador, the Philippines, Panama, Indonesia, Honduras, and other less-developed countries have been destroyed by fish farming (FAO, 1995; Miller, 1999; Naylor et al., 2000). For sustainable development, FAO (2010b) has provided technical guidelines on the ecosystem approach to aquaculture, a strategy on the integration of science, policy, and management.

2.5 Mechanical Functions

2.5.1 Tool for Industrial Operations

Through extreme pressure and speed, water can even be used as a tool to remove the bark of trees (hydraulic debarker) and cut rocks or other materials, including metal, glass, ceramic tile, marble, granite, wood, plastic, and stainless steel. In construction, water jets with high speed and pressure are used to excavate mountains and cut tunnels during massive construction projects. Using hydraulic principles, hydraulic jacks and levels can raise or lower heavy loads precisely with a small force. Also, water clocks and steam engines are powered by water. Water jet cleaning machines can do jobs on hard-to-reach and/or hard-to-scrub locations.

The Bureau of Mines has designed a water jet perforator, which issues a high-velocity water jet, to penetrate nonmetallic well casings for the purpose of completing or stimulating in situ uranium-leaching wells (Savanick and Krawza, 1981). In situ uranium leaching is a mining method in which wells are drilled from the surface to the mineralized rock. An oxidizing leaching solution is injected through these wells into the uraniferous rock. The solution dissolves the uranium minerals, and the uraniferous solution is subsequently drawn into another well. It is then pumped to a processing plant where the uranium is extracted and precipitated as uranium oxide.

Properly channeled water with high pressure has a cutting power much sharper and effective than knifes, forceps, scissors, and scalpels. Technique on water knifes or jet scalpels has been rapidly developed in recent decades and tested in a variety of medical surgeries. Applications of water jet cutters on medical surgery may include the following:

- Stone destruction in the common bile duct
- Tattoo removal, without damaging the skin
- Kidney surgery, reducing the risk of bleeding
- Cataract operations
- Liver tumor operations
- Liver cancer operations

Based on the aforementioned experiences, surgery using jet cutters leads to less loss of blood, is faster in operation, is easier on the patients, results in better healing, is less costly, does no damage to nerves, has greater precision, and leaves no scares (Jian and Sun, 2001).

2.5.2 Material for Musical Instruments

Benjamin Franklin developed a set of musical glass bowls horizontally nested on an iron spindle. The sound was produced by touching the rims of the bowls with moistened fingers. He called his instrument "armonica" or "glass harmonica" (Hopkin, 1996).

The *hydraulophone*, also known as *water organ*, is a water-based musical instrument invented by Steve Mann in the 1980s (Mann et al., 2006). It uses a pipe with an array of holes as the keyboard and is fed with pressurized water to jet out through each hole. These water jets function like a key on a keyboard instrument; each hole (water jet) corresponds to one note. Users can play the instrument by direct interfering, such as touching, diverting, restricting, or obstructing, with the water jets that come out of the holes.

2.6 Political Functions

2.6.1 Cause of International Conflicts

Conflicts over water can be attributed to four important factors: (1) water is essential in life and critical in economical growth, (2) water supply and demand are often imbalanced, (3) freshwater on the land moves constantly from mountains to plains and eventually to sea; water resources are not stable over time, and (4) a river or lake may be used by more than one country (7 for the Amazon, 10 for the Nile, and 12 for the Danube). As a result, disputes and conflicts over water resources have occurred at the international, regional, and local levels; some even have been developed into military maneuvers and bloodshed events. Gleick (2004) listed 126 worldwide conflict events relevant to water resources throughout history; many involved force and violation.

2.6.1.1 Potential Conflicts

A river that flows through more than one country often develops competition for water resources between upstream and downstream countries. The potential in conflicts is great, especially when population densities, cultural development, water consumption rates, and economic growths are different between the upstream and downstream countries. If water allocation between upstream and downstream countries cannot be settled through negotiations and cooperation, it may develop into diplomatic problems, economic sanctions, or even military actions.

2.6.1.1.1 Euphrates and Tigris

The Euphrates and Tigris are two great rivers originated, only 30 km apart, high in the mountains of Turkey and flow southeasterly together into the Persian Gulf. The Euphrates is 2700 km long with the upper 40% in Turkey, the middle 25% in Syria, and the lower 35% in Iraq. Lying northeast of the Euphrates, the Tigris is 1900 km long with the upper 20% in Turkey, 78% in Iraq, and only 2% in Syria. The upper courses of these two rivers are at elevations of 2000–3000 m above sea level, carry heavy suspended particles (as much as 3×10^6 t in a single day), and are responsible for the great deposits of alluvium in the Mesopotamian Plain (Hillel, 1994).

Turkey launched the Southeast Anatolia Project (GAP) in 1977 to construct 22 dams, 19 hydroelectric power stations, and extensive irrigation works on the headwaters of both the Euphrates and Tigris for developing 10% of the country bordering Syria and Iraq. The project causes declines in water quantity and quality downstream and may exacerbate tension between Turkey, Syria, and Iraq—possibly resulting in violent conflict. It is frequently cited as a water-development scheme that may lead to future instability in the Middle East (Harris, 2002).

2.6.1.1.2 Golan Heights

The Golan Heights is not only a strategic area from a military point of view; it is also a water-rich area (the headwaters of the Jordan River) of the region controlled by Israel since the 1967 "Six-Day War." With a population that is only about 50% more than Jordan's, Israel uses nearly twice as much water from the Jordan and Yarmuk rivers. However, in the 1996 peace negotiation between Israel and Syria, Israel refused to give up water from the Golan Heights. Israeli Foreign Minister Ehud Barak told a closed-door parliamentary session, "The Syrians know that the waters of the Galilee and of the Jordan are for exclusive use by us" (U.S. Water News, 1996). Disputes about water rights must be settled before any peace agreement in the Middle East can be achieved.

2.6.1.1.3 Nile River

The conflict on the water allocation of the Nile River is more severe than those in the two Middle East regions described earlier. The White Nile originates in the highlands of Rwanda and Burundi; flows into the huge basin of Lake Victoria in Tanzania; travels northward through Kenya, Uganda, Zaire, Ethiopia, Sudan, and Egypt; and then empties into the Mediterranean Sea. At Khartoum, Sudan, the Blue Nile, originating from Lake Tana in Ethiopia, merges with the White Nile to form a total drainage area of 3.11×10^6 km². The river sprawls from about lat. 5° S to lat. 31.5° N and covers about one-tenth of the African continent. Table 2.4 lists the 10 countries sharing the Nile River basin, areas of contribution to the basin, and the hydrologic data.

Although the Nile (6484 km) is the second longest river in the world, its annual discharge is only a small proportion of that of other major rivers, for example, 3% of the Amazon, 14% of the Mississippi, and 43% of the Danube (Said, 1981). The low volume of waters shared by 10 countries makes allocation of the waters of the Nile River crucial to downstream areas. Egypt's water supply is almost totally derived from the Nile. The entire 1530 km of the Nile within the borders of Egypt receives not a single tributary. A new dam is being built in Uganda on the Victoria Nile and there are plans for several in Ethiopia (Schultz, 2011). Such water development projects in upstream countries can cause water stress for countries in downstream areas, could potentially create instability in the region, and even develop into military confrontations.

2.6.1.2 Iraq–Iran War, 1980–1988

Iraq and Iran (Persia prior to 1935) are two side-by-side Middle East countries. They share a border of about 1458 km between the tri-point with Turkey on Kuh-e Dalanper and the

TABLE 2.4

Nile River Basin Countries, Areas, and Hydrologic Data

| Country | Total Area of the Country | | Total Area in the Basin | | Inflow or Outflow | | Basin Rainfall (mm/year) |
	10^3 km²	% in Basin	10^3 km²	% of Basin	In (km³/year)	Out (km³/year)	
Burundi	27.8	47.5	13.3	0.4	0.00	1.50	1110
Rwanda	26.3	75.5	19.9	0.6	1.50	7.00	1105
Tanzania	945.1	8.9	84.2	2.7	7.00	10.70	1015
Kenya	580.4	8.0	46.2	1.5	0.00	8.40	1260
DC Congo	2344.9	0.9	22.1	9.7	0.00	1.50	1245
Uganda	235.9	98.1	231.4	7.5	28.70	37.00	1140
Ethiopia	1100.0	33.2	365.1	11.7	0.00	80.10	1125
Eritrea	121.9	20.4	24.9	0.8	0.00	2.20	520
Sudan	2505.8	79.0	1978.5	63.6	117.10	55.50	500
Egypt	1001.5	32.6	326.8	10.5	55.50	Rest to Sea	15
The Basin			3112.4	100.0			615

Source: Data from FAO, *Irrigation potential in Africa: A basin approach*, FAO Land and Water Bulletin 4, Land and Water Development Division, the UN Food and Agriculture Organization, Rome, Italy, 1997.

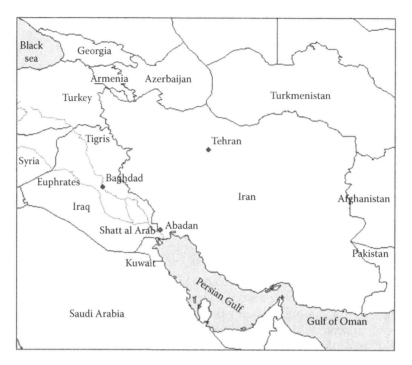

FIGURE 2.3
Shatt al Arab, a 200 km waterway on Iran and Iraq borders.

terminal point of the Shatt al Arab in the Persian Gulf. Iraq lies on the southwestern side of the border, occupying the greater part of the ancient land of Mesopotamia—the plain between the Tigris and Euphrates rivers. Mesopotamia was the location of the Garden of Eden in the Biblical time, and Abraham was born in Ur, an old city about 190 miles southeast of Baghdad today. Iran is on the northeast of Iraq within the Alpine–Himalayan mountain system. The Shatt al Arab, a 200 km waterway combining Euphrates and Tigris rivers together and flowing into the Persian Gulf, is the south-most border between Iraq and Iran (Figure 2.3). It was the territorial dispute over this 200 km waterway, along with historical, political, and religious reasons, that caused the two countries to engage in a war that lasted 8 years in the 1980s.

In January 1979, Ayatollah Khomeini led a revolution in Iran and toppled the Shah's regime. The revolution caused chaos in Iran and weakened Iran's military. Iraq seized the opportunity to reclaim full control of the river by military forces. It launched air strikes on 10 Iranian military airports and crossed 5 army divisions into Iran on September 22, 1980. The war lasted 8 years with a cease-fire sponsored by the UN Security Council 598 Resolutions on August 20, 1988. It was one of the longest and most destructive wars of the twentieth century, with 600,000 Iranian and 400,000 Iraqis dead and an estimated total economic loss of $1.2 billion. Both sides made no significant territorial or political gains by the war; the fundamental issues on the waterway remained unsolved at the end of the war.

2.6.2 Issue of Regional Stability

Lagash and Umma, two ancient Mesopotamian cities, were in dispute over water as early as 4500 BC (Clarke, 1993). In the United States today, competition over the most precious and most wasted resource—water—is steadily rising among regions, states, cities, farmers, industries, Native Americans, and the federal government. Today, lawyers, lobbyists, and politicians dispute over water in courtrooms and legislatures, using arguments based on history, law, economics, considerations of fairness, environmental impacts, and issues of survival. The impact of the disputes is profound.

One example of water confrontation at the state level is Arizona versus California on water allocation of Colorado River in 1934. The Colorado River is a major life-sustaining source of water for drinking, irrigation, and other uses by people in the arid southwestern United States. It supplies water to over 25 million people and helps to irrigate 1.42 million ha of farmland. Originating in the Rocky Mountain National Park on the west of the Continental Divide, the Colorado River flows southwesterly toward the Gulf of California and the Pacific Ocean. It routes through Colorado, Utah, Arizona, Nevada, California, and Mexico with a length of 2330 km. It is one of the siltiest and wildest rivers in the world and is also the most legislated, most debated, most litigated, and most managed river in the United States.

In the late nineteenth century, California was the only state in the Colorado River basin with a substantial population that had withdrawn water along the Colorado River for irrigation, while other states were still largely uninhabited and were not much offended by California. Since California runoff contributes hardly at all to the Colorado River, the irrigation use of water from the river did not represent an equal share of water on one hand and could reduce water available for other states on the other hand. It was largely the California's Imperial Valley irrigation projects in the beginning years of the twentieth century that triggered the negotiations among the basin states on water allocations along the Colorado River and finally reached agreement on "The Colorado River Compact" in November 1922.

The Imperial Valley, or called the Dead Valley, is a flat and dry valley across the border of California and Mexico with a rainfall of about 6 cm/year, the lowest in the United States. Between 1901 and 1908, a channel was cut from the river at six different times to irrigate the Imperial Valley. However, these channels were either silted up or got flooded. The Colorado River-dependent California farmers turned to a dam as a solution. Dams could mitigate flooding, reduce channel siltation, and provide water for irrigation in dry years. It was imperative that negotiations with other basin states be carried out if a dam was to be built along the Colorado River.

Arizona was most affected if California built a dam as proposed. Through meeting with California, negotiating with the Secretary of the Interior, and filing law suit in the Supreme Court, Arizona could not get a favorable contract on water allocations. Two months before the rejection of the Supreme Court in 1934, Arizona governor Benjamin B. Moeur ordered 100 Arizona National Guards and one ferryboat to prevent the erection of Parker Dam, a proposed structure 240 km downstream from Hoover Dam. This appears to be the first water-related confrontation at the state level that involved military forces in U.S. history. The Justice Department asked the Supreme Court for an injunction against Arizona. The court denied the request because the construction of Parker Dam had never specifically been authorized. The Congress remedied that oversight 4 months later and Arizona finally had to agree to the construction of Parker Dam (Hundley, 1975).

2.6.3 Drive in Local Confrontations

2.6.3.1 *Uprising in Cochabamba (Bolivia), 2000*

The uprising in Cochabamba, Bolivia in 2000 is an example of people against an enterprise giant and government for the life-threatening raise of water tariffs. Bolivia is the poorest country in South America with nearly two-thirds of the population living below the

22   *Forest Hydrology: An Introduction to Water and Forests*

poverty line. In the name of economic efficiency, the World Bank and the International Monetary Fund (IMF) have pushed the Bolivian governments to sell its public enterprises to international investors as a precondition for national loans.

2.6.3.1.1 *Water Prices/Privatization*

In September 1999, the Bolivia government granted a 40-year contract to Aguas de Tunari to run the water supply of Cochabamba. Aguas de Tunari is a subsidiary of the San Francisco-based engineering giant, Bechtel Corp. The company got the rights not only to supply water to a network of municipalities but also to industrial, agricultural, and residential uses in all of Cochabamba Province. All rights on surface water and ground water were submitted to the private company. In addition, the Bolivian government guaranteed the company a 15% rate of return on capital and linked water tariffs to the consumer index in the United States. It passed Law 2029 (the Drinking and Sanitation Law), legalizing the privatization of drinking water services (Sridhar, 2003).

Immediately after the takeover of the water rights and water systems, Aguas raised water tariffs by between 30% and 300% in Cochabamba in December 1999. Many Bolivian families earned as little as $100/month; they had to pay $20 and higher, or more than one-fifth of their monthly income, for the new water bills. These increases were catastrophic, far beyond what the city's many poor families could afford. Moreover, people had traditionally considered ponds, lakes, and other local sources of water the property of their communities. Now they were not allowed to access water from these sources. Water had become a commodity owned exclusively by a private monopoly.

2.6.3.1.2 *Uprising*

Being shocked by the deal "behind closed doors" and unable to survive under the burden of the water price hikes, the people staged protests in Cochabamba in January 2000. The protesters effectively shut down the city for four consecutive days and left the city standing still—no cars, no buses, no air flights, or any transport in or out of the city. After negotiations, the government formally agreed to review the water company's contract and the water law if all protests were suspended. The public leaders gave the government 3 weeks to work out the price and water rights problems (Schultz, 2000).

The government did not keep its words and no change in water prices was made in February. People then refused to pay the bills to Bechtel, and the company threatened to shut off the water supplies in response. The dispatched soldiers and people converted every block leading to the city plaza into a mini-battle ground, resulting in 175 injuries and a 17-year being shot to death by the army.

The strikes then quickly spread from Cochabamba through the nation. President Banzer placed Bolivia under martial law, and the public uprising and strikes turned the city of 800,000 in turmoil for 4 months. Finally, President Banzer announced the termination of the water contract between Bolivia and Aguas del Tunari on April 10, 2000. Consequently, the company left the country and filed a "request for arbitration" before the International Centre for Settlement of Investment Disputes (ICSID), a branch of the World Bank, in November 2001. It sought a compensation of $25 million from the Bolivian government for calling off the contract (Olivera and Lewis, 2004).

The Bechtel versus Bolivia case gained attention from individuals and organizations around the world, including the United States. Its importance and implications on water as a human right are far beyond Bolivia. More than 300 organizations from 43 countries sent an International Citizens Petition demanding the case be transparent and open to citizen participation in 2003. Activists have engaged in campaigns to pressure Bechtel to drop the case against Bolivia. On January 19, 2006, Bechtel signed an agreement to abandon the ICSID case for a token payment of 2 bolivianos ($0.30).

2.6.3.2 Pastoralists versus Farmers in Kenya, 2005

The fighting and killing between pastoralists and farmers around the Narok, Kajiado, and Nakuru districts in Kenya in 2005 was an example of people against people triggered by disputes on water resources. Water conflicts between tribes are common in Africa (Omosa, 2005).

Maasai and Kikuyu are two of the three major ethnical groups (the other is Kamba) in Kenya. The Maasai are seminomadic pastoralists with cattle as a major sign of their wealth, while the Kikuyu rely heavily on agriculture with a terrific reputation for money management.

Water in a small river called Ewaso Kedong is shared by the upstream Kikuyu community for farming and the downstream Maasai community for livestock. Disputes over access to water and pasture between the Maasai and Kikuyu communities have frequently occurred since the 1960s. The tension was always high during droughts in Ewaso Kedong River, the only water source for some 30,000 Maasai and their cattle. The 2004 short rainy season had received less than 20% of normal rainfall for the region. This severe drought caused a drastic reduction of water in Ewaso Kedong River and led to clashes between the Maasai and Kikuyu communities in late January–February, 2005, the most deadly worldwide water conflict in recent years.

Clashes erupted after Maasai pastoralists destroyed water pipes and disconnected electricity from the home of a Kikuyu leader who had diverted waters of the Ewaso Kedong River to irrigate his crops, causing a shortage of water downstream for animals. Maasai claimed that small irrigation dams built by upstream neighbors along the river had complicated the drought situation for themselves and their livestock. Also, their livestock had been denied access to water points.

Angered by the move, the leader's neighbors and supporters beat up a local Maasai chief, set his vehicle on fire, and accused him of colluding with his relatives to terrorize their neighbors. In retaliation, the pastoralists attacked anybody from the Kikuyu communities, resulting in fierce fighting among individuals. Warriors from both sides fought fiercely with swords, machetes, spears, bows and arrows, or clubs. The fighting over water over a 2 week period resulted in at least 17 people killed, 30 injured, 40 homes burned, and thousands of people displaced. Destroyed water pipes and uprooted power poles were scattered all over the violence areas. Business stood still as traders ran for safety and schools were closed following the weekend's bloody clashes.

2.7 Military Functions

2.7.1 Weapon against Enemies

Water has been used as a weapon for achieving certain military objectives throughout history (Gleick, 2004). Inundating cities, military camps, warehouses, and other strategically important sites can block the enemy's military operations and activities, drown enemies, and damage military materials and supplies. The story of Moses and his followers passing through the Red Sea to escape Pharaoh's pursuing army in the Bible is familiar to

Christians and other cultures. There were many military operations that had used water as a weapon against enemies; only two operations are described here, one in Europe and the other in Asia.

2.7.1.1 France–Dutch War, 1672–1678

In the seventeenth century, the Dutch became the leading commercial and maritime power in the world, dominating overseas trade and enjoying a trade surplus with all its European neighbor countries. King Louis XIV of France ruled France and launched the war with the Dutch from 1672 to 1678, an attempt to end Dutch competition with French trade and to extend the French empire.

Before the war, Louis XIV worked diplomatically to isolate the Dutch. He obtained the support of Charles II of England by a secret treaty and allied himself with Sweden and several German states. England was promised subsidies in case France entered into war with the Dutch Republic (Price, 1998).

The French army invaded the Netherlands suddenly in May 1672. The military maneuver was initially very successful and soon the French army overran the southern provinces of the Netherlands. Greatly outnumbering the Dutch force, the French army continued to advance northward toward Amsterdam. There would have been little chance for the Dutch force to defend Amsterdam because of imbalance in the army. Instead, the Dutch opened the sluice gates at Muiden (in the north near Amsterdam) on June 20 and inundated an area of land called the "waterline." The area stretched from Muiden to Gorcum (on the River Waal), a distance of about 55 km. The flooding took about 2 weeks because the summer months of 1672 were unusually dry. The land remained underwater during the winter of 1672–1673 and the advance of France was stopped. France's army finally withdrew from the waterline to positions on the River IJssel during the summer of 1673 when the invasion of Cologne by allies of the Dutch threatened to turn the French flank.

Before the flooding, the Dutch army had melted away in the face of French attacks and cities had surrendered without even token defense. The technique of flooding the land to hold off superior force had also been used repeatedly by the Dutch in their 80 years (1568–1648) of war against the Spanish (Summerville, 2006).

2.7.1.2 Sino–Japanese War, 1937–1944

2.7.1.2.1 Background

Beginning in the late nineteenth century, Japan embarked on an aggressive foreign policy and expansionism. The Japanese ambition of domination over China and the Far East was intensified by a planned provocative event at Lu-Gou-Qiao (盧溝橋), an outskirt bridge of Beijing in Hebei Province, on July 7, 1937. By the end of May 1938, Japanese forces had advanced to Kaifeng (開封), Henan Province, an old capital city in ancient China, located on the south side of the Yellow River. About 60 km due west of Kaifeng is Zhengzhou (郑州), a key junction center of four major railroad routes in China and is only a few kilometers south of the Yellow River. If Zhengzhou was lost, then the Japanese could unite the three active battle fields in the east, north, and west into a dynamic system for military maneuvers.

The Japanese forces were much better equipped in terms of weapon quantity and quality, training, and readiness than the Chinese forces. China could not effectively fight against the aggression of Japanese forces at that time. The generalissimo Chiang-Kai-Shek decided to use water from the Yellow River instead of soldiers to confront and block the westward advancement of Japanese.

2.7.1.2.2 *Breaching the Yellow River's Dikes, 1938*

On May 31, 1938, Japanese soldiers had already reached the outskirts of Kaifeng. The dikes on the south side of the Yellow River at Zhaokuo (趙口) were ordered breached by Chiang-Kai-Shek on June 5, but the flow was small and the breached sites were silted again by dirt and debris eroded from the dikes. At this time, Japanese soldiers had already entered Kaifeng City.

A new sinuous site at Huayangkou (花園口) was quickly selected and breached to 50 m wide on June 8. The next day, June 9, rains unusually occurred all day along the Huang He (Yellow) River basin, causing water in the river to rise rapidly. The rising flow ran southeasterly through the breached site at Huayangkou, reopened up the silted site at Zhaokuo again, and flooded 44 counties in three provinces—Henan, Anfui, and Jiangsu over 15,000 km² (Shu and Finlayson, 1993; Larry, 2001).

2.7.1.2.3 *Impacts*

The flooding effectively disturbed the Japan's military plans; it forced Japanese to abandon the occupied counties and retreat from the areas. There were about 7000 Japanese soldiers who did not retreat in time from the flooding as they either drowned or were stranded. The military plans to advance westward along the Longhai railroad tracks were abandoned. Instead, they had to switch their major forces from the Yellow River to the Yangtze River or from north to south. China gained time to assemble forces and supplies to defend the Japanese invasion along the Yangtze River. It turned the war into a long-term confrontation.

Although the Japanese threats in the north were temporarily relieved, China had paid a tremendous price on the flooding. The breaching dikes sent the entire bulk of the Yellow River pouring to the southeast, inundated more than 15,000 km² in the Yellow-Huai Plain, drowned 500,000 people, and left 4 million people homeless. Epidemic was widely spread afterward. The Plain remained flooded for 7 years—about 0.8 million ha of fertile farmlands were still waterlogged when Japan surrendered in 1945 (Clark, 1982; Shu and Finlayson, 1993; Lary, 2001).

2.7.2 Target of Terrorist's Attack

As terrorism is becoming a serious threat to our society, the safety of water supply systems is potentially vulnerable. If successful, water sabotage could disastrously impact vast populations. Cutoff of water supplies could be devastating to fire departments, energy generation, and other water-dependent operations.

2.7.2.1 *Water Infrastructure Systems*

In a broad sense, water infrastructure systems refer to all basic facilities, equipment, and installations for water utilization, including surface and groundwater sources of treated and untreated water, publicly and privately owned operations, storage and delivery systems, and domestic and nondomestic services. In the Unites States, these systems comprise more than 75,000 dams and reservoirs, thousands of miles of water pipes, aqueducts, and sewer lines, 168,000 public drinking water facilities, and about 16,000 publicly owned wastewater treatment plants (Copeland and Cody, 2003). These are the systems that store, deliver, or treat water for water supply, flood control, irrigation, energy generation, aquatic life, recreation, fire protection, and public uses.

2.7.2.2 Attack Measures

In 1999, water supplies and wells in Kosovo were contaminated by Serbs; bomb blast destroyed the main waterline in Lusaka, Zambia, cutting off water to the city and its 3 million inhabitants (U.S. Water News, 2003). President George W. Bush mentioned in his 2002 State of the Union address that U.S. soldiers found diagrams of U.S. public water facilities in Afghanistan. In late 2002, federal officials arrested two suspects in Denver with documents describing how to poison U.S. water supplies. One of the suspects pled guilty of aiding the Taliban and was released in 2004 because of providing substantial assistance to the government (Public Citizen, 2005).

Terrorist attacks on water systems may use (1) biological agents to contaminate water, such as viruses, cysts, *Escherichia coli*, or anthrax spores; (2) chemicals to poison water, such as synthetic organic compounds and radioactive materials; (3) physical measures to damage dams, water companies, pumping stations, and treatment facilities; (4) cyber attacks to computer operation systems; and (5) power disruption to paralyze telecommunications and electronic control systems. All these attacks could be catastrophic.

2.7.3 Tactical and Strategic Model of War

The way that water acts was taught by Sun Tsu in his classic Chinese military treatise— The *Art of War*, to be learned as a model for formulating tactics and strategies in wars.

Sun Tsu, also known as Sun Wu, was a military general in the 6th BC during the Spring and Autumn Period of China. His classical book, *Sun Tsu's Art of War* (Sawyer, 1996), was the oldest and most influential writings on strategic thought in the world. The book has been adopted as a text in many schools of military academy around the world. Theories and techniques stated in the book have also been applied to fields other than military such as politics, diplomacy, business, and playfield, etc. Its wisdom has never been more highly regarded or used.

The *Art of War* covers 13 chapters; Chapter 6 deals with "Illusionism and Reality." Sun Tsu taught in that chapter that military operations and tactics should pattern the way that water acts. In nature, water runs down the ground from high to low and penetrates at where the ground is weak. Also, water changes its course and shape according to the nature of the ground. Accordingly, the actions of army should try to avoid the strong points and strike at what is weak. A commander-in-chief should modify his tactics in accordance with the conditions of the enemy. Just like water retains no constant shape, there are no constant conditions in military operation either. The conditions and strengths should not be kept constant.

2.7.4 Torture against Suspects

Water has been used as a means to torture suspects and prisoners throughout history, especially during the period, Roman against Christians. Even in recent times, water torture is still a common method in police and security sectors for interrogation, punishment, intimidation, retaliation, or obtaining information. Water torture, including water boarding, forced indigestion, dunking, water trials, water dungeons, water whipping, etc., often causes severe pain and suffering, both physical and mental, yet it leaves no external marks on victims (Walker, 1973).

Torture is an aggravated and deliberate form of cruel, inhuman, and degrading treatment or punishment upon individuals. It should be abandoned by all countries and all governments at all levels.

References

Berry, G.W., Grim, P.J., and Ikelman, J.A., 1980, *Thermal Springs List for the U.S.*, U.S. National Oceanic and Atmospheric Administration, National Geophysics and Solar-Terrestrial Data Center, Boulder, CO.

Bloom, A.L., 1978, *Geomorphology—A Systematic Analysis of Late Cenezoic Landforms*, Prentice Hall, New York.

Changnon, S.A., Jr. and Jones, D.M.A., 1972, Review of the influences of the Great Lakes on weather, *Water Resour. Res.*, 8, 360–371.

Clark, C., 1982, *Flood*, Time-Life, Chicago, IL.

Clarke, R., 1993, *Water: The International Crisis*, MIT Press, Cambridge, MA.

Copeland, C. and Cody, B., 2003, Terrorism and security issues facing the water infrastructure sector, CRS Report for Congress, RS21026, Congressional Research Service, The Library of Congress, Washington, DC.

Dahl, T.E., 1990, Wetlands losses in the U.S. 1780s to 1980s, U.S. Fish and Wildlife Service, Washington, DC.

Dahl, T.E. and Johnson, C.E., 1991, Wetlands status and trends in the conterminous U.S., mid-1970s–1980s, U.S. Fish and Wildlife Service, Washington, DC.

DeKorne, J.B., 1992, *The Hydroponic Hot House*, Loompanics Unlimited, Port Townsend, WA.

Dvorchak, R.J., 1993, *The Flood of '1993*, Wieser & Wieser, New York.

FAO, 1993, *Technical manual: Popular hydroponic gardens*, Regional Office for the Latin America and the Caribbean, Santiago, Chile.

FAO, 1995, Review of the state of world fishery resources: Aquaculture, FAO Fisheries Circular No. 886, Food and Agriculture Organization of the United Nations, Rome, Italy.

FAO, 1997, Irrigation potential in Africa: A basin approach, FAO Land and Water Bulletin 4, Land and Water Development Division, the UN Food and Agriculture Organization, Rome, Italy.

FAO, 2010a, *The State of World Fisheries and Aquaculture 2010*, FAO Fisheries and Aquaculture Department, Rome, Italy.

FAO, 2010b, Aquaculture development: 4. Ecosystem approach to aquaculture, *FAO Technical Guidelines for Responsible Fisheries*, No. 5, Suppl. 4, the UN Food and Agriculture Organization, Rome, Italy, 53 pp.

Floyd, C., 1990, *America's Great Disasters*, Mallard Press, New York.

Fowler, G., 1991, *Crazy Water*, Texas Christian University Press, Fort Worth, TX.

Gleick, P.H., 2004, Environment and security: Water conflict chronology version 2004–2005, *The World's Water, 2004–2005*, Island Press, Washington, DC, pp. 234–255.

Harris, L.M., 2002, Water and conflict geographies of the Southeastern Anatolia Project, *Soc. Nat. Resour.*, 15, 743–757.

Haynes, W.M. (Editor-in-Chief), 2011, *Handbook of Chemistry and Physics*, 91st edn., CRC Press, New York.

Hillel, D., 1994, *Rivers of Eden, the Struggle for Water and the Quest for Peace in the Middle East*, Oxford University Press, Oxford, U.K.

Hopkin, B., 1996, *Musical Instrument Design, Practical Information for Instrument Making*, Sharp Press, Tucson, AZ.

Hundley, N., Jr., 1975, *Water and the West, the Colorado River Compact and the Politics of Water in the American West*, University of California Press, Berkley, CA.

Jian, Y. and Sun, J., 2001, The development of the water jet scalpel with air pressure, *J. Fluids Eng.*, 123(2), 246–248.

Kenny, J.F. et al., 2009, Estimated use of water in the United States in 2005, U.S. Geological survey Circular 1344.

Lary, D., 2001, Drowned earth: The strategic breaching of the Yellow River dyke, 1938, *Water in History*, 8(2), 191–207.

Lee, J., 1981, *Commercial Catfish Farming*, Interstate Printer & Publishers, Inc., Danville, IL.

Mann, S. et al., 2006, The hydraulophone: Instrumentation for tactile feedback from water fountain fluid streams as a new multimedia interface, *IEEE International Conference on Multimedia and Expo*, July 2006, Toronto, Canada, pp. 409–412.

Marieb, E. and Hoehn, K., 2007, *Human Anatomy and Physiology*, 7th edn., Pearson Education, San Francisco, CA.

Miller, G.T., Jr., 1999, *Environmental Science: Working with the Earth*, 7th edn., Wadsworth, Belmont, CA.

Muryn, M., 1995, *Water Magic*, Simon & Schuster, New York.

Naylor, R.L. et al., 2000, Effect of aquaculture on world fish supplies, *Nature*, 405(29), 1017–1024.

NOAA, 2007, *Fisheries of the United States 2005*, National Marine Fisheries Service, National Oceanic and Atmospheric Administration, Current Fishery Statistics No. 2005, Silver Spring, MD.

Olivera, O. and Lewis, T., 2004, *Cochabamba!: Water War in Bolivia*, South End Press, Cambridge, MA.

Omosa, E.K., 2005, The impact of water conflicts on pastoral livelihood: The case of Wajir District in Kenya, International Institute for Sustainable Development, available at: http://www.iisd.org, Winnipeg, Manitoba, Canada, 18 pp.

Pielke, R.A. et al., 2008, Normalized hurricane damage of the United States: 1900–2005, *Nat. Hazards Rev.*, 9(1), 19–41.

Price, J.L., 1998, *The Dutch Republic in the Seventeenth Century*, St. Martin's Press, New York.

Public Citizen, 2005, Water unsecured: Public drinking water is vulnerable to terrorist attack, available at: http://www.citizen.org/document/water.pdf

Resh, H.M., 1989, *Hydroponics Food Production*, Woodbridge Press, Santa Barbara, CA.

Richardson, C.J., 1994, Ecological functions and human values in wetlands: A framework for assessing forestry impacts, *Wetlands*, 14, 1–9.

Said, R., 1981, *The Geological Evolution of the River Nile*, Springer-Verlag, Heidelberg, Germany.

Savanick, G.A. and Krawza, W.G., 1981, Water jet perforation, a new method for completing and stimulating in situ leaching wells, Report of Investigations 8569, U.S. Department of the Interior, Washington, DC.

Sawyer, R.D., Translator, 1994, *Sun Tsu's Art of War*, Westview Press, Boulder, CO.

Schultz, J., 2000, A war over water, The Democracy Center On-Line 2/4/00 Syndicated, Pacific News Service, available at: http://www.democracy.org/waterwar/

Schultz, R., 2011, Personal notes, Iowa State University, Ames, IA.

Shu, L. and Finlayson, B., 1993, Flood management on the lower Yellow River: Hydrological and geomorphological perspectives, *Sedimentary Geol.*, 85(1–4), 285–296.

Sridhar, V., 2003, Water war of Bolivia, *Frontline*, India's National Magazine from the Publishers THE HINDU, 20(03), February 1–14, 2003, available at: http://www.flonnet.com/fl2003/stories/20030222140002060000.htm/

Summerville, J.P., 2006, Personal communication: Dutch–France war in 1672, Department of History, University of Wisconsin, Madison, WI.

Ting, J.C., 1989, Conflict analysis of allocating freshwater inflows to bays and estuaries—A case study on Choke Canyon Dam and Reservoir, Texas, Doctoral dissertation, College of Forestry, Stephen F. Austin State University, Nacogdoches, TX.

UNESCO, 1992, Water and health, IHP Humid Tropics Programme Series No. 3, Division of Water Sciences, Paris, France.

U.S. Department of Energy, 2007, Methane hydrate: Future energy within our grasp, available at: http://fossil.energy.gov/programs/oilgas/publications/methane_hydrates/MHydrate_overview_06-2007.pdf

U.S. Fish and Wildlife Service, 2002, National wetlands inventory: A strategy for the 21st century, available at: http://library.fws.gov/Wetlands/21stcentury.pdf/

U.S. National Weather Service, 1975, *Climatological Data Annual Summary–California*, vol. 79, issue 13, Table 2; *Iowa*, vol. 86, issue 13, Table 2; *Nebraska*, vol. 80, issue 13, Table 2, National Climatic Data Center, Asheville, NC.

U.S. Water News, 1996, Israel refuses to give up a drop of water from Golan Heights, *U.S. Water News*, 12, 2.
U.S. Water News, 2003, Water used as military weapon and target during armed conflicts, *U.S. Water News*, 20, 4, 11.
Walker, P.N., 1973, *Punishment—An Illustrated History*, Arco Publishing Co., Inc., New York.
World Resources Institute, the United Nations, and the World Bank, 1996, *World Resources, a Guide to the Global Environment, 1996–1997*, Oxford University Press, Oxford, U.K.

3

Science of Water

Water has been a topic for systematic measurement and scientific study since the seventeenth century. Earlier, humans experienced water with reverence, superstition, folklore, fear, and speculation on the one hand and on the other hand showed many ingenious ways of utilizing water. We shall first overview significant events in water management and development in history and then introduce the disciplines of water science.

3.1 Water in History

The history of water resources management is an integral part of the history of civilization (Bennett, 1939; Frank, 1955). The floodplains along China's Yellow River, Egypt's Nile River, India's Ganges, and Babylon's Tigris and Euphrates cradled the four oldest civilizations of the world. Yet history has taught us how the depletion of soil and water resources caused by wars among nations and tribes has caused the fall of cities, countries, and even civilizations. An ancient Chinese proverb, "To rule the mountains is to rule the waters," reveals the general principle learned from these events—that upstream watershed management is necessary for control of downstream flooding.

3.1.1 Asia

Early civilizations worked strenuously and continuously to reduce flooding at higher latitudes, where precipitation is greater than evaporation, and to develop and conserve water resources at lower latitudes, where evaporation exceeds precipitation. In China, Emperor Yau (about 2880 BC) put a man named Yu in charge of flood control in north China. Because of Yu's success on taming the Yellow River with dams, dikes, diversion ditches, and other drainage works, he was chosen to succeed Emperor Shun as the ruler of China and is known as "Yu the Great."

From old documents and recent archeological evidence, we know that the ancient Chinese had the skills for digging wells for drinking water as early as 6000 or 7000 years ago (Young, 1985). A well excavated at Ho-Moo-Du, Yee-Yau in the Tse-Kiang Province of China is believed to have been built during the New Stone Ages. The well was cased by four rows of logs with a squared frame attached to the logs at the top of the well. Also excavated were more than 60 tile wells southwest of Beijing that are believed to have been built around 600 BC for drinking and irrigation. Figure 3.1 is a mechanical water wheel used during the Tang dynasty (AD 618–907) for agricultural irrigation and domestic water supply.

In central China, the Tukianguien irrigation system, built by Lee Bien about 2200 years ago, was an ingenious multipurpose water resources project (Wang, 1983). The project diverted the flow of the Ming River, a tumultuous stream emerging from the Tibetan

FIGURE 3.1
Ancient hydraulic wheel and a pedal irrigator from the Tang dynasty (AD 618–907).

plateau, through a series of dams and dikes on the main river where it first enters the broad plain from a mountain canyon. It employed bamboo frames weighted down by rocks to dam or divert the water, a spillway to adjust flow volume in the river, and periodic dredging activity when sediment built to a certain level, as indicated by stone marks in the channel. The project irrigated about 200,000 ha of fertile soils, reducing greatly the heavy toll on life and property resulting from spring and summer floods. As a result, the Ming River basin became the most productive region of China. In turn, the river provided the necessary resources to enable Chin Shihuang (the "First Chin Emperor") to overthrow the Chow dynasty and its feudal lords, making China a unified country. Even today, the system is still the major agricultural water supply in the region.

Another great water project in Asia is the Grand Canal of China, the longest waterway system of the world. The canal integrates the peoples and economies of the north, south, inland, and coastal areas into a single political–economic entity. It was the economic artery and a political arm over six major dynasties, from the Sui to the Ching. It had two major periods of development. The first period began in AD 584 during the Sui dynasty, connecting and expanding upon preexisting waterways to form a two-branch water system. One branch extended northeasterly from the Yellow River at Luoyang, the eastern capital in central China, to today's Beijing, about 800 km in length. The other branch extended southeasterly from Luoyang through the Huai River basin to the Yangtze River region, also about 800 km. The second period of development started in AD 1283 during the Yuan Dynasty, stretching the canal from the capital city Beijing in the north to Hangzhou in the south with a total length of about 1100 km (Chen et al., 1992). Variations in climate, geology, topography, and elevation between the north and south are great, and a few major rivers cross all the regions. The development and construction of the canal system reflects the ingenuity and technological skill of the Chinese.

Lo Lan was a small country located on the western bank of Lop Nor (a salt lake now dried), Sinkiang between 300 BC and AD 400. It was a trading center in the middle of the silk routes between China and countries in the Near East during the Han dynasty period. The area was covered with beautiful forests, fertile soils, and thick pastures nourished by flowing rivers and springs. However, the country disappeared after AD 400, and the area is dry and barren today, leaving a series of questions. Recent archaeological work revealed that Lo Lan had a full set of laws regarding taxes, water conservancy, lands, hunting, forest protection, criminals, and property. The forest-protection law set the fine for eradicating a living tree at one horse and for chopping a tree's limbs at one head of cattle. The importance of forests for the control of soil erosion, sand-dune stabilization, and protection of stream siltation was well recognized (Southern Chinese Daily News, 1997). It was probably the earliest law regarding forests and environment in the world.

3.1.2 Middle East

In drier and lower latitudes, the people of Assyria, Babylon, Egypt, and Israel began construction of water supply and drainage systems about 5000 years ago. Egyptians had core drilling in rock for wells as early as 3000 BC. The world's oldest known dam was built by the Egyptians in about 3000 BC to store drinking and irrigation water. Perhaps it also controlled flooding waters. The dam was a rock-fill structure, about 108 m long and with a crest of about 12 m above the riverbed. It failed soon after construction. However, a Roman reservoir in Jordan (Jacob's well) was so well-built that it holds water today (Frank, 1955). Small earth and masonry dams built by the Roman empire were used not only to store runoff water but also to raise the general water level in streams (The United Nations Environmental Program, 1983).

The oldest known streamflow records in the world are the markings of the flood stages of the Nile River carved in cliffs between Semneh and Kumneh around 2000 BC (Boyer, 1964). Mentions of the annual flooding of the Nile date back to between 3000 and 3500 BC. In the lower reach of the Nile River, annual floods deposit rich sediment along the floodplains, making the land agriculturally productive. Thus, the crop yields in the Nile valley are dependent upon the annual flood of the river, and the flood stage in the Nile each year was a criterion for taxation in early Egypt. The Egyptian pharaoh Menes developed a flood-control system for the Nile River more than 3000 years ago that included at least 20 recording stations (Grant, 1992). These stations used a crude staff gage—referred to today as the Nilometer—to measure the water levels, probably the oldest hydrometric in the world.

In and around the third millennium BC, well-planned city drainage and water supply systems, public toilets, and baths were constructed with burnt bricks along the Indus River in Pakistan. Most houses in the capital city Mohenjo Daro, discovered in the early twentieth century, were supplied with water from household wells as deep as 25 m. The average distance between wells was about 35 m in an area of at least 300 ha. Artificial elevation was constructed on the residential areas to survive the threat of the river's annual inundation (Jansen, 1999).

Records on central water supply and wastewater disposal date back about 5000 years to Nippur of Sumeria. The Sanskrit medical lore of India and Egyptian wall inscriptions taught that foul water can be purified by boiling, exposure to sunlight, filtering through charcoal, and cooling in a tile vessel. These teachings were verified by the English philosopher Sir Francis Bacon, who in 1627 published experiments on the purification of water by filtration, boiling, distillation, and clarification by coagulation. Bacon stated that purifying water tends to improve health and increase the pleasure of the eye (Viessman and Hammer, 1985). The chemical treatment of water reportedly started as early as 1500 BC in which the Egyptians used chemical alum to cause suspended particles to settle out of water (EPA, 2000).

3.1.3 Europe

Early Europe made many important contributions to water utilization and management. The collection of water by rain harvesting dates back to prehistoric times in Europe. Designs have been found in the ruins of the palace of Knossos (1700 BC), the center of Minoan Crete. Small courts were built between the western and eastern wings of the palace. Besides providing light for the lower floors, these courts collected rainwater, which was drained out for ritual activities and air conditioning. Rainwater also was collected from the roof, channeled to small cisterns, and stored for various purposes (The United Nations Environmental Program, 1983).

Waterwheels for irrigation, water supply, and milling corn were used in ancient Greece and Rome. However, due to the scarcity of cheap slave and animal labor, the application of water power did not become widespread until the twelfth century.

Concerns about water and the environment were documented in the thirteenth century. Louis VI of France promulgated "The Decree of Waters and Forests" in 1215, considered to be the earliest written document in the West concerning the relation between waters and forests (Kittredge, 1948). Before the French revolution of 1789, Bernardin de Saint Pieer (1737–1814) published a book, *The Studies of Nature*, describing the positive impact of forests on rain and streamflow in Mauritius (Andréassian, 2004). Analysis of such a relationship was addressed again in the 1830s, when systematic streamflow measurements became available. Dr. Heinrich Berghaus, a German hydrographer, found that the streamflows in the Elbe and Oder rivers gradually decreased from 1778 to 1835. He attributed the decrease of streamflows to the destruction of forests, cultivation of the soil, and the draining of swamps. In 1873, an Austrian hydrographer, Gustave Wex, attributed the diminishment of water in the Rhine, Elbe, Oder, Vistula, and Danube rivers to the decrease in precipitation, which, in turn, was due to forest destruction (Zon, 1927). Wex's works generated a great interest in the relation between waters and forests in Europe in the nineteenth century and were well accepted by many prominent scientists, although his interpretations are no longer considered acceptable in the light of modern hydrology.

In 1837, a special commission set up to investigate the causes of decreasing river discharge in Russia concluded that it was due to deforestation and expanding agricultural acreage. Later, G. Wex blamed the severe droughts occurring through most of Europe in 1870 on the effects of expanding agriculture (Molchanov, 1963). In 1860, France began a reforestation program in a region where some 320,000 ha of farmland had been seriously damaged as a result of clearcutting on the headwaters of streams. The planting program resulted in complete control of damage from 163 torrents, 31 of which were considered hopeless about half a century earlier. It was then concluded that forest cover is one of the most effective means of erosion control and that the best place to control streamflow is at the headwaters of streams.

A German scientist, Krutsch, started the first investigation of the effects of forest upon precipitation under a pine stand in 1863. Over a period of 16 months, he found that throughfall was merely 9% when rainfall was very light (not over 0.5 mm), and it might amount to 80%–90% in the case of strong showers (Molchanov, 1963; Friedrich, 1967).

There was also a great interest in the forest and floods relationship in the last part of the nineteenth century in Europe. The first scientific watershed research on the effects of forest upon streamflow regularity was initiated by the Swiss Central Experimental Station in 1890 (Zon, 1927). Two small watersheds of similar topography, geological formation, soil, and latitude, but different in forest cover (one was 98% forested while the other was 30%), were selected for the study. Rainfall, snowfall, runoff, and temperature data for these two watersheds were collected carefully using recording or nonrecording gages. The preliminary results based on 11 years of data showed the following:

1. The deforested watershed carried 30%–50% more water per unit of area during the high flood periods.
2. The low flows were higher in the forested watershed.
3. Total annual discharges of the two watersheds were about the same. The stream-flow discharge in the forested watershed was more uniform than that of the deforested watershed.

3.1.4 The United States

America inherited many of the European concepts about water and forests. Many conservationists claimed that deforestation caused floods in wet seasons and made streams and creeks drier in the summer. Conversely, the existence of forests could prevent floods because tree–root systems held soil in place, and the soil stored moisture. Thus, streamflow benefited in the dry seasons from the slow release of stored water and from the increase of precipitation in the forested area.

The forest–streamflow hypothesis and consequent conservation movement inevitably generated disputes, conflicts, and controversies in the United States (Walker, 1983; Pisani, 1992). The disputes extended from the nineteenth century to the early twentieth century. Hiram M. Chittenden, an officer in the U.S. Army Corps of Engineers, presented a paper titled "Forests and reservoirs in their relation to streamflow with particular reference to navigable streams" at the 1908 annual meeting of the American Society of Civil Engineers. He stated that forests did not prevent floods and that the main benefit of forests was to protect soil from erosion. Willis L. Moore, chief of the Weather Bureau, issued a pamphlet in 1910 entitled "A report on the influence of forests on climate and on floods," which aimed to show that forests were an insignificant factor in floods. This opposition was forcefully attacked by the chief of the U.S. Forest Service, Gifford Pinchot, who stated in 1910, "The connection between forests and rivers is like that between father and son. No forests, no rivers" (Sartz, 1983). The heated debate and nationwide campaign of the Forest Service led to the passage of the Weeks Act in 1911 (Douglas and Hoover, 1987). The act authorized federal purchases of forestlands in the headwaters of navigable streams and federal matching funds for approved state agencies to protect forested watersheds of navigable streams.

Although there was no agreement with respect to the relationship between forests and floods, many realized that the deficiencies in scientific information and experimental data for justification needed to be addressed. The first experimental watershed project of the United States, similar to that of the Swiss study, was established in 1910 at Wagon Wheel

Gap, headwaters of the Rio Grande in Colorado, and continued until 1926. The project, a joint effort between the Forest Service and the Weather Bureau, was designed to compare streamflows from a denuded watershed and an undisturbed, forested watershed. Similar watershed experiments included the San Gabriel River basin in Southern California (1917); San Dimas, California (1935); White Hollow Watersheds, Tennessee (1934); Coweeta, North Carolina (1933); Fraser Experimental Forest, Colorado (1937); H. J. Andrews Watersheds, Oregon (1948); Hubbard Brook, New Hampshire (1955); Parsons, West Virginia (1955); Oxford, Mississippi (1957); and many others. Today, more than 400 major experimental watersheds are located in 51 different areas throughout the conterminous United States (see Figure 16.1 in Chapter 16).

The concerns of the U.S. Congress about stream-water quality were manifested by the passage of the Rivers and Harbors Act in 1899. The act prohibited the discharge of refuse into waterways or deposits of materials on the bank of any navigable waters. It was the first legislation to include nonpoint sources of water pollution in a federal program. However, it was not until the end of World War II that the concerns became more widespread and a series of laws were passed to regulate national water-quality conditions caused by various land uses and industrial activities. Major water-pollution acts include the Federal Water Pollution Control Act of 1948 (PL80-845) and Amendments of 1956 (PL84-660) and of 1961 (PL87-88), Water Quality Act of 1965 (PL89-234), Clear Water Restoration Act of 1966 (PL89-753), Water Quality Improvement Act of 1970 (PL91-224), Federal Water Pollution Control Act Amendments of 1972 (PL92-500), Clear Water Act of 1977 (PL95-217), and Water Quality Act of 1987.

3.2 Hydrosciences

3.2.1 Hydrology

The scientific study of water, called hydrology, is a branch of earth science. It is concerned with the problems of water on the earth. Such problems may involve water quantity and quality, interrelations between water and the environment, and the impact of human activity on the occurrence, circulation, and distribution of water. Hydrology looks for the causes and effects of these problems, predicts water-related events and problems, and studies the adjustment, management, and operation of water and water resources to benefit society and the environment.

Human interest in water may be as old as our civilization. The source of water in streams and springs was a puzzling problem that was the subject of much speculation and controversy until comparatively recent times. King Solomon, who lived nearly 1000 years before Christ, wrote in the Old Testament that all streams flow into the sea but the sea is never full. The Greek philosophers, such as Thales (about 650 BC), Plato (427–347 BC), and Aristotle (about 384–322 BC), developed ideas that springs and streams are supplied from the ocean, driven into rocks by winds, and elevated in the mountains, by rock pressure, by vacuum produced by the flows of springs, by pressure exerted on the sea, or by the virtue of the heavens.

The French physicist Pierre Perrault (1608–1680) quantitatively demonstrated that rainfall in the Seine River is sufficient to account for discharge by the river, conceiving the concept of the hydrological cycle. Soon after Perrault, the English astronomer Edmund Halley

(1656–1742) made observations on the rate of evaporation. He demonstrated that evaporation from the Mediterranean Sea was ample to supply the quantity of water returned to that sea by rivers (Meinzer, 1942). These scientists unveiled our modern concept of the hydrologic cycle and initiated quantitative studies of modern hydrology.

3.2.2 Disciplines in Hydrology

Water can be found in solid, liquid, and gaseous states at common earth temperatures. The presence of water, changes in water from one state to another, and water translocation and storage serve environmental, biological, and sociological functions. Hydrology embraces such a large and diversified field that no one can study it all. The broad science of hydrology, more properly called hydroscience, is further broken down into disciplines and specifications that can be grouped into bodies of water, land-use conditions, and interdisciplines of water and land.

Bodies of Water	Land-Use Conditions	Interdisciplines
Potamology	Rangeland hydrology	Geomorphology
Limnology	Agriculture hydrology	Paleohydrology
Cryology	Forest hydrology	Engineering hydrology
Oceanography	Urban hydrology	Watershed management
Glaciology	Wetland hydrology	Hydrobiology
Hydrometeorology	Desert hydrology	
Hydrogeology		
Hydrometry		

3.2.2.1 On Bodies of Water

Disciplines focusing on the body of waters include the following:

Potamology. The study of surface streams may emphasize stream dynamics and morphology, the fluvial processes (Knighton, 1984; Rosgen, 1996), hydraulic characteristics, transport capacity (Morisawa, 1968, 1985), or physical habitat, management, and classification (Gordon et al., 1993; Petts and Amoros, 1996). Streams have to be controlled in times of flood; kept open for transportation; regulated for agricultural, industrial, and municipal usage; elevated for electrical energy; and managed for environmental integrity, leisure enjoyment, and biological habitats.

Limnology. The study of life and phenomena of lakes and ponds—of the functional relationships and productivity of freshwater communities as related to their physical, chemical, and biotic environment—is called limnology (Lampert and Sommer, 1997). Study topics may include physical, chemical, and biological properties of lakes; element cycles; the distribution, origins, and forms of lakes; biotic communities such as phytoplankton and zooplankton, fish, and benthic animals; the ontogeny (successional development) of lake ecosystems; the structure and productivity of aquatic ecosystems; and interrelations between water quality and biotic communities (Kalff, 2003; Wetzel and Likens, 2010).

Cryology. Also called snow hydrology, cryology is the study of snow and ice. Topics of study may include the occurrence and distribution of snow; measurements, physics, and properties of snow cover, snowmelt, and runoff; snow and ice on lakes;

avalanches; snow on buildings, highways, and airports; snowpack management; and recreation (DeWalle and Rango, 2008; Singh, 2010).

Oceanography. The study of oceans and their phenomena is called oceanography. About 97% of the total waters of the earth are confined in the oceans, which in turn cover about 71% of the earth's surface area. Oceans provide ecological niches for more than 250,000 marine plant and animal species, which serve as foods for humans and other organisms. They are also a source of salts and minerals for human utilization; a sink for industrial, domestic, and instream disposals; a storage for solar-energy dissipation; and a medium for moderating the earth's climate. Oceanography can be divided into three branches: physical, chemical, and biological oceanography (Longhurst, 1998). Physical oceanography deals with the physical properties and processes of seas and oceans and is of the most concern to hydrologists. Topics in this study include properties of sea water, ocean currents, waves and tides, sea levels, temperature variation and distribution, thermal interactions between the ocean and the atmosphere, topography of the ocean floor, sediment deposition, ocean precipitation and evaporation, and the drift of icebergs (Knauss, 2005; Trujillo and Thruman, 2010).

Glaciology. A glacier is a body of ice originating on land by the recrystallization of snow or other solid precipitation; it presents a slow transfer of mass by creeping from region to region. The study of ice and glaciers in all aspects is termed glaciology. The scope may include formation of ice in the ground, glacier classification, runoff, movement, climatic effects and changes, and hydrologic problems related to glaciers (Hubbard and Glasser, 2005; Singh, 2010).

Hydrometeorology. Meteorology deals with the movement of water in the atmosphere, while hydrology is concerned with the distribution and occurrence of water on and under the earth's surface. The hydrologic cycle is a concern common to both sciences. Thus, the application of meteorology to hydrological problems such as developing water resources and flood control is called hydrometeorology (or hyetology, precipitation hydrology). Major topics may include the estimation of probable maximum precipitation, storm models, temporal and spatial variations of storms, storm transposition, water resource planning, design and development, river forecasting, and accuracy and representativeness of precipitation measurements (Upadhyay, 1996; Rakhecha and Singh, 2009; Sene, 2009).

Hydrogeology. More than 22% of all freshwaters on Earth are confined under the ground, while surface water in lakes and rivers makes up only about 0.36%. Currently, about 80% of water withdrawals in the United States come from streams and lakes. Groundwater is then logically seen as a major resource for easing water-shortage problems. In fact, groundwater is even more desirable than surface water due to (1) a lack of pathogenic organisms in general, (2) constancy of temperature and chemical composition, (3) absence of turbidity and color, (4) no effects of short droughts on supplies, and (5) difficulty of radiochemical and biological contamination. However, groundwater development may be difficult in some areas due to its costs, low permeability, land subsidence problems, and great content in dissolved solids. The science that studies groundwater occurrence, distribution, and movement is called hydrogeology or groundwater hydrology. The exploration of groundwater, the effects of geological environment on groundwater chemistry and mode of migration, and groundwater contamination are subjects of interest to hydrogeologists (Hiscock, 2005; Domenico and Schwartz, 2008).

Hydrometry. The science of water measurements is called hydrometry. More than 3000 years ago the Egyptian pharaoh Menes developed the first staff gage, now called the Nilometer, to monitor the water level of the Nile River (Stevens, 1987). The Nilometer was marked in graduations on walls to obtain a visual reading of the water level on site (Kolupaila, 1960). Today, water is measured in terms of water level, velocity, discharge, and depth, and it can be measured manually, semiautomatically, and automatically with radio and satellite communication systems and computer data storage and manipulation capability (USGS, 1985; Herschy, 1999; Boiten, 2008).

3.2.2.2 On Land-Use Conditions

Disciplines concerned with land-surface conditions include

Rangeland hydrology. Land on Earth may be classified into five categories: nonproductive land (about 15%), forestland (30%), rangeland (40%), cropland (10%), and urban-industrial land (5%). Rangelands are natural grasslands, savannas, shrublands, most deserts, tundra, alpine communities, coastal marshes, and wet meadows. Rangelands are often intermingled with other types of land and are distributed from sea level to above the timberline. These lands are more suitable for management by ecological principles than for management by economic principles. Rangeland vegetation is predominantly short, consisting of broad-leafed plants such as grasses, forbs, and shrubs. Rangeland provides a variety of natural resources including forage, livestock, fish and wildlife, minerals, recreation, and water beneficial to people in both tangible and intangible aspects. The study of hydrologic principles as applied to rangeland ecosystems is called rangeland hydrology. Topics associated with rangeland hydrology may include vegetation management in relation to water loss and conservation; grazing impact on surface runoff, soil erosion, stream sedimentation, and water quality; snowpack management; and water harvesting (Branson et al., 1981; American Society of Agricultural Engineers, 1988).

Agricultural hydrology. Water use for agricultural irrigation and raising livestock is the greatest single use in the United States, accounting for 41.8% of the total water utilized in 1985; thermoelectric, industrial, and domestic uses of water accounted for 38.7%, 9.1%, and 10.4%, respectively (Paulson et al., 1988). The study of application of hydrologic principles to agricultural development, production, and management is called agricultural hydrology (agrohydrology). Land drainage, irrigation, water harvesting, water conservation, soil erosion and sedimentation, water quantity, and quality of surface and groundwater are some major topics and concerns in agricultural hydrology (van Hoorn, 1988).

Forest hydrology. With respect to its height, density, and thickness of crown canopy, fluffy forest floor, spread root system, and wide horizontal distribution and vertical coverage, forest is the most distinguished type of vegetation on the earth. Generally, forests prosper in regions where the minimum net radiation is 20,000 langley/year and the minimum precipitation is 500 mm/year. Those areas are the major sources of our surface water for domestic and industrial use. In the United States, for example, forests occupy about 30% of the total territory, yet this 30% of land area produces about 60% of total surface runoff. Harvesting and other activities in the forest area will inevitably disturb forest canopies and floors, consequently affecting the quantity, quality, and timing of water resources. The geographical significance of forestland, along with the uniqueness of forest ecosystems, makes the study of water in forested areas a specialized scientific field. The study of water in forest areas was largely covered under the discipline of forest influences in the first half of this century. All effects of natural vegetation on climate, water, and soil are under the scope of forest influences. By the middle of the century, Kittredge (1948) suggested that the water phases of forest influences be called forest hydrology. Forest hydrology is a study of hydrology in forestland. It is concerned with forest and forest activity in relation to all phases of water— an interdisciplinary science that brings forest and hydrology together. Thus, all the influences of forest cover along with forest management and activity on precipitation, streamflow, evapotranspiration, soil water, floods, drought, soil erosion, stream sediment, nutrient losses, and water quality are within the scope of forest hydrology. Forest hydrology provides the basic knowledge, principles, scientific evidence, and justification for managing water resources in forested watersheds. Topics in forest hydrology are discussed in detail by Penman (1963), Monke (1971), Lee (1980), and Black (1996). Studies extending issues in forest–water relations to plant–water relations may be called "eco-hydrology" (Baird and Wilby, 1999) or "environmental hydrology" (Ward and Trimble, 2004).

Urban hydrology. The physical environment of urban and industrial areas is completely different from that of forests, agricultural lands, and rangeland. Many observations and studies have shown greater precipitation in and around major urban areas than in the surrounding countryside due to the enormous number of condensation nuclei produced by human activities and atmospheric instability associated with the heat island generated by the city (Sayok and Chang, 1991). The impervious surfaces created by urban development may cause an increase in flooding, soil erosion, stream sedimentation, and pollution of land and water bodies (Lazaro, 1990; Akan and Houghtalen, 2003). The study and analysis of hydrology in urban areas and the hydrologic problems associated with urbanization fall within the scope of urban hydrology (Bedient et al., 2007).

Wetland hydrology. Wetlands are areas where the water table is at or near the surface of the land for at least a consecutive period of time or is covered by shallow water up to 2 m (6 ft) deep. The abundant water in the soils makes soil properties significantly different from those in the uplands, and the soil is suitable for growth of certain plant species. Five major wetland systems have been recognized: marine, estuarine, lacustrine, riverine, and palustrine. Marine and estuarine systems are coastal wetlands, such as tidal marshes and mangrove swamps. Lacustrine and riverine wetlands are associated with lakes and rivers, respectively. The last system includes marshes, swamps, and bogs (Niering, 1987). Thus, wetlands are transitional zones between aquatic and terrestrial ecosystems, all rich in plants and animals. Water is the primary factor that controls the environment, plants, animals, and soils. Wetlands are no longer considered wastelands as they were thought to be in the past. They are not only habitats of a variety of plant and animal species but also important natural flood-control mechanisms, nature's water-purification plants, nutrient and food suppliers to aquatic organisms, a major contributor to groundwater recharge, and a buffer zone for shoreline erosion (Chang, 1987). Wetland hydrology studies vegetation and flooding, hydrologic characteristics of wetlands, streamflow, and sediments. It examines the impact of development projects on wetland ecosystems, groundwater fluctuation, water quality, nutrient removal and transformation, wetland construction and restoration, wetland management, soil characteristics, wetlands delineation and classification, erosion control, etc. (Marble, 1992; Lyon, 1993; U.S. Environmental Protection Agency, 1993; Mitsch and Gosselink, 2000).

Desert hydrology. Although arid and semiarid lands cover about 30% of the world's land area, only about 15% of the world's population lives in these regions. However, the regions are of strategic and economic importance because they contain over one-half of the precious and semiprecious minerals and most of the oil and natural gas in the world. Here rainfall is little, highly variable, and infrequent. Coupling rainfall scarcity with great rates of soil and water evaporation makes water deficiency a general phenomenon and characteristic across the region. A vast area depends on springs and groundwater for its water supply. Desert hydrology studies the characteristics and processes of the hydrology cycle in arid and semiarid areas, springs, and groundwater resources. It also examines water conservation and harvesting, the adaptation of plants to the environment, hydro-climate changes, desertification, and water quality (Beaumont, 1993; Shahin, 1996).

3.2.2.3 On Interdisciplinary Studies

Interdisciplinary areas of study include:

Geomorphology. Geomorphology is concerned with landforms and drainage characteristics created by running water and other physical processes. The study of drainage basins and channel networks was largely qualitative and deductive prior to the 1950s. Dr. Arthur N. Strahler and his Columbia University associates have made great contributions to transform the science from descriptive into quantitative study. Geomorphology provides an invaluable basis for assessing the potential of land for development, for land-use planning, and for environmental management. Since the behavior of streamflow is highly affected by watershed topographic characteristics, the quantitative description of drainage basins and channel networks enables hydrologists to evaluate the spatial variations of streamflow, precipitation, temperature, snow distribution, and other hydroclimatic variables, and consequently to produce hydrologic models and simulations (Leopold, 2006; Gregory, 2010; Huggett, 2011).

Paleohydrology. Many hydraulic structures are designed to meet climatic and hydrologic conditions 50–100 years in the future. The estimation of these future events is based on data collected in the past. The accuracy of these estimates is affected greatly by data availability and representativeness of the data observations. If a short period of hydrologic data recorded in a relatively dry period were used in the analysis, the designed flood for a structure might well be too small to accommodate large floods in wet periods. The physical and financial success of such a project might be seriously jeopardized. Paleohydrology is the study of hydrologic conditions in ancient times. It may be useful in understanding the changes in precipitation, temperature, stream levels, and water balance from the past to modern times, defining the physical laws that govern the fluctuations of hydrologic conditions, and studying environmental change on continents (Gregory et al., 1996; Benito et al., 1998). It is related to paleoclimatology (Cronin, 1999). Techniques used may include investigations of tree rings (dendrochronology), glacial fluctuations, fossils, pollen deposition in bogs (palynology), sediment cores in deep seas and lakes, and other geological evidence.

Engineering hydrology. Hydrological information is essential in the design, operation, and management of flood-control works, irrigation systems, water-supply projects, storm-runoff drainage, erosion controls, highway culverts, and many other hydraulic structures. Engineering hydrology is the study of hydrologic characteristics of a watershed or a drainage system required to solve these engineering problems. The study may include frequency analysis of hydrologic events, depth–area–duration analysis, estimation of evapotranspiration, hydrograph analysis, streamflow routing, rain–runoff relations, streamflow simulation, estimation of soil erosion and stream sedimentation, and risk analysis (Linsley et al., 1982; Ponce, 1994; Soliman, 2010).

Watershed management. All land enclosed by a continuous hydrologic drainage divide and lying up-slope from a specific section on a stream is a watershed. Thus, every piece of land belongs to one watershed or another, depending on the reference section in question, and in any watershed all surface water is drained out through that particular section in the channel. The watershed, also called a catchment or drainage basin, is the unit of land area that the hydrologist and watershed manager consider for study or management. It is analogous to the silviculturist's stand, or the forest manager's compartment. However, a stand or compartment may be more-or-less an artificial unit, while watersheds have natural boundaries.

Watersheds have been employed as social and economic units for community development and conservation of natural resources including water, soils, forests, wildlife, and others. Because of the steadily increasing demands for water in our modern society and the fact that much of the water for agriculture, industry, recreation, and domestic use has its source in forested land, watershed management has become increasingly important to foresters. Thus, watershed management is management of all natural resources including forests, land, wildlife, recreation, and minerals within a watershed for the protection and production of water resources while maintaining environmental stability.

Accordingly, watershed management is water oriented; it is primarily concerned with water resources and related problems. Since soil erosion may greatly affect water quality, flood damage, land deterioration, environmental aesthetics, and many other factors, the soil stabilization and prevention of soil erosion is also one of the most important challenges in protecting water resources, an area that is generally covered under the discipline of soil conservation. Soil conservation is soil oriented, and ignorance of water movement causes severe soil erosion and environmental problems. Thus watershed management and soil conservation share the same concerns in some areas. In reality, watershed management involves much more complex problems and broader tasks than soil conservation does. It deals with water as well as soil problems, along with land planning and resource management activities in upstream forested land, while soil conservation deals with soil problems relating to land productivity in downstream agricultural regions. In a sense, watershed management is integrated resources management for the production and protection of watershed water resources and should properly be called "integrated watershed management."

Watershed management is based on principles in forest hydrology and the knowledge in general hydrology, meteorology, climatology, ecology, soils and geology, engineering, land planning, environmental regulation, and social science. It is essential that the atmospheric, plant, soil, and water systems of the watershed—and frequently the impact of human activity on the complex system—be considered simultaneously. Ignorance of any one variable may make watershed projects costly and ineffective. Detailed discussions of watershed management are given by Satterlund and Adams (1992), Ffolliott et al. (2003), DeBarry (2004), and Heathcote (2009).

Hydrobiology. Aquatic plants can be grouped into six categories based upon size, shape, and growth habits: plankton algae, filamentous algae, submersed weeds, emerged weeds, marginal weeds, and floating weeds. Extensive infestations of these plants create problems in water for recreation, irrigation, flood control, and navigation; increase water loss due to transpiration; increase organism-transmitted diseases; and affect physical properties. Topics such as the characteristics of aquatic plants, plant distribution, plant growth and development, aquatic plants and environment, and control and management of these plants are interesting to hydrobiologists (Gangstad, 1986; Riemer, 1993; Caffrey, 2000).

Other disciplines, such as medical hydrology (water in relation to human health) and hydroecology (water and its environmental and physiological functions), also have been mentioned in the reports and studies.

References

Akan, A.O. and Houghtalen, R.J., 2003, *Urban Hydrology, Hydraulic, and Stormwater Quality: Engineering Applications and Computer Modeling*, John Wiley & Sons, New York.

American Society of Agricultural Engineers, 1988, Modeling agricultural, forest, and rangeland hydrology, *Proceedings of the 1988 International Symposium*, December 12–13, 1988, ASAE Publ. 07-88, St. Joseph, MI.

Andréassian, V., 2004, Waters and forests: From historical controversy to scientific debate, *J. Hydrol.*, 29, 1–27.

Baird, A.J. and Wilby, R.L., 1999, *Eco-Hydrology*, Routledge, New York.

Beaumont, P., 1993, *Drylands: Environmental Management and Development*, Routledge, New York.

Bedient, P.B., Huber, W.C., and Vieux, B.E., 2007, *Hydrology and Floodplain Analysis*, 4th edn., Prentice Hall, New York.

Benito, G., Baker, V.R., and Gregory, K.J., Eds., 1998, *Paleohydrology and Environmental Change*, John Wiley & Sons, New York.

Bennett, H.H., 1939, *Soil Conservation*, McGraw-Hill, New York.

Black, P.E., 1996, *Watershed Hydrology*, 2nd edn., Ann Arbor Press, Chelsea, MI.

Boiten, W., 2008, *Hydrometry: A Comprehensive Introduction to the Measurement of Flow in Open Channels*, 3rd edn., Taylor & Francis Group, Boca Raton, FL.

Boyer, M.C., 1964, Streamflow measurement, *Handbook of Applied Hydrology*, Chow, V.T., Ed., McGraw-Hill, New York.

Branson, F.A. et al., 1981, *Rangeland Hydrology*, Kendall/Hunt, Dubuque, IA.

Caffrey, J.M., Ed., 2000, *Biology, Ecology and Management of Aquatic Plants*, Kluwer, Dordrecht, the Netherlands.

Chang, M., 1987, Monitoring bottomland hardwoods in response to hydrologic changes below dams, *Bottomland Hardwoods in Texas*, McMahan, C.A. and Frye, R.G., Eds., Texas Parks and Wildlife Department, San Angelo, TX, pp. 162–164.

Chen, M.T. et al., 1992, *Brief Histories of the Chinese Sciences and Technology*, Ming-Wen Book Co., Taipei, Taiwan (in Chinese).

Cronin, T.M., 1999, *Principles of Paleoclimatology*, Columbia University Press, Irvington, NY.

DeBarry, P.A., 2004, *Watersheds: Processes, Assessment, and Management*, John Wiley & Sons, New York.

DeWalle, D.R. and Rango, A., 2008, *Principles of Snow Hydrology*, Cambridge University Press, Cambridge, U.K.

Domenico, P.A. and Schwartz, F.W., 2008, *Physical and Chemical Hydrogeology*, 2nd edn., Wiley, New York.

Douglas, J.E. and Hoover, M.J., 1987, History of Coweeta, *Forest Hydrology and Ecology at Coweeta*, Swank, W.T. and Crossley, D.A., Jr., Eds., Springer-Verlag, Heidelberg, Germany.

EPA, 2000, The history of drinking water treatment, U.S. Environmental Protection Agency, EPA-816-F-00-006, Washington, DC.

Ffolliott, P.F. et al., 2003, *Hydrology and the Management of Watersheds*, 3rd edn., Iowa State University Press, Ames, IA.

Frank, B., 1955, The story of water as the story of man, *Water, the Yearbook of Agriculture*, USDA, Washington, DC.

Friedrich, W., 1967, Forest hydrology research in Germany, *International Symposium on Forest Hydrology*, Sopper, W.E. and Lull, H.W., Eds., Pergamon Press, Elmsford, NY, pp. 45–47.

Gangstad, E.O., 1986, *Freshwater Vegetation Management*, Thomas Publications, Fresno, CA.

Gordon, N.D., McMahon, T.A., and Finlayson, B.L., 1993, *Stream Hydrology*, John Wiley & Sons, New York.

Grant, D.M., 1992, *ISCO Open Channel Flow Measurement Handbook*, 3rd edn., ISCO, Inc., Lincoln, NE.

Gregory, K.J., 2010, *The Earth's Land Surface: Land forms and Processes in Geomorphology*, Sage Publications Ltd, London, U.K.

Gregory, K.J., Starkel, L., and Baker, V.R., Eds., 1996, *Global Continental Paleohydrology*, John Wiley & Sons, New York.

Heathcote, I.W., 2009, *Integrated Watershed Management: Principles and Practices*, 2nd edn., John Wiley & Sons, New York.

Herschy, R.W., Ed., 1999, *Hydrometry: Principles and Practices*, 2nd edn., John Wiley & Sons, New York.

Hiscock, K.M., 2005, *Hydrogeology: Principles and Practices*, Wiley-Blackwell, New York.

Hubbard, B. and Glasser, N.F., 2005, *Field Techniques in Glaciology and Glacial Geomorphology*, John Wiley & Sons, New York.

Huggett, R.J., 2011, *Fundamentals of Geomorphology*, 3rd edn., Routledge, New York.

Jansen, M.R.N., 1999, Mohenjo Daro and the River Indus, *The Indus River—Biodiversity, Resources, and Humankind*, Meadows, A. and Meadows, P.S., Eds., Oxford University Press, Oxford, U.K.

Kalff, J., 2003, *Limnology*, 2nd edn., Prentice Hall, Upper Saddle River, NJ.

Kittredge, J., 1948, *Forest Influences*, McGraw-Hill, New York.

Knauss, J.A., 2005, *Introduction to Physical Oceanography*, Waveland Press, Prospect Heights, IL.

Knighton, D., 1984, *Fluvial Forms and Processes*, Edward Arnold, London, U.K.

Kolupaila, S., 1960, Early history of hydrometry in the United States, *Proc. Am. Soc. Civil Eng., J. Hydraul. Div.*, 86, 1–50.

Lampert, W. and Sommer, U., 1997, *Limnoecology: The Ecology of Lakes and Streams*, Oxford University Press, Oxford, U.K.

Lazaro, T.R., 1990, *Urban Hydrology—A Multidisciplinary Perspective*, Technomic, Lancaster, PA.

Lee, R., 1980, *Forest Hydrology*, Columbia University Press, New York.

Leopold, L.B., 2006, *A View of River*, Harvard University Press, Cambridge, MA.

Leupold & Stevens, Inc., 1987, *Stevens Water Resources Data Book*, 4th edn, Leupold & Stevens, Inc., Beaverton, OR.

Linsley, R.K., Jr., Kohler, M.A., and Paulhus, J.L.H., 1982, *Hydrology for Engineers*, 3rd edn., McGraw-Hill, New York.

Longhurst, A.R., 1998, *Ecological Geography of the Sea*, Academic Press, New York.

Lyon, J.G., 1993, *Practical Handbook for Wetland Identification and Delineation*, Lewis Publishers, Boca Raton, FL.

Marble, A.D., 1992, *A Guide to Wetland Functional Design*, Lewis Publishers, Boca Raton, FL.

Meinzer, O.E., 1942, *Hydrology*, Dover, New York.

Mitsch, W.J. and Gosselink, J.G., 2000, *Wetlands*, 3rd edn., John Wiley & Sons, New York.

Molchanov, A.A., 1963, *The Hydrological Role of Forests*, Israel Program for Scientific Translations, U.S. Department of Agriculture, Washington, DC.

Monke, E.J., Ed., 1971, Biological effects in the hydrological cycle, *Proceedings of the Third International Seminar for Hydrology Professors*, Department of Agricultural Engineering, Purdue University, West Lafayette, IN.

Morisawa, M., 1968, *Streams, Their Dynamics and Morphology*, McGraw-Hill, New York.

Morisawa, M., 1985, *Rivers*, Longman, New York.

Niering, W.A., 1987, *Wetlands*, Alfred A. Knopf, New York.

Paulson, R.W., Chase, E.B., and Carr, J.E., 1988, Water supply and use in the United States: U.S. geological survey national water summary 1987, *Water-Use Data for Water Resources Management*, Waterstone, M. and Burt, R.J., Eds., American Water Resources Association, TPS-88-2, Bethesda, MD.

Penman, H.L., 1963, *Vegetation and Hydrology*, Commonwealth Agricultural Bureaux, Farnham Royal, Buks, England.

Petts, G.E. and Amoros, C., Eds., 1996, *Fluvial Hydrosystems*, Chapman Hall, London, U.K.

Pisani, D.J., 1992, *To Reclaim a Divided West: Water, Law, and Public Policy, 1848–1902*, University of New Mexico Press, Albuquerque, NM.

Ponce, V.M., 1994, *Engineering Hydrology: Principles and Practices*, Prentice Hall, New York.

Rakhecha, P. and Singh, V.P., 2009, *Applied Hydrometeorology*, Springer, New York.

Riemer, D.N., 1993, *Introduction to Freshwater Vegetation*, Krieger Publishing Co., Melbourne, FL.

Rosgen, D., 1996, *Applied River Morphology*, Wildland Hydrology, Pagosa Springs, CO.

Sartz, R.S., 1983, Watershed management, *Encyclopedia of American Forest and Conservation History*, Vol. 2, Davis, R.C., Ed., MacMillan, New York, pp. 680–584.

Satterlund, D.R. and Adams, P.W., 1992, *Wildland Watershed Management*, John Wiley & Sons, New York.

Sayok, A.K. and Chang, M., 1991, Rainfall in and around the city of Nacogdoches, Texas, *Texas J. Sci.*, 43, 173–178.

Sene, K., 2009, *Hydrometeorology: Forecasting and Applications*, Springer, New York.

Shahin, M., 1996, *Hydrology and Scarcity of Water Resources in Arab Region*, A.A. Balkema, Rotterdam, the Netherlands.

Singh, P., 2010, *Snow and Glacier Hydrology*, Springer, New York.

Soliman, M.M., 2010, *Engineering Hydrology of Arid and Semiarid Regions*, CRC Press, Boca Raton, FL.

Southern Chinese Daily News, 1997, Exploring the secrecy of Lop Nor: Archaeologists revealed the life styles of Lo Lanian (Editorial), *Southern Chinese Daily News*, December 28, 1997 (in Chinese).

Trujillo, A.P. and Thruman, H.V., 2010, *Essentials of Oceanography*, 10th edn., Prentice Hall, New York.

United Nations Environmental Program, 1983, *Rain and Stormwater Harvesting in Rural Areas*, Tycooly International Publishing Ltd., Dublin, Ireland.

Upadhyay, D.S., 1996, *Cold Climate Hydrometeorology*, John Wiley & Sons, New York.

U.S. Environmental Protection Agency, 1993, *Created and Natural Wetlands for Controlling Nonpoint Source Pollution*, Office of Wetlands, Oceans, and Watersheds, Washington, DC.

USGS, 1985, *Techniques of Water-Resources Investigations of the United States Geological Survey*, U.S. Government Printing Office, Alexandria, VA.

van Hoorn, J.W., 1988, Agrohydrology—recent developments, *Proceedings of the Symposium Agrohydrology*, Wageningen, the Netherlands, September 29–October 1, 1987, Elsevier, New York.

Viessman, W., Jr. and Hammer, M.J., 1985, *Water Supply and Pollution Control*, Harper & Row, New York.

Walker, L.C., 1983, Forest influences, *Encyclopedia of American Forest and Conservation History*, Vol. I, Davis, R.C., Ed., Macmillan and Free Press, New York, pp. 215–217.

Wang, P., 1983, A brief history of engineering technology in China, *The History of Science and Technology in China*, Vol. II, Wu, L.C., Ed., Natural Science and Culture Publishing Co., Taipei, Taiwan, pp. 255–262 (in Chinese).

Ward, A.D. and Trimble, S.W., 2004, *Environmental Hydrology*, 2nd edn., Lewis Publishers, Boca Raton, FL.

Wetzel, R.G. and Likens, G.E., 2010, *Limnological Analyses*, 3rd edn., Springer-Verlag, Heidelberg, Germany.

Young, L.S., 1985, How did our ancestors utilize groundwater? *International Daily News*, March 2 (in Chinese).

Zon, R., 1927, *Forests and Water in the Light of Scientific Investigation*, U.S. Forest Service, Washington, DC.

4

Properties of Water

Water is the most vital and ubiquitous substance on Earth. It has been described as a miracle of nature, the mirror of science, and the blood of life. Many properties of water are used as references and standards for properties of other substances. This chapter examines the properties of water from four perspectives: physical, hydraulic, chemical, and biological.

4.1 Physical Properties

4.1.1 Three States of Water

Water has three states: solid, liquid, and gaseous. They all exist at common Earth temperatures. At standard atmosphere (760 mmHg) and at about 0°C, all three states of water are in equilibrium when the partial pressure of water vapor is 4.6 mmHg. Changes from one state to another depend on water temperature and atmospheric pressure. Increases in air pressure will lower the freezing point and raise the boiling point of water. When atmospheric pressure is below 760 mmHg and temperature is lower than 0°C, water can change from the solid state directly into the vapor state, a phenomenon called sublimation (Figure 4.1).

Changes in the state of water do not alter its chemical properties but do affect the cohesion of water molecules and water density (Table 4.1). Water molecules are rigidly bound to one another, but the space between molecules and their rigidity differ among the three states. The arrangement of the molecules in water vapor is much looser than that in the solid or liquid states. Water molecules in the liquid state are the closest among the three states. This makes ice's density less than, and its volume larger than, liquid water's, a property unique among the solid states of substances. Water vapor is a much poorer heat conductor than is liquid water. The thermal conductivity of ice is about 141 times greater than that of water vapor (Table 4.2).

4.1.2 Latent Heat of Water

Any change in the state of water involves a tremendous amount of heat transfer. At 20°C, it takes 586 cal of heat to convert 1 g of water from the liquid state to the vapor state. The heat required is reduced to about 541 cal/g when water temperature is at 100°C, the boiling point. Water temperature does not increase beyond the boiling point even if heat

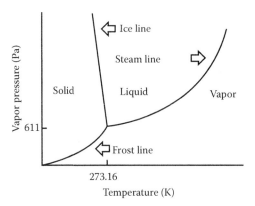

FIGURE 4.1
Triple point of water.

TABLE 4.1

Density (g/cm³) of Water in Solid, Liquid, and Gaseous States at Selected Temperatures

State of Water	Temperatures (°C)					
	0	4	10	20	30	40
Solid	0.91600					
Liquid	0.99984	0.99997	0.99970	0.99821	0.99565	0.99222
Gaseous[a]	4.845×10^{-6}	6.36×10^{-6}	6.94×10^{-6}	17.30×10^{-6}	30.38×10^{-6}	51.19×10^{-6}

[a] Saturation over water.

TABLE 4.2

Specific Heat and Thermal Conductivity of Water Vapor, Liquid Water, and Ice at 0°C

Property	Water Vapor	Liquid Water	Ice
Specific heat (cal/g/°C)	0.44	1.00	0.50
Thermal conductivity (cal/cm/s/°C)	$0.038 \ (10^{-3})$	$1.32 \ (10^{-3})$	$5.35 \ (10^{-3})$

is continually added. The added heat is not retained in the water but is carried off by the vapor through the boiling process. This is the so-called latent heat of vaporization (L_v), which can be computed by the following equation:

$$L_v = 597 - 0.564 \ (T) \tag{4.1}$$

where
  T is water temperature (°C)
  L_v is expressed in calories per gram of water

The L_v decreases at the rate of about 1% per 10°C.

Similarly, the heat required to change 1 g of water from the solid state into the liquid state without change in temperature, referred to as the latent heat of fusion (L_f), is a function of surface temperature T, in degrees Celsius:

$$L_f = 80 + 0.564 \ (T) \tag{4.2}$$

where L_f is expressed in calories per gram. Thus, to melt 1 g of ice at 0°C requires 80 cal, only 13.4% of the latent heat of vaporization at the same temperature. The process of sublimation requires a latent heat equivalent to the sum of L_v and L_f, or 677 cal/g.

Vaporization, sublimation, and fusion represent an energy sink of surfaces because heat is absorbed from the environment to conduct these processes. Conversely, the same amounts of heat will be liberated if water vapor is condensed into dew (a process called condensation), liquid water is frozen into ice (crystallization), or water vapor is directly deposited as frost (frostilization). Therefore, we do not feel cold during the snowfall but do feel cold when the snow is melting.

4.1.3 Saturation Vapor Pressure

Water vapor, like other gases, exerts a partial pressure in the air. This is called the vapor pressure of the atmosphere (e). The maximum amount of water vapor that can be held in the atmosphere is dependent upon air temperature (Figure 4.2). When the amount of water vapor in the air reaches the maximum at a specific temperature, the air is considered saturated and the pressure exerted by the water vapor is called the saturation vapor pressure, e_s. The difference between e_s and e, or ($e_s - e$), is the saturation vapor deficit. For temperatures common in the biosphere, saturation vapor pressure e_s in millibars can be estimated by (Tabata, 1973)

$$\ln e_s = 21.382 - \frac{5347.5}{T_a} \tag{4.3}$$

where
  T_a is air temperature (K)
  ln is the natural logarithm

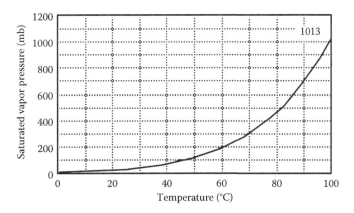

FIGURE 4.2
Saturation vapor pressure as a function of air temperature.

Vapor pressure e can be converted into vapor density ρ_v (or absolute humidity) by

$$\rho_v = \frac{217e}{T_a} \tag{4.4}$$

where
  ρ_v is in grams per cubic meter
  e is in millibars
  T_a is in Kelvin

Vapor density is greater when it is near moist ground, water, or canopy surfaces and decreases with height.

4.1.4 Vapor Diffusion

Water vapor, like other gases, diffuses in the air in response to a gradient in density or partial pressure of water molecules. Under air free of any convective motion and turbulence, the flux of water vapor q_v is proportional to the gradient of vapor density (ρ_v) over the distance z by

$$q_v = -D_v \frac{d\rho_v}{dz} \tag{4.5}$$

where
  q_v is expressed in grams per square centimeter per second
  D_v is the diffusion coefficient of water vapor (cm^2/s)
  z is expressed in centimeters
  the negative sign shows that the diffusion takes place in the direction of decreasing density

The D_v increases with temperature T due to the increase of molecular activity but decreases with the atmospheric pressure P due to the greater frequency of molecular collision by

$$D_v = \frac{210 + 1.5T_a}{P} \tag{4.6}$$

where
  T_a is in degrees Celsius
  P is in millibars

For example, a standard atmosphere (1013 mbar) air parcel at 20°C has a D_v of 0.237 cm^2/s.

4.1.5 Heat Capacity

Heat is a form of energy, a quantity, and a measure of the total kinetic energy of all molecules of a subject. Temperature is an indicator of hot and cold, a scale, and a measure of the mean kinetic energy per molecule of a subject. A reservoir of water may have a lower temperature than the land and still have a greater content of heat because of its greater mass and greater specific heat.

    The heat required to raise the temperature of 1 g of any substance 1°C is called the specific heat of that substance. The amount of heat required to raise the temperature of 1 g of

TABLE 4.3

Some Physical and Hydraulic Properties of Water, Alcohol, and Mercury

Properties	Ethyl Alcohol	Water	Mercury
Density at 20°C (g/cm³)	0.789	0.988	13.55
Volume expansion (10^{-3}/°C)	1.4	0.21	0.18
Latent heat of vaporization (cal/g at 20°C)	220	586	70.4[a]
Specific heat (cal/g/°C)	0.58	1.00	0.033
Surface tension (dyn/cm at 20°C)	22.75	73.05	484
Heat conductivity (cal/s/cm/°C at 20°C)	0.00040	0.00143	0.0198
Viscosity, centipoise at 20°C	1.200	1.002	1.554

[a] At boiling point 356.58°C.

water in 1°C is 1 cal. Thus, the specific heat of water is 1 cal/°C/g, and the specific heat of all other substances is expressed in proportion to that of water (Table 4.3). The heat capacity of a substance is the total heat contained in that substance as determined by its specific heat, temperature, and mass:

$$HC = (c)(T)(v)(\rho) \tag{4.7}$$

where
  c is the specific heat (cal/°C/g)
  T is the temperature (°C)
  v is the volume of the substance (cm³)
  ρ is the density (g/cm³)
  HC is the heat capacity (cal)

Thus, the heat capacity for 1 cm³ of ice ($c = 0.5$ cal/°C/g) with $T = 1$°C is 0.050 cal or

$$HC = (0.50 \text{ cal/°C/g})(1°C)(1 \text{ cm}^3)(0.1 \text{ g/cm}) = 0.050 \text{ cal} \tag{4.8}$$

The heat capacity for 1 cm³ of water at $T = 1$°C is 1 cal, or about 20 times greater than that of ice. Note that 1 cal = 4.19 J = 4.19×10^7 ergs = 1/252 British thermal units (Btu) = 3.09 ft-lb.

The specific heat of seawater decreases with increasing ocean salinity. At a temperature of 17.5°C and a pressure of 1 atm, the specific heat is 0.968 cal/°C/g at 1% salinity, 0.951 cal/°C/g at 2% salinity, and 0.926 cal/°C/g at 4% salinity (List, 1971). For soils, the specific heat ranges from 0.2 to 0.6 cal/°C/g, depending on moisture content, porosity, and temperature.

4.1.6 Thermal Conductivity

The thermal conductivity (k) is defined by

$$H = -k\left(\frac{dT}{dZ}\right) \tag{4.9}$$

where
  H is the rate of heat conduction per unit area
  dT/dZ is the temperature gradient (°C/cm)

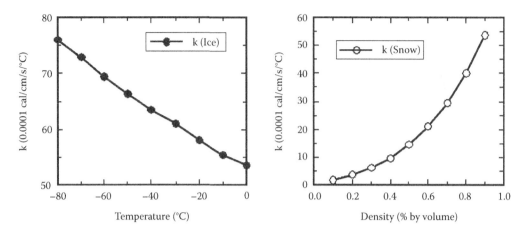

FIGURE 4.3
Thermal conductivity (k) for ice and snow. (From List, R.J., *Smithsonian Meteorological Tables*, 6th rev. edn., Smithsonian Institute Press, Washington, DC, 1971.)

The thermal conductivity (k) increases with increasing temperature. At 20°C and under 1 atm, the k value is 0.00143 cal/cm/s/°C for liquid water and 0.0000614 cal/cm/s/°C for air. In other words, the thermal conductivity of water is about 23 times greater than that of air. The thermal conductivity is even greater for ice. It is 0.00535 cal/cm/s/°C at 0°C and increases to 0.00635 cal/cm/s/°C at −40°C (Figure 4.3).

Snow is a poor heat conductor and consequently a good heat insulator. The average thermal conductivity of fresh snow with a density of 0.1 g/cm³ is about 0.00018 cal/cm/s/°C or about 14% of that for water at 0°C. This makes winter soil temperatures under snow cover less extreme than those in the open.

4.2 Hydraulic Properties

Hydraulic properties of water refer to the behavior of water in the states of motionlessness (hydrostatics) and pure motion (hydrokinetics) and to forces involved in motion (hydrodynamics). The following are some parameters describing the properties of water as a fluid. Other properties of water in various states, surfaces, and conveyors can be found in books dealing with fluid mechanics, hydraulics, or fluid dynamics.

4.2.1 Density

The density of matter is defined as the mass per unit of volume. In equation form, it may be expressed by

$$\rho = \frac{m}{v} \tag{4.10}$$

where
  m is the mass (g)
  v is the volume (cm³)
  ρ is expressed in g/cm³

The maximum density of water is at 4°C ($\rho = 0.99997\,g/cm^3$) and, consequently, dense water sinks to the bottom. This unique property is responsible for thermal stratification of oceans, large rivers, and deep lakes, along with the floating ice at the surface. The density of water in solid, liquid, and gaseous states for a few selected temperatures is given in Table 4.1.

4.2.2 Pressure

Pressure is defined as weight per unit area. Since the weight (W) of an object is the downward force exerted by that object due to gravitation, or

$$W = mg \tag{4.11}$$

where
  g is the acceleration due to gravity
  m is the mass of the object, pressure is more properly defined as the normal force exerted by the object per unit area

For a water column of cross section A and height h, the total pressure is equal to

$$P = \frac{W}{A} = \frac{\rho g h A}{A} = h\rho g \tag{4.12}$$

The total pressure of water exerted on the bottom of a container depends on the height of the water, not on the shape of the container or the quantity of water (Figure 4.4a). In the hydrostatic state, water pressure acts in all directions and is perpendicular to the wall with which water is in contact.

Since the pressure in a fluid depends only on depth, any increase in pressure at the surface of one end must be transmitted to every point in the fluid at the other end. This principle is applied in hydraulic jacks and car lifts to produce a large force with a small force (Figure 4.4b).

The product of ρ and g in Equation 4.12 is called the specific weight of water, and if the equation is rearranged to divide total pressure by the specific weight of water, or

$$\frac{P}{\rho g} = h \tag{4.13}$$

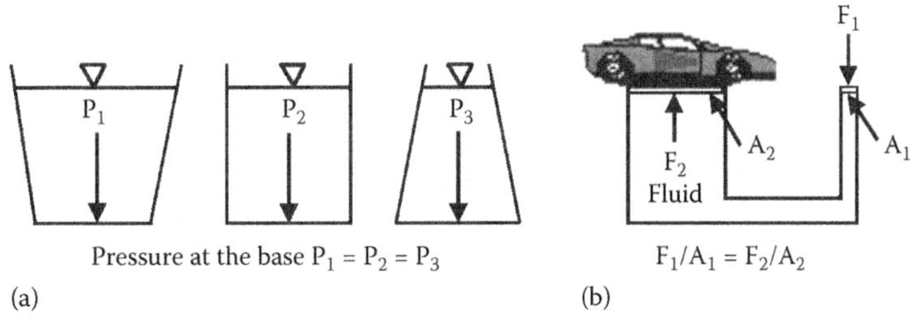

Pressure at the base $P_1 = P_2 = P_3$ $\qquad$ $F_1/A_1 = F_2/A_2$

(a) $\qquad\qquad\qquad\qquad\qquad$ (b)

FIGURE 4.4
(a) Pressure exerted by a column of water is determined by its height, not by the shape of the container. (b) The principle of car lift.

then the h is termed *pressure head*. The pressure head is the *hydraulic head* (H) if the water is at hydrostatic state and the reference datum is at the base of the water column, or H = h.

If the water is in motion, then the total hydraulic head H is

$$H = \frac{V^2}{2g} + \frac{P}{\rho g} + z \tag{4.14}$$

where V is flow velocity. The equation shows that total hydraulic head in a steady flow system is the sum of velocity head ($V^2/2g$), pressure head ($P/\rho g$), and position head (z). The position head is the height between the point in question and the reference datum; the pressure head is due to the water column above that specific point; and the velocity head is the height of water surface created by motion of water. Thus, the flow of water that has risen to a height h meters (i.e., the velocity head) above its surface by a vertical obstacle has a velocity $V^2 = 2gh$, or

$$V = (2gh)^{0.5} \tag{4.15}$$

If h is in meters (g = 9.80 m/s²), Equation 4.15 can be simplified to

$$V = 4.427(h)^{0.5} \tag{4.16}$$

to give V in meters per second. The velocity head can be used to estimate streamflow velocity in small creeks.

The hydraulic head pertains to the energy status of water and is analogous to potential in electrical flow or to temperature in heat conduction. It can be used to describe energy levels for flow transitions in open channels. Figure 4.5 shows that the total energy level at a point in the upstream is equal to the total energy level at a downstream point plus the head loss (in the mass of water) that occurred over the reach. The energy conservation along a streamline is expressed by the Bernoulli equation. A reduction in the cross-sectional area of the lower reach, due to either rise of floor or contraction of banks, results in increased flow velocity and decreased surface elevation.

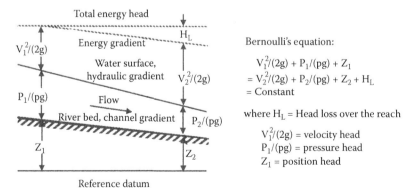

FIGURE 4.5
Energy level along a flowing stream.

4.2.3 Buoyancy

A solid object of volume v submerged in water displaces an equal volume of water, consequently causing the water level to rise to accommodate the displaced volume. If ρ_o and ρ_w are the densities of the submerged object and water, respectively, then the weight (W) of the object after submersion is

$$W = v\rho_o - v\rho_w \tag{4.17}$$

Equation 4.17 can be rearranged to solve for the density of the submerged object, ρ_o, by

$$\rho_o = \frac{W + v\rho_w}{v} \tag{4.18}$$

If ρ_o is smaller than ρ_w or the density of a submerged object is less than the density of water (or other fluids), then the object floats. This means that the hydrostatic upthrust is greater than the downward pressure of the object.

This is the so-called *Archimedes' Principle*, which states that the loss in weight of an object placed in a fluid equals the weight of the fluid displaced by the object. The principle can be applied, as an example, to design floating lysimeters, devices for measuring percolation, leaching, and evapotranspiration (ET) from a column of soil under controlled conditions (Chang et al., 1997).

A floating lysimeter consists of a soil tank floating in a larger outer tank filled with water (or other fluid). The two containers are buried under the ground to imitate natural conditions. Changes in weight of the soil container due to ET are detected by changes in water level in the outer tank due to buoyancy of the water.

To float a soil tank of diameter d_1 and height H_1 in a water tank with its rim at least h ($H_1 > h$) above water surface, the total volume of water displaced (v_1) is

$$v_1 = \frac{(d_1^2)(\pi)(H_1 - h)}{4} \tag{4.19}$$

and the weight of water displaced (W) by the soil tank can be as much as

$$W = (v_1)(\rho_w) \tag{4.20}$$

where ρ_w is water density. If the diameter and height of the outer water tank are d_2 and H_2, respectively, then the volume of water required (v_2) in the water tank must be

$$v_2 = \frac{[(d_2^2)(H_2 - h) - (d_1^2)(H_1 - h)](\pi)}{4} \tag{4.21}$$

in order to keep the soil tank flush with the water tank. To meet the aforementioned conditions, the total weight of the soil tank, which is the sum of soil, tank, and other bedding materials, must be equal to W in Equation 4.20.

A loss of water of depth δ_s in the soil tank has a weight (w) equal to

$$w = (\delta_s)(A_s)(\rho_w) \tag{4.22}$$

where A_s is the surface area of the soil tank. This loss in depth in the soil tank will cause a change of water level δ_w in the water tank equal to

$$\delta_w = \frac{w}{(A_s + A_w)\rho_w} \tag{4.23}$$

where A_w is the area of the annular water surface in the water tank. Since the lysimeter is installed under the ground, the gain or loss of water in the water tank is detected through a reservoir connected to the water tank. In this case, the loss of water δ_s in the soil tank will cause a change in water level in the reservoir equal to

$$\delta_w = \frac{(\delta_s)(A_s)}{(A_s + A_w + A_r)\rho_w} \tag{4.24}$$

where A_r is the surface area of the reservoir.

4.2.4 Surface Tension and Capillary Rise

The force exerted by molecules to bind two different substances in contact together is adhesion, while the force holding molecules of the same substance together is cohesion. For example, when a drop of liquid is placed on a solid surface, the liquid tends to spread out over and wet the surface if the adhesive force between the two substances is greater than the cohesion of the liquid, such as water. On the other hand, the liquid will remain in the drop form if the adhesive force between the two is smaller than the cohesion of the liquid, as is the case with mercury.

The cohesion of molecules creates a phenomenon at the surface of a liquid called surface tension. When a liquid is in contact with the wall of a vessel, the edge of the liquid tends to form a considerable curvature due to the inward pull of intermolecular forces. At liquid–air interfaces, the attraction of molecules to each other at the liquid surface is greater than in the air above. This results in an inward force at the liquid surface, which acts as though it has a thin membrane stretched over it. The surface is under tension and contraction. The force causes the liquid surface to become as small as possible and causes fog drops, raindrops, and soap bubbles to assume spherical shapes as they fall from the air.

When a fine glass tube is placed in water, the water surface rises inside the glass tube. First, adhesion causes water to wet and spread over the glass tube. Then, cohesion causes water to creep up along the glass wall. The rise will stop when the vertical component of the surface tension is equal to the weight of the water raised. This is called capillary action. The height of rise h for any liquid in a capillary tube can be calculated by

$$h = \frac{2T}{r\rho g} \tag{4.25}$$

where
   T is the surface tension
   r is the radius of the tube
   ρ is the density of the liquid
   g is the force of gravity

For water, T = 73.05 dyn/cm at 20°C (Table 4.3). Equation 4.25 is reduced to give h in cm per unit area in the simple expression

$$h = \frac{0.15}{r} \quad (4.26)$$

Surface tension has remarkable effects on water behavior in soils. Water has a T value (surface tension) higher than that of most common liquids. This causes water to be absorbed strongly by capillary attraction of porous materials, which are capable of holding water for various purposes.

4.2.5 Viscosity

All fluids exhibit, more or less, a property of resistance to changes in shape (or to flow) under the action of external forces. Highly viscous liquids approach solid conditions. This is due to the cohesiveness of the molecules in the fluid. According to Newton, the shear stress τ at a point within a fluid is proportional to the velocity gradient dv/dz at that point, or

$$\tau = \mu \left(\frac{dv}{dz} \right) \quad (4.27)$$

where μ (in dyn-s/cm² or lb-s/ft²) is the constant of proportionality known as *dynamic* or *absolute viscosity*. Under ordinary conditions, the viscosity of liquids is only affected by temperature, not by pressure. As temperature increases, gas molecular momentum increases and viscosity decreases. For water, a simple way to calculate μ by temperature is (Russell, 1963)

$$\mu = \frac{0.01799}{1 + 0.03368(T) + 0.000221(T^2)} \quad (4.28)$$

where
T is temperature (°C)
μ is in dynes-second per square centimeter (or g/s/cm)

One dyn-s/cm² (or 0.1 N-s/m²) is equal to one poise, the unit of viscosity in the metric system (named in honor of Poiseuille). One centipoise (cp) is equal to 0.01 poise. Values of the dynamic viscosity for water, alcohol, and mercury are given in Table 4.3. There is no other inorganic liquid with a viscosity as low as that of water, and freshwater has a lower viscosity than saltwater does. Many organic liquids, such as ether, acetone, chloroform, and benzene, have a lower viscosity than water does at room temperature. As water temperature decreases, viscosity increases, which causes water to flow more slowly and requires more work for wind to produce surface waves, for fish to swim, and for athletes to row a boat.

A parameter often used in fluid motion analyses is the *kinematic viscosity* ν, which is the ratio of the dynamic viscosity μ to the density of the fluid ρ, or

$$\nu = \frac{\mu}{\rho} = \frac{(dyn\text{-}s)/cm^2}{g/cm^3} = \frac{dyn\text{-}s\text{-}cm}{g} = cm^2/s \quad (4.29)$$

4.2.6 Reynolds Number

The state of flow in an open channel is often described as laminar flow or turbulent flow. When velocity exceeds a critical value, the flow becomes turbulent, chaotic, and disorderly. Otherwise, the flow is smooth, calm, and undisturbed. Changes of flow from the laminar to the turbulent state depend on the ratio between inertial forces and viscous forces. Inertia tends to maintain its speed along a straight line, promoting turbulent conditions, while viscosity dampens disturbances, promoting laminar conditions. The Reynolds number (N_R) expresses this ratio as follows:

$$N_R = \frac{(V)(R)}{v} = \frac{(\rho)(V)(R)}{\mu} \text{ [Dimensionless]} \quad (4.30)$$

where
  V is the mean velocity of flow (m/s)
  R is the hydraulic radius (m)
  v is the kinematic viscosity $= (\mu)/(\rho)$ (m²/s)
  ρ is the fluid density (kg/m³)
  μ is the dynamic viscosity (N-s/m²)

When $N_R \leq 500$, viscous forces are dominating and the flow is laminar. When $N_R \geq 2000$, the inertial forces are dominating and the flow is turbulent. The flow in an open channel can be either laminar or partly turbulent within the transitional range of $N_R = 500$–2000 (Chow, 1959). The Reynolds number may range from 0.00001 for a bacterium swimming at 0.01 mm/s to 3,000,000,000 for a large whale swimming at 10 m/s (Vogel, 1994). The Reynolds number is highly affected by the size and speed of organisms.

The Reynolds number is of great biological significance, since it affects aquatic species distribution and activity, as well as the transport of nutrients, energy, and gases to an organism (Gordon et al., 1993; Vogel, 1994). In streamflow measurements with a precalibrated gauging device, maintaining a laminar flow in the approach section is a must for satisfactory readings.

4.2.7 Shear Stress

The forces that moving water exerts on a wetted surface are the tractive force, drag force, and shearing force. They are the components of the weight of water (w) in the direction of flow divided by the area (A) over which they act. Dividing the shearing force by the area over which it acts gives the shear stress. This is the force that causes flow resistance along the channel boundary. The stress at which the channel material moves is known as the critical stress. A greater value of shear stress than critical stress will cause the channel to be unstable. The shear stress is calculated by

$$\tau = \frac{w(\sin\alpha)}{A} = \frac{\rho g V(\sin\alpha)}{A} \quad (4.31)$$

$$\tau = \gamma R S$$

where
  τ is the shear stress (N/m²)
  w is the weight of streamflow $= \rho g V$

ρ is the water density (1000 kg/m^3 at 4°C)

g is the acceleration due to gravity (9.80 m/s^2)

V is the volume of streamflow (m^3)

α is the channel slope (deg), for small α, sin $\alpha \cong$ tan $\alpha = S$

A is the channel bed area (m^2)

R is the hydraulic radius (m) = channel cross-sectional area (m^2)/length of wetted perimeter (m)

γ is the specific weight of water = ρg (= 9800 N/m^3)

S is the slope gradient (dimensionless)

The shear stress has been applied extensively to studies in soil erosion, sediment transport, and channel-bank stabilization. More discussion is given in Chapters 12 and 14.

4.2.8 Stream Power

Another important index for describing the erosive capacity of streams is stream power. Power is the amount of work done per unit of time, and work is the product of force applied in the direction of the displacement and the distance of the displacement. In mathematical form,

$$\text{Work (w, N-m} = \text{ Joule)} = \text{Force (F, Newton)} \times \text{Distance (D, m)} \tag{4.32}$$

$$\text{Power } (\omega, \text{J/s)} = \text{Work/Time} = \text{N-m/s} = (F)(V) \tag{4.33}$$

Note that 1 N of force = 1 kg-m/s^2. Expressing stream power per unit of streambed area ω_A, then

$$\omega_A \text{ (N-m/s} = W/m^2) = \left(\frac{F}{A}\right)V = (\tau)(V) \tag{4.34}$$

where

A is streambed area (m^2)

τ is shear stress (N/m^2)

Thus, as velocity increases, stream power increases, resulting in a more erosive flow. A flash flood in steep, mountainous areas can generate values of stream power much greater than those of major rivers.

Stream power can also be expressed as the rate of potential energy over a unit length of stream channel (Bagnold, 1966):

$$\omega_L = \rho g Q S \tag{4.35}$$

where

Q is the whole discharge of the stream (m^3/s)

S is the energy slope of the channel (dimensionless)

ρ and g are defined in Equation 4.31

ω_L is the stream power in unit length (N-m/s^3 or W/m)

The derivation of ω_L is given in Chapter 14, Equation 14.14.

4.3 Chemical Properties

4.3.1 Water Molecule

A water molecule is made of one electronegative oxygen (O^{2-}) atom and two much smaller electropositive hydrogen (H^+) atoms. Each atom consists of electrons in negative charge and a nucleus in positive charge. Electrons circle around the central nucleus. The number of electrons in an atom (the atomic number) varies among elements. An oxygen atom has eight electrons and a hydrogen atom has one electron.

There are two types of chemical bonds in water: covalent bonds and hydrogen bonds. Each H atom in a water molecule is attached to the O atom by a single covalent bond. Thus, the three atoms are held together by two covalent bonds. Oxygen is the second most electronegative element. As a result, the two covalent bonds in water are polar and are structured in a V shape of 105°. The polarization creates a partial negative charge on the oxygen atom and a partial positive charge on each hydrogen atom, which are responsible for many of water's unique properties, such as its being an excellent solvent.

Hydrogen bonds are intermolecular bonds, or bonds formed by electrostatic attraction between atoms in different molecules. Each oxygen atom in water can form two hydrogen bonds, one through each of the unbonded pairs of electrons. Thus, each water molecule is linked to others through intermolecular hydrogen bonds to form a three-dimensional aggregate of water molecules (Buchan, 1996). The bonds give water a much larger and heavier molecule and explain in part its relatively higher melting point, boiling point, heat of fusion, and heat of vaporization.

4.3.2 Formation of Water

Water is formed when hydrogen is burned in air. However, the ignition will cause an explosion if it occurs in hydrogen mixed with air or oxygen:

$$2H_2 + O_2 \rightarrow 2H_2O + 115.6\,\text{kcal}$$

Other reactions that produce water include the process of respiration in the cells and organs of living plants:

$$C_6H_{12}O_6 + 6O_2 \rightarrow 6CO_2 + 6H_2O + 673\,\text{kcal}$$

neutralization of acid and base, for example,

$$H_2SO_4 + 2NaOH \rightarrow Na_2SO_4 + 2H_2O$$

and combustion of hydrogen-containing materials, for example,

$$CH_4 + 2O_2 \rightarrow CO_2 + 2H_2O + 192\,\text{kcal}$$

4.3.3 Chemical Reactions

Water molecules are very stable with respect to heat. Water decomposes into its elements to the extent of about 1% at temperatures up to 2000°C. However, water reacts with many substances, including metals, nonmetals, and metal and nonmetal oxides, or it may become a part of the crystalline structure of compounds known as hydrates through the process of hydration (Hein, 1990).

The reaction of water with metals may take place at cold temperatures, such as in calcium sinks in water, to form calcium hydroxides and liberate a gentle stream of hydrogen:

$$Ca\ (solid) + 2H_2O\ (liquid) \rightarrow H_2 \uparrow + Ca(OH)_2$$

or at high temperatures, such as zinc and water, to form oxidized zinc:

$$Zn\ (solid) + H_2O\ (steam) \rightarrow H_2 \uparrow + ZnO\ (solid)$$

Metal oxides react with water to form bases known as basic anhydrides, such as

$$CaO\ (solid) + H_2O \rightarrow Ca(OH)_2\ (calcium\ hydroxide)$$

Nonmetal oxides known as acid anhydrides react with water to form acids, such as

$$CO_2\ (gas) + H_2O \times H_2CO_3\ (carbonic\ acid)$$

Hydrolysis and oxidation are two chemical processes of great importance in the environment. Hydrolysis is the reaction of an ion or mineral with water in which the water molecule is split into H^+ and OH^- ions, while an increase in the oxidation number of an atom as a result of losing electrons is called oxidation. For example, the hydrolysis of the mineral olivine ($MgFeSiO_4$) produces sepentine and ferrous oxide (FeO), which may be immediately oxidized to ferric oxide (geothite):

$$Hydrolysis:\ 3MgFeSiO_4 + 2H_2O \rightarrow H_4Mg_3Si_2O_9 + SiO_2 + 3FeO$$

$$Oxidation:\ 4FeO + O_2 + 2H_2O \rightarrow 4FeOOH\ (geothite)$$

Hydration is the attachment of water to minerals, such as

$$Al_2O_3 + 3H_2O \rightarrow Al_2O_3 \cdot 3H_2O$$

4.4 Biological Properties

4.4.1 Temperature

4.4.1.1 Measures

Temperature is a scale to measure hotness and coldness of water or other substances. Four temperature scales are in general use today: Centigrade (°C), Fahrenheit (°F), Absolute (K),

TABLE 4.4

Freezing and Boiling Points of Water, Alcohol, and Mercury

Substance	Celsius (°C)	Fahrenheit (°F)	Absolute (K)	Rankine (°R)
Water				
Freezing	0	32	273.16	491.69
Boiling	100	212	373.16	671.69
Alcohol				
Freezing	−117.3	−179.14	155.86	280.55
Boiling	78.4	173.12	351.56	632.81
Mercury				
Freezing	−38.87	−37.97	234.29	421.82
Boiling	356.58	673.84	629.74	1133.53

and Rankine (°R). Water freezes at 0°C and boils at 100°C (Table 4.4). Conversions among these scales are

$$°C = (°F - 32)\frac{5}{9} = K - 273.16 \tag{4.36}$$

$$°F = °C\left(\frac{9}{5}\right) + 32 = \frac{9}{5}(K - 273.16) + 32 \tag{4.37}$$

$$K = °C + 273.16 = (°F - 32)\frac{5}{9} + 273.16 \tag{4.38}$$

$$°R = °F + 459.69 = \left(\frac{9}{5}\right)(K - 273.16) + 491.69 \tag{4.39}$$

The main sources of heat for streams and reservoirs are solar radiation (R_s), longwave radiation from the atmosphere (R_a), the energy advected into the body of water by precipitation and inflow (R_i), and the conducted heat from the ground (R_c). However, heat also can be lost from water to the environment through surface reflection of solar radiation (R_r), longwave radiation, or sensible heat to the atmosphere (R_w), and latent heat of vaporization (R_v). The difference between heat gain and heat loss is balanced by a positive or negative term of heat storage, R_{st}, by the following equation:

$$(R_s + R_a + R_i + R_c) - (R_r + R_w + R_v) \pm R_{st_s} = 0 \tag{4.40}$$

Thus, water temperature will increase or decrease in response to the positive or negative value of R_{st}. In small forested streams, the exposure of the stream channel to direct solar radiation and wind movement due to clearcutting of riparian vegetation may cause an increase in stream temperature adversely affecting aquatic ecosystems. Increases in stream temperature may take more than 10 years to recover to pre-clearcutting levels (Moore et al., 2005), due to regrowth of riparian vegetation. Accordingly, mean and maximum daily water temperatures in afforested streams were reported to be significantly lower than those of open moorland streams in northern England (Brown et al., 2010).

Stream and lake waters are often used in industrial cooling processes. Hot water discharged into streams from a power plant may raise the temperature of water enough to produce major changes in aquatic communities. The effect of thermal discharge from power plants on stream temperature can be calculated by the following mixing equation:

$$T_d = \frac{(T_p)(Q_p) + (T_u)(Q_u)}{Q_p + Q_u} \tag{4.41}$$

where
  T is temperature
  Q is flow rate

  Subscripts p, u, and d refer to power plant, upstream, and downstream, respectively

4.4.1.2 Significance

Water temperature is a primary factor affecting physical and chemical properties of water, including the rate of chemical reaction, gas solubility, decomposition of organic matter, latent heat, saturation vapor pressure, changes in the state of water, water density, viscosity, heat conduction, velocity of particle settlement, water circulation, and others. It also regulates biological activities in the aquatic environment. Cool water is always better in quality than warm water because of a higher concentration of dissolved oxygen (DO) and lower rates of microbial activities, organic matter decomposition, and chemical reaction.

The internal temperature for most aquatic animals follows closely with water temperature. As a rule of thumb, an increase in temperature of 10°C doubles the metabolic rate of cold-blooded organisms and the rate of chemical reactions. The optimum temperature range for aquatic organisms varies with species and the life stage of each species. Changes in temperature of only a few degrees can adversely affect the production, growth, development, and survival of an organism. The optimum temperature range for most salmon species is about 12°C–14°C and the lethal level for adults is about 20°C–25°C, depending on duration and rate of the temperature increase. Some salmon eggs and juveniles may be killed at 13.5°C (MacDonald et al., 1991).

4.4.2 Dissolved Oxygen

4.4.2.1 Measures

DO refers to the amount of oxygen dissolved in water and is expressed in parts per million or milligrams per liter. The concentration of DO in stream water is determined by the oxygen-holding ability, oxygen depletion, and oxygen replenishment of the stream. Oxygen dissolves in water through diffusion at the interface between water and air. The solubility of oxygen in water is inversely proportional to water temperature and increases with increasing atmospheric pressure (Figure 4.6). At one standard atmospheric pressure (760 mmHg), the saturated concentration of DO in water is about 14.6 mg/L at 0°C and gradually decreases to 7.6 mg/L at 30°C (Vesilind and Peirce, 1983). For actual saturation values at pressures other than 760 mmHg, DO can be approximated by the equation:

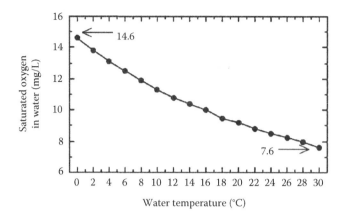

FIGURE 4.6
Saturated concentration of oxygen as a function of water temperature.

$$DO_p = DO_s \left(\frac{P}{760} \right) \qquad (4.42)$$

where
  DO_p is the saturation concentration at atmospheric pressure P
  DO_s is the value reported at one standard atmosphere

Oxygen in streams is depleted by respiration of aquatic plants and the biochemical oxygen demand (BOD) of substances in the stream. The BOD refers to the chemical oxidation of dissolved, suspended, or deposited organic materials in streams and the decomposition of these materials by aquatic microorganisms. The depletion of oxygen is replenished by photosynthesis of aquatic plants and the dissolution of oxygen from the atmosphere. Increasing water temperature not only reduces the dissolution of oxygen in water but also increases the rate of BOD. Since oxygen is transferred to the stream by the diffusion process, stream conditions that maximize the contact between the atmosphere and water, such as turbulent flows, will increase the dissolution level. Thus, a slow-moving stream is usually low in DO because of lower reaeration rates, higher water temperatures, and greater values of BOD. For these reasons, many forested streams of northern Louisiana have summer DO concentrations below the 5 mg/L criterion recommended by EPA (Ice and Sugden, 2003).

4.4.2.2 Significance

The concentration of DO in the stream is critical to the character and productivity of biological communities. Aquatic organisms depend on oxygen for survival, growth, and development. The requirements for DO levels are different among species, below which they will not reproduce, feed, or survive. Generally large fish have a greater need for oxygen than a small fish of the same species (Reardon and Thibert-Plante, 2010). In the northwest United States, for example, the instream DO level should be at least 11 mg/L for embryos and larvae of salmonids. A drop of DO level to 9 mg/L will cause slight production impairment, while production is severely impaired at a DO level of 7 mg/L (EPA, 1986).

4.4.3 pH

Water molecules (HOH) are normally dissociated into hydrogen ions (H^+) and hydroxyl ions (OH^-). If more hydrogen ions than hydroxyl ions are present, then the water is acidic. Conversely, an excess of hydroxyl ions indicates a basic solution. Water or a solution with equal concentrations of hydrogen and hydroxyl ions is called "neutral."

4.4.3.1 Measures

The pH of a solution is a measure of hydrogen–ion concentration in moles (molecular weights in grams) per liter. It is defined as the negative logarithm of the concentration of H^+ ions, or

$$pH = \frac{1}{\log H^+} = -\log H^+ \tag{4.43}$$

Thus, every unit change in pH represents a 10-change in the concentration of H^+ and OH^- ions. The product of H^+ and OH^- concentrations is a constant, 10^{-14} mol/L. If the concentration of H^+ is 10^{-5}, then the concentration of OH^- must be equal to $10^{-14}/10^{-5}$, or 10^{-9}. A concentration of 10^{-5} (pH 5) is greater than a concentration of 10^{-9}, and the solution is acidic. Thus, the lower the pH value, the higher the acidity. For a neutral solution, the H^+ concentration is 10^{-7} and the pH is 7.

The pH for pure water at 25°C is 7. However, many solutes, solid or gaseous, may enter into the water. The reaction or dissociation of these solutes in water may produce H^+ or OH^- ions, resulting in changes of pH. For example, the reaction of dissolved carbon dioxide in water produces carbonic acid (H_2CO_3). Then the H_2CO_3 is further dissociated into H^+ and HCO_3^-. In 2007, the concentration of carbon dioxide in the atmosphere is 384 ppm by volume, and rainwater pH at 25°C and 1 atm is calculated to be 5.62 (Rogan et al., 2009).

4.4.3.2 Significance

Surface waters of very low to moderate pH are generally derived from contact with volcanic gases containing hydrogen sulfide, hydrochloric acid, and other substances associated with oxidizing sulfide minerals such as pyrite and organic acids from decaying vegetation. Waters associated with sodium-carbonate-bicarbonate have high to moderately high pH. The pH values are generally higher for waters from limestone than waters from clay-rich sediments. Rivers in the United States with no effects of pollution generally have a pH ranging from 6.5 to 8.5, while pH in groundwater ranges from about 6.0 to 8.5. Unusual values of pH as high as 12.0 and as low as 1.9 have been observed in water from springs in the United States (Hem, 1985).

The solubility of many metal compounds and the chemical equilibrium in water are affected by pH. It is important in chemical treatment and corrosion control. Low pH is very damaging to fish and other aquatic organisms because of their sensitivity to pH changes. The effects of acid rain and acid-mine drainage on the aquatic ecosystem in streams and reservoirs are well documented. pH must be controlled and monitored at an optimum range for microorganism activity in wastewater treatment.

4.4.4 Conductivity

The ability of a substance to conduct an electric current is termed conductivity or electric conductivity (EC). It is the reciprocal of resistance.

4.4.4.1 Measures

Since the unit for resistance is ohms, the unit for conductivity is therefore the inverse of ohms or mhos. Conductivity is expressed in terms of conductance per unit length, or mho/cm, but, it is usually given in μmho/cm to account for the small values for natural waters: $1\,mho/cm = 10^6\,\mu mho/cm$. In System International d'Unites (SI), electrical conductivity is expressed in S/m (Siemen per meter), and $1\,S/m = 10\,mmho/cm$ (millimho per centimeter), or $1\,\mu S/cm = 1\,\mu mho/cm$.

Because the electric current flow increases with temperature, EC values are usually standardized to 25°C and the values are technically referred to as specific EC. The specific EC for pure water not in contact with the atmosphere is about 0.05 μmho/cm. Such water is difficult to produce because of the dissolution of carbon dioxide. Normal distilled water or water passed through a deionizing exchange unit usually has a conductivity of at least 1 μmho/cm. Seawater has a concentration of dissolved solids of about 35,000 mg/L; thus, its conductivity is about 50,000 μmho/cm.

4.4.4.2 Significance

The EC of water is dependent upon the concentration of dissolved ions and water temperature. Changes in water conductivity are less sensitive at higher temperatures (Light et al., 2005). As the concentration of dissolved ions in water increases, the ability of water to conduct electric current increases. Thus, stream ecologists commonly find that using electroshock to stun fish for monitoring their abundance and distribution is difficult if water is too soft, meaning that the EC is low.

Conductivity is often used as a parameter to determine the suitability of water for a specific purpose such as food preparation, agricultural irrigation, or water supply; to test the result of wastewater treatment; or to control water quality and other manufacturing processes. It is especially fundamental for ultrapure water required in many critical applications, such as chip fabrication for semiconductors, intravenous solutions for pharmaceuticals, etc. In soils, conductivity is higher on clays and lower on sands. Many soil properties such as texture, subsoil characteristics, salinity, cation exchange capacity, level of organic matter, and drainage conditions can be related to EC.

4.4.5 Sediment

Stream sediments are particles, usually inorganic, released from land surfaces or stream banks by the action of raindrops, gravity, animals, overland flows, and streamflows. They can be in suspension (no contact with streambed), saltation (bouncing along streambed), and bed-load (rolling along streambed). No streams are free from sediments.

4.4.5.1 Measures

Expressed in parts per million or milligrams per liter, the concentration of sediment is affected by environmental conditions of the watershed, such as geology, soil, topography, climate, vegetation cover, and land-use activity. Sometimes, the amount of stream sediment is expressed in terms of turbidity. Turbidity, an optical measure, can be measured in terms of Jackson Turbidity Units (JTU; one such unit is the turbidity produced by 1 ppm of silica in distilled water), Nephelometric Turbidity Units (NTU, based on a USEPA approved and stable polymetric suspension as a standard), or Formazin Turbidity Units (FTU, based on a Formazin polymer standard, $C_2H_4N_2$). The JTU is a visual method, based on the light path through a suspension causing the image of the flame of a standard candle to disappear. The method is no longer in standard use. It was removed from "Standard Methods for the Examination of Water and Wastewater" by the American Public Health Association in their 17th edition in 1989.

Turbidity measures the clarity of water as affected by light attenuation of suspended particles (Davies-Colley and Smith, 2001). Since the optical property of water samples is greatly affected by the type, size, shape, configuration, color, origin of particles, concentration, the wavelength of light, and the angle of measurement, there is no direct relationship between NTU or FTU readings and JTU readings. The correlation between these JTU readings and the weight of sediments per unit volume of water is valid only through local calibration.

Sometimes sediment concentration is weighed by flow volume to obtain sediment loss in mass per unit time and per unit watershed area. It is also referred to as sediment yield or sediment load. Generally, watershed area has a negative effect on sediment loss. This may be attributable to the fact that a small watershed is generally associated with less watershed storage, greater rainfall intensity and depth, steeper mean watershed slope, and greater chance for the entire watershed to be covered by a storm.

4.4.5.2 Significance

Sediment concentration has tremendous impacts on the physical and chemical properties of water and consequently affects living things in the aquatic environment. Physically, sediments affect water turbidity, odor, taste, temperature regime, and abrasiveness; reduce light penetration; increase solar absorption at the water surface; deplete reservoir capacity; and clog stream channels. Recreational and aesthetic values of streams can be impaired and turbines in hydroelectric plants can be damaged because of sediment.

Sediments can carry many elements and compounds that may interact with one another in water, consequently affecting water chemistry and quality. The nature and content of these compounds and elements resemble the origin of soil particles plus the residuals from applications of fertilizers, insecticides, herbicides, industrial wastes, fallout from air pollution, and decomposition of organic matter. Concentration of sediment in streams is often highly correlated with nutrient concentration and can be used to estimate nutrient losses in streams (Chang et al., 1983; Granillo et al., 1985).

The effects of stream sediment on aquatic ecosystems can be direct through its presence or indirect through interactions with the physical and chemical environment of the stream. Streams with sandy beds have the lowest species diversity and aquatic productivity. The habitat space for small fish, invertebrates, and other organisms can be reduced if the interstices between coarse particles are filled by fine sediments (MacDonald et al., 1991). Photosynthesis of aquatic plants can be inhibited because of the reduction of light penetration. This may lead to declines in foods along the aquatic food chain. In addition, increases in absorption of solar radiation at the water surface may raise water temperature, which will lead to declines in DO for aquatic life consumption.

4.4.6 Chlorophyll *a*

Chlorophyll is the green photosynthetic pigment that enables plants, including phytoplankton (microscopic plants floating in the water such as algae and blue-green algae), to transform sunlight energy into chemical energy (carbohydrates) through photosynthesis processes. Chlorophyll absorbs light in the red and blue-violet portions of the visible spectrum; the green portion of light is not absorbed but reflected, giving chlorophyll its characteristic color. There are several types of chlorophyll, including chlorophyll *a*, *b*, *c*, and *d*, slightly different in their molecular structure and constituents. Chlorophyll *a* is the most important, making up about 75% of the chlorophyll in green plants and being common to all phytoplankton.

4.4.6.1 Measures

Since chlorophyll *a* is the most important and common to phytoplankton, measuring chlorophyll *a* concentrations in water is a surrogate for an actual measurement of phytoplankton (algae) biomass. Phytoplankton form the base of food web for fish and bottom dwelling invertebrates such as clams and worms, the measurement of chlorophyll *a* is thus also a measure of photosynthesis and a measure of the primary food source of aquatic food webs. It is widely used as an indicator of the production level of these microscopic plants (e.g., primary production or trophic state) and an indicator of water quality, especially in lakes, reservoirs, and bay estuaries and associated tributary water bodies (Boyer et al., 2009).

Expressed in µg/L, the concentrations of chlorophyll *a* can be determined by in situ measurements, laboratory analyses, remote sensing, or satellite measurements (Yacobi and Gitelson, 2000; Pinto et al., 2001; Thomas and Brickly, 2006; Knefelkamp et al., 2007). It may involve instruments such as spectrophotometer, fluorometer, and high-pressure liquid chromatography.

4.4.6.2 Significance

The distribution and concentration of phytoplankton (chlorophyll *a*) is a major issue of water quality and aquatic ecology. Excess concentrations of phytoplankton are often introduced by anthropogenic inputs of plant nutrients (particularly phosphorus). These excess(uneaten) algae sink to the bottom, consumed by bacteria and decompose, which can deplete oxygen from the water column and cause nuisance conditions in surface waters. Chlorophyll *a* can reduce light penetration in shallow-water habitats, which in turn has a direct impact on submerged plants. On the other hand, too little chlorophyll *a* would mean insufficient "fish food" available to fuel the food web. Phytoplankton assemblages dominated by single species represent poor food quality, or even produce toxins, which can impair the animals that feed directly on them. Thus, chlorophyll *a* biomass is the primary regulator of clarity, color of water, and DO levels in many lakes and reservoirs. Deforestation has reported to cause increase in chlorophyll *a* concentrations in small boreal headwater lakes in Canada (Steedman, 2000).

References

Bagnold, R.A., 1966, An approach to the sediment transport problem from general physics, U.S. Geological Survey Professional Paper, 422-I.

Boyer, J.N. et al., 2009, Phytoplankton bloom status: Chlorophyll a biomass as an indicator of water quality conditions in the southern estuaries of Florida, USA, *Ecol. Indicators*, 9(6), Supplement 1, S56–S67.

Brown, L.E. et al., 2010, A comparison of stream water temperature regimes from open and afforested moorland, Yorkshire Dales, northern England, *Hydrol. Proc.*, 24(2), 3206–3218.

Buchan, G.D., 1996, Ode to H_2O, *J. Soil Water Conserv.*, 51, 467–470.

Chang, M., McCullough, J.D., and Granillo, A.B., 1983, Effects of land use and topography on some water quality variables in forested East Texas, *Water Resour. Bull.*, 19, 191–196.

Chang, M. et al., 1997, Evapotranspiration of herbaceous mimosa (*Mimosa strigillosa*), a new drought-resistant species in the southeastern United States, *Resour. Conserv. Recycl.*, 21, 175–184.

Chow, V.T., 1959, *Open Channel Hydraulics*, McGraw-Hill, New York.

Davies-Colley, R.J. and Smith, D.G., 2001, Turbidity, suspended sediment, and water clarity: A review, *J. Am. Water Resour. Assoc.*, 37, 1085–1101.

EPA, 1986, Ambient water quality criteria for dissolved oxygen, U.S. Environmental Protection Agency, Office of Water Regulations and Standards, Washington, DC.

Gordon, N.D., McMahon, T.A., and Finlayson, B.L., 1993, *Stream Hydrology, an Introduction for Ecologists*, John Wiley & Sons, New York.

Granillo, A.B., Chang, M., and Rashin, E.B., 1985, Correlation between suspended sediment and some water quality parameters in small streams of forested East Texas, *Texas J. Sci.*, 37, 227–234.

Hein, M., 1990, *Foundations of College Chemistry*, Brooks/Cole, Pacific Grove, CA.

Hem, J.D., 1985, *Study and Interpretation of the Chemical Characteristics of Natural Water*, 3rd edn., U.S. Geological Survey, Water Supply Paper 2254, Washington, DC.

Ice, G. and Sugden, B., 2003, Summer dissolved oxygen concentrations in forested streams of northern Louisiana, *South. J. Appl. For.*, 27(2), 92–99.

Knefelkamp, B., Carstens, K., and Wiltshire, K.H., 2007, Comparison of different filter types on chlorophyll a retention and nutrient measurements, *J. Exp. Marine Biol. Ecol.*, 345, 61–70.

Light, T.S. et al., 2005, The fundamental conductivity and resistivity of water, *Electrochem. Solid-State Lett.*, 8(1), E16–E19.

List, R.J., 1971, *Smithsonian Meteorological Tables*, 6th rev. edn., Smithsonian Institute Press, Washington, DC.

MacDonald, L.H., Smart, A.W., and Wisssmar, R.C., 1991. Monitoring guidelines to evaluate effects of forestry activities on streams in the Pacific Northwest and Alaska, EPA/910/91–001, U.S. Environmental Protection Agency, Region 10, Seattle, WA.

Moore, R.D., Spittlehouse, D.L., and Story, A., 2005, Riprian microclimate and stream temperature response to forest harvesting: a review, *JAWRA, American Water Resources Association*, 41(4), 813–834.

Pinto, A.M., Von Sperling, E., and Moreira, R.M., 2001, Chlorophyll a determination via continuous measurement of plankton fluorescence methodology development, *Water Res.*, 35(16), 3977–3981.

Reardon, E.E. and Thibert-Plante, X., 2010, Optimal offspring size influenced by the interaction between dissolved oxygen and predation pressure, *Evol. Ecol. Res.*, 12, 377–387.

Rogan, R.A.J. et al., 2009, Changes in rainwater pH associated with increasing atmospheris carbon dioxide after the industrial revolution, *Water Air Soil Pollut.*, 196(1/4), 263–271.

Russell, G.R., 1963, *Hydraulics*, Holt, Rinehart and Winston, Inc., New York.

Steedman, R.J., 2000, Effects of experimental clearcut logging on water quality in three boreal forest lake trout (*Salvelinus namaycush*) lakes, *Can. J. Fish. Aquat. Sci.*, 57, 92–96.

Tabata, S., 1973, A simple but accurate formula for the saturation vapor pressure over liquid water, *J. Appl. Meteorol.*, 12, 1410–1411.

Thomas, A.C. and Brickly, P., 2006, Satellite measurements of chlorophyll distribution during spring 2005 in the California current, *Geophys. Res. Lett.*, 33, L22S05, 5, doi: 10.1029/2006GL026588.

Vesilind, P.A. and Peirce, J.J., 1983, *Environmental Pollution and Control*, Ann Arbor Science, Ann Arbor, MI.

Vogel, S., 1994, *Life in Moving Fluids, the Physical Biology of Flow*, Princeton University Press, Princeton, NJ.

Yacobi, Y.Z. and Gitelson, A.A., 2000, Simultaneous remote measurement of chlorophyll and total seston in productive inland waters, *Verh. Internat. Verein. Limnol.*, 27, 2983–2986.

5

Water Distribution

Earth is the only planet in the solar system where water appears in solid, liquid, and vapor states at common temperatures. The existence of water on Earth has much to do with Earth's size and distance from the Sun. Hydrogen (90%), helium (9%), and oxygen are the three most abundant elements in the universe, followed by neon, nitrogen, and carbon (Cox, 1989). Helium and neon are solitary elements. Two hydrogen atoms and one oxygen atom are easily bound to become water (H_2O), three hydrogens join one nitrogen to become ammonia (NH_3), and four hydrogens join one carbon to become methane (CH_4). The melting points of H_2O, NH_3, and CH_4 are 0°C, −78°C, and −183°C, respectively, while the boiling points are 100°C, −33°C, and −162°C.

All the energy used and exchanged in the atmosphere and biosphere of Earth originates from the Sun. Solar radiation decreases inversely with the square of the distance. The distance between Earth and the Sun makes the average Earth temperature around 27°C, enabling water to stay in the liquid state. The mean distances from the Sun to Mercury and to Jupiter are, respectively, 0.37 and 5.2 times the distance from the Sun to Earth (Figure 5.1). The surface temperatures at the visible sunlight surfaces of Venus and Jupiter are approximately 430°C and −120°C, respectively. Water cannot be maintained in the liquid state at such temperatures.

A planet's gravity is largely affected by its size. Large planets have greater gravity and are able to attract light volatile elements and molecules. If a planet is too small, it cannot attract volatile elements. Then, there are no gases, just nonvolatilized matter such as iron, rocks, and sulfates around the planet. Earth's size and appropriate distance from the Sun enable it to keep water in the liquid state. Water could not exist in all three states on Earth's two nearest neighbors, Venus and Mars, under their current conditions. The surface temperature of Venus is too high for ice or liquid water. As for Mars, although small amounts of water vapor and ice crystals have been detected, the surface temperature (polar caps at −38°C to −66°C) is too cold to keep water liquid. From Mars out to Pluto, all water, if any, is frozen all the time.

5.1 Globe

The total volume of water on Earth is about 1.348×10^9 km³ (Table 5.1), enough to cover Earth's surface to a depth of 2.7 km. Water is present in oceans, lakes, and rivers; underground; at the ground surface; in the atmosphere; and in organic materials, minerals, and rocks. The quantity of water is discussed in two categories: saline water and freshwater.

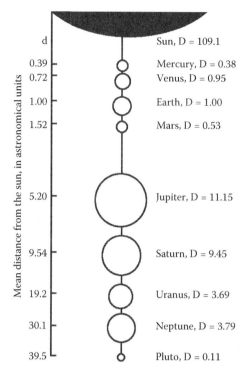

FIGURE 5.1

Mean distance to the Sun (d, between Earth and Sun = 1.0), along with relative diameter (D, Earth diameter = 1.0), for each of the nine planets. (Data from Dormand, J.R. and Woolfson, M.M.: *The Origin of the Solar System: The Capture Theory*, John Wiley & Sons, New York, 1989.)

TABLE 5.1

Water around Earth

Category	Volume		Percentage of	
	km³	acre-ft	Total Water	Freshwater
Saline water	1.348×10^9	1.093×10^{15}	97.398	—
Freshwater	3.602×10^7	2.920×10^{13}	2.602	100.00
Ice, glaciers	2.782×10^7	2.255×10^{13}	2.010	77.23
Groundwater	7.996×10^6	6.482×10^{12}	0.578	22.20
Soil moisture	6.123×10^4	4.964×10^{10}	0.0044	0.17
Lakes	1.261×10^5	1.022×10^{11}	0.009	0.35
Rivers; hydrated and organic materials	3.602×10^3	2.920×10^9	0.0003	0.01
Atmosphere	1.300×10^4	1.054×10^{10}	0.0009	0.04
Total	1.384×10^9	1.122×10^{15}	100.000	—

Source: Adapted from Baumgartner, A. and Reichel, E., *The World Water Balance, Mean Annual Global, Continental and Maritime Precipitation, Evaporation, and Runoff*, R. Oldenbourg, Munich, 1975.

5.1.1 Saline Water

Water is saline if it contains more than 1000 mg of dissolved solids per liter or one part per thousand (1 ppt) by weight. More than 99.97% of saline water is in the ocean, and only a very small fraction is in saltwater lakes.

Ocean water contains dissolved salts and minerals with a concentration of about 35 ppt. These dissolved solids are cumulatively delivered from rivers and are derived from the weathering of rocks, erosion, or human activities. Ocean evaporation also leaves salt behind, increasing salt concentration in the ocean. Dissolved solids in water increase water density, reduce light penetration, and lower the freezing point and temperature of maximum water density.

Horizontally, saline water covers 362×10^6 km^2 or 70.8% of Earth's surface. It makes up 60.6% of the total area in the Northern Hemisphere and 81.0% in the Southern Hemisphere. Geographically, the ocean is recently re-delineated by five names: Pacific, Atlantic, Indian, Arctic, and Southern. The Southern Ocean was named by the International Hydrographic Organization in 2000, covering the southern portions of the Atlantic, Indian, and Pacific oceans. It completely surrounds Antarctica and extends from the coast of Antarctica north to 60° south latitude, forming the fourth largest of the world's five oceans. The Southern Ocean is a unique oceanic system important to studies of the Southern Circulations, El Nino occurrence, and global warming.

The Pacific Ocean is the biggest (660×10^6 km^3, about 49% of the water on Earth) and largest (162×10^6 km^2, or 45% of the total) one of the five. Also, the greatest depth of the oceans, measured at 10,911 m, is located in the Marianas Trench of the Pacific Ocean. The Arctic Ocean is the shallowest (1205 m), the least (19×10^6 km^3), and the smallest (16×10^6 km^2) of the five (Table 5.2).

The total volume of saline water is estimated to be 1.348×10^9 km^3, and 57% of it is in the Southern Hemisphere (Table 5.1; Baumgartner and Reichel, 1975). This volume represents 97.4% of the total water in the world; the other 2.6% is freshwater contained in rivers, lakes, grounds, polar regions, organic matter, and the atmosphere. If Earth had a uniform surface, the water in the ocean could cover the ground to a depth of 2643 m.

TABLE 5.2

Dimensions of the World's Oceans

Name	Area		Volume		Depth (m)	
	10^2 km	% of Total	10^3 km	% of Total	Average	Maximum
Arctic Ocean	15,558	4.3	18,750	1.4	1205	5,567
Atlantic Ocean	85,333	23.5	310,411	23.3	3646	8,484
Indian Ocean	70,560	19.5	264,000	19.8	3741	7,906
Pacific Ocean	161,760	44.7	660,000	49.4	4080	10,803[a]
Southern Ocean	21,960	6.1	71,899	5.4	3270	7,075
South China Sea	6,963	1.9	9,880	0.7	1419	7,352
Total	361,900	100.0	1,334,841	100.0	3688	10,803

Source: Eakins, B.W. and Sharman, G.F., Volumes of the World's Oceans from ETOPO12, NOAA National Geophysical Data Center, Boulder, CO, 2010.

[a] Deepest ocean depth is in the Marianas Trench, measured at 10,911 m. Maximum depths from ETOPO1 (ETOPO1 is a 1 arcmin digital representation of Earth's solid surface that integrates land topography and ocean bathymetry from numerous global and regional datasets.) are not expected to exactly match the known measured maximum depths as ETOPO1 represents average depths over ~4 km^2 areas.

The salty Caspian Sea is the largest lake in the world with a surface area of about 3.71×10^5 km^2, larger than the second Lake Superior by 4.5 times (Table 5.3). Its volume of water, 0.75×10^5 km^3, is about 75% of the total volume of all saltwater lakes in the world. Some groundwater and springs too are saline.

5.1.2 Freshwater

The total volume of freshwater of Earth is about 36.0×10^6 km^3, or less than 3% of the total volume of saline water on Earth, including oceans and saline lakes (Table 5.1). Of the total freshwater, about 77.23% is locked up as icecaps and glaciers in the polar regions and alpine areas, and 22.2% or 8.0×10^6 km^3 is confined in underground aquifers. Water in freshwater lakes and rivers, about 0.13×10^6 km^3, accounts for only 0.36% of the total. Major lakes and rivers of the world are given in Tables 5.3 and 5.4, respectively.

5.1.2.1 Icecaps and Glaciers

Ice (water in the solid state) is formed in polar regions and in mountains at higher elevation. The total volume is equivalent to 27.8×10^6 km^3, the largest entity of all freshwater in the world. In Antarctic regions, icecaps extend to an area of about 15.54×10^6 km^2, and the total volume is about 25×10^6 km^3 (Nace, 1984). In other words, about 90% of all existing solid water is in the Antarctic. Icecaps in Greenland cover an area of 1.23×10^6 km^2, with an average depth of 1.52 km. The total volume of water on Greenland is about 2.62×10^6 km^3, or 9.5% of solid water. Alpine areas, continental patches, and the Arctic regions account for 0.208×10^6 km^3 of water, only a small fraction of that contained in the Greenland and the Antarctic icecaps.

5.1.2.2 Water under the Ground

Beneath the ground, water can be divided into two distinct zones. Immediately below the ground surface is the zone of aeration. Here, both water and air are present in the ground. Under the *zone of aeration* is a layer in most areas of the ground where sands, gravel, and bedrock are saturated with water. This is the *zone of saturation*. The boundary between these two zones is the groundwater table. A shallow, wet layer immediately above the groundwater table to which water can rise up slightly due to capillary force is called the *capillary fringe*. Water in the surface layer where plants' root systems can reach is called soil water. Water present between soil water and the groundwater table is *vadose* or *gravitational water* (Figure 5.2). A typical content of soil water is about 25% by volume, or 0.25 g/cm^3. The total volume of soil water and vadose water of Earth is around 61×10^3 km^3.

Water in the zone of saturation is called either confined or unconfined groundwater. If groundwater is under pressure due to the weight of the overlying impervious stratum and the hydrostatic head, it is called *confined groundwater*. The permeable geological formations that permit appreciable water to move through them are known as *aquifers*.

Below the zone of saturation, waters have been imprisoned in the interior or chemically bound with rock since the formation of Earth. These are *juvenile* or *internal waters* that do not originate from precipitation or surface runoff.

Since water under the ground is held in pores, voids, interstices, or spaces, and in fractures in and between rocks, the occurrence of groundwater must be within the top of Earth's crust where pore spaces exist. Areas with unconsolidated sand and gravels of alluvial, glacial, lacustrine, and deltaic origin; sedimentary rocks such as limestones,

TABLE 5.3

Thirty Largest Natural Lakes of the World

Lake	Country	Surface Area (km²)	Maximum Depth (m)
Caspian Sea	Russia, EP	371,000 (1)	995 (3)
Lake Superior	United States and Canada, NA	83,300 (2)	307 (29)
Lake Victoria	Uganda, Kenya, and Tanzania, AF	68,800 (3)	80
Lake Aral	Russia, AA	66,458 (4)	—
Lake Huron	United States and Canada, NA	59,570 (5)	223 (54)
Michigan	United States, NA	57,016 (6)	265 (42)
Lake Tanganyika	Zaire, Burundi, Zambia, and Tanzania, AF	34,000 (7)	1,435 (2)
Great Bear Lake	Canada, NA	31,792 (8)	445 (17)
Lake Baikal	Russia, AA	31,500 (9)	1,620 (1)
Lake Nyasa (Lake Malawi)	Malawi, Tanzania, and Mozambique, AF	30,500 (10)	706 (4)
Great Slave Lake	Canada, NA	28,438 (11)	614 (6)
Lake Erie	United States and Canada, NA	25,719 (12)	64
Lake Winnipeg	Canada, NA	24,530 (13)	28
Lake Ontario	United States and Canada, NA	18,760 (14)	225 (53)
Lake Ladoga	Russia, EP	18,400 (15)	225 (51)
Lake Balkhash	Russia, AA	17,000–19,000 (16)	26.5
Lake Chad	Niger, Chad, Nigeria, and Cameroon, AF	12,000–26,000 (17)	4–11
Lake Maracaibo	Venezuela, SA	14,343 (18)	250 (47)
Lake Onega	Russia, EP	9,600 (19)	124
Lake Eyre	Australia, AS	0–15,000 (20)	—
Lake Titicaca	Bolivia, SA	8,100 (21)	304 (31)
Lake Athabaska	Canada, NA	8,080 (22)	60
Lake Gairdner	Australia, AS	7,000 (23)	—
Reindeer Lake	Canada, NA	6,330 (24)	—
Issyk Kul	Russia, AA	6,200 (25)	702 (5)
Lake Resaieh (Urmia)	Iran, AA	3,900–5,930 (26)	—
Lake Torrens	Australia, AS	5,776 (27)	—
Vanern	Sweden, EP	5,546 (28)	89
Lake Winnipegosis	Canada, NA	5,447 (29)	12
Lake Mobuto-Sese-Seko (Lake Albert)	Zaire and Uganda, AF	5,300 (30)	48

Source: Adapted from Czaya, E., *Rivers of the World*, Van Nostrand Reinhold, New York, 1981.
Notes: AA=Asia; NA=North America; SA=South and Central America; AF=Africa; EP=Europe; AS=Australia. Figures in parentheses indicate rank in the world.

TABLE 5.4

Thirty Largest Rivers of the World

River	Country at River Mouth	Drainage Area (1000 km²)	Maximum Length (km)	Average Annual Discharge at Mouth (m³/s)
Amazon (with Tocantins)	Brazil	7,180 (1)	6,516 (1) 6,400	180,000 (1)
Congo	Zaire and Angola	3,822 (2)	4,700 (8)	42,000 (2)
Mississippi–Missouri	United States	3,221 (3)	6,019 (3)	17,545 (8)
Ob-Irtysh	Russia	2,975 (4)	5,570 (5)	12,600 (13)
Nile	Egypt	2,881 (5)	6,484 (2)	1,584
Rio de la Plata	Argentina and Uruguay	2,650 (6)	4,700 (9) 6,650	19,500 (7)
Yenisei	Russia	2,605 (7)	5,550 (6)	19,600 (6)
Lena	Russia	2,490 (8)	4,270 (12)	16,400 (9)
Niger	Nigeria	2,092 (9)	4,030 (14)	5,700 (23)
Yangtze Kiang	China	1,970 (10)	5,800 (4) 6,300	35,000 (3)
Amur	Russia	1,855 (11)	4,510 (10)	12,500 (14)
Mackenzie-Peace	Canada	1,805 (12)	4,250 (13)	7,500 (19)
Volga	Russia	1,380 (13)	3,688 (15)	8,000 (17)
Zambesi	Mozambique	1,330 (14)	2,660 (28)	2,500 (37)
Orinoco	Venezuela	1,086 (15)	2,500 (34)	28,000 (4)
Ganges	Bangladesh	1,073 (16)	2,700 (27)	15,000 (11)
Nelson	Canada	1,072 (17)	2,575 (31)	2,300
Murray-Darling	Australia	1,072 (18)	2,570 (32)	391
St. Lawrence River	Canada	1,030 (19)	3,100 (20)	10,400 (16)
Tarim-Khotan	China	1,000 (20)	2,000 (40)	—
Indus	Pakistan	960 (21)	3,180 (19)	3,850 (27)
Brahmaputra	Bangladesh	938 (22)	2,900 (22)	20,000 (5)
Yukon	United States (Alaska)	855 (23)	3,185 (18)	7,000 (20)
Orange	South Africa and Namibia	850 (24)	1,860	—
Shatt-al-Arab	Iraq	808 (25)	2,900 (24)	856
Danube	Romania	805 (26)	2,850 (26)	6,450 (22)
Okavango	Botswana	800 (27)	1,600	—
Mekong	Vietnam	795 (28)	4,500 (11)	15,900 (10)
Hwang Ho	China	745 (29)	4,845 (7)	1,365
Chari	Cameroon and Chad	700 (30)	1,400	—

Source: Adapted from Czaya, E., *Rivers of the World*, Van Nostrand Reinhold, New York, 1981.
Note: Figures in parentheses indicate rank in the world.

dolomites, sandstones, and conglomerates; or volcanic rocks with pores and fractures are the most common aquifers and abundant in groundwater. The depth can go to at least 800 m, with a typical porosity of about 4% (Mather, 1984). Total water in this layer is estimated to be around 4×10^6 km³, a volume about 3000 times greater than the volume of water in all rivers.

Groundwater can go down below 800 m to about 3–4.5 km deep. However, rocks and formations here are tight, and their porosity is generally around 1% or less. Although the

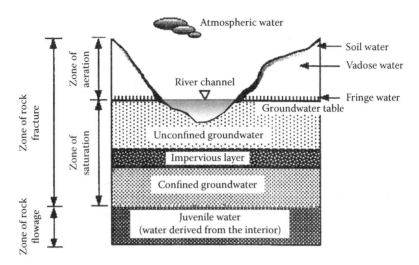

FIGURE 5.2
Vertical distribution of water on Earth.

volume of water in this layer is about equal to that in the overlying layer, recovery of the water may not be economically feasible. The total volume of water below the groundwater table is around 8×10^6 km³.

5.1.2.3 Lakes

The total surface area of freshwater lakes in the world is about 8.45×10^5 km², while the area of saltwater lakes is about 6.91×10^5 km².

The total volume of water aggregated in freshwater lakes is about 1.3×10^5 km³, greater than the volume of water in rivers by as much as 100 times. Most of this surface water is retained in a few relatively large lakes in Africa, Asia, and North America. Lake Baikal in Asiatic Russia (north of Mongolia) contains 0.258×10^5 km³ of water, more than the total volume of water in the five North American Great Lakes by 0.03×10^5 km³. The total volume of freshwater in Lakes Baikal (Siberia), Tanganyika (Africa), Nyasa (Africa), Superior (United States–Canada), and Great Bear (Canada) is about 0.74×10^5 km³, equivalent to almost 60% of the total. Lake Baikal is the greatest and the deepest (1620 m) single body of freshwater in the world, while the Caspian Sea is the biggest (0.75×10^5 km³) and the largest (0.37×10^5 km²) lake of the world, almost 12 times larger than that of Lake Baikal (Table 5.3).

5.1.2.4 Rivers

The volume of water stored in rivers is more difficult to estimate than that stored in lakes. The estimates, besides length, require information on average width and depth for each river and the sum of estimated water for all streams on Earth. There are uncountable numbers of rivers, streams, creeks, and brooks on Earth, with widths varying from a few meters to many tens of kilometers. For example, in the United States, there are about 10,000 rivers with lengths greater than 40 km each and a combined length of 512,000 km. The total length of all sizes of streams and rivers in the United States lies at about 5.76 million km (Palmer, 1996).

Total water in rivers is about 1.23×10^3 km^3 (L'vovich, 1979; Nace, 1984), sufficient to cover Earth's surface to a depth of just 2.4 mm. This is about one-hundredth of that in freshwater lakes and one-tenth of that in the atmosphere. Of all the rivers in the world, the Amazon River contains the greatest volume, accounting for 20% of the total. The 30 largest (drainage area), longest, and biggest (discharge) rivers of the world are given in Table 5.4. The ranks of these rivers differ among authors because of different scales of maps used in measurements and disagreement over river sources.

5.1.2.5 Atmospheric Water

Earth is embraced by a layer of water vapor with a density up to 4% by volume (3% by weight) near the surface and decreasing to about 3–6 ppm by volume at 10–12 km above the ground. The total volume of water vapor in the atmosphere at any given time is around 1.3×10^4 km^3, sufficient to cover Earth's surface to a depth of about 25 mm if all were converted into the liquid state.

5.2 United States

The total volume of water in the continental United States is around 146×10^3 km^3, of which 86% or 126×10^3 km^3 is under the ground and 13% or 19×10^3 km^3 is in freshwater lakes. Waters in stream channels, glaciers, saltwater lakes, the atmosphere, and the soil root-zone sum to 0.995×10^3 km^3, or 0.68% of the total (Table 5.5).

5.2.1 Rivers

Although the country has a network of rivers 5.86×10^6 km long (excluding Alaska), the total volume of water in rivers is only 50 km^3, or about 0.263% of the volume of water

TABLE 5.5

Distribution of Water in the Continental United States

Categories	Volume km^3	%	Annual Circulation (km^3/year)	Replacement Period (years)
Groundwater				
<800 m deep	63.0×10^3	43.20	310	>2,000
>800 m deep	63.0×10^3	43.20	6.2	>10,000
Lakes				
Freshwater	19.0×10^3	13.0	190	100
Saltwater	0.058×10^3	0.04	5.7	>10
Soil moisture (0.35 m deep)	0.63×10^3	0.43	3100	0.2
Rivers	0.05×10^3	0.03	1900	<0.03
Atmosphere	0.19×10^3	0.13	6200	>0.03
Glaciers	0.067×10^3	0.05	1.6	>40

Source: Federal Council for Science and Technology, Report by an ad hoc panel on hydrology and scientific hydrology, Washington, DC, 1962.

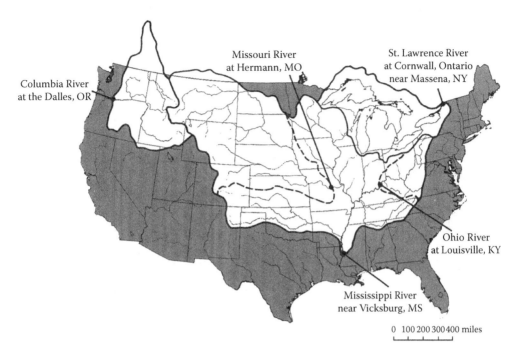

FIGURE 5.3
Five major rivers in the United States.

in freshwater lakes. The Mississippi River, the Missouri River, the Ohio River, the St. Lawrence River, and the Columbia River are the five major rivers in the United States (Figure 5.3). They have a total drainage area of more than half of the conterminous United States. Some statistical data on the streamflow of these five rivers are given in Table 5.6.

The Missouri River flows into the Mississippi River north of St. Louis, Missouri, making the river system the third longest (6019 km), with the third largest (3221 km^2) drainage basin in the world. However, the average annual discharge of the river system is only 17,545 m^3/s, less than 10% of that of the Amazon River. The Amazon is a tropical giant, with its main channel swaying along the south side of the Equator. Rainfall within the Amazon basin ranges from 1500 to 3500 mm/year. The Mississippi–Missouri River is in the temperate region, with rainfall ranging from 500 to 1650 mm/year, much less than the Amazon River.

5.2.2 Lakes

The five Great Lakes of North America—Superior, Michigan, Huron, Erie, and Ontario (Table 5.7)—are the most important inland freshwater bodies of the United States and Canada. Spanning more than 1200 km in width, they contain a total volume of 22.7×10^3 km^3, about 18% of the total water in lakes and six times greater than that in rivers worldwide.

There were 2654 man-made reservoirs with storage capacity of 5000 acre-ft (6.165×10^6 m^3) or greater in the United States. The total volume of storage was 480×10^6 acre-ft or 592 km^3. In addition, there were at least 50,000 reservoirs with capacities between 50 and 5000 acre-ft and about 2 million ponds with capacities of less than 50 acre-ft (der Leeden et al., 1990).

TABLE 5.6

Statistics on Streamflow for the Five Biggest Rivers in the United States

Stream Location	Drainage Area (km²)	Mean (m³/s)	Maximum (m³/s)	Minimum (m³/s)
Mississippi River at Vicksburg, MI	2.909×10^6	16,332	27,761	8889
(Up to 1983, 55 years)		16,377	58,864	2813
Missouri River at Hermann, MO	1.342×10^6	2,233	12,611	193
(Up to 1983, 86 years)		2,265	19,131	119
St. Lawrence at Cornwall, Ontario, near Massena, NY	0.765×10^6	6,874	9,896	4379
(Up to 1983, 123 years)		6,885	9,962	3934
Columbia River at The Dalles, OR	0.607×10^6	5,473	19,003	1051
(Up to 1983, 105 years)		5,476	35,092	342
Ohio River at Louisville, KY	0.233×10^6	3,284	16,858	140
(Up to 1983, 55 years)		3,274	31,413	59

Source: Saboe, C.W., The big five, some facts and figures on our nation's largest rivers, U.S. Geological Survey, General Interest Publications, 1984.
Note: Maximum and minimum discharges are daily averages.

TABLE 5.7

Physical Features of the Great Lakes of North America

Features	Superior	Michigan	Huron	Erie	Ontario
Drainage area (km²)	49,300	45,600	51,700	30,140	24,720
Water area (km²)	82,100	57,800	59,600	25,700	18,960
Volume (km³)	12,100	4,920	3,540	484	1,640
Mean depth (m)	147	85	59	19	86
Maximum depth (m)	406	282	229	64	244
Length (km)	563	494	332	388	311
Width (km)	257	190	245	92	85

Source: Government of Canada and U.S. Environmental Protection Agency, *The Great Lakes, an Environmental Atlas and Resources Book*, 3rd edn., The Great Lakes National Program Office, Chicago, IL, 1995.

As of 1998, the total surface area of reservoirs, lakes, and ponds in the United States was 168,760 km² (EPA, 2009).

5.3 Hydrologic Cycle

Water continues to change its state in response to changes in temperature and atmospheric pressure or to flow to a new location due to gravitational effect. Such movements and changes can occur in the atmosphere (precipitation process), on the ground (runoff process), and in the interface between ground and the atmosphere (evapotranspiration process). Thus, precipitation, runoff, and evapotranspiration are the major components of the hydrologic cycle. The circulation of water on Earth through the atmosphere, land, and ocean forms the hydrologic cycle. It is a continuous process with no beginning or end (Figure 5.4).

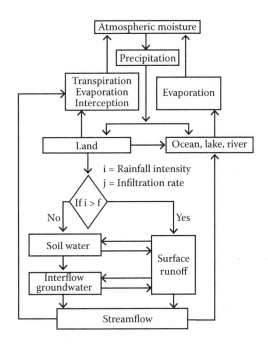

FIGURE 5.4
Hydrologic cycle.

5.3.1 Hydrologic Processes

5.3.1.1 Precipitation

Precipitation is water delivered to Earth from the atmosphere in solid or liquid states. It is the major source of water to a watershed system including streams, springs, soil moisture, groundwater, and vegetation. However, not all rains or snows from a storm event will reach the ground. Some may be vaporized before reaching the ground, and some may be intercepted by vegetation canopies and litters.

For precipitation to occur, water vapor in the atmosphere has to be cooled down below the dew point through certain temperature-cooling mechanisms. This makes water vapor condense into liquid or solid states around condensation nuclei. The formed water droplets are usually very small and drift in the air. If these water droplets grow large enough, through coalescence processes, to be pulled down by gravity, then the water precipitates. However, the total volume of water vapor in the air at a given time is equivalent to about 25 mm (1 in.) of liquid water in depth. Thus, a sufficient amount of water vapor must be supplied and converged at the stormy area in order to have an observed rainfall greater than 25 mm.

Before reaching the ground, part of precipitation will be intercepted and retained by the aerial portions of vegetation and ground litter. The intercepted precipitation will be either vaporized back to the air or dripped to the ground. This process, called forest interception, can affect precipitation disposition, soil moisture distribution, and the impact energy of raindrops on the soil. The amount of interception loss is determined by storm and forest characteristics.

After precipitation has reached the ground, it can enter into the soil to become soil moisture or groundwater. Some of it can run through land surfaces and the soil profile to become streamflow or is lost to the air again through evaporation and transpiration.

5.3.1.2 Evapotranspiration

Liquid water is vaporized to the air whenever there is a vapor pressure gradient between the surface and the ambient atmosphere. Vaporization can occur over surfaces such as soil, water bodies, streams, snowpack, and vegetation. If the active surface of vaporization is leaves (through stomata or cuticle) or bark (through lenticels) and water is conducted through the root system, it is called transpiration. Since separation of transpiration and evaporation is rather difficult in a vegetated watershed, evapotranspiration is a collective term used to describe total loss of water to the air in vapor state. Thus, evapotranspiration is a negative item in the watershed hydrologic balance. Every 1 cm of water lost to the air requires about 600 cal/cm^2 of energy. It represents an energy sink of the environment.

5.3.1.3 Runoff

Precipitation that reaches the soil surface can be entirely or partly absorbed by the soil in the process of infiltration. The infiltrated water can be lost to the air through evapotranspiration, be retained in the soil as soil moisture storage, become groundwater through percolation, or run laterally in the soil profile as interflow or subsurface runoff to reach stream channels. The rate of infiltration processes largely depends on precipitation intensity and soil properties.

When precipitation intensity is greater than the infiltration rate of the soil, or precipitation is greater than the soil water-holding capacity, the excess water will run over the ground surface as overland flow or surface runoff to the nearest stream channel. Some of the overland flow can be trapped in concavities of the surface as depression storage. The combination of surface and subsurface runoff is called direct runoff. It is an influent stream if streamflow comes solely from direct runoff. Such streams flow only during storms, after storms, and in wet seasons.

Under the ground, the profile is divided into two zones: the zone of aeration overlying the zone of saturation. The zone of aeration provides soil moisture, air, nutrients, and sites for plants and animals. The zone of saturation provides a natural reservoir to feed springs, streams, and wells. A stream is effluent if its channel intersects with a groundwater table in which water flows all year round.

Runoff is considered the residual of the hydrologic system in a drainage basin. Precipitation is first to satisfy the watershed storage and evapotranspiration demands before overland runoff can occur. The runoff process involves translocation, water storage, and the change of the state of water. Since its occurrence is in response to watershed climate, topography, vegetation, soil, human activity, and streamflow quantity and quality provide an effective indicator for watershed management conditions.

5.3.2 Hydrologic Budget

In a hydrologic system, the law of conservation applies, and water input must be equal to water output plus storage. The hydrologic budget seeks a quantitative water balance between the input and output components of a watershed system.

5.3.2.1 Globe

For Earth as a whole, there is neither gain nor loss in runoff, and what goes up (vaporization, V) must be equal to what comes down (precipitation, P_t), or

$$P_t = V \tag{5.1}$$

The estimated total precipitation on Earth varies among investigators. According to Baumgartner and Reichel (1975), it is about 973 mm/year, and the average annual precipitation is about 746 mm for the land versus 1066 mm for the oceans. Since the area of land is only about 41.35% of the oceans, the ratio of precipitation volume falling on land to that on the oceans is about 1:3.5.

Equation 5.1 is not valid for all individual localities on Earth. Over some regions of the ocean, P_t exceeds vaporization V, whereas in other regions V exceeds P_t. Overall, V is greater than P_t over the ocean, and the extra water required maintaining sea level, nearly unchanged with time, comes from runoff (RO) of the land:

$$V = P_t + RO \quad \text{or} \quad V - (P_t + RO) = 0 \tag{5.2}$$

For land, the overall precipitation is greater than vaporization (V) and the difference appears as runoff to supplement the vaporization need in the ocean, or

$$P_t - V = RO \tag{5.3}$$

Note that V is evaporation in oceans and evapotranspiration on land. Mean annual balances of P_t, V, and RO for land, ocean, and the globe are given in Table 5.8.

5.3.2.2 The United States

Annual precipitation for the 48 contiguous states is about 76 cm over the entire surface. About 55 cm, or 72.4%, of the precipitation is vaporized back to the air, leaving 21 cm, or 27.6%, to become runoff in rivers and streams. Spatial variations of precipitation and streamflow are given in Figures 5.5 and 5.6, respectively.

The inflow of water vapor in the United States is generally from the Pacific Ocean and the Gulf of Mexico and the outflow northward and eastward to the Atlantic Ocean. However, on the routes of water vapor between inflow and outflow, some water vapor is lost to the ground as precipitation, and some new water vapor joins the routes as vaporization from the ground. Thus, the water vapor balance of the atmosphere is

$$I_v + V - P_t = O_v \tag{5.4}$$

where I_v and O_v are the inflow and outflow of water vapor, respectively. Court (1974) showed that the total inflow of water vapor is about 122 cm/year, or a rate of 0.344 cm/day. Thus, the outflow must be $122 + 55 - 76 = 101$ cm/year, or 0.277 cm/day. The mean vapor

TABLE 5.8

Annual Water Balance of Earth Calculated by Different Investigators

Land			Ocean			Globe	
P_t	V	RO	P_t	V	RO	$P_t = V$	Reference
745	476	269	1066	1177	111	973	Baumgartner and Reichel (1975)
800	485	315	1270	1400	130	1130	UNESCO (1978)
765	489	276	1140	1254	114	1030	L'vovich (1979)

Note: All figures are in mm/year.

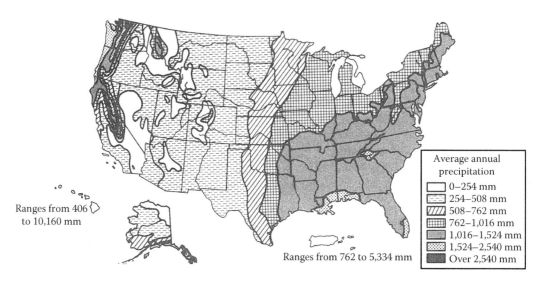

FIGURE 5.5
Average annual precipitation in the United States. (From U.S. Forest Service, An assessment of the forest and range situation in the United States, Forest Resources Report 22, Washington, DC, 1982.)

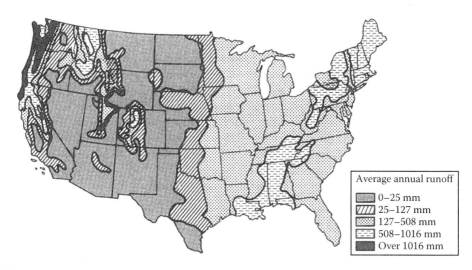

FIGURE 5.6
Average annual streamflow in the United States. (From U.S. Forest Service, An assessment of the forest and range situation in the United States, Forest Resources Report 22, Washington, DC, 1982.)

content of the atmosphere over the United States is about 1.7 cm. An inflow rate of 0.334 cm/day would have a residence time of about 5–6 days to replace the existing water vapor.

5.3.2.3 Watersheds

Equation 5.3 is frequently employed to balance a long-term water budget for a watershed. For watersheds of small size, the boundary for surface water may not be in agreement with the boundary for groundwater. This can make watershed leakage a serious problem because subsurface water can flow out to or flow in from neighboring watersheds. Also, small watersheds may have unsteady soil moisture storage, especially during short periods of time. Under these circumstances, watershed hydrologic balance should take this form:

$$P_t - RO - V \pm \delta S \pm \delta L = 0 \tag{5.5}$$

where δS and δL are, respectively, changes in soil moisture storage and watershed leakage. Note that the variables δS or δL can be positive or negative. If gauging stations are carefully constructed with no occurrence of seepage, such as in many experimental watersheds, and there is no subsurface inflow and outflow caused by physiographic features, $\delta L = 0$ and Equation 5.5 can be employed to solve for vaporization loss V by

$$V = P_t - RO \pm \delta S \tag{5.6}$$

Since there is so much variation in the geology, soil, topography, vegetation, and climate within a watershed, many hydrologic studies have been conducted on small plots that are more uniform in environmental conditions. The small plot watershed approach makes replication of treatments possible. During a dry period, $P_t = 0$; no free water is standing in the soil profile of small plot watersheds. Thus it is logical to assume that there is no surface runoff, no downward water percolation, and no lateral movement of subsurface water in or out of the plot. In this case, Equation 5.6 becomes

$$V = \delta S \tag{5.7}$$

or changes in soil moisture storage can be used to estimate V of the plot. If the soil surface is sealed to prevent soil evaporation, then δS is a measure of plant transpiration.

5.3.3 Energy Budget

The hydrologic cycle described earlier is a continuous change of the state of water between Earth and the atmosphere. A tremendous amount of energy is involved in every component and process of this cycle. Without this supply of energy, there would be no hydrologic cycle and no climate changes.

The energy used in the hydrologic cycle is mainly determined by four components—incoming solar radiation (R_s), outgoing solar reflection (R_r), incoming atmospheric radiation (R_a), and outgoing terrestrial radiation (R_g). The sum of these four components of energy is termed net radiation (R_n). Thus, the energy budget is to track the quantity of energy gains and losses at a given location. In equation form, net radiation (R_n) appears

$$R_n = R_s - R_r + R_a - R_g \tag{5.8}$$

and can be positive or negative, depending on geographic location, season of the year, time of day, and surface conditions. The observed net radiation is further dissipated into latent heat (L_h), sensible heat (S_h), and conductive heat (C_h), or

$$R_n = L_h + S_h + C_h \tag{5.9}$$

For long-term averages, C_h is considered constant; then, R_n is dissipated into either L_h or S_h. Table 5.8 shows the average net radiation to be 49 kcal/cm^2/year for land, 91 kcal/cm^2/year for ocean, and 79 kcal/cm^2/year for Earth as a whole. Of the total net radiation of

Earth, about $67 kcal/cm^2/year$ or 84.5% is used for vaporization. The average temperature of Earth is about 300°K. The latent heat of vaporization corresponding to the 300°K is 582 cal for every cubic centimeter of water. In other words, the $67 kcal/cm^2/year$ of L_h is sufficient to vaporize about 1151 mm of water every year (see Table 5.8).

The heat budgets of different continents and oceans are given in Table 5.9. In humid areas where water is not a limiting factor for vaporization, L_h takes a greater proportion than S_h does. On the other hand, in arid areas such as land at 20°–30° north and south latitudes, water shortage causes S_h to use a greater proportion of R_n than does L_h (Table 5.10).

TABLE 5.9

Heat Balance, in $kcal/cm^2/year$, of Continents and Oceans

	Continent						Ocean		
Component	EUP	ASA	AFA	NAM	SAM	AUA	ATL	PAC	IND
R_n	39	47	68	40	70	70	90	96	97
L_h	28	24	31	24	51	29	79	88	84
S_h	11	23	37	16	19	41	9	10	7

Source: UNESCO, World Water Balance and Water Resources of the Earth, UNESCO Press, Paris, 1978.

Notes: EUP = Europe; ASA = Asia; AFA = Africa; NAM = North America; SAM = South America; AUA = Australia; ATL = Atlantic; PAC = Pacific; IND = Indian; R_n = Net radiation; L_h = Latent heat; A_h = Sensible heat.

TABLE 5.10

Heat Balance, in $kcal/cm^2/year$, of Earth's Surface and the Atmosphere

	Land			Ocean				Earth as a Whole				Atmosphere	
Latitude	R_n	L_h	S_h	R_n	L_h	S_h	F_o	R_n	L_h	S_h	F_o	R_a	L_r
70–60 N	20	16	4	22	26	23	−27	21	19	10	−8	−77	34
60–50	30	23	7	40	47	19	−26	34	33	12	−11	−66	51
50–40	45	25	20	63	65	15	−17	54	44	18	−8	−66	51
40–30	60	23	37	90	96	15	−21	77	65	24	−12	−76	46
30–20	69	19	50	111	106	7	−2	95	73	23	−1	−90	39
20–10	71	32	39	123	117	7	−1	109	94	15	0	−91	70
10–00	72	57	15	126	106	7	13	114	95	9	10	−84	128
00–10 S	72	61	11	128	97	6	25	115	88	7	20	−81	101
10–20	73	45	28	123	113	9	1	115	98	13	0	−84	74
20–30	70	28	42	110	108	11	−9	101	90	18	−7	−81	61
30–40	62	29	33	93	86	11	−4	89	79	14	−4	−78	68
40–50	41	22	19	72	38	4	20	71	47	4	20	−70	77
50–60	31	22	9	45	34	9	2	45	34	9	2	−63	89
Overall	49	27	22	91	82	9	0	79	67	12	0	−79	67

Source: UNESCO, World Water Balance and Water Resources of the Earth, UNESCO Press, Paris, 1978. With permission.

Note: R_n = net radiation balance at Earth's surface; L_h = latent heat; S_h = sensible heat; F_o = redistribution of heat by sea currents; R_a = radiation balance of the atmosphere; L_r = heat input from condensation in the atmosphere. The quantities R_n, R_a, and L_r are taken as positive when they represent a heat input, and other quantities as positive when they represent an expenditure of heat.

References

Baumgartner, A. and Reichel, E., 1975, *The World Water Balance, Mean Annual Global, Continental and Maritime Precipitation, Evaporation, and Runoff*, R. Oldenbourg, Munich, Germany.

Court, A., 1974, Water balance estimates for the United States, *Weatherwise*, December 1974, 252–255, 259.

Cox, P.A., 1989, *The Elements, Their Origin, Abundance, and Distribution*, Oxford University Press, Oxford, U.K.

Czaya, E., 1981, *Rivers of the World*, Van Nostrand Reinhold, New York.

der Leeden, F.V., Troise, F.L., and Todd, D.K., 1990, *The Water Encyclopedia*, Lewis Publishers, Chelsea, MI.

Dormand, J.R. and Woolfson, M.M., 1989, *The Origin of the Solar System: The Capture Theory*, John Wiley & Sons, New York.

Eakins, B.W. and Sharman, G.F., 2010, Volumes of the World's Oceans from ETOPO12, NOAA National Geophysical Data Center, Boulder, CO.

EPA, 2009, National water quality inventory: Report to congress, 2004 reporting cycle, EPA 841-R-08-001, U.S. Environmental Protection Agency, Office of Water, Washington, DC.

Federal Council for Science and Technology, 1962, Report by an ad hoc panel on hydrology and scientific hydrology, Washington, DC.

Government of Canada and U.S. Environmental Protection Agency, 1995, *The Great Lakes, an Environmental Atlas and Resources Book*, 3rd edn., The Great Lakes National Program Office, Chicago, IL.

L'vovich, M.I., 1979, *World Water Resources and Their Future*, Litho Crafters, Chelsea, MI.

Mather, J.R., 1984, *Water Resources, Distribution, Use, and Management*, John Wiley & Sons, New York.

Nace, R., 1984, Water of the world, 1984-421-618/107, U.S. Geological Survey, U.S. Government Printing Office, Washington, DC.

Palmer, T., 1996, *America by Rivers*, Island Press, Washington, DC.

Saboe, C.W., 1984, The big five, some facts and figures on our nation's largest rivers, U.S. Geological Survey, General Interest Publications, Washington, DC.

UNESCO, 1978, *World Water Balance and Water Resources of the Earth*, UNESCO Press, Paris, France.

U.S. Forest Service, 1982, An assessment of the forest and range situation in the United States, Forest Resources Report 22, Washington, DC.

6

Water Resource Problems

Water resource problems are problems of water quantity, quality, and timing. Some regions may have too much water (flooding), while others may have too little water (drought). Water may not occur at the right time and in the right place (timing), or water may not be clean enough for drinking and other uses (water pollution). Moreover, the complexity of water usage in our modern society and the requirement of water for economic development make the rights to capture water and water allocation important issues in water resources (water rights). No region in the world is immune from these problems, which stem from the uneven distribution of water, water mobility, imbalance in supply and demand, steady increases in population, economic growth, poor environmental management, and lack of concern by the public.

6.1 Water Demand and Supply

6.1.1 Water Demand

Water demand is often described as either water withdrawal or water consumption. The term "withdrawal" refers to pumping or taking water from a source for various uses or storage. Water consumption or consumptive use of water refers to the part of withdrawn water that is vaporized to the air; consumed by humans, plants, or livestock; or incorporated into products and not discharged into a supply source for potential reuse.

6.1.1.1 World

Total freshwater withdrawals of the world increased with time, from 1.589 km^3/day in 1900 to about 3.742 km^3/day in 1950, and 10.416 km^3/day in 2000 (Gleick, 2000; World Resources Institute, 2005). The withdrawal rate increased about 0.043 km^3/day/year (15.740 km^3/year) between 1900 and 1950 and about 0.133 km^3/day/year (48.720 km^3/year) between 1950 and 2000. Overall, total freshwater withdrawals in 2000 were 6.57 times greater than those in 1900 (Table 6.1 and Figure 6.1).

The increases in water withdrawal are reflections of population growth, agricultural expansion, and industrial development and were most rapid after World War II. The world population increased 3.81 times during the 100 year period, from about 1.6 billion in 1900 and 2.5 billion in 1950, to 6.1 billion in 2000 (World Resources Institute et al., 2003). Total water withdrawals on a per capita basis were 0.991 m^3/person/day in 1900 and increased to 1.734 m^3/person/day in 2000; North America was the greatest of all regions, about 3.5 times greater than those of South America in 2000.

Agriculture is the biggest consumer despite a continuing fall in its proportion of total water withdrawals from 91% (525 km^3/year) in 1900 to 70% (2661 km^3/year) in 2000.

TABLE 6.1

The World's Freshwater Resources and Total Water Withdrawals in 2000

Region	IRWR[a]	ARWR[b] km³	ARWR[b] m³/Person	Total km³	Per Capital m³/Person	Agri.	Industry	Domestic
Asia w/o M. East	11,192	14,582	4,079	2148	631	81	12	7
Europe	6,591	7,771	10,655	400	581	33	52	15
M. East/N. Africa	518	657	1,505	325	807	86	6	8
Sub-Saharan Africa	3,901	5,463	6,322	113	173	88	4	9
N. America	6,271	6,574	19,992	525	1663	38	48	14
C. Am/Caribbean	1,190	1,259	6,924	101	603	75	6	18
S. America	12,380	17,274	47,044	164	474	68	12	19
Oceania	1,693	1,693	54,637	26	900	72	10	18
World	43,219	55,273	8,549	3802	633	70	20	10

Source: Extracted from the World Resources Institute, Freshwater Resources 2005, in EarthTrends Data Tables: Freshwater Resources, available at: http://www.earthtrends.wri.org/pdf_library/data_tables/wat2_2005.pdf, 2005.

[a] IRWR: Internal renewable water resources, including the average annual flow of rivers and the discharge of groundwater generated from endogenous precipitation—precipitation occurring within a country's boarder.

[b] ARWA: Actual renewable water resources, the sum of the IRWR and ERWR (external renewable water resources). ERWR are the portion of the country's RWR which is not generated within the country.

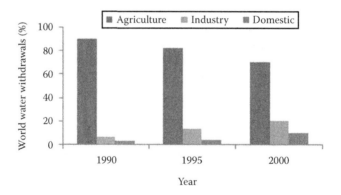

FIGURE 6.1

Percent of world water withdrawals for three water-use categories in 1900, 1950, and 2000. (Data from Shiklomanov, I.A., Assessment of water resources and water availability in the World, Report for the Comprehensive Assessment of the Freshwater Resources of the World, United Nations, Data archive on CD-ROM from the State Hydrological Institute, St. Petersburg, Russia, 1998; World Resources Institute, Freshwater resources 2005, in EarthTrends Data Tables: Freshwater Resources, available at: http://www.earthtrends.wri.org/pdf_library/data_tables/wat2_2005.pdf, 2005.)

Industry used 7% in 1900 and 20% in 2000 (Figure 6.1). Of the major regions around the world, withdrawals in Asia, the biggest water consumer and the fastest-growing region in the world, made up 56% of the world total in 2000 (Table 6.1). Rosegrant and Cai (2002) estimated the total water withdrawals in developing countries to be 74% of the world total in 2025.

6.1.1.2 The United States

The rates of total withdrawals in the United States were much greater than the average withdrawal rate of the world. In 1900, the U.S. population was 76 million, only 4.75% of the world population (Gleick, 2004). Total withdrawals, however, were 0.153 km^3/day, almost 10% of the world total. This represented a per capita withdrawal of 2.019 m^3/person/day, about double that of the world average. In 2000 the U.S. population was 4.67% (285 million) of the world total, but total withdrawals had increased to 1.544 km^3/day (Huston et al., 2004), or 14.8% of the world total. This represented a per capita withdrawal of 5.418 m^3/person/day, about 3.172 times greater than the world average (Table 6.2).

TABLE 6.2

Trends of Estimated Water Use (in km^3/Day) in the United States, 1950–2005[a]

Water Use	1950[b]	1960[c]	1970[d]	1980[e]	1990[e]	2000[e]	2005
				Year			
Population, ×10^6	150.7	179.3	205.9	229.6	252.3	285.3	300.7
Offstream use:							
Total withdrawals	0.6813	1.0220	1.4005	1.6654[f]	1.5443	1.5443	1.5519
Public supply	0.0530	0.0795	0.1022	0.1287	0.1457	0.1635	0.1673
Rural uses	0.0136	0.0136	0.0170	0.0212	0.0299	0.0343	0.0226
Irrigation	0.3369	0.4164	0.4921	0.5678	0.5185	0.5185	0.4845
Industrial:							
Thermal power	0.1514	0.3785	0.6435	0.7949	0.7381	0.7381	0.7608
Others	0.1400	0.1438	0.1779	0.1703	0.1132	()[g]	()[g]
Source of water:							
Ground	0.1287	0.1893	0.2574	0.3142[b]	0.3005	0.3153	0.3013
Fresh	()[g]	0.0015	0.0038	0.0034	0.0046	0.0048	0.0114
Saline							
Surface	0.5299	0.7192	0.9463	1.0977	0.9803	0.9917	1.0220
Fresh	0.0379	0.1173	0.2006	0.2687	0.2581	0.2309	0.2195
Saline treated wastewater	()[g]	0.0023	0.0019	0.0019	0.0028	()[g]	()[g]
Consumptive use	()[g]	0.2309	0.3293[h]	0.3785[h]	0.3558[h]	()[g]	()[g]
Instream use							
Hydro power	4.1635	7.5700	10.5980	12.4905	12.4527	()[g]	()[g]

Sources: Huston, S.S. et al., Estimated use of water in the United States in 2000, U.S. Geological Survey Circular 1263, Washington, DC, 2004; Kenny, J.F. et al., Estimated use of water in the United States in 2005, U.S. Geological Survey Circular 1344, Washington, DC, 2009.

[a] Data for 1980–2000 were modified in Kenny et al.'s (2009) report.
[b] 48 States and District of Colombia.
[c] 50 States and District of Colombia.
[d] 50 States and District of Colombia and Puerto Rico.
[e] 50 States and District of Colombia, Puerto Rico, and Virgin Islands.
[f] Revised.
[g] Data not available.
[h] Freshwater only.

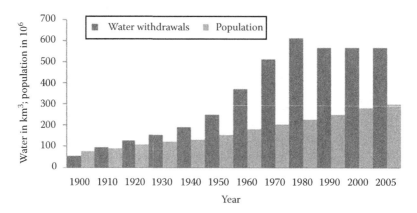

FIGURE 6.2
Total water withdrawals and population in the United States, 1900–2005. (Data from Gleick, P.H., *The World's Water, 2004–2005,* Island Press, Washington, DC, 2004; Kenny, J.F. et al., Estimated use of water in the United States in 2005, U.S. Geological Survey Circular 1344, Washington, DC, 2009.)

Unlike worldwide withdrawals, total withdrawals in the United States did not increase steadily during the period of record. U.S. withdrawals increased from 0.153 km³/day in 1900 to a peak of 1.665 km³/day in 1980 (Figure 6.2). The increasing rate was 0.019 km³/day/year (6.925 km³/year) or 12.4% per year during the 1900–1980 period. However, the population growth rate during the same period was 2.5%, only about one-fifth of the growth rate for water withdrawals.

Despite the continuous growth in population, industry, and agriculture, water withdrawals declined from 1.665 km³/day in 1980 to 1.544 km³/day in 2000 and 1.552 km³/day in 2005, about 93% of the peak 1980 level (Kenny et al., 2009). Agricultural irrigation and industry are the two most important categories in water demand. They make up more than 80% of total water withdrawals in the U.S. public water supply; rural domestic and livestock account for only small fractions of the total demand. The declines in water withdrawals in the 1980s and 1990s seem to reflect improvements in the efficiency of water use in industry and agriculture. Withdrawals for irrigation and thermal power industry in 2005 were, respectively, at 85.3% and 95.7% of levels in 1980, while public water supply and rural domestic water uses continued to grow (Table 6.2).

6.1.1.3 Projections for the Future

Past water usage provides insight for estimating water demands in the future. Such estimation is necessary for water-resources management and planning. However, factors that affect water demand such as population growth, per capita income, energy consumption, agricultural expansion, economic development, and advances in technology are very complex. Projections for the future demand for water are therefore extremely difficult to make and far from precise. Thus, a short-term projection is more reliable than a long-term one, and constant revisions are necessary as new information and technology become available.

Projections of both water withdrawals and consumption by continental regions for three time frames (2000, 2010, and 2015) are given in Table 6.3. These were the assessments for the United Nations made by Shiklomanov (1998). Total worldwide freshwater withdrawals in 2000 were projected at 3927 km³/year, about 7.1% of the actual renewable freshwater resources (Table 6.1), and about 59.3% of the withdrawals were consumptive.

TABLE 6.3

Projections of Total Freshwater Withdrawals and Consumption in Different Continents of the World and the Historical Estimates of Total Freshwater Withdrawals in 2000

Continent	Projection of Withdrawals			Projection of Consumptions			History In 2000
	2000	2010	2025	2000	2010	2025	
Africa, w/M. East	230	270	331	169	190	216	438
Asia, w/o M. East	2245	2483	3104	1603	1721	1971	2148
Europe	534	578	619	191	202	217	400
North America	705	744	786	243	255	269	525
South America	180	213	257	104	112	122	164
Australia/Oceania	32.6	35.6	39.6	18.9	21.0	23.1	127
World	3927	4324	5137	2329	2501	2818	3802

Sources: Data from Shiklomanov, I.A., Assessment of water resources and water availability in the world, Report for the Comprehensive Assessment of Freshwater Resources of the World, United Nations, Data archive on CD-ROM from the State Hydrological Institute, St. Petersburg, Russia, 1998; World Resources Institute, Freshwater Resources 2005, in EarthTrends Data Tables: Freshwater Resources, available at http://www.earthtrends.wri.org/pdf_library/data_tables/wat2_2005.pdf, 2005.
Note: All figures are in km^3/year.

The projection was only 3% deviated from the historical use estimated by the World Resources Institute (2005). Withdrawals are greatest in Asia (excluding Middle East), projected at 57% of the global total in 2000 and 14.4% of its actual freshwater resources. Australia and Oceania are the least, only a fraction, both of global withdrawals and its own resources.

In the United States, total freshwater withdrawals projected by the U.S. Forest Service (1989) were 532, 585, 637, 665, and 727 km^3/year for the years 2000, 2010, 2020, 2030, and 2040, respectively. The actual renewable freshwater resources in the United States are about 3069 km^3 (World Resources Institute, 2005).

6.1.2 Water Supply

6.1.2.1 World

Total land runoff of the world is 266 mm/year (Baumgartner and Reichel, 1975) or 39.162×10^3 km^3/year. This is renewable water that is available for management. However, not all of this water can be utilized due to geographical location or lack of suitable sites for capture.

Total freshwater withdrawals were 3802 km^3 or 633 m^3/person in 2000. The world's water resources are finite, but the demand for water steadily increases with increasing population and economic development. A 2% annual population growth rate would bring the world's population from 6.1 billion in 2000 to 44 billion by the end of the year 2100. This population requires 28×10^3 km^3 of water to meet water withdrawal standards at the 2000 levels, equivalent to about 50% of total renewable water resources of the world. Rosegrant and Cai (2002) estimated the total global water withdrawals to be 10% of total renewable water resources in 2025.

Water supply is an issue of global quantity and distribution. Annual rainfall ranges from 4000 mm at some stations in Taiwan to sometimes nothing at all in the Sahara Desert of Libya.

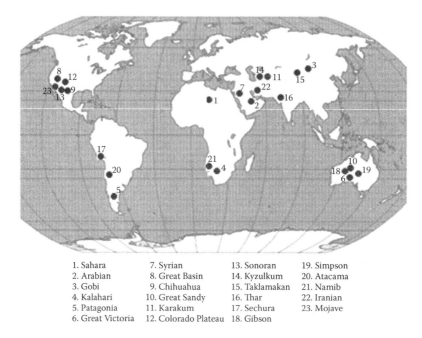

1. Sahara	7. Syrian	13. Sonoran	19. Simpson
2. Arabian	8. Great Basin	14. Kyzulkum	20. Atacama
3. Gobi	9. Chihuahua	15. Taklamakan	21. Namib
4. Kalahari	10. Great Sandy	16. Thar	22. Iranian
5. Patagonia	11. Karakum	17. Sechura	23. Mojave
6. Great Victoria	12. Colorado Plateau	18. Gibson	

FIGURE 6.3
Distribution of major deserts of the world. (Blank map provided by http://studentsfriend.com; Information from Goudie, A.S., *Great Warm Deserts of the world—Landscapes and Evolution*, Oxford University Press, Oxford, U.K., 2003.)

The uneven distribution of precipitation and consequently runoff around the world made water scarce for 132 million people in 1990 and a projected 1.06–2.43 billion people by 2050 (World Resources Institute et al., 1996). About one-fifth of the land on Earth is classified as arid or desert; these lands, when combined with semiarid lands, make up about one-third of the total land area (Robinson, 1988). Deserts are found in North Africa, southwest Asia, Central Eurasia, the Pacific coasts of Peru and Chile, Argentina, the Great Plains and the Southwest of the United States, the Atlantic coasts of Namibia and South Africa, and most parts of Australia (Figure 6.3). There, the annual precipitation is less than 254 mm, evaporation is greater than precipitation, and water deficit prevails. Water supplies in these regions depend on groundwater, imported water, or other alternatives. Water scarcity in these regions may force residents to adjust lifestyles and cause limitations in economic development.

Since water supplies are finite and water consumption increases with time, a measure to assess pressures on water supplies called the *water stress index* has been developed. It represents the annual renewable water resources per person available to meet needs for production and domestic use. An index of $1000 \, m^3$/person/year has been proposed as a benchmark for moderately developed countries in arid regions. Countries with a water stress index below that benchmark are most likely to experience chronic water scarcity sufficient to impede development and harm human health. By this measure, some 20 countries including Kuwait ($75 \, m^3$/person), Israel ($461 \, m^3$/person), and Singapore ($222 \, m^3$/person) already suffered from water scarcity in 1990 (World Resources Institute et al., 1996).

UNEP (1999) reported that about one-third of the world's population already suffers from moderate to high water stress in which water consumption exceeds 10% of the renewable freshwater supply. If the present consumption pattern continues, two out of three persons on Earth will live in a water-stressed condition by 2025.

6.1.2.2 The United States

About 80% of water withdrawals in the United States come from streams and reservoirs (surface water), and the other 20% from water deep under the ground. The surface water is of course replenished by precipitation (P_t). However, surface runoff is only a portion of P_t. A great deal of P_t is lost to the atmosphere through evapotranspiration (ET; or vaporization, V). Thus, the amount of water supply in a region is dependent upon the magnitudes of P_t and ET. In areas where annual P_t is less than annual ET, water deficit prevails and the water supply depends largely on groundwater. On the other hand, water surplus occurs if annual P_t is greater than annual ET. In 2005, for example, groundwater contributed about 61% of total water withdrawals in Nebraska, a semiarid state in the mid-central United States. However, groundwater represented only 6% of total withdrawals in Pennsylvania (Kenny et al., 2009), a humid state in the East where P_t exceeds ET.

Annual precipitation in the continental United States is about 16.05 km³/day, and about 11.42 km³/day is lost to the air through evapotranspiration. This leaves 4.63 km³/day of water (the manageable supply) to percolate to groundwater or run directly over the land surface to streams that eventually drain into the oceans.

Of the 1.55 km³/day total water withdrawals in 2005, about 1.24 km³/day was surface water (Table 6.2). This was about 27% of annual runoff. The remaining 73% or 3.39 km³/day was available to supply additional growth in water demand. The projected surface water withdrawals in 2040, according to the U.S. Forest Service, are 1.55 km³/day. U.S. water supplies should be adequate to meet this demand.

Although overall the United States seems to have plenty of water to meet demand in the future, a large region of the United States has experienced water shortages every year due to uneven distribution of population, industry, agriculture, and available water. A deficit of water exists where potential evapotranspiration is greater than precipitation or demand for water is greater than supply. These areas lie generally west of the 96th meridian and east of the Rocky Mountains. Here annual precipitation is generally less than 60 cm, and surface runoff of less than 10 cm occurs only during or immediately after storms, except in areas in higher mountains. Table 6.4 shows that the 17 western states have 33%, or 544 km³/year, of the total runoff in the contiguous United States, but the West withdrew 44% or 240 km³ of the nation's streamflow in 1990. The greatest demand for water in the West is for agricultural irrigation, and in the East it is for industrial purposes. These uses, respectively, make up 82% of total off-stream withdrawals in each region.

The precipitation regime of a region is affected by:

1. Position relative to air-mass movement and moisture content of the air
2. Proximity to major pressure systems
3. Elevation and other topographic characteristics

The United States is located in a region of Earth that is influenced by the westerlies (prevailing winds blowing from west to east). Major tracks of storms and weather fronts move from the Pacific Ocean and the Gulf of Mexico to the Atlantic Ocean. These cause the windward (western) slopes of the Rocky Mountains and the Northern Pacific coastal plains to

TABLE 6.4

Total Water Withdrawals, Population, and Per Capita Use of
Water between the Eastern (31 States) and Western (17 States)
United States in 1990 and Their Normal Runoff

Items	Western States Volume	%	Eastern States Volume	%	Total Volume
Water withdrawals, km³/day					
Surface water					
Fresh	0.4158	41.4	0.5810	58.6	0.9968
Saline	0.0548	25.6	0.1583	74.4	0.2132
Sum	0.4707	30.9	0.7393	69.1	1.2100
Groundwater					
Fresh	0.1833	64.0	0.1031	36.0	0.2864
Saline	0.0037	94.9	0.0002	05.1	0.0039
Sum	0.1870	64.4	0.1033	35.6	0.2903
Total					
Fresh	0.5991	59.9	0.6841	40.1	1.2832
Saline	0.0586	27.0	0.1585	73.0	0.2171
Sum	0.6577	43.8	0.8426	56.2	1.5003
Per capita, m³/day					
Fresh	7.1353		3.8529		4.9088
Total withdrawals	7.8431		4.7455		5.7391
Population, 1000s	83,853	32.1	177,563	67.9	261,416
Runoff (1931–1960), km³/day	1.4898	32.8	3.0522	67.2	4.5420

Sources: Data for water withdrawals and population were calculated
from Solley, W.B. et al., Estimated use of water in the United
States in 1995, U.S. Geological Survey Circular 1200,
Washington, DC, 1998; Runoff data were calculated from
Murray, C.R., Estimated use of water in the United States, 1965,
U.S. Geological Survey Circular 556, Washington, DC, 1968.

Note: Percents in rows total to 100.

record the greatest precipitation in the United States. These areas are followed by the Gulf Coast around the mouth of the Mississippi River, southeast Florida around West Palm Beach, and the high mountain plateau of western North Carolina. Annual precipitation for the community of Forks on the western side of the Olympic Mountains, Washington, DC is about 3000 mm, but it drops to about 450 mm at Port Townsend, Washington, only 120 km away on the leeward (eastern) side of the same mountains.

The Intermountain region and the Southwest are in a rain-shadow created by the Cascade Range and Coast Ranges to the west. The Coast Ranges block the influx of warm moist-air masses moving from the Pacific. Also, moist-air masses from the Gulf of Mexico and the Atlantic gradually lose moisture as they travel inland. These factors cause the Southwest and the Intermountain regions to be the driest in the nation. Normal annual precipitation is only about 100 mm at Yuma, an Indian reservation near the border of Arizona, California, and Mexico. The seasonal precipitation distribution for a few selected locations in the United States is shown in Figure 6.4.

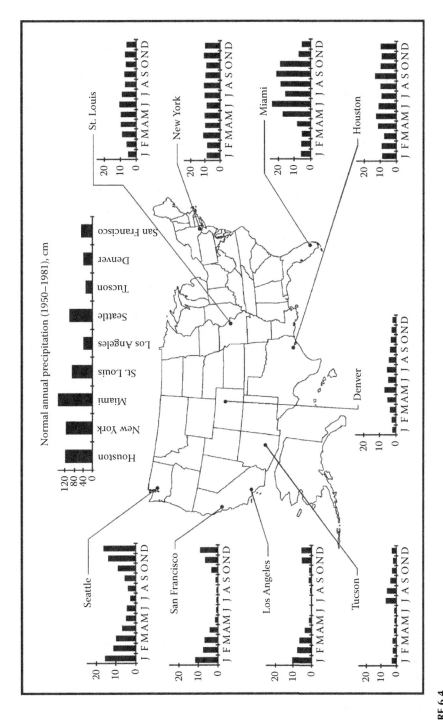

FIGURE 6.4
Seasonal precipitation distribution for selected locations in the United States.

6.2 Water Quantity

6.2.1 Drought

When water demand is greater than supply, water shortage occurs. If the shortage extends over a significant period of time, it is a drought. However, water shortage can apply to agriculture, to the water level in the rivers and reservoirs, or to human comfort as a result of lack of rainfall, high temperature, and low humidity. Thus, drought has been defined through meteorological, agricultural, hydrological, or even socioeconomic points of view. It means different things to different people and varies from region to region. In equatorial areas, a week without rain is a drought, while such a period can extend to 2 years in Libya.

Unlike floods, drought occurs slowly, grows gradually, and acts silently. There are no exciting violent occurrences or immediate disruptions of life and social order associated with drought. The damages of drought are chronic. But once it becomes noticeable, it may have already caused tremendous losses in agriculture, reductions in economic growth, increases in food prices, and problems of infestation associated with insects, fungi, bacteria, and viruses, which in turn may shift farming practices and lifestyles. Drought is not just a natural phenomenon that only affects crops or water supplies. The effects of drought cover all aspects of life and are much more profound and widespread than those of floods. In the United States, average losses from drought have been estimated at $6–$8 billion/year and exceeded $20 billion in 2002 (Wilhite and Buchanan-Smith, 2005).

No panacea can be used with economic feasibility to fight effectively against drought. Although elimination of drought is unattainable, numerous measures have been developed to cope with shortages of water and reduce drought damage. These measures either augment water supply or conserve water use and work better when applied in concert. They include the following:

Water Augmentation	Water Conservation
Rainmaking	Household conservation
Water translocation	Evaporation reduction
Desalination	Irrigation efficiency
Water reclamation	Vegetation control
Rain harvesting (water storage)	Genetic engineering
Iceberg harvesting	Drought forecast
	Price control

6.2.1.1 Rainmaking

Artificially induced rain or snow, or *precipitation enhancement*, would be an ideal solution to droughts if precipitation timing, location, and amounts could be controlled. Rainmaking is done by seeding clouds with crystals of dry ice, silver iodide (AgI), salts, or even clay particles when rain clouds are present in the sky. These seeding aerosols are used to enhance condensation activities and can be launched by airplane, artillery, or rockets. Precipitation enhancement (PTE) is often used to mitigate drought damage and control forest fires in China and has been reported to increase precipitation up to 55%. In 1997 alone, PTE was conducted through 360 airplane flights over 18 provinces in northern

China (Hu, 2005). Experience in the United States shows that such activities could increase precipitation on the order of 10%–50% in some regions, with no significant increases or even decreases in precipitation in other regions. Much of the uncertainty is due to the nature of precipitation variability.

Besides the uncertainty problem, the fact that rain clouds are a requirement makes rainmaking inapplicable in very dry areas where such clouds are rarely available. In addition, using these clouds may create legal disputes over their ownership. As water is removed from clouds in one area, it could deprive neighboring areas of water. Large-scale application can cause undesirable changes of regional or even global weather patterns. A cloud-seeding experiment conducted near Rapid City, South Dakota, in the summer of 1972 was allegedly responsible for the subsequent 356 mm of rain and flooding that followed (Owen, 1985). The usefulness of rainmaking is limited by a lack of control over the volume, intensity, and distribution of resulting precipitation.

6.2.1.2 Water Translocation

The distribution of freshwater supplies in nature is very uneven geographically. Cities are expanding, becoming so large that water supplies are not sufficient for their demands; agricultural irrigation is always needed in arid areas but water is in limited supply. A conventional solution to the water shortage is to transport water from an area of surplus to an area in need through pipelines, channels, canals, trucks, etc., depending on distance and quantity.

Libya is located in North Africa with the Mediterranean Sea to the north, Egypt and Sudan to the east, and Chad and Niger to the south. About 95% of the country is desert, and the average annual rainfall is 26 mm with only 7% of the land exceeding 100 mm/year. The arid climate leads Libya to rely on water stored in reservoirs during wet years and in underground aquifers for its water supply.

In 1986 Libya initiated the "Great Manmade River Project" to pump water out of aquifers deep beneath the Sahara desert formed 20,000–40,000 years ago. Some 1300 wells and 4000 km of concrete pipeline (4 m in diameter × 7.5 m in length), buried 6–7 m under the ground, will stretch from the desert to cities along the Mediterranean (Borrell, 1986; El-Geriani et al., 1998). The project consists of five phases with a total volume of water $2 \, km^3$/year to be transported for a period of at least 50 years. It is expected to meet the demands for water for the heavy population along the coast and to irrigate nearly 200,000 ha of agricultural land. The first and second phases were inaugurated in 1991 and 1996, respectively. The first water reached Tripoli in August 1996 and Gharyan in September 2007. Tapping the deep aquifers has raised concerns that smaller pockets of water close to the surface and scattered oases may disappear. As a result, large areas of desert may be depopulated, and the traditional way of life of its nomadic inhabitants destroyed. Because of low rainfall and high evaporation, natural recharge of groundwater is practically nonexistent. Continuing abstractions of groundwater may cause drawdown of the peizometric surfaces, making it difficult to keep yields sustainable (Abdelrhem et al., 2008).

Perhaps the most complex and expensive water-translocation project in the United States is California's aqueduct system. About 70% of California's potentially usable water originates from the northern third of the state, while 77% of the demand comes from the semiarid southern two-thirds of the state. A dam of 235 m height blocks water in the Feather River in the North to form Lake Oroville, with a capacity of $4.32 \times 10^9 \, m^3$ and depth of 213 m. The water is then alternately translocated toward Los Angeles, San Diego, and the San Fernando Valley through 21 dams, 22 pumping plants, and 1046 km of canals, tunnels, and pipelines (California Department of Water Resources, 1973). The project has been criticized for its excessive costs in construction and energy, losses in scenic beauty and free-flowing wild rivers, and destruction of fish and wildlife habitats (Owen, 1985).

New large-scale water-transfer projects are increasingly difficult to establish because of the following reasons:

1. Political and environmental issues can create lengthy delays.
2. The costs are extremely high.
3. The construction time is too long.
4. Smaller projects or other approaches with fewer environmental-impact issues are more attractive to policymakers.
5. Environmental groups insist on greater water use efficiency prior to construction.

6.2.1.3 Desalination

Saline water includes brackish water (1,000–4,000 mg/L of dissolved solids), salted water (4,000–18,000 mg/L), seawater (18,000–35,000 mg/L), and brine water (more than 35,000 mg/L). With seawater accounting for more than 97% of the world's total water budget, converting saline water into usable water seems to be a logical alternative for augmentation of water supply. Technologically, the conversion can be done by the use of distillation processes or reverse osmosis. Commercially, the techniques are not yet in widespread use because of energy requirements, costs, and environmental constraints such as waste disposal problems, heat, and air pollutants. Desalted water costs about $1 per 1000 gal (3.785 m^3) compared to $0.25–$0.40 per 1000 gal for tap water. In 1999, the desalination capacity of the world was about 0.021 km^3/day (0.15% of water consumption), and 75% of the capacity was in 10 countries (Gleick, 2000). As new energy efficient technologies continue to improve, desalinization may well become a major option for increasing water supplies in coastal regions as well as interior areas in abundance of brackish water (Bell, 2005).

6.2.1.4 Water Reclamation

Sewage wastes have long been used to irrigate agricultural crops and, more recently, forested lands as a provider of both moisture and fertilizers in areas where water supplies are limited (Stulp, 1995). Although many people are concerned about aesthetic effects and about the possible transmission of viruses and other contaminating organisms, waste effluents in fact can be treated to yield a quality suitable for many uses, including body contact. Such processed sewage water has already been used for a variety of functions, including irrigation, steel cooling, groundwater recharging, aquaculture, creation of constructed wetlands, streamflow augmentation to enhance salmon runs and to improve water quality, recreational lakes, and domestic and industrial supplies.

As new water resources and transfer projects become more difficult to develop, reclaimed water provides an economical source to ease our water supply constraints. It already has become a safe, reliable, and drought-proof source of nonpotable water for arid states such as California, Arizona, Nevada, and Florida. In 1992, the state of Washington passed the Reclaimed Water Act (codified as RCW 90.46) to encourage beneficial use of reclaimed water for various applications. The reclamation and reuse system created by the city of Yelm, Washington, is a success story (Skillings, 1999). The system totally eliminates the discharge of wastewater into the Nisqually River to protect water quality and salmon habitat and the city's groundwater resources, and it provides water for public uses such as ground watering, vehicle washing, and a wetland park.

6.2.1.5 Rain Harvesting

In rain harvesting, rainwater is captured to increase water availability for crop production, gardening, household, livestock, or other uses. Methods include trapping more runoff through land-surface treatments; collecting rainwater from roofs; increasing snow accumulation through snow fences, terrain modification, and forest management practices; and catching fog and cloud drips in coastal areas or at tops of mountains by installing artificial screens. Some techniques for harvesting rainwater from roofs and ground catchments were described by the United Nations Environment Programme (1983).

Land surfaces can be treated to produce more runoff water by digging ditches to catch water from hillsides and directing the water to a pond for various uses. The ditches must be on soils with low water permeability; otherwise the soil needs to be treated with sodium salts or spread with water-repellent compounds such as asphalt, paraffin, latex, or silicones. One can use large sheets of plastic or Agri-fabric, a nonwoven engineering geotextile made from 100% Trevira polyester, to cover the entire runoff-producing area and the storage reservoir. To reduce evaporation losses, the reservoir should be deep rather than large in surface area. Rainwater can also be harvested from house roofs using gutters, drains, or troughs and stored in tanks (cisterns). These are all ancient practices and can be applied in areas where average annual rainfall is as low as 50–80 mm. In the U.S. Virgin Islands roof rainwater-catchment systems are quite common, and local ordinances specify catchment areas and storage volumes (Sharpe, 2001). Systems like these have been promoted in urban areas to provide water for landscape maintenance. Several publications are available that provide methods for calculating necessary roof areas and storage volumes based on rainfall amounts (Young and Sharpe, 1989).

In alpine areas, snowfall is the major type of precipitation and the dominating factor in the yield and timing of streamflow. Intensive management of alpine forests and snowfields offers excellent opportunities for increasing snow accumulation and delay of snowmelt, and consequently augmenting water yields. Along coastal areas or fog-prone regions, reversing the effect of rainfall interception can be observed due to condensation of fog or cloud moisture on the forest canopy, dripping to the ground as additional rainwater (Goodman, 1985; see part II-B, Chapter 8). Arranging lattices and nets above woods in Japan has collected as much as 20 times more water on the windward than on the leeward side (Oberlander, 1956; Cavelier et al., 1996). Under an old-growth Douglas-fir forest near Portland, Oregon, fog drip could add an additional 882 mm of water to total precipitation during a year (Harr, 1982). Accordingly, loss of fog drip due to patch clearcutting of the watershed could cause a significant decrease of streamflow in the summer (Ingwersen, 1985).

A fog-collection system of 75 polypropylene nets was installed to capture water drops at Chungungo, Chile. The trapped water was directed into troughs where the water was chlorinated. It was then fed into a gravity system that descended to the village through a 6.5 km pipeline. Before the system was installed, each villager lived on 13 L of water per day delivered by trucks once a week. Now, the fog-collection system furnishes each resident 53 L of water a day at half of the previous cost (*U.S. Water News*, 1994; Olivier, 2002).

The barrel-shaped baobab trees (*Adansonia digitata*) of tropical Africa have trunks up to 9 m in circumference (Figure 6.5). Some tribesmen hollow out the trunks of these trees for houses and commercial stores or for water storage. In Sudan, people dig a moat around a

FIGURE 6.5
Two baobab trees, *Adansonia digitata* of Tanzania on left and *Adansonia grandidieri* of Madagascar on right. The trunk of baobab trees in Africa is often hallowed out for storage of water, up to $4\,m^3$. (Courtesy of Wikimedia Commons, http://en.wikipedia.org, a freely licensed media file repository.)

baobab tree for rain harvesting. The harvested rain is then bucketed, as much as $4\,m^3$, into the hollow trunk for storage.

6.2.1.6 Iceberg Harvesting

More than 77% of the world's freshwater is locked up in ice, mostly in Antarctica and in the Arctic. If the ice could be melted and drained into reservoirs, there would be enough water to supply our needs for more than 7000 years at the present withdrawal rates. Icebergs in Antarctica are broad and flat, resembling giant float tables or ice islands. It may be economically feasible to tow Antarctic icebergs to Southern California, South America, Saudi Arabia, Australia, or other coastal places using oceangoing tugs aided by the favorable Humboldt Currents. These icebergs would then be melted and piped inland for domestic, industrial, or agricultural purposes. The idea received considerable attention when the *First International Conference and Workshop on Iceberg Utilization* was held at Iowa State University in Ames, Iowa, in 1978 (Husseiny, 1978). However, some problems exist regarding the utilization of icebergs:

1. The ownership of icebergs
2. The environmental effects on coastal water temperature, fish reproduction, and migration as well as on climatic patterns
3. The hazards that might be created if chunks of icebergs were broken off in international shipping lanes

It is unlikely that this intriguing idea will be implemented in the near future. But as water shortages continue to grow and the operational technology is further developed, icebergs could provide a new water resource that will meet our water demands for a long time.

6.2.1.7 Household Conservation

About 30%–50% of the water used in the United States is wasted unnecessarily. Water conservation means using water wisely, increasing efficiency, and reducing waste. Thus, household conservation involves the use of technologies, practices, policies, and education to reduce per capita use of water by people in drinking, cooking, bathing, toilet flushing, lawn watering, fire protection, swimming pools, car washing, laundering, and washing dishes. Much of the water around individual homes is wasted due to poor habits, lack of concern, or inefficient plumbing fixtures. Slow drips from faucets can waste as much as 650 L of water each day or 240 m^3 each year. The Texas Water Development Board (1994) has shown how household water consumption can be reduced in a brochure called Forty-Nine Water Saving Tips. Frequently, city ordinances are required to help govern plumbing codes, lot sizes, drainage grades and slopes, or outdoor usage, which can reduce the quantity of water used. Nationally, legislation was passed in 1992 requiring water-efficient plumbing fixtures (Energy Policy Act of 1992).

6.2.1.8 Evaporation Reduction

Annual lake evaporation in the United States, excluding Alaska and Hawaii, ranges from about 500 mm in the north of Maine to 2000 mm in the southwest of Texas (U.S. Weather Bureau, 1959). Saving 10 mm of water from a reservoir of 50,000 ha in size is sufficient to supply domestic water to 50,000 people for more than 5 months. Reservoir evaporation can be reduced by reducing the evaporation surface, by the application of mechanical covers to the surface, or by spreading the surface with thin films of chemicals. The surface-film method is the simplest and the most economical for evaporation reduction. Chemicals used for this purpose must be insoluble, nontoxic, pervious to oxygen and carbon dioxide, penetrable by raindrops, flexible with motion of the water surface, and inexpensive. Depending on wind effects, type of chemical, and the conditions under which it is used, normal evaporation can be reduced as much as one-third by efficient use of monofilms.

6.2.1.9 Irrigation Efficiency

Water loss due to inefficient agricultural irrigation is another important item in water conservation. As much as 50% of the irrigated water is not used by crops due to evaporation loss or seepage. Improving the efficiency of existing irrigation practices can increase crop yields, reduce weed growth, and cut irrigation water by 50%–60%. The methods include:

1. Better scheduling of water application (based on real-time soil moisture and meteorological conditions)
2. Using closed conduits in water-conveyance systems
3. Using trickle or drip irrigation systems to apply water directly into the roots of plants rather than flooding the entire area
4. Reducing irrigation water runoff by using contour cultivation and terracing, and storing runoff water for use in farm ponds or small lakes
5. Covering the soil with mulches

6.2.1.10 Vegetation Control

Because of surface area, deep root systems, more available energy, and wind effects, water losses from vegetated watersheds are greater than if the same area were bare ground. Plant size, canopy density, roots, and leaf characteristics vary among species, resulting in variable transpiration losses. Thus, vegetation management, especially on phreatophytes and forested watersheds, provides promising opportunities to reduce water loss and consequently augment water yield.

Phreatophytes are water-loving plants whose roots penetrate into the groundwater table or spread widely for possible sources of water. They represent the most serious problem in water management in semiarid and arid regions. In Texas, one such plant, mesquite, grows in drainage ways with annual rainfall of 15 cm extending to beyond 75 cm. Its roots penetrate the soil more than 5 m deep and spread as much as 15 m from its trunk, transpiring a luxurious amount of water to the air. Another plant, saltcedar, grows extensively in 15 of the 17 western states. It forms an almost-continuous band of vegetation along many rivers and streams. Saltcedar not only consumes a large quantity of water (possibly reaching 2700 mm/year) but also chocks river beds with vegetation and sediment, thereby increasing flooded areas and peak flows. The elimination of saltcedar growth followed by grassland restoration might result in a collective net water saving of 300 mm/year (Texas Water Development Board, 1968). Other species of such water-wasting plants include cottonwood, elm, post oak, black oak, live oak, shin oak, whitebrush, sagebrush, sassafras, cactus, juniper, yaupon, retama, and willow.

Forests transpire more water than other types of vegetation do. Water yield can thus be increased by manipulation of forests without increasing surface runoff and soil erosion. The method is to reduce transpiration by:

1. Converting vegetation from species that use much water to species that require less water
2. Clearcutting or partial cutting of the forest
3. Reducing forest densities
4. Using antitranspirants

Many management methods, including conversions of chaparral to grass, tree to grass, tree to shrub, and hardwood to conifer, or forest clearcutting, patch cutting, understory cutting, and thinning, have been tested in small watersheds for their effects of streamflows in many areas around the world (Reynolds and Thompson, 1988). The results show an increase in streamflow ranging from 30 to 540 mm/year in the United States, depending on annual precipitation, vegetation, management, and physiographic factors.

Since most of the vegetation-management studies were conducted in small watersheds, it is uncertain how much water can be increased from large watersheds and when and how that water will be released. Upstream forest manipulation results in increases of water yield downstream. Frequently, the landowners and beneficiaries are different people, especially when large watersheds are involved. Until the legal problems are solved, and objective models for evaluating input and output components of water augmentation projects over a broad range of environmental conditions have been developed, validated, and accepted by the courts, the approach, if carried out, will inevitably be restricted to public lands. Private forest owners will not base their management strategies on water production. Moreover, the method is inapplicable in areas where annual rainfall is less than 500 mm or topography is so rugged and fragile that clearcutting would create severe soil-erosion problems.

6.2.1.11 Genetic Engineering

Through research and experiments in genetics, it may be possible to breed plants that require considerably less water or grow well in saline soils. Once these salt-resistant and drought-resistant crops are developed, seawater can be used for irrigation, and vast areas of arid or semiarid regions can become productive. A new saltwater barley developed by two scientists from the University of California at Davis grew well on a tiny windswept beach with a yield of 1500 kg/ha (Owen, 1985). There are 1.8×10^6 ha of once-prime agricultural land in California that are worthless now due to the salinization caused by improper irrigation practices or seawater intrusion. The new barley could be expected to do well in this type of soil.

6.2.1.12 Drought Forecast

Low-flow information is essential for reservoir design, the management of water-supply systems, and the study of the waste-assimilative capacity of streams. A common criterion for these purposes is the lowest 7 day, 10 year flow (Chang and Boyer, 1977). Information on the probability of occurrence of droughts with various severity and duration during a single year or during any specific period of years can be statistically analyzed using historical records of precipitation and streamflow. The severity of droughts can be measured by parameters such as deficiencies in rainfall and streamflow, declines in soil moisture and groundwater level, persistence of rainless days, various drought indices, and the storage required to meet prescribed withdrawals. Analysis of past records indicates that the sequence of dry years is not random, and a prolonged drought in the Western United States occurs about every 22 years. Large-scale studies on the distribution, occurrence, severity, and duration of droughts may provide clearer insight and more reliable information to help prepare for problems in the future, including addressing them with conservation programs and water storage.

6.2.1.13 Price Control

Water has been priced at reproduction costs in the past. However, most systems charge a minimum price for a given quantity of water with a decreasing price for additional use, while others charge a minimum price with an increasing rate for additional use. Other water systems charge a uniform rate that is independent of use. Innovative increasing block rates have been adopted in many municipalities to encourage conservation (Young et al., 1983). When water-supply savings are added to sewer and energy costs, consumer savings can be substantial.

6.2.2 Floods

An inundation or overflow of the floodplain of a river, which causes substantial loss of life, personal property, public facilities, or agricultural productivity, is called a flood. Floods are the number one killer among natural disasters. A flood that occurred along the Yellow River (Huang Ho) in China in October 1887 surged through 1500 towns and villages and drowned between 900,000 and 2.5 million people (Clark, 1982). The death toll outnumbered that of any other major natural disaster known today, including the earthquake of Tangshan, China, in July 1976 (800,000 lives), the cyclones of Bangladesh in November 1970 (300,000 lives), and the tornadoes of Missouri, Illinois, and Indiana on March 18, 1925 (689 lives).

6.2.2.1 Forms and Causes

Floods occur in various forms: flash floods, river floods, urban floods, coastal floods, and terminal lake floods. Each type has features different from the others. Knowing the characteristics of these floods can help us prepare for the threat and reduce the damage to a minimum level. These forms and their causes are discussed in great detail in Chapter 14.

6.2.2.2 Nature of Floods

Floods can be measured for their elevation, area inundation, peak discharge, volume of flow and duration. The size of these variables depends upon the rate, duration, distribution and direction of the storm as well as the weather and the condition of the ground on which it falls. Thunderstorm activity and intense storms are predominantly warm-season phenomena in the United States. However, no specific intensity of storm or rate of flow will cause flooding. What may be a flood in one section of a river may be well-controlled flow in another section.

6.2.2.2.1 The Impact of Season

The flood-producing potential of intense storms is much greater during winter and in urban areas. When unexpected storm intensities occur during cold seasons, especially during late winter, streamflow rates may be maximum, and where there is no canopy interception or transpiration in hardwood forests, evaporation is negligible, and the soil possibly frozen or highly saturated, the results are frequently disastrous. Also, when an abnormally warm temperature, which can melt much of the snowpack, is followed by an intense storm in the cold season, flooding results are frequently disastrous.

6.2.2.2.2 The Impact of Urbanization

In urban areas, paved streets prevent rainwater from entering into soils, and the removal of trees in urban areas reduces transpiration loss of water from the soil. Consequently, run-off in urbanized areas can be two to six times over what would occur on natural terrain.

6.2.2.2.3 The Impact of Watershed Size

The hydrological behavior of watersheds varies with basin sizes. In small tributaries or upstream areas, the average slope steepness is greater, the average rainfall intensity is higher, the soil depth is shallower, the watercourse is shorter and steeper, and the variables of climate may be more extreme than in large watersheds. Thus, small watersheds are very sensitive to high-intensity rainfalls of short duration and can be greatly affected by land use. Floods in small watersheds are characteristically flash floods of short duration and high peak flows per unit area. On large watersheds, the effect of channel flow or basin storage is the dominant factor in flood hydrographs. Thus, if flood flows from several small tributaries arrive at the main stream at about the same time, a major flood can occur. Major floods do not occur in direct response to storm intensity. They occur gradually and drain out slowly.

6.2.2.3 Flood Damages

People can be drowned in floodwaters or indirectly killed by flood-induced fire, diseases, animal attack, or even snake bite. Similarly, property damages can be direct, indirect, or intangible. Direct losses apply to those items that can be determined in cash value, while indirect losses are others, such as property depreciation and loss of a business or job, caused by the effects of flooding. Intangible losses include damages to environmental aesthetics, environmental hazards, loss of ecological habitat, migration of population, or crime activity that cannot be assessed by cash value. The assessment of flood losses is not easy. For example, flooding can cause substantial soil erosion and consequently soil deposition on crop fields, streets, buildings, reservoirs and many other locations. In Bangladesh, the raging floodwaters in the summer of 1998 submerged tens of thousands of wells across the country, making the well water undrinkable. The government reported nearly 194,000 cases of diarrhea and at least 160 deaths. Thus, the damages may include loss of crops, costs due to the cleanout of soil deposition, depletion of land productivity, degradation of water quality, reduction of reservoir capacity, and impairment of landscape aesthetics.

River flooding can deliver a large quantity of freshwater to the ocean, dilute the ocean's salinity level, and consequently affect marine biology. For example, the 1998 summer flood that occurred along the middle and lower Yangtze River was the largest recorded in China in the twentieth century. It lasted more than 60 days, drowned more than 1300 people, and caused a direct economic loss exceeding 177 billion Chinese yuan. The floodwaters poured into the Yellow Sea, drifted to Cheju Island, South Korea, and reduced the island's salinity level from 3.1% to 2.5%. It was reported that the island's marine shellfish farming had already been reduced by 30% during the Yangtze River's flood of August 1996.

Flood damages can also cause the disruption of communication and transportation systems as telephone lines go down, bridges are washed out, and highways and railroad tracks are torn apart. Broken power lines and ruptured gas mains can cause fire, rats and dead animals can spread diseases, and floodwaters can be contaminated from sewers, animal feedlots, and other sources. If flooded areas are not treated properly and promptly, epidemic diseases can prevail. After the livestock and crops are wiped out, famine can menace the land. Many of these effects can take months or even years to assess fully.

6.2.2.3.1 Flood Damages Increase with Time

Flood damages in the Nation have increased with time and relatively high damage years are becoming more common in recent decades. Based on the NWS flood data 1926–2001, the annual increasing rate of change in flood damages of the Nation is estimated at 3.45% (Cartwright, 2005). Fluctuations of the NWS flood damage data 1903–2009, adjusted to 2007 value, are plotted against year in Figure 6.6. Increases in flood damages may attribute to climate variability and climate change, population growth and urban development, increasing property values, and violations of flood management guidelines.

As population increases exponentially with time, so does utilization of and development in floodplain areas. The concentration of population and intensive development along the

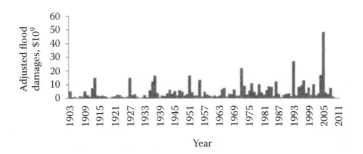

FIGURE 6.6
Flood damages, adjusted to 2007 dollars, in the United States, 1903–2009. (Data from NWS Hydrologic Information Center, Flood losses, compilation of flood loss statistics, available at: http://www.nws.noaa.gov/oh/hic/flood_stats/Flood_loss_time_series.htm, accessed on January 24, 2011.)

floodplain make property values higher and the potential flood damage greater. Moreover, the utilization of groundwater to meet water demands for population increase often causes a subsequent land subsidence, resulting in more areas that are subject to flooding. In the past, flash floods in the Eastern United States were often associated with storms of about 10 cm or greater (Maddox and Chappel, 1979). Now, floods occur in many areas with a storm of only 5 cm or greater.

6.2.2.3.2 *Losses per Unit Area Decrease with Increasing Watershed Size*

Streamflow phenomena and behavior are quite different between large watersheds and small watersheds due to their environmental conditions. A large watershed usually implies a condition in which watershed slope, rainfall intensity, flow velocity per unit area, and the chances of getting the entire watershed covered by storms are less, soil is deeper, and watershed storage is greater. Small watersheds are very sensitive to land-use conditions and to storms of high intensity and short duration. This means that the increase of streamflow in response to storm rainfall is more rapid for small or bare watersheds than for large or forested watersheds. Stream sedimentation and flow volume per unit area decrease with increasing watershed area, as do flood damages.

6.2.2.3.3 *Forested Watersheds Are Damaged Less*

Forests are the most distinguished types of vegetation on Earth. Their large and thick canopies, fluffy forest floor, and deep root systems impose tremendous effects on water, soil, and nutrient movements within the watershed. For example, forested watersheds transpire a great deal of water to the air, cut down the amount of rainwater to reach mineral soils, increase soil water-holding capacity, and reduce overland flow velocity and soil-erosion losses (Anderson et al., 1976). These factors make flood damages in forested watersheds less than those in agricultural or urban watersheds. Moreover, the lesser population and development in forested watersheds make the potential damage lower than in other watersheds.

6.2.2.3.4 *Inundation Can Be a Blessing*

A flood carries a tremendous amount of water, energy, soil particles, nutrients, and other substances. Geomorphologic, hydrologic, and ecological processes dynamically affect the release of this energy and these materials in river systems. Frequently, flooding results in the formation of new channels, floodplains, riffles, pools, sandbars, oxbows, islands, sloughs, deltas, backwaters, and other formations, depending on the stream's energy level and its dissipation. These geomorphologic features along stream channels can support a wide array of new habitats that stimulate the growth and reproduction of living organisms. Thus, from an ecological point of view, flooding has positive impacts on natural systems. The 1993 Mississippi flood was reported to enhance the number and types of many kinds of plants and animals. It was an ecological boon.

The lower reach of the Nile River is flooded every summer by tropical storms and melting snows from the Abyssinian plateau. When the floodwaters begin to recede in October every year, they deposit black sediment rich in nutrients to nourish soils along the floodplain. The Egyptians developed agricultural systems in compliance with the periodic inundation of the Nile, which made the growth of crops such as cotton, grain, and sugar cane possible in the otherwise arid sands. Thus, the greatest problems for farmers along the Nile occurred not when it flooded, but in the years when it did not flood. No wonder the Nile has been called the "lifeblood" of Egypt. These fertile soils, through periodic inundation, conceived one of the oldest and richest civilizations in the world. The Aswan Dam now prevents this flooding.

6.2.2.4 Flood Control and Management

We cannot prevent all floods, but we can control some of the lesser ones and reduce the magnitude and destructiveness of others through flood-control measures and floodplain management. In general, flood-control measures use techniques that retain as much stormwater and surface runoff in the soil (or in reservoirs) as the soil water-holding capacity is able to retain, and drain the excess water out of the watershed as fast as possible. The floodplain-management approach can involve flood forecasting to prepare for floods ahead of time, floodplain zoning to relocate valuable property and life to areas less subject to flood hazards and flood insurance to spread individuals' flood losses among participating majorities. For an effective flood-prevention program over a drainage area, none of these approaches should be neglected.

6.2.2.4.1 Flood-Control Measures

Flood-control measures can be grouped into four categories:

1. Flood-control reservoirs
2. Channel modification (dikes and levees)
3. Soil and vegetation treatments
4. Land drainage

The main function of flood-storage and retarding reservoirs is to reduce peak discharge but not flood volume. However, reduction of peak flows by small reservoirs diminishes rapidly with distance below the dam. Thus, small reservoirs cannot fully replace protective works downstream, nor can downstream measures contribute protection to headwater floods. Some dams are built to raise the level of the stream so that flows can be diverted into canals for irrigation of lowlands or to spreading grounds for groundwater recharge. Most are built for multiple purposes, such as hydroelectric power, recreation, water supply, irrigation, and flood control.

Despite the value of reservoirs in flood control and other uses, big dams have drawn much criticism. Some charge that:

1. Dams cost too much and take too long to build.
2. Prime agricultural land and scenic sites above the dam are lost due to flooding, and wetlands and bottomlands below the dam suffer flow reduction.
3. Dams can result in saltwater intrusion to freshwater aquifers in coastal areas and alter marine environments due to reduction of instream flow to bay estuaries.
4. The natural habitats of endangered species are destroyed, and the upstream migration of adult salmon is blocked.
5. The reduction of periodic flooding below the dam can result in soil-salinization problems in arid and semiarid regions.
6. Many existing reservoirs have siltation problems that seriously deplete their storage capacity and over time shorten their useful life.

Soil and vegetation treatments are designed to increase infiltration rate, soil water-holding capacity, surface detention, depression storage, and roughness coefficient; slow surface-flow velocity; reduce peakflow; and delay time of flow concentration. The most common practices are contouring, strip cropping, no tillage, terracing, vegetated watercourses, crop rotation, seasonal cover crops, green manure, mulching, gully control, reforestation, and grazing control. The choice among these practices depends largely on topographic characteristics, soil conditions, vegetation types, climatic factors, and other variables affecting the hydrologic behavior of the watershed. It is generally agreed that the effects of these treatments on the reduction of flow rates and delayed time of concentration are most effective in small watersheds and for storms of low intensity and short duration. No data have been published on the effects of such treatments over areas larger than 5–10 km². Land treatments can be effective for watersheds up to 50–100 km², while for areas over 1000 km² or for storms of high intensity and long duration, they would be of lesser importance (Chang and Watters, 1984).

Once storm water can no longer be held in the soil or stored in the reservoirs, the excess of water will run into a stream. Damage is minimized if the flow moves rapidly enough to reduce depth and duration of flooding. This can be accomplished through channel modifications such as:

1. Deepening or widening channels to increase flow capacity
2. Removing vegetation, debris, and sandbars from the watercourse, lowering the water level at the outlet, channel straightening, or setting levees and embankments to increase flow velocity
3. Placing revetments and jetties on the shoreline or vegetation planting on the bank to reduce channel erosion

Although channelization can benefit the flood-control program of upstream areas, it usually leads to increases in the amount of flooding, sedimentation, and bank erosion in downstream areas. The habitats of certain fish along with plant and animal species may be eliminated due to the drainage of wetlands as a result of channelization. Moreover, a winding stream possesses the aesthetic value of a natural area, while a straight and open ditch is a degradation of such beauty. Thus, the conflicts between upstream and downstream need to be weighed carefully before any action is implemented.

The main purposes of land drainage are (Shaw, 1983):

1. To prevent waterlogging in agricultural land
2. To minimize flood damage to agricultural crops and urban areas
3. To confine rivers in their channels
4. To dispose of surface water from urban areas
5. To prevent seawater from penetrating into land areas

Methods used in land drainage include land smoothing, drainage ditches, tile drainage, and drainage wells. Water spreading over a gentle slope can divert some of the extra water from flooding streams, consequently recharging groundwater, rehabilitating rangeland, and enhancing forage and crop production.

Frequently, drainage projects may have reduced local flood damages but have also created drainage and flood problems below the project area. Many flooded river valleys, marshlands, tidewater swamps, and bottomlands are natural habitats for waterfowl and other wildlife species, and spawning and nursery areas for economically important aquatic life. These wetlands in a watershed also attenuate flood peaks and storm flows by temporarily storing surface flows (Carter et al., 1978; Novitzki, 1978). Not only are the natural habitats and aquatic life lost if these wetlands are drained, but also the flood volumes and frequency are increased.

6.2.2.4.2 Flood Forecasting

Flood forecasts are a modern technology used for predicting the occurrence of floods in advance so that flood damage can be mitigated to a minimum level. The forecasts usually cover flood regions, the height of flood crest, the date and time when the river is expected to overflow its banks, and the date and time when the flow in the river is expected to recede within its banks. They give time for authorities to remove people, livestock, and movable goods from the floodplain; to make reinforcements of bridges, levees, and other flood-control structures; and to prepare food, water, and medicine for emergency purposes. Reservoir operation for flood control and water supply is more efficient if reliable river forecasts are available.

River-flood forecasts for the United States are prepared by 13 National Weather Service (NWS) river-forecast centers and disseminated by 218 NWS weather-forecast offices to the public for appropriate preparations. However, the NWS cannot prepare flood forecasts alone. A great deal of support and coordination from agencies at federal, state, and local levels and private agencies is required for data collection, storage, and release. These agencies include the U.S. Geological Survey (USGS), the U.S. Army Corps of Engineers (COE), the U.S. Bureau of Reclamation (USBR), the U.S. Natural Resources Conservation Service (NRCS), state water resources departments, and many others.

The USGS is a major cooperator in the nation's flood-warning system. Chartered in 1879 by Congress, the USGS operates and maintains a cooperative program with other federal, state and local agencies on 7292 stream-gauging stations or more than 85% of the total throughout the nation. These continuous streamflow records and the stage–discharge relationship at each gauging station are used by the NWS for river-flood forecasting.

Besides the USGS streamflow data, the NWS also needs data on precipitation, temperature, snowpack, sunshine duration, soils, vegetation, and topography collected from its own weather-observation networks or provided by agencies such as the U.S. Forest Service, COE, USBR, and NRCS. These data are applied to specially calibrated river-runoff models to predict the discharges and stages at about 4000 points along the nation's major rivers.

Procedures for an effective flash-flood warning system include:

1. Accurate determination of regions in which a severe storm is developing
2. Observation, collection, and assimilation of hydrologic data
3. The prediction of runoff volumes and peakflow by means of developed rainfall-runoff models

If the expected storm event is big enough to produce serious flooding in a given region, a special weather statement, such as a river (flash) flood warning or watch, is issued to state and local agencies and to the general public through the news media.

6.2.2.4.3 Floodplain Zoning

Zoning, administered by a zoning commission or planning board and enforced by the courts, was originally designed to classify or protect a community's land-use pattern.

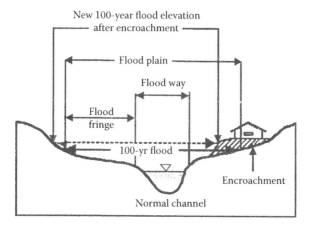

FIGURE 6.7
A 100-year floodplain.

Typically, a community is divided into zones of agriculture, industry, business, multifamily residences, and single-family residences, and the uses of land within the community will follow the zoning ordinances. Someone who desires to put a piece of land into a nonconforming use may apply to the zoning commission, and the application is considered, with public input, on a case-by-case basis.

Zoning now is used as a tool for environmental improvement and for protection of the public from hazards. Wetlands such as swamps, bogs, marshes, ponds, and creeks; forested areas; and rugged landscapes may be classified as protected natural areas and hence not open to development. An area zoned floodplain (Figure 6.7) may be banned for certain land uses, or certain building codes may be imposed for flood protection. Local, state, and federal governments bear many of the costs of flood damage incurred in a classified floodplain area, such as rescue, cleanup, building restoration, and construction of flood-control measures. A desirable and low-cost insurance is available to participating communities under the National Flood Insurance Program of 1968, provided that the residents have taken active measures to prevent further development in the floodplain. Part of the insurance is paid for by the federal government and is used to entice local communities to adopt the idea of floodplain zoning.

On May 24, 1977, President Jimmy Carter issued Executive Order #11988 for floodplain management (CEQ, 1978). It recognizes the natural and beneficial values of floodplains to our citizens, and seeks to avoid possible long- and short-term adverse impacts associated with the occupancy and modification of floodplains and to avoid direct or indirect support of floodplain development wherever there is a practicable alternative. The executive order directs every federal agency to provide leadership and take action to reduce the risk of flood loss, and to restore and preserve the special values served by floodplains.

6.2.2.4.4 *Flood Insurance*

The development of economically and aesthetically attractive floodplains makes all properties in those areas subject to periodic flood damage. Flood insurance is an overall community program of corrective measures for restoration of such damage. However, because of the high-risk potential and frequent ignorance of construction techniques that reduce flood damage to new or remodeled buildings, flood insurance is available only through federal programs.

The National Flood Insurance Act of 1968 (PL90–448) established the National Flood Insurance Program (NFIP) under the administration of the Federal Insurance Administration (FIA) and the Mitigation Directorate (MD) within the Federal Emergency Management Agency (FEMA). The program enables property owners in participating communities to buy flood insurance at reasonable rates. To participate, a community must:

1. Make efforts to reduce flood losses through more comprehensive floodplain management
2. Require new buildings to be elevated or floodproofed up to or above the 100 year flood level

Floodplain management may include zoning, building codes, subdivision development, and special-purpose floodplain ordinances (FEMA, 1997).

The NFIP was broadened and modified with the passage of the Flood Disaster Protection Act of 1973 and the National Flood Insurance Reform Act of 1994. Commercial insurance companies can sell and service flood insurance in their names under the FIA's Write Your Own (WYO) program. WYO was intended to increase the geographic distribution of the NFIP policies and to improve service to the policyholders. As of October, 1996, about 90% of insurance companies had signed arrangements with FIA under the WYO program.

Federal flood insurance coverage is available to all owners of insurable property in a community participating in the NFIP. The insurable property refers to a building and its contents that are principally above ground and not entirely over water. Buildings entirely over water or principally below ground, animals, crops, and all objects in the open are not insurable.

Flood insurance under the NFIP is not available within communities that do not participate in the NFIP. A participating community that does not enforce its floodplain management ordinances can be placed on probation or suspended from the program. The probation may last up to 1 year, and an additional $50 charge is added to the premium for each policy for at least 1 year after the probation period begins.

6.2.2.5 Disaster Aids

The federal government also provides assistance programs that compensate disaster victims for economic damages suffered from flooding. FEMA coordinates most of the direct assistance related to nonagricultural losses while the U.S. Department of Agriculture (USDA) provides direct compensation for damages to crops. Both programs are available only in counties where the president has declared a disaster.

FEMA administers the Hazard Mitigation Grant Program for compensation of damages to local governments and the Individual and Family Assistance Program for direct aid in temporary housing, counseling, and property loss to flood victims. FEMA distributed over $5 billion in direct assistance to presidentially declared disaster areas between 1965 and 1989 and $27.6 billion between 1989 and 1993.

The USDA provides two programs to compensate for crop damages: the Federal Crop Insurance Program and Agricultural Disaster Assistance. The Federal Crop Insurance Program is implemented through designated insurance companies and covers 51 crops including wheat, cotton, sorghum, barley, corn, and rice. Farmers pay premiums up front and get compensation when crop losses are incurred. The Agricultural Disaster Assistance program is a noninsured assistance program that applies only to presidentially declared disaster areas. If qualified, each individual can get assistance up to 60% of total losses or $50,000.

The declaration of a disaster starts at the county level. A county committee gathers data on weather, crops and production to support the disaster conditions. The data are then sent to a committee at the state level. The state committee, after reviewing recommendations from local committees, determines the disaster areas and requests a declaration from Washington, District of Columbia. Once an area is declared a disaster area by the president, the local government and residents are qualified to apply for assistance from USDA or FEMA.

6.3 Water Quality

Due to its solubility and its role as a habitat for aquatic life, all water in nature contains organic matter, inorganic matter, and dissolved gases derived from its environment, from human activities, from the atmosphere, or from living organisms. The concentration of these substances and their biological, physical, and chemical effects are the basic criteria in the determination of water quality. In practice, water is considered polluted if it is not suitable for intended uses such as drinking, recreation, irrigation, industry, or aquatic life. The U.S. Environmental Protection Agency (EPA, 1976, 1998) has established national water-quality standards for various uses.

Water pollution is probably the most common hazard to people in much of the world. About 20% of the world's population lacks access to safe drinking water, and about 50% lacks adequate sanitation (UNEP, 1999). In the United States, the EPA indicated that 95% of the 246 hydrologic drainage basins in the United States were affected by water pollution in 1977 (CEQ, 1978). Under Section 305(b) of the Clean Water Act, the states assessed water-quality conditions of 17% of total river-miles, 42% of total lake-acres, and 78% of estuarine waters in 1994. The results showed that 64% of river-miles, 63% of lake-acres, and 63% of estuarine waters assessed were found to fully support uses designated for them (Bucks, 1997). This was a great improvement in the 20 year period. However, about 60% of the impairments in rivers and 50% in lakes came from agriculture. Groundwater is generally assumed safe and pollution free, counting on the filtering effect of Earth on moving water. However, some groundwater aquifers are occasionally contaminated due to intrusion of seawater, sewer leakage, waste discharged to the ground, and other contaminants.

Polluted water affects human health and aquatic life and it compounds the problems of water scarcity. It affects the health of about 120 million people and contributes to the death of about 15 million children under the age of 5 year every year (WMO, 1992). Many water pollution issues are discussed in Chapter 11; this section only addresses general approaches on water-quality control and introduces major water-quality laws and regulations.

6.3.1 Water-Quality Control

Controlling water quality is a tedious and complicated process. It involves scientific research, technological development, resources management, economic analysis, political movements, law enforcement, public participation, and education. An effective water-quality program requires technical and financial assistance from governments, cooperation and support from industries and farmers, and awareness and participation from the public.

In dealing with water-quality problems, one should keep in mind that prevention is cheaper than reclamation and that the nation's pollutants from nonpoint sources are much greater than those from point sources. Thus, protecting and preventing deterioration of streams with good water quality should be the first consideration. Proper and improving land management is an important element in the nation's water-quality programs. Control of water-quality programs includes the following tasks.

6.3.1.1 Monitoring and Research

The environmental effects of pollutants and their tolerance levels for various uses need to be studied. The basic knowledge for each pollutant in the aquatic environment such as its biological magnification; the residence time in soil, air, water, plants, animals, and people; any synergistic interaction with other pollutants; and any adverse effects on ecosystems, aquatic life, physical environment, and human health all need to be fully understood. The tolerance level of each pollutant needs to be determined, and water-quality standards for various uses need to be established.

Monitoring is basic to research and management. Water-quality levels and sources of pollutants need to be continuously monitored at local, regional, national, and global levels. Pollution sources need to be identified. The long-term water-quality conditions from undisturbed forested areas can serve as an important basis for aquatic water-quality standards for the region. In addition to these chemical and physical measurements, ecological and biological changes in our ecosystems need to be measured. The ecological measurements should be evaluated with respect to short- and long-range effects of water pollution.

6.3.1.2 Public Education

Awareness on the part of the general public of water-pollution problems is an important element of water-pollution-control programs. It can be accomplished through public schools, TV programs, workshops, and town meetings.

6.3.1.3 Prevention and Treatment

Nonpoint sources of water pollution can be greatly reduced by preventing pollutants, mainly sediment and nutrients, from running off to creeks by application of the "best management practices." Point sources of water pollution, such as industries and cities, can be addressed first by use of "best practical technology" and later by improved "best available technology." Land management practices such as mulching, terracing, contour cropping, strip cropping, and planting cover crops greatly increase soil infiltration, reduce overland runoff, and consequently cut down sediment delivered to streams. Traditionally, the effluent from paper and pulp plants impairs the water quality of streams, but a new technique using a dry-barking process removes bark from poplar and aspen trees without water, and consequently no discharge of pollutants to streams occurs (Owen, 1985). Many industries have developed effective technology either to reduce the amount of discharged wastes or to convert their wastes into commercially valuable byproducts.

Polluted water can undergo various levels of purification and treatment, depending on the sophistication of the plant and the degree of purity desired. Treatment of liquid wastes may involve mechanical (primary), biological (secondary), and specialized chemical and physical (tertiary) processes. Disinfection is also carried out in the final stages of sewage treatment to remove water coloration and to kill bacteria and viruses.

6.3.1.4 Management

Standards for the design, construction, operation, and maintenance of sewer systems and treatment plants need to be established. State and federal governments may need to provide financial assistance to municipalities in water-pollution-abatement programs. It is also important to develop different water-quality standards for various uses. Institutional activities of water-pollution control among governments at different levels and between countries need to be coordinated properly. National water-pollution policies need to be established as guidelines for controlling various programs.

6.3.1.5 Legislation and Enforcement

Policies and regulations on water quality need to be established legislatively at the federal, state, and local levels. Legislatures provide authorities, standards, finance, guidelines, responsibilities, and assessment for control and management programs on water quality. Industries and individuals are fined or penalized if they do not comply with environmental laws and regulations. Severe violations of such laws can cost them their operating licenses, and companies may even be shut down through legal action.

6.3.2 Water-Quality Laws and Regulations

Two laws are of paramount importance in the control of water-quality conditions in the United States: the Rivers and Harbors Act (1899) and the Clean Water Act (1977). The 1899 Act flatly prohibits discharging refuse matter of any kind from a ship or shore installation into navigable waters in the United States. Depositing material of any kind in any place on the bank of any navigable waters that can be washed off into the waters is also prohibited. The U.S. Army Corps of Engineers was authorized to require permits before anything can be dumped, deposited or constructed in any navigable U.S. waterway.

The Clean Water Act (CWA), formally known as the Federal Water Pollution Control Act Amendments (1972) (PL 92–500) and originally enacted in 1948 (PL 80–845), is the principal law governing pollution in the nation's streams, lakes, and estuaries. It set an ambitious interim goal of making all U.S. waters safe for fish, shellfish, wildlife, and people by 1983 and a national goal of eliminating all discharges into U.S. waters by 1985. The EPA, the federal agency charged with the programmatic responsibilities for the nation's environmental quality, was required by PL92–500 to establish a system of national effluent standards. Each state was required to establish regulations for nonpoint sources of water pollution (Section 208). The Army Corps of Engineers was authorized to regulate the discharge of dredged and fill material into navigable waters (Section 404). Both point and nonpoint sources were considered threats to the nation's water-quality conditions and were controlled simultaneously. Water-quality standards are published pursuant to Section 304 of PL92–500 and provide guidance for states and tribes to use in adopting water-quality standards.

Congress made certain important amendments in 1977 (PL 95–217), 1981 (PL 97–117), and 1987 (PL 100–4). A new section (319) entitled "Nonpoint Source Management Programs" to improve Section 208, "Nonpoint Source Planning Process," was added to the law in 1987. In addition to the existing industrial and municipal stormwater-discharge programs known as the National Pollutant Discharge Elimination System (NPDES), it also established a program for controlling toxic-pollutant discharges. Overall, the law is a comprehensive and complex water-quality-management law; it establishes water-quality policies, regulations, permits, requirements and deadlines on water-quality standards, quality assessments, reduction of discharges from point and nonpoint sources, water-quality certification, enforcement, financial assistance, and nationwide corporations.

Five primary provisions of the law have impacts on forestry and other related natural-resources activities. They are Sections 208 (nonpoint pollution planning), 303 (total maximum daily load, TMDL), 304 (water-quality standards), 319 (water pollution assessment and management programs), and 404 (wetlands and dredging activities). Traditionally, nonpoint sources of water pollution are managed through the two nonregulatory Sections 208 and 319. Recent developments show that the TMDL process is becoming a powerful tool to control point and nonpoint sources of activities (Dubensky and Ice, 1997).

6.4 Water Rights

Increasing competition and conflicting interests in water make water rights a legal, and some-times a political, problem in water resources. The problem is triggered by the following facts:

1. Water is necessary for survival and development.
2. There is not always enough water of the right quality in the right place at the right time.
3. Water in a river is often shared by many countries and by users of different withdrawal intensities, such as farmers in upper reaches, manufacturers or rec-reationalists in middle reaches, city planners downstream, and fishermen in the bay estuaries.

An absence of clearly defined water rights and rules of liability can cause hateful actions among individuals and communities or between nations.

The fact that water is a moving, returnable, and critical resource with various tangible and intangible values makes water allocation necessarily different from the traditional concepts of private ownership. Absolute ownership may not be appropriate. All allocations among uses and users should be considered in terms of survivability, productivity, environmental integrity, and fairness.

Generally, waters on Earth can be grouped into five distinct categories in accordance with their geographic locations: atmospheric, ocean, diffused, surface, and groundwater. No legal rights have yet been established with respect to uses of clouds for artificial rain. It might become a problem, however, as cloud-seeding techniques advance to greater effectiveness. Ocean water is generally regarded as a universal commodity. It can be used for navigation, fishing, recreation, and other purposes, subject to national and international laws. The diffused, surface, and groundwater rights are of great concern to watershed management.

6.4.1 Diffused Water

Water on the ground surface not in connection with a stream channel is considered dif-fused water. Diffused waters include bogs, fallen rain, fallen snow, springs, seepage, water detached from subsiding floods not forming a part of a water course or lake, and any water that has not yet reached a recognized water course, lake, or pond. Thus, springs that run directly into a stream are not considered diffused water.

Diffused water is generally regarded as the property of the landowner. It allows landowners an absolute right to any use of diffused waters on their lands, including damming, storing, selling, or even preventing it from flowing to adjoining lands. The unlimited right to capture diffused water may affect downstream water flow and groundwater level, especially in arid and semiarid regions. Because virtually no states attempt to regulate landowner uses of diffused water, no legal action can be taken to stop this capture.

6.4.2 Surface Water

A body of water flowing in a well-defined channel or watercourse is regarded as surface water. It is the water that is found in lakes, ponds, rivers, creeks, streams, and springs. Water contained in a surface depression as a result of rainfall or snowmelt is generally not considered surface water. Allocations for surface water are generally guided by three doctrines of water law: riparian, prior appropriation (or Colorado), and hybrid (or California).

6.4.2.1 Riparian Rights

Lands bordering a watercourse give the landowners special rights to make use of water in the channel, provided such use is reasonable relative to all other users. These rights include domestic and household uses, and water for livestock, navigation, power generation, fishing, recreation and other purposes. Generally, the landowner has no right to use water if his land does not adjoin a stream course. If there is insufficient water to satisfy all reasonable riparian needs, then uses of water must be reduced in proportion to the land size. The doctrine of riparian right is applied mainly in the eastern half of the United States.

6.4.2.2 Prior Appropriation

The riparian-rights doctrine is inadequate in the arid West, where water is generally scarce and less dependable than in the humid East. Under the prior-appropriation doctrine, both riparian and nonriparian owners have specific rights to divert water from streams or other bodies of water as long as their use does not conflict with prior claims. The theme of water rights is first-come, first-served with priority determined by a recording system similar to that used for land transactions. Thus, water rights depend on usage, not land ownership. Prior appropriation gives an exclusive right to the first appropriator. The rights of later ones are conditioned by the prior rights of those who have preceded. However, all rights are conditioned upon beneficial use as a protection to later appropriators against the wasteful use by those with earlier rights. The appropriation rules are exercised in nine western states: Alaska, Arizona, Colorado, Idaho, Montana, Nevada, New Mexico, Utah, and Wyoming.

6.4.2.3 Hybrid Doctrine

Ten other states—California, Kansas, Mississippi, Nebraska, North Dakota, Oklahoma, Oregon, South Dakota, Texas, and Washington—adopt portions of the riparian and the prior-appropriation doctrines. In general, this modified doctrine restricts the rights of the riparian owner while recognizing appropriation rights to reasonable amounts of water used for beneficial purposes. The limitations of riparian rights include:

1. Reasonable use
2. Elimination of unused riparian rights by statute

Thus, a riparian cannot prevent an appropriation to a nonriparian unless interference with the riparian's reasonable use of water can be proven.

6.4.2.4 Reserved-Rights Doctrine

The United States is entitled to claim water rights for instream flows on federal lands or in Indian reservations. In March 1891, Congress passed the Forest Reserve Act (or Creative Act) allowing the president of the United States to set aside forest reserves, whether of commercial value or not, to be inaccessible to the public for any purpose. The forest reserves were further specified under the Organic Act of 1897 to provide a continuous supply of timber and secure favorable conditions of water flow. In the 1963 Arizona vs. California case, the court ruled that a sufficient reserve of water accompanied the U.S. lands when they were reserved for a particular purpose (Gordon, 1995). Under this doctrine, lands such as national forests, national parks, national wildlife refuges, and the wild and scenic rivers that are reserved from the public domain for a particular purpose may be claimed for water rights to carry out that purpose (Lamb and Doerksen, 1990). Thus, reserved water rights on federal lands do not exist by statute. They must be identified and quantified by court procedures.

In another case, the Colorado Supreme Court concluded in 1987 that the federal government could claim water rights based on the Organic Act and that each claim must be properly examined or determined for its:

1. Documents reserving the land from the public domain and the Organic Act
2. Precise purposes
3. Importance of water for such purposes
4. Precise quantity of water required

On the basis of the Organic Act, the Department of Justice, representing the Forest Service and acting on behalf of the United States, filed claims for federal reserved water rights in 1976 to keep certain amounts of water for the Arapaho, Pike, Roosevelt, and San Isabel National Forests within Water Division 1 in Colorado to protect stream channels and timber. These claimed water rights were for three purposes:

1. Firefighting, unlimited amount
2. Administrative sites, not more than 10 acre-ft ($12,335\,m^3$) per site per year and not more than one site per 100,000 acres (40,470 ha) of national forest
3. Instream flows for channel maintenance, 50% of average annual runoff

However, these claims were challenged by the State of Colorado and water conservancy districts in northern Colorado that divert water from national forests. The opposition claimed that the United States did not need water rights because it had other mechanisms for controlling diversions such as special-use permits, and that such claims of water rights would injure other water users, especially during the critical spring runoff period.

The case went to trial in 1990 in District Court, Water Division 1, of the State of Colorado. Closing arguments were made in March 1992 and the court decision and order were on February 12, 1993. The judge recognized the reserved water rights of the United States, but the applicant (the United States) failed to show that the claims were necessary to preserve timber or to secure favorable water flows, and failed to establish the minimum quantity needed to ensure that purpose. Thus, the court granted water rights for firefighting and administrative sites to the United States and suggested the Forest Service use its special-use permitting authority to control water diversion in lieu of obtaining water rights on instream flows (Gordon, 1995).

6.4.3 Groundwater

Either flowing water in an underground channel (aquifer) or diffused percolating water is considered groundwater. Allocating rights and obligations in groundwater among the states is based on either the concept of property ownership or the notion of a shared public resource, or both (Getches, 1984). Thus, a state may:

1. Recognize the ownership of groundwater by the overlying landowner
2. Limit the use of groundwater to a reasonable level
3. Provide a special protection for prior users
4. Manage groundwater resources as public property with rights created under a permit program

Groundwater is a part of the hydrological system. Any pumping activities at one site would affect the water table and water yield in nearby locations. Thus, absolute ownership and uncontrolled use of groundwater are no longer feasible. Many states have adopted the concepts of reasonable use and correlative rights to protect all landowners against the inadvertent activities of a few others.

References

Abdelrhem, I.M., Rashid, K., and Ismail, A., 2008, Integrated groundwater management for Great Man-Made River project in Libya, *Eur. J. Sci. Res.*, 22(4), 562–569.

Anderson, H.W., Hoover, M.D., and Reinhart, K.G., 1976, Forest and water: Effects of forest management on floods, sedimentation, and water supply, General Technical Report PSW-18, U.S. Forest Service, Pacific Southwest Forest and Range Experiment Station, Berkeley, CA.

Baumgartner, A. and Reichel, E., 1975, *The World Water Balance, Mean Annual Global, Continental and Maritime Precipitation, Evaporation, and Runoff*, R. Oldenbourg, Munich, Germany.

Bell, T.C., 2005, Desalination should solve interior U.S. water supply problems, *U.S. Water News*, 22(10), 9.

Borrell, J., 1986, A plan to make the desert gush, *Time*, September 29, 1986.

Bucks, D.A., 1997, Competition for water use—An agricultural perspective, *Environ. Prof.*, 19, 33–34.

California Department of Water Resources, 1973, The California Water Plan, Sacramento, CA.

Carter, V. et al., 1978, Water resources and wetlands, *Wetland Functions and Values: The State of Our Understanding*, Greeson, P.E., Clark, J.R., and Clark, J.E., Eds., American Water Resources Association, Middleburg, VA, pp. 344–375.

Cartwright, L., 2005, An examination of flood damage data trends in the United States, *J. Contemp. Water Res. Educ.*, 130(1), Article 4, 20–25, available at: http://opensiuc.lib.siu.edu/jcwre/vol130/iss1/4

Cavelier, J., Solis, D., and Jaramillo, M.A., 1996, Fog interception in montane forests across the central Cordillera of Panama, *J. Trop. Ecol.*, 12, 357–369.

CEQ, 1978, *Environmental Quality, The 9th Annual Report*, Council on Environmental Quality, Government Printing Office, Washington, DC.

Chang, M. and Boyer, D.G., 1977, Estimates of low flows using watershed and climatic parameters, *Water Resour. Res.*, 13, 997–1001.

Chang, M. and Watters, S.P., 1984, Forests and other factors associated with streamflows in east Texas, *Water Resour. Bull.*, 20, 713–720.

Clark, C., 1982, *Flood*, Time-Life Books, Alexandria, VA.

Clean Water Act, 1977, 33 USC (United States Code) 1251, H.R.3199.

Dubensky, M. and Ice, G., 1997, Common issues and challenges in Clean Water Act programs—Natural resource policy and research perspectives, *Environ. Prof.*, 19, 58–61.

El-Geriani, A.M., Essamin, O., and Loucks, D.P., 1998, Water from the desert: Minimizing costs of meeting Libya's water demands, *Interfaces*, 28(6), 23–35.

EPA, 1976, *Quality Criteria for Water*, Government Printing Office, Washington, DC.

EPA, 1998, Current Drinking Water Standards, Office of Ground Water and Drinking Water, available at: http://www. epa.gov/OGWDW/wot/appa.html

Federal Water Pollution Control Act Amendments, 1972, 33 USC (United States Code), 1151, 70 stat. 498, 84 stat. 91, S.2770.

FEMA, 1997, Answers to Questions about the National Flood Insurance Program, Federal Emergency Management Agency, Washington, DC.

Getches, D.H., 1984, *Water Law*, West Publishing, St. Paul, MN.

Gleick, P.H., 2000, *The World's Water, 2000–2001*, Island Press, Washington, DC.

Gleick, P.H., 2004, *The World's Water, 2004–2005*, Island Press, Washington, DC.

Goodman, J., 1985, The collection of fog drip, *Water Resour. Res.*, 21, 392–394.

Gordon, N., 1995, Summary of technical testimony in the Colorado water division 1 trial, General Technical Report RM-GTR-270, U.S. Forest Service, Rocky Mountain Forest and Range Experiment Station, Fort Collins, CO.

Goudie, A.S., 2003, *Great Warm Deserts of the world—Landscapes and Evolution*, Oxford University Press, Oxford, UK.

Harr, R.D., 1982, Fog drip in the Bull Run municipal watershed, Oregon, *Water Resour. Bull.*, 18, 785–789.

Hu, Z., 2005, Precipitation enhancement in China: A developing technology for obtaining additional water from sky, Chinese Academy of Meteorological Sciences, available at http:// www.lanl. gov/projects/chinawater/documents/huzhijing.pdf

Husseiny, A.A., Ed., 1978, *Iceberg Utilization*, Pergamon Press, New York.

Huston, S.S. et al., 2004, Estimated use of water in the United States in 2000, U.S. Geological Survey Circular 1263, Washington, DC.

Ingwersen, J.B., 1985, Fog drip, water yield, and timber harvesting in the Bull Run municipal watershed, Oregon, *Water Resour. Bull.*, 21, 469–473.

Kenny, J.F. et al., 2009, Estimated use of water in the United States in 2005, U.S. Geological Survey Circular 1344, Washington, DC.

Lamb, B.L. and Doerksen, H.R., 1990, Stream water use in the United States—Water laws and methods for determining flow requirements, *National Water Summary 1987—Hydrologic Events and Water Supply and Use*, Carr, J.E. et al., Ed., Water Supply Paper 2350, U.S. Geological Survey, Reston, VA, pp. 109–116.

Maddox, R.A. and Chappel, C.F., 1979, Flash flood defenses, *Water Spectr.*, 11, 1–8.

Murray, C.R., 1968, Estimated use of water in the United States, 1965, U.S. Geological Survey Circular 556, Washington, DC.

Novitzki, R.P., 1978, Hydrologic characteristics of Wisconsin's wetlands and their influences on floods, streamflow, and sediment, *Wetland Functions and Values: The State of Our Understanding*, Greeson, P.E., Clark, J.R., and Clark, J.E., Eds., American Water Resources Association, Middleburg, VA, pp. 377–388.

NWS Hydrologic Information Center, 2011, Flood losses, compilation of flood loss statistics, available at: http://www.nws.noaa.gov/oh/hic/flood_stats/Flood_loss_time_series.htm, accessed on January 24, 2011.

Oberlander, G.T., 1956, Summer fog precipitation on the San Francisco Peninsula, *Ecol.*, 37, 851–852.

Olivier, J., 2002, Fog-water harvesting along the West Coast of South Africa: A feasibility study, *Water SA*, 28(4), 349–360.

Owen, O.S., 1985, *Natural Resources Conservation*, 4th edn., MacMillan, New York.

Reynolds, R.C. and Thompson, F.B., Eds., 1988, *Forest, Climate, and Hydrology*, United Nations University Press, Tokyo, Japan.

Rivers and Harbors Act, 1899, 33 USC (United States Code), 401–413.

Robinson, D.F., Ed., 1988, *Living on the Earth*, National Geographical Society, Washington, DC.

Rosegrant, M.W. and Cai, X., 2002, Global water demand and supply projections, part 2: Results and prospects to 2025, *Water Int.*, 27, 170–182.

Sharpe, W.E., 2001, personal communication.

Shaw, E.M., 1983, *Hydrology in Practice*, Van Nostrand Reinhold, New York.

Shiklomanov, I.A., 1998, Assessment of water resources and water availability in the World, Report for the Comprehensive Assessment of the Freshwater Resources of the World, United Nations, Data archive on CD-ROM from the State Hydrological Institute, St. Petersburg, Russia.

Skillings, T.E., 1999, Water reuse: The wave of the future, *Watershed Management to Protect Declining Species*, Sakrison, R. and Sturtevant, P., Eds., American Water Resources Association, Middleburg, VA, December 1999, pp. 481–484.

Solley, W.B., Pierce, R.R., and Perlman, H.A., 1998, Estimated use of water in the United States in 1995, U.S. Geological Survey Circular 1200, Washington, DC.

Stulp, J.R., 1995, Social, political, and educational factors involved in facilitating municipal waste utilization, *Agriculture Utilization of Urban and Industrial By-Products*, Karlen, D.L. et al., Eds., American Society of Agronomics, Special Publication No. 58, Madison, WI, pp. 1–10.

Texas Water Development Board, 1968, The Texas Water Plan, Austin, TX.

Texas Water Development Board, 1994, Forty-Nine Water Saving Tips, Austin, TX.

UNEP, 1999, Global Environment Outlook—2000, United Nations Environment Programme, Nairobi, Kenya.

United Nations Environment Programme, 1983, *Rain and Stormwater Harvesting in Rural Areas*, Tycooly International, Dublin, Ireland.

U.S. Forest Service, 1989, RPA assessment of the forest and rangeland situation in the United States, 1989, USDA Forest Service Forest Resource Report No. 26, Washington, DC.

U.S. Water News, 1994, Chilean fog collection system furnishes daily baths, *U.S. Water News*, 11, 2.

U.S. Weather Bureau, 1959, Evaporation maps for the United States, Technical Paper No. 37, U.S. Department of Commerce, Washington, DC.

Wilhite, D.A. and Buchanan-Smith, M., 2005, Drought as hazard: understanding the nature and social context, *Drought and Water Crises*, White, D.A., Ed., Taylor & Francis Group, Boca Raton, FL, pp. 1–29.

WMO, 1992, *International Conference on Water and the Environment: Development Issues for the 21st Century*, January 26–31, 1992, Dublin, Ireland.

World Resources Institute, 2005, Freshwater resources 2005, in EarthTrends Data Tables: Freshwater Resources, available at: http://www.earthtrends.wri.org/pdf_library/data_tables/wat2_2005.pdf

World Resources Institute et al., 1996, *World Resources, 1996–97*, Oxford University Press, Oxford, U.K.

World Resources Institute et al., 2003, *World Resources, 2002–2004*, World Resources Institute, Washington, DC.

Young, C.E., Kinsley, K.R., and Sharpe, W.E., 1983, Impact on residential water consumption of an increasing rate structure, *Water Resour. Bull.*, 19, 81–86.

Young, E.S. and Sharpe, W.E., 1989, Rainwater cisterns—Design, construction, and water treatment, college of agriculture, Special Circular 277, Penn State University, University Park, PA.

7

Characteristic Forests

Forests have existed on Earth for 350 million years; they reached a peak about 270 million to 220 million years ago during the Carboniferous period (Burch et al., 1976). Today, forests cover about one-third of the Earth's land surface. They are the most distinguished type of vegetation community and provide many resources and environmental functions that far exceed those of other vegetation covers. Accordingly, forests have always played a vital role in the survival, development, and growth of human society since prehistoric times. Maintaining healthy forests helps improve environmental quality.

7.1 Natural Resource

A forest is a community of somewhat dense growth of vegetation, dominated by trees and other woody plants, that supports an array of microbes and wildlife and occupies an area large enough to have a distinguished microclimate. Thus, forests are composed not only of many large trees but also of smaller plants with multiple canopy levels, and are home to many animals and other living organisms. They include land, streams, and climate in a shared and dynamic ecological system. Each forest is uniquely different from others because of differences in species composition, age, soils, wildlife, and microclimatic conditions.

There are no clear boundaries on how small an area can be called a "forest." An area of plant community that is not large enough or dense enough to create an environment and microclimate significantly different from those of the surrounding areas probably should be considered woodland rather than a forest. If the trees in a stand are artificially regenerated and professionally cultivated for economic gains, they are called a plantation. The Food and Agriculture Organization of the United Nations terms forest all lands with trees of 7 m or taller and a minimum crown cover of 10% in developing countries or 20% in developed regions. Detailed definitions of forest, deforestation, afforestation, and reforestation are given by Lund (2011).

7.1.1 Characteristics

As a natural resource, forests possess a few important characteristics. First, natural resources are generally divided into either renewable or nonrenewable, depending on whether or not their uses can be carried on indefinitely. A nonrenewable resource, like coal and minerals, will be depleted in time if used continuously. If used properly, forests act like a renewable resource, such as water and solar energy, which cannot be exhausted. However, some types of use can destroy the original forest environment and exhaust the forest and its associated resources. In this case, forests behave as a nonrenewable resource.

For example, many nutrients in tropical rain forests are stored in the biomass above the ground and recycled directly between litter and plants. There, soils are fully weathered, rich in iron and aluminum oxides; most nutrients are retained in the organic layers. The high productivity in tropical forests is due to high temperature, long growing season, and rapid recycling of the nutrients, not to the nutrient levels in the mineral soils. Thus, forest clearcutting causes rapid decomposition of the organic matter and extensive leaching of nutrients, particularly the common cations K^+, Ca^{2+}, and Mg^{2+} and the organic phosphate reserves (Medina and Cuevas, 1997). As a result, the clearcut sites may take centuries, or may never be able, to regenerate a forest similar to the original, and loss or even extinction of many plants from the forest occurs.

Accordingly, forests are in a position between renewable and nonrenewable natural resources, depending on how well they are treated. They can be classified as "potentially renewable" resources (Mather, 1990)—resources that cannot be taken for granted. Abusive use of forest resources has resulted in the degradation of the environment, the conversion of fields into deserts, and the disappearance of civilizations.

Second, a forest not only produces wood, fibres, foods, fuels, and medicinal materials; it also provides an environment that affects soil, wildlife, water, and atmosphere. Thus, forest resources should not be referred to as timber resources only. Soil, water, plant, wild-life, fish, livestock, and recreation resources are all associated with the forest. Without the forest or if the forest is improperly used, all the associated resources will be degraded, damaged, or destroyed.

Third, the forest is considered a "common property" in that any forest activities have on-site as well as off-site effects. In other words, the effects can expand from the activity forest (site-scale) to adjacent areas (local-scale), distance areas (regional-scale), and remote areas (global-scale), depending on the areal coverage and intensity of the activity, weather conditions, and the general circulation of the atmosphere. In Indonesia, the forest fires of summer 1997 covered over 1 million ha of forests, causing haze and smoke in Singapore, Malaysia, Brunei, and the surrounding region for weeks. They released an estimated $220–290 \times 10^6$ tons of CO_2 to the atmosphere, about half of the U.K.'s annual emissions.

Fourth, forest growth and development are exposed to a series of threats from nature and human activities. Such threats can come from wildfires, tornadoes, ice storms, severe drought, prolonged floods, tsunamis, insect infestation, diseases, and air pol-lution. Many of them come from nature, while others are derived from unbalanced nutrients and foods in the ecosystem or from the results of industrial activity. In most cases, they are beyond the control of forest managers, a risk and uncertainty that need to be recognized.

Finally, the term *resources* refers to attributes of materials that lead to exchanges and trade-offs, not the materials themselves (Duerr, 1979). In that sense, resources are a man-made concept in that their value depends on how humans need them. Thus, resources, including forest resources, are always changing because of culture and lifestyles. Traditionally, forests were a resource for lumber. Today, their environmental settings and functions are at least as important as timber production. On fragile lands, steep slopes, or sites critical to the environment, however, forests are grown primarily for environmental protection; monetary income is not of primary consideration.

7.1.2 Forest Trees

There are 60,000–70,000 species of trees belonging to either Gymnospermae (conebear-ing plants, commonly called softwoods) or Angiospermae (flowering plants, commonly

TABLE 7.1

Biological and Environmental Functions of Canopy, Stem, and Root System

Component	Biological Functions	Environmental Functions
Canopy	1. Photosynthesis	1. Carbon sinks and pools
	2. Transpiration	2. Affecting soil moisture and infiltration
	3. Respiration	3. Intercepting advective energy
	4. Reproduction	4. Shading
	5. Gas exchange	5. Aesthetic value
	6. Assimilation of foods	6. Shelters for birds and insects
	7. Food storage	7. Barriers to wind movement
		8. Precipitation interception
		9. Affecting solar radiation dissipation
		10. Fog and cloud condensation
		11. Reducing raindrop impact to soil
		12. Snow accumulation and snowmelt
Stem	1. Transport water and nutrients	1. Mechanical barriers to wind and water
	2. Support for canopies	2. Tree rings as an index for past hydroclimatic conditions
	3. Storage of materials	
	4. Photosynthesis, if green	3. Mechanical support and nutrient provider to vines, mosses, and lichens
	5. Use for plant regeneration	
	6. Transpiration	
Roots	1. Absorption of water and nutrients	1. Reinforcement of soils
	2. Transport of water and nutrients	2. Increased soil permeability
	3. Anchoring of plant	3. Improvement of soil structure
	4. Storage of materials	4. Addition of soil organic matter when decayed
	5. Use for plant regeneration	5. Slow down overland runoff, such as buttress roots
	6. Respiration	6. Depletion of soil moisture content
	7. Nitrogen fixation, if legumes	

called hardwoods). All of them contain vascular or conducting tissues and are grouped under the seed vascular plants, to be distinguished from seedless vascular plants or non-vascular plants of the plant kingdom. Seed vascular plants are shrub- or tree-like, woody or herbaceous, upright or creeping, perennial or annual, and photosynthetic, living in a wide variety of environments. Morphologically, a tree consists of three components: a root system in the soil, a foliar canopy in the air, and stems in between to connect roots and canopies. The three plant components perform many physiological functions required for growth and development as well as environmental functions important to human society. They are summarized in Table 7.1. The height (depth), size, shape, and biomass of the three components differ among species (Figure 7.1). For example, redwoods (*Sequoia sempervirens*) in northern California can grow to a height over 100 m and a diameter more than 6 m. The lifespan of some oaks is as long as 1500 years, while bristlecone pines can be over 4500 years old.

7.1.2.1 Canopies

The canopy is the most distinguished part of a plant. It is composed of leaves, flowers, fruits, buds, and stalks (petiole) supported by small branches and shoots. In the

FIGURE 7.1
The shapes of canopies for a few species in the United States. (Compiled from The National Audubon Society, Inc., *The Audubon Society Field Guide to North America Trees*, Alfred A. Knopf, New York, 1992.)

presence of light, leaves convert solar energy into chemical-bond energy (carbohydrates) and release oxygen by using carbon dioxide in the air and water in the plant through the mechanism of chlorophyll, a process called photosynthesis. The process is important biologically because it manufactures food for plants. Environmentally, it serves as carbon sinks and pools, a major sector affecting the carbon cycle of Earth and global warming.

The water required in photosynthesis is obtained from the soil through root systems. However, a great deal of water transmitted from roots to canopies is transpired to the air through stomata. A mature forest can transpire as much as 1000 tons of water to produce

1 ton of wood. Some plants, such as willow and saltcedar, because of large canopies or extensive root systems, can transpire many times more water than others.

Since photosynthesis and transpiration are conducted through leaves, the difference in the quantity of leaves among species is commonly expressed in terms of *leaf area index* (LAI), or total surface of leaf per unit land area in m^2/m^2. LAI varies greatly among species and within species due to differences in site quality, age, stand composition, density, and season. Species growing in cool and arid climates usually have small LAI and reach maximum LAI at later ages than those growing in warm and wet environments.

LAI is about 3.1 for aspens in the central Rocky Mountains (Kaufmann et al., 1982); 5.3 for a 10 year old stand of eastern white pine (Swank and Schreuder, 1974); 4.2–9.6 for tropical regenerating forests near Manaus, Brazilian Amazonia (Honzák et al., 1996); and 10.2 for Norway spruce in Wisconsin (Gower and Son, 1992). For a forest stand with multiple stories, LAI can go much higher than that of a single species. The LAI is 42 for mature stands of western hemlock and silver fir in the western Oregon Cascades (Gholz et al., 1976) and 40–50 for a mature stand with 90% spruce and 10% fir in the central Rocky Mountains (Kaufmann et al., 1982).

The maximum LAI was reported to be about 20.0 at age 40 for a well-stocked lodgepole pine (*Pinus contorta*) stand in Wyoming (Long and Smith, 1992) and 17.3 at age 12 for an eastern white pine (*P. strobes*) stand in North Carolina (Vose and Swank, 1990). Many biological processes and environmental functions such as transpiration, canopy interception, light penetration, heat exchange, carbon fluxes, respiration, photosynthesis, and wind movements are highly correlated with LAI.

7.1.2.2 Root Systems

The root systems of woody plants are composed of taproots, lateral roots, and fibrous roots. Taproots are located directly under the stems, grow rapidly downward, and penetrate deeply in the soil. Lateral roots are those that grow outward from the rootstock to all different directions, acting as an anchorage of the plant and allowing the plant to reach water and minerals in a sphere much greater than its occupied site. The term rootstock refers to the underground section immediately below the stem where the taproot and lateral roots develop. Fibrous roots are adventitious, massive, fine, threadlike, and short lived. They spread out below the soil surface and expose the plant to greater sources of water and nutrients.

7.1.2.2.1 Amounts

Standing root biomass varies from 0.2 to 5.0 kg/m^2 across the world's terrestrial biomass (Richardson and Dohna, 2003). In temperate coniferous forests, standing root biomass is about 20% that of standing above ground (Jackson et al., 1996) and 70% or more of the roots tend to concentrate in the A horizon (Farrish, 1991). Based on 475 soil profiles located at 209 locations throughout the world, Schenk and Jackson (2002) found that over 90% of these profiles had 50% or more of all roots in the upper 0.3 m and 95% in the upper 2 m. The roots of *Boscia albitrunca* and *Acacia erioloba* have been found at depths of 68 and 60 m, respectively, under the ground (Canadell et al., 1996). A big spreading oak tree can have roots totaling many hundreds of kilometers. Living roots of mesquite, a phreatophyte growing on rangeland in the Southwest, United States were found 53.3 m below the original surface of an open-pit mine in Arizona (Phillips, 1963) and commonly spread more than 15 m wide from a single tree.

Root area index (RAI), the ratio between total root surface area and ground surface area in m^2/m^2, is at least comparable to LAI in all terrestrial systems. The average RAI of fine roots is 11.6 in temperate coniferous forests and 79.1 in temperate grassland (Jackson et al., 1997). In the Harvard Black Rock Forest, New York, the average surface coverage of the root systems was about 4.5 times greater than the ground surface covered by the canopies for 25 mature hardwood trees. A sprout clump of chestnut oak, 17 years old and 10 m in height, had a root system that covered 57 m^2 surface area and was about 41 times greater than the ground area covered by its canopies (Stout, 1956). Differences in the depth, surface coverage, and density of root systems are reflections of species and a variety of environmental factors such as soil texture, depth, moisture content, nutrients, groundwater table, and plant competition. Sprouting shrub species tend to have deeper roots than nonsprouting species do (Keeley and Keeley, 1998).

7.1.2.2.2 Impacts

Root systems are an important factor affecting soil properties, surface hydrology, and slope stability. Large roots can anchor the plant and soil mantle to the substrate, while fine roots, fungal mycelia, and decomposed organic matter can contribute to the formation of stable aggregates for subsurface soils. The improvement of soil granulation may be due to:

1. The pressure exerted by growing roots
2. The dehydration of the soil around root systems
3. The secretion of substances produced by roots and accompanying bacteria that bind and cement soil aggregates

Roots grow and die every year. The dead roots, root secretions, and fallen litter, through microbial activities, become important parts of organic matter in the soil. They provide a physical environment suitable for a vast array of plants and animals in various sizes. The large population of macro- and microorganisms in and around the soil promotes the decomposition of organic matter, the mixing of organic matter and inorganic surface soil, the creation of soil organic horizon, and the improvement of the soil's physical and chemical properties. As a result, soil porosity, infiltration capacity, soil water-holding capacity, and plant transpiration are increased, and surface runoff is reduced.

Plants enhance soil stability against downslope movement due to the reduction of soil water content through transpiration and the mechanical reinforcement of the root system (Waldron and Dakessian, 1982). Roots increase the cohesive and frictional components of the soil's shear strength, which is a force counter to the force causing the soil to move downslope. The increase in soil shear strength is proportional to the fresh weight of roots per unit volume of soil. Trees usually exert a greater strength than grasses do, and the removal of trees causes a decrease in root tensile strength due to the decay of root systems. Thus, forested vegetation is more likely to provide better protection against bank erosion (Kleinfelder et al., 1992; Wynn and Mostaghimi, 2006); forest removal has caused many landslides in New Zealand and the northwestern United States (Sidle et al., 1985). (Also see Chapter 13, Section 13.2.5 and Chapter 14, Section 14.4.1.)

7.1.2.3 Stems

Stems serve as a transport path of fluids between roots and canopies as well as storage for nutrients. Forest trees usually have a single stem; its environmental functions are far fewer than those of the canopy and the root. Stems can impose mechanical disruptions to wind and water movements; however, the effectiveness is largely dependent upon the diameter and density of stems in the area. Some of the environmental functions of canopies are affected by stem height. Since the stem accounts for about 80% or more of the above-ground biomass of a single tree, carbon storage is perhaps the most important environmental function of stems.

The total above- and below-ground biomass contained in the world's forests in 2010 amounted to 600 Gt (10^9 tons) (FAO, 2010). Using the worldwide weighted root-shoot ratio of 0.24, the global above-ground biomass in forests is 456 Gt and the total stemwood biomass is about 80% of that or 365 Gt. Assuming 1 ton of organic matter is equivalent to 0.5 tons of organic carbon (Brown et al., 1989), the world's forest stems and trunks contain 183 Gt of carbon. Using the factor 3.67 to convert organic carbon to carbon dioxide (Department of Energy, 1992), the global stemwood biomass would retain CO_2 as much as 672 Gt. Estimated global CO_2 emissions by fossil fuels and cement manufactures for the 1990s were 23.5 Gt/year (IPCC, 2007).

7.1.3 Forest Distribution

Plants require energy and water for survival and development. No plants can thrive where monthly air temperature is below freezing year round. Thus, temperature and rainfall are the two major determinants that control the distribution of forests (Figure 7.2), while the variation in topographic conditions affects forest types in a given region. In general, forests occur in areas where the annual precipitation is greater than 38–50 cm and the frost-free period is at least 14–16 weeks long (Buell, 1949). The minimum net solar radiation

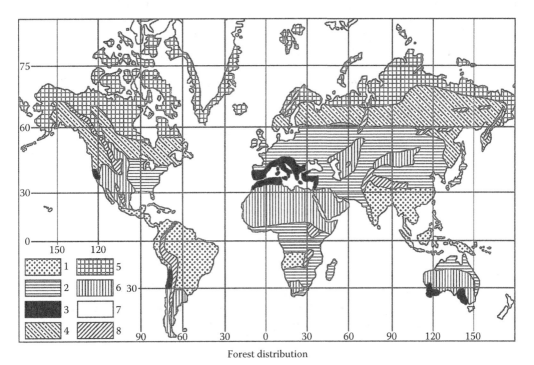

Forest distribution
1. Tropical forest, 2. Temperate forest, 3. Chaparral/mediterranean, 4. Boreal forest, 5. Tundra, 6. Desert/semidesert, 7. Ice, 8. Mixed mountains

FIGURE 7.2
The world's major vegetation biomes.

required for a forest is about 20,000 cal/cm^2/year, or 27 W/m^2 (Reifsnyder and Lull, 1965), which is in regions around lat. 60°–65°, depending on cloud conditions.

In middle latitudes where energy is not an issue, rainfall is the main control on lowland forests and is responsible for the occurrence of deserts and steppe zones that separate tropical and temperate forests. The situation is different in higher latitudes where temperature determines the northern latitudinal limits of forests. The isotherm of 10°C for the warmest month has long been considered the approximate north bound of forests. Regions with mean July temperature of 10°C correspond very well with the 20 kcal/cm^2/year net radiation line. North of this line is the tundra, where low-growing and densely matted arctic plants dominate the region and no forest trees are present.

7.1.3.1 Horizontal Distribution

On the latitudinal scale are three zones of great precipitation. The greatest zone occurs in the equatorial areas, and two secondary zones are in the high latitudes around lat. 50° N and lat. 50° S. The precipitation patterns cause the majority of Earth's forests to be distributed in two regions, one in the tropical areas and the other in the Northern Hemisphere roughly between 35° and 65° latitudes. According to the temperature regimes, forests are broadly grouped into three major types: tropical, temperate, and boreal.

7.1.3.1.1 Tropical Forests

The tropical regions are bounded roughly between the Tropic of Cancer (lat. 23.5° N) and the Tropic of Capricorn (lat. 23.5° S), where the mean temperature of the coldest month is at least 18°C, and precipitation tends to decrease outward from the equatorial region. Accordingly, there is a vegetation gradient from tropical rain forests (evergreen) in the equatorial zone to tropical seasonal forests, tropical scrub forests, and tropical savannah forests.

Tropical rain forests are found in the Amazon Basin, Southeast Asia, and central Africa. Here the climate is hot and damp throughout the year, plant growth is profuse, the diversity of tree species is high, plant litter deposition is very rapid, and soils are old, weathered, and well leached. Outward from the equatorial areas is the tropical seasonal forest where dry seasons are clear and pronounced. Tropical seasonal forests are well distributed in the northern part of South America, from northeast India to Burma and northern Australia, and from Angola to Tanzania in south-central Africa. The forest is simpler and lower than either a semievergreen or deciduous rain forest. Tropical deciduous forests, also known as monsoon forests, exist in areas where there is a distinguished dry season in the year, all the arboreal species are deciduous, and the ground is bare for long or short periods of time.

Tropical scrub forests occur in regions of light rainfall, bordering wet forests in the south and deserts in the north. They run roughly along lat. 10°–14° N in north-central Africa, South Africa, and southern Australia. Small-leaved evergreen vegetation dominates the forest; winters are warm and wet, and summers are long and dry. Where the dry season lasts on the order of 6 months, a mixture of trees and grasslands called a savanna forest occurs. Savanna forests, dominated by grasses and sedges with open stands of well-spaced trees, are widespread in Africa and in southern Brazil.

7.1.3.1.2 Temperate Forests

Forests in the Temperate Zone can be grouped into five types: deciduous, coniferous, broad-leaved evergreen, mixed evergreen, and Mediterranean. Here, winter is distinct, growing seasons are long, and precipitation exceeds evapotranspiration. Climate changes from season to season and within each season as well. Temperate deciduous broad-leaved forests are distributed in regions with 4–6 frost-free months and precipitation either relatively even throughout the year or higher in the summer. They occur in western and central Europe, eastern North America, and northeastern Asia. Coniferous forests are found mainly in North America, stretching from Minnesota to New England, much of the coastal plains from New Jersey to east Texas, and the Pacific coast from Alaska to British Columbia, Washington, Oregon, and northern California. The winter is mild and precipitation in winter is greater than in summer in these areas.

In areas where winter is mild and wet, summer is warm and dry, and annual rainfall is less than 1000 mm in general, forests tend to be dominated by mixed evergreen coniferous and broad-leaved species, the so-called Mediterranean forests. They can be found in Mediterranean basins, western and southwestern North America, temperate Asia, and Australia. However, west of the Mediterranean climate and along the coastal areas from northern California to southeast Alaska, their precipitation remains concentrated in the winter but reaches from 1500 to over 2000 mm/year. Here, redwoods, Douglas fir, hemlock, and Sitka spruce are mixed with evergreen hardwood species. These forests are some of the most long-lived and productive in the world, longer in life and production than many tropical rain forests (Perry, 1994).

Evergreen broad-leaved forests occur mainly in southern China, the lower slopes of the Himalayas, southern Japan, the southeastern United States, much of southeast and southwest Australia, New Zealand, and the southern coast of Chile. All these areas have mild temperatures, frost-free winters, and relatively high precipitation (>1500 mm/year).

7.1.3.1.3 Boreal Forests

North of the temperate forest where winter temperature is too cold and growing seasons are too short to support hardwood species, cold-tolerant conifers known as the boreal forests become dominant. Typically, the boreal forests develop in areas with about 6 months of below-freezing temperatures, 50–125 days of the growing season, 1–3 months of mean summer temperature above 10°C, and 35–50 cm of annual precipitation. They are distributed in two bands, one in North America from Alaska to Newfoundland and the other in Eurasia from the Atlantic coast of Scandinavia to the Pacific coast of Siberia, roughly between lat. 50° N and lat. 60° N.

The boreal forests are much simpler and more uniform than forests in other regions are. Species of spruce, fir, larch, and pine are numerically dominant throughout most of the forests. Contrary to the 4 years for temperate deciduous forests and 0.7 years for tropic forests, the mean residence time of organic matter on the forest floor for boreal forests is 350 years (Mather, 1990). Thus, most of nutrients in the boreal forests, unlike those in tropical forests, are contained in the litter, not in the vegetation. The boreal coniferous forests are by far the most extensive remaining forests outside of the tropical forests.

7.1.3.2 Vertical Distribution

In the free air—air not in contact with the ground—there is a fairly uniform decrease of temperature with increasing elevation. The rate of temperature decrease is about, on the average, 0.65°C/100 m. This is called the normal lapse rate or vertical temperature gradient. On the horizontal scale, the temperature gradient between lat. 20° N and lat. 80° N varies from −0.5 to −1.0°C/1° latitude. This means that a mountain at lat. 20° N where its mean base surface annual temperature is 25.3°C would have a mean annual temperature

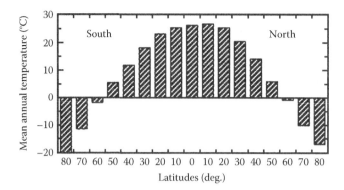

FIGURE 7.3
Mean annual air temperature of the world, by degree zones. (Data from Landsberg, H., *Physical Climatology*, Gray Printing Co., DuBoise, PA, 1968.)

of 5.8°C at 3000 m, a thermal climate equivalent to the sea-level temperature at lat. 50° N (Figure 7.3).

Also, due to orographic effects, precipitation is greater on windward slopes and at higher elevations of the mountain. However, the greater precipitation at higher elevations occurs only up to a certain level. Above this level, the air humidity is too low, precipitation declines, and the environmental harshness increases with height. Therefore, precipitation and temperature variation cause changes in forest types along an elevation gradient in mountainous areas.

At a sufficient height, harsh environmental conditions such as short growing seasons, frigid air and soil temperatures, low available water, high wind speed, and deficient carbon dioxide can cause the abrupt termination of a closed forest stand. Beyond this forest limit, or the so-called *timberline*, tundra and alpine vegetation predominate. Forest types below the timberline start with conifers, mixed conifers and hardwoods, hardwoods, shrubs, grasslands or deserts, and tropical rain forests, depending on latitude, height of the mountain, and distance to the ocean. This vertical distribution of vegetation can be uniquely illustrated by the slopes of the central Sierra Nevada in California (Figure 7.4). Rising to more than 3965 m, over a dozen distinct plant zones have been identified between the base and the mountain peak.

The height of timberline is 900 m at Haines, Alaska (lat. 60° N); 1850–1900 m at Garibaldi Park, British Columbia (lat. 50° N); 3300 m at the Sierra Nevada mountains, California (lat. 38° N); 4500 m in southeast Sinkiang, China (lat. 38° N); and up to 4900 m in northern Chile (lat. 19° S) (Kimmins, 2003). It tends to increase in elevation from polar regions toward the equator, but the greatest height occurs in warm-temperate belts, around 25°–30° (Spurr and Barnes, 1973). In the Northern Hemisphere, the rate of decline between 40° and 55° is about 100 m/1° latitude (Peet, 1988). Southern exposures and areas with maritime climate have a timberline higher than northern exposures and continental climate.

7.1.3.3 Forest Areas

In 2010, forests covered about 4.03×10^9 ha or 30% of Earth's land surface (Table 7.2). Russia has the greatest area of forests, 809×10^6 ha or 20% of the global total, followed by Brazil (12.9%), Canada (7.6%), the United States (7.5%), and China (5.1%) (FAO, 2010). About 50% and 45% of the total areas in South America and Europe, respectively, are occupied by

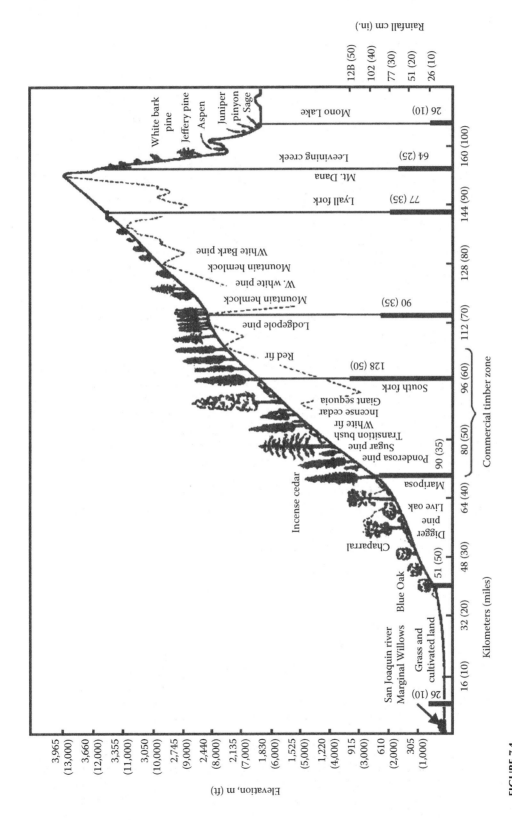

FIGURE 7.4

Vertical distribution of major vegetation types along the slopes of the central Sierra Nevada in California. (After Hughes, R.O. and Dunning, D., *Yearbook of Agriculture*, U.S. Department of Agriculture, Washington, DC, 1949.)

TABLE 7.2

Total Land Areas and Inland Water in 2010, Forest Areas in 1990 through 2010 by Continent

Continent	Land Area (10⁶ ha)	Inland Water (10⁶ ha)	Forest Area (10⁶ ha)			
			1990	2000	2005	2010
Africa	3,031.5	57.5	749.2	708.6	691.5	674.4
Asia	3,182.5	91.0	576.1	570.2	584.0	592.5
Europe	2,306.3	91.5	989.5	998.2	1,001.2	1,005.0
Oceania	856.1	7.0	199.7	198.4	196.8	191.4
North America	2,199.2	138.2	676.8	677.1	677.8	679.0
South America	1,783.0	36.8	946.5	904.3	882.3	864.4
Central America/ Caribbean	75.6	1.7	31.6	28.4	27.5	26.4
World	13,434.2	423.7	4,168.4	4,085.2	4,061.0	4,033.1

Source: Extracted from FAO, Global Forest Assessment 2010, Main Report, FAO Forestry Paper 163, Rome, Italy, 2010, 340 pp.

forests, while forests only make up 2% of land uses in the Middle East and North Africa regions. Percent of three major land uses in 2002 in each continental regions of the world is given in Figure 7.5.

In the United States, forests covered about 296.3×10^6 ha or 32.3% of the land in 1990 and 304.0×10^6 or 33.1% in 2010 (FAO, 2010). This was an increase in forest areas 0.13% per year during the 20 year period. For the conterminous United States, forest as a percent of the total land area, however, is reduced to 21.1% or 165.6×10^6 ha. The acreage of land uses in eight different categories in the conterminous United States is given in Figure 7.6.

Concerns about forest alteration in the tropics have greatly increased in recent years. According to the FAO (2010), the world lost 135.3×10^6 ha, or −3.3%, of its forest cover between 1990 and 2010. Losses in forest cover in seven tropical countries made up 85% of the total, or 5.75×10^6 ha/year during the 20 year period. Brazil had the greatest loss (about 2.77×10^6 ha/year), followed by Indonesia (1.21×10^6 ha/year). Together, they accounted for about 59% of the global total.

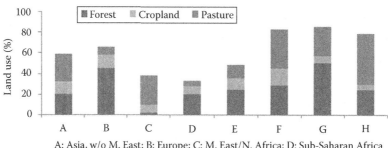

A: Asia, w/o M. East; B: Europe; C: M. East/N. Africa; D: Sub-Saharan Africa
E: N. America; F: C. America/Caribbean; G: S. America; H: Oceania

FIGURE 7.5

Percent of three major land uses in 2002 in each continental regions of the world. (Data from World Resources Institute, Land use and human settlements, in Earth Trends Data Tables: Forests, Grasslands and Drylands, http://www.earthtrends.wri.org/pdf_library/data_tables/forl_2005.pdf, 2005.)

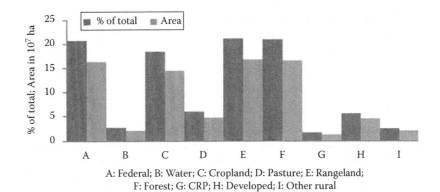

A: Federal; B: Water; C: Cropland; D: Pasture; E: Rangeland;
F: Forest; G: CRP; H: Developed; I: Other rural

FIGURE 7.6
Acreage of land uses (in 10^6 ha and % total) in the conterminous United States. (Data from U.S. Department of Agriculture, Summary Report: 2007 National Resources Inventory, NRCS, Washington, DC, and Center for Survey Statistics and Methodology, Iowa State University, Ames, IA, 123 pp., 2009.)

7.2 Environmental Functions

The physical environment of Earth is composed of three phases: the atmosphere, the hydrosphere, and the lithosphere. Collectively, they form the so-called geosphere of Earth. The interface of these three phases is life, or the biosphere, and the combination of the geosphere and the biosphere is called the ecosphere (Figure 7.7). Forests are the largest system in the biosphere and provide numerous significant impacts to the environment.

FIGURE 7.7
The atmosphere, hydrosphere, lithosphere, and biosphere forming Earth's ecosphere.

The actions of a tree's canopy, root systems, and stem on the environment are outlined in Table 7.1. The integrated actions of all canopies, stems, root systems, and litter floor of the forest system acting upon the environment are significant functions of the forest. These functions can be hydrological, climatological, mechanical, biological, and societal.

7.2.1 Hydrological

Forests affect both water quantity and quality. First, the amount of precipitation that reaches the mineral soil is reduced by canopy interception. Then, a great amount of soil moisture is transpired to the air through the roots–stem–leaf system. Finally, the root systems, organic matter, and litter floor increase the infiltration rate and soil moisture-holding capacity. Combining these three processes makes overland runoff smaller, runoff timing longer, and water yield lower in forested watersheds than those in non-forested watersheds (see Chapter 10, Section 10.3).

A reduced amount of runoff carries less sediment and elements to the stream. This reduced runoff, combined with the shielding and shading effects of canopies, the binding effect of root systems, and the screening effect of forest floor, makes streamflow from forested watersheds have less sediment, lower dissolved elements, cooler temperature, and higher dissolved oxygen (see Chapter 10, Section 10.4).

7.2.2 Climatological

The Earth's surface conditions, such as topography, water, land and vegetation, play an important role in the state and properties of the atmosphere near the ground. These conditions modify water and energy exchanges between Earth and the atmosphere and affect the local and regional patterns of the atmospheric circulation. This is especially true in the forest because of its height, canopy density and depth, and area coverage.

Solar radiation is the major energy source that affects the climate of Earth. In a forest, canopies usually receive more incoming solar radiation than do pasturelands or the bare ground because of their dark color and great roughness, but only a fraction of the received energy is transmitted to the ground surface. On the other hand, the emission of longwave radiation from the ground surface to the sky is reduced due to the shielding effect of forest canopies and less wind movement in the heat-transfer processes. Thus, the forest can cause net radiation to be greater and air and soil temperatures to be cooler in summer and warmer in winter. A long-term study at six locations in the Amazon basin showed that replacing forest by pasture reduced the net radiation at the surface by 11% (Culf et al., 1996). In east Texas, the mean annual air and soil temperatures were lower and the amplitudes of fluctuation were smaller in an undisturbed loblolly and shortleaf pine forest than in a nearby cleared site by 0.5°C–1.0°C (Chang et al., 1994).

The most significant effect of forests on precipitation is canopy interception, which reduces the net amount reaching the soil and delays snowmelt. However, whether or not forests increase the quantity of local precipitation has been a matter of conjecture and debate for decades. Some conservationists, noting generally higher precipitation in forested regions, have suggested that meteorological droughts are a result of forest cutting and that reforestation enhances local precipitation. Others have argued, on the basis of physical processes, that forests result from abundant and frequent rains, but they do not increase the amount of precipitation in the area. Because of the temporal and spatial variability of precipitation, the accuracy limit of modern precipitation measurements, and the scale of a forest involved, it is difficult to resolve the arguments through field observations.

Any evidence for a forest influence on precipitation must be weighted with theoretical support and justification (see Chapter 8, Section 8.4).

Perhaps the largest-scale climatological functions of forests are carbon sequestration. Carbon dioxide is a greenhouse gas; the steady increase in CO_2 concentrations in the atmosphere in recent decades is widely considered as the main cause of global warming. About one-half of the CO_2 that is uptaken by plants through photosynthesis is returned to the atmosphere by respiration and the other one-half is stored as carbon in plants (Pidwirny, 2010). The above-ground carbon storage is about 315 Mt of CO_2/ha in forests and is only 56 Mt of CO_2/ha in grasses and crops (Gorte and Sheikh, 2010). Thus, agroforestry, afforestation, and reforestation are often suggested as a mitigation measure for CO_2 concentrations in the atmosphere (Schoeneberger, 2009), more in Section 7.5.

7.2.3 Mechanical

Many of the hydrologic and climatological functions mentioned earlier are due to the mechanical performance of roots, canopy, and litter floor. However, a more recognized mechanical function is the forest's effects on water and wind erosion. Forests are the most efficient means of controlling soil water erosion by:

1. Reducing overland runoff through canopy interception and transpiration
2. Increasing soil porosity through the organic horizon and root systems
3. Slowing overland flow velocity through litter coverage
4. Reducing the terminal velocity of raindrops through canopy interception
5. Enhancing soil aggregates and binding through root reinforcement

As a result, the soil-erosion rate from a forested watershed can be one-thousandth of that of a bare-ground watershed (Chang et al., 1982). Forest clearing along a stream bank often causes severe channel erosion and even stream-bank collapse.

Like raindrops and streamflow, wind has the energy to detach and transport soil particles. The detachment, rate of soil movement, and transport capacity grow with the second, third, and fifth powers of wind's drag velocity (WMO, 1964). The drag (friction) velocity describes the erosive power of wind. It is a function of surface shearing stress and air density and can be assumed to be about one-tenth of the mean wind velocity (Sutton, 1953). Wind speed is reduced behind a shelterbelt. The retardation of wind speed depends on wind direction, the density and height of canopy, distance from the windbreak, and height above the ground (WMO, 1981). For dense foliage perpendicular to the prevailing wind, wind speed at 1.0 tree's height (H) immediately behind the windbreak can be reduced by about 40% (Figure 7.8). The reduction of the erosive power of wind coupled with the trap of soil particles by canopies makes shelterbelts the most efficient, economic, and relatively permanent control of wind erosion and sand-dune stabilization.

In addition to protecting against water and wind erosion, a forest can also protect against avalanches if it occupies the zone of potential occurrence areas. The shadow cast by shelterbelts along with the retardation of wind speed has a great impact on soil evaporation and soil moisture conservation.

7.2.4 Biological

Forests provide habitats for an array of fauna and flora that live and properly develop in a particular environment. They not only are the genetic pools for life on Earth but also play

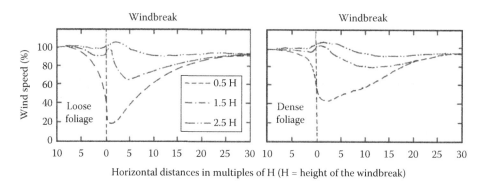

FIGURE 7.8
Change of wind speed at different levels above the surface behind a shelterbelt of height H. (Adapted from Van Eimern, J. et al., 1964, Windbreaks and shelterbelts, WMO-No. 147.TP.70., Technical Note No. 59, Geneva, Switzerland.)

a crucial role in the stability and viability of the biosphere. The biosphere is a large-scale life-support system with all components mutually interactive in a state of equilibrium. Large-scale alterations of forests or destruction of habitats can cause the loss of species of potential importance to human health, medicine, food production, and other uses. Most important, a chain reaction and the cumulative effects, if not corrected in time, may lead to a collapse of food chains and to a biological disaster.

There are about 1,360,000 species of identified animals on Earth. About 90% of non-human primates, 40% of birds of prey, and 90% of insects live in tropical forests (Mader, 1998). Many species are listed as endangered due in part to the loss of their habitats in forests. Some of the many endangered species include spotted owl, California condor, red-cockaded woodpecker, Indian tiger, black rhino (Africa), African elephant, peregrine falcon (North and South America), wood thrush (migratory North American songbirds), and hyacinth macaw (exotic birds in Brazil).

Plants have contributed to the development of 25%–50% of all prescription drugs used in the United States, either directly or indirectly by providing biochemical models (Swerdlow, 2000). Instances of effective disease remission include childhood leukemia by the Madagascar rosy periwinkle (*Catharanthus roseus*), ovarian cancer by Pacific yew (*Taxus brevifolia*), dementia by ginkgo (*Ginkgo biloba*), malaria by qing hao (*Artemisia annua*), tumor and cancer activity by happytrees (*Camptotheca acuminata*), and many others. In China, herbal medicine has been practiced for more than 5000 years. Li Shih Tseng wrote the book *Outlines of Medicinal Plants* in 1595 (the Ming Dynasty). It listed 1,097 drugs extracted from 512 plant species and 11,096 prescriptions for botanical remedies or pharmaceutical treatments. Some utilized plants' roots and bark, while others used leaves, flowers, shoots, nuts, fruits, fruit skins, stems, and sap. The U.S. National Cancer Institute has identified 3000 plants as potential sources for cancer-fighting chemicals.

7.2.5 Societal

Cool and shaded forests, with their vast area, wilderness setting, and fresh air, are ideal places for people to relax from tension, pressure, and hectic activities. Green canopies with blue sky and white clouds are a natural beauty to our eyes, and the environmental tranquility relaxes our ears and minds. Thus, many recreation and leisure-time activities—picnicking, sightseeing, bird watching, hiking, camping, and canoeing—take place in forest areas. Psychologists have even used "forest bath," a mental bath in the forest environment, as a treatment for persons with depression. Retreating from hectic cities to wilderness has become a part of the lifestyle in modern societies.

The presence of forests can benefit the health of people living in their vicinity. This may be attributable to the continuous replenishment of oxygen and the reduction of dust and air pollutants in forested areas. City and traffic noise can be attenuated and unsightly environments can be blocked by a greenbelt of trees.

Wilderness areas serve as field laboratories for various research and educational purposes. They provide opportunities for the public to study and appreciate forest ecosystems, ecological processes, habitats, landscapes, and natural-resources conservation.

7.3 Functional Forests

Although forests perform many environmental functions, some functions may be more significant than others for a particular forest because of its location and environment. Thus, a forest may be intentionally managed to maximize one or two specific functions, with little attention paid to others. Such a forest can be described as a *functional forest*.

Functional forests can be grouped into four categories: production forests, protection forests, preservation forests, and public forests. The functions that a forest is managed to perform are mostly a matter of site and environmental conditions. Frequently, however, the ownership, economic constraints, and prospective value of the forest play an important role in determining management objectives. Also, many environmental regulations and governmental policies provide guidelines for managing forest resources. For example, the Multiple-Use and Sustained Yield Act of 1960 requires that national forests be established and administered for outdoor recreation, range, timber, watershed, wildlife, and fish purposes. The National Forest Management Act of 1976 further strengthened multiple-use practices in the national forests. Multiple uses are less emphasized in private and industrial forests.

7.3.1 Production Forests

The main purpose of production forests is to obtain financial profit from the forest by producing timber, pulpwood, fuels, wildlife, forest and agricultural byproducts, livestock, and recreation services. In the United States, production forests are owned and operated mainly by private individuals (57%) and industry (15%), while only 28% are national and public forests (MacCleery, 1993). Traditionally, the primary concern of industrial and private owners has been timber production. Only in recent decades have other products and services become a serious consideration in maximizing forest incomes. Today, hunting for a fee is available in many private forest clubs throughout the nation, and many ranches and cabins nestled in forested areas are operated for families to spend a weekend in the forest for relaxation, enjoyment, and outdoor activities.

In rural areas of the tropical regions, a sizeable population lives in and around forests. They grow crops in the forest for food and harvest branch and litter for fuel. India has begun a social program in which people plant and grow trees in their back yards and community woodlots for fuel and other purposes. Many feasibility studies have shown that power stations could be operated and the liquid fuel ethanol could be produced based on trees growing in "energy plantations" (Fung, 1982). Forest grazing is also a common practice that often damages trees, destroys litter floor, and compacts soils. These exploitative agroforestry practices interrupt nutrient cycling in the forest, increase soil and water erosion, deplete land productivity, and eventually lead to the abandonment of forests for production. In northeastern India alone, about 2.6×10^6 people depend on shifting cultivation, affecting about 2.7×10^6 ha of forestland (Shafi, 1992).

In the pursuit of maximum economic gain from a forest, exploitative uses of its resources should be avoided. Best-management practices should be incorporated in all forestry activities so that land productivity can be maintained and water quality is not impaired. The U.S. Forest Service adopted ecosystem management as a policy in June 1992. The ecosystem-management approach is to use ecological principles and the best available science and technology in managing natural resources. It recognizes soil, water, vegetation, wildlife, society, and the economy as integrated factors in the formulation of management objectives. Experience from the national forests can serve as guidelines to managing private forests.

7.3.2 Protection Forests

On rough terrain, steep slopes, stream banks, water-resource areas, wind-prone regions, or potential landslide sites, forests are often established to reduce soil erosion, increase sand stability, improve water quality, retain reservoir capacity, mitigate flood damage, and attenuate air pollution. Forests are also managed to protect habitats for birds, fish, and other animals such as pandas in China and tigers in India. As noted Sakals et al. (2006), forests are often managed to provide protection functions associated with floods, debris floods, debris flows, snow avalanches, and rockfells. Protection forests emphasize environmental functions; economic income is insignificant or even totally ignored. In India, protection forests make up about 23.4% or 15×10^6 ha of the nation's forests (Lal, 1992).

Since protection forests are there to protect a specific site and environmental condition, the species used are more restrictive, and management activities need to ensure the sustainability of the forest. For example, salt spray, high wind speed with abrasive sands, low fertility, high temperature, and extremely wet and dry soil conditions are the general characteristics of coastal sand dunes. Because of their adaptability, species such as sand live oak (*Quercus virginiana*) and southern waxmyrtle (*Myrica cerifera*) in the south Atlantic coast and casuarina (*Casuarina equisetifolia*) in Taiwan are often planted to protect against wind erosion and sand-dune movement in coastal areas. Protection forests are usually in areas sensitive to environmental problems; clearcutting, grazing, cropping, and litter harvesting should not be practiced. A clearcut in these sensitive areas would make artificial regeneration very difficult or take too long to establish. It can consequently make the destruction of forests in the protected area devastating. Thus, legal enforcement may be required to prevent protection forests from damage by cultivation, harvesting, grazing, and other impairing activities.

Special forests commonly managed for environmental protection include:

- Headwater resource (reservoir) protection forests
- Coastal protection forests
- Windbreaks
- Sand-stabilization forests
- Riparian forests

- Habitat-protection forests
- Avalanche-protection forests
- Soil-erosion-prevention forests
- Landslide-protection forests
- Swamp-prevention forests
- Noise-reduction belts (greenbelts)

In fact, all forests can be considered protective in view of their function as sinks of atmospheric carbon dioxide, a gas linked to global warming (see Section 7.5).

7.3.3 Preservation Forests

It has been estimated that about 30% of Earth's land surface has been devegetated since farming began. Between 1990 and 2010, about 115×10^6 ha of the highly biodiversified tropical forests in seven tropical countries were lost due to deforestation, accounted for 85% of the global total. Nigeria lost 47.5% of forests, followed by 20.3% of Indonesia, 19.5% of Tanzania, and 19% of Myanmar (FAO, 2010). Loss of forests leads to the loss of many plant and animal species from the genetic and pharmaceutical pools of the world. The nature and ecosystems of managed forests are different from those of the virgin forests. Impacts on the hydrological cycle, soil and nutrient losses, and climate changes are highly significant and well documented, especially in the Amazon basin.

In the United States, Congress has established two systems for preserving forests and related areas in natural and unimpaired condition. One is the National Wilderness Preservation System authorized under the Wilderness Act of 1964 and the other is the National Wild and Scenic Rivers System authorized under the National Wild and Scenic Rivers Act of 1968. A wilderness is an area of undeveloped land retaining its primeval character and influence and an area with its ecosystems untrammeled by people. About 4% of United States land area is preserved under the wilderness system, most of it in Alaska, the West, and Florida. The preserved areas can be used only for recreational activities such as hunting, fishing, camping, and rafting, and for scientific studies and educational programs. They are free from any development and management, even for disease and insect salvages. Harvesting, logging, road construction, mining, damming, mechanized transport, and other commercial activities are prohibited.

The Scenic Rivers System allows river segments with outstanding recreational, scenic, geological, wildlife, ecological, historical, or cultural values to be protected from development and alteration. They are used for enjoyment and recreational activities such as camping, swimming, fishing, and hunting. The river segments under the systems may not be dammed, diverted, straightened, widened, dredged, or filled. Currently, river segments totaling 18,400 km in length, or 0.32% of the 5.6×10^6 km of rivers in the United States, are protected under the system.

7.3.4 Public Forests

Forests that are developed and managed for the public to enjoy are referred to as public forests. They may include parks, botanic gardens, zoos, and wildlife refuges. The majority of public forests are national, state, and city parks. In 1872, the federal government set aside over 8×10^6 ha of forest as Yellowstone National Park in northwestern Wyoming, the first national park of the world. Today there are 54 major national parks in the United States and over 1100 national parks of at least 1000 ha each in more than 120 countries around the world (Miller, 1999).

The national parks are dedicated to preserving vegetation, wildlife, natural wonders, cultural heritage, and historical sites for people's pleasure and health. They also provide educational programs through forest trails, on-site written explanations, and viable environmental interpretation as well as a natural setting for ecological research and natural-resources studies. All preservation, development, and management take place through integral plans administered through the National Park Service.

7.4 Threats to Forests

The most serious threats to forests, such as deforestation, grazing, forest fires, and air pollution, come from human activities. Naturally induced insects, disease, tornadoes, volcanic eruptions, and winter storms can cause substantial damage to forests, forest health, and forest sustainability. Weakened trees created by overcrowding, age, or other agents are less resistant to insect and disease attacks. Populations of insects and disease-causing organisms build up easily in these weakened trees and then spread to the entire forest. In many cases, poor forest management can cause insects and disease threats and climate change may cause severe outbreaks of native pests. A severe winter storm may cause a massive wind-throw of timber in forested areas. Unless active salvage actions are taken, potential for insect infestations and wildfire outbreaks is great.

7.4.1 Deforestation and Grazing

Because of population and economic pressures, deforestation and grazing caused a great loss of tropical forests in the lower latitudes. Tropical forests were lost in most cases to shifting cultivation, about 6.04×10^6 ha/year in 10 tropical countries between 2000 and 2010. In the middle latitudes, however, reforestation and conservation programs increased temperate forests and other woodlands by 4.41×10^6 ha/year in 10 countries in the 2000s (FAO, 2010). Thus, deforestation and forest grazing are more a regional and global issue in the lower latitudes and a local issue in the middle latitudes. Changes in global forest areas during the 1990–2010 period are given in Table 7.2.

In the United States, about 50% of all forestlands are grazed by domestic livestock, ranging from 41% in the Northeast to 83% in the West (Troeh et al., 1999). Grazing causes damage to tree seedlings, sprouts, and roots; destroys organic floor and soil structure; and reduces soil porosity and infiltration capacity. The average water erosion on grazed forestlands is about 5.2 versus 1.6 t/ha/year on nongrazed forestlands in the United States (National Research Council, 1986).

Besides the decline of forest resources, soil erosion, and water quality degradation, deforestation and other land use activities breakdown large forest into smaller ones. Fragmentation of forests may lead to changes in ecological processes, reduction in biological diversity, increases in wildfire susceptibility and tree mortality, alterations of plant and animal species composition, and easier access to interior forest. *Forest fragmentation* and *edge effects* have been identified as one of the most pervasive and deleterious processes in the tropics today (Broadbent et al., 2008). In the United States, forests were fragmented with about 44% being within 90 m, 62% within 150 m, and <1% being more than 1230 m from the forest edge (Tkacz et al., 2008).

TABLE 7.3

Average Annual Burned Areas and Number of Fires for Every 400,000 ha Protected Forests Managed by the U.S. Forest Service (USFS) from 1940 to 2000

USFS Region	Burned Area (ha)	Total Fires (Number)	Lightning Causes		Human Causes	
			Area (ha)	Number	Area (ha)	Number
Northern	560.09	38.70	387.11	32.16	172.98	6.54
Rocky Mountain	298.52	23.06	135.28	14.11	163.23	8.95
Southwestern	709.46	94.03	439.73	73.10	269.73	20.94
Intermountain	762.06	26.45	589.05	19.71	173.01	6.73
Pacific Southwest	1895.40	69.23	688.99	42.81	1206.41	26.42
Pacific Northwest	509.80	57.40	401.88	37.79	107.92	19.61
Southern	1233.71	108.71	77.12	12.65	1156.51	96.07
Eastern	339.83	42.12	11.84	2.17	327.99	39.95
Alaska	4.21	1.08	0.01	0.03	4.20	1.05

Source: Reconstructed from Stephens, S.L., *Int. J. Wildland Fire*, 14, 213–222, www.publish.csiro. au/journals/ijwf, 2005.

7.4.2 Forest Fires

Like precipitation and wind, wildfire is a natural phenomenon, but it occurs less frequently than other weather events do. It ranges from once in 3 years for long-leaf pine communities in the southeastern United States to once in 230–1000 years for mixed forests in New Brunswick (Kimmins, 2003), depending on vegetation types and regions. Since plant adaptations to fire differ among species, fire plays an important role in the characteristics of an ecosystem.

However, many regions have more forest fires caused by human activities than those that are naturally induced. California has experienced both the highest relative area burned and the largest human-caused burned areas (relative) during 1940–2000 in the Unites States However, the number of human-caused fires was largest in the Southern Region. Lightning strikes are stochastic, making it difficult for fire manager to forecast (Table 7.3).

From the management point of view, fires can cause on-site and off-site as well as detrimental and beneficial effects to soils, water, nutrients, vegetation, and wildlife. The intensity and duration of these effects depend on the type of vegetation, the severity and frequency of fires, the type of burning (ground, surface, or crown fires), season, slope, aspect, soil texture, and climatic conditions. Virtually all terrestrial ecosystems have been affected by fire at one time or another. Foresters consider forest fires a great threat because they can destroy the forest and its protective functions, reduce the value of existing timbers, induce insect or disease infestation, damage recreational and scenic value, and cause forest regeneration to take a long time. The effects of fire on watersheds are a major concern of watershed managers.

7.4.3 Air Pollution

Air pollution is the concentration of certain chemicals or particles in the air at levels that can cause harmful impacts to humans, plants, animals, materials, soils, and water. Pollutants originate from natural events such as ocean splash, wind erosion, forest fires, and volcanic eruptions, or from human activities such as fuel emissions and industry. These original pollutants can react with the chemicals and moisture already in the air to form induced pollutants. Air pollutants, including dry and wet deposition of acids, the emissions of acidifying sulfur and nitrogen compounds, ozone, and heavy metals, present a serious long-term threat to forest health and productivity in the northeastern United States, central Europe, the eastern Canadian provinces, and southern China. It is estimated that 49% of global forests will be exposed to tropospheric O_3 with concentrations at damaging levels by 2100 (Perce and Ferretti, 2004). For forest strategic and operational decision makings, monitoring networks, such as those adapted in Sweden (Wulff et al., 2011) are required to provide long-term forest assessments on health conditions along with damaging agents and effects.

Acute vegetation damage caused by smelters, power plants, and other large point sources of air pollution has been reported frequently. However, the most widespread effects of air pollution on the forest are probably due to acid deposition. Pure water has a pH of 7.0; the pH of uncontaminated rain is 5.6. Industrial emissions such as nitrogen oxides and sulfur oxides are transformed chemically into acids with moisture in the atmosphere and fall to the ground as acid rain, snow, and fog or dry acid particles with pH values lower than the reference level. In the Hubbard Brook Experimental Forest, New Hampshire, precipitation pH has been recorded as low as 3.0 (Likens et al., 1977).

Acid deposition can adversely affect forest vegetation either directly by damaging protective surface structures (cuticles) of the canopy or indirectly through the acceleration of soil acidification. Damage to the cuticular layer can lead to malfunction of guard cells, alteration of leaf- and root-exudation processes, interference with reproduction, water stress, and leaching of minerals from the canopy. Soil acidification can lead to leaching of basic nutrient ions, alterations of nutrient availability, slowdown of microbiological processes, reduction of microbial populations and variety, and increases in the level of ion toxicity to plants. Combining these effects on soils and plants can ultimately result in leaf coloration and abscission, and in reduction of forest growth, productivity, and species diversity. The former West Germany employed a loss of tree foliage by 11% or more as the main criterion in air-pollution damage inventories (Huettl, 1989).

In Europe, between 3.5 and 4 million ha of trees showed injury linked to air pollution in the early 1980s (Postel, 1984). The former West Germany estimated forest damage by air pollution at 0.562×10^6 ha in 1982 and increased the estimate to 2.545×10^6 ha in 1983; this represented 34% of the nation's forests. The damage further increased to 50% in 1985 and 54% in fall 1986, of which 19% was marked by foliage losses greater than 25% (Huettl, 1989).

In the United States, injury and mortality of pine has been most extensive in California, of red spruce in the Appalachian Mountains, of yellow pines in the Southeast, of eastern white pines in the East, and of sugar maples in the Northeast and Canada. Some of the damage began as early as in the late 1940s. In view of the complexity of the forest ecosystems, MacKenzie and El-Ashry (1989) stated that these injuries and declines probably were a collective reflection of multiple stresses, such as acid deposition, ozone, and drought, rather than of a single cause.

7.4.4 Biotic Infestation

Herbivorous and saprophagous insects and pathogenic fungi are important components in forest ecosystems. They play significant roles as consumers and decomposers in the energy flow and nutrient cycles of the forest. When a forest is under stress conditions, due to human activities, climate change, or other nature events, the populations and activities of root disease fungi and bark beetles are subsequently increased, resulting in defoliation, growth retardation, or even mortality of the entire forest. Infested forests are often poor in lumber quantity and quality, and are more susceptible to fires and wind throws (Jenkins et al., 2008). The potential U.S. loss from forest insects and forest pathogens is estimated to exceed 4.2×10^9/year (Pimentel et al., 2000), and the risk of insect and disease mortality was estimated to be 23.5×10^6 ha of forested land in the U.S. in 2006 (Holsten et al., 2008). Bark beetles and pathogenic fungi are considered one of the most destructive of the many threats to forest productivity.

7.5 Forests and Climate Change

Evidence for global climate change includes increases in surface and ocean temperatures, melting of glaciers, snow and ice caps, sea-level rises, dry cloud forests, wildfires, and severe weather conditions. This is especially true of recent decades in which many devastating floods, hurricanes, droughts, earthquake, volcanic activities, and weather irregularities have occurred, but had not been previously observed. Forests have also been affected in respect to the reduction of snowpack amounts and duration, the advances of peak snowmelt runoff, the increases of wildfire size and severity, and the outbreaks of bark beetles at unprecedented levels in the western United States in the 2000s (Jones et al., 2009).

These phenomena are largely considered to be associated with the "global warming" caused by the anthropogenic emissions of greenhouse gases to the atmosphere. Natural events, such as solar radiation outputs and aerosols produced by volcano eruptions also can affect global temperatures, but only greenhouse gases can satisfactorily explain the observed warming trends on Earth. Strategies on climate change mitigation have focused on reducing emissions of greenhouse gases, especially CO_2, produced by fossil fuel combustions and industrial activities. Deforestation and shifting agriculture in developing countries are a significant source of anthropogenic CO_2 emissions, consequently creating great concern about their impact on climate change.

7.5.1 Global Warming

Global warming refers to the continuous increase on the average ambient air and surface sea temperatures, especially in the last three decades. The increasing trend is expected to continue if no significant mitigation measures, such as green energy and reforestation, etc., are adopted worldwide. Under various scenarios, climate models project a likely rise of the global surface temperature by 1.1°C–6.4°C during the twenty-first century (IPCC, 2007).

7.5.1.1 Global Surface Temperatures

7.5.1.1.1 Ambient Air Temperatures

Around the time of the industrial revolution, circa 1750, global air temperature started to rise with its most significant increases from the 1950s. Based on the most available data in 2007, ICPP (2007) reports that 11 of the last 12 year temperatures (1995–2006) rank among the 12 warmest years since 1850, the year that instrumental temperature records became available. The 100 year linear increase in temperatures was 0.74°C (0.56°C–0.92°C) for 1906–2005, and 0.60°C (0.4°C–0.8°C) for 1901–2000. The linear warming trend over the last 50 year was 0.13°C/decade, nearly twice the trend for the last 100 year (0.07°C/decade). The total temperature increase over the past 150 years was about 0.76°C. Figure 7.9 shows the annual (1880–2008) average global land and sea combined temperature anomalies with reference period to the twentieth century average (land 8.5°C, sea 16.1°C, and land + sea 13.9°C).

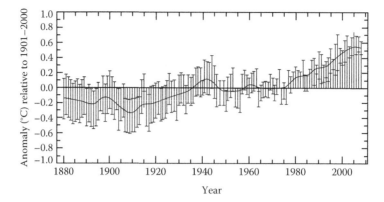

FIGURE 7.9
The annual average global surface and sea combined temperature anomalies with reference period to the twentieth century average (13.9°C), 1880–2008. (Adapted from U.S. National Climatic Center, Global Surface Temperature Anomalies, Available at: http://www.ncdc.noaa.gov/cmb-fag/anomalies.php, accessed on August 18, 2011.)

7.5.1.1.2 Sea Surface Temperatures

Warming trends have been observed on the sea surface temperatures since 1850, and especially for recent periods. The linear warming between 1850 and 2004 was 0.52°C ± 0.19°C (95% confidence interval) for the globe, 0.59°C ± 0.20°C for the Northern Hemisphere, 0.46°C ± 0.29°C for the Southern Hemisphere and 0.35°C ± 0.35°C for the Indian Ocean. However, excluding the first 50 years, the linear warming trends of the four regions for the later period, 1901–2004, were warmer than those for the entire period, 1850–2004, by 0.07°C–0.37°C. They were, in that same sequence, 0.68°C, 0.66°C, 0.68°C, 0.58°C, respectively. If differences in temperatures due to data uncertainties were filtered and corrected for these regions, the increasing trends would range from 0.56°C to 0.71°C for the 1850–2004 period and from 0.63°C to 0.75°C for the 1901–2004 period (Rayner et al., 2006). Thus, the warming trends on the sea surface temperatures and the ambient air temperatures are apparently in the same magnitude. (Note that the aforementioned analyses are based on a gridded dataset of global sea surface temperatures since 1850 maintained as HadSST2 by the Hadley Centre for Climate Prediction and Research, Met Office, at Exeter, United Kingdom) The increases in surface temperatures can cause thermal expansion of the sea and was estimated at 0.42 ± 0.12 mm/year for 1961–2003 and 1.6 ± 0.5 mm/year for 1993–2003. Total sea-level rise was estimated at 1.8 ± 0.5 mm/year for 1961–2003, 3.1 ± 0.7 mm/year for 1993–2003, and 0.17 ± 0.05 m for the twentieth century (IPCC, 2007).

7.5.1.2 Greenhouse Effect

Besides N_2 (78%) and O_2 (21%), the atmosphere also contains some highly variable gases such as H_2O (0%–4%), CO_2 (0.038%), CH_4, O_3, N_2O, SO_2 and some industrial gases such as chlorofluorocarbons (CFCs), hydrofluorocarbons (HFCs), and perfluorocarbons (PFCs), etc. Nitrogen and oxygen are stable gases, essential to human life but having little effect on weather and other atmospheric processes. Those variable elements in the atmosphere, of less than 1%, impose a greater impact on both short-term and long-term climate on Earth. They allow solar (short-wave) radiation to penetrate through the atmosphere, but shield and absorb the long-wave (infrared) radiation emitted from the ground surface. This process prevents excessive heat lost from the ground to space and the heat absorbed by the atmosphere is re-emitted back to the ground, consequently keeping the Earth warmer than what it would be. It is similar to the heat trapped in a greenhouse or an automobile, a process called the "greenhouse effect" and those elements are referred to as the "greenhouse gasses."

The warming impact of greenhouse gases depends on their warming power and the lifetime in the atmosphere. Water vapor is the most abundant greenhouse gas, contributing 95% of the total. Locally, it may be variable in response to the air temperature, but globally the concentration of water vapor remains constant. The lifetime of water vapor in the atmosphere is about 10 days, not responsible for the global warming.

Excluding water vapor, the contributions of CO_2, CH_4, N_2O, and CFCs to the total greenhouses gases in the atmosphere are 73.4%, 7.1%, 19.0%, and 1.4%, respectively. CO_2 not only abundantly outnumbers the others in quantity, it also has a long atmospheric lifetime of 50–200 years. Its large volume and long atmospheric lifetime make CO_2 the most prominent greenhouse gas. A molecule of CH_4 has about 21 times greater effect on warming than a molecule of CO_2 over a 100 year period, but its lifetime (12 years) is short and its concentration in the atmosphere is small. Nitrous oxide is 300 times more powerful than CO_2, but its concentration is also much lower than CO_2 (IPCC, 2007). None of these gases adds as much warmth to the atmosphere as CO_2 does.

7.5.1.3 Atmospheric CO_2 Concentrations

Concentrations of CO_2 in the atmosphere were 280 ppm by volume in the beginning of the industrial revolution, 295 ppm in 1900, 310 ppm in 1950, 370 ppm in 2000, and 390 ppm for 2010. This steady increase in CO_2 in the atmosphere, especially in the last 60 years, is of primarily anthropogenic origins. They include fossil fuel combustion, industrial production processes, farming practices and changes in land use, deforestation and population growth. These increases in CO_2 concentrations in the atmosphere, especially in the last 60 years, correspond very well with the increases in air temperature, and it is thought to be responsible for the climate change. In 2000, energy and industrial processes accounted for about 75% of CO_2 emissions, and about 22% was derived from land use change and forests (Anderson et al., 2008).

7.5.2 Carbon Stocks

Carbon is a common constituent of all organic matter, important to all life processes. It exists on Earth as: (1) carbon dioxide in the atmosphere, (2) organic matter in organisms and soils, (3) fossil fuels and sedimentary rock deposits in the lithosphere, and (4) dissolved carbon dioxide in the oceans and calcium carbonate shells in marine organisms. Of all the simple carbon products, CO_2 is by far the most abundant and most influential to the environment. Estimated major stores of carbon on Earth are given in Table 7.4 (Pidwirny, 2010). On the global average, carbon stored as organic matter in soils and in wetlands is more than that in vegetation above the ground by 4 and up to 15 times, respectively (Table 7.5).

In tropical forests, carbon contained in vegetation is double the level than that in other forests and four times more than the global vegetation average (Table 7.5). However, tropical forest soils contain only average levels of carbon. When tropical forests are converted to cultivated land or pasture, the loss of carbon from plants to the atmosphere is about 90%–100% of their initial carbon stocks and 12%–25% of the initial stocks in soils (Houghton, 2005). This is due to the warm and humid conditions that accelerate the rapid decomposition of organic matter, and the high and intense rainfall that causes organic matter/minerals to be rapidly moved out from soils.

TABLE 7.4

Estimated Major Stores of Carbon on Earth

Sink	Quantity of C in Billion (10⁹) Mt
Atmosphere	578 (as of 1700) to 766 (as of 1999)
Terrestrial plants	540–610
Soil organic matter	1,500–1,600
Fossil fuel deposits	4,000
Ocean	38,000–40,000
Marine sediments/sedimentary rocks	66,000,000–100,000,000

Source: Pidwirny, M., Carbon cycle, in *Encyclopedia of Earth*, Cleveland, C.J., Ed., Environmental Information Coalition, National Council for Science and the Environment, Washington, DC, First published in the *Encyclopedia of Earth*, May 31, 2010; Retrieved on November 3, 2010 at http://www.eoearth.org/article/Carbon_cycle

TABLE 7.5

Average Carbon Stocks (in Mt of CO_2/ha) for Various Biomes on Earth

Biome	Area (10⁹ ha)	Plant Carbon	Soil Carbon	Total Carbon
Tropical forests	1.76	442	450	892
Temperate forests	1.04	208	352	561
Boreal forests	1.37	236	1260	1490
Tundra	0.95	23	467	490
Croplands	1.60	7	293	300
Tropical savannas	2.25	108	430	538
Temperate grasslands	1.25	26	865	892
Desert/semi-desert lands	4.55	6	154	160
Wetlands	0.35	157	2357	2514
Weighted average or total	15.12	113	488	601

Source: Gorte, R.W. and Sheikh, P.A., Deforestation and climate change, CRS Report for Congress, R41144, Congress Research Service (www.crs.gov), Washington, DC, 41 pp., 2010.

Note: 1 ton C = 3.67 tons CO_2.

7.5.2.1 Carbon Cycle

Carbon is a part of the atmosphere, soil, rocks, oceans, and all living things. It exists as CO_2 in the atmosphere, $CaCO_3$ in rocks and corals, coal, petroleum and natural gas deep under the ground, dead organic matter and humus in soils, HCO_3^{-1} when dissolving in water, and organic compounds in plants and animals. The movement of carbon, in its many forms, between the biosphere, atmosphere, oceans and geosphere is described as the *carbon cycle*.

Carbon in the form of CO_2 in the atmosphere is assimilated into organic compounds by plants and dissolved in oceans through diffusion. Some of the plant's organic compounds are passed down to heterotrophic animals through consumption and some stored as carbon in soils. Dead plants and animals slowly decomposed into organic materials, buried deep under the ground or in ocean floors, and formed fossil fuels through high temperature, pressure and bacterial processes over millions of years. Thus, carbon stored in the

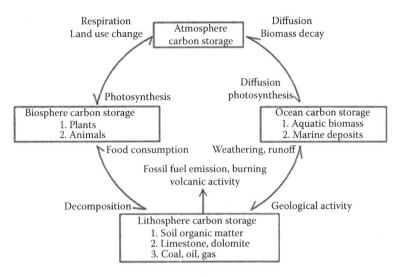

FIGURE 7.10
The carbon cycle.

lithosphere can be in inorganic (fossil fuels like coal, oil, and natural gas, oil shale, and carbonate-based sedimentary deposits) and organic (litter, organic matter, humus) forms.

The dissolved CO_2 in seawater can remain in the ocean or be converted into carbonate (CO_3^{-2}) or bicarbonate (HCO_3^-). Combining these substances with calcium (Ca^{+2}) produces calcium carbonate ($CaCO_3$), which is used to produce shells and other body parts by organisms, such as coral and clams. When they die, the shells and body remains of these organisms sink to the ocean floor and accumulate as carbonate deposits.

Carbon is released as CO_2 to the atmosphere by respiration of plants and animals, by decomposition of organic matter, or by emissions of fossil fuel combustions. Without the burning of fossil fuels, CO_2 concentrations in the atmosphere are relatively stable over time. The movement of carbon on Earth is illustrated in Figure 7.10.

7.5.2.2 Forest Carbon Sequestration

Trees absorb CO_2 and energy (sunlight, solar radiation) from the atmosphere and convert, while releasing oxygen, carbohydrates (carbon) into leaves, trunks and roots. This food production process is called "photosynthesis." In respiration, trees reversely absorb oxygen from the atmosphere, break down carbohydrates and release CO_2. The difference in products between photosynthesis and respiration is called the "net primary production," responsible for tree growth.

Trees/plants absorb more CO_2 than they release, and the trapped CO_2 is stored as carbon in the biomass (foliage, branches, trunks, and roots) and soils. This is the so-called carbon sequestration. The carbon storage in a forest is about 50% of its biomass. In Oregon (USA), carbon storage in forests ranges from 1 to 60 kg C/m^2 ground, depending on stand age and biomass (Tuyl et al., 2005). When cleared, the stored C may be lost to the atmosphere, more quickly through burning and slowly through decay. It may take 100 years or more with some wood products. Deforestation also reduces CO_2 uptake by trees. In the oak and maple forests of northern Iran, a study showed CO_2 uptake was about 6 t/ha/year (Khademi et al., 2009). Total CO_2 uptake by terrestrial ecosystem flux is about $123 \pm 8 \times 10^9$ tons of C/year, and tropical forests and savannahs account for about 60% of the total

(Beer et al., 2010). About one-half of the CO_2 that is uptaken by plants by photosynthesis is returned to the atmosphere through respiration and the other one-half is stored as carbon in plants (Conzalez-Meler et al., 2004; Pidwirny, 2010).

The forest growing stocks on Earth are estimated at 384×10^9 m³ and the biomass of forests, excluding dead wood, litter, and soil, contains about 240×10^9 Mt of carbon (FAO, 2009). Tropical forests are rich in biomass and carbon storage. Two tropical countries, Brazil (49×10^9 tons of C) and the Democratic Republic of Congo (23×10^9 tons of C) combined together make up 30% of the total carbon stocks in forests.

7.5.3 Impacts of Deforestation

Total forest areas on Earth were 4033×10^6 ha in 2010, or about 30% of the land surface. Global forests that were converted to other uses or lost through natural causes were estimated at 13×10^6 ha/year for the 2000–2010 period, and 16×10^6 ha/year for the previous decade, 1990–2000 (FAO, 2010). However, due to both a decrease in the deforestation rate and an increase in the area of afforestation/reforestation and the natural expansion of existing forests, net global losses in forest areas were substantially reduced to at 5.211×10^6 ha/year between 2000 and 2010, and at 8.323×10^6 ha/year between 1990 and 2000 and (Table 7.2). Two tropical countries, Brazil and Indonesia, contributed 60.4% for the period 2000–2010, and 57.8% of the total loss for the 1990–2000 period.

The impact of deforestation on climate is discussed on three different perspectives: site and local-, regional-, and global-scales.

7.5.3.1 On Site- and Local-Scale

Forests are characterized with big size, great surface roughness, dense, thick and dark canopies, and deep root systems. These unique features allow for forests to significantly modify local climates through shading and blocking effects on the dissipation of solar radiation (thermal regimes), advective heat and moisture transfers, wind movement, precipitation deposition, snow accumulation and snowmelt and overland flow. Net radiation over a dense forest canopy is usually diminished at the forest floor by at least an order of magnitude during clear days and is increased slightly at night (Lee, 1978). Daytime net radiation at forest sites is always higher than that at grasslands, up to about 25% in all seasons in Germany (Stiller et al., 2005).

Differences in air and soil temperatures between open and forested sites are complicatedly affected by a number of environmental factors including forest characters, slope, aspect, latitude, soil, wind speed, cloudiness, and precipitation. Generally, maximum temperatures are lower and minimum temperatures are higher in forest than in open. Also, air temperatures at night are usually lower in forest than in open or grassland, regardless of season. The largest differences in temperature between cut and uncut sites in central Massachusetts occurred in the afternoons of the summer season (Brooks and Kyker-Snowman, 2008). In humid East Texas, the mean annual air and soil temperatures were lower and the amplitudes of fluctuation were smaller in an undisturbed loblolly–shortleaf pine forest than in a nearby cleared site by 0.5°C–1.0°C (Chang et al., 1994). In northern Wisconsin, differences in growing season (May–September) mean maximum air temperatures between the clear-cut and forest sites were 5.7°C in the cool year and 4.7°C in the warm year, while the difference for mean minimum air temperatures was −3.3°C in the cool year and −3.2°C in the warm year (Potter et al., 2001). Shelterbelts and shading trees are most beneficial to farms and individual houses. Well-planted deciduous trees can save 10%–50% of home electric bills in summer.

Cutting forests can reduce evapotranspiration loss to the atmosphere by more than 20%. However, the reduced moisture content in the atmosphere is not necessarily transferable into a decrease in local precipitation, except in the tropical regions where internal circulation dominates local climates (see Chapter 8). Snow accumulation in forests can be less than that in nearby clearings by up to 30%–50%. But, snowmelt rate is slower and snowpack duration is longer in forests than it is in openings. Wind speeds below and within forest canopies are several times lower as compared to those in open fields. It is most likely relative humidity is higher in forests than in the open because of lower temperatures.

7.5.3.2 On Regional-Scale

Depending on size and location, impacts of deforestation on climate at the regional scale are more difficult to evaluate, and frequently these impacts spark controversy and debate when it comes to the occurrences of flooding, precipitation, and springs. In tropical regions where great surface heating makes storm activities dominated by convective (internal) circulation, upward fluxes of water vapor could have a direct impact on downward fluxes. Deforestation removes forest canopies, causing an increase in direct solar radiation to the ground (warmer soil surface and ambient air temperatures) and a reduction in evapotranspiration to the air (drier air). The combined effects could result in lower rainfall. A simulation study of potential impacts of tropical deforestation and greenhouse warming on climate shows that the joint climate changes comprise large reduction in evapotranspiration by –180 mm/year and rainfall by –312 mm/year along with an increase in surface temperature by 3.0 K over the Amazon basin. It also shows that rainfall is decreased by –172 mm/year and an increase in surface temperature by 2.1 K over Southeast Asia. The large increase in surface temperature is not solely produced by the increase in CO_2 concentration, but is a joint effect of CO_2 and deforestation (Zhang et al., 2001).

In another impacts, deforestation breakdowns forest coverage, causing an increase in forest edge and the sub-division of large forest into smaller non-contiguous fragments. *Forest fragmentation* means that more forests located in close proximity to forest edges and edge effects may extend into interior forest areas, up to 5–10 km. It may alter microclimates and increase carbon emissions, primarily due to tree mortality.

Forest fires initiated by farmers clearing forest land for cultivation in Indonesia have caused haze and smog to become an annual health and traffic problem for neighboring countries. Visibility of haze sky is down to 200 m in some cases. The 1997/1998 smog originating in Indonesia blanketed 3 million km² in area and affected the health of 75 million people in 6 countries. The haze problem, triggered by forest fires, drought and wind in the dry season from September through November, has been acknowledged by the Association of Southeast Asian Nations (ASEAN) for more than a decade.

In middle and higher latitudes where weather is more affected by general circulations, the impacts of deforestation on regional climates are expected to be small or insignificant as compared to those in the tropics. Precipitable water is mainly brought in through horizontal moisture convergence by large-scale air mass movements. Thus, sources of atmospheric moisture for precipitation are largely of ocean origins, not continents, and a drier air due to deforestation is not likely to cause a significant impact on regional precipitation. However, deforestation can cause surface albedo (reflectivity to incoming solar radiation) to be increased by 10%–30%. Results from experiments with a three-dimensional coupled global carbon-cycle and climate model show that the warming carbon-cycle effects of deforestation are overwhelmed by the net cooling associated with changes in albedo and evapotranspiration. Combining a higher albedo, a lower evapotranspiration, and a greater CO_2 emission together could cause regional-scale deforestation to reduce mean air temperatures by $-1.6°C$ and $-2.1°C$, respectively, for the middle- and higher latitudes of the North Hemisphere in the last decade of 2100 (Bala et al., 2007).

Major floods can occur in forested as well as non-forested watersheds. The effect of forest cutting on flooding is most effective in small watersheds and for small storms. As the size of watersheds, or the size of storms, increases, the effect of deforestation on flooding decreases to an insignificant level (see Chapter 14).

7.5.3.3 Global-Scale

On the global-scale, a major concern of deforestation is its release of CO_2 to the atmosphere, a greenhouse gas largely blamed for potential impacts on the global warming. Deforestation reduces the plant's uptake of CO_2 from the atmosphere through photosynthesis on the one hand, and releases CO_2 to the atmosphere through the combustion and degradation of biomass and organic matter above, on, and under the ground on the other hand. A study showed that CO_2 emissions due to global-scale deforestation could increase air temperature by 1.3 K (Bala et al., 2007).

Using data from Table 7.5, the weighted average carbon stocks in plants of tropical, temperate and boreal forests is 316 Mt of CO_2/ha. A global deforestation of 13×10^6 ha/year during the 2000–2010 period would cause plants to release 4.1×10^9 Mt of CO_2/year to the atmosphere, comparative to 4.4×10^9 Mt of CO_2 (or 1.2×10^9 Mt of C)/year estimated for 2008 by van der Werf et al. (2009). Estimated CO_2 emissions by IPCC (Denman et al., 2007) for the 1990s were 5.89×10^9 Mt of CO_2/year for changes in land use and 23.5×10^9 Mt of CO_2/year for the emission from fossil fuels and cement manufactures.

The carbon cycle in ecosystems, very complex and difficult to measure with current techniques, is not fully understood for its processes and variations under various climatic conditions. Recent studies suggest that the availability of water, often more important than temperature, plays a decisive role for the carbon cycle in ecosystems. Over 40% of Earth's vegetated surface, plants photosynthesize more when the supply of water increases, and less during droughts. However, the amount of CO_2 uptake in temperate grasslands and shrub lands is much strongly dependent upon water supply than that in the tropical rainforest. Respiration rate does not even double as previously thought when the temperatures increases by 10°C (Beer et al., 2010; Mahacha et al., 2010). This makes accurate estimates of CO_2 emissions from ecosystems extremely difficult.

Terrestrial vegetation takes up about 120 Gt of C/year from the atmosphere through photosynthesis and roughly half of the assimilated CO_2 is released back to the atmosphere by plant respiration. The terrestrial vegetation-atmosphere fluxes of CO_2 far exceed anthropogenic inputs of CO_2 to the atmosphere each year; consequently, a small change in vegetation respiration could have a significant impact on the annual increment of CO_2 in the atmosphere. Literature has indicated that plant respiration rates are often reduced when plants are exposed to high CO_2. As such, doubling the current atmospheric CO_2 concentration could reduce 15%–20% plant respiration and increase the net sink capacity of global carbon by 3–4 Gt/year (Drake et al., 1999). However, recent studies have shown that leaf respiration on average will not be greatly changed by increasing atmospheric CO_2 (Gonzalez-Meler et al., 2004). The result makes the role of plant respiration in augmenting the sink capacity of terrestrial ecosystems stated earlier uncertain.

Besides increasing CO_2 concentration in the atmosphere, deforestation also causes an increase in surface *albedo* and a decrease in evapotranspiration and surface roughness as stated previously. A higher albedo at deforested sites will lead both net radiation and latent heat to be lower (see Figure 9.4). Also, destruction of forest canopies greatly reduces the effective (leaf) surfaces for transpiration, and moisture deep in the soil is difficult to lose through evaporation. Combining these two will reduce evapotranspiration to the atmosphere which can trigger changes in atmospheric moisture content, cloudiness, and the lapse rate of air temperatures. An albedo effect could lead to cooling; an evapotranspiration effect could lead to warming and decreases in precipitation. The balance between these processes varies with latitude as the albedo effect being stronger in temperate and boreal zones of the Northern Hemisphere (a cooling effect) and the ET effect stronger in the tropics (a warming effect) (Davin and de Noblet-Ducoudré, 2010).

Simulation studies on the global replacement of forests by grassland have shown that combining the albedo and ET effects could cause a decrease in air temperature $-0.4\,K$ by Gibbard et al. (2005), $-1.6\,K$ by Bala et al. (2007), and $-1.0\,K$ by Davin and de Noblet-Ducoudré (2010).

References

Anderson, J., Fergusson, M., and Valsecchi, C., 2008, An overview of global greenhouse emissions and emission reduction scenarios for the future, Policy Department Of Economics and Science Policy, European Parliament, IP/A/CLIM/NT/2007-07, Brussels, Belgium, 28 pp.

Bala, G. et al., 2007, Combined climate and carbon cycle effects of large-scale deforestation, *Proc. Natl. Acad. Sci. USA*, 104(16), 6550–6555.

Beer, C. et al., 2010, Terrestrial gross carbon dioxide uptake: Global distribution and covariance with climate, *Science*, 239(5993), 834–838.

Broadbent, E.N. et al., 2008, Forest fragmentation and edge effects from deforestation and selective logging in the Brazilian Amazon, *Biol. Conserv.*, 141, 1745–1757.

Brooks, R.T. and Kyker-Snowman, T.D., 2008, Forest floor temperature and relative humidity following timber harvesting in southern New England, USA, *For. Ecol. Manage.*, 254, 65–73.

Brown, S., Gillespie, A.J.R., and Lugo, A.E., 1989, Biomass estimation methods for tropical forests with applications to forest inventory data, *For. Sci.*, 35, 881–902.

Buell, J.H., 1949, Trees living together: The community of trees, *Trees, Yearbook of Agriculture*, USDA, Washington, DC, pp. 103–108.

Burch, W.R., Jr., Alan, F., and Hermann, R.K., 1976, *Forest and Future Resource Conflicts*, Department of Printing, Oregon State University, Corvallis, OR.

Canadell, J. et al., 1996, Maximum rooting depth of vegetation types at the global scale, *Oecologia*, 108, 583–595.

Chang, M. et al., 1994, Air and soil temperature under three forest conditions in East Texas, *Texas J. Sci.*, 46, 143–155.

Chang, M., Roth, F.A., II, and Hunt, E.V., Jr., 1982, Sediment production under various forest-site conditions, *Recent Developments in the Explanation and Prediction of Erosion and Sediment Yield*, *Proceedings of the Exeter Symposium*, Walling, D.E., Ed., IAHS, Exeter, U.K., pp. 13–22.

Culf, A.D. et al., 1996, Radiation, temperature and humidity over forest and pasture in Amazonia, *Amazonian Deforestation and Climate*, Gash, J.H.C. et al., Eds., John Wiley & Sons, New York, pp. 175–191.

Davin, E.L. and de Noblet-Ducoudré, N., 2010, Climatic impact of global-scale deforestation: Radiative versus nonradiative processes, *J. Clim.*, 23, 97–112.

Denman, K.L. et al., 2007, Couplings between changes in the climate system and biogeochemistry, *Climate Change 2007: The Physical Science Basis. Contribution of Working Group I to the Fourth Assessment Report of the Intergovernmental Panel on Climate Change*, Solomon, S. et al., Eds., Cambridge University Press, Cambridge, U.K., pp. 501–587.

Department of Energy, 1992, Sector-Specific Issues and Reporting Methodologies Supporting the General Guidelines for the Voluntary Reporting of the Greenhouse Gases under Section 1605(b) of the Energy Policy Act of 1992.

Duerr, W.A., 1979, American forest resource management, *Forest Resource Management, Decision-Making Principles and Cases*, Duerr, W.A. et al., Eds., W.B. Saunders, Philadelphia, PA, pp. 9–20.

Drake B.G. et al., 1999, Does elevated CO_2 inhibit plant mitochondrial respiration in green plants? *Plant Cell Environ.*, 22, 649–657.

FAO, 2009, *State of the World's Forests, Food and Agriculture*, Organization of the UN, Rome, Italy, p. 152. ·

FAO, 2010, Global forest resources assessment 2010: Main report, FAO Forestry Paper 163, Rome, Italy, 340 pp.

Farrish, K.W., 1991, Spatial and temporal fine-root distribution in three Louisiana forest soils, *Soil Sci. Soc. Am. J.*, 55, 1752–1757.

Fung, P.Y.H., 1982, Wood energy prospects, *Energy from Forest Biomass*, Smith, W.R., Ed., Academic Press, New York, pp. 155–170.

Gholz, H.L., Fitz, F.K., and Waring, R.H., 1976, Leaf area differences associated with old-growth forest communities in the western Oregon Cascades, *Can. J. For. Res.*, 6, 49–57.

Gibbard, S. et al., 2005, Climate effects of global cover change, *Geophys. Res. Lett.*, 32, L23705, 2005.

Gonzalez-Meler, M.A., Taneva, L., and Trueman, R.J., 2004, Plant respiration and elevated atmospheric CO_2 concentration: Cellular responses and global significance, *Ann. Bot.*, 94(5), 647–656.

Gorte, R.W. and Sheikh, P.A., 2010, Deforestation and climate change, CRS Report for Congress, R41144, Congress Research Service (http://www.crs.gov), Washington, DC, p. 41.

Gower, S.T. and Son, Y., 1992, Differences in soil and leaf litterfall nitrogen dynamics for five forest plantations, *Soil Sci. Soc. Am. J.*, 56, 1959–1966.

Holsten, E. et al., 2008, *Insects and Diseases of Alaskan Forests*, R10-TP-140, USDA Forest Service, Washington, DC.

Honzák, M. et al., 1996, Estimation of leaf area index and total biomass of tropical regenerating forests: Comparison of methodologies, *Amazonian Deforestation and Climate*, Gash, J.H.C. et al., Eds., John Wiley & Sons, New York, pp. 366–381.

Houghton, R.A., 2005, Tropical deforestation as a source of greenhouse gas emission, *Tropical Deforestation and Climate Change*, Moutinho, P. and Schwartzman, S., Eds., Amazon Institute for Environmental Research, Belem, Brazil, pp. 13–21.

Huettl, R.F., 1989, "New types" of forest damages in central Europe, *Air Pollution's Toll on Forest and Crops*, MacKenzie, J.J. and El-Ashry, M.T., Eds., Yale University Press, New Haven, CT, pp. 22–74.

Hughes, R.O. and Dunning, D., 1949, Pine forests of California, *Yearbook of Agriculture*, U.S. Department of Agriculture, Washington, DC, pp. 352–358.

IPCC, 2007, *Climate Change 2007: The Physical Science Basis. Contribution of Working Group I to the Fourth Assessment Report of the Intergovernmental Panel on Climate Change*, Solomon, S. et al., Eds., Cambridge University Press, Cambridge, U.K.

Jackson R.B. et al., 1996, A global analysis of root distribution for terrestrial biomass, *Oecologia*, 108, 389–411.

Jackson, R.B., Money, H.A., and Schulze, E.-D., 1997, A global budget for fine root biomass, surface area, and nutrient contents, *Proc. Natl. Acad. Sci. USA*, 94, 7362–7366.

Jenkins, M.J. et al., 2008, Review: Bark beetles, fuels, fires and implications for forest management in the Intermountain West, *For. Ecol. Manage.*, 254, 16–34.

Jones, J.A. et al., 2009, Hydrologic effects of a changing forested landscape—Challenges for the hydrological sciences, *Hydrol. Process.*, 23, 2699–2704.

Kaufmann, M.R., Edminster, C.E., and Troendle, C.A., 1982, Leaf area determination for subalpine tree species in the Central Rocky Mountains, Research Paper RM-238, USDA Forest Service.

Keeley, J.E. and Keeley, S.C., 1998, Chaparral, *North American Terrestrial Vegetation*, Barbour, M.G. and Billings, W.D., Eds., Cambridge University Press, Cambridge, MA, pp. 163–207.

Khademi, A., Babaei, S., and Mataji, A., 2009, A study on productivity of carbon storage and CO_2 uptake in the biomass and soil of coppice stand, *Am. J. Environ. Sci.*, 5(3), 346–351.

Kimmins, J.P., 2003, *Forest Ecology, a Foundation for Sustainable Management*, 3rd edn., Prentice Hall, New York.

Kleinfelder, D. et al., 1992, Unconfined compressive strength of some streambank soils with herbaceous roots, *Soil Sci. Soc. Am. J.*, 56, 1920–1925.

Lal, J.B., 1992, Conservation and sustainable use of India's forest resources, *Forest Ecosystems of the World*, Shafi, M. and Raza, M., Eds., Rawat Publications, New Delhi, India, pp. 27–35.

Landsberg, H., 1968, *Physical Climatology*, Gray Printing Co., DuBois, PA.

Lee, R., 1978, *Forest Microclimatology*, Columbia University Press, New York.

Likens, G.E. et al., 1977, *Biogeochemistry*, Springer-Verlag, Heidelberg, Germany.

Long, J.N. and Smith, F.W., 1992, Volume increment in *Pinus contorta* var. latifolia: The influence of stand development and crown dynamics, *For. Ecol. Manage.*, 53, 53–64.

Lund, H.G., 2011, Definitions of forest, deforestation, afforestation, and reforestation, available at: http://home.comcast.net/~gyde/index.html/

MacCleery, D.W., 1993, American forests, a history of resiliency and recovery, FS-540, USDA-Forest Service.

MacKenzie, J.J. and El-Ashry, M.T., 1989, Tree and crop injury: A summary of the evidence, *Air Pollution's Toll on Forest and Crops*, MacKenzie, J.J. and El-Ashry, M.T., Eds., Yale University Press, New Haven, CT, pp. 1–21.

Mader, S.S., 1998, *Biology*, McGraw-Hill, New York.

Mahacha, M.D., et al., 2010, Global convergence in the temperature sensitivity by respiration at ecosystem level, *Science*, 239, 5593, 838–840.

Mather, A.S., 1990, *Global Forest Resources*, Timber Press, Portland, OR.

Medina, E. and Cuevas, E., 1997, Biomass production and accumulation in nutrient-limited rain forests: Implications for responses to global change, *Amazonian Deforestation and Climate*, Gash, J.H.C. et al., Eds., John Wiley & Sons, New York, pp. 221–239.

Miller, G.T., Jr., 1999, *Environmental Science, Working with the Earth*, Wadsworth, New York.

National Audubon Society, Inc., 1992, *The Audubon Society Field Guide to North America Trees*, Alfred A. Knopf, New York.

National Research Council, 1986, *Soil Conservation: Assessing the National Resources Inventory*, Vol. 1, The National Academies Press, Washington, DC, p. 7.

Peet, R.K., 1988, Forests of the rocky mountains, *North American Terrestrial Vegetation*, Barbour, M.G. and Billings, W.D., Eds., Cambridge University Press, Cambridge, MA, pp. 63–101.

Perce, K.E. and Ferretti, M., 2004, Air pollution and forest health: Toward new monitoring concepts, *Environ. Pollut.*, 130, 113–125.

Perry, D.A., 1994, *Forest Ecosystems*, Johns Hopkins University Press, Baltimore, MD.

Phillips, W.S., 1963, Depth of roots in soil, *Ecology*, 44, 424.

Pidwirny, M., 2010, Carbon cycle, in *Encyclopedia of Earth*, Cleveland, C.J., Ed., Environmental Information Coalition, National Council for Science and the Environment, Washington, DC, First published in the *Encyclopedia of Earth*, May 31, 2010; Retrieved on November 3, 2010 at http://www.eoearth.org/article/Carboncycle

Pimentel, D. et al., 2000, Environmental and economic costs of nonindigenous species in the United States, *BioScience* 10, 53–65.

Postel, S., 1984, Air pollution, acid rain, and the future of forests, Worldwatch Paper 58, Worldwatch Institute, Washington, DC.

Potter, B.E., Teclaw, R.M., and Zasada, J.C., 2001, The impact of forest structure on near-ground temperatures during two years of contrasting temperature extremes, *Agric. For. Meteorol.*, 106, 331–336.

Rayner, N.A. et al., 2006, Improved analyses of changes and uncertainties in sea surface temperature measured in situ since the mid-nineteen century: The HadSST2 dataset, *J. Clim.*, 19, 446–469.

Reifsnyder, W.E. and Lull, H.W., 1965, Radiant energy in relation to forests, Technical Bulletin No. 1344, USDA Forest Service, Washington, DC.

Richardson, A.D. and Dohna, H.Z., 2003, Predicting root biomass from branching patterns of Douglas-fir root system, *Oikos*, 100, 96–104.

Sakals, M.E. et al., 2006, The role of forests in reducing hydrogeomorphic hazards, *For. Snow Landsc. Res.*, 80(1), 11–22.

Schenk, H.J. and Jackson, R.B., 2002, The global biogeography of roots, *Ecol. Monogr.*, 72, 311–328.

Schoeneberger, M.M., 2009, Agroforestry: Working trees for sequestrating carbon on agricultural lands, *Agrofor. Syst.*, 75, 27–37.

Shafi, M., 1992, Utilization and conservation of forests in India with special reference to social forestry, *Forest Ecosystems of the World*, Shafi, M. and Raza, M., Eds., Rawat Publications, New Delhi, India, pp. 21–26.

Sidle, R.C., Pearce, A.J., and O'Loughlin, C.L., 1985, *Hillslope Stability and Land Use*, Water Resources Monograph Series 11, American Geophysical Union, Washington, DC.

Spurr, S.H. and Barnes, B.V., 1973, *Forest Ecology*, 2nd edn., Ronald, New York.

Stephens, S.L., 2005, Forest fire causes and extent on United States Forest Service lands, *Int. J. Wildland Fire*, 14, 213–222, http://www.publish.csiro.au/journals/ijwf

Stiller, B. et al., 2005, Continuous measurements of the energy budget components at a pine forest and at a grassland site, *Meteorol. Z.*, 14(2), 1–6.

Stout, B.B., 1956, Studies of the root systems of deciduous trees, Black Rock Forest Bulletin No. 15, Harvard University, Cambridge, MA.

Sutton, O.G., 1953, *Micrometeorology*, McGraw-Hill, New York.

Swank, W.T. and Schreuder, H.T., 1974, Comparison of three methods of estimating surface and biomass for a forest of young eastern white pine, *For. Sci.*, 20, 91–100.

Swerdlow, J.L., 2000, Medicines in nature, *Natl. Geogr.*, April 2000, 98–117.

Tkacz, B. et al., 2008, Forest health conditions in North America, *Environ. Pollut.*, 155, 409–425.

Troeh, F.R., Hobbs, J.A., and Donahue, R.L., 1999, *Soil and Water Conservation: Productivity and Environmental Protection*, Prentice Hall, New York.

Tuly, S.V. et al., 2005, Variability in the net primary production and carbon storage in biomass across Oregon forests—An assessment integrating data from forest inventories, intensive sites, and remote sensing, *For. Ecol. Manage.*, 209, 273–291.

U.S. Department of Agriculture, 2009, Summary Report: 2007 National Resources Inventory, Natural Resources Conservation Service, Washington, DC, and Center for Survey Statistics and Methodology, Iowa State University, Ames, IA, p. 123, available at: http://www.nrcs.usda.gov/technical/NRI/2007/2007_NRI_Summary.pdf

U.S. National Climatic Center, Global Surface Temperature Anomalies, available at: http://www.ncdc.noaa.gov/cmb-fag/anomalies.php, accessed on August 18, 2011.

van der Werf, G.R. et al., 2009, CO_2 emissions from forest loss, *Nat. Geosci.*, 2, 737–738.

Van Eimern, J. et al., 1964, Windbreaks and shelterbelts, WMO Technical Note No. 59.

Vose, J.M. and Swank, W.T., 1990, Assessing seasonal leaf area dynamics and vertical leaf area distribution in eastern white pine (*Pinus strobus* L.) with a portable light meter, *Tree Physiol.*, 7, 125–134.

Waldron, L.J. and Dakessian, S., 1982, Effects of grass, legumes, and tree roots on soil shearing resistance, *Soil Sci. Soc. Am. J.*, 46, 894–899.

WMO (World Meteorological Organization), 1964, Windbreaks and Shelterbelts, WMO-No. 147. TP.70, Technical Note No. 59, Geneva, Switzerland.

WMO, 1981, Meteorological aspects of the utilization of wind as an energy source, WMO-No. 575, Technical Note No. 175, Geneva, Switzerland.

World Resources Institute, 2005, Land use and human settlements, in Earth Trends Data Tables: Forests, Grasslands and Drylands, available at: http://www.earthtrends.wri.org/pdf_library/data_tables/for1_2005.pdf

Wulff, S. et al., 2011, Adapting forest health assessments to changing perspectives on threats—A case example from Sweden, *Environ. Monit. Assess.*, 184(4), 2453–2464.

Wynn, T. and Mostaghimi, S., 2006, The effects of vegetation and soil type on streambank erosion, southwestern Virginia, USA, *J. Am. Water Resour. Assoc.*, 42(1), 69–82.

Zhang, H., Henderson-Sellers, A., and McGuffie, K., 2001, The compounding effects of tropical deforestation and greenhouse warming on climate, *Clim. Change*, 49, 309–338.

8

Forests and Precipitation

Precipitation in the forms of rain and snow is the major input to a watershed hydrologic system. Its occurrence, distribution, amount, intensity, and duration affect streamflow, soil moisture, soil erosion, nutrient losses, and distribution of plant species. Precipitation in the forms of sleet, frost, dew, and hail, due to its lower occurrence and small quantity, is less important to hydrology. Hydrologic studies and watershed research are mainly concerned with rain and snow.

8.1 Precipitation Processes

8.1.1 Atmospheric Moisture

The atmosphere is a mixture of many gaseous, liquid, and solid substances. Water, up to 4% by volume, is one of the constituents and can be in solid, liquid, and gaseous states in the same region. For a storm to occur, sufficient water has to be present in the atmosphere.

8.1.1.1 Magnitude

Water is always present in the atmosphere even in areas above desert or during severe drought conditions. The amount of moisture in the atmosphere at any one time is about 1.233×10^4 km^3 or about 0.035% of the total freshwater of Earth. If all of the moisture in the air fell as rain, it would cover Earth's surface to a depth of about 25 mm. The average annual precipitation on Earth is about 1000 mm. This means that the resident period of water vapor in the atmosphere is about 10 days.

The source of renewed water in the atmosphere comes from vaporization in the ocean and at land surfaces, including lakes, soils, and vegetation. Plant transpiration is often more important than soil evaporation. However, evaporation from the ocean is the major source of moisture for precipitation due to large surface area and great evaporation rate. Water loss per unit area from the land is about 40% of that from the ocean, but it is only 17% in terms of total volume.

Although oceans are the major supply of water for precipitation, nearness to the ocean does not necessarily lead to abundant precipitation. This is evidenced by the fact that many subtropical islands and deserts have edges around the ocean but are low in rainfall. The general air mass movement, orographic effects, and distance to the storm tracks are major factors affecting the precipitation climate of a region.

8.1.1.2 Measures

A variety of measures have been employed to describe the water content of the atmosphere. Some of them are explained as follows:

Specific humidity, q, is the ratio of the mass of water vapor (M_v) in a sample of moist air to the total mass of the sample (sum of mass of dry air M_d and mass of water vapor), or

$$q \text{ (in g/kg)} = \frac{M_v}{M_v + M_d} \cong 622 \frac{e}{p_a} \tag{8.1}$$

where e and p_a are actual vapor pressure and total air pressure, both in millibars, respectively.

Absolute humidity or vapor density, V_d, is defined as the mass of water vapor per unit volume of air V, or

$$V_d \text{ (in g/m}^3) = \frac{M_v}{V} \tag{8.2}$$

Relative humidity, RH, is the ratio between actual (e) and saturated (e_s) vapor pressures at a given temperature, or

$$RH \text{ (in \%)} = \left(\frac{e}{e_s}\right) 100 \tag{8.3}$$

Saturation deficit, SD, is the difference between saturation and actual vapor pressure, both in millibars, at the ambient temperature, or

$$SD \text{ (in mbar)} = e_s - e = e_s(1 - RH) \tag{8.4}$$

which is a measure of evaporation potential of the air.

Dew point is the temperature to which a given parcel of air mass must be cooled at constant pressure and constant vapor content in order for saturation to occur. If the air is cooled below the dew point, the excess of water vapor in the air is condensed, which forms clouds. This is the initial mechanism required in precipitation processes; however, for a precipitation to occur, additional mechanisms are required.

8.1.2 Precipitation Formation

The total amount of water vapor in the atmosphere at a given time is about 25 mm. This amount of water vapor can supply a moderate rainfall for about 2–3 h. However, many intense storms have intensity greater than 100 mm/h. For an observed rate of storm precipitation, four conditions are required:

- Mechanisms to cool the air temperature below dew point
- Small particles and nuclei to enhance the condensation of water vapor
- Growth of water droplets large enough to be pulled down by gravitation
- The convergence of atmospheric moisture to the stormy region

8.1.2.1 Cooling Mechanisms

Air temperature can be cooled due to one or more of the following causes:

Adiabatic cooling due to convective heating or orographic lifting
Frontal cooling due to mixing of two air masses different in physical properties

Contact cooling due to a colder surface

Radiation cooling due to the loss of heat at the ground surface

When a parcel of air is lifted up due to convective heating or mountain barriers, the reduced atmospheric pressure at higher elevations causes the volume to expand. The volume expansion consumes the internal heat energy and results in reduction of air temperature. Adiabatic cooling and frontal cooling can produce large-scale and significant amounts of precipitation, while contact and radiation cooling produces small condensation such as dew, frost, and fog.

8.1.2.2 Condensation Nuclei

When the air temperature is cooled down below the dew point, water vapor content in the atmosphere is then greater than the maximum capacity that the atmosphere can hold. The excess of water vapor is then condensed around small condensation nuclei about 0.1–10 µm in diameter to form water droplets or ice crystals. It is very difficult for condensation to occur in pure air, and nuclei of marine origin are more effective than other sources. However, they are usually not a limiting factor in precipitation formation.

8.1.2.3 Growth of Water Droplets

Water droplets or ice crystals formed in the air are small and drifting. They will not fall out as rain or snow unless these water particles grow to a size large enough to be pulled down by gravitation. The growth of these particles may be due to collisions with one another when they drift in the air. They can grow at the expense of other particles due to evaporation and condensation because saturation vapor pressure can differ among water droplets and ice crystals. This process is called coalescence.

8.1.2.4 Moisture Convergence

Water vapor content in the atmosphere is about 25 mm at any given time. Thus, additional water must be supplied from the surrounding areas dominated with high pressure to the stormy areas dominated with low pressure in order for a storm to remain at a constant rate or even increase during a storm system. The greater the atmospheric pressure gradient between the high- and low-pressure areas, the greater the storm intensity.

8.1.3 Precipitation Types

Precipitation is often categorized into three types in accordance with cooling mechanisms that generate vertical lifting and formation of precipitation.

8.1.3.1 Orographic Precipitation

When an air mass is lifted up mechanically by mountain barriers, the reduced atmospheric pressure at higher elevations causes cooling along with condensation by expansion. It results in greater precipitation (Figure 8.1). Thus, precipitation is greater at higher elevations on the windward slope of mountainous areas. The annual increasing rate, the so-called *precipitation gradient*, varies with 125–167 mm/100 m in the coastal range of Washington, 67–83 mm/100 m in the Sierra Nevada (Spurr, 1964), 138 mm/100 m in northern Colorado

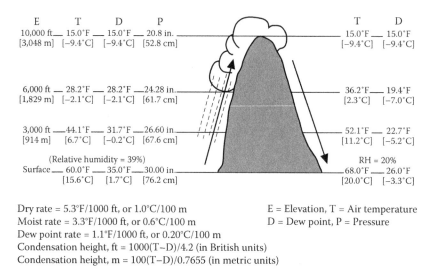

E   T   D   P   T   D
10,000 ft— 15.0°F — 15.0°F — 20.8 in. ——————————   15.0°F — 15.0°F
[3,048 m] [−9.4°C] [−9.4°C] [52.8 cm]   [−9.4°C] [−9.4°C]

6,000 ft — 28.2°F — 28.2°F —24.28 in.—   36.2°F— 19.4°F
[1,829 m] [−2.1°C] [−2.1°C] [61.7 cm]   [2.3°C] [−7.0°C]

3,000 ft —44.1°F— 31.7°F —26.60 in.—   52.1°F — 22.7°F
[914 m]   [6.7°C] [−0.2°C] [67.6 cm]   [11.2°C] [−5.2°C]

(Relative humidity = 39%)   RH = 20%
Surface— 60.0°F — 35.0°F—30.00 in.—   68.0°F — 26.0°F
         [15.6°C] [1.7°C] [76.2 cm]   [20.0°C] [−3.3°C]

Dry rate = 5.3°F/1000 ft, or 1.0°C/100 m   E = Elevation, T = Air temperature
Moist rate = 3.3°F/1000 ft, or 0.6°C/100 m   D = Dew point, P = Pressure
Dew point rate = 1.1°F/1000 ft, or 0.20°C/100 m
Condensation height, ft = 1000(T−D)/4.2 (in British units)
Condensation height, m = 100(T−D)/0.7655 (in metric units)

FIGURE 8.1
Orographic effects on precipitation in Washington state. (After Small, R.T., *Weatherwise*, 204–207, 1966.)

(Daniels, 2007), and 49 mm/100 m on the windward slopes of the Wasatch Range, Utah (Fan and Duffy, 1993). However, there is an upper limit on the increase in precipitation with elevation. The zone of maximum precipitation is at about 2135 m at the Alps, 1525 m in northern California, and 2440 m in the southern Sierra Nevada. Beyond this level, precipitation decreases.

8.1.3.2 Convective Precipitation

Unequal heating between different surfaces, such as plowed field versus forest or land versus water, or the increase of water vapor content in the air due to evapotranspiration in hot summer afternoons, can make air unstable. Heated air over the hot surface expands, becomes lighter, and begins to rise. The unstable air continues to rise and is replaced by cool air from the surroundings. This can cause pronounced vertical movements, adiabatic cooling, condensation, and precipitation. Convective storms are spotty with intensity ranging from light to cloudbursts (100 mm/h or more).

A good example of unequal surface heating is precipitation in the Great Lakes area. The amount of precipitation and frequency of thunderstorm and hailstorm activity over the Great Lakes and their downwind areas tend to decrease in summer and increase in winter and fall (Changnon and Jones, 1972). Apparently, this is because lake water in the fall and winter is warmer than the overlying air. Precipitation over the Great Lakes and their downwind area is enhanced when moisture and heat are added from the lakes to increase atmospheric instability. On the other hand, because the lakes are cooler than the overlying air in summer and tend to stabilize the atmosphere, no precipitation is induced.

8.1.3.3 Cyclonic Precipitation

Cyclonic precipitation can cover a large area over a long duration. It can be either frontal or non-frontal. *Fronts* are boundaries that separate masses of air having significantly different physical properties in terms of humidity, temperature, pressure, and motion. If the moving warm air masses are pushed upward by a cold air mass, it is a *cold front*. If the cold air retreats, warm air pushing over it produces a *warm front*. When the boundary does not move, the front becomes stationary. Fronts usually bring bad weather.

Cold fronts usually move faster, the frontal surfaces are steeper, their upward movements are more rapid, and precipitation rates are much greater than those of warm fronts. Non-frontal precipitation results from air lifting through horizontal convergence of the inflow from high-pressure areas into low-pressure areas.

The precipitation types mentioned earlier can have intensities from near zero to over 100 mm/h. Rainfall intensities and their corresponding drop sizes and terminal velocities are given in Table 8.1 and Figure 8.2.

TABLE 8.1

Rainfall Intensity, Drop Diameter, and Terminal Velocity

Popular Name	Intensity (mm/h)	Drop Diameter (mm)	Terminal Velocity (m/s)
Fog	Trace	0.01	0.003
Mist	0.05	0.1	0.25
Drizzle	0.25	0.2	0.75
Light rain	1.00	0.45	2.00
Moderate rain	4.00	1.0	4.00
Heavy rain	15.00	1.5	5.00
Excessive rain	40.00	2.1	6.00
Cloudburst	100.00	3.0	8.00

Source: Humphreys, W.J., *Physics of the Air*, Dover, New York, 1964.

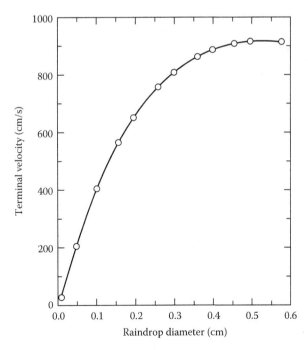

FIGURE 8.2
Terminal velocity as a function of raindrop diameters.

8.2 Forest Interception

Forest canopies stand up in the air, serving as a barrier against precipitation reaching the ground. A portion of precipitation is inevitably intercepted by the canopy (*canopy interception*), flows along the stem to the ground surface (*stemflow*), drips from the foliage and branches or passes through canopy openings to the ground (*throughfall*), or is further intercepted by forest floor (*litter interception*). These processes cause a reduction in precipitation quantity and a redistribution of precipitation toward the soil.

8.2.1 Interception Components

Quantitatively, total *forest interception* (I_F) is the sum of canopy interception (I_C) and litter interception (I_L; Figure 8.3), or

$$I_F = I_C + I_L \tag{8.5}$$

The precipitation that actually reaches the mineral soil is called effective precipitation (P_E). It is the difference between gross precipitation in the open (P_G) and total forest interception (I_F), or

$$P_E = P_G - I_F \tag{8.6}$$

The amount of precipitation that reaches the forest floor or the sum of throughfall (P_{TH}) and stemflow (P_S) is termed net precipitation (P_N), or

$$P_N = P_{TH} + P_S \tag{8.7}$$

In this case, canopy interception can be estimated by the difference between gross precipitation and net precipitation (Figure 8.3), or

$$I_C = P_G - P_N = P_G - \left(P_{TH} + P_S\right) \tag{8.8}$$

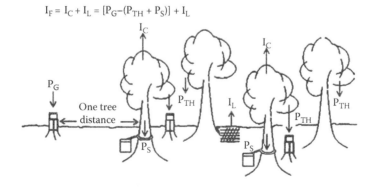

FIGURE 8.3
Forest interception components, canopy interception (I_C), litter interception (I_L), throughfall (P_{TH}), stemflow (P_S), gross precipitation (P_G), and total forest interception (I_F).

Forest interception is an important event in the hydrologic cycle because of its effects on rainfall deposition, soil moisture distribution, snow accumulation and snowmelt, wind movement, heat dissipation, and impact energy of raindrops on soil erosion (particle detachment).

8.2.1.1 Canopy Interception

Approximately 10%–25% of annual precipitation is lost by canopy interception, depending on evaporation power of the air, storm characteristics, and vegetation. In field observations, canopy interception is frequently expressed in this empirical form:

$$I_C = a + b\, P_G \tag{8.9}$$

where a and b are regression coefficients. However, a can be considered the *canopy storage*, which is the amount of water retained on the canopy when rainfall and throughfall have ceased, and b is equivalent to the average evaporation rate in fractions of precipitation during the storm.

During the beginning of a storm, water storage in the canopy increases with time and gradually levels off at maximum canopy storage capacity as storm duration is prolonged or intensity increases. Thus, Equation 8.10 can be rewritten to reflect the interception process by (Linsley et al., 1975)

$$I_C = (S_C + E\, t)\left[1 - \exp(-kP_G)\right] \tag{8.10}$$

where
S_C is the canopy storage capacity
E is the evaporation rate
t is the storm duration
k is a constant equal to $1/(S_C + E\, t)$

For a very small storm, I_C approaches P_G, or dI_C/dP_G will approach unity when rainfall P_G is near zero. Evaporation of canopy-intercepted water is affected by available energy, vapor pressure gradient, aerodynamic resistance of the leaves, and air characteristics (Stewart, 1977; Xiao et al., 2000). When the whole canopy is wet, the Monteith–Penman equation (Monteith and Unsworth, 1990) can be used to estimate the potential evaporation rate of the intercepted water. If the canopy moisture level is less than saturation, then the potential evaporation is only proportional to the percentage of wet canopy (Rutter and Morton, 1977).

A juniper interception study was conducted at ten locations in seven counties stretching across the Edwards Aquifer recharge area in south-central Texas. All of the rainfall in storms of <6.4 mm was intercepted by juniper canopies and evaporated to the atmosphere. About 88% of rainfall events were <12.7 mm and 48.2% of that annual rainfall was lost due to the juniper interception (Owens and Lyons, 2002).

8.2.1.1.1 Canopy Storage

The capacity of canopy storage is affected by a variety of factors such as leaf quantity, orientation, and arrangement; leaf and bark surface characteristics; and wind movement. Temperature also has effects on the storage capacity by modifying water viscosity and surface tension (Satterlund and Adams, 1992). This effect may be too small to be significant or within a magnitude less than measurement errors.

Generally, canopy storage capacity is on the order of about 0.2–2 mm (Gash et al., 1995; Xiao et al., 2000), but it may go as high as 4.3 mm in temperate rainforests (Link et al., 2004) and 8.3 mm in tropical rainforests (Herwitz, 1985). An extensive review in the American literature showed that the storage capacity ranges from 0.3 to 6.6 mm for conifers, 0.03–2.0 mm for hardwoods, 0.3–2.0 mm for shrubs, and 1.0–1.5 mm for grasses (Zinke, 1967). The overall average storage capacity for 27 worldwide forests was 0.69 ± 0.07 mm (Liu, 1998).

8.2.1.1.2 Evaporation Rate

The evaporation rate of intercepted water is about 0.1–0.5 mm/h (Calder and Wright, 1986; Gash et al., 1995), depending not only on meteorological conditions but also on the actual canopy storage. When a canopy is fully saturated, the actual evaporation is equivalent to the potential rate and is driven by advective energy rather than by net radiation (Teklehaimanot and Jarvis, 1991). In a rainforest of central Kalimantan, Indonesia, advective energy accounted for 0.38 mm/h of the 0.50 mm/h of evaporation, whereas radiative energy accounted for only 0.13 mm/h (Asdak, 2003).

8.2.1.1.3 Leaf Area Index

Canopy interception is highly affected by *leaf area index* (LAI), the ratio of total leaf surface area of vegetation to the covered areas in m^2/m^2. LAI, a measure of forest structure and canopy density, plays an important role in energy, gas, and water interactions between vegetation and the atmosphere. The composition and structure of plant communities are often affected by CO_2 and climate (Williams et al., 2008), or by topography (Spadavecchia et al., 2008), through regulating leaf production and species competition. Generally, the tall, mature, and evergreen conifer species have the highest LAI value, followed by deciduous hardwoods, shrubs, forbs, and grasses. Forest species generally have a LAI of around 5–10, but values had been reported to be as high as 40–50 for mature spruce–fir mixed forests in the Rocky Mountains (see Chapter 7, Section 7.1.2). Vose et al. (1994) summarized LAI studies for nine pine species at various ages and stand densities around the world. They showed that the smallest values were for Scotch pine (*Pinus sylvestris*), ranging from 2.8 to 8.5, while the largest were for Monterey pine (*Pinus radiata*), ranging from 10.2 to 32.0. Another worldwide analysis used about 1000 published estimates of LAI from nearly 400 field sites for 15 biome/land cover classes over the period 1932–2000 (Scurlock et al., 2001). The results showed that mean LAI ranged from 1.31 ± 0.85 (standard deviation) for desert biomes to 8.72 ± 4.32 for tree plantations, and the maximum LAI of all forests is 47.

Typically, the storage volume of water per unit area of leaves is 0.1–0.4 mm/m², and may go as much as 1.0 mm/m² (Link et al., 2004).

8.2.1.2 Litter Interception

Litter interception is much smaller than canopy interception. The amount is largely dependent upon the thickness of litter, water-holding capacity, the frequency of wetting, and evaporation rate. Studies have shown that it is only a few millimeters in depth in most cases and can be up to 11 mm during a storm. Generally, about 1%–5% of annual precipitation (Helvey and Patric, 1965) and less than 50 mm/year are lost to litter interception. However, it has been reported to be as high as 17% of gross precipitation under pole-size (30–70 years) ponderosa pine stands (Alden, 1968) and 94 mm/year for shortleaf pine (Rusk, 1969). Evaporation of litter interception under a beech forest was measured to be 34% of throughfall in Luxembourg (Gerrits et al., 2006).

Although litter reduces the quantity of precipitation actually reaching the mineral soil, most importantly, it also affects the velocity of overland flow, which allows more time for soils to absorb runoff water. It also protects the ground surface from direct impact of raindrop energy and wind energy, as well as shades the soil surface, which in turn can reduce soil evaporation. Thus, the conservation of soil moisture due to evaporation reduction for a period of time can outweigh litter interception of the forest floor. Runoff and sediment production in areas covered by litter are much lower than in areas with no cover of litter.

Unger and Parker (1976) showed that a 1 cm thick floor of crop litter can reduce potential evaporation to 46% and can further reduce it to about 17% and 6% if the litter floor is increased to 3 and 5 cm in depth, respectively. A litter floor of 3 cm is common in forest stands. The soil moisture conservation of a 3 cm litter floor in a 5 day period can reach 20 mm of water, much greater than the 1%–5% of rainfall that can be lost due to litter interception in one or two storms.

8.2.1.3 Throughfall and Stemflow

The total of throughfall and stemflow, termed net precipitation, is subject to litter interception before reaching the mineral soil. Throughfall is highly variable among forests and within a forest. Drips from certain points in a stand can cause more throughfall than the total rainfall in the open (Roth and Chang, 1981). Average throughfall of a stand is determined by factors such as species, age, density, season, and storm characteristics. Weighted throughfall and stemflow for a few species in North America are given in Table 8.2. Average throughfall was reported as low as 12% of gross precipitation for a 25 year old Douglas-fir forest in southern Washington, United States (Pypker et al., 2005).

The greatest amount of stemflow reported in the literature is 12% of gross precipitation in a loblolly pine stand in South Carolina (Swank et al., 1972). This may be due to the sharp angles between branches and trunks. In an upland hardwood forest of western Georgia, stemflow was reported as little as 0.54% of gross precipitation (Bryant et al., 2005). For most species, about 2%–5% of gross precipitation flows to the ground along stems. Although it is small in quantity, it may be of ecological importance because rainwater flows directly into the rooting zone of the tree.

TABLE 8.2

Weighted Equations for Estimating Throughfall and Stemflow as a Function of Gross Precipitation for a Few Species in North America

Species	Throughfall	Stemflow
Red pine	$P_{TH} = 0.87P_G - 1.02$	$P_S = 0.02P_G$
Loblolly pine	$P_{TH} = 0.80P_G - 0.25$	$P_S = 0.08P_G - 0.51$
Shortleaf pine	$P_{TH} = 0.88P_G - 1.27$	$P_S = 0.03P_G$
Ponderosa pine	$P_{TH} = 0.89P_G - 1.27$	$P_S = 0.04P_G - 0.25$
Eastern white pine	$P_{TH} = 0.85P_G - 1.02$	$P_S = 0.06P_G - 0.25$
Spruce–fir–hemlock	$P_{TH} = 0.77P_G - 1.27$	$P_S = 0.02P_G$

Source: Helvey, J.D., A summary of rainfall interception by certain conifers in North America, in *Biological Effects of the Hydrological Cycle, Proceedings of the 3rd International Seminar on Hydrology Professors*, Lafayette, IN, pp. 103–113, 1971.

8.2.1.4 Total Interception

Precipitation intercepted by forest canopies and litter floor, collectively termed *total* or *forest interception* (I_F), is lost to the air by evaporation. In forest environments, transpiration is the largest term of water loss to the air, followed by forest interception and soil evaporation (Licata et al., 2011; also see Table 9.10). The annual loss of forest interception is about 15%–40% of P_G, depending on species, plant and stand characteristics, and storm conditions. Studies have shown that annual interception losses reach 40%–60% of P_G for dark fir and spruce forests in Russia (Shiklomanov and Krestovsky, 1988) and 45% for *Cupressus sempervirens* in Iran (Hashemi, 2011). Forest interception may go as much as 100% of P_G in small storms. For eastern white pine stands 10, 35, and 60 years old in western North Carolina, Helvey (1967) showed that total interception loss increased with maturity and could be estimated empirically from gross precipitation for a given season (P_G) and number of storms (N) in that season:

$$I_F = 1.27\,N + 0.08\,P_G\,(\text{for 10 year stand})$$

$$I_F = 1.27\,N + 0.12\,P_G\,(\text{for 35 year stand}) \tag{8.11}$$

$$I_F = 1.52\,N + 0.18\,P_G\,(\text{for 60 year stand})$$

8.2.2 Related Events

Other events relevant to canopy interception exist. The following events are important to forest hydrologists.

8.2.2.1 Canopy Deposition

Along coastal areas, on top of mountains, or at sites prone to the occurrence of fog, the contact of moist, warm air mass, fog, and cloud with dry, cold forest canopies causes deposition of water vapor on the foliage. The deposited water then drips from the foliage or runs down the stems to the ground as an additional water supply. This process, reversing the effect of canopy interception, is often referred to as horizontal precipitation, occult precipitation, negative precipitation, fog precipitation, canopy condensation, cloud drip, or fog drip in literature.

8.2.2.1.1 Amount

Canopy deposition is most significant on top of mountains and in coastal areas. It decreases with increasing distance from the ocean. The type, density, and size of foliage; slope position; and moisture content, duration, and frequency of fog (clouds) also play important roles in canopy deposition. Big trees, or those located in the forest margins, on the windward direction, capture more droplets (González, 2000). Trees with needle-type leaves, such as pine, redwood, and fir, are by far the more effective collectors of moisture (Goodman, 1985). Because of shorter and smaller canopies, brush and chaparrals collect relatively less fog drip than trees. Cloud interception tends to be higher in tropical regions than in temperate regions (Hamilton et al., 1995).

   Canopy deposition can range from less than 10 to more than 100% of the annual rainfall (Table 8.3). In San Francisco, California, summer (May to October) is dry and most rainfall

TABLE 8.3

Fog Deposition Observed at Various Locations around the World

Location	Elevation (m)	Forest	Duration	Rainfall (mm)	Fog Drip (mm)	Reference
Hawaii						
Maui, windward	1951	Hawaiian cloud forest	2 years	3761/year	1212/year	Giambelluca et al. (2011)
Maui, leeward	1219	Hawaiian dry forest	2 years	1139/year	166/year	
Mauna Loa	2500	Mamane/naio forests	2 years	1039/year	706/year	Juvick and Ekern (1978)
California, USA						
Eel River Valley	670	Douglas fir	07–09/71	37	83–425	Azevedo and Morgan (1974)
San Francisco	550	Monofilament	05–09/82	600/year	252–804	Goodman (1985)
Regua	47–191	Redwood	3 years	1315/year	447/year	Dawson (1998)
Puerto Rico: Pico del Este	1010	Elfin cloud forests	44 days	280/44 days	93/44 days	Holwerda et al. (2006)
Venezuela: Cerro Santa Ana	815	Elfin	12 months	1630	518	Cavelier and Goldstein (1989)
Oregon: Portland	1000	Douglas fir	1 year	1739	387	Harr (1982)
Colombia: Tambito	1450	Montane	1 year	6233	561	González (2000)
Macuira	865	Elfin	1 year	853	796	Cavelier and Goldstein (1989)
Bavaria: Waldstein	775	Spruce forests	10 months	1236	117	Klemm and Wrzesinky (2007)
Costa Rica: Monteverda	1500	Lower montane c. forest	1 year	3191	886	Clark et al. (1998)
Guatemala: Minas	2550	Epiphytes/ferns	44 weeks	2559	203	Holder (2003)
Malaysia: G. Brinchang	2031	Tropical montane	10 months	1530	135	Kumaran and Ainuddin (2005)
Mexica: Totutla	1330	Oaks, etc.	1 year	3010	944	Vogelmann (1973)
Panama: Cordillera	1100	Montane forest	1 year	1495	2295	Cavelier et al. (1996)
Peru: Cordillera Yanachaga	2815	Montane forest	36 weeks	2753/year	21	Gomez-Peralta et al. (2008)
Spain: Canary Island	1270	Myrtle/conifers	1 year	611	110	Katata et al. (2009)
Taiwan: Yuanyang L.	1650	Yellow cypress forest	1 year	3280	328	Chang et al. (2006)

occurs in the winter. Fog drip under tan oak (*Lithocarpus densiflorus*) over a period of about 1 month (July 20–August 28) in the Santa Cruz Mountains, California, totaled 1494 mm of water, higher than the annual precipitation for the area (Oberlander, 1956). At Serranía de Macuria, Colombia, annual fog deposition totaled 796 mm, as compared to 853 mm of annual rainfall observed in the open (Cavelier and Goldstein, 1989).

8.2.2.1.2 Impacts

The deposition of fog, cloud, and moist air masses on forest canopies, or other artificial canopy simulators such as vertical nylon nets, provides additional water supply to soil moisture and streamflow during fog-prevailing seasons of the year. This deposition causes a subsequent reduction of transpiration loss and keeps plants vigorous. In the Bull Run Municipal Watershed, Oregon, net precipitation under an old-growth Douglas-fir forest totaled 1739 mm during a 40 week period, 387 mm of water more than in adjacent clear-cut areas. Converting these data to an annual basis and adjusting interception losses, fog drip could have added 882 mm of water or 41% to annual precipitation in the open. Consequently, the removal of 25% of the forest in two small forested watersheds caused a small (20 mm) but significant reduction of streamflow (Harr, 1982), a result contradicting many watershed studies that claim timber harvesting increases water yield. Further analyses indicated that the decrease in streamflow after timber harvesting was detectable only in June and July, the season when the occurrence of fog is thickest and frequent and precipitation is lowest of the year. The decrease in streamflow began recovering to pretreatment level after 5–6 years due to the regrowth of trees (Ingwersen, 1985).

In the Sierra de las Minas Biosphere Reserve, Guatemala, fog precipitation can contribute 19% of the hydrological inputs to the water budget of the cloud forest in the dry season. The conversion of cloud forest to agricultural land may decrease water resources in local communities (Holder, 2006). Moisture input to the Redwood Forests from fog can constitute between 30% and 75% of the annual water budget (Parsons, 1960). A hydrogen isotopic analysis showed that as much as 8%–34% and 6%–100% of water used by the coastal redwood and the understory vegetation, respectively, came from fog precipitation (Dawson, 1993, 1996). Thus, water input from fog could reduce plant moisture stress, enhance growth, or even affect the development and composition of vegetation community due to additional nutrients in the drip (Azevedo and Morgan, 1974; Cavelier et al., 1997).

In addition, low clouds and prevailing fog create an environment that blocks solar radiation, reduces heat loss from the ground, stabilizes temperature fluctuations, lowers vapor pressure deficit, and raises atmospheric humidity. All these lead to a lower rate of evaporation and transpiration. Fog precipitation can help plants maintain vigor through direct absorption during dry periods and reduce the inflammability of litter in the fire-prone season. In northern Chile, fog deposition is the major water supply to desert plants (Westbeld et al., 2009).

8.2.2.1.3 Application

Fog drip can be artificially utilized as an augmentation for water supply. The first fog-collection project for supplies of water is believed to have been implemented at Mariepskop in Mpumalanga, South Africa, during 1969/1970 (Schutte, 1971, cited by Olivier, 2002). Two large fog screens, 28.0 m × 3.6 m each constructed from a plastic mesh, were erected at right angles to each other and to the prevailing winds. During a 15-month period (October 1969–December 1971), the screens collected an average of 31,000 L of water per month, or about 11 L/m^2/day. When yields for only foggy days were taken into account, the mean was 23,395 L/month, almost 800 L/day. During the entire period, fog/cloud drips exceeded rainfall by a factor of 4.6 and was up to 17 times greater during certain foggy months.

The collected water was used as an interim measure to supply water to the South African Air Force personnel at the Mariepskop radar station.

Chungungo is a small desert coast village in Chile. A system of 50 polypropylene nets (4 m × 12 m), resembling black volleyball nets, was installed by the National Catholic University of Chile to trap fog that shrouds nearby mountains on most days. The "fog trappers" feed troughs where water is chlorinated. It is then fed into a gravity system that causes the water to descend to the village through a 6.5 km pipeline. Originally, 330 villagers were dependent on a once-a-week truck with 13 L/day/person of water. Now, the fog-collection system furnishes 11,000 L/day on average or 33 L/day to each person at half the cost of the trucked-in water. Fog collection for water supply has also been collected in other countries such as Peru, Ecuador, and Oman, and a number of other countries (Schemenauer and Cereceda, 1994).

8.2.2.2 Transpiration Reduction

Canopy interception is conventionally considered a loss in the hydrologic budget. However, intercepted water appears as a thin layer on a leaf's surface, and the diffusion resistance of water on the leaf is smaller than that of water in the leaf. This makes the evaporation rate of intercepted water greater than the transpiration rate from dry leaves under similar environmental conditions. During the evaporation process of intercepted water, the energy available for transpiration is reduced. Water in stomata has no access to the atmosphere or to heat energy as long as the leaves are sealed by intercepted water. Therefore, the intercepted water has to be lost first before transpiration can occur, and a reduction in transpiration is expected (Chang, 1977).

Results in studies of interception-induced transpiration reduction vary due to species and environmental conditions. Reported results were 100% reduction for grass vegetation (Burgy and Pomeroy, 1958), 20%–40% for a hardwood forest (Singh and Szeics, 1979), 9% for 6 year old to 7 year old ponderosa pine (Thorud, 1967), and 6% for Colorado blue spruce and Austrian pine (Harr, 1966). In England, Stewart (1977) showed that the average rate of evaporation of intercepted precipitation in a Scots pine stand was three times the average rate of transpiration, an indication of partial compensation for the subsequent suppression of transpiration loss.

The transpiration reduction of canopy-intercepted water can offset the water loss of sprinkler irrigation in agriculture. Sprinkler irrigation efficiency is declined due to evaporation of (1) the applied water that is intercepted by the crop, (2) the wind-drifted water droplets that never reach the ground, and (3) the ponded water at soil surface that is never used by crops. However, irrigation and evaporation of the intercepted water can sharply reduce canopy temperature and within-canopy vapor pressure deficit, and consequently lower transpiration rates. Corn evaporation (by weighing lysimeter) and transpiration (by heat balance method) rates were determined for two plots with one kept wet (application of sprinkler irrigation) and the other dry (no irrigation) at Bushland, Texas, United States. For a daytime impact irrigation application of 15.28 mm, the intercepted water on foliage caused a subsequent reduction of transpiration from 4.99 to 3.40 mm or about 10% of the irrigated water. For an application of 21.4 mm, the reduction was about 5% (Tolk et al., 1995).

8.2.2.3 Mechanical Barriers

Forest canopies also provide mechanical barriers to the transfer process of many other substances. These mechanical effects reduce wind speed in the forest, trap suspended

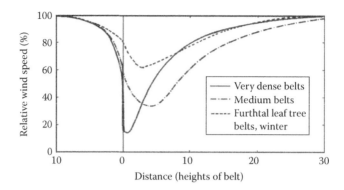

FIGURE 8.4
Relative wind speed around shelterbelts of different degrees of penetrability. (After Naegeli, W., *Mitteil. Schweiz. Anstalt Forestl. Versuchw.*, 24, 659, 1946.)

particles and dusts, redistribute energy budget at the ground, and slow down the terminal velocity of raindrops. Other effects in the forest environment include the increase of relative humidity, the suppression of soil evaporation, the amelioration of the environment for livestock and wildlife, the reduction of costs for home cooling and heating, and the enhancement of the environment for people.

By use of the mechanical barrier, a few rows of trees and shrubs, in a proper design and arrangement, can be used as a shelterbelt and windbreak to protect the leeward side from wind and heat effects (Caborn, 1964; WMO, 1964; Cornelis and Gabriels, 2005). Shelterbelts are often established along coasts, in wind-prone areas, beside the edge of deserts, in arid and semiarid regions, and on the west or southwest sides of farms and villages. Wind speed can be reduced by shelterbelts, depending on the density, height, width, distance from the forest, and wind direction, to about 15%–60% on the leeward side at a distance about one to eight times the height of trees (Figure 8.4). They are also managed for increasing snow accumulation and delay of snowmelt for water augmentation (Gray and Male, 1981).

8.2.2.4 Canopy Wettability

Some species have a higher degree of leaf water repellency, meaning less canopy wettability than others. Leaf water repellency can be photographically measured by calculating the contact angle between a droplet of water and a leaf surface, generally ranging from 89° to 150° (Holder, 2007). Leaf surfaces are considered non-wettable (repellent) with contact angles >130°, and wettable (non-repellent) with contact angles <110° (Smith and McClean, 1989). A comparison study using 12 cloud forest species from the Sierra de las Minas, Guatemala, 12 species from tropical dry forests in Chiquimula, Guatemala, and 12 species from foothills-grassland vegetation in Colorado, United States, Holder (2007) found that leaf water repellency was greater for species in the drier sites of Colorado and Chiquimula than for those in the cloud forest of the Sierra de las Minas, and it was significantly greater on the abaxial surface than on the adaxial surface for 22 species

Differences in leaf water repellency among species are due to leaf structure, surface chemical properties, inclination angle, age, and environmental conditions. The persistence of water droplets on leaf surfaces can inhibit photosynthetic carbon exchange, reduce plant transpiration, and increase canopy storage. Species with water-repellent leaf surfaces may have lower retention of water and dust particles, consequently causing an increase in the quantities of throughfall, stemflow, and fog deposition. This may result in greater hydrological inputs beneath the canopy, an important adaptation event in water-stressed environments. However, the significance of these impacts is yet to be determined.

8.3 Snow Accumulation and Snowmelt

Snow is often referred to as "delayed precipitation" because of its coverage on the ground and gradual release of water until as late as the spring. In areas where snow is the dominant type of precipitation, melting snow is the major source of water supply and a potential cause of spring floods. Many of the snow-prone, water-producing regions are usually covered with forest vegetation. Here, forest-management activities could impose significant effects on snow accumulation and snowmelt and, consequently, water supply and flood alleviation in downstream areas.

8.3.1 Forest Effects

The distribution of snow within forests is significantly affected by timber volume and canopy density, while snowmelt in the open is related to solar radiation, temperature, aspect, wind, and elevation. Thus, dense conifer forests intercept the largest amounts of snow and consequently cause greatest loss from canopy snow sublimation, while thin and leafless hardwoods or open lands catch the least. It is therefore expected that snow accumulation and water equivalent are greater in the open than in the forest (Berris and Harr, 1987) and that they decrease with increasing canopy density (Table 8.4). Also, due in part to the impact of canopy drips from the saturated snow, snowpacks in the forest are usually denser and have higher free water contents than those in the open.

Forest canopy also modifies microclimates within the forest. Generally, it produces lower air and dew-point temperatures, less shortwave radiation, higher relative humidity, and slower wind speed. As a result, energy inputs to snowpacks are consistently greater in the open. This makes snowmelt rate slower and snowpack duration longer in the forest (Haupt, 1979; Ffolliott et al., 1989).

TABLE 8.4

Snow Accumulation (cm) at the Tully Forest of the State University of New York College of Forestry, 1961–1962

Cover Type	Crown Density (%)	Total Accumulation (November 30–April 27)
Open land	0	154.2
Brush hardwoods	2.8	234.4
Northern hardwoods	7.6	184.9
Red pine, thinned	85	163.6
Red pine, dense	93	135.4
Norway spruce, thinned	94	130.3
Norway spruce, dense	96	121.9

Source: Eschner, E.R. and Satterland, R.D., Research Note NE-13, USDA Forest Service, 1963.

TABLE 8.5

Predicted Snowmelts (mm) Based on Various Available Energy versus
Measured Snowmelts in an Old-Growth Forest in the Western
Cascades of Oregon

	Clear-Cut Plot[a]		Forested Plot[b]	
Energy Source	Predicted	Measured	Predicted	Measured
Shortwave radiation	0.3 (1%)		0.0 (0%)	
Longwave radiation	2.9 (13%)		2.8 (25%)	
Sensible heat	7.8 (35%)		2.7 (24%)	
Latent heat	4.8 (22%)		1.8 (16%)	
Rain	6.4 (29%)		3.9 (35%)	
Total	22.2 (100%)	25.1	11.2 (100%)	11.9

Source: Berris, S.N. and Harr, R.D., *Water Resour. Res.*, 23, 135, 1987.
[a] From 2300 hours February 11, 1984 to 0300 hours February 13, 1984.
[b] From 2300 hours February 11, 1984 to 1900 hours February 12, 1984.

A study compared snow accumulation and melt between an old-growth of Douglas-fir
and western hemlock forest and an adjacent clear-cut plot in the H. J. Andrews Experimental
Forest in the western Cascades of Oregon. The results showed the following in open areas:

1. Water equivalents were two to three times greater.
2. Water outflow (rain plus snowmelt) was 21% greater.
3. Total available energy was 40% greater than in the forest.

Differences in various sources of available energy based on predicted snowmelts between
the forested and clear plots are given in Table 8.5 (Berris and Harr, 1987).

Forest cover reduces snow accumulation up to 30%–50% of that in nearby clearings
(López-Moreno and Latron, 2007; Ellis et al., 2010). A 3 year study shows that about 60% of
snowfall was intercepted by forest canopies dominated by Douglas fir–hemlock–pine in
Oregon (Storck et al., 2002). Any forest-management activities that reduce forest canopies
will cause an increase in snow accumulation on the ground and an overall increase in water
yield. Such management activities have included thinning (Goodell, 1952), shelterwood
cutting (Bay, 1958), selection cutting (Anderson and Gleason, 1960), patch or block cutting
(Gary, 1980; Troendle and Leaf, 1981), and strip cutting (Gary, 1979). The increase in snow
accumulation in the forest openings is due to reduction in canopy interception and sublimation, and to the redistribution of snow induced by wind effects. Although forest cuttings
cause an increase in snow accumulation in the opening and in the upwind forest, they may
cause a decrease in the downwind forest too. Clear-cuts in mixed spruce-fir–pine stands
may result in larger increase in snow accumulation than clear-cuts in pine stand. More
snow may accumulate in young forests than in clear-cut sites (Winkler and Roach, 2005).

8.3.2 Site Characteristics

Physiographic characteristics too play an important role in snow accumulation and snowmelt. Snowmelt on a south-facing slope in the northern hemisphere is greater than that on
a north-facing slope (Rosa, 1956).

In northern Idaho, strip clear-cutting in a young–mature stand of mixed conifers
resulted in a 56% initial increase in peak snow water equivalent to a 37% increase at

the end of the 34th year on a northern slope. However, the initial increase on the south slope was only 37% due to greater winter melt and evaporation and sublimation loss (Haupt, 1979).

A snow accumulation study by clear-cutting mature lodgepole pine in blocks of 2, 4, and 8 ha was conducted on the Big Horn Mountain, Wyoming (Berndt, 1965). It showed that the greatest increase in peak snow accumulation was 97 mm of water equivalent for the 4 ha block on the east aspect, but the greatest average increase was 72 mm on the south aspect. Snow packs persisted in the uncut forests about 10–14 days longer than in the clear-cut blocks. These effects were attributed to the reduction in interception loss, snow redistribution due to disruption of wind speed and patterns, and the increase in heat and mechanical energy input to snowpack surfaces. In British Columbia, Canada, a 2 year snow course study showed that elevation, forest cover, and aspect explained 80%–90% of the large-scale variability in snow accumulation in the watershed. Forests accumulated 39% less snow than clear-cuts in 1 year and 27% in the other year (Jost et al., 2007).

8.3.3 Clear-Cut Size

Cutting forests into small openings traps more snow, but snow cover in openings disappears faster. A greater rate of snow disappearance implies more loss of water to the atmosphere, reducing water available for soil moisture recharge. Also, the flood potential is high if snow melts rapidly in late spring. What, then, is the optimum size of forest cutting for proper snowfall management?

Openings of 1 tree-height (H) wide have been suggested to accumulate the greatest depth of snow (Kittredge, 1953; McGurk and Berg, 1987). However, in the Sierra Nevada, Kattelmann (1982) reviewed 70 years of studies and stated that openings about 0.4 ha in size with a solid wall of trees to the south and not more than 2 H wide from south to north will provide the maximum accumulation and delayed melt. In Alberta, the greatest snow accumulation was an opening of about 2 H in size, and the lowest ablation rates were in the 1 H openings (Golding and Swanson, 1978; Bernier and Swanson, 1993).

The optimal forest harvesting for snow redistribution in Colorado and Wyoming is reported to be block or patch cuttings of about 5 H in width, spaced at least 5 H apart. These cuttings should not be more than 50% of the forest at any one time, and openings should be protected from wind (Gary and Troendle, 1982).

As the size of openings increases, so do the duration of exposure and intensity of solar irradiation and wind speed. This will make snow sublimation increase with increasing opening sizes. If forest cuttings do not affect the total amount of snowfall to be disposed in the entire watershed and they cause only snow redistribution, then cutting a forest opening with proper wind and shade protection becomes of utmost importance in snowpack management.

The length of shadow (S) cast by a stand of trees is a function of latitude, aspect, stand height (H), season, and time of the day. It can be calculated by the following equation:

$$S = H \frac{\sin \beta}{\tan \alpha} \tag{8.12}$$

where
  α is solar altitude
  β is the acute angle between the edge of stand and the sun's azimuth

Values of β can be calculated by the absolute difference between stream azimuth and azimuth of the sun. If the absolute difference is greater than 90°, subtract the difference from 180°. The solar altitude α and azimuth z are given by

$$\sin \alpha = \sin \phi \sin \delta + \cos \phi \cos \delta \cos h \tag{8.13}$$

$$\sin z = -\cos \delta \sin h / \cos \alpha \tag{8.14}$$

where
     φ is latitude of the stand
     δ is declination of the sun (various with seasons)
     h is hour angle of the sun or the angular distance from the meridian of the stand, negative
       before solar noon and positive after noon
     z measures eastward from north

Equations 8.13 and 8.14 can be read directly from sun-path diagrams given in *Smithsonian Meteorological Tables* (List, 1971). The sun-path diagram at φ = 40° is given in Figure 8.5.

For a stand on the south with its edge facing to the north, the S/H ratios for four selected days between winter solstice (December 22) and vernal equinox (March 21), for five morning hours and for two latitude locations are given in Figure 8.6. The shadow is longer for a smaller α, or a greater β. It decreases from morning toward the solar noon and

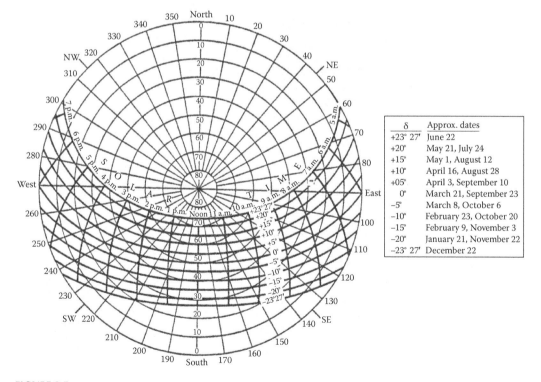

δ	Approx. dates
+23° 27′	June 22
+20°	May 21, July 24
+15°	May 1, August 12
+10°	April 16, August 28
+05°	April 3, September 10
0°	March 21, September 23
−5°	March 8, October 6
−10°	February 23, October 20
−15°	February 9, November 3
−20°	January 21, November 22
−23° 27′	December 22

FIGURE 8.5
The sun-path diagram for lat. 40°N.

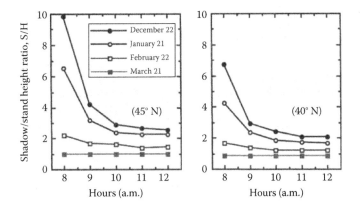

FIGURE 8.6
Shadow length (S)/tree height (H) ratios for a forest stand with its edge facing north at lat. 45°N and lat. 40°N for 4 selected days between winter solstice and vernal equinox.

from winter solstice toward summer solstice (June 21). On March 21, the shadow length is constant throughout the entire solar day (Lee, 1978).

Besides 1–5 H as optimal opening size suggested for trapping snow accumulation on the ground, lengths of shadow cast by a forest stand provide useful information for snowpack management.

8.3.4 Water Yields

Although total snow accumulation in a watershed is not affected by cutting small openings in the forest, many studies have shown that forest cuttings increased water yields. This is attributable to the reduced transpiration losses, the greater concentration of snowmelt water in the openings, and the greater soil moisture that carries over from the previous year (Gary and Troendle, 1982). Increases in water yield ranged from 7% to 111% in the Southwest. They are more pronounced in wet years; no increases or insignificant increases are expected in dry years.

For increasing water yield, 30%–50% of a forest could be put in small clear-cut patches. Trees growing at the base of a slope may use more water than trees upslope. Therefore, harvesting timber in those areas should yield more water to streamflow than cutting in other areas would. Cuttings should provide the longest shadow to shade snow for delayed snowmelt. In this case, openings should face to the north with tall wall trees on the south and upper slopes. If it is a strip cut, the longer sides should be in an east–west direction.

Patches should also be perpendicular to prevailing winds. Frequently, right angles to prevailing winds and east–west aspect are unlikely to be satisfied at the same time. In this case, a compromise should be reached between the two.

If snow is managed to prolong melt as long as possible, strip width can be greater on north slopes than on south slopes. Slope gradient needs to be considered too. Strips can be narrower for a steep slope on the south but wider on the north. Desynchronizing runoff from different parts of watersheds can reduce peak flow during snowmelt. This can be done by having large openings on south slopes to hasten snowmelt and small openings on north slopes to retard snowmelt.

8.4 Do Forests Increase Precipitation?

It is well established, as discussed in the previous sections, that precipitation, deposition, and distribution are greatly affected by forest canopies, but the question of whether forests increase precipitation above the canopy has been a subject of arguments over a long period of time.

8.4.1 Arguments

8.4.1.1 Personal Experience

Based on folklore, experience, visual observation, and imagination, quotations and statements on positive effects were already prevalent in the 1820s (Hazen, 1897). The frequent phenomenon in hilly regions that fog and low-lying clouds suspend near a forest, not over a clearing, led to the idea that forests attract rainfall. Our densest forests are found in areas of the greatest precipitation. No forests form in areas where precipitation is less than 300–400 mm/year. The association of forests with humid environments implies more precipitation, similar to the greater precipitation over the ocean.

The idea that planting trees could modify the local humid environment was highlighted during the major drought of the Midwest in the 1930s. More than 200,000 trees and shrubs were planted in strips of 29,800 km in length stretching from North Dakota to northern Texas between 1935 and 1942. Zon (1927) stated that forests increase both the abundance and frequency of local precipitation to more than 25% in some cases.

8.4.1.2 Meteorological Inductions

The linkage of forests to precipitation can be explained by the following meteorological reasons:

1. Warm air passing over a forest may drop its temperature below dew point by contact cooling, and moisture is condensed with a chance for precipitation (Brown, 1877).
2. Forests add effective height to the mountains, especially those on the crests of hills and the higher windward slopes, resulting in an increase of orographic precipitation by up to 3% (Kittredge, 1948).
3. The roughness of forest canopies may slow down air mass movement, but the air mass behind continues to flow in. As a result, the air mass above the canopy is piled up to a greater depth, initiating vertical turbulence and currents (Rakhmanov, 1963). This can cause condensation due to adiabatic cooling.
4. The occurrence of precipitation requires water vapor in the atmosphere. Forests transpire a greater amount of water to the air than do other surfaces. The additional water transpired to the air will trigger more condensation and precipitation in the forested area.
5. Deforestation can cause an increase in surface albedo, which leads to a decrease in net radiation and a decrease in evaporation, resulting in a decrease in local precipitation (Hahmann and Dickinson, 1997).

8.4.1.3 Field Measurements

The strongest argument on the alleged forest influences in the past has relied on quantitative comparisons between rainfall measured in a forest and in a nearby open field. A network of four rainfall stations within a distance of 9 km was established in France in 1872. The first pair, Bellefontaine (forested), recorded 4% more rainfall than Nancy (open) in 8 years, while in the second pair, about 164 m higher in elevation and hilly, Cinq-Tranchees (forested), recorded 25% more than Amance (open) in 25 years (Hazen, 1897).

In an extensive overview of Russia's literature, Molchanov (1963) stated that forests increase the amount of precipitation up to 10%. Fedorov and Burov (1967) showed that precipitation was 7% greater in a spruce forest than in adjacent open areas in their 15 year study.

In the Copper Basin of eastern Tennessee, a 4 year study showed that rainfall in the forest (1459 mm/year) was as much as 19% greater than in a denuded area and 9% greater than in the grassland (Hursh, 1948). However, average wind speed during the 4 years was 33, 167, and 226 cm/s for the forest, grass, and denuded areas, respectively. Chang and Lee (1974) showed that the greater precipitation in the forested area was due to lower wind speeds. Differences in precipitation among the three surface conditions were insignificant if the wind effects on precipitation were adjusted.

A regional analysis was used to evaluate the relationships between rainfall and percent of forested area in the northeastern regions of Thailand. Based on data collected at 36 stations during a 34 year period, Tangtham and Sutthipibul (1989) found no significant relationships between percent forested area and annual, seasonal, and monthly rainfall or rain days. If the data were expressed in 10 year or longer periods of moving averages, total rainfall tended to significantly decrease while the number of rain days significantly increased with the depletion of percent forested areas. However, interpretations of the results need to be cautious because a period of 34 years may not be long enough to calculate the moving averages of 10 years or longer.

The Thailand study was reexamined again by increasing the record from 34 to 43 years (1951–1994) at 34 stations (11°–17° N latitudes) in a nonparametric Mann–Kendall rank test and a linear regression analysis (Kanae et al., 2001). Significant decreases in precipitation, in association with deforestation, of about 100 mm/month over the study period were detected for the month of September. The decreasing trend was related to occur at the time when the monsoon wind is weak and advective heat transfer is insignificant in this month (September). However, there was also a significant increase in precipitation in and near the northeastern edge of Thailand in July and August.

8.4.2 Counterarguments

Meteorologists and hydrologists have argued that it is mainly the general circulation of the atmosphere and topographic characteristics that affect the horizontal distribution of precipitation on Earth (Figure 8.7), not the forest. In the equatorial regions, excessive solar heating over the oceans makes the air highly unstable and rapidly uprising in response. The uplifting warm and moist air expands due to reduced air pressure, and the expansion causes cooling and, in turn, condensation due to the consumption of internal energy. Therefore, the regions are dominated by low pressures and characterized by frequent local thunderstorm activity and the most abundant rainfall on Earth. As a result, forests are profuse, diverse, and complex. On the other hand, the polar regions are dominated by high pressure, and the air is dry, cold, heavy, and stable. Here precipitation is minimal and the lands are barren. Hazen (1897) stated: "The 'early and latter rain' are experienced in Palestine today just as they were 4000 years ago."

The statement that local transpiration triggers local precipitation is questionable, at least in the temperate zones. Since the atmosphere is in constant motion, the sources of moisture

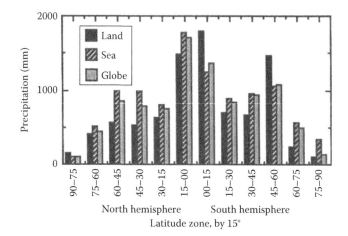

FIGURE 8.7
The horizontal distribution of precipitation for every 15° latitude zone. (*From* Baumgarter, A. and Reichel, E., *The World Water Balance, Mean Annual Global, Continental and Maritime Precipitation, Evaporation, and Runoff,* R. Oldenbourg, Munich, Germany, 1975.)

contributing to precipitation come from oceans hundreds and thousands of kilometers away. In the Mississippi River Basin, only about 10% of the precipitable water comes from the basin vaporization; the remainder is from the oceans. The average water vapor content of the air over deserts may be greater than over a rain forest (Penman, 1963). For the formation of precipitation, water vapor content is not enough. It requires certain mechanisms to enhance the condensation, growth, and convergence of water vapor (see Section 8.1).

On the other hand, deforestation may even increase precipitation. For example, in warm and humid areas, forest clear-cutting creates different surface heating between bare ground and the forest, triggering vertical lifting of the hot air over the clear site. It could result in convective storm activities during summer months. Also, the denuded lands might contribute more dust as condensation nuclei to the air, resulting in more rainfall.

8.4.3 Assessment

It is well-recognized that deforestation causes substantial alterations of energy and water vapor transport between ground and the atmosphere through increases in surface albedo and decreases in both surface roughness and effective transpiration surfaces (leaves). As a result, evapotranspiration is lower (drier air) and sensible heat is higher (warmer air temperature). How significant of a drier air in relation to precipitation may depend on the scale and geographic location of deforestation and is controlled by the atmospheric circulation of the region (see Section 7.5). Note that the best technology in modern precipitation measurements is about 5%; a small increase in precipitation may not be able to verify in the fields.

Thus, reasonable arguments for positive, negative, and no effects of deforestation on the amount of rainfall can be established, but careful examination of the supporting evidence often leads to uncertain conclusions. The main difficulty is due to the great spatial variation of rainfall within a short distance. It is impossible to compare the differences in rainfall for a site with and without forest cover at the same time. Quantitative verification of some arguments may never be attempted due to the scale and costs. This has prompted many investigators to examine the problem through analytical approaches or computer simulations in recent years. One of such studies showed that deforestation in Indochina Peninsula tended to increase rainfall on downwind side and decrease rain on the upwind side, and the effects could far reach East China and Tibetan Plateau through monsoon flows in the summer (Sen et al., 2004).

A review of 94 experimental watersheds around the world showed that a 10% reduction in conifer and eucalypti, deciduous hardwood, and shrub would cause an average increase of water yield by 40, 25, and 10 mm/year, respectively (Bosch and Hewlett, 1982). Thus, cutting all forests on Earth might increase water yield, or reduce transpiration, by a prospective 250 mm. There are 4 billion ha of forests on Earth. A complete deforestation might reduce the water content of the atmosphere by 10^{13} m^3. Since the global annual precipitation is 4.6×10^{14} m^3/year, the prospective reduction in precipitation is about 2%.

A forest adds effective height to the ground and causes an increase in orographic precipitation. Assume a saturated air mass is uplifted to an additional 30 m above the forest. This will make the air temperature over the canopy about 0.24°C cooler than temperature at the ground (using the wet temperature lapse rate with height at −0.6°C/100 m). The additional water vapor condensed due to the decrease in air temperature is about 1.7%, depending on actual air temperature. It is doubtful that an effect of this size is detectable with our standard measurement techniques.

In tropical regions such as the Amazon Basin, where internal circulation is dominant in local climates, deforestation is likely to have a significant impact on local rainfall. Measurements of the isotope oxygen-18 content of precipitation in the Amazon Basin have provided evidence of the large contribution of reevaporated moisture to the basin water balance (Salati et al., 1979). A number of studies showed the actual percent of recycled water accounts for more than 50% of precipitation (Salati et al., 1983; Salati and Nobre, 1991). Using simulation models, the potential impact of deforestation on rainfall varies from −312 mm/year (Zhang et al., 2001) to −600 to −640 mm/year (Salati and Vose, 1983; Shukla et al., 1990) in Amazon basin, and −172 mm/year in Southeast Asia (Zhang et al., 2001). Deforestation also contributes to an increase in the duration of the dry season from 5 to 6 months in Amazon and Cerrado (Costa and Pires, 2010).

Scientists therefore generally accept that deforestation has a significant impact on rainfall in the tropical regions (Bruijnzeel, 1990), but the extent of deforestation and its levels of impacts in other regions require further study. Vertical circulation of transpired water in the temperate zones is probably insignificant. The amount of rainfall profits the forest, not the other way around.

References

Alden, E.F., 1968, Moisture loss and weight of the forest floor under pole-size ponderosa pine stands, *J. For.*, 66, 70–71.

Anderson, H.W. and Gleason, C.H., 1960, Logging and brush removal effects on runoff from snow cover, IASH Pub. No. 51, Merelbeke, Belgium, pp. 478–489.

Asdak, C., 2003, Evaporation of intercepted precipitation in unlogged and logged forest areas of central Kalimantan, Indonesia, *Water Resources Systems—Water Availability and Global Change*, Franks, S., et al., Eds., IAHS Publ. No. 280, Merelbeke, Belgium, pp. 275–281.

Azevedo, J. and Morgan, D.L., 1974, Fog precipitation in coastal California forests, *Ecology*, 55, 1135–1141.

Baumgarter, A. and Reichel, E., 1975, *The World Water Balance, Mean Annual Global, Continental and Maritime Precipitation, Evaporation, and Runoff*, R. Oldenbourg, Munich, Germany.

Bay, R.R., 1958, Cutting methods affect snow accumulation and melt in black spruce stands, Technical Notes 523, USDA Forest Service, Lake States Forest Exp. Sta., St. Paul, MN.

Berndt, H.W., 1965, Snow accumulation and disappearance in lodgepole pine clearcut blocks in Wyoming, *J. For.*, 63, 88–91.

Bernier, P.Y. and Swanson, R.H., 1993, The influence of opening size in the forests of the Alberta Foothills, *Can. J. For. Res.*, 23, 239–244.

Berris, S.N. and Harr, R.D., 1987, Comparative snow accumulation and melt during rainfall in forested and clear cut plots in the western Cascades of Oregon, *Water Resour. Res.*, 23, 135–142.

Bosch, J.M. and Hewlett, J.D., 1982, A review of catchment experiments to determine the effect of vegetation changes on water yield and evapotranspiration, *J. Hydrol.*, 55, 3–23.

Brown, J.C., 1877, *Forest and Moisture*, Simpkin, Marshall & Co., London, U.K.

Bruijnzeel, L.A., 1990, Hydrology of moist tropical forests and effects of conversion: A state of knowledge review, UNESCO International Hydrological Programme, Paris, France.

Bryant, M.L., Bhat, S., and Jacobs, J.M., 2005, Measurements and modeling of throughfall variability for five forest communities in the southeastern US, *J. Hydrol.*, 313, 95–108.

Burgy, R.H. and Pomeroy, C.R., 1958, Intercepted losses in grassy vegetation, *Trans. Am. Geophys. Union*, 39, 1095–1100.

Caborn, J.M., 1964, *Shelterbelts and Windbreaks*, Faber & Faber, London, U.K.

Calder, I.R. and Wright, I.R., 1986, Gamma ray attenuation studies of interception from Sitka spruce: Some evidence for an additional transport mechanism, *Water Resour. Res.*, 22, 409–417.

Cavelier, J. and Goldstein, G., 1989, Mist and fog interception in elfin cloud forests in Colombia and Venezuela, *J. Trop. Ecol.*, 5, 309–322.

Cavelier, J., Jaramillo, M.A., Solis, D., and León, D., 1997, Water balance and nutrient inputs in bulk precipitation in tropical montane cloud forest in Panamá, *J. Hydrol.*, 193, 83–96.

Cavelier, J., Solis, D., and Jaramillo, M.A., 1996, Fog interception in montane forests across the central Cordillera of Panama, *J. Trop. Ecol.*, 12, 357–369.

Chang, M., 1977, Is canopy interception an accurate measure of loss in the hydrologic budget? *Texas J. Sci.*, 18, 339–346.

Chang, M. and Lee, R., 1974, Do forests increase precipitation? *West Virginia For. Notes*, 2, 16–20.

Chang, S.-C. et al., 2006, Quantifying fog water deposition by in situ exposure experiments in a mountainous coniferous forest in Taiwan, *For. Ecol. Manage.*, 224, 11–18.

Changnon, S.A., Jr. and Jones, D.M.A., 1972, Review of the influences of the Great Lakes on weather, *Water Resour. Res.*, 8, 360–371.

Clark, K.L., Nadkarni, N.M., Schaefer, D., and Gholz, H.L., 1998, Atmospheric deposition and net retention of ions by the canopy in a tropical montane forest, Monteverde, Costa Rica, *J. Trop. Ecol.*, 14, 27–45.

Cornelis, W.M. and Gabriels, D., 2005, Optimal windbreak design for wind erosion control, *J. Arid Environ.*, 61(2), 315–332.

Costa, M.H. and Pires, G.F., 2010, Effects of Amazon and Central Brazil deforestation scenarios on the duration of the dry season in the arc of deforestation, *Int. J. Climatol.*, 30, 1970–1979.

Daniels, J.M., 2007, Flood hydrology of the North Platte River headwaters in relation to precipitation variability, *J. Hydrol.*, 344, 70–81.

Dawson, T.E., 1993, Hydraulic lift and water use in plants: Implications for performance, water balance, and plant-plant interactions, *Oecologia*, 95, 565–574.

Dawson, T.E., 1996, The use of fog precipitation by plants in coastal redwood forests, *Proceedings of Coast Redwood Forest Ecology*, Humboldt State University, Arcata, CA, pp. 90–93.

Dawson, T.E., 1998, Fog in the California redwood forests: Ecosystem inputs and use by plants, *Oecologia*, 117, 476–485.

Ellis, C.R. et al., 2010, Simulation of snow accumulation and melt in needleleaf forest environments, *Hydrol. Earth Syst. Sci. Discuss.*, 7, 1033–1072.

Eschner, E.R. and Satterland, R.D., 1963, Research Note NE-13, USDA Forest Service.

Fan, Y. and Duffy, C.J., 1993, Monthly temperature and precipitation fields on a storm-facing mountain font: Statistical structure and empirical parameterization, *Water Resour. Res.*, 29, 4157–4166.

Fedorov, S.F. and Burov, A.S., 1967, Influence of the forest on precipitation, *Soviet Hydrology: Selected Papers*, AGU, Washington, DC, Vol. 3, pp. 217–224.

Ffolliott, P.F., Gotterfried, G.J., and Baker, M.B., Jr., 1989, Water yield from forest snowpack management: Research findings in Arizona and New Mexico, *Water Resour. Res.*, 25, 1999–2007.

Gary, H.L., 1979, Duration of snow accumulation increases after harvesting in lodgepole pine in Wyoming and Colorado, Research Note RM-266, U.S. Forest Service, Fort Collins, CO.

Gary, H.L., 1980, Patch clearcut to manage snow in lodgepole pine, *Watershed Management 1980, Proceedings of ASCE*, Boise, ID, pp. 335–348.

Gary, H.L. and Troendle, C.A., 1982, Snow accumulation and melt under various stand densities in lodgepole pine in Wyoming and Colorado, Research Note RM-417, USDA Forest Service, Rocky Mountain Forest and Range Experiment Station, Fort Collins, CO.

Gash, J.H.L., Lloyd, C.R., and Lachaud, G., 1995, Estimating sparse forest rainfall interception with an analytical model, *J. Hydrol.*, 170, 79–86.

Gerrits, A.M.J. et al., 2006, Measuring forest floor interception in a beech forest in Luxembourg, *Hydrol. Earth Syst. Sci. Discuss.*, 3, 2323–2341.

Giambelluca, T.W. et al., 2011, Canopy water balance of windward and leeward Hawaiian cloud forests on Haleakala, Maui, Hawaii, *Hydrol. Process.*, 25, 438–447.

Golding, D.L. and Swanson, D.H., 1978, Snow accumulation and melt in small forest openings in Alberta, *Can. J. For. Res.*, 8, 380–388.

Gomez-Peralta, D. et al., 2008, Rainfall and cloud-water interception in tropical montane forests in the eastern Andes of central Peru, *For. Ecol. Manage.*, 255, 1315–1325.

González, J., 2000, Monitoring cloud interception in a tropical montane cloud forest of the southwestern Colombian Andes, *Adv. Environ. Monit. Model.*, 1, 97–117.

Goodell, B.C., 1952, Watershed management aspects of thinned young lodgepole pine stands, *J. For.*, 50, 374–378.

Goodman, J., 1985, The collection of fog drip, *Water Resour. Res.*, 21, 392–394.

Gray, D.M. and Male, D.H., 1981, *Handbook of Snow*, Pergamon Press, Elmsford, NY.

Hahmann, A.N. and Dickinson, R.E., 1997, RCCM2-BATS model over tropical South America: Applications to tropical deforestation, *J. Clim.*, 9, 1150–1162.

Hamilton, L.S., Juvik, J., and Scatena, F., 1995, The Puerto Rico tropical cloud forest symposium: Introduction and workshop synthesis, *Ecol. Stud.*, 110, 1–23.

Harr, R.D., 1966, Influence of intercepted water on evapotranspiration from small potted trees, PhD dissertation, Colorado State University, Fort Collins, CO.

Harr, R.D., 1982, Fog drip in the Bull Run Municipal Watershed, Oregon, *Water Resour. Bull.*, 18, 785–788.

Hashemi, S.A., 2011, Comparison of canopy interception loss between trees species, *Am. J. Sci. Res.*, 13, 151–157.

Haupt, H.F., 1979, Effects of timber cutting and revegetation on snow accumulation and melt in North Idaho, Research Paper INT-224, U.S. Forest Service, Fort Collins, CO.

Hazen, H.A., 1897, Forests and rainfall, *Proceedings of the American Forestry Congress, the 7th Annual Meeting*, Washington, DC, pp. 133–139.

Helvey, J.D., 1967, Interception by eastern white pine, *Water Resour. Res.*, 3, 723–729.

Helvey, J.D., 1971, A summary of rainfall interception by certain conifers of North America, *Biological Effects of the Hydrologic Cycle, Proceedings of the 3rd International Seminar on Hydrology Professors*, Lafayette, IN, pp. 103–113.

Helvey, J.D. and Patric, J.J., 1965, Canopy and litter interception of rainfall by hardwoods of eastern United States, *Water Resour. Res.*, 1, 193–206.

Herwitz, S.R., 1985, Interception storage capacities of tropical rainforest canopy trees, *J. Hydrol.*, 77, 237–252.

Holder, C.D., 2003, Fog precipitation in the Sierra de las Minas Biosphere Reserve, Guatemala, *Hydrol. Process.*, 17, 2001–2010.

Holder, C.D., 2006, The hydrological significance of cloud forests in the Sierra de las Minas Biosphere Reserve, Guatemala, *Geoforum*, 37, 82–93.

Holder, C.D., 2007, Leaf water repellency of species in Guatemala and Colorado (USA) and its significance to forest hydrology studies, *J. Hydrol.*, 336, 147–154.

Holwerda, F. et al., 2006, Estimating fog deposition at a Puerto Rican elfin cloud forest site: Comparison of the water budget and eddy covariance methods, *Hydrol. Proc.*, 20, 2669–2692.

Humphreys, W.J., 1964, *Physics of the Air*, Dover, New York.

Hursh, C.R., 1948, Local climate in the Copper Basin of Tennessee as modified by the removal of vegetation, USDA Circular 774.

Ingwersen, J.B., 1985, Fog drip, water yield, and timber harvesting in the Bull Run municipal watershed, Oregon, *Water Resour. Bull.*, 21, 469–473.

Jost, G. et al., 2007, The influence of forest and topography on snow accumulation and melt at the watershed-scale, *J. Hydrol.*, 347, 101–115.

Juvick, J.O. and Ekern, P.C., 1978, A climatology of mountain fog on Mauna Loa, Hawaii Island, Technical Report No. 118, Water Resources Research Center, University of Hawaii at Manoa, Honolulu, HI, 63 pp.

Kanae, S., Oki, T. and Musiake, K., 2001, Impact of deforestation on regional precipitation over the Indochina Peninsula, *Am. Meteorol. Soc.*, 2, 51–70.

Katata, G. et al., 2009, Application of a land surface model that includes fog deposition over a tree heath-laurel forest in Garajonay National Park (La Gomera, Spain), *Estudion en la Zona no Saturada del Suelo*, Vol. IX, Silva, O., et al., Ed., Barcelona, 18 a 20 de Noviembre, 2009, pp. 1–8.

Kattelmann, R.C., 1982, Water yield improvement in the Sierra Nevada snow zone: 1912–1982, *Proceedings of the 50th Western Snow Conference*, Reno, NV, pp. 39–48.

Kittredge, J., 1948, *Forest Influences*, McGraw-Hill, New York.

Kittredge, J., 1953, Influence of forests on snow in the ponderosa pine–sugar pine–fir zone in the Central Sierra Nevada, *Hilgardia*, 21, 1–96.

Klemm, O. and Wrzesinky, T., 2007, Fog deposition fluxes of water and ions to a central mountainous site in central Europe, *Tellus*, B59, 705–714.

Kumaran, S. and Ainuddin, A.N., 2005, Hydrometeorology of tropical montane cloud forests in Malaysia, poster at Universiti Putra Malaysia Research and Innovation Exhibition, March 16–18, 2005, Forestry, Universiti Putra Malaysia, Peninsular, Malaysia.

Lee, R., 1978, *Forest Microclimatology*, Columbia University Press, New York.

Licata, J.A. et al., 2011, Decreased rainfall interception balances increased transpiration in exotic ponderosa pine plantations compared with native cypress stands in Patagonia, Argentina, *Ecohydrology*, 4, 83–93.

Link, T.E., Unsworth, M., and Marks, D., 2004, The dynamics of rainfall interception by a seasonal temperate rainforest, *Agric. For. Meteorol.*, 124, 171–191.

Linsley, R.K., Kohler, M.A., and Paulhaus, J.L.H., 1975, *Hydrology for Engineers*, McGraw-Hill, New York.

List, R.J., 1971, *Smithsonian Meteorological Tables*, 6th rev. edn., Smithsonian Institution Press, Washington, DC.

Liu, S., 1998, Estimation of rainfall storage capacity in the canopies of cypress wetlands and slash pine uplands in North-Central Florida, *J. Hydrol.*, 170, 79–86.

López-Moreno, J.I. and Latron, J., 2007, Influence of canopy density on snow distribution in a temperate mountain range, *Hydrol. Process.*, 22(1), 117–126.

McGurk, B.J., and Berg, N.H., 1987, Snow redistribution: Strip cuts at Yuba Pass, California, *Forest Hydrology and Watershed Management, Proceedings of the Vancouver Symposium*, August 1987, IAHS Publ. No. 167, pp. 285–295.

Molchanov, A.A., 1963, *The Hydrological Role of Forests*, Israel Program for Science Translations, Jerusalem, Office of Technical Services, U.S. Department of Commerce, Washington, DC.

Monteith, J.L. and Unsworth, M.H., 1990, *Principles of Environmental Physics*, Edward Arnold, London, U.K.

Naegeli, W., 1946, Further investigation on wind conditions in the range of shelterbelts, *Mitteil. Schweiz. Anstalt Forstl. Versuchsw.*, 24, 659–737.

Oberlander, G.T., 1956, Summer fog precipitation on the San Francisco Peninsula, *Ecology*, 37, 851–852.

Olivier, J., 2002, Fog-water harvesting along the West Coast of South Africa: A feasibility study, *Water SA*, 28(4), 349–360.

Owens, M.K. and Lyons, R.K., 2002, Evaporation and interception water loss from Juniper Communities on the Edwards Aquifer Recharge Area, UREC-02–028, Texas A&M Agricultural Research and Extension Center at Uvalde, Texas.

Parsons, J.J., 1960, Fog-drip from coastal stratus, with special reference to California, *Weather*, 38, 58–62.

Penman, H.L., 1963, *Vegetation and Hydrology*, Commonwealth Agricultural Bureaux, Technical Communication No. 53. Farnham Royal, Bucks, England.

Pypker, T.G. et al., 2005, The importance of canopy structure in controlling the interception loss of rainfall: Examples from a young and an old-growth Douglas-fir forest, *Agric. For. Meteorol.*, 130, 113–129.

Rakhmanov, V.V., 1963, Dependence of the amount of precipitation on the wooded area in the plains of the European USSR, *Meteorol. Geoastrophys. Abs.*, 14, 712.

Rosa, J.M., 1956, Forest snowmelt and spring floods, *J. For.*, 84, 231–135.

Roth, F.A., II and Chang, M., 1981, Throughfall in planted stands of four southern pine species in East Texas, *Water Resour. Bull.*, 17, 880–885.

Rusk, C.R., 1969, Interception of precipitation by forest canopies and litter, MS thesis, University of Tennessee, Knoxville, TN.

Rutter, A.J. and Morton, A.J., 1977, A predictive model of rainfall interception in forests. III: Sensitivity of the model to stand parameters and meteorological variables, *J. Appl. Ecol.*, 14, 567–588.

Salati, E. et al., 1979, Recycling of water in the Amazon Basin: An isotopic study, *Water Resour. Res.*, 15, 1250–1258.

Salati, E., Lovejoy, T.E., and Vose, P.B., 1983, Precipitation and water recycling in tropical rain forests with special reference to the Amazon Basin, *Environmentalist*, 3, 67–72.

Salati, E., and C. A. Nobre, 1991, Possible climate impacts of tropical deforestation. *Clim. Change*, 19, 177–196.

Salati, E. and Vose, P.B., 1983, Analysis of Amazon hydrology in relation to geoclimatic factors and increased deforestation, *Beitr. Hydrol.*, 9, 11–22.

Satterlund, D.R. and Adams, P.W., 1992, *Wildland Watershed Management*, 2nd edn., John Wiley & Sons, New York.

Schemenauer, R.S. and Cereceda, P., 1994, Fog collection's role in water planning for developing countries, *Nat. Resour. Forum*, 18(2), 91–100.

Schutte, J.M., 1971, Die Onttrekking van Water uit die Lae Wolke op Mariepskop, Technical Note No. 20, Division of Hydrological Research, Department of Water Affairs, Pretoria, South Africa.

Scurlock, J.M.O., Asner, G.P., and Gower, S.T., 2001, Worldwide historical estimates of leaf area index, 1932–2000, ORNL Technical Memorandum TM-2001/268, Oak Ridge National Laboratory, Oak Ridge, TN.

Sen, O.L., Wang, Y., and Wang, B., 2004, Impact of Indochina deforestation on the East Asian summer monsoon, *J. Clim.*, 16, 1366–1380.

Shiklomanov, I.A. and Krestovsky, O.I., 1988, The influence of forests and forest reclamation on streamflow and water balance, *Forests, Climate, and Hydrology*, Reynolds, E.C. and Thompson, F.B., Eds., United Nations University Press, The United Nations University, Tokyo, Japan.

Shukla, J., Nobre, C., and Sellers, P.J., 1990, Amazon deforestation and climatic change, *Science*, 247, 1322–1325.

Singh, B. and Szeics, G., 1979, The effect of intercepted rainfall on the water balance of a hardwood forest, *Water Resour. Res.*, 15, 131–138.

Small, R.T., 1966, Terrain effects on precipitation in Washington State, *Weatherwise*, 19(5), 204–207.

Smith, W.K. and McClean, T.M., 1989, Adaptive relationship between leaf water repellency, stomatal distribution, and gas exchange, *Am. J. Bot.*, 76, 465–469.

Spadavecchia, L. et al., 2008, Topographic controls on the leaf area index and plant functional type of a tundra ecosystem, *J. Ecol.*, 96(6), 1238–1251.

Spurr, S.H., 1964, *Forest Ecology*, Ronald Press, New York.

Stewart, J.B., 1977, Evaporation from the wet canopy of a pine forest, *Water Resour. Res.*, 13, 915–921.

Storck, P., Lettenmaier, D.P., and Bolton, S.M., 2002, Measurement of snow interception and canopy effects on snow accumulation and melt in a mountainous maritime climate, Oregon, United States, *Water Resour. Res.*, 38(11), 1223, 1223–1238.

Swank, W.T., Goebel, N.B., and Helvey, J.D., 1972, Interception loss in loblolly pine stands of the South Carolina piedmont, *J. Soil Water Conserv.*, 27, 160–164.

Tangtham, N. and Sutthipibul, V., 1989, Effects of diminishing forest area on rainfall amount and distribution in North-East Thailand, *Regional Seminar on Tropical Forest Hydrology*, September 4–9, Institute Penyelidikan Perhutanan Malaysia, Kuala Lumpur, Malaysia.

Teklehaimanot, Z. and Jarvis, P.G., 1991, Direct measurement of evaporation of intercepted water from forest canopies, *J. Appl. Ecol.*, 28, 603–618.

Thorud, D.B., 1967, The effect of applied interception rates on potted ponderosa pine, *Water Resour. Res.*, 3, 443–450.

Tolk, J.A. et al., 1995, Role of transpiration suppression by evaporation of intercepted water in improving irrigation efficiency, *Irrig. Sci.*, 16, 89–95.

Troendle, C.A. and Leaf, C.F., 1981, Effects of timber harvest in the snow zone on volume and timing of water yield, *Interior West Watershed Management*, Baumgartner, D.M., Ed., Washington State University, Pullman, Washington, DC, pp. 231–243.

Unger, P.W. and Parker, J.J., 1976, Evaporation reduction from soil with wheat, sorghum, and corn residues, *Soil Sci. Soc. Am. J.*, 40, 936–942.

Vogelmann, H.W., 1973, Fog interception in the cloud forest of eastern Mexico, *Bioscience*, 23, 96–100.

Vose, J.M. et al., 1994, Factors influencing the amount and distribution of leaf area of pine stands, *Ecol. Bull.*, 43, 102–114.

Westbeld, A. et al., 2009, Fog deposition to Tillandsia carpet in the Atacama Desert, *Ann. Geophys.*, 27, 3571–3576.

Williams, J.W., Gonzales, L.M., and Kaplan, J.O., 2008, Leaf area index for northern and eastern North America at the Last Glacial Maximum: A data–model comparison, *Global Ecol. Biogeogr.*, 17(1), 122–134.

Winkler, R. and Roach, J., 2005, Snow accumulation in BC's southern interior forests, *Streamline Watershed Manage. Bull.*, 9(1), 1–5.

WMO, 1964, Windbreaks and Shelterbelts, World Meteorological Organisation, Technical Note No. 59, WMO-No. 147.TP.70, Geneva, Switzerland.

Xiao, Q. et al., 2000, Winter rainfall interception by two mature open-grown trees in Davis, California, *Hydrol. Proc.*, 14, 763–784.

Zhang, H., Henderson-Sellers, A., and McGuffie, K., 2001, The compounding effects of tropical deforestation and greenhouse warming on climate, *Clim. Change*, 49, 309–338.

Zinke, P.J., 1967, Forest interception studies in the United States, *International Symposium on Forest Hydrology*, Sopper, W.E. and Lull, H.W., Eds., Pergamon Press, Elmsford, NY.

Zon, R., 1927, Forests and water in the light of the scientific investigation, U.S. National Waterways Communication, Final Report 1912 (Senate Document No. 469, 62nd Congress, 2nd Session).

9

Forests and Vaporization

Vaporization refers to the change of water from the liquid state into the vapor state. It occurs on water, snow, soil, and vegetation surfaces and is a negative item in the hydrologic balance of a watershed system. The loss of water is commonly termed evaporation when it occurs on water, snow, and soil surfaces; transpiration when it occurs on vegetation surfaces; and evapotranspiration when it refers to the total loss of water in the vapor state. In addition, the process by which snow and ice dissipate due to melting and evaporation is referred to as *ablation*.

9.1 Vaporization Processes

9.1.1 Water Surface

The process of water evaporation involves the transfer not only of mass but of energy. Above absolute zero, water molecules are active and move in different directions. Some move from the water surface to the air and some jump from air into water. The net loss of water molecules of the water is called *evaporation*; the net gain of water molecules is called *condensation*; and the equilibrium between air and water is called *saturation*.

In the narrow layer immediately above the water surface where the air is still (boundary layer), the occurrence of evaporation is analogous to heat conduction. Water molecules diffuse from the surface (water) of greater vapor density to the overlying air of less vapor density. Thus, the net rate of water molecular diffusion is driven primarily by the vapor density gradient between water surface and air, $\rho_s - \rho_a$, which is further modified by the diffusivity of the air and the thickness of the boundary layer in the path of diffusion. Thus, evaporation rate (E) can be expressed by

$$E\,(g/cm^2/s) = -\frac{(\rho_s - \rho_a)(D_v)}{\delta_v} \tag{9.1}$$

where
  D_v is the diffusion coefficient of water vapor in the air (cm^2/s) as estimated by Equation 4.6
  $(\rho_s - \rho_a)$ is the vapor density gradient (g/cm^3)
  δ_v is the thickness of the wind speed-dependent boundary layer (cm)

Since the reciprocal of D_v/δ_v is aerial resistance (r_a) to vapor diffusion in s/cm, Equation 9.1 can be rewritten as

$$E\,(g/cm^2/s) = -\frac{\rho_s - \rho_a}{r_a} \tag{9.2}$$

Values of r_a generally range from 0.1 to 1.0 s/cm. In still air (free convection) at 20°C and 100 kPa (1000 mbar), Campbell (1977) showed that r_a can be estimated by

$$r_a \, (s/cm) = 2.654 \left(\frac{D}{T_s - T_a} \right)^{-0.25} \tag{9.3}$$

where
  D is a characteristic dimension (cm)
  T_s and T_a are the surface and air temperature, respectively (°C)

As evaporation from water surface proceeds, the molecules of water entering the air mass near the surface can cause a substantial increase in r_a, therefore reducing both the vapor density gradient and the vaporization rate. But if the air mass near the surface is renewed continually, as by wind, a higher rate of evaporation can continue indefinitely. In other words, evaporation from a free-water surface is proportional both to the wind and the vapor density (or pressure) difference. In equation form,

$$E = f(u)(e_s - e_a) = (a + bw)(e_s - e_a) \tag{9.4}$$

where
  f(u) is the function of wind speed
  $e_s - e_a$ is the vapor pressure gradient between a water surface and the air
  a and b are constants
  w is the wind speed

The evaporation in Equation 9.4 has appeared in many different forms; one of them shows (Harbeck, 1962)

$$E \, (cm/day) = 0.000169 w_2 (A^{-0.05})(e_s - e_a) \tag{9.5}$$

for estimating reservoir evaporation in the United States, where A is reservoir area up to 120 km², w_2 is wind speed at 2 m above the surface in kilometers per day, and e_s and e_a are in millibars.

The saturated vapor pressure of water, e_s, is very sensitive to changes in temperature. It increases by about 7% for every 1°C increase in temperature between 10°C and 30°C. However, temperature has only a small effect on the vapor pressure of an air mass, e_a, so the difference, $e_s - e_a$, is largely a temperature function.

9.1.2 Bare Soils

Evaporation from a wet, bare-soil surface is similar to that from a water surface and occurs at the potential rate as determined by meteorological factors and by the energy available to change the state of water (often referred to as Stage I). As evaporation proceeds, soil moisture is reduced, the soil profile cannot supply water to sustain atmospheric evaporativity, and the evaporation rate falls steadily below the potential rate. During this falling stage of drying (Stage II), the evaporation rate is primarily determined by the water flow properties within the soil, namely, hydraulic conductivity. When the surface zone becomes so dry that liquid flux of water virtually ceases, water transmission occurs through vapor flux as governed by soil vapor diffusivity. Soil vapor diffusivities are affected by soil-moisture content, temperature, and pressure. The evaporation rate at this vapor-diffusion stage (Stage III) is slow and steady. It can persist for a long period of time until water is supplied to the soil again by precipitation or irrigation.

Soil evaporation modeling during the entire drying process that links the transport of liquid, vapor, and heat is long, complicated, and requires much information on soil properties. The simplest form for most applications is to neglect vapor flux except at the soil surface by (Campbell, 1985).

$$E = E_p \left(\frac{RH_s - RH_a}{1 - RH_a} \right) \tag{9.6}$$

where

E_p is potential evaporation of the soil (the saturated stage)
RH_s is the relative humidity at the soil surface in fraction
RH_a is the relative humidity of the atmosphere

When soil is saturated, $RH_s = 1.00$, and $E = E_p$. Equation 9.6 is sufficient for water budget modeling but fails to provide correct estimates of the evaporation rate at the third stage.

The cumulative evaporation E_c (cm) and the evaporation flux e_f (cm/day) during the first and second stages of drying are also related to soil-moisture contents, square root of time (t, d), and the weighted-mean diffusivity ($\overline{D}$, cm²/day) by these equations (Gardner, 1959; Hillel, 1982):

$$E_c = 2(\theta_i - \theta_f) \left(\frac{\overline{D} t}{\pi} \right)^{0.5} \tag{9.7}$$

$$e_f = (\theta_i - \theta_f) \left(\frac{\overline{D}}{\pi t} \right)^{0.5} \tag{9.8}$$

where θ_i and θ_f are the initial profile wetness and final surface wetness in fractions of volume, respectively. For a soil with $\theta_i = 0.45$ (45%), θ_f (air-dry) = 0.05 (5%), and $\overline{D} = 100$ cm² / day the cumulative E_c at the end of 5 days = $2(0.45 - 0.05)[100 \times 5/3.1416]^{0.5} = 10.09$ cm, and the mean e_f at the fifth day = $0.40[100/(3.1416 \times 5)]^{0.5} = 1.01$ cm/day.

Under the isothermal condition, the vapor flux (q, g/cm²/s) of a soil is affected by vapor-diffusion coefficient and vapor pressure gradient as described by the following equation (Hanks, 1958; Jalota and Prihar, 1998):

$$q = -\alpha\phi \left[\frac{DM}{RT} \right] \left[\frac{P}{P - P_V} \right] \left[\frac{dP_V}{dX} \right] \tag{9.9}$$

where

D is the diffusivity of water vapor into still air (cm²/s)
P is the atmospheric pressure (dyn/cm²) (1 atm = 1.013×10^6 dyn/cm²)
P_V is the water vapor in the soil (dyn/cm²)
R is the gas constant, 8.314×10^7 dyn-cm/K/mol

T is the soil temperature (K)

M is the molecular weight of water vapor (= 18 g/mol)

dP_V/dX is the vapor pressure gradient in a dry layer X ($dyn/cm^2/cm$)

α is the tortuosity factor (taken as 0.87 for silt loam and 0.66 for sandy loam)

ϕ is the volume fraction of air-filled voids

The vapor flux is greater in soils with higher diffusivity, higher vapor pressure, and finer textures but is lesser in soils with deeper drying layers. For a 3-cm layer of dry soil with $\alpha = 0.66$, $\varphi = 0.58$, $D = 0.245 \, cm^2/s$, $T = 293 \, K$, $P = 1023 \times 10^3 \, dyn/cm^2$, and $PV = 16 \times 10^3 \, dyn/cm^2$, the calculated $q = 0.0037 \times 10^{-4} \, g/cm^2/s$, or $0.032 \, g/cm^2/day$ (0.032 cm/day).

Mulches have been used frequently as a management practice for reducing soil evaporation. However, they are effective only during the early stages of drying when climatic factors dominate evaporation rate. During the drying stages, the soil surface begins to dry up and gradually forms a "self-mulch" that is more effective than gravel or other mulches (Hanks and Gardner, 1965). Also, tillage is generally thought to destroy soil structure and break down the capillaries. It causes the tilled layer to dry out quickly, retards liquid movement of water from the underlying untilled layer, and keeps the lower portion of the tilled layer moist. Recent studies showed that the upward capillary flow was not completely cut off by tillage; it was only retarded because of the decrease in vapor diffusivity (Jalota and Prihar, 1998).

9.1.3 Vegetative Surfaces

Water loss from vegetative surfaces, or transpiration, is basically an evaporation process except that its active surface is through stomata or cuticles on leaves and lenticels on bark, and water is conducted through the root system. Thus, compared to water loss per unit area of soil, transpiration has a greater and deeper evaporating sphere. This means that while the rate of soil evaporation is reduced because of the depletion of water content near the surface, transpiration can continue to be active because water deeper in the soil can still be available through plant root systems.

Water potential is a term often used in plant physiology and soil physics to describe the status of water in terms of energy level. A potential is defined as the work required to move a unit mass of water from a reference point of pure free water (the potential $\psi = 0$) to the point in question ($\psi = -$works/mass or ergs/g; tension, suction, or negative pressure). When water potential in the stomata is greater than the air, water in the stomata is vaporized to the air due to vapor pressure gradient, $e_s - e_a$. The losses of water to the air cause the total water potential in the stomata to be less than that in adjacent cells. This causes water in the adjacent cells to move to the stomata and thus form a driving force from the stomata all the way to root tips and surrounding soils. In equation form, transpiration T_r can be expressed by

$$T_r = -3600 \left(\frac{c_p \rho_a}{\gamma L_V} \right) \left(\frac{\delta_e}{r_c + r_a} \right) \tag{9.10}$$

where

c_p is the specific heat of air (cal/g/°C)

ρ_a is the air density (g/cm^3)

γ is the psychrometric constant (from 0.654 to 0.670 mbar/°C), 0°C–30°C

δ_e is the saturation vapor pressure deficit of the air (mbar) or $e_s - e_a$

The denominators r_c and r_a are canopy and aerial resistance (generally <10 s/cm), respectively

L_V is the latent heat of vaporization (cal/g, Equation 4.1)

Constant 3600 converts the result T_r from cm/s into cm/h

Equation 9.10 is basically a calculation of total energy used for transpiration and then converts the total (latent) energy into the depth of water per unit area. The terms $(c_p\rho_a)/(\gamma L_V)$ are relatively constant, making $\delta_e/(r_c+r_a)$ the determining factors of transpiration rate. δ_e is very sensitive to canopy temperature and relative humidity of the air, while r_c and r_a are affected by wind speed and canopy height. Generally, r_c is greater than r_a (Lee, 1980), and r_a decreases with the height of vegetation (Shuttleworth, 1993). Thus, a smaller value of aerial resistance means more efficient water vapor diffusion in the air.

9.1.4 Snow

The vaporization of snow is often referred to as *sublimation*. It occurs when there is a vapor pressure gradient between snowpack surface and the overlying air or when the dew point temperature of the air is lower than that of snow surface and the air is not saturated with water. Snowpack evaporation can be directly measured by the use of evaporation pans of various sizes and composition (Bengtsson, 1980), or estimated by physically based mathematical equations to determine the spatial and temporal variations of snow evaporation for hydrological simulations. One popular model uses the aerodynamic principle to estimate snow evaporation E by (Bernier and Swanson, 1993; Storck et al., 1995)

$$E = \rho_a C_a (e_s - e)\left(\frac{0.622}{P}\right) \tag{9.11}$$

where

E is the evaporation rate (g/cm^2/s)

ρ_a is the density of the air (g/cm^3)

$e_s - e$ is the vapor pressure gradient (kPa) between the snow surface and the air at height z_a

P is the atmospheric pressure (kPa)

C_a is the aerodynamic conductance above the snowpack, estimated by

$$C_a = \frac{[k^2 u_a (1-bR_b)^2]}{[\ln^2(z_a/z_0)]} \tag{9.12}$$

where

k is the von Karman constant (0.41)

u_a is wind speed (cm/s) at height z_a

b is an empirical constant (estimated at five by Dyer, 1974)

z_a is the height at which temperature measurements are taken (cm)

z_0 is the roughness length (cm) as determined by the measurements of the wind profile (Rosenberg et al., 1983) or equal to 0.1 of stand height as given by Rutter et al. (1971)

ln is the natural logarithm

R_b is the bulk Richardson number computed by

$$R_b = \frac{\left[g(T_a - T_s)(z_a - z_0)\right]}{\left[T_k(u_a)^2\right]} \tag{9.13}$$

where
  g is the gravitational constant
  T_a and T_s are the air temperatures (K) at height z_a and at the surface z_0, respectively
  T_k is the average temperature of the air layer between the surface and the height z_a of the instruments (K)

It is common to have an average snow evaporation at 1 mm/day.

9.2 Sources of Energy

At 20°C, vaporization of each gram of water requires about 586 cal of latent heat. The amount of energy available for vaporization is primarily determined by net radiation at the site and how the net radiation is dissipated in the environment.

9.2.1 Net Radiation

Four major components determine net radiation at a given site: incoming solar radiation ($H + I$; shortwave), outgoing reflected solar radiation (R_r; shortwave), incoming sky radiation ($\varepsilon\sigma T_a^4$; longwave), and outgoing terrestrial radiation (σT_s^4; longwave). Solar radiation is the primary source of energy on Earth. Its intensity, which decreases inversely with the square of the distance, at normal incidence outside the atmosphere at the mean solar–Earth distance is about 1.94 cal/cm²/min (or 1.35 kW/m²; 1 cal/cm²/min = 697.3 W/m² = 1 langley/min). This is the so-called *solar constant*. The radiation intensity is further subject to absorption, reflection, and transmission when it passes through the atmosphere. Thus, some solar radiation reaches the ground surface directly (direct radiation, I), some diffuses or scatters from the atmosphere to the ground (diffuse radiation, H), and some is reflected back to the atmosphere when it hits the ground (Figure 9.1). In equation form, net radiation (R_n) can be calculated by

$$R_n = (H + I) - (R_r) + \varepsilon\sigma T_a^4 - \sigma T_s^4 \quad \text{or} \quad R_n = (H + I)(1 - r) + \varepsilon\sigma T_a^4 - \sigma T_s^4 \tag{9.14}$$

where
  $r = R_r/(H + I)$, called the surface albedo
  T_a and T_s are the air and surface temperature in K, respectively
  σ is the Stefan–Boltzmann constant (8.26×10^{-11} cal/cm²/min/K⁴)
  ε is the sky emissivity, a measure of the efficiency of a body in emitting radiation at a specified wavelength and temperature

The Stefan–Boltzmann law states that any object radiates energy at a rate proportional to the fourth power of its absolute temperature. Thus, the terms $\varepsilon\sigma T_a^4$ and σT_s^4 in Equation 9.14 estimate sky radiation and terrestrial radiation, respectively, as a function of temperature and emissivity by the Stefan–Boltzmann law. The emissivity of a blackbody such as the sun is 1.0. It emits radiation at the maximum possible intensity for every wavelength. Earth acts as a blackbody with ε of about 0.97–0.99 in wavelength bands 9–14 μm and its

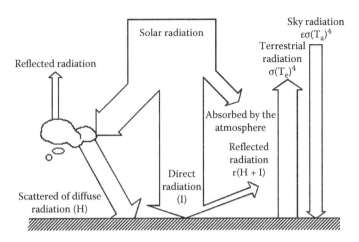

FIGURE 9.1
Radiation balance at the ground surface.

emissivity is assumed to be 1.0. Most objects radiate a fraction of radiation with an emissivity less than unity and are referred to as gray bodies. The ε of air can be estimated by an exponential function of air temperature T_a in degrees Celsius (Idso and Jackson, 1969):

$$\varepsilon = 1 - 0.26 \exp[-7.77(10^{-4})(T_a)^2] \tag{9.15}$$

The solution of Equation 9.15 is given in Figure 9.2.

The surface reflectivity (albedo) to incoming solar radiation varies with the incident angle and surface characteristics such as color, moisture content, and roughness. Snowpacks, depending on maturity, reflect about 70%–90% of solar radiation, while water surface only reflects 5%–15% (Table 9.1). During the daytime, differences in temperature between ground surface and air are small, which makes surface albedo an important factor affecting net radiation. Thus, net radiation over a dark-colored coniferous forest is often greater than over a nearby light-colored hardwood forest or bare ground. At night, there is no incoming solar radiation, and the R_n is determined by the temperature differences between the

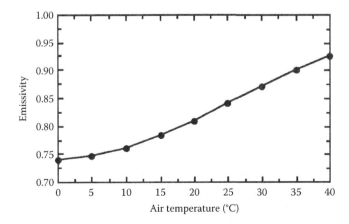

FIGURE 9.2
The solution of Equation 9.15 for estimating air emissivity.

TABLE 9.1

Albedo of Some Natural Surfaces

Surface	Albedo	Surface	Albedo	Surface	Albedo
Snow, fresh	0.80–0.95	Waters	0.03–0.10	Forest, spruce	0.05–0.10
Snow, old	0.45–0.70	Oceans	0.03–0.07	Hardwoods	0.15–0.20
Soil, clay	0.20–0.35	Grass, short	0.15–0.25	Forest, mixed	0.10–0.15
Soil, sandy	0.25–0.45	Grass, tall	0.15–0.20	Crops	0.15–0.25

TABLE 9.2

Radiation Balance (cal/cm²) for a Clear-Cut and a Forested Site Observed on Two Typical Days at Parsons, West Virginia

Date	R_n	(H+I)	r	$\varepsilon\sigma T^4$	σT_s^4
Clear-cut site with dense slash litter and humus					
July 21, 1966	414	658	0.131	326	484
February 22, 1967	53	245	0.624	162	201
Eastern hardwoods, uneven-					
July 21, 1966	425	658	0.202	326	426
February 22, 1967	129	245	0.310	162	202

Source: Adapted from Hornbeck, J.W., *For. Sci.*, 16, 139, 1970.
Note: All radiation components are explained in Equation 9.9.

surface and the air. The radiation balance for a clear-cut and a forested site observed on a typical day in summer and winter at Parsons, West Virginia, is given in Table 9.2.

9.2.2 Energy Dissipation

The net radiation determined by Equation 9.14 for a given site is further dissipated into three major components: sensible heat (S_h), latent heat (L_h), and conductive heat (C_h), or

$$R_n = S_h + L_h + C_h \tag{9.16}$$

The solution of R_n in Equation 9.14 requires measurements of incoming solar radiation H+I (also called *global radiation*), surface reflectivity r, and air and surface temperatures. If measurements of (H+I) are not available, R_n (cal/cm²/day) can be estimated by the empirical equation

$$R_n = R_a(1-r)(0.3+0.5n/N) - \sigma T_a^4\left[0.56-0.09(e_d)^{0.5}\right](0.1+0.9n/N)(1440) \tag{9.17}$$

where
R_a is the daily solar radiation at the top of the atmosphere (cal/cm²/day) (Table 9.3 or see Section 9.2.3)
n/N is the percent of sunshine (Figure 9.3; Table 9.4)
e_d is the actual vapor pressure of the atmosphere in mmHg
Variables r, σ, and T_a are as defined in Equation 9.14
The constant 1440 is to convert the longwave radiation from cal/cm²/min into cal/cm²/day

Equation 9.17 is more reliable for long-term averages.

TABLE 9.3

Total Daily Solar Radiation (cal/cm^2/day) by Latitude Zones at Top of the Atmosphere for Selected Days

Latitude (°)	March 21	April 13	May 6	June 22	July 15	August 8	September 23	October 16	November 8	December 22	January 13	February 4
N												
90		423	772	1077	949	765						
80	155	423	760	1060	980	754	153	7				
70	307	525	749	1012	934	742	303	129	24			24
60	447	635	809	979	929	801	442	273	146	49	73	146
50	575	732	867	989	954	859	568	414	286	176	205	289
40	686	807	910	991	967	901	677	545	429	317	350	434
30	775	865	929	975	960	921	765	663	564	466	494	568
20	841	894	923	935	930	916	831	760	685	605	630	691
10	882	897	893	873	877	886	871	835	789	733	752	795
0	895	873	837	790	800	830	885	886	870	843	855	878
S												
10	882	824	760	687	704	753	871	910	927	933	936	936
20	841	750	660	567	590	654	831	907	959	999	993	968
30	775	654	543	436	463	538	765	877	964	1041	1025	973
40	686	538	413	297	328	409	677	819	944	1059	1032	953
50	575	408	276	165	192	274	568	743	901	1056	1018	909
60	447	269	140	47	68	139	442	644	840	1046	992	847
70	307	127	23			23	303	532	778	1081	998	785
80	155	7					153	429	790	1132	1046	796
90								429	801	1149	1062	809

Source: List, R., *Smithsonian Meteorological Tables*, 6th edn., Smithsonian Institution Press, Washington, DC, 1971.

Note: Based on the solar constant 1.94 cal/cm^2/min.

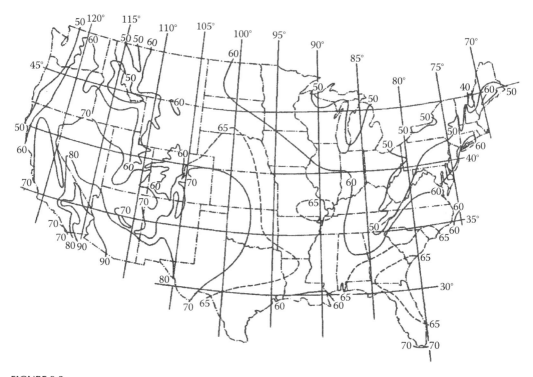

FIGURE 9.3
Annual percentage of possible sunshine in the United States. (From EDS, *Climatic Atlas of the United States*, U.S. Department of Commerce, Washington, DC, 1968.)

9.2.2.1 Sensible Heat

The flow of heat between surface and air by convection is often referred to as sensible heat flux. This is the heat transfer process with mass motions; it affects air temperature. One of the methods to estimate sensible heat flux is to use the aerial resistance approach. Consider the equation

$$S_h = -\frac{(c_p \rho_a)(T_s - T_a)}{r_a} \tag{9.18}$$

where
　c_p and ρ_a have been defined previously
　T_s and T_a are the surface and air temperature, respectively (°C)
　r_a is the *aerial resistance* to the flow of sensible heat S_h

In the case of forced (turbulent) convection for air at 20°C and 1000 mbar (100 kPa), Campbell (1977) showed

$$r_a = 3.07 \left(\frac{D}{W}\right)^{0.5} \tag{9.19}$$

TABLE 9.4

Average Percent Possible Sunshine for a Few Cities in the United States

Location	January	February	March	April	May	June	July	August	September	October	November	December
Montgomery, AL	48	53	59	65	64	64	62	64	62	65	56	50
Phoenix, AZ	78	80	84	88	93	94	85	85	89	88	84	78
San Francisco, CA	56	62	69	73	72	73	66	65	72	70	62	53
Denver, CO	71	70	69	67	65	71	71	72	74	72	65	67
Miami, FL	69	68	75	79	73	74	75	75	75	75	70	67
Boise, ID	39	50	62	68	71	76	87	85	81	69	43	39
Chicago, IL	49	47	51	52	59	68	68	65	58	54	40	45
Shreveport, LA	50	55	57	59	64	71	74	73	69	69	58	52
Reno, NV	65	68	76	81	81	85	92	92	91	83	70	64
New York, NY	51	55	57	59	61	64	65	64	62	61	52	49
Houston, TX	43	48	50	54	58	64	66	65	62	61	49	51
Seattle, WA	28	34	42	47	52	49	63	56	53	37	28	23

Source: Bair, F.E., *Weather of U.S. Cities*, 4th edn., Gale Research, Detroit, MI, 1992.

and for free convection (at 20°C and 1000 mbar), he gave

$$r_a = 2.656 \left(\frac{D}{\delta T} \right)^{0.5}$$ (9.20)

where
 r_a is in seconds per centimeter
 D is a characteristic dimension (the mean width for a leaf) (cm)
 W is the wind speed (cm/s)
 $\delta T = T_s - T_a$ (°C)

The negative sign in Equation 9.18 means that heat fluxes move from the surface toward the air. S_h is positive if heat fluxes move from the air toward the surface.

9.2.2.2 Conductive Heat

The fraction of net radiation that is dissipated into the soil is heat conduction. The transfer process occurs by the contact of soil particles; it does not change the shape of particles and does not require mass motions. Heat conduction into the soil profile is determined by the temperature gradient and the *thermal conductivity*. Soil porosity, moisture content, and organic matter content make thermal conductivity different among soils. A typical dry soil can have a thermal conductivity of $4–8 \times 10^{-4}$ cal/s/cm/°C, while a wet soil can have a thermal conductivity of $3–8 \times 10^{-3}$ cal/s/cm/°C, about ten times greater. In comparison, the thermal conductivity of water at 20°C is 1.43×10^{-3} cal/s/cm/°C. The soil heat flux can be downward or upward depending on the temperature gradient between the surface and the underlying soil.

Changes in soil heat content (cal/cm^2/day) for a given period can be estimated by the following equation:

$$C_h = \frac{(c_s)(d_s)(T_2 - T_1)}{\delta t}$$ (9.21)

where
 c_s is the soil heat capacity (0.24 cal/cm^3/°C for a typical soil)
 d_s is the soil depth (cm)
 T_2 is the soil temperature at the end of the period (°C)
 T_1 is the soil temperature at the beginning of the period (°C)
 δt is the length of the period (days)

Soil temperature and ambient air temperature are closely related to each other. During the daytime and in summer, the soil surface warms up more intensely than does the air. At night and in winter, the surface temperature falls below that of the air. The fluctuation of soil temperature decreases with depth, and the mean soil temperature is higher than the mean air temperature in general. If soil temperatures required for the calculation of soil heat contents are not available, then an empirical equation can be developed to estimate soil temperatures from ambient air temperatures and a time angle in a periodic regression function (Chang et al., 1994):

TABLE 9.5

Coefficients and Statistics of Equation 9.22 for Estimating Soil
Temperatures (°C) at 15 and 60 cm Depths for a Fuller Fine Sandy
Loam in East Texas

Site Conditions	Depth (cm)	b_0	b_1	b_2	b_3	R^2	SEE
Forested	15	0.490	0.417	−4.708	−1.999	0.91	2.1
	60	12.337	0.273	−5.307	−2.634	0.91	2.0
Cleared	15	13.154	0.227	−5.404	−2.914	0.91	1.8
	60	16.717	0.093	−5.357	−3.723	0.94	1.3

Note: R^2 = coefficient of multiple determination; SEE = standard error of estimate, °C.

$$T_s = b_0 + b_1 T_a + b_2 \cos\left(\frac{2\pi t}{365}\right) + b_3 \sin\left(\frac{2\pi t}{365}\right) \qquad (9.22)$$

where
   T_s is the soil temperature for any day of a year t starting with January 1 as 001
   T_a is the corresponding air temperature
   b_0, b_1, b_2, and b_3 are the regression coefficients
   $(2\pi t/365)$ is the daily unit of time angle in the annual cycle (rad)

As an example, the coefficients of Equation 9.22 for estimating soil temperatures (in degrees
Celsius) at two different depths in a Fuller fine sandy loam in the Davy Crockett National
Forest, Texas, are given in Table 9.5.

9.2.2.3 Latent Heat

Unlike convection and conduction, the transfer of latent heat does not cause a direct
change in temperature. Using a term known as the "Bowen ratio" in Equation 9.16 can
solve the amount of latent heat used in a given area. The Bowen ratio β is defined as the
ratio between sensible heat S_h and latent heat L_h, or the ratio of the differences in temperature (°C, $\delta T = T_s - T_a$) and vapor content (mbar, $\delta\theta = \theta_s - \theta_a$) between the surface and the
ambient air:

$$\beta = \frac{S_h}{L_h} = \gamma\left(\frac{\delta T}{\delta\theta}\right) \qquad (9.23)$$

where γ is the psychrometric constant in millibars per degree Celsius. The constant γ, to
account for units, is calculated by

$$\gamma = \frac{c_p P_a}{0.622 L_V} \qquad (9.24)$$

where
   c_p is the specific heat of air at constant pressure (cal/g/°C)
   0.622 is the ratio of the molecular weight of water vapor to that of dry air
   P_a is the atmospheric pressure (mbars)
   L_V is the latent heat of vaporization (cal/g) (Equation 4.1)

Substituting β of Equation 9.23 in Equation 9.16, the total latent heat can be calculated by

$$L_h = \left(\frac{R_n - C_h}{1+\beta}\right) \qquad (9.25)$$

In humid environments, Equation 9.25 can be simplified further by eliminating the variable C_h without committing serious errors if the estimates are for long-term averages such as monthly and annual values. Note that $L_h = (L_V)(E)$ where E = vaporization in centimeters, or $E = L_h/L_V$.

9.2.2.4 Proportionality

The relative dissipation of net radiation into convective, evaporative, and conductive heat varies with water availability, climatic regions, land use, and time periods. Generally, conductive heat, the smallest of the three, is less than 25% of the net radiation in humid areas and is negligible on an annual basis. A comparison of heat dissipation on three surface conditions in Germany showed that net radiation was greatest over the forest, followed by grass and bare ground (Baumgartner, 1967). Only 5%, 17%, and 25% of the net radiation was dissipated into heat conduction for the forest, grass, and bare ground, respectively (Figure 9.4a).

A similar study of energy partitioning for three major land use conditions during four snow-free seasons was conducted in the Wolf Creek Research Basin near Whitehorse, Yukon Territory in western Canada (Granger, 1999). Net radiation averaged to be equivalent to 629.8 mm of water for the boreal forest consisting of mixed spruce, pine, and poplar during the summers (May to September) of 1994–1997, compared to 540.7 mm for the shrub taiga and 463.5 mm for alpine tundra. Although variations in elevation and cloud cover may account for some differences in net radiation, the most significant contributing factor to the differences is the surface albedo, which averaged to 0.09, 0.18, and 0.21 at the forest, taiga, and alpine sites, respectively. Ratios of S_h to L_h were 1.02, 0.91, and 0.72 for the forest, taiga, and alpine, respectively (Figure 9.4b). Although the average precipitation was lowest

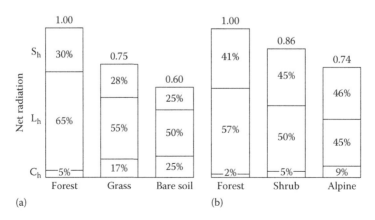

FIGURE 9.4
Relative dissipation of energy for various land use conditions in two humid regions: (a) Germany and (b) Canada.

in the forest (86.8 mm) and highest in the alpine (137.2 mm), the average vaporization was highest in the forest (358.8 mm) and lowest in the alpine (210.7 mm).

In humid environments, a great deal of energy is used for vaporization, causing latent heat to be greater than sensible heat. The situation is reversed in arid or semiarid areas. There the limited vaporization due to the lack of available water causes the conserved latent heat to be used for warming the air. As a result, sensible heat is greater than latent heat. Thus, the Bowen ratio (sensible heat/latent heat) is usually less than 1.0 in humid regions and greater than 1.0 in semiarid regions. In the desert, the ratio can go above 3.0.

At higher latitudes, such as Copenhagen, Denmark, net radiation is negative during winter months and the energy for L_h comes largely from air masses of the warmer ocean. Thus, L_h exceeds R_n in the winter months, and S_h can take up more than 50% of R_n in the summer. At Tempe, Arizona, strong advective energy from the nearby desert areas causes L_h from a fully covered alfalfa field to exceed R_n every month of the year except January and December (Pruitt, 1972).

9.2.3 Extraterrestrial Radiation

As mentioned previously, virtually all energy utilized and exchanged in the biosphere comes from the sun. The radiant energy moves out from the sun in all directions and spreads over the surface of a sphere. At some distance r from the source R_o, the energy per unit area R_s is

$$R_s = \frac{R_o}{4\pi r^2} \tag{9.26}$$

This means that the energy intercepted by surfaces of a given area declines with the square of the distance from its source.

Solar radiation decreases to 1.94 cal/cm^2/min (1.35 kW/m^2) at the top of the atmosphere. This is the average flux at a surface perpendicular to the incoming solar radiation at the mean sun–Earth distance. The daily extraterrestrial radiation R_a in cal/cm^2/day for a given geographical location and time of the year can be estimated by (Shuttleworth, 1993)

$$R_a = 920 d_r \left(\omega_s \sin\phi \sin\delta + \cos\phi \cos\delta \sin\omega_s\right) \tag{9.27}$$

$$\omega_s = \cos^{-1}(-\tan\phi \tan\delta) \tag{9.28}$$

$$\delta = 0.4093 \sin\left(\frac{2\pi J}{365} - 1.405\right) \tag{9.29}$$

$$d_r = 1 + 0.033 \cos(J) \tag{9.30}$$

where
  ω_s is the sunset hour angle (rad)
  d_r is the relative distance between the earth and the sun
  δ is the angle of solar declination (rad)
  J is the Julian day number (January 1, J = 1)
  ϕ is the site latitude in degrees (positive in the Northern Hemisphere and negative in the Southern Hemisphere)

The extraterrestrial radiation undergoes a great deal of change with respect to quantity, quality, and direction when it passes through the atmosphere.

Example 9.1

Calculate R_a for a site at $\phi = 30°N$ on April 15 ($J = 105$).

Solution:

Use the following steps to solve Equation 9.27:

$\phi = 30° = 30°/57.2956°/\text{rad} = 0.5236$ rad.
$J = 105\text{th day} = 105/57.2956\,\text{d/rad} = 1.8326$ rad.
$\delta = 0.4093 \times \sin[2 \times 3.1416 \times 1.8326/365 - 1.405] = 0.16$ rad.
$\omega_s = \cos^{-1}(-\tan\phi \times \tan\delta) = \cos^{-1}(-\tan 0.5236 \times \tan 0.16) = 1.664$ rad.
$d_r = 1 + 0.033 \times \cos(1.8326) = 0.9915$.
$R_a = 920 \times 0.9915 \times (1.664 \times \sin 0.5236 \times \sin 0.16 + \cos 0.5236 \times \cos 0.16 \times \sin 1.664)$
$= 897.45\,\text{cal/cm}^2/\text{day}$.

9.3 Evapotranspiration

Separating vaporization by different surfaces is not necessary in many practical applications. A collective term, evapotranspiration (ET), often is used to describe the total vaporization for a given watershed. Evapotranspiration can be measured only under control conditions; indirect observations or empirical and theoretical models are frequently employed to make the estimates.

9.3.1 Potential Evapotranspiration

The concept of *potential evapotranspiration* (PE) is often used for estimating watershed ET losses. It is, as defined by Thornthwaite (1948) and Penman (1963), the total water loss in the vapor state from the center of an extended surface completely covered by short, even-height green vegetation with no limit on water supply. This definition emphasizes the following:

1. The advective heat exchanges to and from the surroundings are insignificant.
2. The heat flux from canopy to soil is negligible.
3. Water loss is independent of vegetation and soil type.
4. The solely limiting factor of PE is available energy.

Thus, PE is the upper limit of ET under a given climatic condition and cannot exceed free-water evaporation. In practice, PE = ET if the water supply to plants is unlimited or if monthly PE is less than monthly precipitation.

The concept of PE is widely accepted and frequently applied in geographic studies, climate classifications, and water resources research, planning, and management. However, the PE concept is questionable if it is applied to small, forested watersheds. This is because of the tremendous effects on energy availability of forest density, species composition, the physiological and phenological behavior of woody plants, topography, and advective heat exchange. It also is invalid if it is applied to a body of pure water. For example, evaporation from a shallow pond exhibits a bell-shaped seasonal trend similar to air temperature patterns. But, evaporation from a large, deep lake under similar weather conditions can show an inverted bell-shaped pattern due to warmer water temperatures in the cold season (Chang et al., 1976). These limitations imply that PE models should not be applied in a short period of time and not be applied to an isolated field or a small, forested watershed.

9.3.1.1 Thornthwaite (1948) Model

This is one of the most widely used empirical models for estimating PE. It expresses PE as a function of air temperature (T_m) and a temperature-dependent heat index (I). The estimated values (mm/month), requiring adjustments in the length of the day for a 12h period, are given by the following equation:

$$PE = 16\left(\frac{10T_m}{I}\right)^a (K) \tag{9.31}$$

where
  I is the summation of monthly heat index i, or $I = \Sigma i = \Sigma (T_m/5)^{1.514}$
  T_m is the monthly air temperature with m ranging from January to December
  $a = 0.4923 + 0.017\,92(I) - 0.000\,072\,7(I)^2 + 0.000\,000\,675(I)^3$
  K is the adjustment in the length of the day for a 12h period

Some of the K values of Equation 9.31 can be found in Table 9.6 or calculated by

$$K = \frac{H}{12} = \frac{24(\omega_s)/\pi}{12} = \frac{2(\omega_s)}{\pi} \tag{9.32}$$

where
  H is the maximum possible daytime hours
  ω_s is the sunset hour angle in radians (Equation 9.28)

Thornthwaite and Mather (1957) prepared tables and bookkeeping procedures to remove the need for complicated calculations. Computer programs also are available for solving these equations (Black, 1995). The model has been used extensively in watershed research, climate characterization and classification, and irrigation studies. Generally, it works best in water-balance estimates for large areas and for long-term periods; it is not accurate for daily estimates. Success with this method has been more frequent in humid areas. Thornthwaite's climate classification system, based on Equation 9.31, water deficit, and water surplus, has proven especially useful to hydrologists, agricultural engineers, and water resource managers (Chang et al., 1996).

9.3.1.2 Hamon (1963) Method

The Hamon model requires only mean daily air temperature and relative humidity, perhaps the simplest form of PE. It shows

TABLE 9.6

Mean Possible Daytime Duration by Latitude φ in Units of 30 Days and 12h

φ	January	February	March	April	May	June	July	August	September	October	November	December
50 N	0.74	0.78	1.02	1.15	1.33	1.36	1.37	1.25	1.06	0.92	0.76	0.70
40 N	0.84	0.83	1.03	1.11	1.24	1.25	1.27	1.18	1.04	0.96	0.83	0.81
30 N	0.90	0.87	1.03	1.08	1.18	1.17	1.20	1.14	1.03	0.98	0.89	0.88
20 N	0.95	0.90	1.03	1.05	1.13	1.11	1.14	1.11	1.02	1.00	0.93	0.94
10 N	1.00	0.91	1.03	1.03	1.08	1.06	1.08	1.07	1.02	1.02	0.98	0.99
00	1.04	0.94	1.04	1.01	1.04	1.01	1.04	1.04	1.01	1.04	1.01	1.04
10 S	1.08	0.97	1.05	0.99	1.01	0.96	1.00	1.01	1.00	1.06	1.05	1.10
20 S	1.14	1.00	1.05	0.97	0.96	0.91	0.95	0.99	1.00	1.08	1.09	1.15
30 S	1.20	1.03	1.06	0.95	0.92	0.85	0.90	0.96	1.00	1.12	1.14	1.21
40 S	1.27	1.06	1.07	0.93	0.86	0.78	0.84	0.92	1.00	1.15	1.20	1.29
50 S	1.37	1.12	1.08	0.89	0.77	0.67	0.74	0.88	0.99	1.19	1.29	1.41

Source: Adapted from Thornthwaite, C.W. and Mather, J.R., *Instructions and Tables for Computing Potential Evapotranspiration and the Water Balance*, Vol. XIX (3), Laboratory of Climatology, Centerton, NJ, 1957.

$$PE\,(mm/day) = 0.1651(\rho w)(K)(C) \qquad (9.33)$$

where
  ρ_w is the absolute humidity in g/m^3 at the mean air temperature
  K is the length of daytime in units of 12h (Equation 9.32)
  C is the correction factor to adjust for measured values

The absolute humidity or vapor density can be calculated by saturated vapor pressure e_s in millibars and mean air temperature T_a in Kelvin by

$$\rho_w = \frac{217\,(e_s)}{T_a} \qquad (9.34)$$

The inclusion of the "C" factor in Equation 9.33 is required because the calculated values are often lower than those calculated by the Penman equation (Chang, 1975) and measured values (Zhou et al., 2008) in humid regions. In the United States, the correction coefficient, generally higher at lower latitudes, ranges from 1.0 at Hubbard Brook, New Hampshire, to 1.1 in Northern Wisconsin, 1.2 at Coweeta, North Carolina, and 1.3 in Florida (Sun et al., 2002, 2008).

Example 9.2

Use Hamon's equation (Equation 9.33) to estimate PE at $\phi = 30°$ N on April 15 with $T_a = 18°C$.

Solution:

Solving Equation 9.33 requires information on K and ρ_w; they are calculated by

e_s (saturated vapor pressure at T_a, mbar) = exp $[21.382 - 5347.5/(273° + 18°)]$
  = 20.20 millibar
$\rho_w = 217\,(e_s)/T_a = 217 \times 20.20/(273 + 18) = 15.06\,(g/m^3)$
$\delta = 0.16$ rad and $\omega_s = 1.664$ rad $\left(\text{see Example 9.1}\right)$
$K = 2\omega_s/3.1416 = 1.0593$

Using C=1.2, the estimated PE=$0.1651 \times 15.06 \times 1.0593 \times 1.2 = 3.16$ (mm/day).

9.3.1.3 Penman (1948) Method

The Penman method, one of the most popular physically based models, combines available energy and aerodynamic function to estimate PE by

$$PE\,(mm/day) = \frac{[(\delta)(E) + (E_o)(\gamma)]}{\delta + \gamma} \qquad (9.35)$$

where
  δ is the slope of saturation vapor pressure curve with air temperature (mbar/°C)

$$= \frac{4098e_s}{(237.3 + T, °C)^2} \tag{9.36}$$

where

γ is the psychrometer constant (mbar/°C) (Equation 9.24)

E is the net radiation $= 10R_n/L_v$ (mm/day)

R_n is the net radiation (cal/cm^2/day) (Equation 9.14 or 9.17)

L_v is the latent heat of vaporization (cal/cm^3) of water (Equation 4.1)

$E_o = 0.263(e_s - e)(1.0 + 0.00625w_2)$ (mm/day) or empirically derived

e_s is the saturation vapor pressure of the surface (mbar)

e is the actual vapor pressure of the air (mbar)

w_2 is the wind speed at 2 m above the ground (km/day), or as suggested by McCuen, 1998):

$$w_2 = w_o \left(\frac{2}{z_o} \right)^{0.15} \tag{9.37}$$

where

z_o is the height of the anemometer (m)

w_o is the wind speed recorded by the anemometer (km/day)

Equation 9.35 was originally developed for estimating open water evaporation and has been modified to estimate PE and ET for various surface conditions and in different regions. One of the modifications, made by Monteith (1965) and later discussed by Monteith (1985) and Monteith and Unsworth (1990), is to incorporate biologically based canopy and physically based aerodynamic resistance parameters into the Penman's wind function for estimating water vapor diffusion over vegetated surfaces. The Penman–Monteith model is probably the most popular equation for estimating PE; however, the Penman–Monteith model requires climatic data that are not usually available, particularly the wind speed and resistance parameters. This limits its applicability.

Example 9.3

Use Penman's equation (Equation 9.35) to estimate PE at $\phi = 30°N$ on April 15 with $T_a = 18°C$, RH = 80%, Pa = 1013 mbar, n/N = 50%, r = 0.20, and $w_2 = 100$ km/day.

Solution:

For $\phi = 30°N$ on April 15, Example 9.1 has already calculated

$\delta = 0.16$ rad, $\omega_s = 1.664$ rad, $e_s = 20.20$ mbar, and $R_a = 897.45$ cal/cm^2/day.

Others are calculated as follows:

$e = (e_s)(RH) = 20.20$ mbar $(0.80) = 16.16$ mbar (actual vapor pressure)

$e_d = (e)(0.75$ mm Hg/mbar$) = 16.16$ mbar $\times 0.75$ mm Hg/mbar $= 12.12$ mmHg

$L_v = 597 - 0.564(18) = 586.85$ cal/g

$\delta = 4098 (20.20)/(237.3 + 18)^2 = 1.27$ (mbar/°C), (From Equation 9.36)

$\gamma = (0.24$ cal/g/°C$)(1013$ mbar$)/(0.622 \times 586.85$ cal/g$) = 0.67$ mbar/°C (From Equation 9.24)

$R_n = (897.45$ cal/cm^2/day$)(1 - 0.2)(0.3 + 0.5 \times 0.5) - 8.26(10^{-11})$ cal/cm^2/min/K^4 $[(273 + 18)$K$]^4 \times$

$(0.56 - 0.09 \times 12.12^{0.5})(0.1 + 0.9 \times 0.5)(1440$ min/day$) = 279.15$ cal/cm^2/day

$$E = 10 \, mm/cm \, (279.15 \, cal/cm^2/day)/586.85 \, cal/g = 4.76 \, mm/day$$
$$E_o = 0.263(20.20 \, mbar - 16.16 \, mbar)(1 + 0.0065 \times 100 \, km/day) = 1.73 \, (mm/day)$$
$$PE = [(\delta)(E) + (\gamma)(E_o)]/(\delta + \gamma) = [(1.27)(4.76) + (0.67)(1.73)]/(1.27 + 0.67) = 3.71 \, (mm/day)$$

9.3.1.4 Hargreaves et al. (1985) Equation

For monthly PE in mm/day, the method only requires measurements of maximum and minimum temperatures and a tabulated value of extraterrestrial radiation:

$$PE \, (mm/day) = 0.0023 R_a (T_m + 17.8)(T_r^{0.50}) \tag{9.38}$$

where
R_a is the extraterrestrial radiation in equivalent depth of water (mm/day)
T_m is the mean monthly air temperature (°C)

$$= (mean \, monthly \, maximum - mean \, monthly \, minimum)/2$$

T_r is the mean monthly air temperature range (°C)

$$= mean \, monthly \, maximum - mean \, monthly \, minimum$$

Values of R_a expressed in depth of water at 20°C for 10° latitude zones of the world are given in Table 9.7. Average temperature ± 10°C from 20°C will result in an error of R_a by less than 1%. For a specific temperature of R_a, one can multiply the tabulated R_a in Table 9.7 by 586 cal/cm^3 to obtain the average daily extraterrestrial radiation in energy unit for the month (i.e., cal/cm^2/day), then use latent heat of vaporization (Equation 4.1) and actual temperature to convert R_a from energy unit into water unit.

The method has been tested against up to 11 other equations in global analyses. It performs almost as well as the classical Penman combination equation, but requires less parameterization and has significantly lower sensitivity to error in climatic inputs (Hargreaves, 1994; Trabucco et al., 2008).The equation is recommended for general use such as climate characterization, water resource planning, design, and other water adequacy studies.

9.3.2 Actual Evapotranspiration

Potential evapotranspiration is an upper limit to water loss as determined by local weather conditions with no shortage of water supply. When water supply for vaporization is deficient or soil-moisture content is below the field capacity, then vaporization cannot proceed at the potential level. Thus, the actual evapotranspiration (AE) is only a fraction of PE, or

$$AE = \alpha(PE) \tag{9.39}$$

where the empirically derived fraction α is affected not only by soil-moisture content but also by climate and species. It is the so-called *crop coefficient* in agricultural irrigation.

AE can equal or exceed ($\alpha \geq 1.0$) PE in both humid and arid environment where soil is wet or crops are well watered. In a 15-year study at Copenhagen, Denmark, Kristensen (1979) showed that grass AE measured from evapotranspirometers could exceed PE calculated by Penman method by 23% in a single month. Seasonal AE for five surface conditions, measured from 358 to 2800 days, at 16 sites across Florida is given in Table 9.8 (Douglas et al., 2009). The smaller values of AE at forest sites, as compared to citrus or grass, could reflect differences in site conditions and water availability.

TABLE 9.7

Extraterrestrial Radiation Expressed in Millimeter per Day of Water at 20°C

Lat	January	February	March	April	May	June	July	August	September	October	November	December
50N	03.8	06.1	09.4	12.7	15.8	17.1	16.4	14.1	10.9	07.4	04.5	03.2
40	06.4	08.6	11.4	14.3	16.4	17.3	16.7	15.2	12.5	09.6	07.0	05.7
30	08.8	10.7	13.1	15.2	16.5	17.0	16.8	15.7	13.9	11.6	09.5	08.3
20	11.2	12.7	14.4	15.6	16.3	16.4	16.3	15.9	14.8	13.3	11.6	10.7
10	13.2	14.2	15.3	15.7	15.5	15.3	15.3	15.5	15.3	14.7	13.6	12.9
0	15.0	15.5	15.7	15.3	14.4	13.9	14.1	14.8	15.3	15.4	15.1	14.8
10S	16.4	16.3	15.5	14.2	12.8	12.0	12.4	13.5	14.8	15.9	16.2	16.2
20	17.3	16.5	15.0	13.0	11.0	10.0	10.4	12.0	13.9	15.8	17.0	17.4
30	17.8	16.4	14.0	11.3	08.9	07.8	08.1	10.1	12.7	15.3	17.3	18.1
40	17.9	15.7	12.5	09.2	06.6	05.3	05.9	07.9	11.0	14.2	16.9	18.3
50	17.5	14.7	10.9	07.0	04.2	03.1	03.5	05.5	08.9	12.9	16.5	18.2

Source: Extracted from Hargreaves, G.H., *J. Irrig. Drain. Eng.*, 120, 1132, 1994.

TABLE 9.8

Actual Evapotranspiration (mm/Day), by
Seasons, Measured Differently from 358 to 2800
Days at 18 Sites across Florida in the United States

Cover Type	Winter	Spring	Summer	Fall
Forest	1.8	2.8	3.2	2.3
Citrus	1.7	3.8	4.4	2.8
Grass	1.3	3.0	3.7	2.0
Marsh	2.7	4.2	4.4	3.3
Open water	3.3	5.3	4.8	3.8

Source: Data extracted from Douglas, E.M. et al., *J. Hydrol.*,
373, 366, 2009.

9.3.2.1 Hydrologic Balance Approach

Watershed AE is often estimated indirectly by measuring the differences between watershed precipitation (P_t) and runoff (RO). P_t can be reliably estimated through proper sampling, and RO is the only component in the hydrologic cycle that can be directly measured with great accuracy. Apparently, the estimate is valid only for a time interval long enough to ensure that those changes in soil-moisture storage are negligible. If the estimate is for a period, such as a month or season, with significant variation of moisture content, then changes in soil-moisture storage need to be considered in the budget equation solution.

In regions where the boundaries of surface watershed and underground watershed are not coincident, the measured RO is not reliable because of underground water leakage. The problem is more serious in small watersheds and becomes less significant as the size of watershed increases. It is important that the groundwater leakage be detected or the site abandoned for hydrologic studies.

9.3.2.2 Lysimeters

A device that measures the quantity and quality of percolated water, or monitors the water budget, from an isolated block of soil in the field is called a *lysimeter*. Basically, it is a soil tank with growing plants inserted into the ground to simulate natural conditions. Rain or irrigation water percolated through the soil profile is collected at the bottom of the tank or pumped to the surface through installed pipes for analysis. Vaporization of water from the lysimeter is measured directly through gravimetric (weighing lysimeters), volumetric (floating lysimeters), or water-budget (percolation lysimeters) approaches. Since the lysimeter is isolated from the surrounding soil, its hydrologic budget can be precisely monitored and evaluated.

The weighing lysimeter may use mechanical scales through levels and counterbalances (Harrold and Dreibelbis, 1958), pressure devices through hydraulic and pneumatic supports (Fritschen and Simpson, 1982), or strain gages with torsional weigh bars (Kirkham et al., 1984). Although weighing lysimeters are the most complicated and expensive, they are generally considered the most accurate means of direct measurement of AE (Dugas et al., 1985). Their high fabrication costs may limit the number of replicates available for statistical evaluation.

Floating lysimeters consist of a soil tank floating in a larger outer tank filled with water or other liquids, both buried under the ground to imitate natural conditions (Chang et al., 1997). Evapotranspiration of the soil tank is detected, based on the Archimedes' Principle, through changes in liquid level in the outer tank. For accurate measurements, it may require to make adjustments for thermal expansion and evaporation loss of the liquid in the outer tank.

Lysimeters have been widely used in research for testing hypotheses on plant–water relationships and calibrating theoretical and empirical ET models. Their accuracy depends on the sensitivity to and representativeness of the surrounding areas.

9.3.2.3 Soil-Moisture Depletion Approach

Soil-water deficit causes AE to decline with time. The decline will stop when it reaches the wilting point or if additional water, either from precipitation or irrigation, is supplied to the soil. If no free water is standing in the soil profile, then the depletion of soil-moisture content over time can be assumed to be due to AE.

In east Texas, Chang et al. (1983) showed that the average annual soil-moisture content of the whole soil profile (0–1.35 m) of a Woodtell soil was 0.292 g/cm^3 for a mature southern pine forest and 0.492 g/cm^3 for a nearby clear, cultivated plot. The difference in soil-moisture content between cultivated and forested Woodtell soil was as much as 0.2 g/cm^3, or 27 cm in depth for the whole soil profile. Note that the depth of soil profile affects total moisture content, and measurements covering the root zone should be satisfactory in calculating AE for most purposes.

Soil-moisture depletion rates can be described by the following function:

$$\theta_{t+n} = K^n\theta_t \tag{9.40}$$

where
θ_{t+n} is the soil-moisture content (g/cm^3) n days after the initial moisture content θ_t
K is the depletion coefficient

Values of K for the six forest conditions on the Woodtell soil, by seasons and depths, are given in Table 9.9. These values indicate that the depletion rate of soil water, or AE, increases with increasing forest coverage. It will take 35 days for the undisturbed forest, but 62 days for the cultivated plot to deplete initial soil-moisture content of 0.45 g/cm^3 in the surface 30 cm profile to 0.20 in g/cm^3 during the growing season.

TABLE 9.9

K Values of Equation 9.40 for Six Forest Conditions in East Texas

Depth	Forest Conditions					
	a	b	c	d	e	f
Growing season						
0–0.30 m	0.9769	0.9795	0.9796	0.9821	0.9869	0.9869
0–1.35 m	0.9895	0.9878	0.9875	0.9902	0.9876	0.9923
Dormant season						
0–0.30 m	0.9858	0.9937	0.9862	0.9836	0.9924	0.9860
0–1.35 m	0.9926	0.9950	0.9943	0.9938	0.9876	0.9911

Note: a = undisturbed mature southern pine forest; b = 50% thinned forest; c = commercial clear-cutting without site preparation; d = clear-cut and chopped; e = clear-cut and sheared; f = clear-cut and cultivated.

Equation 9.40 can be rewritten as $ET_{t+n} = K^n ET_t$, where ET is daily evapotranspiration and K^n is the decay coefficient of ET as determined by the soil-moisture depletion rate. A similar relationship is also used in Tueling et al.'s (2006) ET and soil-moisture studies. Although soil-moisture depletion rate is a simple method, the estimate of AE is valid only for the period in which no free water is standing in the soil profile and no water is flowing in from or out to surrounding areas. Thus, it may be necessary to wait 1 or 2 days after a storm for all free water to drain out of the area before soil moisture is monitored. This results in an underestimate of AE because the AE during the waiting period is not included. Also, a reliable estimate of AE for the entire watershed is difficult to obtain because of the heterogeneity of the soil profile.

9.4 Forested versus Nonforested

Forests are the most distinguishable and complicated vegetation communities on Earth. Their canopy structures, organic floor, and root systems form a complex environment that significantly affects water and energy transfers between ground and the atmosphere. Numerous studies have shown that ET in forested watersheds is greater than that in bare or nonforested watersheds. Consequently, deforestation causes reduction in AE, and the conserved water will contribute to increases in water yield.

9.4.1 Environmental Conditions

The great loss of water to the atmosphere in forested watersheds is attributable to the great transpiration surface, deep root system, canopy interception loss, more available energy, and wind effects. Evaporation from soils occurs only at the soil surface and is active only in the surface layer. When water at the surface is lost to the air, the lost water must be replenished to the surface from lower soil layers. However, once all free water has drained out through the soil profile, the transport of water to the surface is very slow, usually taking 1 or 2 days. Under this condition, bare-soil evaporation is limited by soil-water content, not by atmospheric factors, and very little soil water below the surface layer—about 20 cm—will ever be lost by evaporation. However, transpiration is not limited to water in the surface layer. The tree's root system can reach water deep in the soil to more than 5 m below the surface and 15 m from trunks. Consequently, a large volume of water in the soil is available for transpiration but not for evaporation.

The total leaf area of a tree is often several times greater than the ground covered by the canopy. Such a ratio is often referred to as *leaf area index* (LAI). The LAI for a typical tree is around 5–7, but it could be higher than 40 for a mature, mixed stand of spruce and fir in the central Rocky Mountains (Kaufmann et al., 1982). *root area index* (RAI), the ratio between total root surface area and ground surface area in m^2/m^2, is at least comparable to LAI in all terrestrial systems (see Chapter 7, Section 7.1.2). A large volume of available water in the soil, coupled with a great area of vaporization surfaces, makes the transpiration rate from trees higher and the duration longer than vaporization from bare soils and short vegetation.

Other factors affecting AE in forested areas, such as net radiation and canopy interception loss, have been discussed in Chapter 9, Section 9.2 and Chapter 8, Section 8.2, respectively.

TABLE 9.10

Components of Vaporization (in % of the Total) for Various Land Uses at a Few Locations around the World

Location	Cover Type	E	I_F	T_r	Reference
Munich, Germany	Forest stands	10	30	60	Baumgartner (1967)
	Meadow	25	25	50	
	Cultivated	45	15	40	
	Bare soil	100			
Oak Ridge, TN	Mixed hardwoods	16	30	54	Oishi et al. (2008)
NOPEX site, Sweden	Mixed conifers	15	20	65	Iritz et al. (1999)
Flanders, Belgium	Forests	10	26	64	Verstraeten et al. (2005)
	Cropland	33		67	

Note: E = soil evaporation; I_F = evaporation of canopy-incepted water; T_r = transpiration.

9.4.2 Forested Watersheds

Transpiration and evaporation in a forested watershed occur at different rates, and evaporation of canopy-intercepted water often exceeds PE for open water surfaces (Shuttleworth, 1993). The AE of a forested watershed is the sum of the following three items:

$$AE = (1-\alpha)E + \alpha(T_r + I_F) \tag{9.41}$$

where
   α is the watershed forest coverage in fractions
   E is the evaporation
   T_r is the forest transpiration
   I_F is the forest interception
   E includes water losses from soil and water surfaces

For a forested watershed with its floor covered by a layer of litter, E only comes from the 1% to 2% water surface of the watershed. It makes up only 5%–10% of the total ET (Szilagyi and Parlange, 1999). The total interception loss I_F is about 10%–40% of annual precipitation, depending on species, atmospheric conditions, and storm characteristics. In humid areas, transpiration from forest stands can make up 60%–65% of the total AE, while evaporation makes up only about 10%–15% (Table 9.10).

The impact of forest coverage on watershed AE can be illustrated by a study using 39 experimental watersheds across the southeastern United States. There, the long-term annual AE in mm/year can be estimated by the linear regression model ($R^2 = 0.82$; Lu et al., 2003):

$$AE = 1098.786 + 0.309(P_t) - 0.289(Elev) - 21.840(Lat) + 1.96(F_o) \tag{9.42}$$

where
   P_t is the watershed mean precipitation (mm/year)
   Elev is the mean watershed elevation (m)
   Lat is the watershed latitude at the outlet (deg)
   F_o is the watershed percent forest cover

The model shows that an increase in forest coverage in the watershed by 10% would cause an increase in AE by 20 mm/year.

9.4.3 Forest Species

Each forest species has its own unique characteristics with respect to height, canopy density, root systems, color, stomatal response to environmental stress, leaf orientation, leaf–sapwood ratio, the length of the growing season, and changes in foliage with seasons. These characteristics affect the amount of available energy in a forest and the mechanism of a plant to regulate water conductance between the evaporative demand of the atmosphere and the water supply in the soil. Large differences in transpiration loss exist among species and within a species due to age and site conditions. In the central Rocky Mountains, ratios of annual transpiration for Engelmann spruce, subalpine fir, lodgepole pine, and aspen stands of equal basal area are 3.2, 2.1, 1.8, and 1, respectively (Kaufmann, 1985). Thus, total water losses to the atmosphere in forested watersheds are greatly affected by species composition in the watershed, which leads to watershed forest manipulation resulting in significant alterations of streamflow quantity and regimes around the world (see Chapter 10, Section 10.3).

A study in SE Arizona showed that cottonwood (*Populus* spp.) stands along a perennial reach had higher leaf–sapwood ratio, LAI, and ground water table (1.1–1.8 m vs. 3.1–3.9 m) as compared to stands along an intermittent reach of the san Pedro River. These plant structural and physiological traits regulated by site conditions cause great differences in transpiration rates. Average transpiration rate in a growing season was 4.41 mm/day at the perennial site and 2.21 mm/day at the intermittent site (Gazal et al., 2006). In Mauna Loa, Hawaii, Kagawa et al. (2009) showed that transpiration varied strongly among forest types even within the same wet tropical climate. Species like *Eucalyptus saligna* and *Fraxinus uhdei* trees, at the stand level, had three- and ninefold higher water use, respectively, than native *Metrosideros polymorpha* trees.

The average daily transpiration rates (T_r, ranging from 0.5–5.9 mm/day) for a few forest species around the world are given in Table 9.11. Although direct comparison of these figures is improper, they do show that species could cause differences in T_r of 1 mm/day for conifers and 5 mm/day for hardwoods. Conifers usually have higher annual T_r than hardwoods do. When there is no water deficiency in the soil, Jones (1997) gives ratios of T_r for various plant species relative to bare ground as 9.0 for larch/fir, 8.5 for spruce, 8.0 for maquis shrub, 7.5 for beech, 5.0 for pine, 3.0 for pasture, and 1.0 for bare ground.

However, based on the T_r (0.4–1.2 mm/day) from different temperate forests conducted by 19 studies in Europe, Roberts (1999) stated that T_r are quite similar among species and from day to day. He argued that there is a strong negative correlation between air humidity deficit and canopy resistance. Physiologically, when water demand in the air is high, stomata tend to close, and when the demand is low, the stomata open. In addition, the T_r for species with higher initial soil-water depletion rates or higher stomatal resistance declines more rapidly as water deficit increases. All these cause differences in T_r to be insignificant among species.

Roberts' (1999) arguments on the atmospheric humidity deficit–leaf resistance describe the general function of stomata, but the responses of stomata to humidity deficit are not necessarily the same among species. The decline in T_r may be faster for those species with higher initial T_r and greater stomatal resistance, but the total transpiration loss is not necessarily the same. In fact, significant differences in T_r among species have been verified

TABLE 9.11

Transpiration Rates for a Few Selected Species around the World

Species	Location	Age (year)	Method	Study Duration	T (mm/day)	Reference
Hardwoods						
Saltcedar	Buckeye, AZ	3–7	Lysimeter	6 years	2.9–5.9	Van Hylckama (1974)
Eucalyptus	New South Wales, Australia	160	Sap flux	02 to 03/77	0.8	Roberts et al. (2001)
		45			1.4	
		14			2.2	
E. Hardwoods	Duke Forest, NC	80–100	Sap flow	4 years	0.9	Oishi et al. (2008)
Sweetgum	Oak Ridge, TN	10	Sap flow	82 days	3.1	Wullschleger and Norby (2001)
Grevillea	Semiarid Kenya	Mature	Sap flow	108 days	1.4	Lott et al. (2003)
Ash	Winchester, United Kingdom	45	Meas/Est.	1990–1991	1.1	Roberts and Rosier (1994)
European beech	Central Slovakia	100	Sap flow	169 days	1.5	Střelcova et al. (2006)
Poplar, clone 1	Swanbourne, United Kingdom	2	Sap flow	91 days, summer	3.5	Allen et al. (1999)
Poplar, clone 2	Swanbourne, United Kingdom	2	Sap flow	91 days, summer	2.3	
Cottonwood	San Pedro River, Arizona	Mature	Sap flow	04 to 10/97	4.8	Schaeffer et al. (2001)
Floodplain forest	Dyje River, Czech Republic	Mature	Sap flow	05 to 09/98	3.3	Šír et al. (2008)
Conifers						
Douglas fir	British Columbia, Canada	7–31	Simulation	04 to 10/59 to 83	1.3	Spittlehouse (1983)
Loblolly pine	Duke Forest, NC	15	Sap flux	3 grow-seasons	1.92	Schäfer et al. (2002)
				Dormant seasons	0.77	
Ponderosa pine	Black Butte, Oregon	40	Sap flux	06 to 08/95 to 96	1.9	Ryan et al. (2000)
		290			1.3	
Norway spruce/	Uppsala, Sweden	50	Sap Flow	Growing seasons	0.9	Cienciala et al. (1997)
Scots pine	Eastern Pyrenees, Spain	60	Sap Flow	120 days, 03/04 each	2.16	Poytatos et al. (2005)
Norway spruce	Fichtelgebirge, Germany	40	Sap flux	04 to 10/95	0.97	Alsheimer et al., (1998)
		140		04 to 10/95	0.51	
Scots pine	East Anglia, United Kingdom	46	Meas/Est.	1 year	1.0	Gash and Stewart (1977)
Amabilis fir	Cascade Mt., Washington, DC	43	Sap flow	28 days, summer	1.7	Martin et al. (1997)

by soil-moisture depletion rates and by vegetation manipulation studies in many parts of the world. The T_r of 0.4–1.2 mm/day for those European forests as summarized by Roberts seem to be significant.

Canopy resistance increases rapidly in the fall as chlorophyll is destroyed and leaves change color. In the northeastern United States, leaves of hickory change color 3–4 weeks earlier than leaves of maple and oak do, which causes T_r to differ among species by several weeks. It is common for canopy resistance to cause differences in daily transpiration by 20% among species (Federer and Lash, 1978). Stomata are the biological valves that regulate water loss in a plant; their mechanisms in response to environmental stress among different species and in relation to transpiration require more study.

References

Allen, S.J., Hall, R.L., and Rosier, P.T.W., 1999, Transpiration by two poplar varieties grown as coppice for biomass production, *Tree Physiol.*, 19, 493–501.

Alsheimer, M. et al., 1998, Temporal and spatial variation in transpiration of Norway spruce stands within a forested catchment of the Fichtelgebirge, Germany, *Ann. Sci. For.*, 55, 103–123.

Bair, F.E., 1992, *Weather of U.S. Cities*, 4th edn., Gale Research, Detroit, MI.

Baumgartner, A., 1967, Energetic bases for differential vaporization from forest and agricultural lands, *International Symposium on Forest Hydrology*, Sopper, W.H. and Lull, H.W., Eds., Pergamon Press, Elmsford, NY, pp. 381–390.

Bengtsson, L., 1980, Evaporation from a snow cover—Review and discussion of measurements, *Nord. Hydrol.*, 11, 221–234.

Bernier, P.Y. and Swanson, R.H., 1993, The influence of opening size on snow evaporation in the forests of the Alberta foothills, *Can. J. For. Res.*, 23, 239–244.

Black, P.E., 1995, *Watershed Hydrology Laboratory Manual*, Pacific Crest Software, Corvallis, OR.

Campbell, G.S., 1977, *An Introduction to Environmental Biophysics*, Springer-Verlag, Heidelberg, Germany.

Campbell, G.S., 1985, *Soil Physics with Basic*, Elsevier Science Publication, New York.

Chang, M., 1975, On the reliability of using the 20-cm evaporation pan to estimate open-water evaporation in Taiwan, *J. Chin. Soil Water Conserv.*, 6, 36–48.

Chang, M., Clendenen, L.D., and Reeves, H.C., 1996, *Characteristics of a Humid Climate*, Center for Applied Studies in Forestry, Stephen F. Austin State University, Nacogdoches, TX.

Chang, M., Lee, R., and Dickerson, W.H., 1976, Adequacy of hydrologic data for application in West Virginia, Bull. #7, WRI-WVU-76–01, Water Research Institute, West Virginia University, Morgantown, WV.

Chang, M. et al., 1983, Soil-moisture regimes as affected by silvicultural treatments in humid East Texas, *Hydrology of Humid Tropical Regions with Particular Reference to the Hydrological Effects of Agricultural and Forestry Practice*, IAHS Publication No. 137, pp. 175–186.

Chang, M. et al., 1994, Air and soil temperatures under three forest conditions in East Texas, *Texas J. Sci.*, 46, 143–155.

Chang, M. et al., 1997, Evapotranspiration of herbaceous mimosa (*Mimosa strigillosa*), a new drought-resistant species in the southwestern United States, *Resour. Conserv. Recycl.*, 21, 175–184.

Cienciala, E. et al., 1997, Canopy transpiration from a boreal forest in Sweden during a dry year, *Agric. For. Meteorol.*, 86, 157–167.

Dyer, A.J., 1974, A review of flux-profile relationships, *Boundary-Layer Meteorol.*, 7, 363–372.

Douglas, E.M. et al., 2009, A comparison of models for estimating potential evapotranspiration for Florida land cover types, *J. Hydrol.*, 373, 366–376.

Dugas, W.A., Bland, W.L., and Arkin, G.F., 1985, Evapotranspiration measurements from different-sized lysimeters, *Advances in Evapotranspiration, Proceedings of the National Conference on Advances in Evapotranspiration*, Chicago, IL, ASAE, Washington, DC, pp. 208–215.

EDS, 1968, *Climatic Atlas of the United States*, U.S. Department of Commerce, Washington, DC.

Federer, C.A. and Lash, D., 1978, Simulated streamflow response to possible differences in transpiration among species of hardwood trees, *Water Resour. Res.*, 14, 1089–1097.

Fritschen, L.J. and Simpson, J., 1982, A pressure transducer for determining atmospheric pressure and evaporation with a hydraulic weighing lysimeter, *Agric. Meteorol.*, 26, 273–278.

Gardner, W.R., 1959, Solutions of the flow equation for the drying of soils and other porous media, *Soil Sci. Soc. Am. Proc.*, 23, 183–187.

Gash, J.H.C. and Stewart, J.B., 1977, The evaporation from Thetford Forest during 1975, *J. Hydrol.*, 35, 385–396.

Gazal, R.M. et al., 2006, Controls on transpiration in a semiarid riparian cottonwood forest, *Agric. For. Meteorol.*, 137, 56–67.

Granger, R.J., 1999, Partitioning of energy during the snow-free season at the Wolf Creek Research Basin, *Proceedings of the Wolf Creek Research Basin: Hydrology, Ecology, Environment*, Pomeroy, J.W. and Granger, R.J., Eds., Whitehorse, Yukon, Canada, March 5–7, 1998, National Water Research Institute, Saskatoon, SK, Canada, pp. 33–43.

Hamon, W.R., 1963, Computation of direct runoff amounts from storm rainfall, *Symposium on Theoretical Studies and Practical Methods of Forecasting the Yield of Rivers for Both Long and Short Terms (Except Floods)*, International Association of Scientific Hydrology, Rennes, France, pp. 25–62.

Hanks, R.J., 1958, Water vapor transfer in dry soils, *Soil Sci. Soc. Am. Proc.*, 22, 372–374.

Hanks, R.J. and Gardner, H.R., 1965, Influence of different diffusivity-water content relations on evaporation of water from soils, *Soil Sci. Soc. Am. Proc.*, 29, 495–498.

Harbeck, G.E., 1962, A practical field technique for measuring reservoir evaporation utilizing mass-transfer theory, U.S. Geological Survey Professional Paper 272-E, pp. 101–105.

Hargreaves, G.H., 1994, Defining and using reference evapotranspiration, *J. Irrig. Drain. Eng.*, 120, 1132–1139.

Hargreaves, G.L., Hargreaves, G.H., and Riley, J.P., 1985, Irrigation water requirements for Senegal River Basin, *J. Irrig. Drain. Eng.*, ASAE, III(3), 265–275.

Harrold, L.L. and Dreibelbis, F.R., 1958, Evaluation of agricultural hydrology by monolith lysimeters, *U.S. Dept Agric. Tech. Bull.* #1179, U.S. Government Printing Office, Washington, DC.

Hillel, D., 1982, *Introduction to Soil Physics*, Academic Press, New York.

Hornbeck, J.W., 1970, The radiant energy budget of clearcut and forest sites in West Virginia, *For. Sci.*, 16, 139–145.

Idso, S.B. and Jackson, R.D., 1969, Thermal radiation from the atmosphere, *J. Geophys. Res.*, 74, 5397–5403.

Iritz, Z. et al., 1999, Test of a modified Shuttleworth-Wallace estimate of boreal forest evaporation, *Agric. For. Meteorol.*, 98–99, 605–619.

Jalota, S.K. and Prihar, S.S., 1998, *Reducing Soil Water Evaporation with Tillage and Straw Mulching*, Iowa State University Press, Ames, IA.

Jones, J.A.A., 1997, *Global Hydrology*, Longman, London, U.K.

Kagawa, A. et al., 2009, Hawaiian native forest conserves water relative to timber plantation: Species and stand traits influence water use, *Ecol. Appl.*, 19(6), 1429–1443.

Kaufmann, M.R., 1985, Annual transpiration in subalpine forests: Large differences among four tree species, *For. Ecol. Manage.*, 13, 235–246.

Kaufmann, M.R., Edminster, C.E., and Troendle, C.A., 1982, Leaf area determination for sub-apline tree species in the central Rocky Mountains, Research Paper RM-238, USDA Forest Service.

Kirkham, R.R., Gee, G.W., and Jones, T.L., 1984, Weighing lysimeters for long-term water balance in investigations at remote sites, *Soil Sci. Am. J.*, 48, 1203–1205.

Kristensen, K.J., 1979, A comparison of some methods for estimation of potential evaporation, *Nord. Hydrol.*, 10, 239–250.

Lee, R., 1980, *Forest Hydrology*, Columbia University Press, New York.

List, R., 1971, *Smithsonian Meteorological Tables*, 6th edn., Smithsonian Institution Press, Washington, DC.

Lott, J.E. et al., 2003, Water use in a *Grevillearobusta*—Maize overstory agroforestry system in semi-arid Kenya, *For. Ecol. Manage.*, 180, 45–59.

Lu, J. et al., 2003, Modeling actual evapotranspiration from forested watersheds across the southeastern United States, *J. Am. Water Resour. Assoc.*, 39, 887–896.

Martin, J.M. et al., 1997, Crown conductance and tree and stand transpiration in a second-growth *Abiesamabilis* forest, *Can. J. For. Res.*, 27, 797–808.

McCuen, R.H., 1998, *Hydrologic Analysis and Design*, Prentice Hall, New York.

Monteith, J.L., 1965, Evaporation and environment, *Symp. Soc. Exp. Biol.*, 19, 205–234.

Monteith, J.L., 1985, Evaporation from land surface: Progress in analysis and prediction since 1948, *Advances in Evapotranspiration, Proceedings of the National Conference on Advances in Evapotranspiration*, Chicago, IL, ASAE, Washington, DC.

Monteith, J.L. and Unsworth, M.H., 1990, *Principles of Environmental Physics*, 2nd edn., Edward Arnold, London, U.K.

Oishi, A.C., Ram Oren, R., and Stoy, P.C., 2008, Estimating components of forest evapotranspiration: A footprint approach for scaling sap flux measurements, *Agric. For. Meteorol.*, 148, 1719–1732.

Penman, H.L., 1948, Natural evaporation from open water, bare-soil polt and grass, *Proc. R. Soc. Lond., A.*, 193, 120–145.

Penman, H.L., 1963, Vegetation and hydrology, Technical Communication No. 53, Commonwealth Agricultural Bureaux, Oxford, U.K.

Poyatos, R., Llorens, P., and Galla, F., 2005, Transpiration of montane *Pinussylvestris L.* and *Quercuspubescens Willd.* forest stands measured with sap flow sensors in NE Spain, *Hydrol. Earth Syst. Sci.*, 9, 493–505.

Pruitt, W.O., 1972, Factors affecting potential evapotranspiration, *Biological Effects in the Hydrological Cycle, Proceedings of the 3rd International Seminar on Hydrology Professors*, UNESCO and Agricultural Experiment Station, Purdue University, West Lafayette, IN, pp. 82–102.

Roberts, J., 1999, Plants and water in forests and woodlands, *Eco-Hydrology*, Baird, A.J. and Wilby, R.L., Eds., Routledge, New York, pp. 181–236.

Roberts, J. and Rosier, P.T.W., 1994, Comparative estimates of transpiration of ash and beech forest at a chalk site in southern Britain, *J. Hydrol.*, 162, 229–245.

Roberts, S., Vertessy, R., and Grayson, R., 2001, Transpiration from *Eucalyptus sieberi* (L. Johnson) forests of different age, *For. Ecol. Manage.*, 143, 153–161.

Rosenberg, N.J., Blad, B.L., and Verma, S.B., 1983, *Microclimate: The Biological Environment*, 2nd edn., John Wiley & Sons, New York, 495 pp.

Rutter, A.J. et al., 1971, A predictive model of rainfall interception in forest, I, derivation of the model from observations in a plantation of Corsican pine, *Agric. Meteorol.*, 9, 367–384.

Ryan, M.G. et al., 2000, Transpiration and whole-tree conductance in ponderosa pine trees of different heights, *Oceologia*, 124, 553–560.

Schaeffer, S.M., Williams, D.G., and Goodrich, D.C., 2001, Transpiration of cottonwood/willow forest estimated from sap flux, *Agric. For. Meteorol.*, 105, 257–270.

Schäfer, K.V.R. et al., 2002, Hydrologic balance in an intact temperate forest ecosystem under ambient and relative atmosphere CO_2 concentration, *Global Change Biol.*, 8, 895–911.

Shuttleworth, W.J., 1993, Evaporation, *Handbook of Hydrology*, Maidment, D.R., Ed., McGraw-Hill, New York, pp. 41–53.

Šir, M. et al., 2008, Measuring and modelling forest transpiration, *XXIVth Conference of the Danubian Countries, IOP Conference Series: Earth and Environmental Science*, 4, 012050, doi:10.1088/1755-1307/4/1/012050.

Spittlehouse, D.L., 1983, Determination of the year-to-year variation in growing season water use of a Douglas-fir stand, *The Forest–Atmosphere Interaction, Proceedings of the Forest Environmental Measurements Conference*, Oak Ridge, TN, Hutchison, B.A. and Hicks, B.B., Eds., D. Reidel Publ. Co., Boston, MA, pp. 235–254.

Storck, P. et al., 1995, Implications of forest practices on downstream flooding, phase II final report, Timber Fish Wildlife Service Report TFW-SH20-96-001, University of Washington, Seattle, WA.

Střelcova, K., Mindaš, J., and Škvarenina, J., 2006, Influence of tree transpiration on mass water balance of mixed mountain forests of the West Carpathians, *Biologia*, Bratislava, 61(Suppl. 19), S305–S310.

Sun, G. et al., 2002, A comparison of the watershed hydrology of coastal forested wetlands and the mountainous uplands in the Southern U.S., *J. Hydrol.*, 263, 92–104.

Sun, G. et al., 2008, Evapotranspiration estimates from eddy covariance towers and hydrologic modeling in managed forests in Northern Wisconsin, USA, *Agric. For. Meteorol.*, 148, 257–267.

Szilagyi, J. and Parlange, M.B., 1999, Defining watershed-scale evaporation using a normalized difference vegetation index, *J. Am. Water Resour. Assoc.*, 35, 1245–1255.

Teuling, A.J. et al., 2006, Observed timescales of evapotranspiration response to soil moisture, *Geophys. Res. Lett.*, 33, L23403, DOI: 10.1029/2006GL028178.

Thornthwaite, C.W., 1948, An approach toward a rational classification of climate, *Geogr. Rev.*, 38, 55–94.

Thornthwaite, C.W. and Mather, J.R., 1957, *Instructions and Tables for Computing Potential Evapotranspiration and the Water Balance*, Vol. XIX (3), Laboratory of Climatology, Centerton, NJ.

Trabucco, A. et al., 2008, Climate change mitigation through afforestation/reforestation: A global analysis of hydrologic impacts with four case studies, *Agric. Ecosyst. Environ.*, 126, 81–97.

Van Hylckama, T.E.A., 1974, Water use by saltcedar as measured by the water budget method, USGS Geological Survey Professional Paper 491-E.

Verstraeten, W.E. et al., 2005, Comparative analysis of the actual evapotranspiration of Jlemish forest and cropland using the soil water balance model WAVE, *Hydrol. Earth Syst. Sci.*, 9, 225–341.

Wullschleger, S.D. and Norby, R.J., 2001, Sap velocity and canopy transpiration in a sweetgum stand exposed to free-air CO_2 enrichment (FACE), *New Phytol.*, 150, 489–498.

Zhou, G. et al., 2008, Estimating forest ecosystem evapotranspiration at multiple temporal scales with a dimension analysis approach, *J. Am. Water Resour. Assoc.*, 44, 208–221.

10

Forests and Streamflow Quantity

A body of water moving over Earth's surface in a network of natural channels such as brooks, creeks, or rivers is called streamflow, a term that is often interchangeable in use with discharge or runoff. Water in stream channels comes from one or all of four components:

- Precipitation intercepted by stream channels
- Overland flow (surface runoff)
- Interflow (subsurface runoff)
- Baseflow (groundwater runoff)

It is the result of the integral effects of thermal, topographic, geological, edaphic, and vegetal factors acting upon storm events in a watershed system. Thus, streamflow is considered the residual of the hydrologic cycle.

10.1 Runoff Generation

Runoff refers to the portion of precipitation running over the land surface or through the soil profile to nearby stream channels. When rainfall intensity is greater than soil-infiltration rate, or rainfall amount exceeds soil-infiltration or percolation capacity, runoff occurs. However, many other environmental factors can either reduce the amount of rainfall reaching the mineral soil or increase the retention of water in the soil profile, consequently affecting the amount of runoff. Those factors are collectively called *watershed storage, retention, or abstraction*. The soil surface does not need to be saturated for overland flow to occur. Watershed storage can include canopy and litter interception, surface detention, surface depression, snowpack, soil moisture, ponds, stream bank storage, channel storage, and even evapotranspiration (ET).

10.1.1 Rainfall Intensity

Rainfall is first reduced by canopy and litter interception before reaching the mineral soil. On bare ground where no water is lost due to canopy and litter interception, a very small amount of water, usually no more than 1 mm, is used to wet the ground surface. This is the so-called *surface detention*. The rainfall that finally reaches the mineral soil is termed *effective rainfall*, P_e. Effective rainfall is affected by rainfall characteristics, vegetation, and climatic conditions; it ranges from 70% to 80% of gross rainfall in forested areas.

Rainfall intensity, expressed in rainfall per unit of time, can be measured by a recording rain gauge. Some hydrologic studies may focus on maximum 1 h intensity, maximum

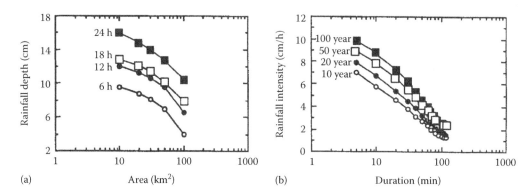

FIGURE 10.1
Typical relationships for (a) rainfall depth–area–duration and (b) rainfall intensity–duration–return periods.

30 min intensity, or intensities of other duration (D) in centimeters per hour. Maximum 1 h intensity is the greatest amount of rainfall in a segment of any consecutive 60 min during the entire storm period. If the maximum consecutive 30 min rainfall (P_t) during a storm is 2.5 cm, then the maximum 30 min rainfall intensity in centimeters per hour is

$$i = \frac{P_t(60 \, \text{min}/\, \text{h})}{D \, \text{min}} = \frac{(2.5 \, \text{cm})(60 \, \text{min}/\, \text{h})}{30 \, \text{min}} = 5.0 \, \text{cm}/\text{h} \qquad (10.1)$$

where i is rainfall intensity in centimeters per hour for a duration D in minutes.

The depth of rainfall generally increases with duration and decreases with the size of area, while maximum rainfall intensity decreases with increasing duration and increases with increasing return period (Figure 10.1). A return period T is the average number of years during which a given storm or hydrologic event X would be expected to equal or exceed value x. Thus, if a 1 h storm (X) with total rainfall (x) of 50 mm or greater is expected to be a 50 year return period, it can be expressed as T $(X \geq x) = 50$ year, and the probability (P_r) of that 1 h storm event occurring in any given year is 2%, or

$$P_r(X \geq x) = \frac{1}{T} \qquad (10.2)$$

$$P_r(1 \, \text{h storm} \geq 50 \, \text{mm}) = \frac{1}{50} = 0.02$$

On the other hand, the probability that event X would not occur in any given year, which is the same as the P_r for a hydrologic event X equal to or less than x, is 98%:

$$1 - P_r(X \geq x) = P_r(X \leq x) = 1 - \left(\frac{1}{T}\right) = 1 - \frac{1}{50} = 0.98 \qquad (10.3)$$

10.1.2 Soil Infiltration

Infiltration refers to the entry of water into the soil, a combined process of capillary attraction and gravitation along with pressure due to water ponding at the ground surface.

The infiltration rate is high at the initial stage and declines with time as soil voids such as animal burrows, root tunnels, interstices, and macro- and micropores gradually fill with water. It will eventually decline to a constant at which the infiltration rate is equal to the rate of water drained out through the soil profile by gravity. This process of draining water to deeper layers is called *percolation*.

Water movement into and through soil profiles is affected by a variety of factors reflecting the surface and subsurface conditions and flow characteristics. Surface conditions such as type of vegetation cover, land management practices, roughness, crusting, cracking, surface temperature, slope, and chemicals have a significant impact on surface ponding, overland flow velocity, and ability of the water to enter the soil. Conditions under the ground can include soil texture, structure, organic matter content, depth, compaction, voids, layering, water content, groundwater table, and root system. These factors affect soil water-holding capacity and ability of water to move. For a given soil, the infiltration in a forest can be many times greater than that over bare ground. Also, infiltration is greater for warmer water containing less sediment.

10.1.2.1 Soil Water

A body of soil consists of solid particles, water solution, and gases. Thus, the total volume of soil V_t is the sum of the volumes of solid particles (V_s), water solution (V_w), and gases (V_a)

$$V_t = V_s + V_w + V_a \tag{10.4}$$

and the sum of V_w and V_a is the total volume of voids or pore spaces. The ratio between pore spaces and total soil volume is *soil porosity*, S_p.

$$S_p = \frac{V_w + V_a}{V_t} \tag{10.5}$$

Coarse soils have larger but less total volume of spaces and consequently smaller porosity, while fine soils have smaller but more spaces and consequently higher porosity. Thus, the rate of water infiltration is higher, movement is faster, and water-holding capacity is smaller for coarse soils. Soil water can be volumetrically expressed by

$$\theta_v = \frac{V_w}{V_t} \tag{10.6}$$

or gravimetrically expressed by

$$\theta_w = \frac{M_w}{M_s} \tag{10.7}$$

where
M_w is the mass of water content
M_s is the mass of dry soil

Converting θ_w into θ_v requires a multiplier, bulk density of the soil d_b, or

$$\theta_v = (\theta_w)(d_b) \tag{10.8}$$

and

$$d_b = \frac{M_s}{V_t} \qquad (10.9)$$

Soils with more pore spaces have lower bulk density. Thus, the bulk density of fine-textured soils is lower than that of coarse-textured soils, and organic soils have a bulk density even lower than that of fine soils, usually less than $1\,g/cm^3$. Prolonged intensive cropping (Brady, 1990), forest cutting, and site preparation (Chang et al., 1994) tend to increase bulk density, especially in the surface layer.

10.1.2.2 Infiltration Capacity

Under a given soil condition, the maximum rate at which a given soil can absorb water is referred to as *infiltration capacity*, f_m. Assuming $f =$ actual infiltration rate and $i =$ rainfall intensity, then $f = f_m$ when $i \geq f_m$. In this case, the difference $i - f_m$ appears to be surface runoff. On the other hand, if $i < f_m$, then $f = i$, and all water enters the soil as soil water or percolation water. No overland flow occurs.

It is often important to know infiltration capacity f_m and total infiltration in runoff generation studies. The many infiltration models that have been developed are described by Skaggs and Khaleel (1982), Rawls and Brakensiek (1983), Gupta (1989), and Rawls et al. (1993). The Green–Ampt infiltration model, the Horton equation, and the SCS runoff curve number (RCN) approach are three popular models with wide application.

10.1.2.2.1 The Green–Ampt Model
For a homogeneous soil profile with (Figure 10.2)

1. An initial water content θ_i
2. A pressure head due to ponding H_o
3. A capillary head at the wetting front H_f
4. A depth to the wetting front L

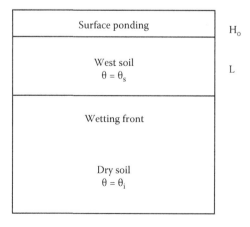

FIGURE 10.2
Soil-moisture profile in the Green–Ampt infiltration model.

infiltration capacity f_m in centimeters per hour can be estimated by the following equation (Green and Ampt, 1911):

$$f_m = \frac{K(H_o + H_f + L)}{L} \tag{10.10}$$

where

$\quad$ K is *saturated hydraulic conductivity* (cm/s)

$\quad$ H_o, H_f, and L are in cm

The sum of H_o, H_f, and L is the *total hydraulic head* H. Since the change in ratio between H and L, or $(\delta H)/(\delta L)$, is the hydraulic gradient, Equation 10.10 is an application of Darcy's law for the velocity of flow V through an unsaturated soil:

$$V = K\left(\frac{\delta H}{\delta L}\right) \tag{10.11}$$

Assuming the ponding head is close to zero and the total infiltration (in centimeters) for a specific period is F where $F = (\theta_s - \theta_i)L$, then Equation 10.10 can be rearranged to become

$$f_m = K\left[1 + \frac{(\theta_s - \theta_i)H_f}{F}\right] \tag{10.12}$$

$$f_m = K\left[1 + \frac{N_s}{F}\right] \tag{10.13}$$

where

$\quad$ θ_s is the saturated soil-water content, a value analogous to the soil porosity but in different units (g/cm^3 vs. percent)

$\quad$ the difference $(\theta_s - \theta_i)$ is the soil-water saturation deficit

$\quad$ N_s is the effective capillary pressure head (in cm) and is equal to $(\theta_s - \theta_i)H_f$

The parameter θ_s can be estimated by (Kidwell et al., 1997)

$$\theta_s = 0.9\left(1 - \frac{d_b}{2.65}\right) \tag{10.14}$$

where

$\quad$ d_b is the soil bulk density (g/cm^3)

$\quad$ 2.65 is the particle density (g/cm^3) for most soils

The integration of Equation 10.13 provides an estimate of cumulative infiltration for a time period t as

$$(K)(t) = F - N_s \ln\left(1 + \frac{F}{N_s}\right) \tag{10.15}$$

In summary, under a steady rainfall condition, infiltration can be treated at three stages of time t, i.e., before surface ponding, at ponding t_p, and after ponding:

1. From the beginning to surface ponding, or $t < t_p$, the actual infiltration rate is equal to the rainfall intensity $(f = i)$ and the cumulative infiltration is equal to cumulative rainfall, $F = \Sigma(f \times t) = \Sigma(i \times t)$.
2. At the time of ponding, or $t = t_p$, the actual infiltration rate is the infiltration capacity and is equal to the rainfall intensity $(f = f_m = i)$. The cumulative infiltration at ponding F_p can be calculated by substituting i for f_m in Equation 10.12 and rearranging the equation to become

$$F_p = \frac{N_s}{[(i/K) - 1]} \tag{10.16}$$

$$t_p = \frac{F_p}{i} \tag{10.17}$$

3. For the time beyond ponding, or $t > t_p$, the infiltration capacity f_m is calculated by Equation 10.13 and the cumulative infiltration can be calculated by the following equation (Mein and Larson, 1971; Skaggs and Khaleel, 1982):

$$K\left(t - t_p + t_p'\right) = F - N_s \ln\left[1 + \frac{F}{N_s}\right] \tag{10.18}$$

where t_p' is the equivalent time to infiltrate volume F_p under initially ponded surface conditions.

Example 10.1

Consider a soil with these input data: $K = 0.2 \, cm/h$, $\theta_s = 0.40$, $\theta_i = 0.15$, $H_f = 16 \, cm$, and $i = 1.20 \, cm/h$. Determine the total infiltration and net runoff of the soil.

1. From Equation 10.16, the cumulative infiltration (F_p) at the time of surface ponding is

$$F_p = \frac{N_s}{(i/K) - 1} = \frac{(\theta_s - \theta_i)H_f}{(i/K) - 1}$$

$$= \frac{16(0.40 - 0.15)}{[(1.20/0.20) - 1]} = 0.80 \, cm$$

2. From Equation 10.17, the ponding time is

$$t_p = \frac{F_p}{I} = \frac{0.80}{1.20} = 0.67 \, h$$

3. From Equation 10.15, the equivalent time to F_p (t_p'—time required to reach F_p under ponding conditions) is

$$(0.20)(t_p') = 0.80 - 16(0.40 - 0.15)\ln\left[1 + \frac{0.8}{16(0.40 - 0.15)}\right]$$

$$t_p' = 0.35\,h$$

4. From Equation 10.18, the cumulative infiltration F after ponding is

$$0.20(t - 0.67 + 0.35) = F - 16(0.40 - 0.15)\ln\left[1 + \frac{F}{16(0.40 - 0.15)}\right]$$

$$t = 0.32 + 5F - 0.20\,\ln(1 + 0.25F) \tag{10.19}$$

A plot of F versus t for a graphical solution of Equation 10.19 is given in Figure 10.3. Values of F for different time t can be read directly from the graph.

5. The overland flow (RO) at each sequential hour is given in Table 10.1.

10.1.2.2.2 Horton's Infiltration Model

Horton (1937) suggested that the infiltration capacity f_m (in mm/h) at time t (h) be described by

$$f_m = f_c + (f_o - f_c)\,e^{-kt} \tag{10.20}$$

where
f_o is the initial infiltration capacity (mm/h)
f_c is the final equilibrium infiltration capacity (approaching constant, mm/h)
k is the recession constant or the decreasing rate (h⁻¹)

Integrating f_m in Equation 10.20 yields the total infiltration F (mm), or

$$F = \int_0^t f_m dt = t(f_c) + \left(\frac{f_o - f_c}{-k}\right)\left(e^{-kt} - 1\right) \tag{10.21}$$

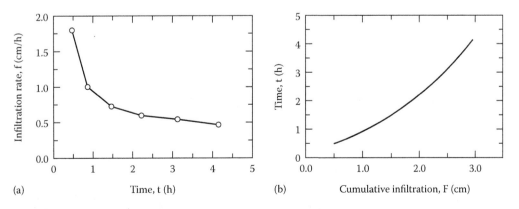

(a)　Time, t (h)　　(b)　Cumulative infiltration, F (cm)

FIGURE 10.3
Graphical solution for Equation 10.19 (b) and infiltration rate versus time (a).

TABLE 10.1

Cumulative Infiltration and Overland Flow Estimated by
the Green–Ampt Approach

Time, t (h)	F (cm)	Δt (h)	ΔF (cm)	PT (cm)	RO (cm)
(1)	(2)	(3)	(4)	(5)	(6)
0.67	0.80				
1.00	1.10	0.33	0.30	0.40	0.10
2.00	1.90	1.00	0.80	1.20	0.40
3.00	2.47	1.00	0.57	1.20	0.63
4.00	2.95	1.00	0.48	1.20	0.72
Total					1.85

Note: (1), time; (2), cumulative infiltration, read from Figure 10.1;
(3), time interval between consecutive times in column 1; (4),
changes between consecutive F in column 2; (5), column
$3 \times i = (\Delta t)(i)$; (6), column 5 − column 4.

Values of f_o and f_c for a given soil can be graphically determined by plotting sequential measurements of infiltration capacity versus time, and the constant k can be calculated by rearranging Equation 10.20, or

$$k = \frac{[\ln(f_o - f_c) - \ln(f_m - f_c)]}{t}$$

(10.22)

Equation 10.20 applies under the condition that rainfall intensity (i, mm/h) is greater than f_m. If $i < f_m$, then $f_m = i$. A soil with $f_o = 88$ mm/h, $f_c = 67$ mm/h, and k = 1.4 h⁻¹, the infiltration capacity at the end of 3 h, $f_m = 67.3$ mm/h, and the cumulative infiltration F = 215.8 mm (Figure 10.4). Note that the actual total infiltration may be less than 215.8 mm if rainfall intensity is not greater than f_m during the entire 3 h period.

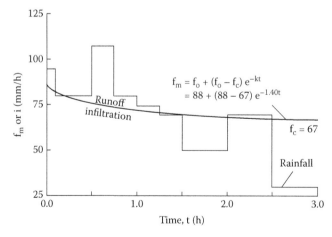

FIGURE 10.4
A Horton's infiltration capacity curve (f_m) and its associated histogram of rainfall intensity (i) versus time (t).

10.1.2.2.3 The SCS RCN Approach

The SCS (Soil Conservation Service, now renamed Natural Resources Conservation Service) estimates watershed abstractions by soil type, soil-moisture condition, land use, and management. The method has wide application in predicting storm runoff. See Section 10.2.3.

10.1.3 Mechanics of Runoff

When effective rainfall exceeds accumulated infiltration, the excess rainfall will run off from the land surface or pond on the surface. At the initial stage, soil-water deficit is relatively high, and a majority of rainfall will enter the soil. Gravity and capillary forces cause the infiltrated water to percolate down to deeper layers or divert laterally from large pores to smaller pores. As pores fill with water, the increase in soil-water content causes a decline in soil-water deficit and consequently slows the infiltration rate. Also, the presence of impermeable or semipermeable rocks or clays in the soil profile can block and slow water movement. Eventually, infiltration will be equal to percolation at a constant rate.

Infiltrated water can become surface runoff again as it flows laterally and downslope or routes to nearby stream channels as subsurface runoff. It is difficult to separate surface and subsurface runoff in hydrologic analysis. The combination of channel rainfall, surface runoff, and subsurface runoff is called direct runoff. Direct runoff implies direct response of streamflow to storm rainfall over a relatively short time frame. Water that flows in stream channels during periods of no rainfall (baseflow) comes from groundwater. The sum of direct runoff and baseflow is total streamflow.

Rangeland watersheds usually have shallow and compact soils with low infiltration capacities, consequently exhibiting a quick streamflow response dominated by surface runoff. On the other hand, forested watersheds are characterized by deep and loose soils, a fluffy floor, complex root systems, big canopies, and high infiltration capacities. As a result, streamflow responses are slow and small in general, and subsurface runoff predominates.

Surface runoff is responsible for producing quick streamflow discharge. However, forested watersheds usually have little surface runoff. Here quick streamflow discharge is largely generated from subsurface flow processes, as described by the variable source area (Hewlett and Troendle, 1975). As a storm proceeds, the increase in watershed storage causes the rise of stream channels and the expansion of saturated areas adjacent to stream channels. These wet and saturated areas contribute water to streamflow directly and efficiently from storm rainfall. The variable source area contributing stormwater increases as storm duration increases but shrinks to original size after the storm stops. At that time, water in underground pores, decayed root channels, animal burrows, and soil tubes will gradually seep out to prolong direct runoff (Cheng, 1988). Interflow in forested watersheds could be more than 90% of rainfall during a calendar quarter (Beasley, 1976).

10.2 Watershed Discharges

The discharge of streamflow varies greatly with watershed and time. It is often characterized by flow rates of various statistical parameters: crest height, volume, duration, frequency, or a combination of these parameters. The fluctuation of streamflow is important to water supply and floodplain management and can be used as an indication of the effectiveness of watershed management conditions.

10.2.1 Hydrograph

Streamflow, either in discharge rate or stage, is often plotted against time for graphical illustrations, studies, and analyses. Such a plotting is called a *hydrograph*, and the area covered under the hydrograph is total runoff over the specified period of time. Several features can be identified in a typical hydrograph (Figure 10.5).

10.2.1.1 Prominent Features

The beginning of storm runoff (a) is the beginning of the rising limb and usually falls in time behind the storm rainfall due to watershed storage. The lag in time is longer for forested than for rangeland watersheds. Large watersheds can take a few days to respond to rainfall.

The rising limb (a–d) represents the increase in watershed discharge. The shape of the rising limb is affected by the watershed characteristics and storm duration, intensity, and distribution.

The crest (d–f) runs from the inflection point on the rising limb to the inflection point on the recession limb, representing the arrival of water from all hydrologically active parts of the watershed to the outlet. This is the highest concentration of storm runoff. The end of the crest indicates the end of direct runoff to the stream outlet.

The recession limb (f–k) represents the contribution of water from watershed storage, including detention storage, interflow, and groundwater runoff. The recession limb, representing the draining-off process, is independent of storm characteristics.

The duration of direct runoff (a–k) is referred to as the time base, the sum of the time (from the beginning) to peak flow, and the time (from the peak flow) to the end of the direct runoff. An intense storm over a small, bare watershed often results in a rapid discharge in which the time to peak flow and time base are short.

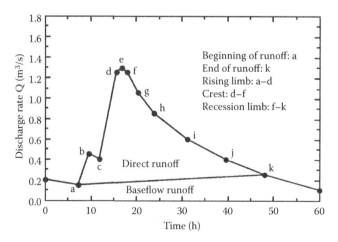

FIGURE 10.5
A typical hydrograph with a separation of direct and baseflow runoff.

10.2.1.2 Direct Runoff versus Baseflow

The area under a hydrograph represents direct runoff and baseflow. Few methods can be used for separating direct runoff from baseflow (Pilgrim and Cordery, 1993); the simplest one is to draw a straight line connecting the beginning of runoff on the rising limb and the end of direct runoff on the recession limb. The area above this line is direct runoff, and under the line is baseflow. The volume of direct runoff can be calculated first by breaking the direct runoff hydrograph into segments of uniform flow rate. Then multiply the average flow rate (m³/s) of each segment by its time interval (s) to obtain the discharge volume (m³) of that segment and sum the discharges of all segments to get the total discharge under the direct runoff hydrograph. Finally, the total discharge volume (m³) is divided by watershed area (m²) and multiplied by 100 to convert it into average depth in centimeter so that it can relate to storm rainfall (Table 10.2). The end of direct runoff can be approximated by (Linsley et al., 1975)

$$N = 0.827 \, A^{0.2} \tag{10.23}$$

where
   N is the time in days from the peak of a hydrograph to the end of direct runoff
   A is watershed area (km²)

10.2.2 Factors Affecting Watershed Discharge

Like rainfall, streamflow fluctuates with time within watersheds in response to watershed climate, topography, and land-use conditions.

TABLE 10.2

Calculations of Direct Runoff for the Hydrograph in Figure 10.5

			Total Discharge		Baseflow		Direct Flow		
Time (h)	Segment	Duration (h)	Reading (m³/s)	Ave. (m³/s)	Reading (m³/s)	Ave. (m³/s)	Ave. (m³/s)	Vol (m³)	Remark
(1)	(2)	(3)	(4)	(5)	(6)	(7)	(8)	(9)	(10)
7.2	A	0	0.15	—	0.150	—	0.00	0.00	BDR
9.2	B	2.4	0.45	0.300	0.165	0.1575	0.1425	1,231.20	
12.0	c	2.4	0.40	0.425	0.178	0.1715	0.2535	2,190.24	
15.6	d	3.6	1.25	0.825	0.189	0.1835	0.6415	8,313.84	
16.8	e	1.2	1.30	1.275	0.196	0.1925	1.0825	4,676.40	Peak
18.0	f	1.2	1.25	1.275	0.205	0.2005	1.0745	4,641.84	
20.4	g	2.4	1.05	1.150	0.214	0.2095	0.9405	8,125.92	
24.0	h	3.6	0.85	0.950	0.223	0.2185	0.7315	9,480.24	
31.2	I	7.2	0.60	0.725	0.232	0.2275	0.4975	12,895.20	
39.6	j	8.4	0.40	0.500	0.241	0.2365	0.2635	7,968.24	
48.0	k	8.4	0.25	0.325	0.250	0.2455	0.0795	2,404.80	EDR
Total								61,927.92	

Note: BDR, the beginning of direct runoff; EDR, the end of direct runoff. Column 5, the average between two proceeding segments in column 4; Column 7, the average between two proceeding segments in column 6; Column 8 = column 5 − column 7; Column 9 = column 8 × column 3 × 3600.

10.2.2.1 Climate

Precipitation is the input in the hydrologic system. Its intensity, depth, duration, type, distribution within the watershed, and direction of storm movement impose direct and significant impacts on streamflow regimes and hydrograph shape. For a given duration, the hydrograph produced from storms of greater intensity will exhibit a higher peak flow and a greater runoff volume. A storm that occurs near the watershed outlet will result in a rapid rise and sharp peak flow. On the other hand, the hydrograph will exhibit a longer time base, lower peak flow, and slower rise to peak if the storm occurs in the upper parts of the watershed. In general, snowmelt tends to exhibit a lower and broader runoff hydrograph than does rainfall.

Lower atmospheric humidity, higher air temperature, solar radiation, and wind speed tend to increase ET and reduce water available for runoff. As a result, streamflow generally decreases with air temperature and increases with rainfall, as illustrated by an empirical equation for estimating annual streamflow (Q, mm) for the 80 km² La Nana Creek watershed in Nacogdoches, east Texas (Chang and Sayok, 1990):

$$Q = -396.16 + 194.23\left(\frac{P_t}{T_a^2}\right) \tag{10.24}$$

where
P_t is annual rainfall (mm)
T_a is annual air temperature (°C)

In areas where snow is the major type of precipitation, a sudden rise in air temperature in early spring often causes rapid melting of snowpacks, which can result in devastating floods (Chang and Lee, 1976).

10.2.2.2 Topography

Watershed topography can be characterized by a variety of parameters such as slope, shape, elevation, size, relief ratio, and stream density and frequency (Chang and Boyer, 1977; Chang, 1982). These parameters affect streamflow and influence the shape of the hydrograph through watershed storage, runoff speed, infiltration, and soil-water content. A watershed with higher elevations implies lower temperature, less ET, greater rainfall, steeper slope, and shallower soil depth, which lead to the production of more runoff. A steep slope can make soil infiltration lower and overland runoff greater and faster. The hydrologic behavior of small watersheds tends to be different from that of large watersheds. A small watershed is very sensitive to high-intensity rainfall of short duration and land use. Thus, overland flow is a dominating factor affecting the peak flow, and the streamflow hydrograph is characterized by a sharp rise to peak flow and a rapid fall in recession. In large watersheds, the average watershed slope, the chances for a single storm to cover the entire watershed, and the average storm depth and intensity are smaller. Thus, the streamflow hydrograph is dominated by watershed and channel storage, and it exhibits a broader crest and longer time base (Figure 10.6).

The positive effects of watershed area A (km²) on streamflow Q (m³/s) lead it to be perhaps the most widely used topographic parameter in hydrologic analyses (Gingras et al., 1994):

$$Q = aA^b \tag{10.25}$$

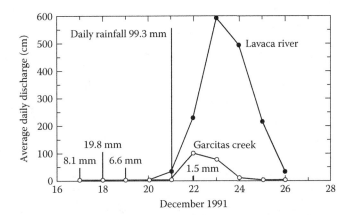

FIGURE 10.6
Hydrographs produced from two side-by-side watersheds in Texas, one large (Lacaca River at Edna, 2116 km²) and one small (Garcitas Creek near Inex, 238 km²); total rainfall was 135.4 mm, covering the entire watersheds.

where a and b are empirical constants. In mountainous regions, however, the predictability of Equation 10.25 can be improved by introducing watershed mean elevation or other topographic parameters into the analysis. For example, the following equation was developed from 36 gaging stations to estimate mean annual flood Q (m³/s) as a function of watershed area A (km²) and watershed mean elevation Elev (m) in New Hampshire and Vermont (Dingman and Palaia, 1999):

$$Q = 0.5154 \left(A^{0.792} \right) \left[10^{0.000833(\text{Elev})} \right] \tag{10.26}$$

The two variables explain about 93% of the flood variation with a standard error of 38%.

10.2.2.3 Land Use

Plant cover, through the unique effects of its canopy and root systems on precipitation interception, infiltration, percolation, surface detention and roughness, transpiration, snow accumulation, and snowmelt, is a very important factor in the hydrologic cycle. Forest cutting usually results in an increase in water yield. The increase is most significant if the cutting is conducted in watersheds with conifer species, followed by hardwoods, chaparrals, and grasses. A forested watershed is expected to produce a hydrograph with lower peak flow, smaller volume of runoff, and broader time base than if the watershed had been cleared, cultivated, and pastured.

Changes in hydrographs due to forests can be illustrated by the comparative hydrographs from two adjacent watersheds (elevations 2880–3536 m) in the Fraser Experimental Forest, Colorado (Figure 10.7). One of the watersheds, Lexen (124 ha), was used as the control to evaluate the effects of forest cutting on streamflow in the treatment watershed, North Fork, Deadhorse Creek (41 ha). The subalpine forest in the treatment watershed was reduced by 36% through patch cut from 1977 to 1978. Prior to the harvest, the annual streamflow for the treatment watershed was 37% of that of the control watershed (15.0 vs. 40.4 cm). After the treatment, the streamflow increased to 55% of the control watershed (23.1 vs. 41.9 cm). The increase in flows due to forest removal appears on the rising limb of the hydrograph, mainly in May (Troendle, 1983).

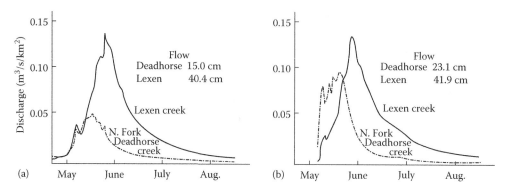

FIGURE 10.7

Comparative hydrographs for North Fork Deadhorse Creek (treatment) and Lexen Creek watersheds in the Fraser Experimental Forests, Colorado: (a) before treatment and (b) after 36% of the subalpine forest in the North Fork Creek was cleared in patches of 1.2 ha each. (Adapted from Troendle, C.A., *Water Resour. Bull.*, 19, 359, 1983.)

In southern pine-dominated East Texas, a regional analysis showed that a reduction of watershed forest area by 10%, other things being equal, would increase annual streamflow about 20 mm (Chang and Waters, 1984). Differences between full forest-cover watersheds and watersheds devoid of vegetation could be as much as 200 mm/year. This difference was much less than that reported in western Oregon (Harr, 1983), greater than that in Arizona (Hibbert, 1983), and about the same as that in West Virginia (Kochenderfer and Aubertin, 1975). For flood flows, the inverse effect of forest can be expressed by the general equation as follows:

$$Q_{d,t} = a\frac{A^b}{F^c}, \quad F > 0\% \tag{10.27}$$

where

$Q_{d,t}$ is the maximum flood discharge (m³ s/day) for a period of d days with a return period of t years
A is watershed area (km²)
F is percent forested area
a, b, and c are coefficients, varying with d, t, and region

The equation shows that an increase in percent forest area causes a decrease in the maximum peak discharge (Figure 10.8). However, the effect is more pronounced for floods of short return periods and for watersheds of small sizes.

10.2.3 Estimation of Streamflow

Streamflow data are usually available only at a few stations or are inadequate in water-resource applications, and many hydrologic projects are too urgent to wait for actual measurements. In any of these cases, streamflow data have to be estimated using techniques such as

1. Hydrograph analysis to estimate runoff volume and peak flow from rainfall
2. Statistical analysis to estimate runoff from environmental factors or other sites

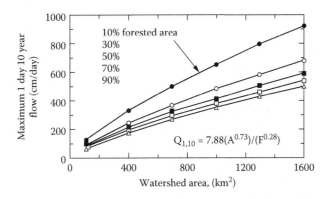

FIGURE 10.8
Maximum 1 day 10 year flow as a function of watershed area and percentage of forest cover. (After Chang, M. and Waters, S.P., *Water Resour. Bull.*, 20, 713, 1984.)

3. Time-series analysis to generate sequential data into the future

4. Hydrologic simulation to generate data based on physical models

Discussion of these techniques is beyond the scope of this book; only a few well-accepted methods are introduced here.

10.2.3.1 SCS RCN Model (Storm Runoff Volume)

The Soil Conservation Service's (1979) curve number approach is one of the most commonly used methods for estimating storm runoff volume. The approach estimates direct runoff Q in inches from storm rainfall P in inches and watershed storage S by

$$Q = \frac{(P - I_a)^2}{(P - I_a + S)} \qquad (10.28)$$

where
I_a is the initial abstraction (in.)
S is the maximum potential difference between P and Q

Both I_a and S are affected by factors such as vegetation, infiltration, depression storage, and antecedent moisture conditions (AMCs). Empirical evidence shows $I_a = 0.2S$; Equation 10.28 becomes

$$Q = \frac{(P - 0.2S)^2}{(P + 0.8S)} \qquad (10.29)$$

The parameter S is defined by

$$S = \left(\frac{1000}{RCN}\right) - 10 \qquad (10.30)$$

where RCN is an arbitrary runoff curve number ranging from 0 to 100. For RCN = 100, S = 0, and P = Q, which is the height potential for causing direct runoff. In other words, there is no watershed storage and all P becomes Q. As RCN approaches 0, S approaches infinity, and all rainfall is stored in the watershed. In this case, there is no occurrence of Q.

The RCN describes the land use, soil-infiltration rate, and soil-moisture conditions prior to the storm event. The SCS has divided soils into four hydrological groups in accordance with their infiltration rates and soil textures (Table 10.3) and defined antecedent moisture conditions (AMCs) into three categories (Table 10.4). The AMC is based on the season and 5 day antecedent precipitation. Once the land use and the hydrologic soil group of the watershed are identified, the RCN can be obtained from Table 10.5 for AMC II, which is an

TABLE 10.3

Hydrologic Soil Groups

Group	Minimum Infiltration Rate (in./h)	Soils
A	0.30–0.45	Deep sand, deep loess, aggregated silts
B	0.15–0.30	Shallow loess, sandy loam
C	0.05–0.15	Clay loam, shallow sandy loam, soils low in organic content, and soils usually high in clay
D	0.00–0.05	Soils that swell significantly when wet, heavy plastic clays, and certain saline soils

Source: Soil Conservation Service, *National Engineering Handbook, Section 4: Hydrology*, U.S. Department of Agriculture, Washington, DC, 1979.

TABLE 10.4

Estimation of Runoff Curve Number for Various AMC

RCN for AMC II	Corresponding RCN		RCN for AMC II	Corresponding RCN	
	AMC I	AMC III		AMC I	AMC III
100	100	100	45	27	65
95	87	99	40	23	60
90	78	98	35	19	55
85	70	97	30	15	50
80	63	94	25	12	45
75	57	91	20	9	39
70	51	85	15	7	33
65	45	83	10	4	26
60	40	79	5	2	17
55	35	75	0	0	0
50	31	70			

Source: Soil Conservation Service, *National Engineering Handbook, Section 4: Hydrology*, U.S. Department of Agriculture, Washington, DC, 1979.

Note: AMC I: Optimum soil-moisture condition from about lower plastic limit to wilting point; the 5-day AMC is less than 0.5 and 1.4 in. for dormant and growing seasons, respectively. AMC II: Average value for annual floods; the 5 day AMC is 0.5–1.1 and 1.4–2.1 in., in that order. AMC III: Heavy rainfall or light rainfall and low temperature within 5 days prior to the given storm; the 5 day AMC is greater than 1.1 and 2.1 in., in that order.

TABLE 10.5

SCS's (1979) Runoff Curve Number for Hydrologic Soil-Cover
Complexes, Antecedent Moisture Condition II

Land Use	Treatment or Practice	Hydrologic Condition	A	B	C	D
Fallow	Straight row	—	77	86	91	94
Row crops	Straight row	Poor	72	81	88	91
	Straight row	Good	67	78	85	89
	Contoured	Poor	70	79	84	88
	Contoured	Good	65	75	82	86
	Contoured and terraced	Poor	66	74	80	82
	Contoured and terraced	Good	62	71	78	81
Small grain	Straight row	Poor	65	76	84	88
		Good	63	75	83	87
	Contoured	Poor	63	74	82	85
		Good	61	73	81	84
	Contoured and terraced	Poor	61	72	79	82
		Good	59	70	78	81
Close-seeded legumes, or rotation meadow	Straight row	Poor	66	77	85	89
	Straight row	Good	58	72	81	85
	Contoured	Poor	64	75	83	85
	Contoured	Good	55	69	78	83
	Contoured and terraced	Poor	63	73	80	83
		Good	51	67	76	80
Pasture or range		Poor	68	79	86	89
		Fair	49	69	79	84
		Good	39	61	74	80
	Contoured	Poor	47	67	81	88
	Contoured	Fair	25	59	75	83
	Contoured	Good	6	35	70	79
Meadow		Good	30	58	71	78
Woods		Poor	45	66	77	78
		Fair	36	60	73	79
		Good	25	55	70	77
Farmsteads		—	59	74	82	86
Roads	Dirt	—	72	82	87	89
	Hard surface	—	74	84	90	92

average value for initial abstraction equal to 0.2S. Using Table 10.4, the RCN for the AMC II condition can be converted into RCN for AMC I or III conditions, if necessary. Solutions of Equation 10.29 are given in Figure 10.9.

The determination of RCN in forested areas is more difficult than that in agricultural areas because of the variation of plant species, ages, heights, canopy coverage, and root systems. Some typical RCN values for forested conditions are given in Table 10.6.

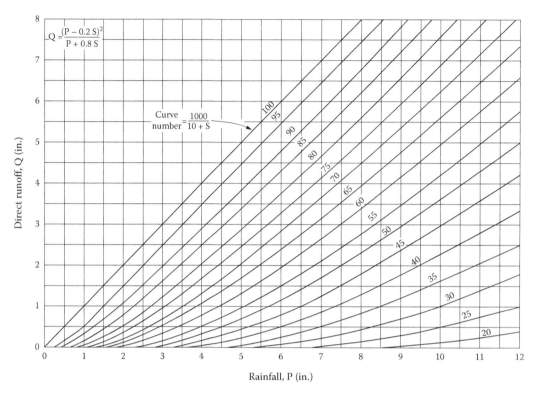

FIGURE 10.9
Solution of SCS runoff equation (Equation 10.29) for rainfall $P = 0\text{--}12$ in.

10.2.3.2 Rational Equation (Storm Peak Flow)

Estimates of peak discharges are required for the design of culverts, drainage works, soil-conservation works, spillways of farm ponds, and small bridges. Although several empirical and hydrographic methods have been developed for estimating peak discharges from storm events, the rational equation is perhaps the most commonly used and simplest one in water-resource applications.

$$Q_p = k\,CIA \tag{10.31}$$

where
   Q_p is peak discharge (ft^3/s) for the rainfall intensity I (in./h)
   A is the watershed area (acre)
   C is runoff coefficient or the ratio of runoff to rainfall
   k is 1.008, or a factor for unit conversion

If I is in millimeters per hour and A is in hectares, then $k = 0.00278$ to give Q_p in cubic meters per second.

   The equation assumes that rainfall continues at a uniform intensity with duration equal to the time of concentration. The time of concentration (t_c) is the time required for water to travel from the remotest part of the watershed to reach the outlet. For uniform rainfall intensity, this would be equal to the time of equilibrium at which the rate of runoff is

TABLE 10.6

Some Typical RCN Values under Forested Conditions

Cover Type	Ground Cover (%)	Hydrologic Soil Group			
		A	B	C	D
Bare	0	77	86	91	94
Fallow	5	76	85	90	93
Shrubland	25	63	77	85	88
Grassland/herbaceous	25	49	69	79	84
Undisturbed forests					
Deciduous and mixed	50	55	55	75	80
Evergreen	50	45	66	77	83
Forests, low severity fires					
Deciduous and mixed	43	59	60	78	82
vergreen	43	49	71	80	85
Shrubland	21	65	79	86	89
Moderate severity fires					
Deciduous and mixed	34	65	65	80	85
Evergreen	34	55	76	82	88
Shrubland	17	68	82	88	90
Forests, high-severity fires					
Deciduous and mixed	25	70	71	83	87
Evergreen	25	60	82	85	90
Shrubland	12	73	88	91	91

Source: Goodrich, D.C. et al., Rapid post-fire hydrologic watershed assessment using the AGWA GIS-based hydrologic modeling too, in *Managing Watersheds for Human and Natural Impacts Engineering, Ecological and Economic Challenges: Proceedings of the 2005 Watershed Management Conference*, Moglen, G.E. Ed., July 19–22, 2005, Williamsburg, VA, American Society of Civil Engineers, Reston, VA, 12 pp., 2005.

equal to the rate of rainfall supply. Accordingly, the size of the watershed should be small enough, up to $10\,\text{km}^2$, to have storm duration equal to t_c. If the duration is less than t_c, then water is not contributed from the entire watershed. On the other hand, if the duration is greater than t_c, then there are no additional areas to contribute water, and the rainfall intensity decreases. This makes Q_p the highest when the duration equals t_c.

In practice, the rainfall intensity is obtained from the rainfall intensity–duration–frequency atlas (Hershfield, 1961) for the location, with its frequency the same as the designed flood and duration equal to t_c. Values of t_c in minutes are often estimated by the Kirpich's (1940) formula:

$$t_c = 0.0078L^{0.77}\,S^{-0.385} \tag{10.32}$$

where
    L is the length of channel from watershed divide to outlet (ft)
    S is the average channel slope (ft/ft)

The constant 0.0078 should be replaced by 3.97 if L is in kilometers and S is in meters per meter. The runoff coefficient C is affected by land use, surface condition, soil, and slope. Some typical C values are given in Table 10.7.

The RCN approach has also been used to estimate the time of concentration t_c, and the calculated t_c is then used to estimate Q_p by the following empirical equations (Schwab et al., 1996; Elliot et al., 2010):

$$t_c = \frac{L^{0.8} \left((1000/\text{RCN}) - 9\right)^{0.7}}{CS^{0.5}} \tag{10.33}$$

$$\log Q_p = 2.51 - 0.7 \log t_c - 0.15 \left(\log t_c\right)^2 + 0.071 \left(\log t_c\right)^3 \tag{10.34}$$

where
  t_c is the time of concentration (h)
  L is the length of watershed (m or ft)
  S is the average watershed gradient (m/m or ft/ft)
  C is the constant 441 for metric, or 1140 for English units
  Q_p is the peak flow (ft^3/s/mile^2/in.) of storm runoff, multiplied by 0.0043 to get metric unitsin m^3/s/km^2/mm of storm runoff

10.2.3.3 Manning Equation (Flood Flows)

When there is no stream gaging station, when the peak discharge of a flood is beyond the highest measured stage-discharge rating curve, or when one is interested in the peak discharge of a previous flood, the Manning equation is often used for estimating steady and uniform flows in open channels. The equation requires measurements of channel depth, cross-sectional area, and water surface slope (or bed slope), with an estimate of roughness coefficient in the following manner:

$$Q_p = (A)(V) = \frac{(A)\left(R^{0.67}S^{0.5}\right)}{n} \tag{10.35}$$

where
  Q_p is the peak discharge (m^3/s)
  A is the channel cross-sectional area (m^2)
  V is the flow velocity (m/s)
  R is the hydraulic radius (m)
  S is the energy slope (m/m)
  n is Manning's roughness coefficient

Dividing the difference in elevation of the high-water marks between upper and lower sections by the reach length in between can approximate the channel's energy slope S. Values of R are calculated by dividing the channel cross-sectional area in square meters by the wetted perimeter of the cross section in meter. The parameter n describes the resistance of a watercourse to flow. It ranges from about 0.009 for closed brass conduits to 0.50 for vegetated, lined channels (Chow, 1959), and has been recorded as high as 4.48

TABLE 10.7

Some Runoff Coefficients (C Values) for Equation 10.31 by Hydrologic Soil Groups (A, B, C, D) and Watershed Slope Range

Land Use	A			B			C			D		
	0%–2%	2%–6%	6%+	0%–2%	2%–6%	6%+	0%–2%	2%–6%	6%+	0%–2%	2%–6%	6%+
Cultivated	0.08[a]	0.13	0.16	0.11	0.15	0.21	0.14	0.19	0.26	0.18	0.23	0.31
	0.14[b]	0.18	0.22	0.16	0.21	0.28	0.20	0.25	0.34	0.24	0.29	0.41
Pasture	0.12	0.20	0.30	0.18	0.28	0.37	0.24	0.34	0.44	0.30	0.40	0.50
	0.15	0.25	0.37	0.23	0.34	0.45	0.30	0.42	0.52	0.37	0.50	0.62
Meadow	0.10	0.16	0.25	0.14	0.22	0.30	0.20	0.28	0.36	0.24	0.30	0.40
	0.14	0.22	0.30	0.20	0.28	0.37	0.26	0.35	0.44	0.30	0.40	0.50
Forest	0.05	0.08	0.11	0.08	0.11	0.14	0.10	0.13	0.16	0.12	0.16	0.20
	0.08	0.11	0.14	0.10	0.14	0.18	0.12	0.16	0.20	0.15	0.20	0.25
Street	0.70	0.71	0.72	0.71	0.72	0.74	0.72	0.73	0.76	0.73	0.75	0.78
	0.76	0.77	0.79	0.80	0.82	0.84	0.84	0.85	0.89	0.89	0.91	0.95
Parking	0.85	0.86	0.87	0.85	0.86	0.87	0.85	0.86	0.87	0.85	0.86	0.87
	0.95	0.96	0.97	0.95	0.96	0.97	0.95	0.96	0.97	0.95	0.96	0.97

Source: Selected Values from McCuen, R.H., *Hydrologic Analysis and Design*, Prentice Hall, New York, 1998.

[a] Runoff coefficients for storm-return periods less than 25 years.
[b] Runoff coefficients for storm-return periods of 25 years or longer.

TABLE 10.8

A Few Values of Manning's Roughness Coefficient for Natural Streams

Type of Stream and Description	Minimum	Normal	Maximum
1. Minor streams (top width at flood stage <30.5 m)			
a. Streams on plain			
Clean, straight, full stage, no rifts, or deep pools	0.025	0.030	0.033
Clean, winding, some pools, and shoals	0.033	0.040	0.045
Sluggish reaches, weedy, deep pools	0.050	0.070	0.080
Very weedy reaches, deep pools, or floodways with heavy stand of timber and underbrush	0.075	0.100	0.150
b. Mountain streams, no vegetation in channel, banks			
Usually deep, trees, and brush along banks			
Usually submerged at high stages			
Bottom: gravels, cobbles, and few boulders	0.030	0.040	0.050
Bottom: cobbles with large boulders	0.040	0.050	0.070
2. Floodplains			
a. Short-grass pasture, no brush	0.025	0.030	0.035
Long-grass pasture, no brush	0.030	0.035	0.050
b. Cultivated areas, no crop	0.020	0.030	0.040
Cultivated areas with mature row crops	0.025	0.035	0.045
c. Scattered brush, heavy weeds	0.035	0.050	0.070
Medium to dense brush, summer	0.070	0.100	0.160
d. Dense willows, summer, straight	0.110	0.150	0.200
Heavy stand of timber, a few trees down, little undergrowth, flood stage below branches	0.080	0.100	0.120

Source: Chow, V.T., *Open-Channel Hydraulics*, McGraw-Hill, New York, 1959.

and 7.14 in channels with extreme vegetation blockages (Green, 2005). There are tables and pictures that give average or typical n-values for various channel conditions. A few values of n compiled by Chow (1959) are given in Table 10.8. Those channel photographs (Barnes, 1967; Arcement and Schneider, 1989; Coon, 1998) for which n values have been computed, along with hydraulic data and particle size, can be used to compare to a site of interest for estimating an n value. Also, equations have been developed to relate n values that were obtained from actual streamflow measurements with the hydraulic data and particle-size characteristics of stream channels (Table 10.9). These equations can be used to estimate n values for sites of similar conditions.

The determination of n is critical to flood estimates. The roughness coefficient is not constant. It varies with time and depth of flows. Vegetation, bank, bed materials, channel configuration, and water temperature may change due to seasons and the magnitude of flow, which in turn impose different resistance to flows. A correction procedure involving the selection of one base value of n along with four adjustment factors (due to surface irregularities, channel configuration cross section, obstructions, and vegetation) and one corrective factor for sinuosity can improve the determination of n values (Cowan, 1956, cited by Arcement and Schneider, 1989). Perhaps the most reliable estimates are obtained by calibration with historical discharge and stage records (Fread, 1993).

In practice, a straight reach that is at least 100 m long and fairly homogeneous must be carefully selected. In other words, the width, depth, flow velocity, channel slope, roughness, bank soil texture, and bed materials must be relatively uniform. Reaches with rapids,

TABLE 10.9

A Few Equations for Estimating Manning Equation's n Values

Equation	Channel Conditions	Reference
$n = 0.113(R)^{1/6}/[1.16 + 2.0 \log(R/d_{84})]$	For high within-bank flows in gravel-bed channels, $d_{50} = 0.6–25$ cm; $S_w < 0.002$.	Limerinos (1970)
$n = 0.104(S_w)^{0.177}$	No significant vegetation in the channel bed, no dominant bed-form features; $d_{50} = 2–15$ cm.	Bray (1979)
$n = 0.32(S_f)^{0.38}(R)^{-0.16}$	$R = 0.15–2.13$ m; $S_f = 0.002–0.09$.	Jarrett (1984)
$n = 0.121(S_w)^{0.18}(R)^{0.08}$	$S_w = 0.0003–0.018$; $R \leq 5.8$ m.	Sauer (1990)
$n = 0.289(R)^{0.14}(R/d_{50})^{-0.44}(R/T)^{0.30}$	$S_w = 0.0003–0.018$; $R \leq 5.8$ m.	Jobson and Froehlich (1988)
$n = 0.217(A)^{-0.173}(S)^{0.267}(S)^{0.156}$	Straight open channels, wide conditions	Dingman and Sharma (1997)
$n = 0.012(F_r)^{-1.028}$	Vegetated channels, density disregarded	Chen et al. (2009)

Source: Cited in British units by Coon, W.F., Estimation of roughness coefficients for natural stream channels with vegetated banks, Water Supply Paper 2441, U.S. Geological Survey, Reston, VA, 1998; Dingman, S.L. and Sharma, K.P., *J. Hydrol.*, 199, 13, 1997; Chen, Y.-C. et al., *Ecol. Eng.*, 35, 1027, 2009.

Note: R = hydraulic radius, m; d_{84} = intermediate particle diameter, m, that equals or exceeds 84% of the particles; S_w = slope of the water surface, m/m; S_f = energy gradient, m/m; d_{50} = median particle diameter, m; T = top width of stream, m; F_r = Froude number, 0.02–0.25.

abrupt falls, excessive channel-width variation, tributary flows, back flows, or submerged flows should be avoided.

10.2.3.4 Water-Balance Approach (Annual Streamflow)

Information on annual streamflow or long-term averages at ungaged watersheds can be estimated from

1. Runoff maps of the nation (McGuinness, 1964; Busby, 1966; U.S. Forest Service, 1982)
2. Empirical equations
3. The water-balance approach

Runoff maps show long-term averages of existing conditions for different regions. They provide only general information for the region in question. Empirical equations estimate streamflow as a function of climatic, topographic, and land-use conditions. They are useful not only for estimating streamflow at ungaged watersheds, but also for providing a guideline for managing water resources of the region. Empirical equations are applicable only for the regions for which they were developed.

The water-balance approach is to solve the difference for streamflow through input, output, and storage components in the hydrologic cycle. First, precipitation (P_t) is used to recharge the soil-moisture deficit ($\theta_s - \theta_i$) and then to supply water required for potential evapotranspiration (PE). The excess of P_t then occurs as streamflow Q. Values of PE can be calculated by the equations given in Section 9.3, and the soil-moisture deficit is calculated for the root zone based on field measurements.

Actual water translocation and movement are much more complicated than the generalized model described earlier. Thus, the water-balance approach is more reliable for estimating streamflow of longer duration, such as monthly and annual. In practice, the water balance should be calculated by monthly sequence to account for soil-moisture deficit from the previous month and then the monthly runoff summed to yield annual runoff. An example of the calculations is given in Table 10.10.

10.3 Deforestation

Forest transpiration is greater than soil evaporation because of the forest's large canopies, litter floor, deep root systems, and greater energy level (see Sections 9.2 and 9.4). Since forests do not increase the amount of precipitation for the area, at least in the temperate regions (see Section 8.4), the greater ET loss to the air must compensate for a reduction of runoff in watersheds. Inversely, forest clear-cut, or converting forests into vegetation of smaller sizes and lower densities, is expected to cause an increase in streamflow.

Numerous investigations and studies on the effects of forest and forest activity on streamflow quantity have been conducted under various forest types and climatic, soil, and topographic conditions. These effects are addressed here on water yield, flow timing, extreme flows, and groundwater.

10.3.1 Water Yield

Cutting forests in small watersheds generally causes an increase in water yield. However, the increase in water yield is greatest the first year after cutting and gradually decreases with time because of the regrowth of vegetation. Such increases in water yield are affected by the intensity of forest cutting, species, amount and timing of precipitation, and soil topographic conditions.

10.3.1.1 Forest-Cutting Intensity

Under a given environmental condition, the increase in water yield is affected by cutting intensity and distribution. Clear-cutting the entire watershed yields the maximum increase, while thinning yields the minimum. For a complete forest clearing, the increase in the first posttreatment year was about 462 mm (39%) at the H.J. Andrews Experimental Forest in Oregon, 427 mm (65%) at Coweeta, North Carolina, and 15 mm (8%) at the Wagon Wheel Gap in Colorado (Table 10.11). However, increases of as much as 470–600 mm/year were reported in the second to fifth posttreatment years in western Oregon (Harr et al., 1982), the greatest ever reported in the United States.

These results provide evidence of forest removal as a potential method for water yield augmentation. However, the augmentation of water yields in the United States is attainable only if

1. Forests cover a significant portion of the entire watershed (Douglass, 1983).
2. Average precipitation exceeds 400–450 mm/year, and is greater than annual ET (Hibbert, 1983).
3. Soil depth is more than about 1 m (Rowe and Reimann, 1961).
4. Forest cover reduction is greater than 20% (Stednick, 1996).

Forests and Streamflow Quantity 241

TABLE 10.10

Calculating Runoff (All in cm) Using Water-Balance Approach

Components	January	February	March	April	May	June	July	August	September	October	November	December	Year
P_t	8.9	9.9	9.6	8.1	8.4	11.9	13.7	13.5	10.2	7.1	5.8	9.1	116.2
ET[a]	1.3	1.5	3.3	6.4	11.2	15.0	16.8	15.5	10.9	5.8	2.3	1.5	91.5
$P_t - ET$	7.6	8.4	6.3	1.7	-2.8	-3.1	-3.1	-2.0	-0.9	1.3	3.3	7.6	24.7
θ_i[b]	37.5	37.5	37.5	37.5	37.5	34.7	31.6	28.5	26.5	25.6	26.9	30.2	
$D = \theta_i - \theta_s$	0	0	0	0	0	-2.8	-5.9	-9.0	-11.0	-11.9	-10.6	-7.3	
$R = P_t - ET + D$[c]	7.6	8.4	6.3	1.7	-2.8	-5.9	-9.0	-11.0	-11.9	-10.6	-7.3	0.3	24.3

[a] If potential ET is used in the calculation, the estimated runoff will be lower by a value equal to PE − ET.

[b] θ_i = The initial soil-moisture content of the month = θ_i of the previous month + (P$_t$ − ET) of the previous month. If the answer is greater than θ_s (saturated moisture content), enter θ_s (=37.5 cm) as θ_i.

[c] R = rainfall excess (runoff). There is no excess for months with negative values. Sum the positive values to yield annual total.

TABLE 10.11

Selected Results on the First-Year Increase in Water Yield after Complete Forest Cutting in the United States

Region	Forest Type	Water Increase		Reference
		mm	%	
The East				
Coweeta, NC	Mixed hardwoods	427	65	Hewlett and Hibbert (1961)
Hot Springs, MS	Shortleaf pine	370	38	Ursic (1991)
		116	13	
Hubbard Brook, NH	Mixed hardwoods	343	40	Hornbeck et al. (1970)
Leading Ridge, PA	Mixed hardwoods	137	23	Lynch and Corbett (1990)
Fernow, W. VA	Mixed hardwoods	130	19	Reinhart and Trimble (1962)
Bear Creek, AL	Pine–hardwoods	297	60	Betson (1979)
The Northwest				
H.J. Andrews, OR	Douglas fir	462	39	Rothacher (1970)
H.J. Andrews, OR	Douglas fir	420	27	Harr et al. (1982)
Coyote Creek, OR	Douglas fir and pine	360	39	Harr et al. (1979)
The North				
Marcell Experimental Forest, MN	Aspen–birch	81	39	Hornbeck et al. (1993)
The Rocky Mountains				
Fool Creek, CO	Alpine and subalpine	94	36	Troendle and King (1985)
Wagon Wheel Gap, CO	Bristlecone pine	15	8	Bates and Henry (1928)
The Southwest				
Beaver Creek, AZ	Ponderosa pine	99	63	Baker (1986)

Since most watershed studies have been conducted in headwater areas, how much the downstream users can enjoy the upstream increase is an addressable question.

While there is great potential to augment water yield by 10%–65% through forest cutting, this process can impose adverse effects, such as nutrient losses, soil erosion, lower water quality, and less aesthetic environments. Watershed managers select a less intensive cutting or a small area of clear-cutting to mitigate the environmental problems that might be incurred by large-scale harvesting (Gottfried, 1983; Johnston, 1984). Small openings in the watershed can also benefit timber and wildlife resources (Folliott and Thorud, 1977), are effective BMPs for controlling nonpoint sources of water pollution (Lynch and Corbett, 1990), and can improve stream habitats, consequently increasing the primary productivity of aquatic biota in headwater streams (Bureau of Land Management, 1987; Moseley et al., 2008).

The increase in water yield produced by partial cuttings is less than that produced by clear-cutting and depends on the intensity and distribution of the cutting. In the eastern United States, increases in water yield for harvest of hardwoods can be estimated by the following equations (Douglass, 1983):

$$Y_1 = 0.00224 \left(\frac{BA}{PI} \right)^{1.4462} \tag{10.36}$$

$$D = 1.5(Y_1) \tag{10.37}$$

$$Y_i = Y_1 + b \log(i) \tag{10.38}$$

where
  Y_1 is the first-year increase in inches for the harvest of hardwoods
  BA is the percent basal area cut of the watershed
  PI is the annual potential insolation index (10^{-6} cal/cm^2) for the watershed calculated by
    methods of Lee (1963) or Swift (1976)
  D is the duration of treatment effect (years)
  Y_i is the yield increase in inches for the ith year after treatment
  b is a coefficient derived by solving Equation 10.38 for the year when i = D and $Y_i = 0$

For example, completely clear-cutting an eastern hardwood watershed (BA = 100) in the Appalachian Highland of West Virginia with PI = 0.3 could increase water yield in the first posttreatment year 9.97 in. The increase would decline to 4.18 in. (25.3 cm) in the 5th year after treatment and to its pretreatment level in the 15th year. If the watershed is cut by 50% of its basal area (BA = 50), then the first-year increase is reduced to 3.66 in., and it takes 6 years for water yield to return to its pretreatment level.

In the West, a 67% selection cut in Caspar Creek dominated by redwood and Douglas fir in California caused an average annual increase in water yield of 94 mm or 15% (Keppeler and Ziemer, 1990). A 25% patch cut in Brownie Creek, Utah, dominated by lodgepole pine, increased annual streamflow by 147 mm or 52% (Burton, 1997). Cutting by single-tree selection or thinning does not produce large increases in water yield (Troendle and King, 1987), unless the density is reduced to a basal area of 6.9 m^2/ha, as it was in Beaver Creek, Arizona (Baker, 1986), or of 9.2 m^2/ha, as in Workman Creek, Arizona (Rich and Gottfried, 1976).

10.3.1.2 Species

Plant transpiration rate is different among species due to characteristic canopy, height, and root systems. Generally, conifers have greater total leaf-surface areas and retain foliage all year round; they intercept and transpire more water than hardwoods do. Chaparrals are deep-rooted evergreen shrubs with a large biomass above ground. Transpiration and interception losses are greater for chaparrals than for grasses of shallower root systems and longer dormancy period. The lower stem, smoother surface, and light-colored canopy of grasses cause the reflection of solar radiation to be greater, the energy absorption to be lower, and consequently the water consumption to be less than those of woody plants. In other words, deep-rooted evergreen species with large and dark canopies transpire a great deal more water every year than shallow-rooted deciduous species with small and light-colored canopies do. Thus, transpiration rate in general is greatest for conifers, followed by hardwoods, chaparrals, and grasses. Converting species composition in watersheds from higher transpiration rate to lower transpiration rate increases water yield.

Many watershed observations have provided evidence on increases in water yield through manipulation of species composition (Table 10.12). In the West, water yields were increased up to 150 mm/year by converting chaparrals into grasses. In North Carolina, one watershed had a 25% yield loss after conversion of hardwoods to white pines. Riparian

TABLE 10.12

A Few Studies on the Effects of Species Conversion and Brush Control on Average Water Yield
(in mm or %) in the United States and at Other Locations

Location	Precipitation (mm/year)	Original Vegetation	Conversion Treatments	Effects on Water Yield (mm or %)		Reference
The United States						
Beaver Creek, AZ	460	Pinyon, etc.	2,4-D to forbs	+157%	8 years	Baker (1984)
Whitespar, AZ	660	Chaparral	Herbicides control	+69 mm	7 years	Davis (1993)
Mingus, AZ	480	Chaparral	To grass by fire	+10 mm	5 years	Hibbert et al. (1982)
Three Bar, AZ	673	Chaparral	To grass by fire	+148 mm	18 years	Hibbert et al. (1982)
Mendocino, CA	920	Oaks	2,4-D to grass	+39%	10 years	Pitt et al. (1978)
Placer County, CA	620	Oaks, forb	2,4-D to clovers	+113 mm	3 years	Lewis (1968)
Boco Mt., CO	259	Sagebrush	Plow to wheatgrass	−9 mm	6 years	Lusby (1979)
Coweeta, NC	1854	Hardwoods	Cut to fescue	+78 mm	5 years	Hibbert (1969)
Coweeta, NC	1930	Hardwoods	Cut to white pine	−64 mm	4 years	Swank and Miner (1968)
Riesel, TX	889	Mesquite	Killed by chemicals	+24 mm	8 years	Richardson et al. (1979)
Sane Creek, WY	572	Sagebrush	Controlled by 2,4-D	+11 mm	11 years	Sturges (1994)
Others						
Lisdale, Australia	756	Eucalypts	To *P. radiate*	−31.3 mm	9 years	Putuhena and Cordery (2000)
Marianza, Ecuador	939	Grassland	To *P. patula*	−242 mm	14 months	Buytaert et al. (2007)
Otago, New Zealand	1000	Pasture	To *P. radiate*	−173 mm	8 years	Smith (1987)
Westfalia, S. Africa	1611	Grassland	To eucalypts	−322 mm	8 years	Scott and Smith (1997)
La Corona, Uruguay	1162	Pasture	To pine	−50 mm	5 years	Chescheir et al. (2009)

vegetation and *phreatophytes* with root systems extending into the groundwater table or capillary fringe transpire much more water than do mesophytes growing on uplands. However, the treatment of phreatophytes is effective only if

1. The water supply exceeds ET loss after treatments.
2. The water table is within reach of the vegetation root systems.
3. The soils are of sufficient depth to permit the reduction in ET if the deep-rooted vegetation is eliminated.

Some suggest that streamflow can be increased with relatively little work by eliminating riparian vegetation through mechanical eradication or transpiration suppression (Ingebo, 1971). However, such treatments need to be done with care due to possible soil erosion and water quality problems and in compliance with environmental regulations. However, riparian vegetation management is more effective in the West than in the East (Lynch, 2001).

10.3.1.3 Precipitation

Forest manipulation for water augmentation is more effective in humid regions and during wet years. Also, precipitation enhances rapid growth of vegetation, and the duration of treatment effects in humid regions is shorter than that in arid regions. In the chaparral country of California and Arizona, precipitation is the major factor in water-yield response to species conversion. In areas where average annual precipitation is less than 400 mm, there is no potential for increasing water; and for areas with precipitation between 400 and 500 mm, the increase is likely to be marginal. On western rangelands of the United States, the mean annual increase in water yield Q in millimeters was related to mean annual precipitation P in millimeters by (Hibbert, 1983)

$$Q = -100 + 0.26(P_t) \tag{10.39}$$

in which P_t has to be >385 mm/year for any positive forest treatment effects on Q. At the H.J. Andrews Experimental Forest in western Oregon, Harr (1983) showed that increases in water yield due to harvest of Douglas-fir forest in watershed HJA-1 could be estimated by

$$Q_i = 308 + 0.87(P_t) - 18.1(Y_i) \tag{10.40}$$

where
  Q_i is the predicted annual increase of water (mm) in the ith year after harvest
  P_t is the annual precipitation (mm)
  Y_i is the ith year after harvesting

Equations 10.39 and 10.40 show that the precipitation effect on yield increase is much greater in the Northwest than in the Southwest of the United States. Water yield increases about 1 mm for each 1 mm increase in precipitation at watershed HJA-1 and for each 4 mm increase in precipitation in the Southwest. This difference is probably associated with

1. Climatic factors such as temperature and monthly distribution of precipitation
2. Soil topographic factors such as soil depth and the gradient, aspect, and shape of slopes
3. Vegetal factors such as species, age, and treatments

10.3.1.4 Soil Topographic Conditions

Hydrological responses to forest harvests vary among watersheds due to the type and depth of soils along with steepness and orientation of the watershed. Soils of deep and fine textures have a much greater water-holding capacity than soils of shallow and coarse textures do; consequently, they have a greater potential for yield increase. Water yield could not be appreciably increased in soils with a depth less than about 1 m. Also, forest openings in upper slopes cause increase in water yield smaller than those in lower slopes or close to stream channels.

Slope aspect affects solar radiation, precipitation, and wind speed and consequently affects soil and air temperatures, snow accumulation, snowmelt, ET, and vegetation type and growth. In the Northern Hemisphere, forest transpiration is generally greater in northern than in southern slopes because of denser vegetative cover and deeper soils (Bethlahmy, 1973). Actual observations showed that clear-cutting on south-facing slopes caused only about a one-third increase of that measured on north-facing slopes in Idaho (Cline et al., 1977) and at Coweeta, North Carolina (Douglass, 1983). Similarly, west-facing forests use more water than those do on east-facing slopes.

10.3.2 Timing of Streamflow

Timing refers to the occurrence of streamflow in a given time scale. Cutting forests can cause increases in water yield. Does the increase occur in growing seasons or in dormant seasons, in a few months or over the entire year? It is of great concern to hydrologists and resource managers.

10.3.2.1 Seasonal Distribution

The seasonal distribution of water yield associated with forest management is not uniform throughout the year. Since increases in water yield caused by forest cutting are attributable to the reduction in canopy interception and transpiration, a large proportion of increased flow must occur during the growing season. However, superimposed on this vegetal effect are precipitation type and seasonal distribution.

In areas where precipitation is dominated by rainfall with no significant distribution pattern throughout the year, such as the eastern United States, most of the flow increases occur in summer and fall. The impact of forest cutting on water yield in winter and early spring is insignificant when both cut and uncut watersheds are saturated and ET is minimum. For an eastern hardwood forest that was cut annually during a 7 year period, Swank and Johnson (1994) showed that 76% of the mean monthly increase in runoff occurred during the low-flow period, July through December.

In the Rocky Mountain region, where precipitation is dominated by snow and 70%–80% of the total annual runoff occurs during the summer months, forest clearing reduces ET and enhances snow accumulation and snowmelt. Thus, the greatest increase in water yield occurs during the snowmelt period. The classic study on the Fool Creek Watershed in Colorado showed that increases in runoff were primarily on the rising side of the annual hydrograph, with May being the greatest, accounting for more than 50% of the estimated annual increase (Troendle, 1983). However, a 25% harvest of lodgepole pine forest on a large watershed (2145 ha) in northeastern Utah gave a different perspective. The cutting caused an average increase in runoff of 52% (14.7 cm) per year, but increases occurred primarily on the recession limb of the annual hydrograph (May through August), with little or no change in winter (Burton, 1997). Hydrologic storage is greater in large watersheds, which can have implications for its timing on flow increases that are different from those in small watersheds.

In areas where summer is dry and the majority of precipitation occurs in fall and winter, the greatest proportion of water-yield increase due to forest cutting occurs during the rainy season. For example, about 80% of annual precipitation in western Washington and Oregon occurs between October and March, and major increases occur between

December and March, when there is plenty of streamflow and no agricultural demand for water. Although large in relative terms, summer increases are small in absolute value when water stress is high. Thus, without storage facilities, water increases in winter seem to be of little economic benefit downstream when water is most needed (Harr et al., 1979).

10.3.2.2 Treatment Duration

Increases following forest cutting are greatest in the first or second year and decline thereafter due to the regrowth of vegetation. The duration has been estimated to be as long as 80 years for the Fool Creek watershed of the Fraser Experimental Forest, Colorado (Troendle and King, 1985), and only 4 years on Leading Ridge III in Pennsylvania (Lynch and Corbett, 1990). Generally, it is shorter in humid regions and in species with deep root systems. Typically, it returns to precutting level in about 7 or more years in the United States.

The Coranderrk project, started in 1958, is the longest running paired watershed study in Australia. Clear-cutting a watershed there predominately with mountain ash (*Eucalyptus regnans*) shows increases in water yield in the first 3 posttreatment years, reaching 300 mm/year at peak. The annual flow then declined to pre-logging level at year 8 and continued for 34 years with no sign of recovery (Bren et al., 2010). This behavior is common to all eucalypts in Australia (Kuczera, 1987; Cornish and Vertessy, 2001). Dr. D.E. Leaman (Personal Communication) of Australia suspects that it may take 250–400 years for background levels of streamflow to be produced.

10.3.2.3 Flow Duration

Flow duration is often depicted by flow-duration curves. A flow-duration curve is a cumulative curve of the percentage time that a given flow magnitude is equaled or exceeded. Thus, the percent time of occurrence is greater for flows of lower magnitude than for flows of higher magnitude.

In West Virginia, forest harvest increased the magnitude of both high and low flows for all frequencies of occurrence (Figure 10.10). Low flows were substantially augmented in both wet and dry years (Reinhart et al., 1963). Conversely, the frequency of both high and low flows was significantly reduced when two watersheds were converted from eastern hardwoods to white pine plantation in Coweeta, North Carolina (Swank and Vose, 1994). By age 24, on one of the watersheds, high flows at 1% and 5% of time were reduced by 33% and 52%, respectively. Flows for other percentages of time were reduced 37%–60% below predicted values (Figure 10.10).

Forty percent of the aforementioned Fool Creek Watershed (290 ha) in the Fraser Experimental Forest, Colorado, was strip clear-cut during 1954–1956. Total seasonal flow (1956–1971) increased 40%, peak flow increased 20%, and most detectable changes occurred in May. The increases appeared to be most for flow durations in the range from 80% to 120% of bankfull; the highest flow durations were not affected (Troendle and Olsen, 1994).

10.3.3 Flow Extremes

For quite some time, forests have been speculatively claimed as streamflow moderators by making high flows lower and low flows higher. Numerous watershed observations and studies in the last century have provided many new perspectives on the speculation of forest–water relationships. If forest clearings do increase water yield in watersheds,

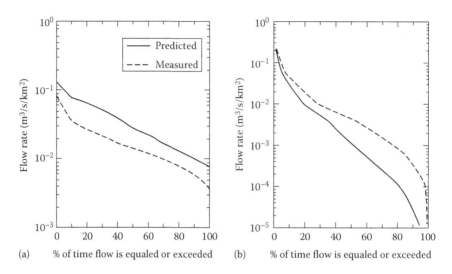

FIGURE 10.10
Flow-duration curves: (a) predicted for a mature hardwood forest versus measured from a watershed with a 24 year old white-pine plantation in North Carolina and (b) predicted versus measured for a clear-cut watershed at Parsons, West Virginia.

as discussed in the previous sections, then the increase must be attributable to quick flows, low flows, or both. Timber harvests have been shown to cause low flows higher in magnitude, shorter in duration, and less frequent in occurrence (Troendle, 1970; Harr et al., 1982; Swank and Vose, 1994). However, the effect on quick (storm) flows is highly variable. It can be an increase (Burton, 1997), no change (Hewlett and Hibbert, 1961), and even a decrease (Harr and McCorison, 1979), depending on the severity of soil disturbance, cutting intensity and distribution, size and distribution of storms, antecedent soil-moisture condition, and precipitation type.

10.3.3.1 Peak Flows

For a given region, peak flows are generally lower in watersheds with a greater percentage of forested area (Reich, 1972; Chang and Waters, 1984). Thus, cutting forests can cause an increase in peak flows. Reported increases include 15% at Coweeta, North Carolina (Swank and Johnson, 1994); 35%–122% in coastal Oregon (Harr, 1976); 10%–89% in the Horse Creek Watersheds, Idaho (King, 1989); 66% in Brownie Creek, Utah (Burton, 1997); 180% in Upper Bear Creek, Alabama (Betson, 1979); and 340%–880% on Leading Ridge-2, Pennsylvania (Lynch and Corbett, unpublished report). The causes can be attributable to more saturated soils, soil compaction, road construction, and differences in snow accumulation and snowmelt. However, peak flows can also be decreased or remain unchanged after logging (Thomas and Megahan, 1998), depending on cutting intensity, treatment location in the watershed, regrowth of vegetation, and climatic patterns. It is generally agreed that as long as the forest floor remains intact, the effect of clear-cutting on peak flows could be significantly reduced (Harr et al., 1979).

The increase in peak flows occurred generally in the growing season, and quick flows were a small portion of the total water increases. For example, of the average increase of 280 mm/year in the Hubbard Brook Experimental Forest, New Hampshire, only 99 mm or 35.4% occurred as quick flows (Hornbeck et al., 1970). In other words, most of the streamflow increase was in the form of baseflow. Storm-runoff events occurring after fall recharge of soil moisture and before the start of spring snowmelt were unchanged by clearing. This is a desirable effect because streamflow increase in the form of baseflow is of more value than quick flow. Also, the occurrence of peak flows in the growing season poses less danger than if the peak flows occur during the seasons when streamflow is high.

Although low peak flows occur in forested watersheds and high peak flows result from forest cutting, this does not imply that forests can be relied on for flooding prevention. It is true that forests, due to the great amount of transpiration loss and the increase in soil water-holding capacity, are the most effective types of vegetation in reducing streamflow. Flood damage in forested areas is also the smallest among all natural surface conditions due to forests' control of soil erosion, landslides, and channel stability. However, when the soil is saturated, any additional rainfalls will run off from the watershed whether it is forested or nonforested. Thus, forests can reduce peak flows for storms of short duration and low intensity but cannot prevent the occurrence of floods that are produced from storms of high intensity and long duration over a large area. See more discussions on forests and flooding in Chapter 14.

10.3.3.2 Low Flows

The increase in water yield due to timber harvest results basically from increases at low flow levels, especially during the growing season. This has been reported in many regions in the United States, including Minnesota (Verry, 1972), New Hampshire (Hornbeck, 1973), New York (Satterlund and Eschner, 1965), North Carolina (Swank and Vose, 1994), Pennsylvania (Lynch et al., 1980), western Oregon (Harr et al., 1982), and West Virginia (Patric and Reinhart, 1971), and in many other countries such as United Kingdom (Johnson, 1998), South Africa (Scott and Lesch, 1997), Australia (Best et al., 2003), and France (Cosandey et al., 2005). For example, timber harvest resulted in fewer low-flow days in both clear-cut and shelterwood cut watersheds in the H.J. Andrews Experimental Forest, Oregon. In 1977, by far the driest year on record in western Oregon, the clear-cut watershed had only 8 low-flow days instead of the 143 days predicted by the pre-logging relationship. The shelterwood watershed had only 2 low-flow days instead of the 135 days predicted by the calibration equation (Harr et al., 1982). However, in fog-prone regions, such as the coast range of Oregon, deforestation could cause a decrease in low flows due to the reduction of fog deposition (Ingwersen, 1985).

10.3.4 Groundwater

Groundwater is the water percolated from precipitation and accumulated under the surface over time to form a relatively permanent zone of saturation. Thus, groundwater can flow through a column of soil with a velocity governed by the hydraulic conductivity (k) of the soil or rock strata and the difference in pressure head ($h_1 - h_2$), and is inversely proportional to the length of the column (L):

$$V = \frac{k(h_1 - h_2)}{L} \tag{10.41}$$

where
   V is the velocity (cm/s)
   k is in centimeters per second
   h and L are in centimeters

Equation 10.41 is called Darcy's law. The value k refers to the ability of the column material to transmit water and is known as *permeability*, and the item $(h_1 - h_2)/L$ is the hydraulic gradient, which determines the flow occurrence and direction. Permeability is greatly affected by the size and configuration of pores in a medium. Sandy materials transmit water more efficiently than clays do.

Groundwater level has been the subject of hectic discussion in forest hydrology. It is defined as the level at which groundwater stands in an open hole under natural potential, expressed either as the distance below the ground surface or a topographic elevation above a standard datum such as mean sea level. Both increase and decrease of water table following deforestation have been reported in many parts of the world. The interactions of soil texture and profile, geological characteristics, local topography, groundwater depth, precipitation type and seasonal distribution, forest type, root system, and other factors may have caused the forest effects to be inconsistent among studies.

10.3.4.1 Groundwater in Steep Terrain

Springs, perennial streams, and water in wells are evidence of the existence of groundwater in mountainous regions. The subsurface stratigraphy in these regions is more complicated than that of level terrain, which can create a perched groundwater table (isolated zone of saturation) in the soil profile or disrupt the groundwater table. Groundwater in these areas can also be deep under the ground. The ability of forests in these areas to sustain springs and streamflow—and the disappearance of springs after deforestation—have been described in the literature (Zon, 1927; Molchanov, 1963; Pritchett, 1979). On the other hand, groundwater table rises of 20–150 cm in response to forest clear-cutting or partial cutting were reported in a number of studies (Smerdon et al., 2009). Rises in groundwater table may trigger preferential flow in headwater watersheds or in shallow groundwater areas. A higher groundwater table under forests than in the open may reflect specific site conditions optimal for infiltration of precipitation. In steep topography, open areas may have greater evaporation and runoff, reducing water entering the ground.

In areas where groundwater is shallow, the most immediate effect of forests is the access of their root systems to the water table. Here groundwater is subject to diurnal and seasonal fluctuations in response to transpiration losses (Groeneveld and Or, 1994). The diurnal pattern of fluctuations is most distinct near small headwater streams; it can be used as an estimate of ET rate over time. In areas where soils are frozen and precipitation is dominated by snow, forests can increase snow accumulation and delay snowmelt in the spring. This frequently results in an increase of precipitation infiltration and consequently raises the groundwater table to above those groundwater tables in the open (Pereira, 1973).

10.3.4.2 Groundwater in Level Terrain

In areas of gently level terrain where the subsurface conditions are more uniform and the horizontal movement of groundwater is slow, forests lower the groundwater table. The forest–groundwater relation is consistent in many studies, including those in South Africa (Wicht, 1949), Italy (Wilde et al., 1953), Denmark (Holstener-Jorgensen, 1967), Finland (Heikurainen, 1967), the United States (Urie, 1977; Williams and Lipscomb, 1981; Bliss and Comerford, 2002), Australia (Borg et al., 1988), Canada (Pothier et al., 2003), and other countries. Those studies showed that the fluctuations in groundwater tables reflected the differences in transpiration rate caused by seasons, time of day, hardwoods versus conifers, clear-cutting intensities, and land-use conditions.

In low-lying swampy areas where groundwater is excessive, lowering the groundwater table by forest transpiration can be beneficial to economic use of the site. However, water-loving plants and trees (phreatophytes) along streams and canals can be detrimental to the water supply in arid and semiarid areas. While clearing phreatophytes is an approach to conserving water resources in arid areas, cutting forests in humid low-lying areas such as the southeast coastal plains can bring groundwater to the surface horizons (Xu et al., 1999). This then turns the whole area into a waterlogged site, creating an anaerobic condition that is harmful to root development, retardatory to tree growth, and very detrimental to forest regeneration.

For example, more than 120,000 ha of somewhat poorly drained saline soils support mature loblolly (*Pinustaeda*) and shortleaf (*Pinus echinata*) pines in central east Texas. Artificial pine regeneration on the clear-cut sites of these soils is extremely difficult, with failure in three attempts in some cases. Transpiration and interception reduction raises the groundwater level, which may have brought salt along with it to near the surface. Thus, seedling mortality may have been caused by high water tables, toxic salt concentrations, nutrient imbalance, or combinations of all three (Chang et al., 1994).

In Western Australia, completely converting a forested watershed into agriculture moved the groundwater table upward more than 2.7 m/year in a 4 year period. This was equivalent to an increase in recharge of 6%–12% of rainfall (Peck and Williamson, 1987). Wilde et al. (1953) referred to the experience of Trappist monks at the Fontana, near Rome, Italy, on malarial swamplands. Planting a deep-rooted, fast-growing species of eucalyptus lowered the groundwater table 1 m and greatly reduced the mosquito population.

10.4 Forest Fires

Fires in forests or wild lands can be ignited either by human or natural forces. Intense wildfires can destroy forest vegetation, shift plant species and types, enhance exotic plants, spread ash, mineralize the forest duff layer, destroy wildlife habitats, degrade air quality, affect soil physical, chemical, and biological properties, and cause tremendous impacts to watershed water resources.

Studies on wildfires that occurred between 1970 and 2003 in western United States show that large wildfire (>400 ha) activity increased suddenly in the middle 1980s, with greater burned areas (>6.5 times), higher frequency (4 times), longer duration (7.5 vs. 37.1 days), and longer wildfire seasons (averaged 78 days) (Westerling et al., 2006). Concerns about uncontrolled wildfire impacts to the environment include, among others, the acceleration of worldwide desertification (Neary, 2009), water quantity and quality, stream habitats, air quality, and carbon release.

10.4.1 Nature of Forest Fires

Forest fires, like other weather events, are natural phenomenon of less frequency. They can be ignited by natural forces such as lightning, earthquakes, erupting volcanoes, and drought. The occurrence interval of forest fires, ranging from a few years to a few hundreds of years, depends on vegetation communities and region. Typically, as vegetation communities mature, combustible matter accumulates more rapidly than decomposition rates, increasing fuel loads and catastrophic wildfire probabilities. In these ecosystems, frequent low-intensity fires keep fuel loads at levels that prevent such conflagrations.

Forest fires can be categorized as *ground fires* (on the humus layer of the forest floor down to the forest soil, not appreciably above the surface), *surface fires* (on forest undergrowth and surface litter), and *crown fires* (burning through the tops of trees or shrubs). The surface fire is low intensity and the most common type of fires, leaving trees intact, while the crown or canopy fire is of high intensity, which jumps along the top of trees, is spread by wind, and can destroy entire forests. Wildfires are sometimes characterized by large burn areas (can go beyond 1,000,000 ha), fast spreading (advancing 400 ha of eucalyptus forests in 30 min in Australia), spotting, sweeping over various directions in very short periods of time, extremely high temperatures, and causing rapid destruction. Sullivan et al. (2003) noted that in Australian wildfires flame tip temperature can be 300°C, with peak temperatures around 927°C.

Wildfires are estimated to burn 20 million ha of forest vegetation each year. The majority of forest fires have been of anthropogenic origin, about 40%–90% of the total (Tishkov, 2004). However, the threat of wildfires varies among different ecosystems.

10.4.1.1 Mediterranean Forests and Shrubs

The dry summer conditions of Mediterranean climates found around the Mediterranean Sea, coastal California, and Southern Australia are regions of the world that are highly prone to wildfire. Wildfires in these regions are rarely ignited naturally, mostly induced by campfires, cigarettes, debris burning, and arson. Analyzing of spatial variation in fire danger during episodic wind events in coastal southern California, Moritz et al. (2010) found fire frequency and ignition probabilities are more associated with factors other than fire weather conditions such as human population and road densities. But, once fires are ignited, the process of fire growth tends to be weather-driven. In particular, vegetation in these regions is highly diverse and fragmented, making fire prevention and control complex. The oil base of eucalypt forests in Australia makes forest fire devastating once ignited.

10.4.1.2 Boreal Forests

Continental climate conditions such as high winds, seasonal drought, and frequent lightning along with highly combustible forests and peat make forest fire a periodic phenomenon of the Boreal ecosystem. Subsequently, fire regulates major functions and processes of these ecosystems, and the forests are able to adapt to their prefire status.

10.4.1.3 Tropical Rainforests

The humid environment makes natural wildfires a very rare and small event. Frequent thunderstorm activities preclude forest ignition by lightning. Large-scale fires are most often associated with human activities such as slash-and-burn agriculture practices. Forest logging and burning for agricultural activities triggered forest fires, burning a total of 9.7 million ha over all major islands of Indonesia during the 1997/1998 dry season and these fires generated between 22% and 33% of global CO_2 emissions that year (Kluser et al., 2004).

10.4.1.4 Subtropical Grass and Shrubs

The subtropical savannahs are extremely flammable during the dry season. Thus, fire is an essential factor for the stability of the ecosystems and lightning ignited most fires in the past. Human activities have now become the major cause of fire in these regions. This is because fire is an inexpensive and effective tool for land clearing, improving the palatability of pasture, increasing hunting visibility, eliminating parasites, and getting rid of insects and undesirable weeds/species.

10.4.2 Impacts on Soil

During a fire, temperatures in the canopy of chaparral brush can reach over 1100°C and temperatures can reach about 850°C at the litter–soil interface (DeBano, 2000). After burning, vegetation cover is combusted, organic matter is consumed, soil pore spaces are clogged with plant ash, oxidized elements are leached out, and soil particles are subject to erosion. All these processes can subsequently alter the physical and chemical properties of soils. Also, burning can cause the volatilization of organic compounds with subsequent condensation on the topsoil surfaces and cooler underlying soil layers, inducing soil water repellency. The fire-induced water repellency (hydrophobicity) of soils, generally confined to a few centimeters or decimeters below the soil surface, prevents dry soils from getting wet or water from entering into the soil. Postfire soil water repellency can be broken down in a few years or destroyed at high temperatures (DeBano, 1981; Larsen, 2009) (Figure 10.11).

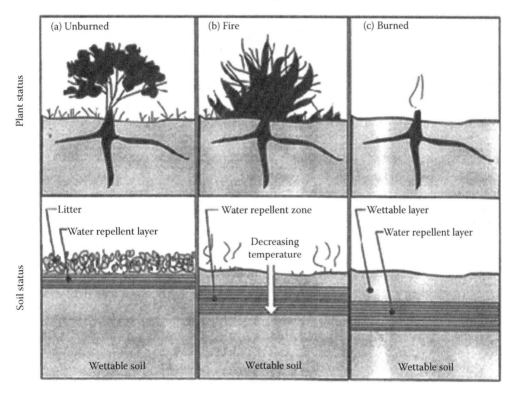

FIGURE 10.11
Soil water repellency found in unburned litter, duff, and mineral soil layers immediately beneath the shrub plants (a), hydrophobic substances are vaporized, moving downward along temperature gradients during fires (b), and a water repellent layer is present below and parallel to the soil surface after fires (c). (Adapted from DeBano, L.F., Water repellent soils: A state-of-the-art, General Technical Report PS W-46, USDA Forest Service, Berkeley, CA, 1981.)

The impact on soil properties can be short term, long term, or permanent, depending on type of vegetation, soil texture, severity, duration and frequency of fires, and postfire climatic conditions. Impacts are usually more pronounced on fine-textured soils and in arid areas. If postfire weather conditions allow plant succession and recolonization in the burned area to proceed promptly, the prefire level of most properties can be recovered and even enhanced. Fires with low or moderate severity, such as prescribed burning used in forest management, can eliminate undesirable species, promote renovation of the dominant vegetation, and increase soil-available nutrients and pH values (Certini, 2005).

Physically, fire usually causes an increase in bulk density and a decrease in hydraulic conductivity and porosity in soils. The increase in bulk density and decrease in porosity are a result of the collapse of the organo-mineral aggregates and the clogging of soil pores by ash or freed clay minerals. Postfire saturated hydraulic conductivity of forest soils has reported reduction of 50% in a few studies (Ekinci, 2006). Since soil organic matter is consumed by fire, soil structure stability can be decreased (Certini, 2005), which makes soil more susceptible to raindrop and surface runoff erosion. All these impacts tend to lower infiltration rate and soil water-holding capacity, thus accelerating surface runoff and enhancing soil erosion.

Low severity fires have very little effect on soil chemical properties (Hatten et al., 2005) and increases in soil pH, available nutrients, and base cations (Certini, 2005). Under high severity of fires, however, much of organic matter is transformed into CO_2 and water, and remaining nutrients are considerably lost through volatilization or mineralized as oxides. Thus, burning gasifies C, N, and P, but leaves Ca, Mg, and K behind, which can be transported by surface runoff. The loss of organic matter can cause a subsequent reduction in cation-exchange capacity. All of these impacts are controlled by fire weather conditions, fuel load, type and moisture content, soil type, and topography.

10.4.3 Impacts on Runoff

Fire destroys forest canopies and understory vegetation; its impacts on water quantity, largely through reduction of transpiration and interception are similar to that of "forest cutting," discussed in the previous section (Section 10.3.1). However, wildfire creates soil water repellency and converts vegetation into ash capable of clogging soil pores; makes soils with physical properties that impedes infiltration and increases overland flow, a great alteration on watershed runoff processes. This leads to the impact of fire-cleared forests on runoff with magnitudes and scale to be greater than that of mechanically cleared forests.

Large increases in peak flow and runoff volume along with decreases in flow timing are common in the first and second years after fires. Compared to magnitudes in unburned watersheds, peak flows increased 6–23 times in the Santa Anita chaparral watershed, southern California burned by the Monrovia Peak fire in late 1953 (Nasseri, 1989), 10–1000 times in the mixed coniferous watersheds near Los Alamos National Lab., New Mexico by the Cerro Grande fire in 2000 (Gallaher and Koch, 2004), and 90–2350 times in the coniferous Stermer Ridge watershed, north-central Arizona by the Rodeo-Chediski wildfire in 2002 (Gottfried et al., 2003).

Volume of runoff in the first 1–2 years after high-severity fires can increase by one to two orders of magnitude over values expected under unburned conditions in southwestern United States (Nasseri, 1989; Gallaher and Koch, 2004). Studies on a wildfire-burned eucalypt-forested watershed in northeastern Victoria, Australia showed an increase in annual streamflow of 41% for the immediate postfire year and decreases in flow to prefire level in

4 years (Lane et al., 2010). In British Columbia, Canada where 50% of annual precipitation occurs as snow and more than 80% of streamflow runs through the April–August period—a watershed with 60% of its coniferous forests severely burned increased the average seasonal water yield by 24% over a 4 year period, caused higher and earlier annual peak flows, and advanced the half-flow volume 17 days earlier as compared to unburned watershed (Cheng, 1980). The removal of forest by fires tends to lead to a higher rate of snowmelt, a shorter snowmelt period, and less snow evaporation and sublimation, and results in greater concentration of snowpack runoff. Also, the dark burned materials have lower albedo values, thus absorbing more solar radiation, and adding more energy to accelerate snowmelt.

Variations on the impacts of wildfire shown previously are largely controlled by fire severity, section of the watershed burned, fuel loads, soil and topography, and postfire weather conditions. A severe wildfire burns the entire watershed and when shortly followed by a severe storm can generate tremendous amount of runoff to stream channels and cause a rapid rise of peak flows. Things could be different if the wildfire or storm were to occur only in upper sides of the watershed.

10.5 Afforestation

The term afforestation refers to the artificial establishment of forest on lands that previously did not grow forest within living memory, while reforestation refers to the artificial establishment of forest on lands that grew forest before (Lund, 2011). The conversion of natural grasslands to plantations is common in the Southern Hemisphere. It has also been proposed as a mitigation strategy for global warming due to its carbon sink capacity. There are some discussions that allow for developed countries to offset part of their obligations on CO_2 emission reduction by purchasing "carbon credits" from afforestation/reforestation programs in developing countries. However, converting grasslands or shrubs to plantations has potential to alter the hydrologic processes in watersheds and should be explicitly addressed, especially in water-shortage areas, before such forest carbon sequestration programs are implemented.

10.5.1 Water Yield

Afforestation introduces large forest canopies and deep root systems to the planted areas, causing great losses of forest transpiration and canopy interception. It is therefore expected to result in a reduction of water yield, along with lower low and peak flows. However, the water yield reductions should not be estimated as a reversed process of "Forest Cutting" as discussed in Section 10.3.1 of this chapter. They are not necessarily opposite in treatment effects due to differences in background land uses and soil conditions. Response in water yield following afforestation is expected to be slower than that of deforestation because vegetation takes time to develop.

Water yield reductions of afforestation are generally affected by species, differences in vegetative characteristics between the original and afforested species, age of plantation, site conditions, and rainfall. Afforestation of eucalypts usually reduces runoff more than that of pine plantations (Scott and Lesch, 1997). Such reductions increase with the percent of watershed afforested, annual precipitation, and the plantation's age for at least 20 years. However, expressing the reduction in water yield in percent, the reduction is greater in drier areas and during drier years (Farley et al., 2005; Trabucco et al., 2008). Reductions are not detectable if afforestation covers less than 15%–20% of the watershed area (Zhang et al., 2007).

The maximum reductions in water yield following the afforestation of grassland watersheds were 470 mm/year for eucalypts and 205 mm/year for pine plantations at Mokobulaan, South Africa (Scott and Lesch, 1997), 289 mm/year for *Pinus radiata* at Rotorua, New Zealand (Fahey, 1994), and 242 mm/year for *P. radiata* in the Andean highlands of Ecuador (Buytaert et al., 2007). Most experiments suggest that a 20%–50% reduction in water yield can be expected for the afforestation of grassland watersheds with eucalyptus or pine plantations.

A review of watershed studies from 94 experiments showed that, on the average, pine and eucalypt plantations cause a 40 mm/year decrease in runoff for a 10% increase of forest cover in grassland watersheds (Bosch and Hewlett, 1982). Figure 10.12b is a curve on relative annual water yield reduction in mm for grassed watersheds totally afforested as a function of annual precipitation. It was derived from a worldwide study of forest and grassland runoff using data from 250 watersheds (Zhang et al., 2001, 2007). According to the curve, for a water yield reduction of 400 mm/year estimated by Bosch and Hewlett (1982), the watershed should be located in areas with annual precipitation around 1750 mm.

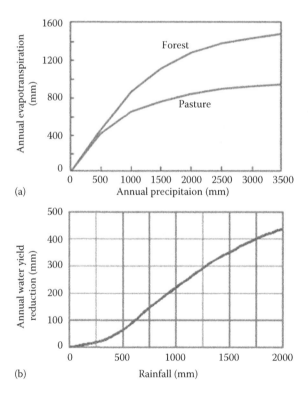

FIGURE 10.12
Annual ET as a function of annual precipitation for forested and pastured watersheds (a), and annual water yield reduction due to afforestation of pastured watersheds as a function of annual precipitation (b), both were derived from 250 experimental watersheds in 28 countries around the world by Zhang et al. (2001). (Adapted Zhang, L. et al., Afforestation in a catchment context: Understanding the impacts on water yield and salinity, Industry Report 1/07, ISBN 1876810 09 2, eWater CRC, Melbourne, Victoria, Australia, 2007. With permission.)

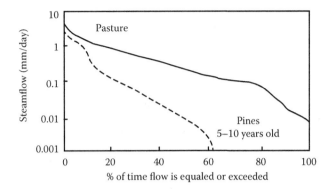

FIGURE 10.13
Flow-duration curves for two small watersheds near Tumut, New South Whales. (Adapted Zhang, L. et al., Afforestation in a catchment context: Understanding the impacts on water yield and salinity, Industry Report 1/07, ISBN 1876810 09 2, eWater CRC, Melbourne, Victoria, Australia, 2007. With permission.)

10.5.2 Streamflow Timing

Depending on species, soils, precipitation, topography, and areal coverage of afforestation, the reduction in water yield usually takes at least 3 years to become significant, at a time when the plant's canopies and root systems are significantly developed. For example, the reduction in water yield of *P. radiate* was not significant for the first 4 years at Mokobulaan, South Africa (Scott and Lesch, 1997) and 6 years in east Otago, New Zealand (Fahey, 1994), but at age 9 the plantation streamflow in southeastern Australia reached the midway between the grassland curve and the forest curve depicted in Figure 10.12a (Bren et al., 2006). The afforestation of *Eucalyptus grandis* starts to reduce streamflow significantly at age 3 in South Africa.

Reductions in water yield may continue for 20–30 years, at a time that the forest is fully developed, growth is most vigorous, ET is highest, and runoff reaches a minimum. As the forest begins to age, ET declines and runoff begins to increase. Generally, the reductions are most significant during dry seasons. It could even cause stream channels to dry up completely in such seasons, a great impact to water supplies.

The difference in streamflow between pine forest and pasture watersheds can be depicted using flow-duration curves from two nearby small catchments with great similarity in respect to aspects, topography, and climate near Tumut, New South Wales (Figure 10.13). For daily low flows equaled or exceeded 0.01 mm, the time of occurrences was 60% for the pine forest watershed and 100% for the pastured. But, for daily high flows equaled or exceeded 1.00 mm, the time of occurrences was about 10% for the pined versus 12% for the pastured watersheds. These duration curves confirm that the afforestation impact of pine on streamflow is greater for low flows than for high flows.

References

Arcement, G.J., Jr. and Schneider, V.R., 1989, Guide for selecting manning's roughness coefficients for natural channels and flood plains, Water-Supply Paper 2339, U.S. Geological Survey, Reston, VA.

Baker, M.B., Jr., 1984, Changes in streamflow in an herbicide-treated pinoy-juniper watershed in Arizona, *Water Resour. Res.*, 20, 1639–1646.

Baker, M.B., Jr., 1986, Effects of Ponderosa pine treatments on water yield in Arizona, *Water Resour. Res.*, 22, 67–73.

Barnes, H.H., Jr., 1967, Roughness characteristics of natural channels, Paper 1849, U.S. Geological Survey, Reston, VA.

Bates, C.G. and Henry, A.J., 1928, Forest and streamflow experiments at wagon wheel gap, Colorado, U.S. Weather Bureau, No. 30, Washington, DC.

Beasley, R.S., 1976, Contribution of subsurface flow from the upper slopes of forested watersheds to channel flow, *J. Soil Sci. Soc. Am.*, 40, 955–957.

Best, A. et al., 2003, A critical review of paired catchment studies with reference to seasonal flows and climatic variability, CSIRO Land and Water Technical Report 25/23, Canberra, Australian, Capital Territory, Australia.

Bethlahmy, N., 1973, Water yield, annual peaks, and exposure in mountainous terrain, *J. Hydrol.*, 20, 155–169.

Betson, R.P., 1979, The effects of clear cutting practices on upper bear creek watersheds, Alabama, Report No. WR28–1–550–101, Tennessee Valley Authority, Division of Water Resources, Knoxville, TN.

Bliss, C.M. and Comerford, N.B., 2002, Forest harvesting influence on water table dynamics in a Florida flatwoods landscape, *Soil Sci. Soc. Am. J.*, 66, 1344–1349.

Borg, H., Stoneman, G.L., and Ward, C.G., 1988, The effect of logging and regeneration on ground-water, streamflow, and stream salinity in the southern forest of western Australia, *J. Hydrol.*, 99, 253–270.

Bosch, J. M. and Hewlett, J.D., 1982, A review of catchment experiments to determine the effect of vegetation changes on water yield and evapotranspiration, *J. Hydrol.*, 55, 3–23.

Brady, N.C., 1990, *The Nature and Properties of Soils*, MacMillan, New York.

Bray, D.I., 1979, Estimating average velocity in gravel-bed rivers, *J. Hydraulic Div.*, 105, 1103–1122.

Bren, L., Lane, P., and Hepworth, G., 2010, Long-term water use of native eucalyptus forest after logging and regeneration: The Coranderrk experiment, *J. Hydrol.*, 384, 52–64.

Bren, L., Lane, P., and McGuire, D., 2006, An empirical, comparative model of changes in annual water yield associated with pine plantations in southern Australia, *Australian For.*, 69(4), 275–284.

Bureau of Land Management, 1987, Assessment of water conditions and management opportunities in support of Riparian values: BLM San Pedro River properties, Arizona, Project Completion Report, USDI, BLM Service Center, Denver, CO.

Burton, T.A., 1997, Effects of basin-scale timber harvest on water yield and peak streamflow, *J. Am. Water Resour. Assoc.*, 33, 1187–1196.

Busby, M.W., 1966, *Annual Runoff of the Conterminous United States*, USGS Hydrologic Investigations, Atlas HA-212, Washington, DC.

Buytaert, W. et al., 2007, The effects of afforestation and cultivation on water yield in the Andean Paramo, *For. Ecol. Manage.*, 251(1–2), 22–30.

Certini, G., 2005, Effects of fire on properties of forest soils: A review, *Oecologia*, 143, 1–10.

Chang, M., 1982, *Laboratory Notes: Forest Hydrology*, Center for Applied Studies in Forestry, College of Forestry, Stephen F. Austin State University, Nacogdoches, TX.

Chang, M. and Boyer, D.G., 1977, Estimates of low flows using watershed and climatic parameters, *Water Resour. Res.*, 13, 997–1001.

Chang, M. and Lee, R., 1976, Adequacy of hydrologic data for applications in West Virginia, Bulletin 7, Water Resource Institute, West Virginia University, Morgantown, WV.

Chang, M. and Sayok, A.K., 1990, Hydrologic responses to urbanization in forested La Nana Creek watershed, *Tropical Hydrology and Caribbean Water Resources*, American Water Resource Association, Herndon, VA, pp. 131–140.

Chang, M., Sayok, A.K., and Watterston, K.G., 1994, Clearcutting and shearing on a saline soil in East Texas: Impact on soil physical properties, *Proceedings of the 8th Biennial Southern Silvicultural Research Conference*, U.S. Forest Service Southern Research Station, Asheville, NC, pp. 209–214.

Chang, M. and Waters, S.P., 1984, Forests and other factors associated with streamflows in east Texas, *Water Resour. Bull.*, 20, 713–719.

Chen, Y.-C. et al., 2009, Retardance coefficient of vegetated channels estimated by the Froude number, *Ecol. Eng.*, 35, 1027–1035.

Cheng, J.D., 1980, Hydrologic effects of a severe forest fire, *Symposium on Watershed Management*, Boise, ID, ASCE, Reston, VA, pp. 240–251.

Cheng, J.D., 1988, Subsurface stormflows in the highly permeable forested watersheds of southwestern British Columbia, *J. Contam. Hydrol.*, 3, 171–191.

Chow, V.T., 1959, *Open-Channel Hydraulics*, McGraw-Hill, New York.

Cline, R.G., Haupt, H.F., and Campbell, G.S., 1977, Potential water yield response following clearcut harvesting on north and south slopes in Northern Idaho, Research Paper INT-191, U.S. Forest Service, Intermountain Forest and Range Experiment Station, Ogden, UT.

Coon, W.F., 1998, Estimation of roughness coefficients for natural stream channels with vegetated banks, Water Supply Paper 2441, U.S. Geological Survey, Reston, VA.

Cornish, P.M. and Vertessy, R.A., 2001, Forest age-induced changes in evapotranspiration and water yield in a eucalypt forest, *J. Hydrol.*, 242, 43–63.

Cosandey, C. et al., 2005, The hydrological impact of the Mediterranean forest: A review of French research, *J. Hydrol.*, 301, 235–249.

Cowan, W.L., 1956, Estimating hydraulic roughness coefficients, *Agric. Eng.*, 37, 473–475.

Davis, E.A., 1993, Chaparral control in mosaic pattern increased streamflow and mitigated nitrate loss in Arizona, *Water Resour. Bull.*, 29, 391–399.

DeBano, L.F., 1981, Water repellent soils: A state-of-the-art, General Technical Report PS W-46, USDA Forest Service, Berkeley, CA.

DeBano L.F., 2000, The role of fire and soil heating on water repellence in wildland environments: A review, *J. Hydrol.*, 231, 195–206.

Dingman, S.L. and Palaia, K.J., 1999, Comparison of models for estimating flood quantiles in New Hampshire and Vermont, *J. Am. Water Resour. Assoc.*, 35, 1233–1243.

Dingman, S.L. and Sharma, K.P., 1997, Statistical development and validation of discharge equations for natural channels, *J. Hydrol.*, 199, 13–35.

Douglass, J.E., 1983, The potential for water yield augmentation from forest management in the eastern United States, *Water Resour. Bull.*, 19, 351–358.

Ekinci, H., 2006, Effect of forest fire on some physical, chemical and biological properties of soil in Çanakkale, Turkey, *Int. J. Agric. Biol.*, 8(1), 102–106.

Elliot, W. et al., 2010, Tools for analysis, Chapter 13, *Cumulative Watershed Effects of Fuel Management in the Western United States*, Elliot, W.J., Miller, I.S., and Audin, L. Eds., General Technical Report RMRS-GTR-231, U.S. Department of Agriculture, Forest Service, Rocky Mountain Research Station, Fort Collins, CO, pp. 245–276.

Fahey, B., 1994, The effect of plantation forestry on water yield in New Zealand, *New Zealand For.*, 39(3), 18–23.

Farley, K.A., Jobbagy, E.G., and Jackson, R.B., 2005, Effects of afforestation on water yield: A global synthesis with implications for policy, *Glob. Change Biol.*, 11, 1565–1576.

Folliott, P.F. and Thorud, D.B., 1977, Water resources and multiple-use forestry in the Southwest, *J. For.*, 75, 469–472.

Fread, D.L., 1993, Flow routing, *Handbook of Hydrology*, Maidment, D.R., Ed., McGraw-Hill, New York, pp. 10, 1–36.

Gallaher, B.M. and R.J. Koch, 2004, Cerro Grande fire impacts to water quality and stream flow near Los Alamos national laboratory: Results of four years of monitoring, Los Alamos National Laboratory Report LA-14177, Los Alamos, NM.

Gingras, D., Adamowski, K., and Pilon, P.J., 1994, Regional flood equations for the provinces of Ontario and Quebec, *Water Resour. Bull.*, 30, 55–67.

Goodrich, D.C. et al., 2005, Rapid post-fire hydrologic watershed assessment using the AGWA GIS-based hydrologic modeling too, *Managing Watersheds for Human and Natural Impacts Engineering, Ecological and Economic Challenges: Proceedings of the 2005 Watershed Management Conference*, Moglen, G.E. Ed., July 19–22, 2005, Williamsburg, VA, American Society of Civil Engineers, Reston, VA, 12 pp.

Gottfried, G.J., 1983, Stand changes on a southwestern mixed conifer watershed after timber harvesting, *J. For.*, 81, 311–316.

Gottfried, G.J. et al., 2003, Impacts of wildfires on hydrologic processes in forest ecosystems: Two case studies, *Proceedings of the First Interagency Conference on Research on the Watersheds*, October 27–30, 2003, Benson, AZ., U.S. Department of Agriculture, Agricultural Research Service, Tucson, AZ, pp. 668–673.

Green, J.C., 2005, Modeling flow resistance in vegetated streams: Review and development of new theory, *Hydrol. Process.*, 19, 1245–1259.

Green, R.E. and Ampt, G.A., 1911, Studies on soil physics: 1. Flow of air and water through soils, *J. Agric. Sci.*, 4, 1–24.

Groeneveld, D.P. and Or, D., 1994, Water table reduced shrub-herbaceous ecotone: Hydrological management implications, *Water Resour. Bull.*, 30, 911–920.

Gupta, R.S., 1989, *Hydrology and Hydraulic Systems*, Waveland Press, Prospect Heights, IL.

Harr, R.D., 1976, Forest practices and streamflow in western Oregon, General Technical Report PNW-49, USDA Forest Service, Pacific Northwest Forest and Range Experiment Station, Portland, OR.

Harr, R.D., 1983, Potential for augmenting water yield through forest practices in western Washington and western Oregon, *Water Resour. Bull.*, 19, 383–394.

Harr, R.D., Fredriksen, R.L., and Rothacher, J., 1979, Changes in streamflow following timber harvest in southwest Oregon, Forest Service Research Paper PNW-249, Pacific Northwest Forest and Range Experiment Station, Portland, OR.

Harr, R.D., Levno, A., and Mersereau, R., 1982, Streamflow changes after logging 130-year-old Douglas fir in two small watersheds, *Water Resour. Res.*, 18, 637–644.

Harr, R.D. and McCorison, F.M., 1979, Initial effects of clearcut logging on size and timing of peak flows in a small watershed in western Oregon, *Water Resour. Res.*, 15, 90–94.

Hatten, J. et al., 2005, A comparison of soil properties after contemporary wildfire and fire suppression, *For. Ecol. Manage.*, 220, 227–241.

Heikurainen, L., 1967, Effect of cutting on the groundwater level on drained peatlands, *International Symposium on Forest Hydrology*, Sopper, W.E. and Lull, H.W., Eds., Pergamon Press, New York, pp. 345–354.

Hershfield, D.N., 1961, *Rainfall Frequency Atlas of the United States*, U.S. Weather Bureau, Silver Spring, MD.

Hewlett, J.D. and Hibbert, A.R., 1961, Increases in water yield after several types of forest cutting, *Int. Assoc. Sci. Hydrol. Bull.*, 6, 5–17.

Hewlett, J.D. and Troendle, C.A., 1975, Nonpoint and diffuse water resources: A variable source area problem, *Watershed Management Proceedings*, ASCE, New York, pp. 21–46.

Hibbert, A.R., 1969, Water yield changes after converting a forest catchment to grass, *Water Resour. Res.*, 5, 634–640.

Hibbert, A.R., 1983, Water yield improvement potential by vegetation management on western rangelands, *Water Resour. Bull.*, 19, 375–381.

Hibbert, A.R., Davis, E.A., and Knipe, O.D., 1982, Water yield changes resulting from treatment of Arizona chaparral, *Proceedings of Symposium on Dynamic and Management of Mediterranean-Type Ecosystems*, June 22–26, 1981, San Diego, CA, U.S. Forest Service, Pacific SW Forest and Range Experiment Station, Berkeley, CA, pp. 382–389.

Holstener-Jorgensen, H., 1967, Influence of forest management and drainage on groundwater fluctuations, *International Symposium on Forest Hydrology*, Sopper, W.E. and Lull, H.W., Eds., Pergamon Press, New York, pp. 325–333.

Hornbeck, J.W., 1973, Storm flow from hardwood-forested and cleared watersheds in New Hampshire, *Water Resour. Res.*, 9, 346–354.

Hornbeck, J.W., Pierce, R.S., and Federer, C.A., 1970, Streamflow changes after forest clearing in New England, *Water Resour. Res.*, 6, 1124–1132.

Hornbeck, J.W. et al., 1993, Long-term impacts of forest treatments on water yield: A summary for northeastern USA, *J. Hydrol.*, 150, 323–344.

Horton, R.E., 1937, Determination of infiltration capacity for large drainage basins, *Trans. Am. Geophys. Union*, 18, 371–385.

Ingebo, P.A., 1971, Suppression of channel-side chaparral cover increases streamflow, *J. Soil Water Conserv.*, 26, 79–81.

Ingwersen, J.B., 1985, Fog drip, water yield, and timber harvesting in the Bull Run municipal watershed, Oregon, *Water Resour. Bull.*, 21(3), 469–473.

Jarrett, R.D., 1984, Hydraulics of high-gradient streams, *J. Hydraul. Eng.*, 110(HY11), 1519–1539.

Jobson, H.E. and Froehlich, D.C., 1988, Basic hydraulic principles of open-channel flow, U.S. Geological Survey Open-File Report 88–707, Reston, VA.

Johnson, R., 1998, The forest cycle and low river flows: A review of UK and international studies, *For. Ecol. Manage.*, 109, 1–7.

Johnston, R.A., 1984, Effects of small aspen clearcuts on water yield and water quality, U.S. Forest Service Research Paper INT-333, Intermountain Forest and Range Experiment Station, Ogden, UT.

Keppeler, E.T. and Ziemer, R.R., 1990, Logging effects on streamflow: Water yield and summer low flows at Caspar Creek in northwestern California, *Water Resour. Res.*, 26, 1669–1679.

Kidwell, M.R., Weitz, M.A., and Guertin, D.P., 1997, Estimation of Green–Ampt effective hydraulic conductivity for rangelands, *J. Range Manage.*, 50, 290–299.

King, J.G., 1989, Streamflow response to road building and harvesting: A comparison with the clear-cut area procedure, General Technical Report INT-401, U.S. Forest Service, Intermountain Forest and Range Experiment Station, Ogden, UT.

Kirpich, Z.P., 1940, Time of concentration of small agricultural watersheds, *Civil Eng.*, 10, 362.

Kluser, T. et al., 2004, Wildland fires, a double impact on the planet, *Environment Albert Bulletin*, UNEP, GRIP/Europe, Nairobi, Kenya, 4 pp., http://www.preventionweb.net/files/1146_ewfire.en.pdf

Kochenderfer, J.M. and Aubertin, G.M., 1975, Effects of management practices on water quality and quantity: Fernow experimental forest, west Virginia, *Municipal Watershed Management, Symposium Proceedings*, General Technical Report NE-13, U.S. Forest Service, Upper Darby, PA, pp. 14–24.

Kuczera, G., 1987, Prediction of water yield reductions following a bushfire in ash-mixed species eucalypt forest, *J. Hydrol.*, 94, 215–236.

Lane, P.N.J. et al., 2010, Modelling the long term water impact of wildfire and other forest disturbance in Eucalypt forests, *J. Environ. Model. Softw.*, 25(4), 467–478.

Larsen, I.J., 2009, Causes of post-fire runoff and erosion: Water repellency, cover, or soil sealing? *Soil Sci. Soc. Am. J.*, 73(4), 1393–1407.

Lee, R., 1963, Evaluation of solar beam irradiation as a climatic parameter of mountain watersheds, Hydrology Paper 2, Colorado State University, Fort Collins, CO.

Lewis, D.C., 1968, Annual hydrologic response to watershed conversion from oak woodland to annual grassland, *Water Resour. Res.*, 4, 59–72.

Limerinos, J.T., 1970, Determination of the manning coefficient from measured bed roughness in natural channels, Water Supply Paper 1898-B, U.S. Geological Survey, Reston, VA.

Linsley, R.K., Jr., Kohler, M.A., and Paulhus, J.L.H., 1975, *Hydrology for Engineers*, McGraw-Hill, New York.

Lund, H.G. (Coordinator), 2011, Definitions of Forest, Deforestation, Afforestation, and Reforestation. [Online] Gainesville, VA: Forest Information Services. Available from the World Wide Web: http://home.comcast.net/~gyde/DEFpaper.htm, Misc. pagination.

Lusby, G.C., 1979, *Effects of Converting Sagebrush Cover to Grass on The Hydrology of Small Watersheds at Boco Mountain, Colorado* (*Hydrologic Effects of Land Use*), U.S. Geological Survey Water Supply Paper, Reston, VA.

Lynch, J.A., 2001, personal communication.

Lynch, J.A. and Corbett, E.S., 1990, Evaluation of best management practices for controlling nonpoint pollution from silvicultural operations, *Water Resour. Bull.*, 26, 41–52.

Lynch, J.A. and Corbett, E.S., unpublished report, Increasing summer storm peakflows following progressive forest clearcutting, School of Forest Resources, Pennsylvania State University, University Park, PA.

Lynch, J.A., Corbett, E.S., and Sopper, W.E., 1980, Impact of forest cover removal on water quality, Research Paper 23–604, Pennsylvania State University, Institute for Research on Land and Water Resources, University Park, PA.

McCuen, R.H., 1998, *Hydrologic Analysis and Design*, Prentice Hall, New York.

McGuinness, C.L., 1964, *Generalized Map Showing Annual Runoff and Productive Aquifers in the Conterminous United States*, Atlas HA-194, USGS Hydrologic Investigations, Reston, VA.

Mein, R.G. and Larson, C.L., 1971, Modeling infiltration component of the rainfall-runoff process, Bulletin 43, Water Resources Research Center, University of Minnesota, Minneapolis, MN.

Molchanov, A.A., 1963, *The Hydrological Role of Forests*, Forestry Institute of the USSR Academy of Sciences, Moscow, Russia (Trans. from Russian, OTS63–11089).

Moritz, M. A. et al., 2010, Spatial variation in extreme winds predicts large wildfire locations in chaparral ecosystems, *Geophys. Res. Lett.*, AGU, 37, L04801. DOI: 10.1029/2009GL041735.

Moseley, K.R. et al., 2008, Long-term partial cutting impacts on *Desmognathus salamander* abundance in West Virginia headwater streams, *For. Ecol. Manage.*, 254, 300–307.

Nasseri, I., 1989, Frequency of floods from a burned Chaparral watershed, USDA Forest Service General Technical Report PSW-109, 00, Berkeley, CA, pp. 68–71.

Neary, D.G., 2009, Post-wildland fire desertification: Can rehabilitation treatments make a difference? *Fire Ecol.* 5(1), 129–144.

Patric, J.H. and Reinhart, K.G., 1971, Hydrologic effects of deforesting two mountain watersheds in West Virginia, *Water Resour. Res.*, 7, 1182–1188.

Peck, A.J. and Williamson, D.R., 1987, Effects of forest clearing on groundwater, *J. Hydrol.*, 94, 47–65.

Pereira, H.C., 1973, *Land Use and Water Resources*, Cambridge University Press, Cambridge, U.K.

Pilgrim, M.P. and Cordery, I., 1993, Flood runoff, *Handbook of Hydrology*, Maidment, D.R., Ed., McGraw-Hill, New York, pp. 9.1–9.42.

Pitt, M.D., Burgy, R.H., and Heady, H.F., 1978, Influences of brush conversion and weather patterns on runoff from a northern California watershed, *J. Range Manage.*, 31, 23–27.

Pothier, D., Prevost, M., and Auger, I., 2003, Using the shelterwood method to mitigate water table rise after forest harvesting, *For. Ecol. Manage.*, 179, 573–583.

Pritchett, W.L., 1979, *Properties and Management of Forest Soils*, John Wiley & Sons, New York.

Putuhena, W.M. and Cordery, I., 2000, Some hydrological effects of changing forest cover from eucalypts to *Pinus radiate*, *Agric. For. Meteorol.*, 100, 59–72.

Rawls, W.J. and Brakensiek, D.L., 1983, A procedure to predict Green and Ampt infiltration parameters, *Proceedings of the American Society of Agricultural Engineers on Advances in Infiltration*, Chicago, IL, 5.1–5.51

Rawls, W.J. et al., 1993, Infiltration and soil water movement, *Handbook of Hydrology*, Maidment, D.R., Ed., McGraw-Hill, New York, 5.1–5.51.

Reich, B.M., 1972, The influence of percentage forest on design floods, *National Symposium on Watershed in Transition*, American Water Resources Association, Urbana, IL, pp. 335–340.

Reinhart, K.G., Eschner, A.R., and Trimble, G.R., Jr., 1963, Effect on streamflow of four forest practices in the mountains of west Virginia, Research Paper NE-1, U.S. Forest Service, Northeastern Forest Experiment Station, Upper Darby, PA.

Reinhart, K.J. and Trimble, G.R., 1962, Forest cutting and increased water yield, *J. Am. Water Works Assoc.*, 54, 1464–1472.

Rich, L.R. and Gottfried, G.J., 1976, Water yields resulting from treatments on the Workman Creek experimental watersheds in central Arizona, *Water Resour. Res.*, 12, 1053–1060.

Richardson, C.W., Burnett, E., and Bovey, R.W., 1979, Hydrologic effects of brush control on Texas rangeland, *Trans. ASAE*, 22(2), 0315–0319.

Rothacher, J., 1970, Increases in water yield following clear-cut logging in the Pacific West, *Water Resour. Res.*, 6, 653–658.

Rowe, P.B. and Reimann, L.F., 1961, Water use by brush, grass, and grass-forb vegetation, *J. For.*, 58, 175–181.

Satterlund, D.R. and Eschner, A.R., 1965, The surface geometry of a closed conifer forest in relation to losses of intercepted snow, USDA Forest Service Research Paper NE-34, NE Forest Experiment Station, Upper Darby, PA.

Sauer, V.B., 1990, Written communication, U.S. Geological Survey, Reston, VA, cited in Coon (1998).

Schwab, G.O., Fangmeier, D.D., and Elliot, W.J., 1996, *Soil and Water Management Systems*, 4th edn., John Wiley & Sons, New York, 371 pp.

Scott, D.F. and Lesch, W., 1997, Streamflow responses to afforestation with *Eucalyptus grandis* and *Pinus patula* and to felling in the Mokobulaan experimental catchments, South Africa, *J. Hydrol.*, 199, 360–377.

Scott, D.F. and Smith, R.E., 1997, Preliminary empirical models to predict reductions in total and low flows resulting from afforestation, *Water SA*, 23(2), 135–140.

Skaggs, R.W. and Khaleel, R., 1982, Infiltration, *Hydrologic Modeling of Small Watersheds*, Hann, C.T., Johnson, H.P., and Brakensiek, D.L., Eds., ASAE, St. Joseph, MI, pp. 121–166.

Smerdon, B.D., Redding, T.E., and Beckers, J., 2009, An overview of the effects of forest management on groundwater hydrology, *B.C. J. Ecosyst. Manage.*, 10(1), 22–44.

Smith, P.J.T., 1987, Variation of water yield from catchments under introduced pasture grass and exotic forest, east Otago, *J. Hydrol. (NZ)*, 26(2), 175–184.

Soil Conservation Service, 1979, *National Engineering Handbook, Section 4: Hydrology*, U.S. Department of Agriculture, Washington, DC.

Stednick, J.D., 1996, Monitoring the effects of timber harvest on annual water yield, *J. Hydrol.*, 176, 79–95.

Sturges, D.L., 1994, High-elevation watershed response to sagebrush control in southcentral Wyoming, Research Paper RM-318, U.S. Forest Service, Rocky Mountain Forest and Range Experiment Station, Fort Collins, CO.

Sullivan, A.L., Ellis, P.F., and Knight, I.K., 2003, A review of radiant heat flux models used in bushfire applications, *Int. J. Wildland Fire*, 12, 101–110.

Swank, W.T. and Johnson, C.E., 1994, Small catchment research in the evaluation and development of forest management practices, *Biogeochemistry of Small Catchments: A Tool for Environmental Research*, Moldan, M. and Cerny, J., Eds., John Wiley & Sons, New York, pp. 383–408.

Swank, W.T. and Miner, N.H., 1968, Conversion of hardwood-covered watershed to white pine reduces water yield, *Water Resour. Res.*, 4, 947–954.

Swank, W.T. and Vose, J.M., 1994, Long-term hydrologic and stream chemistry responses of southern Appalachian catchments following conversion from mixed hardwoods to white pine, *Hydrologie kleiner Einzugsgebiete: Gedenkchrift Hans M. Keller*, Landolt, R., Ed., Schweizerische Gesellschaft fur Hydrologie und Limnologie, Bern, Switzerland, pp. 164–172.

Swift, L.W., Jr., 1976, Computational algorithm for solar radiation on mountain slopes, *Water Resour. Res.*, 12, 108–112.

Thomas, R.B. and Megahan, W.F., 1998, Peak flow responses to clear-cutting and roads in small and large basins, western Cascades, Oregon: A second opinion, *Water Resour. Res.*, 34, 3393–3403.

Tishkov, A.A., 2004, Forest fires and dynamics of forest cover, *Natural Disasters*, Kotlyakov, V.M., Ed., the Encyclopedia of Life Support Systems (EOLSS) Developed under the Auspices of the UNESCO, Eolss Publishers, Oxford, U.K.

Trabucco, A. et al., 2008, Climate change mitigation through afforestation/reforestation: A global analysis of hydrologic impacts with four case studies, *Agric. Ecosyst. Environ.*, 126, 81–97.

Troendle, C.A., 1970, The flow interval method for analyzing timber harvesting effects on streamflow regimen, *Water Resour. Res.*, 6, 328–332.

Troendle, C.A., 1983, The potential for water yield augmentation from forest management in the Rocky Mountain region, *Water Resour. Bull.*, 19, 359–373.

Troendle, C.A. and King, R.M., 1985, The effect of timber harvest on the Fool Creek watershed, 30 years later, *Water Resour. Res.*, 21, 1915–1922.

Troendle, C.A. and King, R.M., 1987, The effects of partial and clearcutting on streamflow at Deadhorse Creek, Colorado, *J. Hydrol.*, 90, 145–157.

Troendle, C.A. and Olsen, W.K., 1994, Potential effects of timber harvest and water management on stream dynamics and sediment transport, *Sustainable Ecological Systems: Implementing an Ecological Approach to Land Management*, Covington, W.W. and DeBnao, L.F., Eds., General Technical Report RM-247, Forest Service Rocky Mountain Forest and Range Experiment Station, Fort Collins, CO, pp. 34–41.

Urie, D.H., 1977, Ground water differences on pine and hardwood forests of the Udell experimental forest in Michigan, Research Paper NC-145, USDA Forest Service North Central Forest Experiment Station, St. Paul, MN.

Ursic, S.J., 1991, Hydrologic effects of two methods of harvesting mature southern pine, *Water Resour. Bull.*, 27, 303–315.

U.S. Forest Service, 1982, An assessment of the forest and range situation in the United States, Forest Resource Report 22, Washington, DC.

Verry, E.S., 1972, Effect of an aspen clearcutting on water yield and quality in northern Minnesota, *National Symposium on Watersheds in Transition*, Fort Collins, CO, American Water Resources Association, Minneapolis, MN, pp. 276–284.

Westerling, A. L. et al., 2006, Warming and earlier spring increase western U.S. forest wildfire activity, *Science*, 313, 940–943.

Wicht, C.L., 1949, Forestry and water supplies in South Africa, Department of Agriculture, South Africa, Bulletin 33.

Wilde, S.A. et al., 1953, Influence of forest cover on the state of the groundwater table, *Soil Sci. Soc. Proc.*, 17, 65–67.

Williams, T.M. and Lipscomb, D.J., 1981, Water table rise after cutting on coastal plain soils, *South. J. Appl. For.*, 5, 46–48.

Xu, Y.J. et al., 1999, Recovery of hydroperiod after timber-harvesting in a forested wetland, *Proceedings of the 10th Biennial Southern Silvicultural Research Conference*, Shreveport, LA, U.S. Forest Service Southern Research Station, Asheville, NC, pp. 282–287.

Zhang, L., Dawes, W.R., and Walker, G.R., 2001, The response of mean annual evapotranspiration to vegetation changes at catchment scale, *Water Resour. Res.*, 37, 701–708.

Zhang, L. et al., 2007, Afforestation in a catchment context: Understanding the impacts on water yield and salinity, Industry Report 1/07, eWater CRC, Melbourne, Victoria, Australia, 60 pp.

Zon, R., 1927, Forests and water in the light of scientific investigation, USDA Forest Service, Lake States Forest Experiment Station 106, St. Paul, MN.

11

Forests and Streamflow Quality

Forested watersheds are generally at high elevations and in remote areas with little development and few disturbances. Forest canopies, floor, and root systems also provide the best mechanisms in nature to protect the watershed from accelerated soil erosion, landslides, and nutrient losses. Coupling these conditions with cooler water temperatures, due either to the higher elevation or the shading effect of riparian vegetation, makes the quality of water flowing through forested watersheds better than in other surface conditions.

Since forests are subject to use and management for their various resources, activities in forested areas can destroy the equilibrium of the ecosystem and adversely affect water quality. About 80% of the Nation's freshwater resources originate from forested watersheds (EPA, 2000a).Thus, maintaining a good water quality in watersheds with forest activities is a challenge to foresters.

11.1 Water Pollutants

All substances that make water unsuitable for intended uses are water pollutants. They can be classified into eight categories.

11.1.1 Sediments

Stream sediments are soil and mineral particles, usually inorganic, but in part organic or composites of the two, released from the land by surface runoff, streamflow, wind, melting glaciers, raindrop impact, animals, gravity, or avalanche. The amount of sediment delivered to waters in the United States each year exceeded 4×10^9 tons in the 1960s, a quantity greater than the total sewage load by 500–700 times (Glymph and Carlson, 1966). Today, it is still the largest single water pollutant in the nation (EPA, 1997, 2000b).

Physically, the presence of sediment can affect water turbidity, light penetration, energy exchange, taste, odor, temperature, and abrasiveness. When deposited, it can deplete reservoir capacity, clog stream channels and drainage ditches, alter aquatic habitat, suffocate fish eggs and bottom-dwelling organisms, form alluvium, and increase flooding and flood damages. Also, excessive siltation interferes with drinking-water-treatment processes and recreational use of rivers. Chemically, many plant nutrients, mineral elements, organic chemicals, fertilizers, insecticides, herbicides, and radioactive substances can attach to the soil particles to be carried to streams with sediment. These substances are responsible for degradation of water quality, lake eutrophication, and harmful effects on aquatic life and human health. Biologically, aquatic ecosystems can be affected directly by the physical presence of sediment or indirectly by the interaction of sediment with the physical and chemical environment of the stream.

11.1.2 Heat

Excessive heat caused by the summer sun, the reduction of flow volume, the removal of riparian plants, and the discharge of heated water from power plants and other industries can be a serious pollution problem. Higher water temperature causes lower viscosity, greater sediment-falling velocity, increased chemical reaction, reduction in dissolved oxygen, and increased evaporation. Undesirable blue-green algae and other destructive microorganisms can reproduce or be introduced in warm water due to their tolerance of higher temperatures. Not only are blue-green algae undesirable as aquatic animal food, they also give off toxic substances and cause a series of problems in water quality. A change in water temperature by a few degrees can adversely affect the survival, migration, spawning, and reproduction of fish.

11.1.3 Oxygen-Demanding Wastes

Oxygen-demanding wastes include domestic sewage, animal manure, and some industrial wastes of largely carbonaceous organic material that can be decomposed by microorganisms to carbon dioxide, water, some basic ions (such as NO_3^-, PO_4^{2-}, SO_4^{2-}), and energy. The decomposition of organic material in water is carried out by bacteria at the expense of dissolved oxygen. Since the maximum level of dissolved oxygen in water is about 15 mg/L, the activity of bacteria can deplete oxygen concentration. This can cause detrimental effects to other aquatic organisms if sufficient organic matter and other conditions are favorable for multiple expansion and bacteria activity. Large amounts of sewage or other oxygen-demanding wastes from industry or agriculture can deplete the dissolved oxygen level in the water to septic conditions.

A common parameter to describe the oxygen demand of waste is the 5 day biochemical oxygen demand (BOD). It measures the rate at which oxygen is used rather than the level of oxygen in the water or in certain specific pollutants. A very low value of BOD implies either that the water is clean or that the bacteria are inactive. In either case, the dissolved oxygen level is usually high.

11.1.4 Plant Nutrients

Frequently, plant nutrients such as nitrogen and phosphorus are added to reservoirs by dissolving in runoff water or attaching to eroding soil particles. Nutrients may come from:

- Soil and water erosion
- Agricultural fertilizers
- Domestic sewage
- Livestock wastes
- Decomposition of plant residuals
- Phosphate detergents

Nutrients stimulate the growth of aquatic plants, which in turn interfere with water use, produce disgusting odors when they decay, and add BOD to the water. Waters excessively enriched with nutrients are referred to as eutrophic. They have high plant populations and appear turbid and greenish in color. In eutrophic waters, algae and aquatic weeds bloom

TABLE 11.1

Estimated National Background Concentrations for Five Nutrients in U.S. Streams and Groundwater

Water Source	Ammonia		Nitrate		Total Nitrogen		Ortho-Phosphorus		Total Phosphorus	
	Sites	mg/L	Sites	mg/L	Sites	mg/L	Sites	mg/L	Sites	mg/L
Streams	89	0.025	108	0.24	88	0.58	89	0.010	84	0.034
Groundwater	177	0.100	419	1.00			160	0.030		

Source: Dubrovsky, N.M. et al., The quality of our nation's waters—Nutrients in the nation's streams and groundwater, 1992–2004, National Water-Quality Assessment Program, USGS Circular 1350, Reston, VA, 2010.

excessively, destroy aesthetic qualities, and interfere with fishing, sports, navigation, irrigation, and the generation of hydroelectric power.

In a 10 year study on stream nutrients across the United States, Dubrovsky et al. (2010) found that all five study nutrients (nitrate, ammonia, total nitrogen [TN], ortho-phosphorus, and total phosphorus [TP]) exceeded background concentrations at more than 90% of 190 sample streams draining agricultural and urban watersheds. The median concentrations of TP and TN were at 0.25 and 4.0 mg/L, respectively, about six to seven times greater than background concentrations. The background concentrations for the five nutrients are given in Table 11.1. They are the nutrient concentrations present in streams and groundwater as a result of purely natural processes.

11.1.5 Disease-Causing Agents

Waste discharges from cities, slaughtering plants, animal feedlots, or ships can carry bacteria, viruses, or other microorganisms capable of producing disease in humans and animals. These microorganisms are responsible for such diseases as cholera, typhoid fever, dysentery, polio, and infectious hepatitis. However, waters contain numerous kinds of pathogenic bacteria, and identification of each one is both time-consuming and expensive for routine pollution tests. Coliform bacteria are relatively harmless microorganisms that live in the gut of animals and are discharged with feces in sewage-contaminated waters. Their existence and density are reliable indicators of the adequacy of treatment for the reduction of pathogens in water. Thus, the most probable number of coliform bacteria in water samples has become a standard parameter for the determination of pathogenic pollution. A high count of coliforms indicates the probability of the presence of a high number of disease-causing microorganisms in the water. The maximum for drinking water is 1 per 100 mL, and for swimming 200 per 100 mL.

11.1.6 Inorganic Chemicals and Minerals

Many acids, salts, heavy metals, and toxic pollutants from industrial and municipal discharges affect water quality adversely in more than 70% of the large river basins in the United States (CEQ, 1978). Acid mine drainage (AMD) arises from strip-mining and underground-mining areas when iron pyrites (FeS_2) are oxidized into ferric hydroxide [$Fe(OH)_3$] and sulfuric acid (H_2SO_4) through water, air, and bacteria. AMD affects a total length of more than 17,600 km of U.S. streams (ReVelle and ReVelle, 1984). It is characterized by highly acidic water and stream bottoms coated with ferric hydroxide (yellow color), creating an environment toxic to aquatic life.

Mercury, cadmium, arsenic, zinc, and lead are some examples of other inorganic chemicals contaminating rivers and water supplies, toxic to aquatic species and harmful to human health. Chlorine has been applied to destroy disease-causing bacteria in drinking water, but it can combine with organic compounds in the water to form chlororganic compounds such as chloroform, carbon tetrachloride, and trihalomethanes (THMS). These chemicals have been found to cause cancer in rats and mice and may have some effects in people (Owen, 1985). Concerns on the content of mercury (Hg) in fish and areas with fish Hg concentrations potentially harmful to human health are wide-spread among federal, state, and local governments (Chalmers et al., 2011). In 2006, a total of 5,739,747 lake hectares and 1,420,687 river kilometers were under fish consumption advisory for mercury, an increase of 8% for lakes and 15% for rivers under advisory between 2004 and 2006 (EPA, 2007).

Many different pollutants are present in the stream environment. The presence of certain particulate matter can increase the toxicity of other pollutants, a function usually referred to as *potentiation*. In other cases the combined effects of two or more pollutants are more severe or different in quality from each individually, a phenomenon called *synergism*. If the combined effects of two pollutants are less severe than those of each individual pollutant, the interaction is called *antagonism*. All these interactions make the effects of water pollution extremely difficult to study.

11.1.7 Synthetic Organic Chemicals

Detergents, plastics, oil, septic-tank cleaners, phenols, DDT and other pesticides, and many other organic compounds are products of modern industrial technology. Some of these chemicals are slowly degradable pollutants, while others are non-degradable. The slowly degradable chemicals remain in water for a long period of time but are eventually broken down to harmless levels by natural processes. Non-degradable pollutants, such as plastics, cannot be broken down in nature and must be controlled either by removing them through mechanical treatment or by preventing them from entering the environment.

Synthetic organic chemicals can disrupt aquatic ecosystems; damage the economic, recreational, and aesthetic value of lakes and rivers; impair the taste and odor of water; enhance growth of algae and aquatic weeds; and poison fish, shellfish, predatory birds, and mammals. They can be toxic to humans, induce birth and genetic defects, or cause cancer. Since many new compounds can be introduced each year without testing of their environmental effects, it is possible that they will have caused chronic damage to human health and to aquatic life long before scientists realize their effects.

Compounds of pesticides, herbicides, fertilizers, and other synthetic chemicals often enter into bodies of water through agricultural applications and urbanization. A 10 year assessment by the National Water-Quality Assessment Program of the USGS showed that at least one pesticide was detected in water from all streams studied. Pesticide compounds were detected with 94%–97% of the time in water from streams with agricultural, urban, or mixed-land-use watersheds. However, their occurrences in streams and groundwater are seldom at concentrations likely to affect human, but may have effects on aquatic life or fish-eating wildlife in many streams (Gilliom et al., 2007).

11.1.8 Radioactive Substances

The ability of certain materials to emit the proton, gamma rays and electrons by their nuclei is known as the radioactivity. Radioactive substances can come from radioactive rocks and soils, from uranium mining and processing, from nuclear power plants, from nuclear weapon tests, or from leakage of radioactive instruments and laboratories. Concentrations of radioactive substances in drinking water above the maximum contaminant levels can cause chronic effects on human health such as cancer, kidney damage, and genetic defects. In rivers, lakes, and oceans, they can be absorbed by aquatic plants, seaweeds, and organisms at the bottom of the food chain. Such plants and organisms are then eater by small fish, which is eaten by bigger fish and eventually the whole food chain is contaminated.

The Cerro Grande forest fires around the Los Alamos National Lab, New Mexico in 2000 caused an increase of concentrations in down-streams 55 times for plutonium-239 and 240, 25 times for total uranium, and 20 times for cesium-137 in the 2–3 post-fire years. The increases in cesium-137 and uranium were predominantly associated with the initial flush of fallout derived cesium-137 in ash from the burned areas into the streams, while the plutonium-239 and 240 increases were due mostly to the large runoff events that accelerated the erosion of LANL-contaminated sediments (Gallaher and Koch, 2004).

11.2 Sources of Water Pollution

For management purposes, surface water pollution is generally identified as point source or nonpoint source.

11.2.1 Point Sources

Those wastes and pollutants that are discharged from identifiable locations are described as point sources of water pollution. Point sources can come from municipal sewage, industrial plants, animal feedlots, combined storm runoff and sewer lines, offshore oil-well or tanker accidents, and power plants. Thus, point-source pollution is generally created by human activities and is discharged into receiving water through "discernible, confined, and discrete conveyances including but not limited to any pipe, ditch, channel, tunnel, conduit, well, discrete fissure, container rolling stock, concentrated animal feeding operation, or vessel or other floating craft," as defined by the Federal Water Pollution Control Act Amendments of 1972 (PL 92–500). When discharges from a point source exceed effluent limitations, the discharges must be treated and treatments must be upgraded by use of the best available technology, or the discharges terminated.

Thus, the source of point pollution is identifiable, its route is continuous, its generation is anthropogenic, and its control is technological.

11.2.2 Nonpoint Sources

If pollutants come from the land surface or roofs (Chang et al., 2004), their sources are difficult to identify, their routes are intermittent and diffused, and their occurrences are associated with storms and surface runoff, then they are nonpoint sources of water pollution. They include:

- Sediment from natural ground surfaces, forest activities, agricultural production, construction, mining, grazing
- Chemicals from applications of fertilizers, pesticides, herbicides, saline water irrigation

- Urban and roof storm runoff
- Acids and minerals drained from mines
- Acid deposition
- Untraceable oils and other hazardous materials

Generally, nonpoint sources of water pollution cannot be corrected using technology and treatments similar to those employed for point sources of water pollution. Instead, "best management practices (BMPs)" to land and vegetation must be applied to all human activities and to all areas that have the potential to generate water pollution. The term BMP is a vague and localized approach, depending on soil, topography, vegetation, and climate of the area. For example, using mulches alone to reduce soil erosion may be a good BMP in location A but might not be effective enough in location B. In other words, BMP is site specific, is subjective, and is artistic.

Groundwater can be contaminated with organic chemicals through a number of point and nonpoint sources. They include waste lagoons; groundwater recharge activities; leaks from sewers, septic tanks, or gasoline-storage tanks; injection wells of waste disposal and oil fields; drainage from landfill and mining operations; seawater intrusion; chemicals applied to soils; and pumping wells.

11.3 Water Quality Determination

Stream water quality refers to the levels of the physical, chemical, and biological properties of water that can affect aquatic environments and intended water usage. These properties are induced by the concentrations of various substances dissolved in the water and their interactions with the natural environment. The concentration levels of those constituents are minimal and stable when the water–soil–vegetation–atmosphere system within the watershed is in equilibrium. However, the levels can be greatly altered and water quality significantly impaired when human activities or natural actions disturb the watershed equilibrium. The degradation of water quality caused by human activities is often referred to as water pollution, water not suitable for intended uses.

Of the eight groups of water pollutants discussed in Section I, sediment, which is considered by far the nation's largest single water pollutant, is discussed in great detail in Chapter 12. This section is more concerned with water chemistry and other pollutants.

11.3.1 Input–Output Budgets

Like hydrological budgets, elements in a watershed system are also involved with input, output, and storage. Under stable conditions, the input, output, and storage components of elements are in equilibrium; vegetation maintains vitality and healthy growth; and element concentrations in the stream are less variable. Any alteration of this equilibrium could adversely affect the land productivity and water quality in streams.

11.3.1.1 Budget Components

Elements in a stream are the output components in the watershed chemical budget system. Elements can be deposited to the watershed with precipitation, released from soils through weathering processes, mineralized from organic matter by biological activity, and leached from canopies with throughfall and stemflow. These are the input components in available phase. On the other hand, the available elements in the ground can be consumed biologically, volatilized and lost to evaporation, percolated to the aquifer, leached out of the watershed through surface and subsurface runoff, or stored in soils by ion exchanges, fixation, and mineralization.

Thus, the chemical-nutrient budgets of a watershed's terrestrial-hydrologic system are affected complexly by an array of factors such as precipitation and temperature, type of vegetation, soil and geology, and topography. Elements in precipitation originate from sources including oceanic spray, gaseous and industrial pollutants, terrestrial dusts, and volcanic emissions. Although there is a great variation in water chemistry in precipitation, the input elements from precipitation are an important source for the ecosystem. In the humid Hubbard Brook Experimental Forest, New Hampshire, precipitation input supplies about 66%, 26%, 6%, 5%, 4%, and 1% of the annual uptake of S, N, Mg, Na, Ca, and K, respectively, by green plant biomass (Likens et al., 1977).

11.3.1.2 Variations

The element balance between precipitation inputs and streamflow outputs varies among regions. In the southern coastal plain of Mississippi and Tennessee, the annual inputs of TKN (total Kjeldahl nitrogen), TP, K, HN_4–N, and NO_3–N in rainfall were 3.5–17 times greater than the annual yields of those nutrients in stormflow (Schreiber et al., 1980; McClurkin et al., 1985). However, the annual mean concentrations of Ca, Na, Mg, K, and NO_3–N in precipitation were only 10%–70% of those in streamflow for undisturbed forest watersheds at Hubbard Brook Experimental Forest, New Hampshire (Likens et al., 1977) and 1%–35% on the western side of Olympic National Park, Washington (Edmonds and Blew, 1997). The total ionic strength of stream water was about twice that of precipitation at Hubbard Brook (0.20 vs. 0.10 meq/L) and Coweeta, North Carolina (0.12 vs. 0.06 meq/L; Table 11.2). Measurements at 3200–3700 m elevations in the Snowy Range of southern Wyoming showed that Ca, Na, Mg, and K in precipitation were lower, while NH_4–N and NO_3–N were higher (9–34 times) than those in streams (Reuss et al., 1993). These great variations reflect the complexity of the interactions of vegetation uptake, biological activities, soil and rock weathering processes, and runoff generation upon element movements and storage.

Generally, water quality is expected to be better for streams in colder regions because higher temperature reduces dissolved oxygen level, increases chemical reactions, accelerates organic matter decomposition, stimulates microorganism activities, and encourages growth of algae and aquatic plants. Coniferous litter tends to be low in metallic cations such as Ca, Mg, and K, which can cause the developed soils to be more acidic (Brady, 1990). Studies of stream chemistry in Coweeta, North Carolina, showed that the pine ecosystems are more nutrient conservative for base cations than hardwood ecosystems are as a result of lower flow discharge and higher nutrient accretion in vegetation (Swank and Vose, 1994). Nitrate–N concentrations and yields tend to be highest in northeastern and mid-Atlantic coastal states, while those of TN are highest in the Southeast and in parts of the upper Midwest. In the Rocky Mountain and Central Plain states, TP is generally the highest in the nation (Clark et al., 2000).

11.3.1.3 Element-Stream Discharge

Water carries particulate matter and elements when it passes through a watershed ecosystem. Thus, elements in water, expressed either in concentrations or in mass per unit area

TABLE 11.2

Weighted Mean Annual Concentrations of Dissolved Substances in Bulk Precipitation and Stream Water for Undisturbed Watersheds at the Hubbard Brook Experimental Forest, New Hampshire (1963–1974) and Coweeta Hydrologic Laboratory, North Carolina (1973–1983)

	Hubbard Brook, WS #1–6, NH				Coweeta, WS #2, NC			
	Precipitation		Stream Water		Precipitation		Streamflow	
Substances	mg/L	μ eq/L	mg/L	μ eq/L	mg/L	μ eq/L	mg/L	μ eq/L
H^+	0.073	72.4	0.012	11.9	0.027	26.64	<0.000	0.2
NH_4^+	0.22[b]	12.2	0.04	2.2	0.095	6.80	0.002	0.2
Ca^{2+}	0.16	7.98	1.65	2.3	0.194	9.70	0.583	29.1
Na^+	0.12	5.22	0.87	37.8	0.170	7.38	1.22	53.2
Mg^{2+}	0.04	3.29	0.38	31.3	0.041	3.35	0.326	26.9
K^+	0.07	1.79	0.23	5.9	0.094	2.41	0.499	12.8
Al^{3+}	—		0.24	26.6	—	—	—	—
SO_4^{2-}	2.9	60.3	6.3	131	1.59	33.1	0.450	9.4
NO_3^-	1.47	23.7	2.01	32.4	0.143	10.2	0.003	0.2
Cl^-	0.47	13.3	0.55	15.5	0.271	7.65	0.662	18.7
PO_4^{3-}	0.008	0.253	0.0023	0.1	0.013	0.42	0.006	0.2
HCO_{3-}	0.006	0.098	0.92	15.1	0.074	1.21	4.97	81.5
Diss. silica			—	4.5	0.030		8.80	
Diss. org. C		2.4	—	1.0				
pH	(4.14)		(4.92)		(4.6)		(6.7)	
Total		102.9(+)		198.0(+)		56.3(+)		122.3(+)
		97.7(−)		194.1(−)		52.6(−)		109.9(−)

Source:　Likens, G.E. et al., *Biogeochemistry of a Forested Ecosystem*, Springer-Verlag, Heidelberg, Germany, 1977 (Hubbard Brook, WS #1–6, NH); Swank, W.T. and Waide, J.B., Characterization of baseline precipitation and stream chemistry and nutrient budgets for control watersheds, in *Forest Hydrology and Ecology at Coweeta*, Swank, W.T. and Crossley, D.A., Jr., Eds., Springer-Verlag, Heidelberg, Germany, pp. 57–79, 1988 (Coweeta, WS #2, NC).

per unit time, are strongly related to the volume of streamflow discharge due to its dilution and leaching effects. The relationships can be properly expressed by a simple linear function for forested, headwater streams. For agricultural or urban watersheds and large drainage areas, power and exponential functions may be more appropriate for describing this relationship (McBroom et al., 1999).

11.3.2　Water Quality Criteria

Substances dissolved in water can alter the physical, chemical, and biological properties of the water and limit its suitability for usage. Established by state and federal governments, water quality criteria are a management approach designed to regulate the discharge of pollutants into receiving streams and to provide desirable requirements for the intended use of the water (Table 11.3). They are based on research and long-term observations on the acute and chronic effects of each pollutant. Thus, a quality standard may need to be modified when new evidence on tolerant levels is substantial and significant.

TABLE 11.3

Water Quality Standards (µg/L) for a Few
Inorganic Elements in the United States

Elements	Fresh Water		Drinking Water
	Acute	Chronic	
Arsenic	340.0	150.0	0.018
Cadmium	2.0	0.25	
Copper	13.0	9.0	1,300.0
Cyanide	22.0	5.2	140.0
Iron			300.0
Lead	65.0	2.5	
Manganese			50.0
Mercury	1.4	0.77	
Nickel	470.0	52.0	610.0
Nitrates			10,000.0
Selenium			170.0
Silver	3.2		
Zinc	120.0	120.0	7,400.0

Source: USEPA, National recommended water quality criteria, EPA-822-R-02-047, Washington, DC, 2002.

The Clean Water Act classifies water uses into five categories:

1. Public water supply
2. Recreation and aesthetic value
3. Aquatic life
4. Agriculture
5. Industries

Each requires a set of water quality criteria different from the others. Generally, the criteria for uses that require large volume are not as rigid as those for uses that require a smaller volume. Water quality criteria for recreation and aquatic life should consider local variations due to differences in climatic, soil, and geological conditions. For example, the minimum level for dissolved oxygen in warm, slow-moving streams in the South should be lower than in the cold, fast-moving streams in the Northwest. Even in the same state, standards for surface water may be different among watersheds (TNRCC, 1995). Accordingly, the USEPA's (2002) recommended water quality criteria for TP, TN, chlorophyll *a*, and Turbidity are different between streams and lakes as well as different among 14 ecoregions across the United States (Table 11.4).

It is also important that the established water quality criteria be for streams and lakes of under purely natural processes without anthropogenic inputs. In other words, criteria must be rational and achievable. Table 11.1 shows that the estimated nation-wide background concentration for TN is 0.58 mg/L. Background concentrations are nutrients present in streams and groundwater as a result of natural processes with low level inputs of nutrients from anthropogenic sources such as atmospheric deposition. The EPA's recommended TN criteria for streams and rivers for 13 ecoregions range from 0.12 to 2.18 mg/L. There are seven

TABLE 11.4

The Recommended EPA Criteria for Total Phosphorus (TP), Total Nitrogen (TN), Chlorophyll *a* (Chl *a*), and Turbidity (Turb) or Secchi (Sec) for Both Lakes and Reservoirs and Rivers and Streams for Each of the Aggregate Nutrient Ecoregions in the United States

Eco-Region	Rivers and Streams				Lakes and Reservoirs			
	TP (μg/L)	TN (mg/L)	Chl *a* (μg/L)	Turb (FTU)	TP (μg/L)	TN (mg/L)	Chl *a* (μg/L)	Sec (m)
I: Willamette and Central Valleys	47.00	0.31	1.80	4.25				
II: Western Forested Mountains	10.00	0.12	1.08	1.30	8.75	0.10	1.90	4.50
III: Xeric West	21.88	0.38.	1.76	2.34	17.00	0.40	3.40	2.70
IV: Great Plains Grass/ Shrublands	23.06	0.56	2.40	4.21	20.00	0.44	2.00[s]	2.00
V: S. Central Cultivated Great Plains	67.00	0.88	3.00	7.83	33.00	0.56	2.30[s]	1.30
VI: Corn Belt/Northern Great Plains	76.25	2.18	2.70	6.30	27.50	0.78	8.59[s]	1.36
VII: Mostly Glaciated Dairy Region	33.00	0.54	1.50	1.70	14.75	0.66	2.53	3.33
VIII: Upper Midwest/ Northeast	10.00	0.38	0.63	1.30	8.00	0.24	2.43	4.93
IX: SE Temperate ForestPlains/Hills	36.56	0.69	0.93[s]	5.70	20.00	0.36	4.93	2.53
X: TX-LA Coastal/MS Alluvial Plains	128.0	0.76	2.10[s]	17.5				
XI: Central/Eastern Forested Uplands	10.00	0.31	1.61[s]	2.30[n]	8.00	0.46	2.79[s]	2.86
XII: Southeastern Coastal Plain	40.00	0.90	0.40[s]	2.30[n]	10.00	0.52	2.60	2.10
XIII: Southern Florida Coastal Plain					17.50	1.27	12.35[T]	0.79
XIV: Eastern Coastal Plain	31.25	0.71	3.75[s]	3.04	8.00	0.32	2.90	4.50

Source: USEPA, National recommended water quality criteria, EPA-822-R-047, Washington, DC, 2002.
Chl *a* = chlorophyll *a* measured by Fluorometric method, unless specified; s = spectrophotometric method; T = trichromatic method; n = NTU (nephelometric turbidity unit); FTU = formazin turbidity unit.

recommended TN criteria lower than the background concentration of 0.58 mg/L. Forest streams without recent management are high in water quality and are assumed by most people to be able to meet the proposed criteria. However, Ice and Binkley (2003) reported that many streams in small forest experimental watersheds exceeded the proposed TN criteria.

11.3.3 Bio-Assessment

Traditionally, stream water quality is determined by collecting water samples from the stream and analyzed in a laboratory for suspected physical and chemical pollutants. This approach is expensive, time-consuming, and sometimes ineffective in detecting nonpoint source pollution problems because of the temporal and spatial variations of pollutants. There are certain types of stream organisms that occur or thrive only under certain water quality conditions. When stream conditions change due to overland flow, or other reasons, the abundance and composition of organisms in the receiving runoff site change as well. Thus, organisms can be used as indicators and criteria for determining stream water quality.

The USEPA (1990) noticed the potential of biological assessment in the 1980s and required states to use biological indicators to accomplish the goals of the Clean Water Act. Accordingly, biological assessment has become one of the standard monitoring tools of water resource protection agencies over the last two decades. Since then, biological criteria and monitoring programs have been created for lakes and streams throughout the United States (Reiss and Brown, 2005).

11.3.3.1 Benthic Invertebrates

Benthic invertebrates, including insects, mollusks (snails and clams), crustaceans (shrimps), and worms, are organisms with no backbones. The word "benthic" means bottom-living. They inhabit tiny spaces between stones, within organic debris, on logs, and by aquatic plants at the bottom of ponds, lakes, and streams. Invertebrates are important in the food chain because they consume other invertebrates or are consumed by fish and birds. Changes in hydrologic regime in streams may affect benthic inverte- brate assemblage composition and structure by changes in population density or taxa richness (Rediske et al., 2009).

Although fish and algae have been used in stream biomonitoring programs, benthic invertebrates are the most commonly used organisms. This is because benthic inverte- brates (Feminella and Flynn, 1999)

- Are numerically abundant in streams
- Have a wide range of sensitivity to pollution
- Are relatively easy to sample and identify
- Have life cycles (months to a few years) intermediate to fish (years to decades) and algae (days to weeks), providing reliable and rapid evidence of water quality conditions
- Characterize effects over a relatively small area (in contrast with fish of travelling over long distances), providing a site-specific indication

11.3.3.2 Applications

In practice, benthic invertebrates are sampled from streams using Scurber or Hess samplers, kick net, D-frame dip nets or rectangular dip nets (Cuffney et al., 1993; Barbour et al., 1999). These samplers and nets are used in streams of less than 1 m in depth. Usually, biotic samples can be collected: (1) upstream and downstream of suspected pollution sites, (2) before and after an environmental event, (3) from suspected and undisturbed reference streams, and (4) along the gradient of a stream reach. Assessments involving the spatial (upstream vs. downstream, or Control vs. Impact) and the temporal (Before vs. After) elements are referred to as the BACI design, while the method looking at suspected versus referenced sites is referred to as Reference Condition Approach (RCA). The assessment along a gradient of degradation in a region requires the development of a multimetric index using the best biotic descriptors for applications to the sites of unknown degradation and is referred to as the Index of Biotic Integrity (IBI) approach (Bailey et al., 2007).

Once samples are collected, the invertebrates are isolated from inhabited substances, identified by their taxonomic group, either in the field or in the laboratory, and counted.

These numerical data and taxonomic groups are then used to quantify the biotic environment by calculating a number of metrics for comparisons and analysis. There are about 50 metrics that have been proposed for bio-assessment investigations (Mandaville, 2002), the basic ones may include:

- Taxa richness (the number of unique types of invertebrates present in a sample)
- Abundance (the total number of invertebrates in a sample)
- Relative abundance (the number of invertebrates in the sample from one species relative to another)
- Species diversity (the distribution of total individual species in the sample)
- EPT (Ephemeroptera, Plecoptera, and Trichoptera) richness (the total number of species within the three most pollution-sensitive aquatic insect orders, mayflies, stoneflies, and caddis flies)
- % EPT (the percentage of the total number of organisms in each sample belonging to the EPT orders)
- % CDF (contribution dominant family, the percentage of the total number of organisms in each sample in the numerically dominant family)

These biotic metrics, or others, are then used to assess water quality conditions at the study sites. Streams with high values of Taxa richness, EPT richness, % EPT, or species diversity are less likely to be polluted than are streams with relatively low values of those metrics in the same region. A shift of stream towards dominance by relatively few taxa (increasing % CDF values) implies environmental stress.

11.3.4 Cumulative Watershed Effects

Frequently a forest is managed for more than one purpose and by adopting more than one activity to reach each purpose. These multiple uses and management activities are not only common in practice but also mandated on federal lands by laws such as the Multiple-Use and Sustained Yield Act of 1960, the National Forest Management Act of 1976, and the Clean Water Act of 1977 (Sidle and Hornbeck, 1991). The effects of multiple activities on streamflow are obviously greater than effects from any single activity.

11.3.4.1 Significance

A watershed impact is seldom influenced by a single activity; it is most likely to be influenced by multiple activities or by a single activity of multiple applications, all superimposed by multiple environmental factors. Cumulative watershed effects (CWEs) refer to the additive, incremental, combined, and interactive impacts created by additional activities through time and space. The Council on Environmental Quality (CEQ) defined a cumulative impact as (CEQ Guidelines, 40CFR 1508.7, issued April 23, 1971):

> … the impact on the environment which results from the incremental impact of the action when added to other past, present, and reasonably foreseeable future actions regardless of what agency (federal or non-federal) or person undertakes such other actions. Cumulative impacts can result from individually minor but collectively significant actions taking place over a period of time.

For example, a cumulative-effects study of land use on water quality along a fifth-order stream in western North Carolina showed that streamflow with low concentrations of chemical and biotic constituents, physical variables, and cations was predominantly associated with forested conditions in upper streams. Most variables were slightly increased downstream, and the increases were pronounced during stormflow stages (Bolstad and Swank, 1997).

In the H.J. Andrews Experimental Forest, Oregon, a 25 year study on landslides and soil erosion showed that the average erosion rate was $2.51\,m^3/ha/year$. Although forest roads cover only 5% of the entire area, they contribute 51% of the annual erosion losses. Landslides and accelerated erosion in clear-cut areas (26% of the total area) accounted for 25% of the erosion, and the other 24% came from the 69% forested area. In other words, the volume of slide material per unit area moved from the roads was 30 times the rate of that in undisturbed forests and about 10 times that in clear-cut areas. In comparison, if a forest fire occurred on 3% of the watershed each year, it would increase erosion only by an estimated $0.05\,m^3/ha/year$ (Mersereau and Dryness, 1972; Swanson and Dryness, 1975).

Concern of CWEs can cover a broad range of issues, but the primary ones are the changes in streamflow quantity and quality, channel morphology, and aquatic life incurred by management activities at the watershed scale. CWEs can be assessed for multiple activities and land uses on a single watershed or for a single activity through upstream–downstream routing with diverse patterns of geology, topography, soil, and land use in the drainage system. They also can be addressed for the present, past, and future with all water, soil, nutrients, and biotic processes and responses in a watershed system interactively considered. CWEs are an advanced means of controlling nonpoint sources of water pollution, an important step on the implementation of the Clean Water Act.

11.3.4.2 Assessments

The National Environmental Policy Act of 1969 requires that federal agencies must present assessments on potential cumulative environmental effects of proposed major land management actions for reviews by the government decision-makers and the public. CWEs may arise from single or multiple activities that cause direct or indirect, on-site or off-site, spatial or temporal, and short-term or long-term impacts.

There are many procedures, methods, models, and tools that have been developed for CWEs assessments. They range from using the users' qualitative observations and professional judgment (CDF, 1991) to indices and accounting procedures such as the Equivalent Clearcut Area (USFS, 1974) and the Equivalent Roaded Area USFS (1988), the process-based induction such as the water resources evaluation of nonpoint silvicultural sources referred to as WRENSS (USFS, 1980), field measurements and judgments (Bevenger and King, 1995; Idaho Dept. of Lands, 2000), and simulation and modeling (Stonesifer, 2007; Elliot et al., 2010a). Many assessment methods and tools are described, discussed, and summarized in works by Reid (1993) and Elliot et al. (2010b).

In assessing cumulative effects, one must consider the past, present and possible future activities that may contribute to watershed disturbance. Often, the magnitude of the effects of the activities tapers off with time because of the re-growth of vegetation and the natural healing of watersheds. Also, the current procedures were developed to address particular issues in particular areas, none can be applied to all land use activities, all watershed-response mechanisms, and all areas. Thus, it is necessary to consider in advance the limitations of those procedures and the targeted issues and needs for the watershed under assessments.

11.4 Forest Practices

Forested watersheds normally yield streamflow with better quality than that from other land uses (Chang et al., 1983; Gravelle et al., 2009). For example, the average TN concentration in forested streams was about 0.850 mg/L in the United States compared with 4.170 mg/L for agricultural watersheds. For TP concentrations, it was 0.014 mg/L for forested watersheds and 0.135 mg/L for agricultural watersheds (Omernik, 1976).

Specific conductance is a parameter to measure the ability of water to conduct electric currents and is dependent upon the total quantity of dissolved ions. It ranges from 3 to 15 mS/m in forested watersheds (Binkley and Brown, 1993) and from 276 to 3855 mS/m for the Lower Arkansas River, an agricultural drainage basin in the Central Plains (Lewis and Brendle, 1998).

The higher water quality in forested watersheds is attributable to:

- More uptake of nutrients and ions by plants
- Fewer runoff and sediment losses
- Lower rates of organic matter decomposition and microbial activity
- Cooler temperature
- Less management

Forest-management activities often remove vegetation, expose the ground to direct environmental impacts, and disturb soils. As a result, the forest functions mentioned in Chapter 7 are largely reduced and water quality is impaired. Thus, the use of BMPs becomes a necessary measure for controlling nonpoint sources of water pollution caused by forest activities (Ice, 2004).

11.4.1 Clear-Cutting

Clear-cutting—felling and removing essentially all trees from a forest—intends to create a site and microclimate for generation of an even-aged forest. It has been the major approach of the silvicultural systems in North America because of its efficiency in wood production and low costs (Kellogg et al., 1996). Since the method is the most intensive among forest-harvesting practices, its potential for causing adverse effects on stream water quality is also the greatest.

Besides stream sediment, clear-cutting can adversely affect stream temperature, concentrations of water chemicals, dissolved oxygen, BOD, specific conductance, pH, fecal coliform and fecal streptococcus, and the aquatic environment (Binkley and Brown, 1993; Bolstad and Swank, 1997). However, these effects are highly variable, depending on the clear-cutting system, species, soils and topography, and climate. A tractor harvest system can create 35% bare soil and 26% compacted soil, while a highlead harvest system can only create 15% and 9%, respectively (Adams, 1999).

Clear-cutting northern hardwoods in New Hampshire caused increases in NO_3^- of over 50-fold (average 15 mg/L) and Al^{+3} by almost 10-fold, and decreases in pH by 0.3 units (Likens et al., 1977). In contrast, clear-cuttings caused only measurable increases in concentrations of NO_3^- (0.015–0.7 mg/L), K^+, and other constituents in hardwood ecosystems of the southern Appalachians. The magnitude of changes in stream constituents was small and unimportant to downstream uses (Swank and Johnson, 1994) and should not adversely affect the sustainability of growth in the successional forest (Swank et al., 2001). The average NO_3^- concentration for a clear-cut watershed of mixed conifer forests in Coyote Creek, Oregon, was 0.1 mg/L, compared with 0.015–0.040 mg/L for the control watershed (Adams and Stack, 1989).

Changes in phosphorus in streams after harvesting are usually insignificant (Binkley and Brown, 1993; Gravelle et al., 2009), and responses of other constituents are likely to be less than that of NO_3^-. However, water temperature has been reported to increase by 16°C (Brown and Krygier, 1970), and dissolved oxygen dropped to 1.7 mg/L (Woods, 1980) in northern California. Impacts become insignificant downstream or as re-growth of vegetation progresses.

Most of water quality problems associated with clear-cutting are caused by overexposure of bare soil, poor design, construction and maintenance of logging roads and skid trails, improper road closure, stream crossings, and lack of Streamside Management Zones. Leaving a riparian buffer strip along stream channels is an effective practice to provide a screen effect for water, sediment, and nutrient movement as well as a shading effect for exposure of stream surface to direct solar radiation. It minimizes increases in element concentrations and water temperature in streams. In steep terrain, a less intensive cutting—such as patches, strips, and tree selection—may be necessary in order to reduce the adverse effects on water quality. Also, it is possible to keep a healthy level of dissolved oxygen if fresh slash is kept out of streams (Ice, 1999).

Thus, with the adoption of BMPs in timber harvesting and log extraction, impacts of clear-cutting on water quality can be kept at insignificant levels (Rashin et al., 2006; Webb et al., 2007; Winkler et al., 2009).

11.4.2 Forest Roads

Forest access systems are essential in forest-management activities, including recreation, production, management, and protection. Road-induced erosion, with its consequent effect on water quality, is a major concern in forestry and has become the focus of intense discussion in recent years (Bengston and Fan, 1999; Gucinski et al., 2001). As much as 90% of all sediment produced from forestland could originate from forest roads and stream crossings (Megahan, 1972; Grace, 1999). There are over 600,000 km of roads in the National Forests, and the recreational use of National Forests has increased more than 18 times since the 1940s (Grace and Clinton, 2007). The pressure imposed to the road protection and maintenance presents a challenge to watershed managers.

The construction of forest roads exposes land surfaces not only to an accelerated erosion process but also to vehicle impacts, especially during the wet season. In hilly terrain, cut and fill slopes account for as much as 50% of the road prism area and can generate 70%–90% of the total soil loss from the road system (Swift, 1984). The magnitude of road erosion has been reported to reach 276 ton/ha/year in the Coastal Plain of Alabama (Grace and Elliot, 2008) and is common to exceed 100 ton/ha/year, while the soil tolerance rates for agricultural lands are about 4–10 ton/ha/year. Forest roads also disrupt watershed drainage patterns and become a convergence for overland flows of greater erosive power. Without proper maintenance, the converged runoff can turn forest roads into drainage channels and dissected gullies.

Stream crossings and road drainage are other major sources of sediment in forested streams. Constructing and maintaining stream crossings are crucial in developing forest road systems. The primary types of stream crossings are bridges, culverts, logs, and fords (stream-bottom crossings). The construction of all of these crossings introduces large amounts of sediment. Fords deliver additional sediment into streams during vehicle crossings and direct sediment-laden runoff from the road into streams. Although fords are the least expensive type of crossing, their impacts on water quality are probably the greatest (Taylor et al., 1999). When bridges and culverts are insufficient to drain floodwaters in streams, or culverts are insufficient to move water from one side to the other of a road prism, the result is often erosion of the crossing fill on the inlet and outlet sides, diversion of streamflow onto the road surface or inboard ditch, or even large landslides (Beschta, 1998; Furniss et al., 1998). In northern California, fill-failures and diversions of road-stream crossings have been found to cause 80% of fluvial hillslope erosion (Best et al., 1995).

Generally, roads built on steep slopes, unstable topography and geology, and floodplains, or in areas with abundant precipitation, have a great potential for sediment production. The mitigation of soil erosion and the protection of water quality from forest road systems are largely dependent upon adequate design, proper water drainage, and routine maintenance. Techniques of erosion control have been discussed by Kochenderfer (1970), the Texas Forestry Association (1989), and Egan (1999).

11.4.3 Forest Fire

The impact of fires on stream water quality is associated with:

- Destructing forest canopies and organic litter
- Heating soils and water
- Accelerating overland flow
- Ash deposition
- Fire control practices

Burning forest canopies and organic litter can significantly reduce interception, transpiration, and infiltration rates; destroy the shading effect of riparian vegetation; increase snow accumulation, storm runoff, and water yield; and release nutrients and metals to streams. Heating soils and water can alter soil physical and chemical properties, induce soil water repellency, and reduce DO concentrations in streams. The acceleration of overland flow can increase nutrients and sediment losses, and alter stream aquatic habitat. The ash deposition and suppression chemicals provide additional constituents to land and water surfaces. Thus, major constituents, nutrients, and suspended sediment in streams are often elevated due to forest fires. The effects are highly variable, can be on-site as well as off-site, and depend on species composition, soil textures, seasons, climatic conditions, geographic location, the severity of fire, and the distribution of fire within the watershed (Tiedemann et al., 1979; Emmerich, 1998). A wildfire can cause temperatures at the soil surface to approach 900°C, and temperatures in water to increase by 15°C (Ice et al., 2004).

11.4.3.1 Burning

When a plant is burned, elements incorporated in the plant are either volatilized from the biotic system, as occurs for N, P, K, Ca, Mg, Cu, Fe, Mn, and Zn, or mineralized and dispersed to become ash, as is the case for Ca, Mg, and K. Metallic elements are oxidized and become more soluble when they react with CO_2 and H_2O, causing increases of bicarbonate salts in soil solution and in streamflow. Increases in NO_3^- concentrations in streams are associated with the nitrification process in the soil as enhanced by a more favorable soil pH (Tiedmann et al., 1979). Many elements are leached out via surface and sediments. Studies of water quality in loblolly-pine watersheds in Tennessee showed that a third or more of TKN and two-thirds of TP exported in stormflow were through sediment (McClurkin et al., 1985).

Suspended sediment and turbidity are often the most dramatic responses of water quality to wildfires. For the Cerro Grande Fire occurred around the Los Alamos Laboratory, New Mexico in 2000, post-fire concentrations of total suspended sediment were 10,000 times larger than perfire at a site in upper Pajarito Canyon (Gallaher and Koch, 2004). However, in British Columbia, the most pronounced effect of a wildfire on water chemistry was on NO_3^- with a maximum increase of 0.87 mg/L (Gluns and Toews, 1989). So was a 4 year study conducted in Montana, United States. Stream NO_3^- concentrations, exceeding 0.80 mg/L, were the greatest response to a wildfire in 2003, increasing up to 10-fold above those in the unburned drainage just prior to the first post-fire snowmelt season (Mast and Clow, 2008). The slow release of snowmelt runoff at higher latitudes may have caused sediment to be less significant. Nevertheless, these increases in NO_3^- concentrations are far less than the federal drinking-water quality standards, 10 mg/L.

Unlike wildfires, prescribed fires are controlled at the ground surface, over a specific time, on a designated area, and to a small scale. Properly executed, prescribed fires have little effect on streamflow since very little of the stand is killed, and the soil organic layer is not completely consumed (Gottfried and DeBano, 1991). Although surface fires are not as destructive as crown fires and have been used as a tool in forest management, a severe surface fire can also kill all trees and understory vegetation and cause serious soil erosion and water quality problems.

Burning also causes the formation of water-repellent layer on the ground surface which can reduce soil wettability and impede soil infiltration. It results in increases in surface runoff, sediment production, and nutrient losses to streams.

11.4.3.2 Fire Control Practices

Fire control practices include fire suppression chemicals, bulldozers and hand clearing. These methods are applied from the air or at the ground level to attack fires or create fire-lines. They all can have impacts on water quality and steam habitats, depending on intensity of the applications, type of chemicals, stream morphology, and wind directions.

Fire retardants have been increasingly in use on fire suppression since their introduction in the 1930s. They are used in direct treatment of burning fuels, in advance treatment of fire-lines, or in control of prescribe burning. Retardants are mixtures of water, selected inorganic salts, clay and gums (thickening additives), bactericides (inhibiting corrosion and spoilage), and dyers (coloring the liquid). The major compounds include chemical elements such as phosphorus, sulfur, antimony, chlorine, bromine, boron, and nitrogen. They have the potential to cause adverse effects on water quality, environment, and human health when applied too heavily or indiscriminately (Kalabokidis, 2000).

Water quality impacts were studied for ground-level applications of an ammonia-based fire retardant to five streams in Oregon, Idaho, and California. The results showed that initial concentrations of ammonia nitrogen ($NH_3 + NH_4$) in water approached fish damage-levels, but no distressed fish were found. The increases in concentrations declined sharply with time and with distance downstream (Norris and Webb, 1989). The application of fire retardants was found to have insignificant impacts on water quality in a study in Arizona (Crouch et al., 2006). It is important that ground-level applications be kept away from water surface and riparian zones.

11.4.4 Mechanical Site Preparation

Following forest harvesting, sites are usually under some degree of preparation for planting, direct seeding, or even natural regeneration. Site preparation can be mechanical, chemical, or by burning, or it may involve a combination of these techniques.

11.4.4.1 Practices

Mechanical site preparation uses tools, machinery, and tractors to treat standing trees, debris, stumps, and ground by disking, furrowing, bedding, chopping, shearing, KG (bulldozer) blading, windrowing, rooting and bulldozing, and cultivation. Frequently, a site can be prepared by more than one method, and the adoption of these methods is dependent upon generation methods, species, stem diameters, soil types, groundwater table, surface drainage, land slopes, and other considerations such as cost and environmental impacts. For example, bedding is often used on excessively wet sites to improve drainage and to elevate seedlings above standing water, but it should not be used in areas where seedlings would have potential for suffering from summer droughts. In sites with too many big stems and stumps to be removed by disking and chopping, a combination of shearing, KG blading, and windrowing are often used to prepare the site for planting.

11.4.4.2 Impacts

Site preparation improves soil moisture conditions; reduces plant competition for water, nutrients, and energy; makes tree planting and other activities easier; and mitigates fire hazard. As a consequence, the survival and growth of seedlings are enhanced, wood production is increased, rotation cycle is shortened and financial return is optimized. However, the heavy equipment and operations of mechanical site preparation also severely disturb soils, expose ground from litter protection, and alter soil physical properties. A saline soil study in east Texas showed that soil bulk density, percent of silt and clay at the surface horizon, soil water content, soil water retention, and height of groundwater table were increased after clear-cutting followed by shearing and KG blading (Chang et al., 1994). These changes in soil properties resulted in increases in overland flow, accelerated soil erosion, and nutrient losses and consequently affected water quality and the aquatic environment (Riekerk, 1983; Muda et al., 1989; Blackburn and Wood, 1990; Sayok et al., 1993).

Six plot-watersheds of 0.02 ha in size, each equipped with a Coshocton N-1 runoff sampler and automatic water-level recorder, were used to study the effects of mechanical site preparation on storm runoff, sediment and nutrient losses from a mature loblolly-pine

TABLE 11.5

Total Net Rainfall, Runoff, Sediment, and Nutrient Losses under Six Forest Conditions from 30 Storms Observed between May 1980 and February 1981 in Forested East Texas

Variables	Treatments[a]					
	a	b	c	d	e	f
Net rainfall (mm)	360	378	430	459	459	459
Runoff (mm)	7.4	17.0	41.7	81.4	137.3	119.0
Sediment (kg/ha)	10.7	17.1	155.8	265.0	3461.6	3423.4
Nutrients (kg/ha)[b]						
K	0.25	0.90	1.85	2.74	6.69	3.85
Na	0.23	0.70	1.57	3.08	8.66	5.21
Ca	0.36	1.55	1.47	3.07	10.83	6.13
Mg	0.08	0.30	0.79	1.48	7.76	6.38
$NO_3 + NO_2$	0.31	1.07	2.30	3.31	7.93	6.63
TKN	0.44	2.22	5.24	3.62	10.86	7.74
SO_4	0.41	1.57	3.30	2.13	12.12	7.23
PO_4	0.20	0.36	1.70	1.90	5.72	3.73
Total	2.33	8.57	18.18	21.23	70.58	46.65

Source: Chang, M. et al., Sediment production under various forest site conditions, in *Recent Developments in the Explanation and Prediction of Erosion and Sediment Yield*, Walling, D.E., Ed., IAHS Publ. No. 137, pp. 13–22, 1982.

[a] Treatments: a = undisturbed mature forest; b = 50% thinning; c = clear-cutting without site preparation; d = clear-cut and chopping; e = clear-cut, sheared, and KG bladed; f = clear-cut, clean till for cultivation.

[b] Covered 23 storms.

plantation in east Texas. The six plots, all on the same type of soil, were under six different treatments:

1. Undisturbed mature forest
2. Fifty percent thinning
3. Clear-cut without site preparation
4. Clear-cut with chopping
5. Clear-cut, sheared, and KG bladed
6. Cultivation

Based on data collected from 30 storms, storm runoff, sediment and total nutrient losses of eight elements increased with increasing intensity of site preparation (Table 11.5). Total sediment and nutrient losses ranged respectively from 10.7 and 2.33 kg/ha for the undisturbed forest to 3462 and 71 kg/ha for the sheared watershed. Chopping and shearing, the two most popular site preparation methods in the South, have about equal site-clearness results, but sediment produced from chopping was less than 7% of that from shearing. Sediment produced from chopping is less than 10% of that produced from any treatment involving shearing and bulldozing in the coastal plain (Balmer et al., 1976).

A chopper usually runs up and down the slope and creates many small furrows perpendicular to hillside slopes by the chopper's wheel blades. These furrows are effective in reducing overland flow velocity, increasing infiltration rate and soil moisture storage, and trapping downward movement of sediment. As a result, the impact of chopping on water quality is significantly less than that of shearing with KG blading.

11.4.5 Grazing

Livestock grazing in forests and woodlots is common in the United States, estimated at about 34% or 70×10^6 ha of all forests (Clason and Sharrow, 2000). Concerns on the impact of grazing to the watershed are triggered by the trampling, browsing and waste discharges (feces) caused by animals. The amount of wet feces along with a number of nutrients produced per 1000 kg of animal live weight per day for four grazing animals are given in Table 11.6. For a mean animal weight of 400 kg, 100-head of cattle will collectively deposit 50 kg of N and 25 kg of P in manure each day on the range (Derlet et al., 2010). Thus, the potential impacts of grazing include: (1) forest vegetation, (2) riparian zones, (3) soils, (4) stream water quantity and quality, and (5) wildlife.

When grazing is poorly managed, it can cause a decrease in canopy and litter cover, soil organic content, plant transpiration, and soil infiltration rate along with an increase in raindrop impact on the soil surface, soil compaction and bulk density. These impacts lead to greater levels of storm runoff, sediment concentration, constituents and nutrients, and pathogen contents in streams. Nonpoint pollution from cattle waste is a serious threat of eutrophication to both surface and ground water sources at both higher and lower elevations (Klott, 2007).

A study with 11 years of grazing near Corvallis, OR, showed that soil in the silvopastures had 13% higher bulk density and 7% lower total porosity than those in adjacent forests (Sharrow, 2007). In UAE, forest lands grazed by Arabian Oryx resulted in lower organic matter content (0.36% vs. 0.78%) and a significant deterioration of species diversity

TABLE 11.6

Mean Fresh Manure and Nutrient Productions (in kg/1000 kg of Animal Live Weight/Day)[a] for Four Grazing Animals

Item	Dairy	Beef	Sheep	Horse
Total manure	86	58	40	51
Urine	26	18	39	10
Total Kjeldahl nitrogen	0.45	0.34	0.42	0.30
Ammonia nitrogen	0.079	0.086		
Total phosphorus	0.094	0.092	0.087	0.071
Ortho-phosphorus	0.061	0.030	0.032	0.019
Potassium	0.029	0.021	0.032	0.025
Total coliforms[b]	1110	63	20	490

Source: Extracted from Hubbard, R.K. et al., *J. Anim. Sci.* 82(E. Suppl.), E255–E263, 2004.

[a] All values are expressed in wet basis.

[b] Mean bacteria colonies per 1000 kg of animal mass multiplied by 1010 colonies per 1000 kg animal/mass divided by kg of total manure per 1000 kg of animal mass multiplied by density (kg/m³) equals colonies per m³ of manure.

than those not grazed (Ksiksi et al., 2006). A quantity review of 54 studies on effects of cattle grazing on rangelands in the western United States conducted between 1945 and 1996 showed that litter biomass (1034 vs. 2573 kg/ha) and soil water infiltration rates (6.07 vs. 9.85 cm/ha) were lower and soil erosion (526 vs. 289 kg/ha) were higher on grazed than those of un-grazed lands (Jones, 2000).

References

Adams, P.W., 1999, Timber harvesting and the soil resource: What does the science show? *Clearcutting in Western Oregon*, Oregon State University College of Forestry, Corvallis, OR, pp. 77–93.

Adams, P.W. and Stack, W.R., 1989, Stream water quality after logging in Southwest Oregon, PNW 87–400, U.S. Forest Service Pacific NW Station, Portland, OR.

Bailey, R.C. et al., 2007, Integrating stream bioassessment and landscape ecology as a tool for land use planning, *Freshwater Biol.*, 52, 908–917.

Balmer, W.E. et al., 1976, Site preparation—Why and how, forest management bulletin, USDA Forest Service, State and Private Forestry, Atlanta, GA.

Barbour, M.T. et al., 1999, *Rapid Bioassessment Protocols For Use in Streams and Wadeable Rivers: Periphyton, Benthic Macroinvertebrates, and Fish*, 2nd edn., EPA 841-B-99-002, U.S. Environmental Protection Agency, Office of Water, Washington, DC, http://www.epa.gov/owow/monitoring/rbp/

Bengston, D.N. and Fan, D.P., 1999, The public debate about roads on the national forests: An analysis of the news media, 1994–98, *J. For.*, 97, 4–10.

Beschta, R.L., 1998, Forest hydrology in the Pacific Northwest: Additional research needs, *J. Am. Water Resour. Assoc.*, 34, 729–741.

Best, D.W. et al., 1995, Role of fluvial hillslope erosion and road construction in the sediment budget of Garret Creek, Humboldt County, California, *Geomorphic Processes and Aquatic Habitat in the Redwood Creek Basin, Northwestern California*, Nolan, K.M., Kelsey, H.M., and Marron, D.C., Eds., U.S. Geological Survey Professional Paper 1454, M1–M9, Reston, VA.

Bevenger, G.S. and King, R.M., 1995, A pebble count procedure for assessing watershed cumulative effects, Research Paper RM-RP-319, U.S. Forest Service, Rocky Mountain Forest and Range Experiment Station, Fort Collins, CO.

Binkley, D. and Brown, T.C., 1993, Management impacts on water quality of forests and rangelands, General Technical Report RM-239, Rocky Mountain Forest and Range Experiment Station, U.S. Forest Service, Fort Collins, CO.

Blackburn, W.H. and Wood, J.C., 1990, Nutrient export in stormflow following forest harvesting and site preparation in East Texas, *J. Environ. Qual.*, 19, 402–408.

Bolstad, P.V. and Swank, W.T., 1997, Cumulative impacts of land use on water quality in a southern Appalachian watershed, *J. Am. Water Resour. Assoc.*, 33, 519–533.

Brady, N.C., 1990, *The Nature and Properties of Soils*, MacMillan, New York.

Brown, G.W. and Krygier, J.T., 1970, Effects of clearcutting on stream temperature, *Water Resour. Res.*, 6, 1133–1140.

CDF, 1991, Guidelines for assessment of cumulative watershed effects, California Department of Forest Fire Protection, Sacramento, CA.

CEQ, 1978, *Environmental Quality, The 9th Annual Report*, Council on Environmental Quality, Government Printing Office, Washington, DC.

Chalmers, A.T. et al., 2011, Mercury trends in fish from rivers and lakes in the United States, 1969–2005, *Environ. Monit. Assess.*, 175, 175–191. DOI: 10.1007/s10661-010-1504-6.

Chang, M., Macullough, J.D., and Granillo, A.B., 1983, Effects of land use and topography on some water quality variables in forested East Texas, *Water Resour. Bull.*, 19, 191–196.

Chang, M., McBroom, M.W., and Beasley, R.S., 2004, Roofing as a source of nonpoint water pollution, *J. Environ. Manage.*, 73, 307–315.

Chang, M., Roth, F.A., II, and Hunt, E.V., Jr., 1982, Sediment production under various forest site conditions, *Recent Developments in the Explanation and Prediction of Erosion and Sediment Yield*, Walling, D.E., Ed., The International Association of Hydrological Sciences, IAHS Publ. No. 137, pp. 13–22.

Chang, M., Sayok, A.K., and Watterston, K.G., 1994, Clearcutting and shearing on a saline soil in East Texas: Impact on soil physical properties, *Proceedings of the 8th Biennial Southern Silvicultural Research Conference*, U.S. Forest Service Southern Research Station, Asheville, NC, pp. 209–214.

Clark, G.M., Muller, D.K., and Mast, M.A., 2000, Nutrient concentrations and yields in undeveloped stream basins of the United States, *J. Am. Water Resour. Assoc.*, 36, 849–860.

Clason, T.R. and Sharrow, S.H., 2000, Silvopastoral practices, *North American Agroforestry: An Integrated Science and Practice*, Garrett, H.E., Rietveld, W.J., and Fisher, R.F., Eds., American Society of Agronomy, Madison, WI, pp. 119–147.

Crouch, R.L. et al., 2006, Post-fire surface water quality: Comparison of fire retardant versus wildfire-related effects, *Chemosphere*, 62(2), 847–89.

Cuffney, T.F., Gurtz, M.E., and Meador, M.R., 1993, Methods for collecting benthic invertebrate samples as part of the National Water Quality Assessment Programs, USGS Open-File Report 93–406, Washington, DC.

Derlet, R.W., Goldman, C.R., and Connor, M.J., 2010, Reducing the impact of summer cattle grazing on water quality in the Sierra Nevada Mountains of California: A proposal, *J. Water Health*, 8, 326–333.

Dubrovsky, N.M. et al., 2010, The quality of our nation's waters—Nutrients in the nation's streams and groundwater, 1992–2004, National Water-Quality Assessment Program, USGS Circular 1350, Reston, VA.

Edmonds, R.L. and Blew, R.D., 1997, Trends in precipitation and stream chemistry in pristine old-growth forest watershed, Olympic National Park, Washington, *J. Am. Water Resour. Assoc.*, 33, 781–793.

Egan, A.F., 1999, Forest roads: Where soil and water don't mix, *J. For.*, 97, 18–22.

Elliot, W.J., Miller, I.S., and Audin, L., Eds., 2010b, Cumulative watershed effects of fuel management in the Western United States, General Technical Report RMRS-GTR-231, USFS, Rocky Mountain Research Station, Fort Collins, CO.

Elliot, W. et al., 2010a, Chapter 12, Tools for analysis, *Cumulative Watershed Effects of Fuel Management in the Western United States*, Elliot, W.J., Miller, I.S., and Audin, L., Eds., General Technical Report RMRS-GTR-231, USFS, Rocky Mountain Research Station, Fort Collins, CO, pp. 246–276.

Emmerich, W.E., 1998, Estimating prescribed burn impacts on surface runoff and water quality in southeastern Arizona, *Proceedings of the Rangeland Management and Water Resources*, Potts, D.F., Ed., AWRA, Middleburg, VA, pp. 149–158.

EPA, 1997, Monitoring guidance for determining the effectiveness of nonpoint source controls, EPA 841-B-96-004, U.S. Environmental Protection Agency, Washington, DC.

EPA, 2000a, Achieving cleaner waters across America: Supporting effective programs to prevent water pollution from forestry operations, EPA 841-F-002, Washington, DC.

EPA, 2000b, The quality of our nation's water, 1998: EPA841-S-00-001, available at http://www.epa.gov/305b/98report/98brochure.pdf

EPA, 2007, Fact sheet: 2005/2006 National Listing of Fish Advisories, EPA-823-07-003, available at: www.epa.gov/waterscience/fish/advisories/2006/tech.pdf

Feminella, J.W. and Flynn, K.M., 1999, Biotic indicators of water quality, Alabama Cooperation Extension System, ANR-1167, Alabama A&M and Auburn University, Auburn, AL.

Furniss, M.J. et al., 1998, Response of road-stream crossings to large flood events in Washington, Oregon, and Northern California, U.S. Forest Service, San Dimas Technology Development Center, San Dimas, CA, 9877 1806-SDTDC, p. 14.

Gallaher, B.M. and Koch, R.J., 2004, Cerro Grande fire impacts to water quality and stream flow near Los Alamos National Laboratory: Results of four years of monitoring, Los Alamos National Laboratory report LA-14177, Los Alamos, NM.

Gilliom, R.J. et al., 2007, The quality of our nation's waters—Pesticides in the nation's streams and ground water, 1992–2001, National Water Quality Assessment Program, USGS Circular 1291, Reston, VA.

Gluns, D.R. and Toews, D.A.A., 1989, Effects of a major wildfire on water quality in southeastern British Columbia, *Proceedings of the Headwaters Hydrology*, Woessner, W.W. and Potts, D.F., Eds., American Water Resources Association, Bethesda, MD, pp. 487–99.

Glymph, L.M. and Carlson, C.M., 1966, Cleaning up our rivers and lakes, American Society Agricultural Engineers, Paper No. pp. 66–74, St. Joseph, MI.

Gottfried, F.J. and DeBano, L.F., 1991, Streamflow and water quality responses to preharvest prescribed burning in an undisturbed Ponderosa Pine watershed, *Effects of Fire in Management of Southwestern Natural Resources*, General Technical Report RM-191, USDA Forest Service, Atlanta, GA.

Grace, J.M., III, 1999, Forest road side slopes and soil conservation techniques, *J. Soil Water Conserv.*, 54, 96–101.

Grace, J.M., III and Clinton, B.D., 2007, Protecting soil and water in forest road management, *Am. Soc. Agric. Biol. Eng.*, 50, 1579–1584.

Grace, J.M., III and Elliot, W.J., 2008, Determining soil erosion from roads in the Coastal Plain of Alabama, *Environmental Connection 08, Proceedings of Conference 39*, February 18–22, 2008, Orlando, FL, International Erosion Control Association, Steamboat Springs, CO.

Gravelle, J.A. et al., 2009, Nutrient concentration dynamics in an inland Pacific Northwest watershed before and after harvest, *For. Ecol. Manage.*, 257, 1663–1675.

Gucinski, H. et al., Eds., 2001, Forest Roads: A synthesis of scientific information, General Technical Report PNW-GTR-509, Forest Service Pacific NW Research Station, Portland, OR.

Hubbard, R.K., Newton, G.L., and Hill, G.M., 2004. Water quality and the grazing animal, *J. Anim. Sci.*, 82(E. Suppl.), E255–E263.

Ice, G., 1999, Streamflow and water quality: What does the science show about clearcutting in western Oregon, *Clearcutting in Western Oregon: What Does the Science Show?* College of Forestry, Oregon State University, Corvallis, OR, pp. 55–74.

Ice, G., 2004, History of innovative best management practice development and its role in addressing water quality limited waterbodies, *J. Environ. Eng.*, 130, 684–689.

Ice, G. and Binkley, D., 2003, Forest streamwater concentrations of nitrogen and phosphorus: A comparison with EPA's proposed water quality criteria, *J. For.*, 10(January/February), 21–28.

Ice, G.G., Neary, D.G., and Adams, P.W., 2004, Effects of wildfire on soils and watershed processes, *J. For.*, 102(2), 16–20.

Idaho Department of Lands, 2000, Forest practices cumulative effects, Process for Idaho, Boise, ID.

Jones, A., 2000, Effects of cattle grazing on North American arid ecosystem: A quantitative review, *West. North Am. Nat.*, 62(2), 155–164.

Kalabokidis, K.D., 2000, Effects of wildfire suppression chemicals on people and the environment—A review, *Global Nest: Int. J.*, 2(2), 129–137.

Kellogg, L.D., Bettinger, P., and Edwards, R.M., 1996, A comparison of logging planning, felling, and skyline yarding costs between clearcutting and five group-selection harvesting methods, *West. J. Appl. For.*, 11, 25–31.

Klott, R.W., 2007, Locating *Escherichia coli* contamination in a rural South Carolina watershed, *J. Environ. Manage.*, 83, 402–408.

Kochenderfer, J.N., 1970, Erosion control on logging roads in the appalachians, Pap NE-158, USDA Forest Service Research, NE Forest Experiment Station, Upper Darby, PA.

Ksiksi, T.S. et al., 2006, Artificial forest ecosystems of the UAE are hot spots for plant species, *World J. Agric. Sci.*, 2(4), 359–366.

Lewis, E.L. and Brendle, D.L., 1998, Relations of streamflow and specific conductance trends to reservoir operations in the lower Arkansas River, Southern Colorado, Water-Resources Investigations Report 97–4239, USGS, Denver, CO.

Likens, G.E. et al., 1977, *Biogeochemistry of a Forested Ecosystem*, Springer-Verlag, Heidelberg, Germany.

Mandaville, S.M., 2002, Benthic macroinvertebrates in freshwaters-Taxa tolerance values, metrics, and protocols, Project H-1, Soil & Water Conservation Society of Metro Halifax, http://www.chebucto.ns.ca/ccn/info/Science/SWCS/H-1/tolerance.pdf

Mast, M.A. and Clow, D.W., 2008, Effects of 2003 wildfires on stream chemistry in Glacier National Park, Montana, *Hydrol. Process.* 22, 5013–5023.

McBroom, M.W., Chang, M., and Cochran, M.C., 1999, Water quality conditions of streams receiving runoff from land-applied poultry litter in East Texas, *Watershed Management to Protect Declining Species, Proceedings of AWRA*, Sakrison, R. and Sturtevant, R., Eds., American Water Resource Association, Bethesda, MD, pp. 549–552.

McClurkin, D.C. et al., 1985, Water quality effects of clearcutting Upper Coastal Plain loblolly pine plantations, *J. Environ. Qual.*, 14, 329–332.

Megahan, W.F., 1972, Logging, erosion, and sedimentation, are they dirty words? *J. For.*, 70, 403–407.

Mersereau, R.C. and Dryness, C.T., 1972, Accelerated mass wasting after logging and slash burning in western Oregon, *J. Soil Water Conserv.*, 27, 112–114.

Muda, A., Chang, M., and Watterston, K.G., 1989, Effects of six forest-site conditions on nutrient losses in East Texas, *Proceedings of the Headwaters Hydrology*, Woessner, W.W. and Potts, D.F., Eds., American Water Resource Association, Bethesda, MD, pp. 55–56.

Norris, L.A. and Webb, W.L., 1989, Effects of fire retardant on water quality, USDA Forest Service General Technical Report PSW-109, pp. 79–86.

Omernik, J.M., 1976, The influence of land use on stream nutrient levels, EPA-600/3-76-014, EPA Environmental Research Laboratory, Corvallis, OR.

Owen, O.S., 1985, *Natural Resources Conservation*, 4th edn., MacMillan, New York.

Rashin, E.B. et al., 2006, Effectiveness of timber harvest practices for controlling sediment related water quality impacts, *J. Am. Water Resour. Assoc.*, 42, 1307–1327.

Rediske, R.R. et al., 2009, Assessment of benthic invertebrate populations in the Muskegon Lake area of concern, prepared for Michigan Department Environmental Quality, Annis Water Resources Institute, Grand Valley S. University, Muskegon, MI.

Reid, L.M., 1993, Research and cumulative watershed effects, USDA Forest Service, General Technical Report PSW-GTE-141, Albany, CA.

Reiss, K.C. and Brown, M.T., 2005, The Florida Wetland Condition Index (FWCI): Developing biological indicators for isolated depressional forested wetlands, Howard T. Odum Center for Wetlands, University of Florida, Gainesville, FL.

Reuss, J.O. et al., 1993, Biogeochemical fluxes in the Glacier Lakes catchments, Research Paper RM314, USDA Forest Service, Rocky Mountain Forest and Range Experiment Station, Fort Collins, CO.

ReVelle, P. and ReVelle, C., 1984, *The Environment, Issues and Choices for Society*, Willard Grant Press, Boston, MA.

Riekerk, H., 1983, Impact of silviculture on flatwood runoff, water quality, and nutrient budget, *Water Resour. Bull.*, 19, 73–79.

Sayok, A.K., Chang, M., and Watterston, K.G., 1993, Forest clearcutting and site preparation on a saline soil in East Texas: Impact of element movement, *Hydrology of Warm Humid Regions, Proceedings of the Yokohama Symposium, International Association of Hydrological Sciences*, Publ. No. 216, pp. 125–133.

Schreiber, J.D., Puffy, P.D., and McClurkin, D.C., 1980, Aqueous- and sediment-phase nitrogen yields from five southern pine watersheds, *Soil Sci. Soc. Am. J.*, 44, 401–407.

Sharrow, S.H., 2007, Soil compaction by grazing livestock in silvopastures as evidence by changes in soil physical properties, *Agrofor. Syst.*, 71, 215–223.

Sidle, R.C. and Hornbeck, J.W., 1991, Cumulative effects: A broader approach to water quality research, *J. Soil Water Conserv.*, 46, 268–271.

Stonesifer, C.S., 2007, Modeling the cumulative effects of forest fire on watershed hydrology: A post-fire application of the distributed hydrology-soil-vegetation model (DHSVM), Unpublished MS thesis, University of Montana, Missoula, MT.

Swank, W.T. and Johnson, C.E., 1994, Small catchment research in the evaluation and development of forest management practices, *Biogeochemistry of Small Catchment: A Tool for Environmental Research*, Moldan, M. and Cerny, J., Eds., John Wiley & Sons, New York, pp. 383–408.

Swank, W.T. and Vose, J.M., 1994, Long-term hydrologic and stream chemistry responses of southern Appalachian catchments following conversion from mixed hardwoods to white pine, *Hydrologie kleiner Einzugsgebiete: Gedenkchrift Hans M. Keller*, Landolt, R., Ed., Schweizerische Gesellschaft fur Hydrologie und Limnologie, Bern, Switzerland, pp. 164–172.

Swank, W.T., Vose, J.M., and Elliott, K.J., 2001, Long-term hydrologic and water quality responses following commercial clearcutting of mixed hardwoods on a southern Appalachian catchment, *For. Ecol. Manage.*, 143, 163–178.

Swank, W.T. and Waide, J.B., 1988, Characterization of baseline precipitation and stream chemistry and nutrient budgets for control watersheds, *Forest Hydrology and Ecology at Coweeta*, Swank, W.T. and Crossley, D.A., Jr., Eds., Springer-Verlag, Heidelberg, Germany, pp. 57–79.

Swanson, F.J. and Dryness, C.T., 1975, Impact of clear-cutting and road construction on soil erosion by landslides in the western Cascades Range, Oregon, *Geology*, 3, 393–396.

Swift, L.W., Jr., 1984, Soil losses from roadbeds and cut and fill slopes in the southern Appalachian Mountains, *South. J. Appl. For.*, 8, 209–215.

Taylor, S.E. et al., 1999, What we know—and don't know—about water quality at stream crossings, *J. For.*, 97, 12–17.

Texas Forestry Association, 1989, *Texas Best Management Practices for Silviculture*, Lufkin, TX.

Tiedemann, A.R. et al., 1979, Effects of fire on water, USDA Forest Service General Technical Report WO-10, Washington, DC.

TNRCC (Texas Natural Resource Conservation Commission), 1995, *Texas Surface Water Quality Standards*, Austin, TX.

USEPA, 1990, Feasibility report on environmental indicators for surface water programs, Office of Water Regulations and Standards and Office of Policy, Planning and Evaluation, Washington, DC.

USEPA, 2002, National recommended water quality criteria, EPA-822-R-02-047, Washington, DC.

USFS, 1974, Forest hydrology Part II—Hydrologic effects of vegetation manipulation, U.S. Department of Agriculture, Forest Service, Missoula, MT, p. 229.

USFS, 1980, *An Approach to Water Resources Evaluation of Non-Point Silvicultural Sources (A Procedural Handbook)*, EPA-600/8-80-012, EPA Environmental Research Laboratory, Athens, GA.

USFS, 1988, Cumulative off-site watershed effects analysis, *USFS Region 5 Soil and Water Conservation Handbook, FSH 2509.22*, USDA Forest Service Region 5, San Francisco, CA.

Webb, A.A., Jarrett, B.W., and Turner, L.M., 2007, Effects of plantation forest harvesting on water quality and quantity: Canobolas State forest, NSW, *Proceedings of the 5th Australian Stream Management Conference, Australian Rivers: Making a Difference*, Wilson, A.L. et al., Eds., Charles Sturt University, Thurgoona, New South Wales, Australia, pp. 443–448.

Winkler, G. et al., 2009, Short-term impact of forest harvesting on water quality and zooplankton in oligotrophic headwater lakes of the Canadian Boreal Shield, *Boreal Environ. Res.*, 14, 323–337.

Woods, P.F., 1980, Dissolved oxygen in intragravel water of three tributaries to Redwood Creek, Humbold County, California, *Water Resour. Bull.*, 16, 105–111.

12

Forests and Stream Sediment

Stream sediment refers to particles that are transported or deposited in stream channels. These particles are derived from rocks, soil, or biological materials in the watershed and are carried to stream channels through runoff, debris flow, and channel and bank erosion. Particles deposited by wind erosion are less significant in humid areas. Sediment impairs the physical, chemical, and biological properties of streams and is the greatest water pollutant in the United States. The fluctuation of sediment concentration and sediment yield are good indicators for the effectiveness of watershed management practices.

12.1 Soil Erosion Processes

Erosion is the movement of soil particles to a new location by raindrop splash, runoff, wind, or other forces. If the erosion is associated with water, it is called water erosion. Likewise, it is called wind erosion if wind movement induces it. Erosion, on one hand, is a natural and geological phenomenon in which the loss of soil is a continuous process and occurs at a normal rate (around 0.3 ton/ha/year). It may, on the other hand, occur at a destructive rate if the surface condition is improperly disturbed by human activities or other causes. This is called "accelerated erosion" and is a discrete event.

Water erosion begins with detachment of soil particles or small aggregates from the bulk of soil mass. The striking force must be greater than resistance forces of the soil to initiate the detachment. Once detached, particles are transported by runoff with a distance controlled by soil properties, topography, runoff energy, and surface conditions. When the flow-carrying capacity of sediment is less than the weight of soil particles, deposition occurs. Thus, water erosion is a three-step process involving detachment, transport, and deposition. A soil particle can get to a nearby creek during a single storm or can take as long as years, decades, or even centuries.

12.1.1 Detachment

The initial soil detachment requires sufficient energy to overcome the cohesive force of soil particles. Once detached, the energy required for transport is much less.

12.1.1.1 Acting Forces

For water erosion, the major forces that cause particles to detach from soil mass are striking raindrops and overland flow. Alternate freezing and thawing enhance rock weathering, but its particle-detachment rate is below the geological erosion level.

12.1.1.1.1 Raindrop (or Flow) Energy

The impact energy of a falling raindrop or flowing water can be calculated by

$$E = \frac{(M)(V_t)^2}{2} \tag{12.1}$$

where
  E is the kinetic energy in ergs (1 erg = force × distance = 1 g × 1 cm/s² × 1 cm = 1 dyn-cm)
  M is the mass of raindrop or flowing water (g)
  V_t is the terminal velocity of raindrop or flow velocity (cm/s)

The terminal velocity of a raindrop increases with increasing raindrop diameter (Figure 8.2), while the velocity of overland flow increases with the depth of the water layer. For a moderate rain of 0.40 cm/h, the average raindrop diameter is about 0.10 cm (Table 8.1), and its terminal velocity is 400 cm/s. It is rare to have an overland flow with speed exceeding 100 cm/s. Thus, the ratio of kinetic energy between raindrops from a moderate storm and an equal mass of overland flow of 100 cm/s is (400/100)², or 16. Apparently, raindrops are very effective for the detachment of soil particles. The impact energy of raindrops was shown to be the major force initiating soil detachment in a rainfall-simulation study of three different soils (Young and Wiersma, 1973).

Generally, the size of raindrops increases with rainfall intensity. In the eastern United States, Laws and Parsons (1943) showed that the median raindrop diameter (D_{50}, cm) can be estimated by the rainfall intensity (I, cm/h) by the following exponential function:

$$D_{50} = 0.188(I)^{0.182} \tag{12.2}$$

However, Hudson (1995) showed that the median raindrop diameter increases with rainfall intensity up to about 8.0–10.0 cm/h, and it is then decreased slightly beyond that intensity. Presumably greater turbulence at greater rainfall intensity makes larger drops unstable and breaks them into smaller drops. Actually, the relationship between median raindrop size and rainfall intensity is different among precipitation types. Therefore, a relationship similar to Equation 12.2 should be developed for major climatic regions. At Hot Springs, western Mississippi, McGregor and Mutchler (1977) showed that only about 6% of the rainfall had intensity greater than 10 cm/h and 70% was 2.54 cm/h or less. Until more relationships are developed for different climatic regions, Equation 12.2 is adopted for most applications in the United States.

Raindrop energy is further modified by storm wind and plant canopy. Wind enhances soil detachment by causing inclination of raindrops, which in turn makes the impact energy greater than it does in still air by

$$E = \frac{[(M)(V_t/\cos\alpha)^2]}{2} \tag{12.3}$$

where ($V_t/\cos\alpha$) is the vectorial velocity and α is the angle of raindrop inclination calculated by

$$\alpha = \tan^{-1}\left(\frac{V_h}{V_t}\right) \tag{12.4}$$

where

V$_h$ is the horizontal wind speed (cm/s)
V$_t$ is already defined in Equation 12.1

Thus, for raindrops of 0.10 cm diameter (V$_t$=400 cm/s), a 200 cm/s wind will cause an inclination of the raindrops of 26.6° from the vertical and make the kinetic energy of the inclined raindrops (1/cos 26.6°)^2 = 1.25 times greater than that in the calm air. Lyles (1977) showed that a storm with winds of 100 cm/s detached soil particles about 2.7 times more than it did in still air, regardless of percent mulch cover.

Plant canopies intercept falling raindrops and make the raindrops falling from the canopy have terminal velocities less than those from free fall. The reduction in fall velocity is dependent upon the height of the canopy. Raindrops falling from a height ≥20 m have a velocity equal to the free fall (or the terminal velocity), but the fall velocity is greatly reduced if the height is less than 20 m (Table 12.1). For a 3 mm raindrop falling from a canopy of 4 m above the ground, its kinetic energy is reduced by 6.68^2/8.06^2 or 69% compared to falling in the open or from a canopy of 20 m or taller.

If the canopy is in direct contact with the soil, the fall velocity is zero because the intercepted raindrops have no remaining fall height to the ground. This means that grasses and litter are more effective in dispersing raindrop energy and consequently reducing soil detachment than plants with canopies high above the ground are. However, the intercepted raindrops could be accumulated on the canopy and drip to the ground with diameters larger than those in the open (Chapman, 1948), although storm intensity is lower under the canopy.

12.1.1.1.2 Storm Energy

The kinetic energy of a storm is a function of storm intensity and duration along with mass, diameter, and velocity of raindrops. Based on Equation 12.2, Wischmeier and Smith (1958) developed the following empirical equation to estimate the kinetic energy of a storm rainfall:

$$E = 210 + 89 \log i \qquad (12.5)$$

where

i is the rainfall intensity (cm/h)
E is the kinetic energy (Mt-m/ha/cm)

TABLE 12.1

Velocities (m/s) of Various Sizes of Raindrops Falling from Different Heights in Still Air

Rainfall Intensity (cm/h)	Drop Diameter (mm)	Drop Fall Height (m)						
		0.25	1.00	2.00	3.00	4.00	6.00	20.00[a]
1.25	2.00	2.89	3.83	4.92	5.55	5.91	6.30	6.58
2.50	2.25	2.93	3.91	5.07	5.74	6.14	6.63	7.02
5.00	2.50	2.96	3.98	5.19	5.89	6.34	6.92	7.41
10.00	3.00	3.00	4.09	5.34	6.14	6.68	7.37	8.06

Source: Laws, J.O., *Trans Am. Geophys. Union*, 22, 709, 1941.
[a] Values in the column are considered terminal velocities.

This means a storm with an intensity of 1 cm/h would have enough energy to lift 210 Mt of soil 1 m high in every hectare of land and in every 1 cm of rainfall. If this storm lasts only 30 min and the total rainfall is 0.50 cm (storm intensity still = 1 cm/h), then the E value is equivalent to 105 Mt of soil 1 m high per hectare. Equation 12.5 applies to $i \leq 7.6$ cm/h. For $i > 7.6$ cm/h, E = 289 ton-m/ha-cm of rain. Other rainfall indices that relate kinetic energy to rainfall intensity or amount can be obtained from McGregor and Mutchler (1977), Morgan (1986), and Lal and Elliot (1994).

Estimating total kinetic energy of a storm requires the sequential progress of rainfall depth versus time, such as observations on a recording rain gage chart. The storm is then broken down into segments of uniform intensity. Calculate the kinetic energy of each segment by the product of Equation 12.5 and the total rainfall in the segment, and finally sum the kinetic energy of all segments to yield the total kinetic energy (see Section 12.2). In forested areas, total kinetic energy needs to be modified by considering percent forest canopy coverage and the height of the canopy in the following form:

$$E_e = E\left[(1-F) + F\left(\frac{M_{th}}{M_o}\right)\left(\frac{V_{th}}{V_o}\right)^2\right] \tag{12.6}$$

where
  E_e is the effective kinetic energy of rainfall
  E is the kinetic energy in the open, calculated by Equation 12.5
  M is the rainfall mass
  V is the raindrop velocity
  F is the fraction of canopy coverage
  Subscripts th and o, respectively, refer to throughfall and gross rainfall in the open

If the forest is taller than 20 m, then $(V_{th}/V_o)^2 = 1.0$ and the velocity ratio is omitted from the equation.

12.1.1.1.3 Freezing Water

Water expands as it freezes. The volume at −10°C is 0.187% greater than at 4°C, the smallest volume of water at any thermal condition. Such volume expansion exerts a pressure equivalent to about 146 kg/cm^2 (Brady, 1990). Thus, alternate freezing and thawing cause disintegration of rock, mineral, and soil aggregates.

12.1.1.1.4 Soil Resistance

Soil particles are bound together due to cohesion, internal friction, and organic colloids. The force that holds particles together is often referred to as shear strength, and the ability of soil structural units to withstand the forces of raindrop impact and surface flow is called detachability. Generally, soil detachability decreases with increasing organic matter content and infiltration rate, and increases with increasing particle size up to about 0.5 mm (Figure 12.1). Organic matter stabilizes soil structure by binding soil particles together. It forms an organo-mineral complex that can resist microbial degradation in soil for many years. The complex determines the duration and effectiveness in stabilizing soil particles (Barry et al., 1991).

For a given volume of soil, the total surface area for small particles is much greater than for large particles, which causes an increase in particle contact and cohesive force and consequently a greater resistance to detachment. However, the weight of particles overcomes

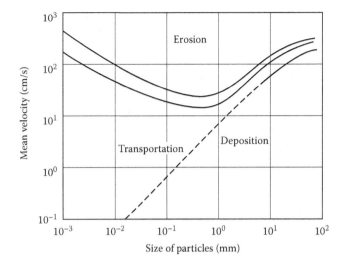

FIGURE 12.1
Erosion, transport, and sedimentation as a function of particle size. (Adapted from Hjulstrom, F., Transportation: Transportation of detritus by moving water, in *Symposium on Recent Marine Sediments*, Trask, P.D., Ed., American Association of Petroleum Geologists, Tulsa, OK, pp. 5–31, 1939.)

the surface attraction if the size is larger than about 0.5 mm. It then requires a great force to move the particles as the size increases. An interrill erosion study conducted on 18 soils with a wide range of textures showed that the silt and silt loam soils were most erodible, the high-clay soils were least erodible, and the loam and sandy loam soils were intermediate (Meyer and Harmon, 1984).

Shear strength is commonly determined by a series of three or four triaxial compression tests at different confining pressures carried to failure stresses in the laboratory and is defined by the Mohr–Coulomb equation (Parker et al., 1980):

$$\tau_c = C + \alpha_n \tan \phi \qquad (12.7)$$

where
τ_c is the shear strength (g/cm^2)
α_n is the effective normal stress (g/cm^2)
C and ϕ are empirical parameters, analogous to soil cohesion (g/cm) and the angle of internal friction, respectively

Cruse and Larson (1977) showed that soil detachment due to raindrop impact was significantly decreased with increasing shear strength in a second-degree polynomial regression function.

12.1.1.2 Detachment Rate

The detachment rate of soils (D) is determined by the relative magnitudes between kinetic energy of raindrops or surface flow (E) and soil shear strength (τ_c), modified by surface and topographic conditions. A number of studies have estimated soil detachment related directly to parameters such as raindrop energy (Bubenzer and Jones, 1971), the product of rainfall kinetic energy and maximum 30 min intensity (Free, 1960), rainfall and soil characteristics (Ellison, 1944), rainfall and water depth (Martinez et al., 1980), and runoff energy and topographic parameters (Meyer and Wischmeier, 1969).

12.1.1.2.1 Interrills

The USDA's (1995) Water Erosion Prediction Project (WEPP) hillside computer model breaks down erosion into interrill erosion and rill erosion. Rill erosion is the erosion that occurs in numerous small channels on hillsides by normal tillage, while erosion that occurs in areas between rills is called interrill erosion, formerly known as sheet erosion. Detachment on interrill areas is due to raindrop impact, and the detached particles are transported to the rills by raindrop impact and surface runoff. Soil detachment in the interrill areas (D_i, kg/s/m^2) is estimated by

$$D_i = (K_i)(I^2)(C)(G)\left(\frac{R_s}{w}\right) \tag{12.8}$$

where
  K_i is interrill erodibility (kg-s/m^4)
  I is effective rainfall intensity (m/s)
  C is the effect of canopy on interrill erosion (0–1)
  G is the effect of ground cover (0–1)
  R_s is the spacing of rills (m)
  w is the computed rill width (m)

Values of K_i for cropland soils with 30% or more of sand are estimated by the equation

$$K_i = 272,800 + 9,210,000 \text{ (vfs)} \tag{12.9}$$

and use the following equation if the sand is less than 30%:

$$K_i = 605,400 - 5,513,000 \text{ (clay)} \tag{12.10}$$

where vfs and clay are the fraction of very fine sand and clay in the surface soil, respectively. The enormous coefficients in Equations 12.9 and 12.10 are due to the large unit of rainfall intensity used in the model. The estimated values from these two equations are then adjusted to account for various site effects, including canopy cover, ground cover, roots, slope, freeze–thaw cycles, and sealing and crusting. Meyer and Harmon (1984) showed that interrill soil erodibilities were negatively correlated with percent dispersed clay (<2 mm), percent water content at 1500 kPa, exchangeable calcium, sum of exchangeable bases, and cation exchange capacity, with correlation coefficients ranging from 0.80 to 0.86.

12.1.1.2.2 Rills

The difference between flow shear stress τ and critical shear strength of soil τ_c, all in force per unit area or Newton/m^2 (1 N/m^2 = 1 kg/m/s^2 = 1 Pa), can be used to estimate soil detachment in the rills, or

$$D_c = K_r(\tau - \tau_c) \tag{12.11}$$

where
  D_c is detachment capacity by rill flow (kg/m²/s)
  K_r is rill erodibility (s/m)

In other words, there is no rill detachment when τ is less than τ_c. For crop soils with 30% or more of sand, the erodibility and the critical shear strength are estimated by

$$K_r = 0.00197 + 0.030 \, (\text{vfs}) + 0.03863 \, e^{-184} \, (\text{OM}) \tag{12.12}$$

$$\tau_c = 2.67 + 6.5 \, (\text{clay}) - 5.8 \, (\text{vfs}) \tag{12.13}$$

where OM is percent organic matter in the surface soil. For crop soils with less than 30% sand,

$$K_r = 0.0069 + 0.134 \, e^{-20} \, (\text{clay}) \tag{12.14}$$

$$\tau_c = 3.5 \tag{12.15}$$

Like K_i, the calculated K_r needs to be adjusted for other soil factors. The flow shear stress τ is calculated by

$$\tau = \gamma R s \tag{12.16}$$

where
  γ is the specific weight of water (Newton/m³ or kg/m²/s²)
  R is the hydraulic radius of rill or substituted by depth of flow (m)
  s is the hydraulic gradient of rill flow

These equations employ the concept that soil detachment on interrill areas is primarily induced by rainfall energy and ignore the minor impact from overland flow. Rainfall intensity is rarely constant during a storm, which causes the soil detachment rate to be a time-dependent function. Unfortunately, the time-dependence effect of changing soil detachment has been lumped with the rainfall parameter in most studies (Foster and Meyer, 1975). On the other hand, detachment in rill areas is primarily induced by concentrated flow and ignores the minor impact from rainfall impact. Once particles are detached from soil mass, the energy required to transport the sediment is much smaller.

12.1.2 Transport

For detached particles to become stream sediment, they require agents to carry them on to new locations and eventually to stream channels. Both rainfall and runoff can transport soil particles.

12.1.2.1 Rainfall Transport

Transport of soil particles by rainfall is due to splash action, and the splash capacity is a function of rainfall amount and intensity, slope gradient, soil characteristics, wind speed, and micro relief. As a rule of thumb, the percentage of total splashed soil that moves downslope equals the percent slope plus 50 (Ekern, 1953). However, the major impact of raindrop energy is for soil detachment, not for transport, in interrill areas. The detached and splashed particles are then transported primarily by broad shallow surface flow to rills (Nearing et al., 1994). There, soil particles from rill and interrill areas are joined together and transported to stream channels by concentrated flow (Young and Wiersma, 1973).

12.1.2.2 Runoff Transport

The maximum amount of sediment that an overland flow can carry, often referred to as transport capacity, depends on its energy level as determined by flow velocity and mass, slope gradient, hydraulic radius, surface roughness, and particle transportability. A turbulent flow is more erosive and carries more particles than does a laminar flow, while ponding of the flow produces rapid deposition. Since the velocity of overland flow increases with depth (Equation 10.35), the transport capacity is greater and transport distance is longer for deep flows. Thus, a reduction in overland flow or discharge, an increase in cross-sectional area, or a change in slope will cause a decrease in transport capacity. As a result, the heavier, larger, and round particles deposit first, while the smaller, lighter, and platy particles remain in suspension. Thus, particles of larger sizes are deposited upstream and smaller sizes downstream.

Silted flows are more erosive than clean ones are. Particles abrade and are crushed against one another as they are transported in the flow. This makes large particles wear away into smaller ones, which are carried farther downstream. In relative terms, shale and sandstone are more easily crushed than are limestone and dolomite, while granite and quartz are less easily crushed (Gottschalk, 1964).

In field investigations, the sediment-carrying capacity of flowing water was found to be approximately proportional to the fifth power of flow velocity (Laursen, 1958), to the 1.5th power of bottom shear stress (Foster and Meyer, 1972), or to stream power, the product of bottom shear stress and flow velocity (Yang, 1972). A more comprehensive equation to estimate total sediment transport capacity (T_c, in mass/width or kg/s · m) of overland flow for a storm is given by Foster (1982):

$$T_c = 138 \, Q \, q \, S^{1.55} C_t \tag{12.17}$$

where
Q is the total discharge/width (m³/s/m)
q is the peak discharge/width (m³/s/m)
S is the sine of slope angle
C_t is a factor reflecting the direct influence of soil cover on the flow's hydraulic forces

The C_t value is 1.0 (maximum) for a bare ground and 0.2 for an 8% slope with light mulch. The coefficient and exponent, depending on particle size, in Equation 12.17 represent conditions typical of most aggregated soils. Since discharge is estimated by the product of flow velocity, width, and depth, Equation 12.17 can be substituted by

$$T_c = 138(V)(q)(d)(S^{1.55})(C_t) \tag{12.18}$$

where
V is flow velocity (m/s)
d is flow depth (m)

12.1.3 Deposition

The deposition of moving particles depends on the relative magnitudes of transport capacity and sediment load (determined by soil detachment). If sediment load is less than transport capacity, then the sediment load moves downslope. If sediment load exceeds transport capacity, then deposition occurs. This can be expressed by

$$D_p = C_d(T_c - D) \tag{12.19}$$

where
D_p is the deposition rate (kg/s/m²)
T_c is the transport capacity (kg/s/m)
D is the sediment load (kg/s/m)
C_d is the first-order reaction coefficient for deposition in per unit width

A small value of C_d means a greater distance in particle transport and consequently less deposition. The value decreases with decreases in particle size and fall velocity, but increases as discharge increases. Foster (1982) gave the following equation to estimate C_d for deposition by overland flow:

$$C_d = 0.5\left(\frac{V_f}{Q}\right) \tag{12.20}$$

where
V_f is the fall velocity (m/s)
Q is the discharge (m³/s/m)

The fall velocity V_f, in centimeters per second, can be estimated by Stokes's law:

$$V_f = \frac{2(\delta_p - \delta)(g)(r^2)}{(9\mu)} \tag{12.21}$$

where
r is the radius of a particle (cm)
g is the acceleration due to gravity
δ_p is the density of the particle
δ is the density of the liquid
μ is the absolute (dynamic) viscosity of the liquid

The equation describes that the fall velocity of particles with a uniform density in a given liquid will increase with the square of the particle radius. For silts in an overland flow with $r = 0.0002$ cm, $\delta_p = 2.65$ g/cm³, μ at 20°C $= 0.01$ g/cm/s (poise), and $\delta = 1.0$ g/cm³, the fall velocity will be 0.00144 cm/s. If the overland flow is 1.0 cm deep in 0.10 m/s velocity, then those particles at the water surface would be carried in the flow for about 694 s and be deposited 69.4 m downslope.

For a typical midwestern soil, the reaction coefficient C_d for clay (2 μm), silt (10 μm), and sand (200 μm) is about 0.009, 0.074, and 22.0/m, respectively (Foster, 1982).

12.2 Watershed Gross Erosion

Soil erosion in a watershed can be categorized into six major sources: interrills, rills, ephemeral gullies, gullies, channels, and landslides. The sum of soil erosion from these six sources is called watershed gross erosion. Only a fraction of the watershed gross erosion will reach stream channels during a storm; the rest is trapped within the watershed subject to actions of other storm events.

12.2.1 Interrill and Rill Erosion

Surface runoff often concentrates in numerous small downslope channels that are uniformly spread across the field. Such small channels can be removed by normal tillage practices, and the soil loss that occurs along these small channels is termed rill erosion. Erosion in areas between rills is called interrill erosion. Severe rill erosion can exceed 448 ton/ha/year (Foster, 1986), while interrill erosion rates as high as 45 ton/ha/year have been measured (Meyer, 1981).

Empirical, factor-based, and physical models have been developed for estimating interrill and rill erosion. Most physical models treat interrill and rill erosion as two different processes, while the empirical and factor-based models combine them in a single estimate.

12.2.1.1 Empirical

The empirical approach relates observed erosion data to environmental factors through statistical analysis. Generally, it is easier to develop models for estimating long-term average soil losses than for single storms and for general areas than for specific areas. Since these models are not derived theoretically, they are applicable only in regions where the studies were conducted and under similar environmental conditions. Thus, the reliability of estimates depends on the representativeness of the data used in the model development and the suitability of the area for model applications.

The first model to estimate soil loss for hillslopes was developed by Zingg (1940) as a function of two topographic parameters.

$$A = f(\tan^{1.48} L^{0.6}) \tag{12.22}$$

where
  θ is the angle of slope gradient
  L is slope length

This pioneer study triggered other studies to relate soil loss to environmental factors, such as climatic conditions, soils, and vegetation cover. Later, using soil loss data collected from 40,000 plot-years under a wide range of rainfall conditions, Musgrave (1947) gave a prediction model that used topographic as well as climatic, edaphic, and cover factors.

$$A = K\,C\,(S^{1.35})(L^{0.35})[(P_{0.5})^{1.75}]\beta \tag{12.23}$$

where
  A is the soil loss (acre-in. or ha-cm)
  K is the inherent soil erodibility (in. or cm)

C is the cover factor

L is the slope length (ft or m)

S is the land slope (%)

$P_{0.5}$ is maximum 30 min rainfall with a 2 year return period (in. or cm)

β is a conversion factor if the equation is expressed in metric units instead of the original British units

The Musgrave equation was used by the Soil Conservation Service (SCS) for about 10 years and was the pioneer version of the universal soil-loss equation (USLE).

12.2.1.2 Factor-Based

A factor-based model contains a number of factors, such as rainfall, soil, topography, vegetation, and management, which may affect soil losses significantly. The effect of each factor on soil loss is separately quantified, based on either observed data or theoretical considerations. Each factor may quantify one or more erosion processes and their interactions. Finally, all factors are multiplied together to give the estimate. Although factor-based models are also largely based on data experience, the final equations are not obtained through statistical analyses. Such models are semiempirical and still have certain limitations. The most popular ones are the USLE (Wischmeier, 1975) and its modified version (USDA, 1995).

12.2.1.2.1 Universal Soil-Loss Equation

The USLE was originally developed by the Agricultural Research Service (ARS) and SCS in the 1950s to estimate average annual interrill and rill erosion from agricultural lands for areas east of the Rocky Mountains. It was extended to cover the entire conterminous United States in the middle of the 1970s. The USLE has been a very valuable and successful tool for land management and planning for nearly 50 years.

The USLE estimates soil loss (A) as the product of rainfall factor (R) and soil factor (K) with modifications of slope steepness factor (S), slope length factor (L), conservation practice factor (P), and cover and management factor (C):

$$A = RKLSCP \tag{12.24}$$

where

A is the estimated soil loss (ton/acre/year [1 ton/acre/year = 2.247 ton/ha/year])

R is the rainfall factor—the energy term to generate soil erosion

For a single storm, the erosive energy is indexed by the product of storm kinetic energy E in ft-ton/acre-in. (1 ft-ton/acre-in. = 0.268 ton/ha-cm) and the maximum 30 min storm intensity (I_{30}) in in./h (or cm/h) divided by 100. For a duration corresponding to A, R is calculated by

$$R = \frac{\left[\sum^{n} \sum^{m} (EI_{30}) \right]}{(100n)} \tag{12.25}$$

where

n refers to the total number of years

m refers to storms in each year

E is storm kinetic energy in British units (ft-ton/acre-in.) calculated by the following equation:

$$E = 916 + 331 \log i, \quad \text{for } i < 3 \text{ in./h} \tag{12.26}$$

where i is in in./h. For i > 3 in./h, E = 1074 ft-ton/acre/in. of rainfall. Equation 12.26 is Equation 12.5 in British units. The conversion of the USLE from British units to SI metric units was given by Foster et al. (1981) in great detail.

The calculation of R for a single storm requires rainfall intensity data (depth vs. time) such as that shown in a recording raingage chart. First, a storm is broken down into segments of uniform intensity. Calculate the E value for each segment first per in. of rainfall using Equation 12.26 (or per cm of rainfall using Equation 12.5), and then multiply the E/in. of rainfall (or E/cm of rainfall) by the total rainfall in inches (or in centimeters) in the segment to obtain the total E value for that segment. Sum the total kinetic energy of each segment to yield the total E value for the entire storm. Finally, multiply the total storm energy E by the maximum 30 min storm intensity in in./h (or in cm/h) and divide the product by 100 to yield the R factor for the storm. The calculation procedure is given in Table 12.2. Average annual R values for the conterminous United States, in hundreds of ft-ton-in./acre-h-year, are given in Figure 12.2 (Wischmeier and Smith, 1978). To convert R in British units to metric units in hundreds of ton-m-cm/ha-h-year or to SI units in MJ-mm/ha-h-year, multiply by the factor 1.735 or 17.02, respectively.

The calculation of R values illustrated earlier is laborious and requires rainfall intensity data that may not be available for the location in question. Many efforts have been made in different parts of the world to estimate R values based on monthly or annual precipitation data (Roose, 1977; Arnoldus, 1980; Simanton and Renard, 1982; Lo et al., 1985). Although reasonable estimates can be obtained by these substitute methods, their application is

TABLE 12.2

Calculations of R Factor of the USLE for Two Single Storms

Date	Time	Duration (min)	Rainfall cm	Rainfall in.	Intensity cm/h	Intensity in./h	E[a] (m-ton/ha) Per cm	E[a] (m-ton/ha) Segment	E[b] (ft-ton/acre) per in.	E[b] (ft-ton/acre) Segment
August 31	2300	0	0.00	0.00	0.00	0.00	0.0	0.0	0.0	0.0
—	2400	60	2.68	1.06	2.68	1.06	248.1	664.9	924.4	979.8
September 1	0130	90	6.23	1.63	4.15	1.63	265.0	1651.0	986.2	2416.3
	0200	30	3.36	2.65	6.72	2.65	283.6	952.9	1056.1	1394.0
	0350	0	0.00	0.00	0.00	0.00	0.0	0.0	0.0	0.0
	0400	10	0.68	0.27	4.08	1.61	264.3	179.7	984.5	265.8
	0500	60	0.13	0.05	0.13	0.05	131.1	17.0	485.4	24.3
	0600	60	0.84	0.33	0.84	0.33	203.3	178.8	756.6	249.7
	1400	480	0.13	0.05	0.02	0.01	58.8	7.6	254.0	12.7
	1500	60	3.28	1.29	3.28	1.29	255.9	893.4	952.6	1228.9
	1745	165	0.32	0.13	0.12	0.05	128.0	41.0	485.4	63.1
Total		1015	17.65	6.95				4524.3		6634.6

Note: In metric units: $I_{30} = 6.72$ cm/h; $R = (EI_{30})/100 = 4524.3 \times 6.72/100 = 304.01$ (m-ton-cm/ha-h). In British units: $I_{30} = 2.65$ in./h; $R = (EI_{30})/100 = 6634.6 \times 2.65/100 = 175.8$ (ft-ton-in./acre-h)

[a] Using Equation 12.5.
[b] Using Equation 12.26.

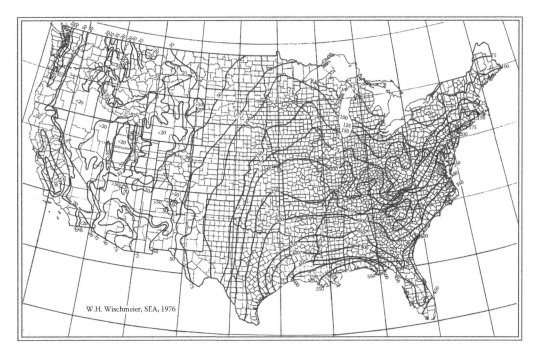

FIGURE 12.2
Values of the R factor (ft-ton-in./acre-h-year) of the USLE for the conterminous United States. (From Wischmeier, W.H. and Smith, D.D., *Predicting Rainfall Erosion Losses, a Guide to Conservation Planning*, Agricultural Handbook No. 537, USDA Agricultural Research Service, Brooksville, FL, 1978.)

limited to the areas where they were developed. Perhaps the largest applications in area are the models Renard and Freimund (1994) developed in the United States. Based on 132 stations throughout the United States (except for locations in western Washington, western Oregon, and northern California), one of the models appeared as

$$R = 0.1922 \, (P_t)^{1.61} \tag{12.27}$$

where
R is in hundreds of ton-m-cm/ha-year-h
P_t is mean annual precipitation (cm)

The equation seems to overestimate R values in the north and the northeast and underestimate those in the other regions.

K, the soil erodibility factor, describes both soil detachability and transportability due to various soil properties such as texture, structure, organic matter, density, compaction, and biological characteristics. These properties in turn affect soil infiltration and percolation, water-holding capacity, and surface runoff, parameters of the utmost importance to soil erosion processes. The K value in Equation 12.24 is the average soil loss per unit value of RLSCP, or

$$K = \frac{A}{(RLSCP)} \tag{12.28}$$

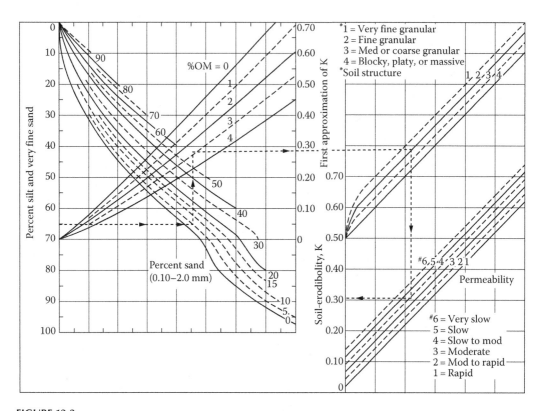

FIGURE 12.3
Nomograph for solving the K value (in British units) of the USLE. (From Wischmeier, W.H. and Smith, D.D., *Predicting Rainfall Erosion Losses, a Guide to Conservation Planning*, Agricultural Handbook No. 537, USDA Agricultural Research Service, Brooksville, FL, 1978.)

Assuming the land is bare, 9% slope, 22.13 m (72.6 ft) long, continuously plowed up and down the hill, and undergoing no conservation practices, all the LSCP values are equal to 1 and the equation becomes K = A/R. Field observations for a wide range of soil types showed that percent silt and very fine sand, percent of sand, percent of organic matter, soil structure, and permeability are most significant to the variation of K. The K value for a particular soil can be obtained from the NRCS' county soil survey report. If the information for the five soil properties is available, it can be read directly from Figure 12.3, or solved by the following equation:

$$100\,K = 2.1M^{1.14}(10^{-4})(12-a)+3.25\,(b-2)+2.5\,(c-3) \tag{12.29}$$

where
M = (silt + very fine sand%) (100 − clay %)
a is the percent organic matter
b is the soil structure code used in soil classification
c is the profile permeability class

Equation 12.29 and Figure 12.3 give K in British units (ton/ac/EI), or 1 K in British units = 0.342 K in metric units (ton/ha/EI in metric units). Values of K typically range from about 0.10–0.45 in British units. Soils with high clay or sand content have lower K values, while soils with high content of silt have higher K values.

The LS factor in Equation 12.24 describes the effects of topography: as the slope length increases, runoff per unit area decreases, but the accumulation (concentration) of runoff increases, resulting in greater soil loss per unit area. On the other hand, steep slopes cause lower infiltration rate, greater surface runoff, higher flow velocity, longer splash distance, and more soil detachment and transport. The relationship between soil loss and slope length (λ) has generally been described with the exponential function, soil loss = $c(\lambda)^a$, where soil loss is in mass or mass/area, λ has units of distance, and c and a are constants. Exponent a, depending on soil loss unit, soil type, and slope, varies greatly from 0 to 1.6 (Truman et al., 2001). In USLE, the two topographic factors are estimated by

$$LS = \left(\frac{\lambda}{22.13}\right)^a \left(0.065 + 0.0454s + 0.0065s^2\right) \qquad (12.30)$$

where
λ is slope length (m)
a is 0.5 for slope $\geq$ 5% and 0.3 for 1%–3%
s is slope steepness (%)

Comparatively, the effect of slope steepness on soil erosion is greater than that of slope length. A 10% error in slope steepness results in about a 20% error in computed soil loss, but a 10% error in slope length gives computed soil loss with 5% error (Renard et al., 1991).

The C factor refers to the ratio of soil loss between an area with specified cover and vegetation management (A_t) and an area with continuous tillage under identical conditions (A_c). Thus, soil-loss observations between treated and tilled plots can be used to evaluate various effects of species, cover density, cropping stage and rotation system, productivity, length of growing season, tillage practices, and residue management by

$$C = \frac{A_t}{A_c} \qquad (12.31)$$

For agricultural land, values of C are given in USDA Agriculture Handbook No. 537 (Wischmeier and Smith, 1965) for various crop rotation systems and for crops at five growing stages. Because the USLE is for average annual soil losses, the C values used should consider long-term changes in the cover and management conditions among years, seasons, and duration. For permanent pasture, range, and idle land, the C values are given in Table 12.3.

USDA Agriculture Handbook No. 537 (Wischmeier and Smith, 1978) also provides a few values of the C factor for forested land under three canopy densities, three litter-coverage groups, and two management statuses. However, these values were largely based on experienced judgment with limited field observations. A set of observed C values, along with other proposed values, under six forest conditions in east Texas is given in Table 12.4. Also, Wischmeier (1975) developed a procedure to estimate the C factor for forested lands. In his method, C values are a product of three forest effects: canopy (type I), ground cover (type II), and root and residues (type III). Later, the three types of effects were expanded into nine subfactors: amount of bare soil, canopy, soil reconsolidation, organic content, fine roots, steps, residual binding effects, on-site storage, and contour tillage, known as the

TABLE 12.3

C Values for Permanent Pasture, Range, and Idle Land

Vegetative Canopy			Cover that Contacts the Soil Surface					
Type and Height[a]	% Cover[b]	Type[c]	0	20	40	60	80	95
No appreciable canopy		G	0.45	0.20	0.10	0.042	0.013	0.003
		W	0.45	0.24	0.15	0.091	0.043	0.011
Tall weeds or short brush with average drop fall height of 6 m	25	G	0.36	0.17	0.09	0.038	0.013	0.003
		W	0.36	0.20	0.13	0.083	0.041	0.011
	50	G	0.26	0.13	0.07	0.035	0.012	0.003
		W	0.26	0.16	0.11	0.076	0.039	0.011
	75	G	0.17	0.10	0.06	0.032	0.011	0.003
		W	0.17	0.12	0.09	0.068	0.038	0.011
Appreciable brush or brushes, with average drop fall height of 2 m	25	G	0.40	0.18	0.09	0.040	0.013	0.003
		W	0.40	0.22	0.14	0.087	0.042	0.011
	50	G	0.34	0.16	0.08	0.038	0.012	0.003
		W	0.34	0.19	0.13	0.082	0.041	0.011
	75	G	0.28	0.14	0.08	0.036	0.012	0.003
		W	0.28	0.17	0.12	0.078	0.040	0.011
Trees, but no appreciable low brush, average drop fall height of 4 m	25	G	0.42	0.19	0.10	0.041	0.013	0.003
		W	0.42	0.23	0.14	0.089	0.042	0.011
	50	G	0.39	0.18	0.09	0.040	0.013	0.003
		W	0.39	0.21	0.14	0.087	0.042	0.011
	75	G	0.36	0.17	0.09	0.039	0.012	0.003
		W	0.36	0.20	0.13	0.084	0.041	0.011

Source: Wischmeier, W.H. and Smith, D.D., *Predicting Rainfall Erosion Losses, a Guide to Conservation Planning,* Agricultural Handbook No. 537, USDA Agricultural Research Service, Brooksville, FL, 1978.

Note: Assume that the vegetation and mulch are randomly distributed over the entire area.

[a] The height is measured as the average fall height of raindrops from the canopy to the ground. Canopy effect is inversely proportional to drop fall height and is negligible if height exceeds 10 m.

[b] Portion of total-area surface that would be hidden from view by canopy in a vertical projection.

[c] G: Cover at surface is grass, grasslike plants, decaying compacted duff, or litter of ≥5 cm deep. W: Cover at surface is mostly broadleaf herbaceous plants (such as weeds with little lateral-root network near the surface) or undecayed residues or both.

TABLE 12.4

Two Proposed and One Observed Sets of C Values of the USLE under Six Forested Site Conditions

Forested Site Conditions[a]						
A	B	C	D	E	F	Reference
0.0001–0.001	0.002–0.004	0.003–0.009	0.03		0.11–0.17	Wischmeier and Smith, 1978
0.001–0.0034		0.0003–0.0100	0.001–0.022	0.004–0.028	0.023–0.068	Dissmeyer and Stump, 1978
0.00014	0.00019	0.00165	0.00325	0.0242	0.097	Chang et al., 1982

[a] A = undisturbed mature forests; B = thinned to 50% density; C = clear-cut, merchantable timber removed but no site preparation; D = clear-cut, roller chopped; E = clear-cut, sheared, root raked, and windrowed; F = clear-cut, clear tilled, and continuous fallow.

USFS procedure (Dissmeyer and Foster, 1984). A proper value is assigned to each of the subfactors, multiplied together to obtain factor C.

The different methods of obtaining C values introduced earlier require field tests for their applicability. One of such tests used C values estimated by three methods: (1) USDA Agricultural Handbook No. 537 (Wischmeier and Smith, 1978), (2) the USFS procedure, and (3) field calibrations (Chang et al., 1982) to test against two annual and 51 single storms of soil-loss data observed under three forest conditions in east Texas (Chang et al., 1992). Soil losses in disturbed forests were found more difficult to estimate than those in undisturbed forests, and the estimates were more accurate for medium- and large-sized storms than for small storms. All three sets of C values provide reasonable estimates for annual as well as single storms for the undisturbed forest. However, the USDA's C values estimated annual and single-storm soil losses 3–51 times greater than those observed at the two disturbed forests (commercial cut and clear-cut with stumps sheared and windrowed). Errors in estimating soil losses by the two other sets of C values for the two disturbed forests ranged from 0.5 to 1.5 times the observed values, much closer to those estimated by the USDA's C values.

P, the conservation practice factor, is the ratio of soil loss with conservation practices to soil loss with up- and downhill cultivation under the same environmental conditions. This is the factor that describes human activities, such as contouring, terracing, and buffer strip cropping, which can modify slope length and steepness, surface roughness, runoff movement, and sediment transport. A land with no conservation practices has a $P = 1.0$, otherwise $P < 1.0$ as recommended in Table 12.5.

Example 12.1

Determine the average annual soil loss on idle land under the conditions given next. If it is required to keep the soil loss under 12 ton/ha/year, what treatments do you recommend to achieve that goal?

Location: Nacogdoches, Texas

Soil: Woodtell series with 18% of clay, 46% of silt plus very fine sand, 5% organic matter, fine granular (structure code 2), and very slow permeability (class 6).

TABLE 12.5

Recommended Values for the P Factor in the USLE

Land Slope (%)	Contouring	Contour Strip Cropping[a]	Terrace and Strip Cropping
1–2	0.60	0.45	0.30
3–5	0.50	0.38	0.25
6–8	0.50	0.38	0.25
9–12	0.60	0.45	0.30
13–16	0.70	0.52	0.35
17–20	0.80	0.60	0.40
21–25	0.90	0.68	0.45

Source: Wischmeier, W.H. and Smith, D.D., *Predicting Rainfall Erosion Losses, a Guide to Conservation Planning*, Agricultural Handbook No. 537, USDA Agricultural Research Service, Brooksville, FL, 1978.

[a] For 4 year rotation of 2 years row crop, winter grain with meadow seeding, and 1 year meadow.

Topography: 100 m long and 15% slope
Land use: Idle land with 25% coverage of brush less than 4 m on average height, 20% of the land surface covered by grass; no conservation practices

Solution:

(a) Determination of the average annual loss by USLE, $A = RKLSCP$:

$R = 400$, from Figure 12.2 in British units

$= 400 \times 1.7 = 680$ (ton-cm/ha-h-year)

$K = 2.1[46 \times (100 - 18)]^{1.14} (10^{-4})(12 - 5) + 3.25(2 - 2) + 2.5(6 - 3)$

from Equation 12.29

$= 0.25$ (in British units)

$= 0.25 \times 0.342 = 0.0855$ (in ton/ha-year-EI in metric units)

$LS = (100/22.13)^{0.5} (0.065 + 0.0454 \times 15 + 0.0065 \times 15^2)$

from Equation 12.30

$= 4.69$

$C = 0.19$, from Table 12.3
$P = 1.0$ (No conservation practices)

$A = 680$ (EI in metric units) $\times 0.0855$ (ton/ha-year-EI in metric units)

$\times 4.69 \times 0.19 \times 1.0$

$= 52$ ton/ha-year

(b) Potential reduction of soil losses:

The tolerated soil loss of 12 ton/ha-year is only 12/52 or 23% of the predicted value. Of the six environmental factors used in the USLE, the most feasible factors for reducing soil losses are C and P. The current C value is 0.19, and it must be reduced to $0.19 \times 0.23 = 0.044$ to achieve that goal. If grass coverage on the land is increased from the current 20% to 60% or greater, then $C = 0.041$ (Table 12.3) and the total soil loss $= 680 \times 0.0855 \times 4.69 \times 0.041 \times 1.0 = 11.18$ (ton/ha-year), which is below the tolerated level.

12.2.1.2.2 *The Revised Universal Soil-Loss Equation*

Although the USLE is a powerful tool that is widely used for estimating soil erosion in the United States and other countries, research and experience have brought new information and improvements to the application of the technology since the 1970s. In late 1987, ARS and SCS began to update the USLE by incorporating new information into the technology given in USDA Agricultural Handbook No. 537 (Wischmeier and Smith, 1978). It results in a revised procedure called revised universal soil-loss equation (RUSLE) and is documented in USDA's Agricultural Handbook No. 703 (Renard et al., 1994, 1997).

Basically, RUSLE is a computerized technology for the determination of individual factors and calculations of the original USLE. Major changes include the following:

- New isoerodent map for the western United States, based on rainfall data from more than 1200 locations
- Corrections for high R factor areas with flat slopes to adjust for splash erosion with raindrops falling on ponded water in the eastern United States
- Development of a seasonal K factor to account for seasonal variation
- Use of a five-subfactor approach to account for prior land use, crop canopy, surface cover, soil moisture, and surface roughness for calculating the C factor
- The capacity to calculate LS factor for slopes of various shapes, and a reduction of LS values by almost half for slopes greater than 20%
- A new P value as the product of P subfactors for individual practices such as contouring, strip cropping, buffer strips, terracing, and subsurface draining and for rangelands

12.2.1.3 Physical

It is impossible to develop an empirical or semiempirical model that can estimate soil loss for all land uses with different management practices, on all topography and soil types, and under all weather conditions. Generally, empirical or semiempirical models were designed to predict long-term soil losses, not to present soil erosion as a dynamic process at different scales of time and space. Physically based models intend to overcome the empirical models' limitations, but no physical models as yet can be satisfactorily applied to all purposes due to the difficulties in quantifying certain physical processes. Most physical models are developed under certain assumptions and for a particular environment. Continuous refinement and calibration are a necessity for improving their applicability. Once validated, a physical model is a powerful and flexible tool for conservation management and planning.

Physical models for estimating upland soil erosion are based on soil detachment and transport processes. Meyer and Wischmeier (1969) developed perhaps the first model by breaking the whole process into four components as follows:

1. Detachment by rainfall
2. Detachment by runoff
3. Transport capacity by rainfall
4. Transport by runoff

If total detached soil is less than transport capacity, then soil is carried downslope and no soil deposition occurs (Figure 12.4). Later, the four processes are applied separately to interrill and rill areas, and the sum of the soil losses from interrills and rills is the total soil loss (Foster and Meyer, 1975). The dynamic of each component can then be described by fundamental hydraulic, hydrologic, meteorological, edaphic, topographic, and vegetative interactions.

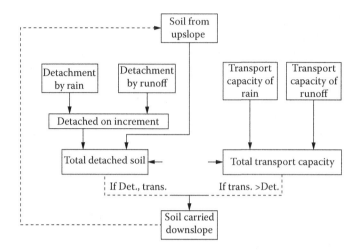

FIGURE 12.4
Simulating the processes of soil water erosion. (From Meyer, L.D. and Wischmeier, W.H., *Trans. ASAE*, 12, 754, 1969.)

Many models, including ANSWERS (Beasley et al., 1980), CREAMS (Knisel, 1980), EPIC (Williams et al., 1983), WEPP (USDA, 1995), and SWAT (Neitsch et al., 2001; Jha et al., 2002; Di Luzio et al., 2004), have been developed for estimating upland soil erosion. Although time scales and space scales along with methods for estimating each process or parameter may be different among models, basic processes and principles remain the same.

12.2.2 Ephemeral Gully Erosion

Overland flow often converges into a few major waterways or swales before leaving the fields. The concentrated flow makes most sediment in the field drain out through these waterways. Erosion from these waterways is called ephemeral erosion, megarill erosion, or concentrated flow erosion, a new type of erosion that was overlooked until the 1980s. Assessment of ephemeral gully erosion rates has shown it to be as high as 274% of interrill and rill erosion rates in the United States (Bennett et al., 2000).

Ephemeral gullies are larger than rills but smaller than gullies (Figure 12.5). Although plowing and tillage may occur across ephemeral gullies, they recur in the same location each year, while rills may change locations from year to year. Flows in rills are characterized by overland runoff, but they are channelized in ephemeral gullies. Characteristics of rill, ephemeral gully, and (classic) gully erosion are given in Table 12.6.

12.2.2.1 Processes

Erosion in ephemeral gullies is initiated and carried by channelized flows. Flow detaches soil particles and transports the sediment load downstream. The detachment rate is proportional to the difference between the shear stress of flow and the soil's critical shear strength. Sediment load in flow is limited either by transport capacity of the flow or by sediment available for transport. Two sources contribute to sediment load, one from interrill and rill erosion from adjacent areas and the other from detachment in upstream

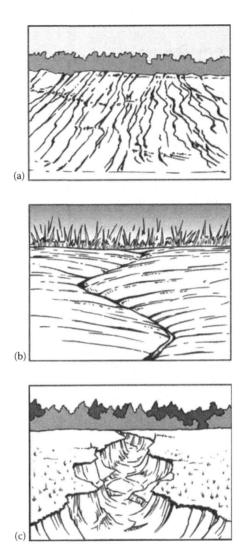

FIGURE 12.5
Interrill and rill erosion (a), ephemeral gully erosion (b), and gully erosion (c).

ephemeral gullies. This causes sediment load to be greater downstream. Transport capacity is greater downstream because of greater discharge.

As the sediment load is greater than the transport capacity of the flow, the fraction of sediment that is greater than transport capacity is deposited to the bottom. In this case, transport capacity is equal to sediment load. Unless transport capacity is increased due to an increase in either discharge or flow velocity, no additional sediment is carried by the flow.

12.2.2.2 Estimates

Ephemeral gully erosion can be estimated based on an empirically based on-site visit procedure (Thorne and Zevenbergen, 1990) or on physically based models (Woodward, 1999). These approaches are described next.

TABLE 12.6

Characteristics of Different Types of Erosion

Interrill and Rill Erosion	Ephemeral Gully Erosion	Gully Erosion
Occurs on smooth side slopes above drainageways	Occurs along shallow drainageways upstream from incised channels or gullies	Generally occurs in well-defined drainageways
May be of any size but are usually smaller than concentrated flow channels	May be of any size but are usually larger than rills and smaller than classic gullies	Usually larger than concentrated flow channel and rills
Flow pattern develops many small disconnected parallel channels that end at concentrated flow channels, terrace channels, or in depositional areas	Usually form a dendritic pattern along water courses beginning where overland flows, including rills, converge. Flow patterns influenced by tillage, rows, terraces, man-made features	Dendritic pattern along natural water courses. May occur in nondendritic patterns in road ditches, terrace or diversion channels, etc.
Cross sections usually are narrow relative to depth	Cross sections usually are wide relative to depth. Sidewalls not well defined	Cross sections usually narrow relative to depth. Sidewalls are deep
	Head-cuts not readily visible; do not become prominent because of tillage	Head-cut prominent. Eroding channel advances upstream
Normally removed by tillage and do not recur in the same place	Temporary feature, usually removed by tillage; recur in the same place	Not removed by tillage
Soil removed in thin layers or shallow channels. Soil profile becomes thinner over entire slope	Soil removed along narrow flow path, to tillage depth if untilled layer is resistant to erosion, or deeper if untilled layer is less resistant	Soil may erode to depth of profile, and can erode into soft bedrock
Low erosion rates not readily visible	Area may or may not be visibly eroding	Erosion readily visible
Detachment and transport by raindrops and flowing water	Detachment and transport by flowing water only	Detachment by flowing water and slumping of unstable banks; transport by flowing water

Source: Laflen, J.M., Ephemeral gully erosion, in *Proceedings of Fourth Federal Interagency Sediment Conference*, March 24–27, 1986, Las Vegas, NV, Vol. I, Sec. 3, pp. 29–37, 1986.

12.2.2.2.1 On-Site Visit Procedure

The field visit method is based on an analysis of field topography including slope, contributing area, and the vertical curvature across the natural drainage way (planform). These topographic parameters are then used to determine a compound topographic index (CTI) as a prediction of the intensity (or stream power) of concentrated surface runoff and the cross-sectional area voided by the ephemeral gully. Thorne and Zevenbergen (1990) reported that ephemeral gullies usually do not form at sites where the planform is less than a critical CTI value; they gave the following procedures:

1. Identify locations of ephemeral gullies, swales (topographic lows), and drainage divides along with each gully head in the field.
2. Based on local topography and size, select representative reaches along each ephemeral gully and flag the intermediate point in each reach. Measure the following topographic parameters from the gully head downstream to each flag (Figure 12.6):

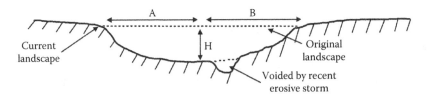

FIGURE 12.6
Field measurements of H, A, and B for ephemeral gully erosion. (Adapted from Thorne, C.R. and Zevenbergen, W.: *Soil Erosion on Agricultural Land*. pp. 447–460. 1990. Copyright Wiley-VCH Verlag GmbH & Co. KGaA.)

    a. Stake height (H, m)
    b. Left swale width (A, m)
    c. Right swale width (B, m)
    d. Local slope (S, m/m)
    e. Upstream area (Area, m²)
    f. Distance downstream from gully or swale head (L, m)

3. Calculate the CTI for each point by

$$CTI = (Area)(S)(PC) \tag{12.32}$$

where PC, the planform curvature of the swale, a control on the convergence of surface runoff and hence the concentration of erosive streampower per unit bed area, is given by

$$PC = \frac{200(H)}{[(A)(B)]} \tag{12.33}$$

4. Calculate the CTIcrit as the average of CTI value for all gully head points. Point CTI values less than the CTIcrit are excluded from further analysis because they would not have an ephemeral gully in an average year.

5. Calculate the gully cross-sectional area for all points with CTI > CTIcrit by the experimentally derived equation:

$$\text{Cross-sectional area, } m^2 = 0.03365(CTI)^{0.25} \tag{12.34}$$

6. Calculate the eroded volume for each reach from the cross-sectional area and the reach length. The cross-sectional area in each reach represents a distance halfway to its upstream and downstream reaches. Summing the eroded volume in each reach yields the annual ephemeral gully erosion for the field. Multiply the volume by soil bulk density to yield total annual soil erosion in mass and then convert the mass into mass per unit area.

The CTI approach can be incorporated into digital terrain analysis tools, such as TOPAZ (Topographic Parameterization) (Martz and Garbrecht, 1998), for rapid and automatic identification of potential locations for ephemeral gully channels at the catchment scale (Parker et al., 2010).

12.2.2.2.2 Physical Models

Physical models use two components relating hydrology and the erosion process. The hydrology component estimates peak discharge and runoff volume by the use of SCS curve number, drainage area, watershed flow length, watershed slope, and 24 h rainfall and standard temporal distribution. The estimated peak flow and discharge volume drive the erosion component, a combination of physical process equations and empirical relationships to estimate the width and depth of the ephemeral gully. Finally, the mass of soil erosion is calculated using the volume voided and the soil bulk density.

A basic equation relating soil detachment from gully boundaries to flow hydraulics (Woodward, 1999) is

$$D_e = K_{ch}(1.35\tau - \tau_c) \qquad (12.35)$$

where
D_e is the detachment rate $(g/m^2/s)$
K_{ch} is the channel erodibility factor $(g/s/N)$
τ is the average shear stress of flowing water (N/m^2)
τ_c is the critical shear strength of soil (N/m^2)
constant 1.35 represents maximum stress at the channel bottom

The amount of detachment is balanced with the sediment transport capacity of the flowing water by the equation

$$\frac{D_e}{D_c} + \frac{D}{T_c} = 1 \qquad (12.36)$$

where
D_e is the detachment rate
D_c is the maximum detachment capacity
D is the sediment
T_c is the sediment transport capacity

The erosion process begins with a gully of initial width W_e. Depending on the duration of runoff, channel slope, and hydraulic roughness, the gully will continue to deepen until an erosion-resistant layer is reached. At that time, further deepening ceases and the channel begins to widen toward the ultimate width, W_u. If the flow ceases after the channel reaches the erosion-resistant layer but before the ultimate width is attained, it results in a width of W between W_e and W_u. Regression equations used to estimate equilibrium gully width (W_e, m), equilibrium erosion rate (E, $g/m^2/s$), and ultimate gully width (W_u, m) are

$$W_e = 2.66 \, (Q^{0.396})(n^{0.387})(S^{-0.16})(\tau_c)^{-0.24} \qquad (12.37)$$

$$W_u = 179 \, (Q^{0.552})(n^{0.556})(S^{0.199})(\tau_c)^{-0.476} \qquad (12.38)$$

$$E = 34.42 \, K_{ch}(Q^{0.811})(n^{0.80})(S^{0.77})(\tau_c)^{-0.48} \qquad (12.39)$$

where

    Q is the peak flow rate (m^3/s)
    n is the Manning's roughness coefficient
    S is the channel slope
    τ_c and K_{ch} are as defined earlier

Procedures used to calculate W and the amount of soil erosion for a single storm as well as the annual average were given in a model and computer program called Ephemeral Gully Erosion Estimator (EGEE) by Watson et al. (1986). The EGEE was later modified to become the Ephemeral Gully Erosion Model (EGEM) to meet the needs of the Natural Resources Conservation Service (Merkel et al., 1988; Woodward, 1999). However, the average annual concept of the model has not yet been validated with field data, and the empirical relationships need to be adjusted as more data from a variety of field conditions become available.

12.2.3 Gully Erosion

When eroded channels reach a size large enough to restrict vehicular access or where it cannot be smoothed over by normal tillage operations, it is termed gully or gully erosion. Gullies are permanent channels unless they are filled with soils by heavy equipment; they are an advanced stage of water erosion. All soils above the bottom in the gully channel have been washed out by the concentration and high velocity of runoff water. If no conservation and control measures are taken, gullies will continue to expand and grow. The results are deeply incised gullies (ephemeral channels) along depressions and discrete landscape along hillside slopes. Deep gullies may extend to the watershed divide and ultimately convert the watershed into badlands. Gullies divide lands into small pieces, completely remove portions of fields from production, and largely reduce the efficiency of farm equipment. Moreover, failure of reservoir–embankment dams and earth spillways often occurs due to head-cutting and gully development.

12.2.3.1 Formation

Gully erosion is developed mechanically by two major processes, down-cutting and head-cutting. Down-cutting causes gully bottoms to be wider and deeper, while head-cutting advances channels into headwater areas and expands tributary nets of the gully.

12.2.3.1.1 Induced by Surface Runoff

The surfaces of hillsides are never uniform due to the variation of land configuration, soil texture, rock distribution, plant root exposure, and debris and branches. This makes runoff different in depth and velocity over a surface. Some runoff may converge in small depressions, forming a small concentrated flow. As the flow in a small surface channel passes an abrupt drop in slope, such as roots or rocks, a small waterfall is created. The energy of a waterfall is much greater than the same amount of water flowing on a uniform slope. As a result, the channel at the foot of the waterfall is deepened, and the banks are caved in from being undermined. Alternate freezing and thawing or landsliding may also result in massive collapse of the exposed gully bank. Thus, the head-cut induced by the sudden change in elevation advances the growth of gully length, while the down-cut (scour) induced by concentrated flow or bank flow makes gullies wider and deeper (Figure 12.7).

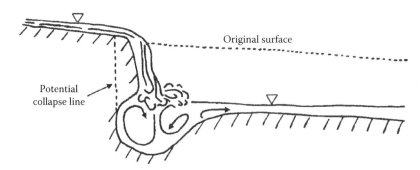

FIGURE 12.7
Formation of gully erosion.

Overgrazing, fire, or other land abuses such as poor road layout and construction often accelerate gully development.

12.2.3.1.2 Induced by Subsurface Runoff

Most water in the forest is removed by subsurface flow through pipes and tunnels formed by macropores. Those pipes and tunnels under the ground can undergo soil erosion due to subsurface runoff. When a forest is cleared, a great amount of water is flushed out of the soil, causing severe erosion to underground tunnels and pipes, the subsidence of ground surface, and the exposure of the pipe and tunnel network as gullies (Morgan, 1986). This problem is especially serious in areas where a sandy soil on the surface overlies a compact, relatively impermeable soil in the profile. Heede (1976) stated that piping soils, which may cause the formation of gullies, have a significantly higher exchangeable sodium percentage (ESP). Sodium enhances clay dispersion, eliminates the macropores, and makes the soils largely impervious to water infiltration.

Chang (1997) reported that severe gully erosion could be developed on a fine sandy loam soil underlain by a sandy clay loam soil by subsurface runoff within a few years due to improper land management. Two 2.8 ha forested tracts lie side-by-side (east–west) on a 4%–8% slope in east Texas. The top soil is fine sandy loam, 75 cm deep with a permeability of about 1.5–5.0 cm/h, underlain by a sandy clay loam at 75–125 cm deep with permeability as low as 0.15–0.015 cm/h. In 1990, the soil on the downslope tract was excavated to convert the entire area into level ground. The excavation exposed a vertical embankment about 1.20–1.50 m high and 275 m long. Following the excavation, a landscape timber retaining wall was installed along the cutoff bank to control the soil erosion. The wall was about 15 cm higher than the ground surface, and neither drainage tiles nor anchors were installed under the ground.

Without the soil excavation and retaining wall, the fine sandy loam soil at the surface enhanced water percolation and the percolated water accumulated at the top layer of the sandy clay loam soil. Now, the retaining wall blocked the surface runoff and turned the runoff water to scour the cutoff bank longitudinally and vertically along the retaining wall. Also, the excavation of soil altered water pressures and water balance under the ground, causing the accumulated water along with soils in the soil profile in the upper slope to gush out through the cutoff bank (Figure 12.8). As a result, severe water erosion appeared as depressions, sinkholes, and hollows under the ground; gullies occurred along the cutoff bank and its upslope areas. Within 5 years, 17 gullies or hollow sites were developed along the retaining wall with the longest site extending 25 m long, 5 m wide, and 1.5 m deep.

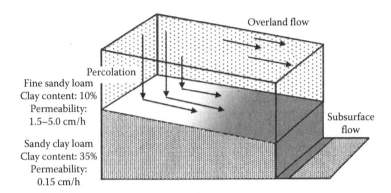

FIGURE 12.8
Soil erosion caused by subsurface runoff. (After Chang, M., Improper soil management causing severe water erosion problems, in *Proceedings of the 28th Conference*, February 25–28, 1997, Nashville, TN, International Erosion Control Association, Reno, NV, pp. 495–501, 1997.)

Down-cutting erosion occurred at about 15–30 cm/year on many sites. The erosion and gully development also demonstrated that downslope activities could impose impacts upslope.

12.2.3.2 Gully Types

In loessial regions and alluvial valleys where soil resistance to erosion is small, U-shaped gullies with vertical walls can develop. On the other hand, a V-shaped gully is more likely if the subsoil is more resistant to erosion. Within a watershed, gully networks can be categorized as continuous, discontinuous, or mixed (Heede, 1978). In a continuous network, many gullies are connected together. The gullies always start high up on the mountainside, beginning with many small rills. They quickly reach gully depth and maintain the depth approximately to the gully mouth. There, depth decreases rapidly in the lowest segment, and the shallower depth in turn widens the channel and causes more sediment deposition.

The discontinuous network is signified by an abrupt head-cut in the gully head, with depth decreasing rapidly downstream. The bottom gradient is gentler than that of the original valley floor. It forms an alluvial fan at the intersection with the valley. A chain of discontinuous gullies can be fused into a single continuous gully network. If the fan is too steep, a head-cut and a new discontinuous gully may form. Uphill advancement of head-cuts may fuse the discontinuous gullies to form or join a continuous gully, but some may still be fully independent to make the entire gully network a mix. Thus, gully head (head-cut) and gully mouth are the two critical locations important to the development of a discontinuous gully.

12.2.3.3 Estimates

Erosion processes in a gully include head-cut advancement, sidewall undercutting, bank collapse, and cleanout of sloughed material. Flow in the gully may produce some erosion, but it may not be significant (Heede, 1975). In many cases, head-cuts and mass wasting of gully banks are the prime erosion processes, not tractive force (shear stress) and power of the streamflow (Piest et al., 1975). Although there is an increasing interest in gully erosion in recent years (Robinson and Hanson, 1996; Bennett and Alonso, 2000), parameters governing the processes of gully growth are still not well defined. In estimating gully erosion, both head-cut erosion and bank erosion need to be considered.

Because of the ambiguity of gully growth process and inadequacy of field observational data, no physical models have been developed yet for predicting gully growth and the generated soil losses. Changes in gully geometry are often used as a basis for quantitative estimates of soil loss in the past. They can be done by comparing old and new aerial photos, doing field inspections, and consulting local people. Once the rates of head-cutting and bank erosion, width, and depth are estimated, the volume of voided soil can be converted into mass by multiplying by an appropriate soil bulk density. However, the estimated past erosion may not be a good indication of future erosion because of differences in gully development stage, climatic conditions, vegetation cover, land management, and other factors.

Some empirical models have been developed for estimating future gully advancement. For example, Beer and Johnson (1963) developed a logarithmic regression model to estimate changes in gully surface area in western Iowa using five physical parameters. These parameters included surface runoff, level terraced area, initial gully length, gully length from upper end to watershed divide, and deviation from normal precipitation. By use of field measurements of 210 gullies scattered around the eastern slopes of the Rocky Mountains, Thompson (1964) showed that the rate of head advance could be estimated using drainage area, slope, precipitation, and soil in a linear regression model, with 77% of its variation explained. The statistical approach for gully growth prediction is valid at best for the watersheds used in the study and maybe for other watersheds of the same region. Variations in soil properties, rainfall intensity, geology and topography, land use and vegetation cover, and gully stages limit the application of these empirical models greatly.

The U.S. Soil Conservation Service (1977) used the following equation to estimate the future rate of gully head-cut advancement:

$$G_h = R_p(A_r^{0.46})(P^{0.20})$$ (12.40)

where
  G_h is the computed future average annual rate of gully head advance for a given reach (m/year)
  R_p is the past average annual rate of gully head advance (m/year)
  A_r is the ratio of the average drainage area of a given upstream reach to the average drainage area of the reach through which the gully has moved
  P is the ratio of the expected long-term average annual rainfall from 24 h rains of 1.27 cm or greater to the average annual rainfall from rainfalls of 1.27 cm or greater for the period in which the existing gully head has moved

Equation 12.40 is a combination of the past growth of gully head and the result of new statistical analysis using Thompson's (1964) field measurement data. Using R_p as the past average annual rate of gully undercut, Equation 12.40 can also be employed to estimate the future rate of gully undercut until equilibrium or base level is established. Beyond this stage, no deepening will occur. Gully widening can be estimated based on the average width–depth ratio of existing gullies within the area. It may use the general relationships observed from a large number of gullies that the width–depth ratio is about 3.00 for cohesive materials and 1.75 for noncohesive materials. The total volume of gully void can be computed once information on head-cut advance, width, and depth is available.

12.2.4 Mass Movements

Slope failures or mass movements refer to the downhill movement of large volumes of soil mass, rocks, and other materials under the direct influences of gravity, such as land creep, landslides, mudflows, debris flows, avalanches, and falling rocks. They can occur anywhere with steep slopes and weak geological structures and often are triggered by intense and prolonged rainfall, snow accumulation and snowmelt, convergence of overland flows and seepage, earthquake, forest harvesting, animal tramping, engineering earthworks, and even thunder. Mass movements can cause death and injury, disrupt highway transportation, damage private property and public facilities, cause loss of productive lands, and degrade the aesthetic environment. The quantity of sediment delivered into rivers and reservoirs by mass movements far exceeds the contributions by interrills, rills, ephemeral gullies, and gullies.

12.2.4.1 Occurrences

Based on limit equilibrium theory, the stability of soil mass on a slope depends on the balance between shear forces (stress, τ), pulling the soil mass sliding down on a critical surface and shear strength along the surface to resist the downward movement (τ_c). When the shear stress is greater than or equal to the shear strength, the mass is unstable and failure occurs. Thus, the ratio between τ_c and τ is often referred to as the factor of safety (FS) for a given slope, or

$$FS = \frac{(\tau_c)}{(\tau)} \tag{12.41}$$

The slope is stable if FS > 1, and instability exists if FS $\leq$ 1.

Neglecting friction, a soil mass on a slope is accelerated down by a force (shear stress) equivalent to the tangential component of its total weight (Figure 12.9), or

$$\tau = (W_s + W_t)\sin(\alpha) \tag{12.42}$$

where
W_s and W_t are soil mass and tree mass, respectively
α is the slope angle (deg)

Shear strength that resists downhill movement is equal to a force due to the normal component of $(W_s + W_t)$ plus the cohesion of soil (C_s) and root systems (C_r), or

$$\tau_c = (C_s + C_r) + (W_s + W_t)\cos(\alpha)\tan(\phi) \tag{12.43}$$

where ϕ is the angle of internal friction of the soil (an expression of the degree of interlocking of individual grains at which, under gravity, sliding would occur). When $\tau_c = \tau$, the angle $\alpha = \phi$. The soil is unstable and gravity would induce movement along the joint plane if $\alpha > \phi$. On the other hand, the slope could stand up to 90° if $\alpha < \phi$.

The factor of safety of a slope decreases with an increase in shear stress τ or a decrease in shear strength τ_c. The τ may be increased as a result of increased water content of the soil mass; snowpack; wind stresses transferred to the soil through root systems; impacts

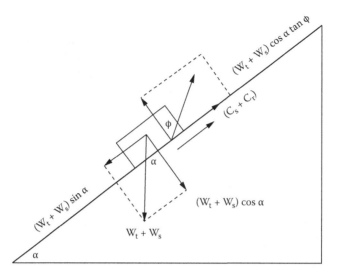

FIGURE 12.9
Forces acting on soil block of a hillside slope.

of people, animals, or transportation; seepage pressure from upslopes; or earthquake. Factors leading to a decrease in shearing resistance can include increase in water content, materials with low internal cohesion, and higher groundwater table as a result of increased precipitation, deforestation, and dam construction. In any case, slope instability is seldom a single-cause event; it usually results from a series of events that lead to mass movements.

Forest vegetation tends to maintain more secure slopes by mechanisms transferred through its canopy and root systems. They include the following:

1. Mechanical reinforcement of root systems that transfers shear stress in the soil to tensile resistance in the roots
2. Depletion of soil moisture by interception, transpiration, snow accumulation, and snowmelt that limits the buildup of positive water pressure
3. Increasing soil organic matter content that strengthens the cohesion of soil
4. Buttressing or soil arching action between trunks and stems that counteracts downslope shear forces

Thus, forest harvesting and related activities on steep terrain often cause slope instability and mass movements. Perhaps the most significant effect of vegetation removal is on root reinforcement. The magnitude of root reinforcement is dependent upon the tensile strength and distribution of roots in the soil column and small roots provides proportionally greater cohesive strength than larger roots. The apparent cohesion provided by roots, varies between 2 and 150 kPa, can represent up to 100% of the cohesive strength of hillslope soils (Hales et al., 2009). Studies showed that landslides were most frequent 4–10 years after logging (Gray and Megahan, 1981). The time of occurrence of maximum landslide hazard seems to be affected by the relative rates of root decay and new root growth. This is because the cohesion decreases significantly with root decay, increasing landslide hazard in deforested sites.

In southwestern Oregon, the rates of mass erosion from debris slides in disturbed areas were on average five times greater than in undisturbed forest sites. Mass erosion was more than 100 times greater in disturbed road rights-of-way areas than on undisturbed natural slopes (Amaranthus et al., 1985). Road construction was also the most likely cause of accelerated landslide activity in the Idaho Batholith area. Combining road construction with logging and forest fires accounted for 88% of total landslides, and the removal of vegetation accounted for 9%. Only 3% of the landslides occurred on slopes with no disturbance (Megahan et al., 1978).

12.2.4.2 Estimates

Because of its size and site specificity, mass erosion is usually estimated based on volumetric measurements made in the field or through aerial-photo interpretations. Modern technologies such as geographic information systems (GIS) and geographic position systems (GPS) can also be employed for large-scale digital mapping and data analysis. Combining these technologies with field reconnaissance and calibration (Rosgen et al., 1980) provides more reliable estimates. Once the voided volume of mass movements is available, multiplying the volume by soil density yields the total mass of landslides.

12.2.5 Channel Erosion

Channel erosion applies to channels in the lower end of headwater tributaries with water flowing all year round, while gully erosion refers to those channels in the upper ends of headwater tributaries with water flowing during or immediately after storms. Channel erosion consists of soil erosion on the stream bank and sediment transport in the stream channel.

12.2.5.1 Stream Bank Erosion

Erosion on stream banks is caused by raindrop splash, surface runoff, the scouring and undercutting of water by the processes of corrosion (chemical reaction of water and rocks), corrasion (grinding action of sediment), cavitation (kinetic and potential of the stream, wave actions), and gravitation. Frequently, stream bank erosion is accelerated by the removal of vegetation, grazing, tillage near the bank, and stream crossing by vehicles and animals. However, the continuous movement (kinetic energy) of water is mainly responsible for the cave-in of banks and washout of sediment, resulting in channel widening and deepening, shifted courses, deposition, and alluvial terraces. The resistance of a bank to cave-in and gravitation drop-off is provided by $W \cos\alpha \tan\varphi$, where W is the weight of bank materials, α is the angle of the bank slope, and φ is the angle of repose of the material. If erosion forces are greater than this erosion resistance, erosion occurs.

12.2.5.2 Sediment Transport

Sediment in streams is transported by suspension (no contact with the stream bed), saltation (bouncing along the stream bed), and bedload movement (rolling along the stream bed). Suspended sediment includes the wash load of fine particles such as silts and clays (<0.06 mm) that are washed from banks and hillsides into streams during runoff events, and the bedload materials that are suspended due to turbulence and high-flow events. Thus, the relative size of particles transported as suspended load or bedload depends on the hydraulic characteristics of the stream and the nature of materials; no clear distinction can be made. The wash load travels at essentially the same speed as the water and will

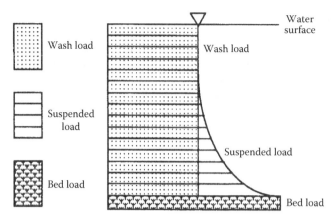

FIGURE 12.10
Total sediment load in a stream profile.

remain suspended in the fluvial process. In practice, total sediment load is separated into two components: suspended load and bedload (Figure 12.10).

The relative percentage contribution of suspended load and bedload to the total sediment load in streams, depending on land use, geology, soils, topography, and channel conditions, is highly variable from time to time, flood to flood, and site to site in a given stream. It was reported that suspended sediment is only 10% of the total load in the Rhine River and 96% of the total load in the Eel River at Scotia, California (Anderson, 1971). The fraction of the total load transported in suspension has been reported at 0.50 in the meltwater river from the Nigardsbreen glacier in Norway (Møen et al., 2010) and varied between 0.0 and 1.0 at the Pitzbach, a mountain stream in Austria (Turowski et al., 2010). Generally, the fraction of suspended load increases with increasing drainage area in a given stream. Suspended sediment is a common subject of measurements, due to its significance in water quality and its useful capacity as an index for total sediment discharge and potential disposition. The annual sediment loads for major rivers of the world are given in Table 12.7.

Basically, a stream will carry particles until its energy (force) is less than that of particle load. In other words, a critical energy level must be reached for channel materials to be suspended, transported, or eroded; otherwise, deposition occurs. Thus, several critical values may exist because of particle sizes and shapes. The difference in energy levels between flow and sediment load in streams can then be used as a parameter for estimating sediment transport rate.

One such parameter commonly used to characterize the energy level in streams is shear stress, more frequently referred to as tractive force. Like pressure, shear stress is force per unit area. However, pressure is the force acting perpendicular to a surface, while shear stress is the force acting parallel along a stream channel. For a given stream section with a slope α in degrees, the shear stress τ (in N/m^2, $1\,N = 1\,kg\text{-}m/s^2$; $1\,N/m^2 = 1\,Pa$) is calculated by

$$\tau = W \sin \frac{\alpha}{A} = \frac{[(\rho)(g)(V)\sin\alpha]}{A} = \rho gRs = \gamma Rs \qquad (12.44)$$

which is discussed and explained in Equation 4.31.

The critical shear stress (τ_c, in N/m^2) is the minimum force that overcomes the submerged weight of the particles and causes particles to leave their original locations. It is affected by

TABLE 12.7

Sediment Loads and Flow Discharges of Selected Large Rivers in the World

River	Country	Area (10^3 km³)	Mean Discharge (m³/s)	Annual Sediment Load (10^6 ton/year)	(ton/km²/year)
Yellow	China	752	1,370	193[a]	256
Ganga	India	955	1,800	1450	1500
Amazon	Brazil	6100	172,000	850	139
Brahmaputra	India	600	12,200	730	1100
Yangtze	China	1807	29,200	480	280
Indus	Pakistan	969	5,500	436	450
Irrawaddy	Burma	430	13,500	300	700
Mississippi	United States	3269	24,000	172[b]	53
Red	Vietnam	120	3,900	130	1100
Congo	Zaire	4014	39,000	72	18
Pearl	China	355	8,000	70	260
Danube	Romania	816	6,200	65	80
Niger	Nigeria	1081	4,900	21	19
Po	Italy	540	1,550	15	280
Ob	Russia	2430	12,200	15	6
Don	Russia	378	830	4.2	11
Rhine	Netherlands	160	2,200	2.8	17
Wesla	Poland	194	950	1.4	7

Source: After UNESCO, *The Water-Related Issues and Problems of the Humid Tropics and Other Warm Humid Regions: A Programme for the Humid Tropics*, Division of Water Sciences, UNESCO, Paris, France, 1991.

[a] Updated from 1.645 Gt/year based on Liu et al. (2008).
[b] Updated from 0.3 Gt/year based on Meade and Moody (2010).

particle size and density and by the flow movement conditions and can be estimated by the following equation (Shields, 1936; cited by Gordon et al., 1993):

$$\tau_c = \theta_c g d (\rho_s - \rho) \tag{12.45}$$

where
  d is a typical sediment size (m)
  g is the acceleration due to gravity
  $(\rho_s - \rho)$ is the difference in densities between sediment and water (kg/m³)
  θ_c is a dimensionless constant, a function of sediment shape, fluid properties, and arrangement of the surface sediment

Shields (1936) showed that θ_c is related to the roughness Reynolds number (R_e), an index to describe the hydraulically laminar or turbulent flow conditions. The value of θ_c decreases from about 0.2–0.08 for R_e between 0.02 and 1.5, within which the flow is considered hydraulically smooth. It remains at about 0.05 for $R_e \geq 40$, the region of hydraulic turbulence. Thus, for a turbulent flow with $\theta_c = 0.05$, $(\rho_s - \rho) = (2650 - 1000)$ or 1650 kg/m³, the calculated $\tau_c = 808.9$(d) in N/m². The particle diameter that can be lifted up in this turbulent flow is about 1.8 times larger than the diameter in a flow that would have been under a laminar condition with $\theta_c = 0.08$.

Sediment-transport capacity can then be estimated as a function of the differences between shear stress and critical shear stress, or $\tau - \tau_c$, with a modification of soil detachment K_d as expressed by

$$E_b = K_d (\tau - \tau_c) \qquad (12.46)$$

where
  E_b is the bank erosion rate (m/s)
  K_d is the bank detachment coefficient (m^3/N/s)
  $\tau - \tau_c$ are in N/m^2

Values of K_d and τ_c can be measured in the field with a jet test apparatus (Hanson and Cook, 2004). Along six streams near Blacksburg, southwest Virginia, Wynn and Mostaghimi (2006) reported K_d ranging from 0.2 to 13.1 cm^3/N/s, and ranging from 1.9 to 4.1 Pa for τ_c. In the more erodible region of northeastern Mississippi, K_d ranges from 0.70 to 105.3 cm^3/N/s, and from 0.01 to 11.89 Pa for τ_c (Ramirez-Avila et al., 2010).

Other parameters that have been used to characterize stream energy level include flow velocity (Hjulstrom, 1939), discharge (Bunte, 1996), and stream power (Bagnold, 1977). In the field, changes in geometry of the stream cross sections and stream banks between recent and old aerial photos can be used to estimate channel-bed erosion and bank erosion. Also, measurements of sediment concentrations at two sites along a reach for periods of time can be used as an estimate for channel-transport capacity. A positive difference in concentration between the lower and the upper sites is due to the channel erosion of that reach, while a negative difference implies deposition and storage.

Erosion pins inserted perpendicularly into the stream bank face is also a common method for field assessments on stream bank erosion. Pins in transects can be installed vertically on stream banks and horizontally along stream channels (Zaimes et al., 2006; Hupp et al., 2009). Pin exposure (bank erosion) and pin burial (deposition) are recorded for assessments at each visit. Also, a method for quantitative prediction of stream bank erosion rates is developed by Rosgen (2001). The model is empirically derived, utilizing a rational estimation and process integration approach. Additional discussions are given in Section 13.2.5.

12.3 Estimation of Sediment Yield

The total quantity of soil particles transported from a source to a downstream point of interest is known as sediment yield. It is the result of complex natural processes involving soil detachment, transport, and deposition, with impacts of human activities. Estimates of sediment yield are needed because it not only affects reservoir capacity, stream water quality, aquatic life, stream habitat, channel morphology, flood stage, and riparian zones, but also is a good indicator for the effectiveness of watershed management conditions.

12.3.1 Sediment Delivery Ratio

Not all eroded particles will reach stream channels during a storm-runoff event; some of them may be trapped in the watershed for an extended period of time. The ratio between

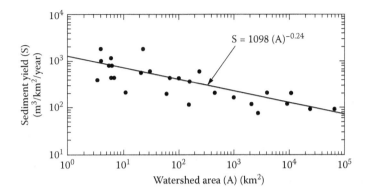

FIGURE 12.11
Sediment yields in m³/km²/year, from reservoir surveys, as a function of watershed area (A) in km² in the semiarid climate of the United States. (Adapted from UNESCO, Sedimentation problems in river basins, Project 5.3 of the International Hydrological Programme, United Nations Educational Scientific and Cultural Organ, Paris, France, 1982.)

sediment yield (S) in a stream or reservoir and the *watershed gross erosion* (ΣG_e) is called sediment delivery ratio (SDR), or

$$SDR = \frac{S}{(\Sigma Ge)} \tag{12.47}$$

Watershed gross erosion is the total erosion from all sources within the watershed, including interrill, rill, ephemeral gully, gully, channel, and landslides. Sediment yields are usually measured from reservoir deposition or total sediment load from stream samples. Each source of erosion can be estimated by available techniques; some of them were introduced in the previous section. However, interrill and rill erosion are the only sources for which well-developed equations are available for the estimates.

SDR represents sediment-transport capacity of a watershed. It has been reported to range from 0.26 to 0.67 in the Blackland Prairie Land Resource Area, Central Texas (Maner, 1962) and from 0.10 to 0.99 in Marshall County, northern Mississippi (Mutchler and Bowie, 1976). Generally, it is small in large watersheds with a high percentage of forest coverage. A large watershed usually has greater watershed storage and smaller values of watershed mean slope, channel gradient, probabilities for storms to cover the entire watershed, and mean rainfall intensity. All these phenomena tend to decrease sediment yield per unit area (Figure 12.11).

Other environmental factors affecting SDR include relief–length ratio, annual runoff, gully density, channel density, main channel length, eroded material, precipitation characteristics, and surface roughness (Renfro, 1975). Based on regional analyses on SDR variations and associated watershed characteristics, a few empirical models have been developed for specific regions (Table 12.8). A generally applicable model for estimating SDR is not available due to the complexity of the processes and the lack of data for assessment (Walling, 1994). Once SDR is determined, sediment yield can be calculated by multiplying SDR with estimated watershed gross erosion.

12.3.2 Sediment-Rating Curve

Sediment concentration in most streams often exhibits a positive relationship with stream discharge. A plot of sediment concentration (mg/L) versus stream discharge (m³/s) over

TABLE 12.8

A Few Empirical Models for Estimating SDRs by Watershed Characteristics

Location	Equation[a]	Reference
The Blackland Prairie, Texas	$\text{Log SDR} = 1.87680 - 0.14191 \log(3.861A)$	Maner (1962)
Marshall County, N. Mississippi	$\text{SDR} = 0.488 - 0.064(A) + 0.01(RO)$	Mutchler and Bowie (1976)
Texas, United States	$\text{SDR} = 1.366(10^{-11}) \, (A^{-0.100}) \, (R/L)^{0.363} \, (CN)^{5.444}$	Williams (1977)
Southern United States	$\text{SDR} = 4.5 - 0.23 \log(3.86A) - 51 \operatorname{colog}(R/L) - 2.79B$	Roehl (1962)
Nacogdoches, Texas	$\text{SDR} = 0.5038 + 0.2779(\rho_d) - 0.0057(F)$	Chang and Wong (1983)
11 watersheds, worldwide	$\text{SDR} = 0.416[A^{0.5}(500 - E^{0.5})(S^2)(P^{-0.5})]^{-0.422}$	Diodato and Grauso (2009)

Sources: Maner, S.B., *Factors Influencing Sediment Delivery Ratios in the Blackland Prairie Land Resource Area*, USDA, Soil Conservation Service, Fort Worth, TX, 1962; Mutchler, C.K. and Bowie, A.J., Effect of land use on sediment delivery ratios, *Proceedings of the Third Federal Inter-Agency Sedimentation Conference*, Denver, CO, pp. I:11–I:21, 1976; Williams, J.R., Sediment delivery ratios determined with sediment and runoff models, in *Erosion and Solid Matter Transport in Inland Waters*, IAHS Publ. No. 122, Wallingford, England, U.K., pp. 168–179, 1977; Roehl, J.E., Sediment source areas, delivery ratios and influencing morphological factors, in *Land Erosion, Proceedings of the Bari Symposium*, International Association of Scientific Hydrology Publ. No. 59, pp. 202–213, 1962; Chang, M. and Wong, K.L., *BeitragezurHydrologie*, 9, 55, 1983; Diodato, N. and Grauso, S., *Environ. Earth Sci.*, 59, 223, 2009.

[a] A, area (km²); RO, runoff (cm); R/L, relief/length ratio; CN, SCS curve number; ρ_d, drainage density; B, bifurcation ratio (ratio of the number of streams of a given order to the number of streams of next higher order); F, percent forest area; E, average basin elevation (m); S, average basin slope (m/m); P, annual basin precipitation (mm).

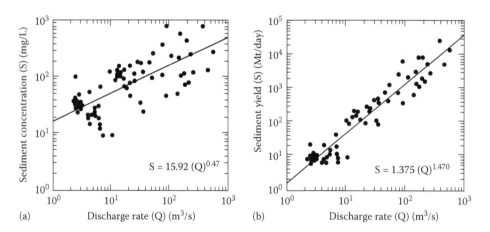

FIGURE 12.12
A sediment rating curve, in concentration (a) or mass (b) versus discharge, for the Iroquois River near Chebanse, Illinois. (From Bhowimik, N.G. et al., Hydraulics of flow and sediment transport in the Kankakee River in Illinois, Illinois State Water Survey, ISWS/RI-98/80, 1980.)

a wide range of data is called a sediment-rating curve (Figure 12.12). Once the sediment-rating curve for a given stream is available, the sediment concentration can be interpreted from the curve for various discharge rates. The total amount of sediment yields can be calculated by breaking down streamflow into many ranges (segments) of small discharges. The sediment yield for each range is the product of sediment concentration, discharge rate, and the frequency of occurrence of that discharge rate (obtained from a flow-duration table or curve), or

TABLE 12.9

Calculation Procedures for Annual Sediment Yield Using Sediment-Rating Curve (Figure 12.12) and a Hypothetical Set of Flow-Duration Data

Streamflow Discharge (m³/s)			Sediment	
Range (1)	Mid-Flow (2)	Frequency (days) (3)[a]	Concentration (mg/L) (4)[b]	Yield (t) (5)[c]
0.000–10.000	5.000	215	33.920	3,150.5
10.001–30.000	20.000	50	65.077	5,622.6
30.001–60.000	45.000	34	95.269	12,593.8
60.001–90.000	75.000	18	121.122	14,127.6
90.001–120.000	105.000	12	141.874	15,444.9
120.001–180.000	150.000	11	167.767	23,916.8
180.001–240.000	210.000	15	196.511	53,482.3
240.001–300.000	270.000	7	221.149	36,112.7
300.001–400.000	350.000	2	249.836	15,110.1
400.001–500.000	450.000	1	281.160	24.3
Total		365		179,585.6

[a] Obtained from a flow-duration table.
[b] Interpreted from Figure 11.11a, or calculated by $S = 15.92(Q)^{0.47}$, where Q is in m³/s.
[c] $0.0864 \times$ Column (2) $\times$ Column (3) $\times$ Column (4).

$$S = \Sigma[0.0864(C_i)(Q_i)(F_i)] \qquad (12.48)$$

where
S is the total sediment yield for all discharge ranges i (segments) (Mt)
C_i is the sediment concentration (mg/L)
Q_i is the median discharge rate of the range (m³/s)
F_i is the frequency of occurrence for each range (days)

The value 0.0864 converts from seconds to days and mg to Mt. The calculation procedure is given in Table 12.9.

Besides graphical plotting, the sediment–discharge relationship can be expressed in the following form:

$$S = aQ^n \qquad (12.49)$$

where
Q is discharge rate
a and n are constants, depending on regions and vegetation cover
S is sediment expressed either in concentration (mg/L) or in yield (ton/day)

Equation 12.49 can be developed for instantaneous, daily, monthly, and annual data of a single stream, or for annual data of many streams covering a large region. The developed equation is more reliable for data of longer duration in a single stream; for large streams; for smaller, homogeneous regions when many streams are involved; and for sediment yield in mass per unit of time (Figure 12.12). This method tends to underestimate high and overestimate low suspended sediment concentrations (Horowitz, 2003).

12.3.3 Modified USLE

Using SDR in conjunction with watershed gross erosion (Equation 12.47) to estimate sediment yield is tedious, and inadequate if one is interested in single storms. Unless already available, developing an SDR model may involve parameters similar to those of the USLE and other models that are used to estimate gross erosion, a duplicate step and time-consuming process. Stream sediment is affected by the carrying capacity and deposition processes of overland flow. However, the storm-energy factor used by the USLE often fails to account for the effective rainfall that generates surface runoff. Also, SDR varies with storms; the assumption of a constant SDR adds another source of error to the estimates.

Williams (1975) showed that the estimate of stream-sediment yield for individual storms could be simplified by using the USLE with its rainfall factor R replaced by a runoff factor. Using 778 storm-runoff events collected from 18 small watersheds, 16 of them near Riesel, Texas, and 2 at Hastings, Nebraska, he modified USLE in the following form:

$$S = a(Q \times q)^{n} \, KLSCP \qquad (12.50)$$

where
  S is storm-sediment yield (Mt) for the whole watershed
  KLSCP are factors of the USLE (Equation 12.24)
  $a(Q \times q)^{n}$ is the runoff factor with Q = runoff volume in m^3
  q is the peak discharge rate (m^3/s)
  a and n are location coefficients

For the areas where the equation was developed, a = 11.8 and n = 0.56. Values of Q and q can be estimated using Equations 10.28 and 10.31 (or Equation 10.35). The procedure should only be used in small watersheds and requires calibrations of a and n if it is not used in Texas and Nebraska, where the equation was originally developed.

12.3.4 Statistical Models

Watershed sediment is affected by an array of parameters in climate, topography, soils, geology, and vegetation. The relative importance and degree of influence of each parameter vary from region to region. With sufficient sediment observations from various watersheds, a statistical model can be developed relating sediment yield to environmental parameters. Numerous models have been developed around the world (Flaxman, 1972; Walling, 1974; Renard, 1980). Parameters used in the predictions include one or a few of the following: watershed area, slope, channel gradient, precipitation, temperature, mean discharge and various return periods, percentage of lakes and forested areas, drainage density, basin form factors, and others. However, these models generally can be applied in the regions where data were collected.

Dendy and Bolton (1976) used sediment data collected from 800 reservoirs with watershed areas of 2.59 km² or larger and developed the following equations to estimate sediment yield S in Mt/km²/year (converted from English units) in the United States:

For Q < 5.0 cm:

$$S = 292Q^{0.46}(1.54 - 0.26 \log A) \tag{12.51}$$

For Q ≥ 5.0 cm:

$$S = 686e^{-0..022Q}(1.54 - 0.26 \log A) \tag{12.52}$$

where
  A is watershed area (km^2)
  Q is annual runoff (cm)

Parameters Q and A explained 75% of the variation of sediment yield S. However, these two equations were developed from average values of grouped data, which provide an approximation for preliminary watershed planning on a regional basis. Application of these equations to predict sediment for individual watersheds, due to local factors, may produce errors as much as 10–100 times larger or smaller than actual sediment yields.

12.3.5 Physical Models

The limitations of statistical models can be overcome if models are developed based on erosion processes and transport mechanisms. Many of these models use the concept that rainfall and runoff energy must be greater than soil resistance in order to detach soil particles. If sediment detachment (supply) is greater than flow transport capacity, then sediment yield is equal to transport capacity. On the other hand, if sediment supply is less than the flow transport capacity, then sediment yield is equal to sediment supply plus detached particles. Thus, two components of models must be developed. One is the hydrology component to predict sequential flow volume and rates from storm precipitation and soil characteristics, and the other is the erosion component to predict soil detachment, transport, and sediment yield using data simulated in the hydrology component. It is important that the developed models be calibrated with field data before application to various management and decision-making processes. Many of such models are summarized by Elliot et al. (2010).

12.4 Vegetation Effects

Vegetation is considered a highly effective, economical, relatively permanent, and almost maintenance-free approach to controlling soil erosion and stream sediment. Vegetation mechanisms that enhance these beneficial effects, plant characteristics that need to be examined for specific conservation purposes, and field observations that provide lessons and evidences of vegetation effects are briefly discussed here.

12.4.1 Vegetation Mechanisms

Forests provide many environmental functions (see Section 7.2). The mechanical and hydrologic functions that modify factors affecting soil erosion processes are of particular importance in controlling soil erosion and stream sedimentation.

12.4.1.1 Mechanical Effects

Soil erosion begins with soil detachment caused by rainfall and runoff energy. Vegetation counterbalances this energy or enhances the resistance of soils to this energy through these mechanisms:

1. Canopies above the ground intercept rainfall and consequently reduce the terminal velocity of raindrop energy. (See Section 8.2 for a discussion of canopy interception, and Section 12.1 for a discussion of terminal velocity.)
2. Litter and canopy at the ground surface reduce raindrop and overland flow energy by shielding the soil and blocking runoff movement.
3. Root systems and organic matter cause an increase in the cohesive and frictional components of soil shear strength, which contributes to soil stability.

The degree of canopy interception's effects depends on canopy density and height. Plants whose canopies are in direct contact with the surface essentially eliminate raindrop kinetic energy because there is no falling height of raindrops from the canopy. These canopies act as blankets on the soil (shielding effects) and are more effective than are canopies high above the ground. On the other hand, canopies and litter increase surface roughness of the ground and slow down runoff velocity (blocking effects), consequently reducing soil erosion.

The effects of litter density on soil erosion can be demonstrated in an intense-storm simulation study (Table 12.10). Straw mulch of only 0.56 ton/ha applied to 3.7 × 10.7 m plots of Fox loam soil at 15% slope, and moderate-to-slow permeability reduced soil losses to less than one-third and flow velocity to about one-half of those from unmulched plots. Another 2 h, 64 mm/h of rainfall simulation study showed that switchgrass strips 72 cm wide and 46 cm high at the bottom of 3.7 m wide × 10.7 m long plots, 8%–16% slopes reduced runoff to 22%–52% and erosion to 53%–63% of similar plots without the hedges (Gilley et al., 2000).

While deep-rooted, woody vegetation is more effective than grass roots for slope stabilization, fine roots of dense grasses and herbaceous vegetation are very beneficial for protection against raindrop and wind erosion. Large roots penetrate the surface horizons to anchor the soil mantle to the substrate; fine roots, fungal mycelia, and decomposed organic matter contribute to the formation of stable aggregates of surface soils. These taproots and lateral roots develop a dense network that interweaves a soil–root system centered on the vicinity of individual trees. Soil mantles are thus stabilized due to the great reinforcement and binding effects of the soil–root system.

The shear strength of soils against erosion is positively related to root concentration (biomass of roots per unit volume of soil in kg/m^3), to root area ratio (the fraction of the

TABLE 12.10

Influence of Straw-Mulch Rate on Flow Velocity and Soil Erosion for Two Soil Conditions

| Mulch Rate (ton/ha) | Fox Loam, Unplowed | | Xenia Silt Loam, Plowed and Disked | |
	Soil Loss (ton/ha)	Velocity (cm/s)	Soil Loss (ton/ha)	Velocity (cm/s)
0	62.3	13.9	32.4	13.1
0.56	20.1	7.1	13.0	6.9
1.12	19.4	6.9	8.2	6.3
2.24	11.5	5.6	3.8	3.4

Source: After Meyer, L.D. et al., *Soil Sci. Soc. Am. Proc.*, 34, 928, 1970.

total cross-sectional area of a soil occupied by roots), and to root tensile strength. Root tensile strengths (the resistance to a force tending to tear roots apart) are about 10–50 MPa for most species. They are higher for roots of smaller diameters (more surface area), growing uphill, and in late fall (the time of strongest growth), and decline with age after tree harvesting. In New Zealand, forest removal often caused landslides on steep slopes along the plane beneath the zone of dense root networks (Sidle et al., 1985).

Generally, conifer roots tend to have a lower tensile strength than the roots of deciduous trees, and shrubs appear to have tensile strengths comparable to that of trees (Gray and Sotir, 1996). This is of particular significance for shallow soils and stream banks, where the growth of trees may accompany an increase in shear stress to the soils due to their greater weight and rigidity. Windthrow and uprooting of trees in high winds can reduce the stability of slopes with shallow soils, stream banks, and levee embankment.

12.4.1.2 Hydrologic Effects

The hydrologic functions of vegetation are very beneficial to the control of soil erosion and stream sediment. They include the following:

1. Interception loss, which reduces net precipitation to reach mineral soils
2. Transpiration, which depletes soil moisture content
3. Litter and organic matter, which slow down flow velocity and prolong time of concentration

All these lead to an increase in infiltration rate and a decrease of sediment-transport capacity in runoff.

Forests reduce erosion and sediment much more than other types of vegetation do because of greater hydrologic impacts induced by their size and density. In a national survey of suspended sediment conducted at 63 stations in Kenya during 1948–1968 (Dunne, 1977), stream sediment was found to be the lowest in watersheds with greater forested area. The survey grouped all watersheds into four land-use types: (1) forest, (2) more forest than agriculture, (3) more agriculture than forest, and (4) grazing dominant. Sediment yield (S, ton/km²/year) for each land use increases with mean annual runoff (Q, mm) by these relationships:

Forests

$$S = 2.670(Q)^{0.38} \tag{12.53}$$

Forest > agricultural

$$S = 0.042(Q)^{1.18} \tag{12.54}$$

Agricultural > forest

$$S = 0.038(Q)^{1.41} \tag{12.55}$$

Grazing dominant

$$S = 0.002(Q)^{2.74} \tag{12.56}$$

In grazed watersheds, vegetation cover and soil structures are likely to be destroyed, soil compaction and bulk density are increased, and infiltration rate is reduced. All these lead to an increase in concentrated flows and more flow energy for detachment and transport. For $Q = 200$ mm/year, the estimated $S = 17.9$, 15.5, 44.5, and 1834.5 in ton/km^2/year, in that order.

12.4.2 Conservation Plants

All seed plants in certain types of trees, shrubs, herbaceous forbs, and grasses are useful for control of soil erosion. However, since each plant grows in a specific environment (habitat) and evolves a unique set of morphological characteristics and life cycle, the effectiveness on soil conservation is not the same among species. Growing plants in an unsuitable environment can affect the plant's survival, growth, and development. Thus, some species are more effective at a particular site, while many others may not be suitable at all. Plants that are particularly useful for control of soil erosion and land reclamation can be referred to as *conservation plants*. The selection of a conservation plant species is based on (1) intended purpose, (2) cause of the erosion problem, (3) site conditions, including soil, topography, and microclimate, (4) plant characteristics, and (5) availability. In other words, the selected plants should have plant characteristics, and be available, that meet the other three considerations, i.e., intended purpose, the erosion problem, and site conditions.

Plant characteristics that need to be examined for possible adoption should cover three aspects: morphological, physiological, and managerial. Some plant characteristics in each of the three aspects are listed in Table 12.11. In general, plants growing on sites similar to the problem area provide a good choice for adoption, and native species are a better choice than exotic species. N-fixation species such as alder and black locust provide additional benefit to soil fertility levels and consequently enhance plant growth. They should be the preference, if available. Frequently, a combination of more than one species such as mixtures of legumes with nonlegumes, pioneer (nurse) species with host plants, cold-season with warm-season grasses, and grass with woody plants provides compensation for the shortages of a single species, which in turn makes the revegetation program more

TABLE 12.11

Plant Characteristics Relevant to Species Selection for Erosion Control

Physiological	Morphological	Managerial
Growth rate, vigorousness	Canopy density	Availability of plant
Tolerance to drought, soil pH,	Shape of plants,	Environmental nuisance
flooding, grazing, and extreme temperatures	individual, clump, creeping, vine, etc.	Harmful effects to man and animals
Regrowth ability	Height of clean stems (height below the canopy)	Plant sources
Method of generation		
Adaptability to the sites	Root systems	
Productivity of forage, litter		
Species competition		
N-fixation ability		
Palatability to wildlife		
Evergreen or deciduous		

effective. However, the selection of species in these mixtures should be done with care to avoid adverse competition for soil moisture and light.

Native species are commonly more preferable than exotic species for erosion control, due to ecological stability, reliable growth, and maintenance costs. A 4 year study in Alabama showed that native species, exotic species, and the wood excelsior erosion mat offered similar benefits on the control of sediment and runoff yield from forest roadside slopes (Grace, 2002). Exotic species vegetation can be preferable for erosion control if application costs can be reduced, the establishment of ground cover is quick and reliable, and they impose no potential threat to the ecological environment.

12.4.3 Forest Activity

Stable forest ecosystems have erosion rates usually less than 0.05 ton/ha/year (Pimentel, 2006). Any forest activities that reduce canopy coverage and disturb forest floor and soils will lead to an increase in gross erosion and sediment yield. The USLE estimates that inter-rill and rill erosion for a clear-cut site, as described by the C factor, can be greater than for a forested site by a factor of more than 1000. The increase in sediment production in a forest activity depends on the degree of forest and soil disturbances, location and proportion of the watershed affected, watershed characteristics, and climatic conditions.

12.4.3.1 Forest Roads

Forest roads are recognized as a major source of erosion in watersheds. Soil erosion from roads has been reported to reach 276 ton/ha/year (Grace and Elliot, 2008), and can account for as much as 90% of all sediment production in forested watersheds (Swift, 1984). This is due to the following reasons:

1. A low permeability of the road surface
2. The alteration of natural surface drainage paths
3. The creation of concentrated overland flow by ditch systems and relief culverts
4. The susceptibility of cut-slopes and fill-slopes to soil erosion
5. The construction and maintenance of stream crossings in developing a forest road system

Damaging roads that result from water-drainage problems can trigger mass movement (Gucinski et al., 2001). Studies have shown that sediment yield is related to road gradient, surface density, the diameter of materials, slope-length factor, percentage of forest cover over the fill, and road density (Burroughs and King, 1989; Brake et al., 1997).

The primary issues in controlling erosion on forest roads are adequate planning, over-land runoff management, topography, and revegetation. Runoff diversion from the road surface or eroding ditch to the forest and a reduction of road gradients are used as primary practices in existing road systems. In humid Alabama, vegetation treatments reduced more than 90% of sediment production and have shown the greatest potential for mitigation of soil erosion on forest road side-slopes (Grace, 2000).

12.4.3.2 Forest Cutting and Site Preparation

Clear-cutting followed by mechanical site preparation is a common practice for enhancing forest regeneration in the southern United States. Because of the intensity of site

TABLE 12.12

Differences in Erosion Rates for Forest Clear-Cutting Followed by
Various Site-Preparation in the Southeastern United States

Treatment	Recovery Time (Years)	Annual Erosion (ton/ha)
Natural	—	0.00–0.05
Logged and roads	3	0.10–0.50
Burned	2	0.05–0.70
Chopped	3	0.05–0.25
Chopped and burned	4	0.15–0.40
Windrowed	4	0.20–0.24
Disked	4	2.50–10.00

Source: Burger, J.A., Physical impacts of harvesting and site preparation on soil, in *Maintaining Forest Site Productivity, Proceedings of the First Regional Technical Conference*, January 27–28, Myrtle Beach, SC, Appalachian Society of American Foresters, Clemson, SC, pp. 3–11, 1983.

disturbance, mechanical site preparation caused soil erosion in the first posttreatment year of as much as 12.8 Mt/ha/year in northern Mississippi (Beasley, 1979). The average erosion rates for forest harvesting followed by site preparation in the southeast United States were estimated to range from 0.05 to 10 ton/ha (Table 12.12). A plot study conducted in east Texas showed that the sediment production generated from two consecutive storms, a 17.65 cm storm with 16 h, 55 min duration followed by a 0.98 cm storm 22 h later, was quite different under six forest conditions (Table 12.13). Under the full forest cover, sediment production was only 1.4 kg/ha, but it increased to 341 kg/ha for the harvested and sheared plot, 244 times higher. The cultivated plot generated 1250 kg/ha sediment, 3.7 times greater than the sheared plot and 893 times greater than the forested plot.

TABLE 12.13

Surface Runoff and Sediment Generated from Two Consecutive Storms
(August 31 to September 2, 1981) under Six Forest Site Conditions near
Nacogdoches, Texas

Variable	Forest Site Conditions[a]					
	A	B	C	D	E	F
Rainfall, cm	14.58	15.32	17.45	18.63	18.63	18.63
Runoff, cm	1.26	1.40	1.45	3.82	5.47	15.49
Runoff/rainfall ratio	0.086	0.091	0.083	0.189	0.294	0.831
Sediment, ton/ha	0.0014	0.0011	0.0051	0.0965	0.3412	1.2504

Source: Chang, M. and Ting, J.C., Applications of the universal soil loss equation to various forest conditions, in *Monitoring to Detect Changes in Water Quality Series, Proceedings of Budapest Symposium*, Lerner, D. Ed., July 1986, IAHS Publication No. 157, Budapest, Hungary, pp. 165–174, 1986.

[a] A, undisturbed mature forest; B, 50% thinned forest; C, commercial cutting without site preparation; D, clear-cut, roller chopped; E, clear-cut, sheared, root raked, slashed, and windrowed; F, clear-cut, tilled, continuous fallow, cultivated up- and downhill.

12.4.3.3 Forest Fires

The effects of forest fire, either naturally induced or prescriptively burned, on erosion and sedimentation depend on the frequency and severity of fires, soil conditions, cover type, topography, and weather conditions of the postfire period. Generally, the effects are greater in the drier West than in the humid East, on fine-textured soils than on sandy soils, for brush lands than for coniferous forests, and in plantations than in mixed forests. A tremendous amount of sediment can be generated if a severe fire that destroys most of the vegetation is followed by high-intensity rainfall. The Cerro Grande wildfire burned nearly17,410 ha of mixed conifer forests near Los Alamos National Laboratory, New Mexico in 2000. Suspended sediment concentrations downstream of the burned areas remained elevated by 10,000 times through 4 years of the postfire period (Gallaher and Koch, 2009). In Colorado, the Bobcat fire burned 4290 ha near Drake and destroyed all forests in the two nearby watersheds designated for postfire studies (Kunze and Stednick, 2006). A large storm generated 950 kg/ha of sediment from the watershed (3.9 km²) without rehabilitation treatments and only 58 kg/ha from the watershed with treatments (2.2 km²).

Prescribed burning is commonly employed as a management tool in forestry to mitigate vegetation competition and fire hazard, and to prepare sites for regeneration. If burning is properly executed, impacts on soil erosion can be insignificant. A 3 year study was conducted to assess the effects of three low-intensity prescribed burns and harvesting on sediment and nutrient losses in watersheds dominated by loblolly pines in the South Carolina Piedmont (Van Lear et al., 1985). During the calibration period, sediment loss from the treatment watersheds was 58 kg/ha/year, about 2.2 times greater than the control watersheds. The sediment loss was increased to 151 kg/ha/year in the first posttreatment year, almost eight times greater than the control watersheds. However, the magnitude and duration of these increases were relatively insignificant when compared with those effects of intensive site preparation. Prescribed burning should not be conducted on fine-textured soils and steep terrain, along stream channels, and in seasons with high potential of fire hazard.

Fire lanes are important potential sources of suspended sediment in streams because they usually are (1) created by bulldozers with severe site disturbances, (2) constructed in urgent circumstances without adequate stream protection measures, and (3) difficult to revegetate due to the removal of top soils. Seeding with application of fertilizers is considered an effective way to resume vegetation cover (Landsberg and Tiedemann, 2000).

12.5 Trends in Sediment Loads of the World's Rivers

Human activity and climate change often impact sediment loads in streams. Currently, there tends to be a decreasing trend on the quantity of sediment transports in rivers around the world. Studies on annual sediment loads at the downstream gaging stations of 145 world rivers showed that 47% of them are decreasing, 5% increasing, and 48% stationary (Walling and Fang, 2003). Thus, some sediment loads given in Table 12.7 may not reflect the current conditions.

12.5.1 Decreasing Trends

The decreasing trend of suspended sediment loads in recent records has been observed in many rivers around the world, including the United States and China. Before 1900, the Mississippi River system transported an estimated 400 Mt/year of sediment from inlands to coastal Louisiana. This transport has declined to 172 Mt/year during the last two decades, 1987–2006 (Meade and Moody, 2010; Figure 12.13). The average suspended sediment load in the Yellow River is frequently cited as 1.6 Gt/year in literature (Shi et al., 2002; Walling, 2008). However, recent analysis on data collected at Lijing, about 40 km from the delta, showed that the average suspended sediment was 0.78 Gt/year for the 1952–2005 period and 0.19 Gt/year for the recent 10 year (1996–2005) period (Liu et al., 2008). In other words, the average suspended sediment in the recent 10 year period was only 25% of that in the 54 year period.

Table 12.14 lists the long-term and recent 10 year averages of suspended sediment loads along with annual runoff for 10 major rivers across China, ranging from the remotest north to the deepest south. They are the furthest downstream stations of the 10 large rivers and cover various climatic conditions in China. All the sediment loads in the recent 10 year records (S_{10}) of the 10 rivers were lower than those of the long-term averages (S_{long}). The S_{10}/S_{long} ratios ranged from 0.01 for the Yongding River at Yanchi to 0.81 for the Qinatang River at Lanxi. Actually, the decrease in sediment loads began in the 1970s and the 1980s in many rivers in China (Xu, 2003; Walling, 2009).

The dramatic decline of suspended stream sediment in China has been attributed to a number of reasons: (1) the erection of dams and reservoirs (Xu et al., 2006), (2) the diversion of water from stream channels and reservoirs for irrigation (Liu et al., 2008), (3) the

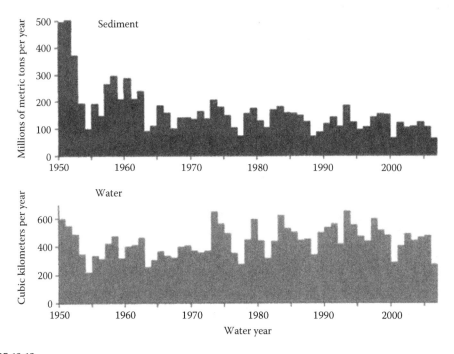

FIGURE 12.13
Annual sediment and water discharge in Mississippi River at Tarbert Landing, Louisiana, 1950–2007. (Reprinted from Meade, R.H. and Moody, J.A., *Hydrol. Processes*, 24, 35, 2010. With permission.)

TABLE 12.14

Long-Term and Recent 10 Year Annual Averages of Annual Runoff (Q_{long}, Q_{10}) and Annual Suspended-Sediment Loads (S_{long}, S_{10}) at the Furthest Downstream Stations of 10 Large Rivers across China

Station/River	Drainage (10^3 km²)	Data	Annual Runoff (L/s/km²)			Annual Sediment (ton/km²/year)		
			Q_{long}	Q_{10}	Q_{10}/Q_{long}	S_{long}	S_{10}	S_{10}/S_{long}
Harbin/Songhua R.	389.8	1955–2005	3.45	2.96	0.86	17.7	12.3	0.74
Tieling/Liaohe R.	120.8	1954–2005	0.80	0.45	0.56	103.5	23.4	0.23
Yanchi/Yongding R.	43.7	1963–2005	0.48	0.24	0.50	3.1	0.0	0.01
Lijing/Yellow R.	752.0	1952–2005	1.32	0.46	0.35	1034.5	256.1	0.25
Bengbu/Huaihe R.	121.3	1950–2005	6.99	6.47	0.93	75.5	42.8	0.57
Datong/Yangtze R.	1,705.4	1950–2005	16.80	17.56	1.05	242.7	164.7	0.68
Lanxi/Qiantang R.	18.2	1977–2005	28.75	28.16	0.98	108.7	87.9	0.81
Zhuqi/Minjing R.	54.5	1950–2005	31.13	32.09	1.03	110.1	42.9	0.39
Boluo/Dongjiang R.	25.3	1954–2005	28.89	27.64	0.96	97.0	57.6	0.59
Gaoyao/Xijing R.	351.5	1957–2005	19.86	20.40	1.03	193.4	134.1	0.69

Source: Liu, C. et al., *Int. J. Sediment Res.*, 23(1), 44, 2008.

sand extraction in stream channels (Chen et al., 2006), (4) the implementation of large-scale soil and water conservation programs (Zhao et al., 1992; Chu et al., 2009), (5) the effort on reforestation (Walling, 2008), and (6) climate change (Walling, 2009). The impact of climate change on sediment fluxes could be as important as human activity. During the period 1970–1997, the overall reduction in sediment load transported by the Lower Yellow River, as suggested by Xu (2003), could be attributed to 55% human impact and 45% reduced precipitation.

Both the total number of dams and the total reservoir storage capacity in China are the greatest in the world. In the Yellow River Basin, there was no hydropower development prior to 1960, but as of 2005, a total of 171 large dams have been constructed (Liu et al., 2008). The Three Gorges Dam on the Yangtze River, the largest hydropower project of the world, started impounding water in 2003. Two downstream stations, Yichang and Datong, measured only 12% and 33% of their 1950–1986 sediment levels, respectively, in 2004 (Xu et al., 2006). Increases in forested areas might also have had an impact on sediment transport. The total forest area in China was 157.1(10^6) ha in 1990 and increased to 206.9(10^6) ha in 2010, an expansion of 2.5(10^6) ha/year or 1.58%/year during the 20 year period (FAO, 2010).

Reported declines of river sediment loads in different parts of the world include River Indus at Kotri, Pakistan, Chao Phraya River at Ban PhaiLom, Thailand, River Danube at Ceatal-Ismail, Romania, and the Colorado River and the Rio Grande River of the United States (Walling, 2008, 2009).

12.5.2 Increasing Trends

In contrast with the decreasing trend discussed in the preceding text, continuous deforestation, abusive agricultural activities, over grazing, mining, road constructions, and urban development can cause substantial fluxes of sediment to stream channels, especially in watersheds with steep terrains, resulting in steady increases in stream suspended sediment over time. The actual impact on the stream sediment depends on the nature of human activity, the proportion and position of the watershed affected, watershed soils and topography, and climatic conditions.

As an example, the Bei-Nan River in Taiwan drains through a mountainous watershed of 1584 km^2 in area characterized by steep slopes, fragile and instable tectonics, intense rainfall, and frequent typhoons. Here, forest clearance and road construction caused the annual sediment load to increase by almost an order of magnitude since the early 1960s (Walling, 2009). However, many watershed disturbances occur in developing countries in which sediment monitoring programs are either nonexistent or insufficient for assessments. This may make the number of streams with increasing trends on sediment loads much more than have been realized.

12.5.3 Implications

Both increasing and decreasing sediment trends have many implications in land and water resources as well as having social and economic dimensions. Attention is usually focused on the detrimental effect of increasing erosion and sediment rates on land productivity, food costs, security and sustainability, reservoir life, water quality, aquatic ecosystems and habitats, and water suitability. Although the decline of sediment loads may improve water quality, aquatic habitats, and navigation, as well as reduce flood problems, the reduction of sediment input to a delta may result in shoreline retreat, loss of wetlands, and increase in coastal flooding. The impoundment of the Three Gorges Dam has caused a substantial decline of sediment in Yangtze River by 110 Mt/year; it also delivers 70 Mt/year of sediment from severe channel erosion to the lower reach of the river (Xu and Milliman, 2009).

When the floodplain along a river does not have the input of fertile sediment from the stream, it may require fertilizers to maintain the land productivity level. The application of fertilizers may impair surface and groundwater quality. In bay estuaries, decreases in sediment inputs and runoff may affect salinity levels and food supplies, consequently affecting the health and productivity of fish and shellfish.

References

Amaranthus, M.O. et al., 1985, Logging and forest roads related to increased debris slides in south western Oregon, *J. For.*, 83, 229–233.

Anderson, H.W., 1971, Relative contributions of sediment from source areas, and transport processes, *Proceedings of Symposium on Forest Land Uses and Stream Environment*, School of Forestry, Oregon State University, Corvallis, OR, pp. 55–63.

Arnoldus, H.M.J., 1980, An approximation of the rainfall factor in the USLE, *Assessment of Erosion*, DeBoodt, M. and Gabriels, D., Eds., John Wiley & Sons, New York, pp. 127–132.

Bagnold, R.B., 1977, Bed load transport by natural rivers, *Water Resour. Res.*, 13, 303–312.

Barry, P.V. et al., 1991, Organic polymers' effect on soil shear strength and detachment by single raindrops, *Soil Sci. Soc. Am. J.*, 55, 799–804.

Beasley, R.S., 1979, Intensive site preparation and sediment losses on steep watersheds in the Gulf coastal plain, *Soil Sci. Soc. Am. J.*, 43, 412–417.

Beasley, D.B., Huggins, L.F., and Monke, E.J., 1980, ANSWER: A model for watershed planning, *Trans. ASAE*, 23, 938–944.

Beer, C.E. and Johnson, H.P., 1963, Factors in gully growth in the deep loess area of western Iowa, *Trans. ASAE*, 6, 237–240.

Bennett, S.J. and Alonso, C.V., 2000, Experiments on headcut growth and migration in concentrated flows typical of upland areas, *Water Resour. Res.*, 36, 1911–1922.

Bennett, S.J. et al., 2000, Characteristics of actively eroding ephemeral gullies in an experimental channel, *Trans. ASAE*, 43, 641–649.

Bhowimik, N.G. et al., 1980, Hydraulics of flow and sediment transport in the Kankakee River in Illinois, Illinois State Water Survey, ISWS/RI-98/80.

Brady, N.C., *The Nature and Properties of Soils*, 10th edn., Macmillan Publ. Co., New York.

Brake, D., Molnau, M., and King, J.G., 1997, Sediment transport distance and culvert spacings on logging roads within the Oregon Coast Mountain Range, Paper No. IM-975018, presented at the *1997 Annual International Meeting*, ASAE, Minneapolis, MN, August 10–14.

Bubenzer, G.D. and Jones, B.A., 1971, Drop size and impact velocity effects on the detachment of soils under simulated rainfall, *Trans. ASAE*, 14, 625–628.

Bunte, K., 1996, Analyses of the temporal variation of coarse bedload transport and its grain size distribution, General Technical Report RM-GTR-288, USDA Rocky Mountain Forest and Range Experiment Station, Fort Collins, CO.

Burger, J.A., 1983, Physical impacts of harvesting and site preparation on soil, *Maintaining Forest Site Productivity, Proceedings of the First Regional Technical Conference*, January 27–28, Myrtle Beach, SC, Appalachian Society of American Foresters, Clemson, SC, pp. 3–11.

Burroughs, E.R., Jr. and King, J.G., 1989. Surface erosion control in roads in granitic soils, *Proceedings of Symposium* Sponsored by *Committee on Watershed Management/Irrigation and Drainage Division*, ASCE Convention, April 30–May 1, Denver, CO.

Chang, M., 1997, Improper soil management causing severe water erosion problems, *Proceedings of the 28th Conference*, February 25–28, 1997, Nashville, TN, International Erosion Control Association, Reno, NV, pp. 495–501.

Chang, M., Roth, F.A., II, and Hunt, E.V., Jr., 1982, Sediment production under various forest-site conditions, *Recent Developments in the Explanation and Prediction of Erosion and Sediment Yield*, Walling, D.E., Ed., IAHS, Washington, DC, pp. 13–22.

Chang, M., Sayok, A.K., and Watterston, K.G., 1992, Comparison of three proposed C-values of the soil loss equation for various forested conditions, *The Environment Is Our Future, Proceedings of Conference XXIII*, February 18–21, Reno, NV, International Erosion Control Association, Steamboat Springs, CO, pp. 241–246.

Chang, M. and Ting, J.C., 1986, Applications of the universal soil loss equation to various forest conditions, *Monitoring to Detect Changes in Water Quality Series, Proceedings of Budapest Symposium*, Lerner, D. Ed., July 1986, IAHS Publication No. 157, Budapest, Hungary, pp. 165–174.

Chang, M. and Wong, K.L., 1983, Effects of land use and watershed topography on sediment delivery ratio in East Texas, *BeitragezurHydrologie*, 9, 55–69.

Chapman, G., 1948, Size of raindrops and their striking force at the soil surface in a red pine plantation, *Trans. Am. Geophys. Union*, 29, 664–670.

Chen, X., Zhou, Q. and Zhang, E., 2006, In-channel sand extraction from the Mid-Lower Yangtze channels and its management problems and challenges, *J. Environ. Plan. Manage.*, 49, 309–320.

Chu, Z.E. et al., 2009, A quantitative assessment of human impacts on decrease in sediment flux from major Chinese rivers entering the western Pacific Ocean, *Geophys. Res. Lett.*, 36, L19603. DOI: 10.1029/2009GL039513.

Cruse, R.M. and Larson, W.E., 1977, Effect of soil shear strength and soil detachment due to raindrop impact, *Soil Sci. Soc. Am. J.*, 41, 777–781.

Dendy, F.E. and Bolton, G.C., 1976, Sediment yield–runoff drainage area relationships in the United States, *J. Soil Water Conserv.*, 31, 264–266.

Di Luzio, M., Srinivasan, R., and Arnold, J.G., 2004, A GIS-coupled hydrological model system for the watershed assessment of agricultural nonpoint and point sources of pollution, *Trans. GIS*, 8(1), 113–136.

Diodato, N. and Grauso, S., 2009, An improved correlation model for sediment delivery ratio assessment, *Environ. Earth Sci.*, 59, 223–231.

Dissmeyer, G.E. and Stump, R.F., 1978, Predicting Erosion Rates for Forest Management Activities and Conditions Sample in the Southeast, USDA Forest Service Southeastern Area State and Private, Atlanta, GA.

Dunne, T., 1977, Studying patterns of soil erosion in Kenya, *FAO Soils Bull.*, 33, 109–122.

Ekern, P.C., 1953, Problems of raindrop impact erosion, *Agric. Eng.*, 34, 23–25, 28.

Elliot, W. et al., 2010, Tools for analysis, *Cumulative Watershed Effects of Fuel Management in the Western United States*, Elliot, W.J., Miller, I.S. and Audin, L., Eds., USDA Forest Service RMRS-GTR-231, Forest Service Rocky Mountain Research Station, Fort Collins, CO, pp. 246–276.

Ellison, W.D., 1944, Studies of raindrop erosion, *Agric. Eng.*, 25, 131–136.

FAO, 2010, Global forest resources assessment 2010: Main report, UN Food and Agriculture Organization, FAO Forestry Paper 163, Rome, Italy, 340 pp.

Flaxman, E.M., 1972, Predicting sediment yield in western United States, *Proc. Am. Soc. Civ. Eng. Hydrol. Div.*, 98, HY12, 2073–2085.

Foster, G.R., 1982, Modeling the erosion process, *Hydrologic Modeling of Small Watersheds*, Haan, C.T., Johnson, H.P., and Brakensiek, D.L., Eds., ASAE, St. Joseph, MI, pp. 295–380.

Foster, G.R., 1986, Understanding ephemeral gully erosion, *Soil Conservation: Assessing the National Resources Inventory*, Vol. 2, National Academic Press, Washington, DC, pp. 90–118.

Foster, G.R. and Meyer, L.D., 1972, A closed-form soil erosion equation for upland areas, *Sedimentation Symposium to Honor Professor Hans Albert Einstein*, Shen, H.W., Ed., Colorado State University, Fort Collins, CO, pp. 12.1–12.19.

Foster, G.R. and Meyer, L.D., 1975, Mathematical simulation of upland erosion by fundamental erosion mechanics, *Present and Prospective Technology for Predicting Sediment Yields and Sources*, USDA Agricultural Research Service, ARS-S-40, Brooksville, FL, pp. 190–207.

Foster, G.R. et al., 1981, Conversion of the universal soil loss equation to SI metric units, *J. Soil Water Conserv.*, 36, 355–359.

Free, G.R., 1960, Erosion characteristics of rainfall, *Agric. Eng.*, 41, 447–449, 455.

Gallaher, B.M. and Koch, R.J., 2009, Water quality and stream flow after the Cerro Grande Fire: A summary, LALP-05-009, Los Alamos National Laboratory, Los Alamos, NM, 40 pp.

Gilley, J.E. et al., 2000, Narrow grass hedge effects on runoff and soil loss, *J. Soil Water Conserv.*, 55, 190–196.

Gordon, N.D., McMahon, T.A., and Finlayson, B.L., 1993, *Stream Hydrology, an Introduction for Ecologists*, John Wiley & Sons, New York.

Gottschalk, L.C., 1964, Sedimentation. Part I: Reservoir sedimentation, *Handbook of Applied Hydrology*, Chow, V.T., Ed., McGraw-Hill, New York, pp. 1–34.

Grace, J.M., III, 2000, Forest road sideslopes and soil conservation techniques, *J. Soil Water Conserv.*, 55, 96–101.

Grace, J.M., III, 2002, Effectiveness of vegetation in erosion control from forest road sideslopes, *Trans. ASAE*, 45, 681–685.

Grace, J.M., III and Elliot, W.J., 2008, Determining soil erosion from roads in the Coastal Plain of Alabama, *Environmental Connection 08; Proceedings of Conference 39*, February 18–22, 2008, Orlando, FL, International Erosion Control Association, Steamboat Springs, CO., 12 pp.

Gray, D.H. and Megahan, W.F., 1981, Forest vegetation removal and slope stability in the Idaho Batholith, U.S. Forest Service, Intermountain Forest and Range Experiment Station, Research Paper INT-271, Ogden, UT.

Gray, D.H. and Sotir, R.B., 1996, *Biotechnical and Soil Bioengineering Slope Stabilization*, John Wiley & Sons, New York.

Gucinski, H. et al., 2001, Forest roads: A synthesis of scientific information, General Technical Report PNW-GTR-509, U.S. Forest Service, Pacific NW Research Station, Portland, OR.

Hales, T.C. et al., 2009, Topographic and ecologic controls on root reinforcement, *J. Geophys. Res.*, 114, F03013, 17 pp. DOI:10.1029/2008JF001168.

Hanson, G. J., and Cook, K. R. 2004. Apparatus, test procedures, and analytical methods to measure soil erodibility in situ. *Appl. Eng. Agric.*, 20(4), 455–462.

Heede, B.H., 1975, Watershed indicators of landform development, *Proceedings of the Hydrology and Water Resources in Arizona and the Southwest*, Vol. 5, Arizona Section, American Water Resources Association and the Hydrology Section, Arizona Academy of Science, pp. 43–46.

Heede, B.H., 1976, *Gully Development and Control: the Status of Our Knowledge*, U.S. Forest Service, Rocky Mountain Forest and Range Experiment Station, Fort Collins, CO.

Heede, B.H., 1978, Designing gully control systems for eroding watersheds, *Environ. Manage.*, 2, 509–522.

Hjulstrom, F., 1939, Transportation: Transportation of detritus by moving water, *Symposium on Recent Marine Sediments*, Trask, P.D., Ed., American Association of Petroleum Geologists, Tulsa, OK, pp. 5–31.

Horowitz, A.J., 2003, An evaluation of sediment rating curves for estimating suspended sediment concentrations for subsequent flux calculations, *Hydrol. Processes*, 17, 3387–3409.

Hudson, N., 1995, *Soil Conservation*, Iowa State University Press, Ames, IA.

Hupp, C.R. et al., 2009, Bank erosion along the dam-regulated lower Roanoke River, North Carolina, *Management and Restoration of Fluvial Systems with Broad Historical Changes and Human Impacts*, James, L.A., Rathburn, S.L., and Whittecar, G.R., Eds., Geological Society of America, Special Paper 451, pp. 97–108.

Knisel, W.G., 1980, CREAMS: A field scale model for chemicals, runoff, and erosion from agricultural management systems, Conservation Research Report No. 26, USDA, Washington, DC.

Kunze, M.D. and Stednick, J.D., 2006, streamflow and suspended sediment yield following the 2000 Bobcat fire, Colorado, *Hydrol. Processes*, 20, 1661–1681.

Laflen, J.M., 1986, Ephemeral gully erosion, *Proceedings of Fourth Federal Interagency Sediment Conference*, March 24–27, 1986, Las Vegas, NV, Vol. I, Sec. 3, pp. 29–37.

Lal, R. and Elliot, W., 1994, Erodibility and erosivity, *Soil Erosion Research Methods*, Lal, R., Ed., Soil and Water Conservation Society, Ankeny, IA, pp. 181–208.

Landsberg, J.D. and Tiedemann, A.R., 2000, Fire management, *Drinking Water from Forests and Grasslands, a Synthesis of the Scientific Literature*, G.E. Dissmeyer, Ed., General Technical Report SRS39, USDA Forest Service, Southern Research Station, Athens, GA.

Laursen, E.M., 1958, Sediment transport mechanics in stable channel designs, *Trans. ASCE*, 123, 195–206.

Laws, J.O., 1941, Measurements of the fall-velocity of water-drops and raindrops, *Trans. Am. Geophys. Union*, 22, 709–721.

Laws, J.O. and Parsons, D.A., 1943, The relation of raindrop size to intensity, *Trans. Am. Geophys. Union*, 24, 452–459.

Liu, C., Sui, J. and Wang, Z.-Y., 2008, Sediment load reduction in Chinese rivers, *Int. J. Sediment Res.*, 23(1), 44–55.

Lo, A. et al., 1985, Effectiveness of EI_{30} as an erosivity index in Hawaii, *Soil Erosion and Conservation*, El-Swaify, S.A., Moldenhauer, W.C., and Lo, A., Eds., Soil Conservation Society of America, Ankeny, IA, pp. 384–392.

Lyles, L., 1977, Soil detachment and aggregate disintegration by wind-driven rain, *Soil Erosion: Prediction and Control*, Soil Conservation Society of America, Ankeny, IA, pp. 152–159.

Megahan, W.F., Day, N.F., and Bliss, T.M., 1978, Landslide occurrences in the western and central northern Rocky Mountain physiographic province in Idaho, *The 5th North American Forest Soils Conference*, Colorado State University, Fort Collins, CO, pp. 116–139.

Maner, S.B., 1962, *Factors Influencing Sediment Delivery Ratios in the Blackland Prairie Land Resource Area*, USDA, Soil Conservation Service, Fort Worth, TX.

Martinez, M.R., Fogel, M.M., and Lane, L.J., 1980, Modeling for upland areas, Paper No. 80-2505, American Society of Agricultural Engineers, St. Joseph, MI.

Martz, L.W. and Garbrecht, J., 1998, The treatment of flat areas and depressions in automated drainage analysis of raster digital elevation models, *Hydrol. Processes*, 12, 843–855.

McGregor, K.C. and Mutchler, C.K., 1977, Status of the R factor in northern Mississippi, *Soil Erosion: Prediction and Control*, Soil Conservation Society of America, Ankeny, IA, pp. 135–142.

Meade, R.H. and Moody, J.A., 2010, Causes for the decline of suspended-sediment discharge in the Mississippi river system, 1940–2007, *Hydrol. Processes*, 24, 35–49.

Merkel, W.H., Woodward, D.W., and Clarke, C.D., 1988, Ephemeral gully erosion model (EGEM), *Model Agricultural, Forest, and Rangeland Hydrology, Proceedings of the 1988 International Symposium*, Chicago, IL, ASAE, St. Joseph, MI, pp. 315–322.

Meyer, L.D., 1981, How rain intensity affects interrill erosion, *Trans. ASAE*, 24, 1472–1475.

Meyer, L.D. and Harmon, W.C., 1984, Susceptibility of agricultural soils to interrill erosion, *Soil Sci. Soc. Am. J.*, 48, 1152–1157.

Meyer, L.D. and Wischmeier, W.H., 1969, Mathematical simulation of the process of soil erosion by water, *Trans. ASAE*, 12, 754–758, 762.

Meyer, L.D., Wischmeier, W.H., and Foster, G.R., 1970, Mulch rates required for erosion control on steep slopes, *Soil Sci. Soc. Am. Proc.*, 34, 928–931.

Møen, K.M. et al., 2010, Bedload Measurement in Rivers Using Passive Acoustic Sensors, USGS Scientific Investigations Report 2010–5091, pp. 336–351.

Morgan, R.P.C., 1986, *Soil Erosion and Conservation*, Longman Scientific & Technical, Hong Kong.

Musgrave, G.W., 1947, The quantitative evaluation of factors in water erosion: A first approximation, *J. Soil Water Conserv.*, 2, 133–138.

Mutchler, C.K. and Bowie, A.J., 1976, Effect of land use on sediment delivery ratios, *Proceedings of the Third Federal Inter-Agency Sedimentation Conference*, Denver, CO, pp. I:11–I:21.

Nearing, M.A., Lane, L.J., and Lopes, V.L., 1994, Modeling soil erosion, *Soil Erosion Research Methods*, Lal, R., Ed., Soil and Water Conservation Society, Ankeny, IA, pp. 127–156.

Parker, C. et al., 2010, Automated mapping of the potential for ephemeral gully formation in agricultural watersheds, *Proceedings 2nd Joint Federal Interagency Conference*, June 27–July 1, 2010, Las Vegas, NV, available at: http://acwi.gov/sos/pubs/2JFIC/Contents/3A_Parker_02_26_10_paper.pdf

Parker, J.C., Amos, D.F., and Sture, S., 1980, Measurements of swelling, hydraulic conductivity, and shear strength in a multistage triaxial test, *Soil Sci. Soc. Am. J.*, 44, 1133–1138.

Piest, R.F., Bradford, J.M., and Spomer, R.G., 1975, Mechanisms of erosion and sediment movement from gullies, *Present and Prospective Technology for Predicting Sediment Yields and Resources*, USDA Agricultural Research Service, ARS-S-40, Brooksville, FL, pp. 162–176.

Pimentel, D., 2006, Soil erosion: A food and environment threat, *Environ. Dev. Sustainability*, 8, 119–137.

Ramirez-Avila, J.J. et al., 2010, Assessment and estimation of streambank erosion rates in the Southeastern Plains Ecoregion of Mississippi, *2nd Joint Federal Interagency Conference*, June 27–July 1, 2010, Las Vegas, NV, U.S. Department of the Interior, Washington, DC, 12 pp.

Renard, K.G., 1980, Estimating erosion and sediment yield from rangeland, *Symposium on Watershed Management 1980*, Boise, ID, Vol. 1, ASAE, New York, pp. 164–175.

Renard, K.G. and Freimund, J.R., 1994, Using monthly precipitation data to estimate the R-factor in the revised USLE, *J. Hydrol.*, 157, 287–306.

Renard, K.G. et al., 1991, RUSLE: Revised universal soil loss equation, *J. Soil Water Conserv.*, 46, 30–33.

Renard, K.G. et al., 1994, The revised universal soil loss equation, *Soil Erosion Research Methods*, 2nd edn., Lal, R., Ed., Soil and Water Conservation Society, Ankeny, IA, pp. 105–124.

Renard, K.G. et al., 1997, *Predicting Soil Erosion by Water: A Guide to Conservation Planning with the Revised Universal Soil Loss Equation (RUSLE)*, Agricultural Handbook No. 703, USDA-ARS, Brooksville, FL.

Renfro, G.W., 1975, Use of erosion equations and sediment delivery ratios for predicting sediment yield, *Proceedings of Sediment-Yield Workshop on Present and Prospective Tech. for Predicting Sediment Yields and Sources*, USDA, ARS-40, Brooksville, FL, pp. 33–45.

Robinson, K.M. and Hanson, G.J., 1996, Gully headcut advance, *Am. Soc. Agric. Eng.*, 39, 33–38.

Roehl, J.E., 1962, Sediment source areas, delivery ratios and influencing morphological factors, *Land Erosion, Proceedings of the Bari Symposium*, International Association of Scientific Hydrology, Publ. No. 59, pp. 202–213.

Roose, E., 1977, Erosion et Ruisselement en Afrique de l'ouest—vingtannees de measures en petites parcelles experimentales, Travaux et Documenys de l'ORSTOM No. 78, ORSTOM, Paris, France.

Rosgen, D.L., 2001. A Practical method of computing streambank erosion rate, *Proceedings of the 7th Federal Interagency Sedimentation Conference II. Stream Restoration*, March 25–29, Reno, NV, U.S. Department of the Interior, Washington, DC, pp. 9–17.

Rosgen, D.L., Knapp, K.L., and Megahan, W.F., 1980, Total potential sediment, *An Approach to Water Resources Evaluation of Non-Point Silvicultural Sources (A Procedural Handbook)*, EPA-600/8-80–012, U.S. Environment Protection Agency, Washington, DC, pp. VI.1–VI.43.

Shi, C., Zhang, D., and You, L., 2002, Changes in sediment yield of the Yellow River basin of China during the Holocene, *Geomorphology*, 46, 267–2893.

Shields, N.D., 1936, Anwendung der Ahnlichkeit Mechanik und der Turbulenzforschung auf die Geschiebelerwegung, Mitt. Preoss Versuchanstaltfür Wasserbau und Schiffau, 26.

Sidle, R.C., Pearce, A.J., and O'Loughlin, C.L., 1985, *Hillslope Stability and Land Use*, Water Resources Monograph Series 11, American Geophysical Union Washington, DC.

Simanton, J.R. and Renard, K.G., 1982, The USLE rainfall factor for southwest U.S. Rangelands, *Proceedings of Workshop on Estimating Erosion and Sediment Yield from Rangelands*, USDR-ARS Agricultural Reviews and Manuals, ARS-W-26, Tucson, AZ, pp. 50–62.

Swift, L.W., Jr., 1984, Soil losses from roadbeds and cut and fill slopes in the southern Appalachian Mountains, *South. Appl. For.*, 8, 209–215.

Thompson, J.R., 1964, Quantitative effect of watershed variables on rate of gully-head advancement, *Trans. ASAE*, 7, 54–55.

Thorne, C.R. and Zevenbergen, W., 1990, Prediction of ephemeral gully erosion on cropland in the southeastern United States, *Soil Erosion on Agricultural Land*, Boardman, J., Foster, I.D.L., and Dearing, J.A., Eds., John Wiley & Sons, New York, pp. 447–460.

Truman, C.C. et al., 2001, Slope length effects on runoff and sediment delivery, *J. Soil Water Conserv.*, 56, 249–256.

Turowski, J.M., Rickenmann, D., and Dadson, S.J., 2010, The partitioning of the total sediment load of a river into suspended load and bedload: A review of empirical data, *Sedimentology*, 57, 1126–1146.

UNESCO, 1982, Sedimentation problems in river basins, Project 5.3 of the International Hydrological Programme, United Nations Educational Scientific and Cultural Organ, Paris, France.

UNESCO, 1991, *The Water-Related Issues and Problems of the Humid Tropics and Other Warm Humid Regions: A Programme for the Humid Tropics*, Division of Water Sciences, UNESCO, Paris, France.

USDA, 1995, USDA Water Erosion Prediction Project (WEPP), User Summary, NSERL Report No. 11, USDA-ARS-MWA, National Soil Erosion Research Laboratory, West Lafayette, IN.

U.S. Soil Conservation Service, 1977, Procedure for determining rates of land damage, land depreciation and volume of sediment produced by gully erosion, *Guidelines for Watershed Management*, FAO, Rome, Italy, pp. 125–142.

Van Lear, D.H. et al., 1985, Sediment and nutrient export in runoff from burned and harvested pine watersheds in the South Carolina Piedmont, *J. Environ. Qual.*, 24, 169–174.

Walling, D.E., 1974, Suspended sediment and solute yields from a small catchment prior to urbanization, *Fluvial Processes in Instrumented Watersheds*, Gregory. K.J. and Walling, D.E., Eds., Institute of British Geographers, London, U.K., pp. 169–192.

Walling, D.E., 1994, Measuring sediment yield from river basins, *Soil Erosion Research Methods*, Lal, R., Ed., Soil and Water Conservation Society, Ankeny, IA, pp. 39–80.

Walling, D.E., 2008, The changing sediment loads of the world's rivers, *Ann. Warsaw. Univ. Life Sci. SGGW Land Reclam.*, 39, 3–20.

Walling, D.E., 2009, The impact of global change on erosion and sediment transport by rivers: Current Progress and future challenges, the UN World Water Assessment Programmes, Sci. Paper, Paris, France, 30 pp.

Walling, D.E. and Fang, D., 2003, Recent trends in the suspended sediment loads of the world's rivers, *Global Planet Change*, 39, 116–126.

Williams, J.R., 1975, Sediment-yield prediction with universal equation using runoff energy factor, *Present and Prospective Technology for Predicting Sediment Yields and Sources*, U.S. Agricultural Research Service, ARS-S-40, Brooksville, FL, pp. 244–252.

Williams, J.R., 1977, Sediment delivery ratios determined with sediment and runoff models, *Erosion and Solid Matter Transport in Inland Waters*, IAHS Publ. No. 122, Wallingford, England, U.K., pp. 168–179.

Williams, J.R., Renard, K.G., and Dyke, P.T., 1983, A new method for assessing the effect of erosion on predictability—The EPIC model, *J. Soil Water Conserv.*, 38, 381–383.

Wischmeier, W.H., 1975, Estimating the soil loss equation's cover and management factor for undisturbed areas, *Present and Prospective Technology for Predicting Sediment Yields and Sources*, USDA Agricultural Research Service, ARS-S-40, Brooksville, FL, ARS-S-40, pp. 118–124.

Wischmeier, W.H. and Smith, D.D., 1958, Rainful energy and its relationship to soil loss, *Trans. Am. Geophys. Union*, 39, 285–291.

Wischmeier, W.H. and Smith, D.D., 1978, *Predicting Rainfall Erosion Losses, a Guide to Conservation Planning*, Agricultural Handbook No. 537, USDA Agricultural Research Service, Brooksville, FL.

Woodward, D.E., 1999, Method to predict cropland ephemeral gully erosion, *CATENA*, 37, 393–399.

Wynn, T. and Mostaghimi, S., 2006, The effects of vegetation and soil type on streambank erosion, southwestern Virginia, USA, *J. Am. Water Resour. Assoc.*, 42(1), 69–82.

Xu, J.X., 2003, Sediment flux to the sea as influenced by changing human activities and precipitation: Example of the Yellow River, China. *Environ. Manage.*, 31, 328–341.

Xu, K. and Milliman, J.D., 2009, Seasonal variations of sediment discharge from the Yangtze River before and after impoundment of the Three Gorges Dam, *Geomorphology*, 104, 276–283.

Xu, K. et al., 2006, Yangtze sediment decline partly from Three Gorges Dam, *EOS, Trans. Am. Geophys. Union*, 87(19), 185, 190.

Yang, C.T., 1972, Unit stream power and sediment transport, *J. Hydraul. Div. Am. Soc. Civ. Eng.*, 98, 1805–1827.

Young, R.A. and Wiersma, J.L., 1973, The role of rainfall impact in soil detachment and transport, *Water Resour. Res.*, 9, 1629–1636.

Zaimes, G.N., Schultz, R.C. and Isenhart, T.M., 2006, riparian land uses and precipitation influences on stream bank erosion in central Iowa, *J. Am. Water Resour. Assoc.*, 42(1), 83–98.

Zhao, W. et al., 1992, Analysis on the variation of sediment yield in the Sanchuanhe river basin in 1980s, *Int. J. Sediment Res.*, 7, 1–19.

Zingg, A.W., 1940, Degree and length of land slopes as it affects soil loss in runoff, *Agric. Eng.*, 21, 59–64.

13

Forests and Stream Habitat

Excluding Alaska, the United States has 5,859,608 km of streams and rivers (USEPA, 2000). Besides lakes and oceans, streams and rivers are also homes to fish, benthic invertebrates, and many other aquatic organisms. The biodiversity, integrity, and biotic productivity of stream organisms are affected by the physical, chemical, and biological properties of stream habitats. Fish are considered a nontimber benefit of forested watersheds. Under ecological management concepts, stream habitats need to be protected and enhanced so that populations of cold- and warm-water fish can be maintained, threatened fish populations can be stabilized, and endangered species can be preserved. Improper forestry activities and management often impose adverse impacts upon stream habitats and in turn upon aquatic life.

13.1 Stream Habitat

In a drainage basin, fingertip tributaries with no branches at the head of the stream are called first-order streams. Two first-order streams join to form a second-order stream; two second-order streams form a third-order stream when they join each other, and so on. Streams in forested watersheds are of lower orders in general (Figure 13.1).

A stream channel in a forested watershed can be characterized not only by flow volume, velocity, depth, width, channel gradient, and substrate but also by pools, riffles, fallen logs (log steps), accumulation of large woody debris (LWD), beaver dams, and cienegas (riparian marshlands). The composition of these physical features along with riparian vegetation, woody debris, temperature, dissolved oxygen, water chemistry, and food availability forms forested stream habitats. They make patterns of stream habitats different from one another. A species resides, adapts, grows, and reproduces in a particular stream habitat and has a tolerance range for a given parameter (e.g., substrate type) of the habitat. Some parameters (e.g., dissolved oxygen) are more critical than others.

13.1.1 Habitat Parameters

Many habitat parameters can affect the biological integrity of streams. Karr and Dudley (1981) grouped those parameters into five primary categories: water quality, flow regime, physical habitat, energy, and biotic interactions. The parameters that relate to flow regime and physical habitat are discussed in the following sections.

13.1.1.1 Streamflow

Running water affects the transport and deposition of instream particles, streambed structure, plankton populations, and the ability of fish to feed (e.g., extraction of nutrients

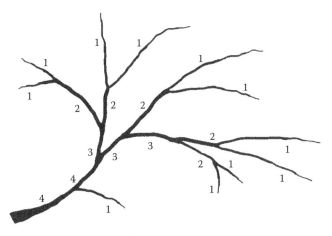

FIGURE 13.1
A fourth-order stream.

in water current). A species must adapt to the flow conditions in which it colonizes. A relatively stable streamflow without extreme floods and droughts is a major factor for productive streams. Extreme high water can destroy available habitats, and fish may not be able to withstand the shear stress of the water. On the other hand, low summer flow could be the major factor limiting the abundance of many species, including coho salmon, anadromous rainbow, and cutthroat trout (Narver, 1971; Hicks et al., 1991a). Many forested headwater streams become intermittent during the summer, which makes fish susceptible to both instream predation and avian predation. Shallow streams, especially those in open-canopy situations, tend to have higher temperatures, greater temperature fluctuation over time, and more energy for photosynthesis. Shade and nutrient input from overhanging riparian vegetation seem to decrease with the width of streams. Generally, large fish tend to prefer deep water while smaller fish live in shallow water. Thus, creating pools and increasing riffle areas are often used as effective approaches for enhancing rearing habitat for salmon and trout (Hall and Baker, 1982; Waters, 1995).

   Although flow velocity and depth are important parameters of stream habitats, a number of regional studies showed that streamflow velocity and depth were not significantly correlated with density, biomass, and catch-per-effort for smallmouth bass (*Micropterus dolomieu*) and spotted bass (*Micropterus punctulatus*) in warmwater streams (McClendon and Rabeni, 1987; Sowa and Rabeni, 1995; Tillma et al., 1998). This suggests that fish have the ability to adapt to a variety of flow conditions or that the microhabitat of these parameters has not been properly assessed.

13.1.1.2 Temperature

Fish are ecothermic (cold-blooded); their internal temperatures closely follow external water temperatures. As a rule of thumb, a rise in temperature of 1°C increases metabolic rates in fish by 10%. Thus, as stream temperatures increase, food demand and dissolved oxygen demand also increase. This is compounded by the fact that warmer water holds less dissolved oxygen (and other gases) than cold water does. Stream temperatures suitable for fish depend on many factors, including species composition and life-history stage. The spawning temperature, optimal temperature for growth, and maximum tolerated temperature for Atlantic salmon (*Salmo salar*) are 0°C–8°C, 13°C–15°C, and 16°C–17°C, respectively. For rainbow trout (*Salmo gairdneri*), they are 4°C–10°C, 20°C–21°C, and 24°C (Moss, 1988). Changes in stream temperature beyond the optimum level can cause disease outbreaks, retard growth, stop migration, and cause death. However, higher summer water temperatures and accompanying lower dissolved oxygen levels seldom impact fish such as spotted bass, largemouth bass, and bluegill sunfish in southern forest streams like they impact salmonids in streams of the Northwest. This is because warm-water fish in the South have evolved mechanisms to cope with these factors (Rainwater, 2001, personal communication). If streamflow in the headwaters is impeded by drought or forest practices and reaches the tolerance limits of temperature and oxygen requirements, the ecosystem will be stressed regardless of the location.

13.1.1.3 Dissolved Oxygen

Oxygen in water is required for respiration in fish and other aquatic animals. It is supplied to the stream from the atmosphere through gas diffusion; the aerating action of wind, currents, and floodwaters; mixing brought about by precipitation; and photosynthesis activities. Low dissolved-oxygen levels can occur in streams due to respiration of aquatic organisms, decomposition of organic matter, and increases in water temperature due to the exposure of the stream surface to direct solar radiation or other reasons. Also, reduced streamflow or a stream surface covered by floating mats of aquatic vegetation or leaves affects dissolved-oxygen levels. Some species such as Atlantic salmon require a cool, highly oxygenated habitat with dissolved oxygen of at least 7.5 mg/L, while a minimum concentration of about 4–4.5 mg/L is sufficient for rainbow trout. Concentrations below 3 mg/L are not satisfactory for many fish in streams (Narver, 1971), but hypoxic conditions (oxygen deficiency in body tissues) for warmwater fish can occur at as low as 2 mg/L (Rutherford, 2001).

13.1.1.4 Sediment

All streams carry particles of various sizes derived either from geologic strata, streamflow erosion, and land-surface runoff, or from anthropogenic sources including agriculture, forestry, grazing, mining, urban development, and other human activities. Transported particles are categorically separated into two groups: suspended load and bedload. Suspended load consists of particles up to 1 mm diameter (coarse sands; Table 13.1), depending on flow-transport capacity. Each stream channel has its own characteristics such as concentrations of suspended sediment and bedload sizes, as do its habitats. Any land-use activities that significantly alter sediment patterns in a stream can damage instream habitats and impact aquatic communities (Henley et al., 2000). Elevated loads of sediment can change channel geometry. Sediment in motion can increase bed scour and bank erosion because the erosive force of the water is increased by the blasting effect of sands.

Physically, sediment in suspended form reduces light penetration and water clarity in streams. The decline of light penetration leads to decreases in photosynthetic rates of the periphyton community (the primary production of a stream) and to the reduction of invertebrate fauna in the stream. This reduces the abundance of food organisms available to fish. Forested headwater streams depend more on energy inputs from streamside zones (e.g., leaf litter and other organic debris) than on primary production unless the canopy is open due to human activities or naturally derived causes. Thus, clear-cutting forests to the stream bank fundamentally changes the nature of energy flow through these ecosystems. Turbid water makes feeding difficult for sight-dependent species such as salmon

TABLE 13.1

USDA's Size Classification of Soil Particles
and Rock Fragments

Rock Fragments		Soil Particles	
Name	Size (mm)	Name	Size (mm)
Spherical	Diameter		
Pebbles (gravel)	2–75	Clay	<0.002
Fine	2–5	Silt	0.05–0.002
Medium	5–20	Sand	0.05–1.0
Coarse	20–75	Very fine	0.05–0.10
Cobbles	75–250	Fine	0.10–0.25
Stones	250–600	Medium	0.25–0.50
Boulders	>600	Coarse	0.50–1.00
Flat	Long	Very coarse	1.00–2.00
Channers	2–150	Gravel	>2.00
Flagstones	150–380		
Stones	380–600		
Boulders	>600		

Source: USDA, *Soil Survey Manual*, USDA-NRCS Agricultural Handbook #18, U.S. Government Printing Office, Washington, DC, 1993.

and trout, all larval fish, basses, sunfish, and most minnows. Also, heat absorption in the surface layer can be increased due to suspended sediment (MacDonald et al., 1991). Biologically, high concentrations of suspended sediment can disrupt fish respiration and cause gill damage (Reynolds et al., 1989).

When sediment is deposited on the streambed, it can coat fish eggs and larvae, seal interstices, and trap fry (for salmonids) in the gravel. These factors cause oxygen deficiencies in gravel beds, inadequate removal of metabolites (free carbon and ammonia, toxic to aquatic organisms), and starvation of fry (only for salmonids). Fine sediment could thereby cause an overall reduction in fry availability for salmonids (Cover et al., 2008). Recommendations of suspended sediment concentrations were less than 30 mg/L for a high level and 100 mg/L for a moderate level of protection. A suspended sediment concentration of 488 mg/L was fatal to juvenile chinook salmon in Washington (Lloyd, 1987). Allowable suspended sediment/turbidity standards for cold water aquatic habitat are different among states in the United States. In Idaho, it requires that turbidity not exceed the background level by more than 50 NTU instantaneously or more than 25 NTU for more than 10 consecutive days. In California, standards for forested lands vary by region from 1 NTU above background to 20% above background (Foltz et al., 2008). Also, Canada attempts to address sediment guidelines by dividing flows into high and low groups, while the United States addresses this by considering flow differences in seasons and the European Union's guidelines are not to apply to exceptional circumstances such as storms and droughts (Biotta and Brazier, 2008).

In addition to the concentration level, the duration of exposure is crucial in a stressed environment. Newcombe and MacDonald (1991) showed that the product of suspended sediment concentration and duration of exposure was a much better indicator of effects ($R^2 = 0.64$) than was suspended sediment alone ($R^2 = 0.14$).

13.1.1.5 Substrate

The substrate, or streambed to hydrologists, of a stream is composed of a variety of particles in various sizes, voids among particles, organic matter, and aquatic animals and plants. It may be predominantly muddy, sandy, stony, or mixed, and may be stable or unstable. In headwater streams, particle sizes are generally larger (except coastal streams in the southeastern United States), channels shallower, cross-sectional areas narrower, slopes greater, and surface rougher than those downstream (Morisawa, 1985). Streams with braided channels are unstable, carry a great amount of bedload, and are severe in bank erosion. Streams with bedrock or bedrock-controlled waterways are stable and have less erosion (Rosgen, 1996). In other words, streambeds dominated by large particles tend to be more stable in general than ones dominated by fine particles.

Substrate is the most biologically active zone in stream profiles. It is a site for egg deposition and incubation, a source for food supplies, shelter for protection from predators and a habitat for aquatic organisms. However, each aquatic species has its own preferred substrate; variations in species abundance and diversity are great among substrates. A mixed substrate of sand, gravel, and large particles (e.g., rubble, cobble, boulder) usually supports the greatest number of species. Lowland streams with silt and clay beds tend to have less diversity of aquatic animals but may be a favorable site for aquatic plants and planktonic organisms (Gordon et al., 1993). Salmonids deposit eggs to a depth of 7.5–38 cm with gravel substrates of clean, mixed, stable gravel of 1–15 cm in diameter, small amounts of fine sediments, and high concentrations of dissolved oxygen (Bratovich et al., 2004). Research suggests that substrates for spawning and incubation of salmonid eggs are in good condition if fine sediment (<1.0 mm) comprising less than 12%–14% of the substrate; and emergence of fry will not significantly affected if sediment of <6.0 mm comprises less than 30% (Kondolf, 2000; Rosser and O'Connor, 2007). Gravel and gravel plus cobble, providing cover for resting and protection of newly hatched larvae, are dominant types of substrate in paddlefish (*Polyodon spathula*) spawning areas (Crance, 1987). In the Southeast, where there is a lack of large substrates such as boulders and bedrock ledges, woody debris becomes an important habitat component for smallmouth bass and spotted bass (Tillma et al., 1998). A study of population characteristics of smallmouth bass and rock bass in Missouri found that large substrate, undercut banks, and aquatic vegetation were the three most important habitat parameters among 32 parameters studied (McClendon and Rabeni, 1987).

13.1.1.6 Pools

Pools are the concave sections along a stream channel where flow velocity is lower, water is deeper, and substrate particles are finer than in adjacent riffles or runs. Thus, when stream channels have no flow in summer, pools can still provide water for fish to survive and serve as sites for escape from intense predation. Deep slow-moving pools are helpful for salmonid survival under harsh winter conditions.

Pools can be created by channel erosion and deposition processes, bank undercut, debris dams, beaver dams, logjams, or other events. They are essential habitats for stream fish, especially larger predatory species such as trout, and a variety of pools is required to provide habitats for different species and age classes of fish. Thus, species abundance and diversity are often related to the area, volume, depth, number, type, and frequency of pools. Land-use and management activities that cause severe soil erosion and stream sedimentation problems can reduce the depth, volume, and size of pools. Creating adequate pools in streams is a major effort in many programs for rehabilitating and enhancing stream habitat (Reeves and Roelofs, 1982; Seehorn, 1992).

Often a pool is followed by a convex streambed section, called a riffle, where current is rapid, water is shallow, and sediment is dominated by coarse particles. Riffles are predominately inhabited by young and small fish. The sequence of pool–riffle along a stream channel provides a great diversity of channel morphology and aquatic habitats.

13.1.1.7 Cover

The shading and sheltering effects of cover are essential to stream salmonids and other species. They can be in the form of streamside or submerged vegetation, undercut banks, logs, rocks, floating debris, turbulence, deep water, or turbidity (Bjornn and Reiser, 1991). Fish use cover for protection from disturbance and predation, feeding stations, food sources, wintering, and resting (Figure 13.2). Streams under overhanging vegetation are cooler in summer and warmer in winter, and have more nutrient input from vegetation. Cooler temperatures under the shaded area imply greater dissolved oxygen, a parameter important to fish activities.

In Kansas streams, root-wad and undercut bank areas were found to be the two most important habitat components affecting the density and biomass of spotted bass (*M. punctulatus*) among the 34 parameters studied (Tillma et al., 1998). Underwater observations in southern Ontario, Canada, noted that both brook trout (*Salvelinus fontinalis*) and brown trout (*Salmo trutta*) preferred holding positions beneath submerged cover structures rather than under above-water structures or uncovered control forms (Cunjak and Power, 1987).

13.1.2 Habitat-Rating Indices

Although many parameters can affect the variation of stream habitats (one study involved 43 parameters; Wang et al., 1998), some are more critical than others for a given species

FIGURE 13.2
Riparian habitat. (Reproduced from Reeves, G.H. and Roelofs, T.D., Rehabilitating and enhancing stream habitat: 2. Field applications, USFS Pacific NW Forest and Range Experiment Station, General Technical Report PNW-140, Portland, OR, 1982.)

or community. Knowing those critical parameters and their impacts on stream biological communities is essential in formulating management, protection, and enhancement programs for aquatic resources. They provide resource managers with adequate information for decision-making without requiring complex data collection in the field. This is an advantage because managers often do not have time and funding to conduct rigorous data collection and analyses. This leads scientists to develop habitat indices or models involving only a few key and sensitive parameters. Once developed, they can be used for (Wang et al., 1998; Brown et al., 2000)

1. Spatial and temporal comparisons and communication with other streams
2. Constructing habitat maps in poorly sampled areas
3. Evaluating impact scenarios of regulatory alternatives
4. Identifying and prioritizing areas for conservation actions
5. Predicting or assessing impacts of environmental change

Habitat indices are simple mathematical expressions to relate habitat quality as a function of one or more environmental parameters. Values of environmental parameters used in mathematical calculations may be based either on assigned ranges given by professional judgment (0–1, or others, depending on how favorable the range is for survival, growth, and reproduction of a given species) or on actual field observations. The derived indices result in a single value to represent habitat quality or conditions for a particular species or a life stage, or an entire fish community at a specified geographical scale. Some developed models include habitat suitability index (HSI) by USFWS (FWS, 1981; Raleigh et al., 1984), rapid bioassessment protocols (RBP) by USEPA (Plafkin et al., 1989), habitat quality index (HQI) in Missouri (Fajen and Wehnes, 1982), qualitative habitat evaluation index (QHEI) in Ohio (Rankin, 1989), fish habitat rating system (SHRS) in Wisconsin (Wang et al., 1998), habitat suitability curve (HSC) in Kansas and Oklahoma (Layher and Maughan, 1987), sediment stress index in British Columbia, Canada (Newcombe and MacDonald, 1991), fish habitat tolerance index (FHTI) in Maryland (Pirhalla, 2004), and an index of reservoir habitat impairment (IRHI) for the contiguous United States (Miranda and Hunt, 2011). Geographic patterns can be discerned from applying the indices and Geographic Information System (GIS) techniques, which provide managers with another tool to help formulate management plans.

13.2 Forest Impacts

Forestry practices have potential impacts on stream habitats and aquatic communities by:

1. Altering the structure and composition of the riparian vegetation
2. Allowing more solar radiation to reach stream surfaces
3. Generating more sediment to stream channels
4. Decreasing the amount of LWD in streams
5. Causing stream bank erosion

Stream-habitat parameters that are involved in the alteration may include supplies of organic matter, stream temperature, the stability of channel morphology, substrates, sediment, plant nutrients, dissolved oxygen, debris and LWD, pools, covers, and others.

13.2.1 On Riparian Vegetation

13.2.1.1 Impacts

The removal of riparian vegetation can affect food supplies of the stream through:

1. Cutting down supplies of leaves, needles, twigs, and wood to streams
2. Shifting the composition of organic matter inputs from the more decay-resistant coniferous material to more decomposable deciduous material
3. Reducing the quantity of insects and animal wastes that drop from canopies

These are of particular importance in small-forested streams, where riparian vegetation virtually covers the water surface and the limited aquatic photosynthesis cannot provide the needed energy base for aquatic invertebrates, bacteria, and fungi. Here, as biologist Dr. Fred L. Rainwater (personal communication December 11, 2001) has stated, a heterotropic (getting energy and nutrients from other organisms)-dominated ecosystem functions in these headwaters. The fish living in this habitat are omnivores (eating both plants and animals) and invertivores (invertebrate eating), and seldom (perhaps a single species) piscivores (fish eating). The reliance on a riparian source of organic particulates declines as the stream width increases and the sunlight strikes the stream surface. It then promotes autotropic (organisms making their own food) dominance over heterotropic. The individual sizes of fish increase, and piscivores become more common. Management objectives need to determine the balance between decreases in organic-matter supply and increases in primary and secondary production in the stream.

Besides a source of food supplies to streams, riparian vegetation provides falling trees, coarse woody debris, shade, and cover important to stream habitats. In the West, California golden trout (*Oncorhynchus mykiss aguabonita*) more often selected undercut bank, aquatic vegetation, and sedge and avoided bare and collapsed banks (Matthews, 1996). Variability of the density and biomass of smallmouth bass and rock bass (*Ambloplites rupestris*) was also associated with aquatic vegetation in Missouri streams (McClendon and Rabeni, 1987). Of the three cover parameters important to trout habitats, overhead riparian vegetation cover explained a greater amount of variation than rubble-boulder-aquatic vegetation and deepwater areas did in trout population size in southeast Wyoming (Wesche et al., 1987).

While most changes in riparian vegetation adversely affect salmonid populations, a few instances showed that fish and invertebrate abundance were increased after opening of the canopy along the stream (Newbold et al., 1980; Murphy and Hall, 1981). Apparently, shading effect, organic matter supply, and photo energy need to be balanced in a given stream. Overall, riparian vegetation is important to:

1. Fish habitat
2. Food chain support
3. Wildlife habitat
4. Thermal cover
5. Stream bank stabilization
6. Water quality protection
7. Channel morphology
8. Flood control

Channels with riparian vegetation are generally narrower, deeper, and simpler than channels without vegetation. In Alberta, Canada, long-term effects of clear-cut harvest in riparian areas increased stream channel width up to 60% in 17–24 years. This consequently generated large volumes of sediment in the streams, raised streambed elevations, and decreased spacing between pools (McCleary et al., 2004).

13.2.1.2 Management

Riparian vegetation is very dynamic and sensitive to changes in management that affect the hydrology of the watershed. Overgrazing, excessive logging, major construction, fire, agricultural activity, or severe weather conditions may disturb or damage the riparian vegetation. Consequently, riparian enhancement or rehabilitation must be implemented in such watersheds to improve riparian quality. These processes intend to restore the riparian and stream habitat to a more productive condition, by means of natural, artificial, or a combination of the two types of measures.

When the riparian vegetation is disturbed or damaged, an important step is to prevent the area from additional human disturbance. Most native species have the ability to adapt themselves to natural processes and disturbance patterns. Many studies have shown that fencing to reduce grazing and browsing pressure can result in rapid and impressive recovery of riparian vegetation. Sound soil and vegetation management in the entire watershed often is the best strategy for a healthy community of riparian vegetation (DeBano and Schmidt, 1989). Restoration is much more difficult if riparian vegetation is completely destroyed. In four hardwood-dominated California streams, riparian restoration, through exclusionary fencing implemented 10–20 years ago, significantly improved channel morphology and fish habitat (Opperman and Merenlender, 2004). There, the treated channels were narrower, long profile elevations were more heterogeneous, frequencies of LWD and debris jams were greater, and late summer water temperatures were lower than those of untreated stream corridors. Thus, watershed condition, geomorphology, stream types, vegetation community types, stream size, velocity, sinuosity and bank slope, uniformity, and stratigraphy should all be considered for the design of revegetation in riparian zones (Carlson et al., 1992).

13.2.2 On Water Temperature

13.2.2.1 Variations

Timber-harvesting operations often expose streams in headwater watersheds to direct solar radiation, a factor that can cause significant changes in stream temperature regimes. These changes include higher daily mean, maximum and minimum temperatures in the summer, greater diurnal temperature fluctuations, and longer durations for temperatures to remain above a threshold value. The effects are most significant in small streams in which streamflows are low and surface areas in relation to flow volumes are large. Studies showed that forest harvesting increased average daily maximum stream temperature 1.2°C–7.2°C in the eastern United States and 0.6°C–8.0°C in the western United States (Stednick, 2000). For summer maximum, increases were reported of more than 10°C in central Pennsylvania and 16°C in western Oregon. However, at higher elevations in northern latitudes, winter stream temperatures may be reduced due to the loss of insulation from riparian vegetation and the increase in outgoing longwave radiation from the stream

(Beschta et al., 1987). Both temperature increases in summer and decreases in winter have negative impacts on biological communities.

13.2.2.2 Impacts

Increases in stream temperature due to the removal of riparian vegetation generally stimulate primary production by microbial communities, algae, and invertebrates (Gregory et al., 1987). At the Coweeta Hydrologic Laboratory, North Carolina, invertebrate taxonomic diversity increased in the stream immediately after clear-cutting of a forested watershed, and benthic invertebrate abundance was three times higher. Biomass and production were two times higher following 16 years in succession than those of the adjacent control watershed (Swank et al., 2001). Higher production levels in green algae and grazer invertebrates suggest higher levels of food available for fish. However, the elevated temperature is accompanied by: (1) an increase in decomposition rates of organic matter by microbial activities that may lead to a deficiency of organic matter in certain periods of time, and (2) a decrease in dissolved oxygen levels that can cause stress, disease, and mortality to fish. Coupling these fast microbial activities with a reduction in organic matter inputs from riparian vegetation may offset the increased primary productivity in streams to a significant degree.

In the Northwest, increased summer temperatures could affect salmonids by altering habitat suitability for rearing, slowing down upstream migration while increasing susceptibility to disease, reducing metabolic efficiency, and shifting the competitive advantage of salmonid over nonsalmonid species (Beschta et al., 1987). Decreases in winter temperature may slow egg development and increase the possibility of surface ice formation (Hicks et al., 1991b).

13.2.2.3 Management

In small-forested streams, convective, conductive, and evaporative heat-transfer processes at the water surface are insignificant due to very low wind movement (Brown, 1980). This means that solar radiation striking the water surface is the major source of heat for small streams. The energy received from the sun will not be dissipated to surrounding areas but will instead be stored in the stream and cause increases in stream temperature. On the other hand, if a zone of riparian vegetation is left to shade the stream surface, it prevents stream surfaces from direct exposure to solar radiation, and the stream temperature is not likely to be affected by clear-cutting.

The length of shadow cast over the stream by riparian vegetation is affected by the height of vegetation, the orientation of stream azimuth in respect to the sun, solar altitude (affected by hour of the day and season), and geographical location. Although not managerial, the shadow cast by the streambank (especially for trench-typed streams) or topography is as effective as vegetation. Equations 8.12 through 8.14 can be used to calculate the length of shadow cast by tree belts or streambank. For estimating the length of shadow required to completely span the stream, or the so-called stream effective width, W_e, the following equations can be used (Currier and Hughes, 1980):

$$W_e = \frac{W_a}{\sin \beta} \tag{13.1}$$

$$\text{Sin}\beta = \sin[180-|\text{ stream azimuth} - \text{sun azimuth }|] \tag{13.2}$$

where

W_a is the actual stream width
β is the acute angle between stream azimuth and azimuth of the sun

At lat. 40° N on August 12, the sun's azimuth at 9 a.m. is 110° (Figure 8.5). If the azimuth of a particular stream is 235° and the actual stream width $W_a = 3\,\text{m}$, then $W_e = (3\,\text{m})/$ $[\sin(180 - |235 - 110|)] = 3.66\,\text{m}$, or it requires a shadow 3.66 m long to cross the stream surface. For actual width to be equal to effective width, it requires $\sin\beta = 1$, or $\beta = 90°$, meaning that the sun and stream are perpendicular. It is the smallest value of W_e. As the acute angle β decreases, W_e increases; W_e will reach infinity when $\beta = 0$, or where the sun and stream lie parallel to each other. Thus, the potential impact of streamside cutting on water temperature is greater for smaller streams because a greater surface area of small streams is more easily shaded by trees.

If additional heat δH (in cal/cm²/min) is received at the stream surface due to the removal of riparian vegetation, its effect on stream temperature can be calculated by (Brown, 1980):

$$\delta t = \frac{0.00017\,(\delta H)(A)}{Q} \tag{13.3}$$

where

δt is the increase in stream temperature (°C)
A is the surface area of the stream (m²)
Q is the stream discharge (m³/s)
0.00017 is a constant to convert discharge in m³/s into g/min

The value δH should represent net radiation in which solar radiation, stream reflection, and the exchange of longwave radiation between the stream and the air should all be considered. However, using solar radiation as the substitute for net radiation, δH was sufficient in several small streams in Oregon (Brown et al., 1971). The stream surface (area A) should include only the portion affected by the removal of riparian vegetation. Procedures for shade measurements in the field are given by Bartholow (1989).

13.2.3 On Sediment

13.2.3.1 Variations and Impacts

Stream habitat can be impaired by additional sediment induced by forest practices such as forest roads, clear-cutting, site preparations, and burning. Annual average sediment concentrations in most forested streams are less than 10 mg/L, while stormflows are up to 100 mg/L (Binkley and Brown, 1993). However, maximum increases in sediment concentrations caused by forest practices have been reported to be 56,000 JTU at Fernow Experimental Watersheds near Parsons, West Virginia and 14,949 mg/L in northern Mississippi in the East, and 7670 mg/L at Alsea Experimental Watersheds near Toledo, Oregon, in the West (Table 13.2). Average increases in the first posttreatment year ranged from 300 to 900 mg/L in general and, as an extreme case, from 2127 mg/L in the control site to over 2800 mg/L in the treatment site in northern Mississippi. The variations reflected

TABLE 13.2

A Few Studies on the Effect of Forest Practices on the Maximum Concentrations of Suspended Sediment (mg/L) or Stream Turbidity (JTU) in the United States

Location	Vegetation	Treatment	Sediment mg/L	Sediment JTU	Measure	Reference
Bull Run, OR	Conifers	25% cut, roads, burned	2,600		Maximum	Harr and Fredriksen (1988)
		Control	2–6		Range	
Alsea, OR	Douglas fir	100% cut, burned	20–7,670		Range	Brown and Krygier (1971)
		Control	32–256		Range	
H/J Andrews Experimental Forest, OR	Douglas fir	25% patch cut/road	1,850		Maximum	Fredriksen (1970)
		Control	8		Maximum	
Entiat Experimental Forest, WA	Firs, pines	Below road construction	497		Peak in July	Fowler et al. (1988)
		Control (above)	4			
Chiachagof Island, AK	Sitka spruce	Logged/ burned	1,268		Maximum	Stednick et al. (1982)
		Control	313		Maximum	
Parsons, WV	Hardwoods	Commercial cutting		56,000	Maximum	Reinhart et al. (1963)
		Control		15	Maximum	
Leading Ridge, PA	Hardwoods	Commercial cutting		550	Maximum	Lynch et al. (1975)
		Control		25	Maximum	
Hot Springs, MS	Shortleaf pine	Cutting/ roads/yarded	6,000		December 1983	Ursic (1991)
			556		5 year mean	
		Control	67		5 year mean	
Northern Mississippi	Loblolly pine	Sheared/ bedded	14,949 1,260		Max. month 1 year mean	Beasley (1979)
Alto, TX	Southern pines	Sheared Control	2,119 90		1 year mean 1 year mean	Blackburn et al. (1986)
Mica Creek, ID	Douglas fir	50% cleared 2.8% roads	1,000 230		Maximum	Karwan et al. (2007)
Upper Pajarito, Los Alamos, NM	Mixed conifers	Control Wildfires, 100% prefire	55,000 5		4 year mean	Gallaher and Koch (2004)

the soil and topographic conditions of the forest activities, weather conditions during the operations, the intensity of soil disturbances, and the presence or absence of prevention measures.

Increased sediment in streams can affect stream habitats in several ways, including (1) altering channel morphology important to aquatic life such as pools, riffles, spawning gravel, flow obstructions (barriers to fish migration), and side channels extended from the stream's edge (Chamberlin et al., 1991), (2) reducing the abundance of food organisms such as benthic invertebrates and periphyton available to the fish (Newcombe and MacDonald, 1991), and (3) degrading gravel quality and strata composition important to spawning and rearing habitats (Waters, 1995). These conditions can cause stress to fish, consequently fish behavior and physiology are negatively affected (Shrimpton et al., 2007). Frequently, the majority of sediment delivered to streams comes from one or a few major storms. The excessive runoff not only carries a great deal of sediment from forest management sites to stream channels but also causes severe bank erosion and bed scouring. As a result, channel forms can be altered, deposited sediment redistributed, the stream bank destabilized, and original habitats destroyed.

13.2.3.2 Management

Many best management practices (BMPs), such as road design and maintenance, harvesting strategies and site management, etc., have been proposed to reduce sediment production from forest activities. Buffer strips are one of the most effective and have become a standard measure in forest operations.

Buffer strips, or streamside management zones (SMZs), are zones of vegetation left along a stream bank to reduce stream sediment and protect water quality from forest practices. A study conducted at 57 stations in 50 northern California streams showed that the diversity of macroinvertebrate communities in streams with buffers ($\geq30\,m$) could not be distinguished from that at control sites but was significantly greater than in streams without buffer strips (Newbold et al., 1980).

The use of SMZs has been adopted by state forestry agencies and private lumber companies as part of their standard guidelines in forestry BMPs. They generate, at least, as cited by Lakel et al. (2010), the following benefits to the riparian environment:

1. Surface runoff velocities are slowed down and runoff volume is reduced by vegetation and litter layers in SMZs; consequently sediment and sediment attached nutrients and chemicals are trapped.

2. Streams are shaded by vegetation and consequently, water temperature is protected against increases.

3. Vegetation in SMZs takes up nutrients and transforms them into organic forms which are less harmful to water quality.

4. Wetting and drying cycles of riparian areas favor transformations of inorganic nitrogen compounds into benign gaseous compounds through aerobic and anaerobic soil processes.

5. The litter and vegetation stabilize stream banks and minimize stream bank erosion.

6. It provides both large and fine debris to streams, creating in-stream habitat and support for the food chain.

Buffer strips require a width greater than the distance sediment can move downslope. Sediment-travel distance is affected mainly by land slope, overland flow velocity and depth, particle size, slope shape, and surface roughness. Predicting the distance of sediment movement is difficult because of the interactions among variables (Seyedbagheri, 1996). Many state agencies have provided guidelines for the width of SMZs (Blinn and Kilgore, 2001). The guidelines generally recommend widths that range from 7.5 to 40 m and wider for watersheds of steeper slopes and for protection of municipal water supplies. In the southern Appalachians, minimum SMZ widths for graded and graveled forest roads without brush barriers on the forest floor can be estimated by the equation (Swift, 1986):

$$SMZ = 13 + 0.42\,S \qquad (13.4)$$

where
  SMZ is in meters
  S is slope in percent

If there are brush barriers on the forest floor, the minimum width of SMZ is reduced to:

$$SMZ = 10 + 0.12\,S \qquad (13.5)$$

For a 30% slope with no brush barriers on the floor, the minimum width of the SMZ is 26 m, while a similar slope with brush barriers reduces the width of the SMZ to 14 m.

Also, a design aid was developed for determining the width of filter strips needed to achieve a given level of trapping efficiencies for sediment or water under a broad range of agricultural site conditions (Dosskey et al., 2008). The aid is a set of curves depicting trapping efficiency versus filter width under seven site conditions. These curves were simulated and developed based on the process-based Vegetative Filter Strip Model (Muñoz-Carpena and Parsons, 2004) with four site parameters—field length, USLE C factor, slope, and soil texture class. Under forested conditions, the C factor is usually much lower than 0.15 and 0.50 used in developing the design aid and the estimated filter widths could be wider than needed.

Buffer strips are subject to damage due to wind effect, logging, fire, or disease. Failures of buffer strips jeopardize streamside-management objectives and watershed-management programs. In the Cascade Mountains of western Oregon, a study on 40 streamside buffer strips showed that wind damage accounted for nearly 94% of timber-volume loss (Steinblums et al., 1984). Thus, an SMZ needs to be wider in wind-prone regions and for species susceptible to windthrow.

13.2.4 On Large Woody Debris

13.2.4.1 Variations

Riparian trees deliver organic matter and debris continuously to the stream throughout their succession stages. The delivery of these materials may through mass wasting, tree mortality, windthrow, bank erosion, and upstream transport. Any woody material in streams with diameter greater than 10 cm and length greater than 1 m is generally considered LWD. The amount of LWD is usually greater in small streams, streams with finer substrates and streams with greater riparian tree density. However, the length and diameter of LWD increase with stream size. Large pieces of woody debris are difficult to flush out by flows, stay in the channel longer, and have greater impacts on stream morphology, flow movement and habitats for small streams compared to large streams. The amount of LWD was found 15–50 pieces/km in coastal Maine (Magilligan et al., 2008), 107–127 pieces/km in coastal British Columbia (Bahuguna et al., 2010), and 340–480 pieces/km in the southern Oregon Coast Range (May and Gresswell, 2003). In the Pacific Northwest, LWD covers as much as 50% of the channel in first- and second-order streams and 25% in third- and fourth-order streams (Sedell et al., 1988). Some coniferous LWD, depending on the transport capacity of streams and the resistance ability to decomposition, breaking, and abrasion, has been in channels of the Pacific Northwest for more than 200 years (Swanson et al., 1976).

The decomposition or depletion of LWD in streams is complex and involves biological, chemical, and physical processes, in addition to species and size. It is most often expressed as a negative exponential decay function:

$$M = M_0(e)^{-kt} \tag{13.6}$$

where
 M_0 is the initial mass, density, or volume of LWD
 M is the quantity of LWD at time t years after deposition
 k is the decay rate calculated through long-term studies

Based on Scherer's (2004) reviews, k values range from 0.01 to 0.03 for western hemlock and sitka spruce in Southeast Alaska, 0.11–0.28 for red oak in North Carolina, and 0.95–1.20 for alder and birch in eastern Québec.

13.2.4.2 Impacts

Traditionally, LWD has been considered a physical barrier to fish migration, a blockage to floodwater movement, a retainer for sediment deposition, a consumer of dissolved oxygen and a supplier of certain compounds toxic to stream biota. These negative impacts led to the removal of LWD from streams and rivers as a general practice through the early 1980s to allow fish migration, to reduce flood hazards, and to improve stream-habitat quality (MacDonald et al., 1991). More recent studies showed that LWD is an important part of the stream morphology–habitat system.

LWD affects channel meandering and bank stability, creates greater variation in stream width and channel gradient, forms pools and waterfalls, provides cover and food supplies for fish and aquatic organisms and regulates the storage and routing of sediment. All of these are important to habitats for aquatic biota. However, these impacts on the landscape are variable in degree due to the position, orientation, size, volume, frequency, and decomposition of LWD in stream channels (Bisson et al., 1987; Sedell et al., 1988; Lemly and Hilderbrand, 2000). Biologically, LWD provides fish populations with diversified habitats, organic matter and energy required in the aquatic food web, cover, and pools as survival shelters during low flows in the summer. A study of 4382 forest stream sites in Sweden found that the abundance of brown trout increased with the amount of LWD up to 8–16 wood pieces per $100\,m^2$ and the size of trout caught was significantly larger at sites with LWD (Degerman et al., 2004).

• Timber harvesting and logging of riparian trees often cause streams to (Bryant, 1980; Bilby, 1984; Carlson et al., 1990; Ralph et al., 1994):

1. Decline in the amount and size of LWD
2. Decrease in pool frequency and size
3. Lose important habitat features
4. Reduce fish-population abundance
5. Increase in bank instability
6. Decline in insect-community diversity
7. Have reduced benthic detritus

In the Pacific Northwest, the removal of riparian forests has reduced sources of LWD to stream channels for several decades, causing LWD abundance in the channel to decline and remain low for 50–100 years after logging (Beechie et al., 2000). It then becomes a technique to artificially increase LWD in streams as a restoration and enhancement measure for aquatic habitats (Clément et al., 2008). It is also a challenge to forest- and fishery-resource managers to decide how much and what kind of LWD is required to add to the stream as an adequate measure for maintaining fish habitats and where and when LWD needs to be removed for increasing floodwater movements.

13.2.5 On Streambank Erosion

13.2.5.1 Occurrences

In small streams, bank erosion can affect stream depth and channel configuration, consequently stream habitats. Bank erosion includes erosion caused by: (1) overland flow, (2) streamflow action, (3) bank drainage (seepage), and (4) mass wasting (bank failure). They occur when the driving forces (shear stress) exceed the resisting forces (shear strength). Thus, if the ratio of shear strength over shear stress is greater than 1.0, the streambank is stable and erosion does not occur.

The shear strength is different among soils due to soil texture and structure, parent material, organic matter content, moisture content and sizes and volume of the root systems. These soil properties affect soil cohesion and the angle of internal friction. Soils with fine particles and high organic matter contents are more cohesive than those with coarse particles and low contents of organic matter, hence more resistant to erosion. The angle of internal friction measures the ability of a soil (or other materials) to withstand a shear stress. Laboratory tests in central Switzerland showed that soils planted with alder (*Alnus incana*) significantly increased the angle of internal friction by 5°, compared to pure soil samples at the same dry unit weight (Graf et al., 2009). Riparian vegetation is thus widely considered as an effective means for reducing streambank erosion or increasing streambank stability.

The plant roots are strong in tension but weak in compression and soils are strong in compression but weak in tension. A combination of soil and roots produces a composite material that is stronger than each individual (Thorne, 1990). The root enforcement, various within and between species, is affected by the size, density, and distribution of the root systems. Many studies show that the root tensile strength (y), ranging from 2 to 125 MPa, and root diameter (x) exhibit a negative power function (i.e., $y = ax^{-b}$). It implies that grasses and shrubs, which possess large number of small, strong roots, could add to forest cover as an optimum measure for streambank stabilization.

Roots generally grow in the upper 30 cm of the soil profile and decline markedly with depth. Very few roots reach a depth below 100 cm. Thus, in streams where bank height is greater than rooting depth, vegetation cannot be expected to increase the shear strength of the entire bank, but the canopy function is still active in protecting streambank from erosion.

13.2.5.2 Vegetation Impacts

The effects of vegetation on streambank erosion are introduced by the reinforcement of soils due to the root system (mechanical effects) and by the reduction of soil moisture content due to canopy interception and transpiration (hydrologic effects). However, the weight of vegetation above the ground produces a surcharge on the streambank, increasing the driving forces which could reduce the soil stability. Also, during rainfall events, canopy interception and stem flow tend to redirect rainwater locally around the stems of plants, and the presence of stems and roots can cause an increase in infiltration rate and capacity of the soil. All these effects can create higher local pore water pressures inside the streambank, consequently reducing bank stability (Simon et al., 2006). Nevertheless, these negative effects are small as compared to those positive effects as evidenced by many field studies showing beneficial impacts of vegetation on the reduction of streambank erosion and bank retreat. In general, the hydrologic effects are as important as the mechanical effects for woody vegetation. But for grasses, the bank stabilization effect is largely mechanical; the hydrologic effects are detrimental because of increases in soil infiltration and moisture content in soil profile during rainfall events (Simon and Collison, 2002). A mixed stand of woody and grass species is most beneficial.

In southern British Columbia, a total of 748 bends in four stream reaches were selected to assess the impact of riparian vegetation on bank erosion following major floods in 1990. Comparisons of pre- and post-flood photos showed that bends without riparian vegetation were likely to have undergone detectable erosion by nearly five times. Major bank erosion was 30 times more prevalent on non-vegetated bends as on vegetated bends (Beeson and Doyle, 1995).

Differences in streambank erodibility also were observed in floodplains between forests and agriculture. Using aerial photos and a numerical model to study river meander migration for the central reach of the Sacramento River in California, United States, Micheli et al. (2004) found that the bank migration rates and erodibility increased roughly 50% in the recent 50 year period (1949–1997) as compared to those in the preceding 50 year period (1896–1949). This was due to the progressive conversion of riparian floodplains from forests to agriculture during the study periods. Agriculture floodplains were 80%–150% more in erodibility than riparian forest floodplains.

If sediment deposition and bank lateral migration are both lower in forested than in non-forested reaches due to the binding effect of the root systems, then the geometry of stream channels could be consequently altered. Many studies have indicated that forested reaches of small streams are wider than contiguous non-forested, grass-bordered reaches of the same streams at different geographical locations (Davies-Colley, 1997; Hession et al., 2003; Allmendinger et al., 2005). Others suggested that the width of stream channels through grassland is generally greater than those through forest (Gregory and Gurnell, 1988; Rosgen, 1996; Marsh et al., 2004). Such discrepancies seem to reflect differences in specific site conditions such as the type and density of vegetation, soil conditions, flow regime, stream size, slopes, geologic setting, disturbance history, and watershed characteristics (Montgomery, 1997). In fact, the effects of vegetation on channel width are scale dependent. In watersheds greater than $10–100\,km^2$, channel widths are narrower for those with thick woody riparian vegetation than for those with herbaceous plants. The widths for forested and non-forested channels are conversed in smaller streams (Anderson et al., 2004). As noted by Eaton and Giles (2008), the effect of vegetation on channel geometry is strongest for smallest streams; it has virtually no effect on the geometry of larger ones.

13.3 Forest Fires

The impact of forest fires on stream habitats and fish populations can be detrimental as well as beneficial, depending on fire intensity, location, season, weather conditions, topography, and types of fish and vegetation. Man-made prescribed fires are of low intensity, at ground level, and under controlled conditions; they are used as a tool in forest management. Fish habitat disruptions resulting from such fires are insignificant in general. This section only discusses the impact of wildfires.

13.3.1 Occurrences

13.3.1.1 Causes

Fires require three elements to burn: fuel, oxygen, and heat. A forest or chaparral has plenty of fuels on, or above, the ground with unlimited supplies of oxygen from the atmosphere. Thus, fires can be easily ignited when sufficient heat is added to the forest. If the fuel is dry, vegetation is vulnerable to fire, and when there are no natural boundaries to stop fires such as streams, small fires can develop, with the acceleration of prevailing winds, into high-intensity wildfires that burn everything standing in their path.

Heat that triggers forest fires can be of human or natural causes. When the temperature reaches 325°C, oxygen can combine with combustible woods (fuels) to ignite a fire. Lightning is the primary natural cause of forest fires. A bolt of lightning can reach temperatures approaching 28,000°C, and trees are frequent conductors of lightning to the ground. In a typical day, the earth receives about eight million lightning strikes. Even as such, only about one out of every five wildfires is naturally induced.

During the periods at which air temperature is high, forest litter is dry, air humidity is low, and a dry spell prevails, fires can start from cigarette butts, matches, campfires, unextinguished charcoals from barbecue grills, sparks from machinery, or even passing trains. Once lit, the anomalous wildfire behaviors are determined by fuel moisture content, species, topography, wind, and precipitation.

13.3.1.2 Frequencies

Fire is a natural component of the Earth's ecosystems, an essential factor in the maintenance of the diversity and stability of plant and animal communities. The average annual acreage and number of wildfires in the United States between 1960 and 2003 were 1.645×10^6 ha/year and 132,400/year, respectively (NIFC, 2005). Also, average annual burned areas and number of fires for the U.S. national forests in different regions between 1940 and 2000 are given in Table 7.3. Those burned acres in the West were found to be strongly dependent on fuel accumulation governed by climatic conditions 10–18 months before the fire season (Westerling et al., 2003), and large wildfires (>400 ha) greatly increased since the middle 1980s—greater burned areas, higher frequency, longer duration, and longer fire seasons (Westerling et al., 2006).

In Europe, the average annual number of forest fires is close to 50,000/year throughout the Mediterranean basins, almost twice as much as during the 1970s (Goldammer, 2002). This is probably due to rapid socioeconomic growth and development of European communities in recent years.

13.3.2 Impacts

13.3.2.1 Fire Activity

The impact of wildfires on stream habitat is seen largely through effects on watershed processes such as riparian vegetation, water temperature, overland flow, sediment, debris, and water quality. Generally, the impacts are greater in small upstream channels, on fine-textured soils, and in arid regions. These impacts are highly affected by the frequency of occurrences. They can be positive and negative.

In southern Oregon, a wildfire reduced shades in streams from 90% to an average of 30% and, consequently, resulted in increases in maximum water temperatures up to 10°C (Amaranthus et al., 1989). Increased water temperatures favor the introduction and proliferation of "warm water" species but are detrimental to "cold water" species already in existence. Fish can also be indirectly affected by the accompanied increases in fish pathogens and algae and decreases in dissolved oxygen.

During combustion, the formation of water-repellent layer can decrease soil wettability and impedes water infiltration (Larsen, 2009). As a result, overland runoff, sediment production, and nutrient losses are all increased. The impact can be serious and devastating when fires are followed by severe storms. An unattended campfire near Pike National Forest ignited a forest fire that blackened 4820 ha of ponderosa pines along Colorado's Front Range. On July 12, 1996, about 2 months after the fire, a storm of 6.35 cm rainfall in 45 min caused a devastating flood on Buffalo Creek. The runoff carried a tremendous amount of sediment to streams that formed a natural sand dam, converted a rocky streambed into a sandy streambed, and deposited in a water treatment plant an amount equal to the plant's total sediment in 13 years of operation (Illg and Illg, 1997). A review on forest fire studies showed that stream sediment yields increased from prefire 3.36 kg/ha/year to postfire 6,220 kg/ha/year in central Texas, and from 1,400 to 110,000 kg/ha/year in southern California (Landsberg and Tiedmann, 2000). The Cerro Grande wildfires near Los Alamos National Laboratory in 2000 caused post-fire suspended sediment concentrations at upper Pajarito Canyon to be 10,000 times larger than pre-fire in 4 years (Gallaher and Koch, 2004).

Dramatic increases in sediment can elevate streambeds, affect channel morphology, and alter pool distribution. Debris and sediment deposits can decrease pool frequency and volume in small streams (Sedell et al., 1988) but promote pool formation in large stream channels (Ice at al., 2004).

Because of low rainfall, slow re-growth of vegetation, and small volume of streamflow, the impact of wildfire on fish is great in the arid West. Three fires in 2002 and two in 2003 in the southwestern United States have caused the loss of all fish in fire-impacted reaches in one stream and a reduction of 70% in total fish numbers in two streams (Rinnie, 2004). Wildfires are considered the most significant risk factor for the sustainability of the endangered Gila trout (*Oncorhynchus gilae*) (Brown et al., 2001) and a limiting factor to native fish survival (Rinnie, 2002).

On the other hand, fires can also be beneficial to stream habitats over the long run. Burton (2005) notes that fine sediments can be removed from stream channels by fire-triggered hydrologic events and coarse substrates are transported to stream channels by debris flows. These processes may well provide the materials needed to maintain productive habitats for fish and other organisms.

13.3.2.2 Fire Suppression

Of the three types of forest fires, crown fires are the most difficult to control, followed by surface fires and ground fires. Controlling and suppressing forest fires requires substantial manpower, heavy equipment, fire line clearing, and materials. The activity of fire suppression may cause additional damage to the watershed and stream habitat, but this impact has received relatively little attention in the past.

Fighting crown fires often calls for aerial bombing with water and/or fire-retardant chemicals. The water may be different in properties than water in burned watersheds, and retardant chemicals may contain substances, such as ammonium phosphate and borate salts, that are toxic to organisms. They can adversely affect stream habitat and kill fish, especially directly applied into stream channels. Fish kills and chemical concentrations reaching levels damaging to fish have been reported in many regions including Alaska, California, Colorado, and Oregon (Norris and Webb, 1989).

13.3.3 Management

Managing forest fires for stream habitat protection needs to cover all three aspects of fire management: pre-fire prevention, on-fire suppression, and post-fire restoration. In formulating such a management plan, watershed conditions and problems need to be revealed through data collection and analysis, management objectives need to be defined in short-term and long-term perspectives, and fire impact and management effectiveness need to be continuously assessed for further improvements. Thus, fire-risk predictions; fire preparedness; prescribed burning; vegetation management; biological, chemical and mechanical means of fuels management; aerial bombing materials; local topography; and stream habitat need to be considered. For example, prescribed fire and mechanical forest thinning are often used as management tools to avoid the detrimental effects of large, high severity wildfires on stream habitats and to provide resource benefits (Arkle and Pilliod, 2010). Detailed field techniques for managing and control of wildfires along with habitat restoration can be found in reports and guidelines given by the U.S. Fish and Wildlife Service (USFWS, 2011).

References

Allmendinger, N.E. et al., 2005, The influence of riparian vegetation on stream width, eastern Pennsylvania, USA, *Geol. Soc. Am. Bull.*, 117, 229–243.

Amaranthus, M., Jubas, H., and Arthur, D., 1989, Stream shading, summer streamflow and maximum water temperature following intense wildfire in headwater streams, *Proceedings of the Symposium on Fire and Watershed Management*, Berg, N.H., Ed., USFS, Pacific SW Forest and Range Experiment Station, General Technical Report PSW-109, Berkeley, CA, pp. 75–78.

Anderson, R.J., Bledsoe, B.P., and Hession, W.C., 2004, Width of streams and rivers in response to vegetation, bank material, and other factors, *J. Am. Water Resour. Assoc.*, 40, 1159–1172.

Arkle, R.S. and Pilliod, D.S., 2010, Prescribed fires as ecological surrogates for wildfires: A stream and riparian Perspective, *For. Ecol. Manage.*, 259, 893–903.

Bahuguna, D., Mitchell, S.J., and Miquelajauregui, Y., 2010, Windthrow and recruitment of large woody debris in riparian stands, *For. Ecol. Manage.*, 259, 2048–2055.

Bartholow, J.M., 1989, Stream temperature investigations: Field and analytic methods, Biological Report 89(17), U.S. Fish and Wildlife Service, National Ecological Research Center, Fort Collins, CO.

Beasley, R.S., 1979, Intensive site preparation and sediment losses on steep watersheds in the Gulf Coastal Plain, *Soil Sci. Soc. Am. J.*, 43, 412–417.

Beechie, T.J. et al., 2000, Modeling recovery rates and pathways for woody debris recruitment in northwestern Washington streams, *N. Am. J. Fish. Manage.*, 20, 436–452.

Beeson, C.E. and Doyle, P.F., 1995, Comparison of bank erosion at vegetated and nonvegetated channel bends, *Water Resour. Bull.*, 31, 983–990.

Beschta, R.L. et al., 1987, Stream temperature and aquatic habitat: Fishery and forestry interactions, *Streamside Management: Forestry and Fishery Interactions*, Salo, E.O. and Cundy, T.W., Eds., University of Washington, Institute of Forest Resources Contribution, Seattle, WA, pp. 191–232.

Bilby, R.E., 1984, Removal of woody debris may affect channel stability, *J. For.*, 82, 609–613.

Binkley, D. and Brown, T.C., 1993, Forest practices as nonpoint sources of pollution in North America, *Water Resour. Bull.*, 29, 729–740.

Biotta, G.S. and Brazier, R.E., 2008, Understanding the influence of suspended solids on water quality and aquatic biota, *Water Res.*, 42, 2849–2861.

Bisson, P.A. et al., 1987, Large woody debris in forested streams in the Pacific Northwest: Past, present, and future, *Streamside Management: Forestry and Fishery Interactions*, Salo, E.O. and Cundy, T.W., Eds., University of Washington, Institute of Forest Resources Contribution No. 57, Seattle, WA, pp. 143–190.

Bjornn, T.C. and Reiser, D.W., 1991, Habitat requirements of salmonids in streams, *Influences of Forest and Rangeland Management on Salmonid Fishes and Their Habitats*, Meehan, W.R., Ed., American Fisheries Society, Bethesda, MD, pp. 83–138.

Blackburn, W.H., Wood, J.C., and DeHaven, M.G., 1986, Storm flow and sediment losses from site-prepared forestland in east Texas, *Water Resour. Res.*, 22, 776–784.

Blinn, C.R., and Kilgore, M.A., 2001, Riparian management practices, a summary of state guidelines, *J. For.*, 99(8), 11–17.

Bratovich, P. et al., 2004, Final report—Evaluation of spawning and incubation substrate suitability for salmonids in the Lower Feather River, Oroville Facilities P-2100 Relicensing, SP-F10, TASK 2A, Department of Water Resources, Sacramento, CA.

Brown, G.W., 1980, *Forest and Water Quality*, O.S.U. Book Stores, Corvallis, OR.

Brown, G.W. and Krygier, T., 1971, Clearcut logging and sediment production in the Oregon Coast Range, *Water Resour. Res.*, 7, 1189–1198.

Brown, G.W., Swank, G.W., and Rothacher, J., 1971, Water temperature in the streamboat drainage, Research Paper PNW-119, USDA Forest Service, Pacific NW Forest Experiment Station, Portland, OR.

Brown, S.K. et al., 2000, Habitat suitability index model for fish and invertebrate species in Casco and Sheepscot Bays, Maine, *N. Am. J. Fish. Manage.*, 20, 408–435.

Brown, D.K. et al., 2001, Catastrophic wildfire and number of populations as factors influencing risk of extinction for Gila trout (*Oncorhynchus gilae*), *West N. Am. Naturalist*, 6(2), 139–148.

Bryant, M.D., 1980, Evolution of large, organic debris after timber harvesting: Maybeso Creek, 1949 to 1978, General Technical Report PNW-101, USDA Forest Service, Portland, OR.

Burton, T.A., 2005, Fish and stream habitat risks from uncharacteristic wildfire: Observations from 17 years of fire-related disturbances on the Boise National Forest, Idaho, *For. Ecol. Manage.*, 211, 140–149.

Carlson, J.Y., Andrus, C.W., and Froehlich, H.W., 1990, Woody debris, channel features, and macroinvertebrates of streams with logged and undisturbed riparian timber in northeastern Oregon, *Can. J. Fish. Aquat. Sci.*, 47, 1103–1111.

Carlson, J.R. et al., 1992, Design criteria for revegetation in riparian zones of the Intermountain area, *Proceedings of the Symposium on Ecology and Management of Riparian Shrub Communities*, USFS Intermountain Research Station, General Technical Report INT-289, Ogden, UT, pp. 145–150.

Chamberlin, T.W., Harr, R.D., and Everest, F.H., 1991, Timber harvesting, silviculture, and watershed processes, *Influences of Forest and Rangeland Management on Salmonid Fishes and Their Habitats*, Meehan, W.R., Ed., American Fisheries Society, Bethesda, MD, pp. 181–205.

Clément, M. et al., 2008, Characterization of the large woody debris in Catamaran Brook, New Brunswick, 1990 to 1997, *Can. Tech. Rep. Fish. Aquat. Sci.*, 2794, 43.

Cover, M.R. et al., 2008, Quantitative linkage among sediment supply, streambed fine sediment, and benthic macroinvertebrates in northern California streams, *J. N. Am. Benthol. Soc.*, 27, 135–149.

Crance, J.H., 1987, Habitat suitability index curves for paddlefish, developed by the Delphi technique, *N. Am. J. Fish. Manage.*, 7, 123–130.

Cunjak, R.A. and Power, G., 1987, Cover use by stream-resident trout in winter: A field experiment, *N. Am. J. Fish. Manage.*, 7, 539–544.

Currier, J.B. and Hughes, D., 1980, Temperature, *An Approach to Water Resources Evaluation of Non-Point Silvicultural Sources (A Procedural Handbook)*, U.S. Environmental Protection Agency, Environmental Research Laboratory, Athens, GA, pp. VII i– VII v, 1–30.

Davies-Colley, R. J., 1997, Stream channels are narrower in pasture than in forest, *NZ J. Marine Freshwater Res.*, 31, 599–608.

DeBano, L.F. and Schmidt, L.J., 1989, Improving southwestern riparian areas through watershed management, *Proceedings of the Symposium on Fire and Watershed Management*, Berg, N.H., Ed., USFS, Pacific SW Forest Range and Experiment Station, General Technical Report PSW-109, Berkeley, CA, pp. 55–62, USFS Rocky Mountain Forest Range and Experiment Station, General Technical Report RM-182, Fort Collins, CO.

Degerman, E. et al., 2004, Large woody debris and brown trout in small forest streams—Towards targets for assessment and management of riparian landscapes, *Ecol. Bull.*, 51, 233–239.

Dosskey, M.G., Helmers, M.J., and Eisenhauer, D.E., 2008, A design aid for determining width of filter strips, *J. Soil Water Conserv.*, 63(4), 232–241.

Eaton, B.C. and Giles, T.R., 2008, Assessing the effect of vegetation-related bank strength on channel morphology and stability in gravel-bed streams using numerical models, *Earth Surf. Process. Landforms*, 34(5), 712–724.

Fajen, O.F. and Wehnes, R.E., 1982, Missouri's method of evaluating stream habitat, *Acquisition and Utilization of Aquatic Habitat Inventory Information*, Armantrout, B., Ed., American Fisheries Society, Western Division, Bethesda, MD, pp. 117–123.

Foltz, R.B., Yanoseka, K.A., and Brown, T.M., 2008, Sediment concentration and turbidity changes during culvert removals, *J. Environ. Manage.*, 87, 329–340.

Fowler, W.B., Anderson, T.D., and Helvey, J.D., 1988, Changes in water quality and climate after forest harvest in Central Washington State, Research Paper PNW-388, U.S. Forest Service, Pacific NW Forest and Range Experiment Station, Portland, OR.

Fredriksen, R.L., 1970, Erosion and sedimentation following road construction and timber harvesting on unstable soils in three small Western Oregon watersheds, Research Paper PNW104, USDA Pacific NW Forest and Range Experiment Station, Portland, OR.

FWS, 1981, Standards for the development of habitat suitability index models for use with the habitat evaluation procedures, Report 103, ESM Release 1–81, U.S. Fish and Wildlife Service, Division of Ecological Services, Washington, DC.

Gallaher, B.M. and Koch, R.J., 2004, Cerro Grande fire impacts to water quality and streamflow near Los Alamos National Laboratory: Results of four years of monitoring, LA-14177, Los Alamos National Laboratory, Los Alamos, NM, 195 pp.

Goldammer, J.G., 2002, Forest fire problems in south East Europe and adjoining regions: Challenges and solutions in the 21st Century, *The International Science Conference, Fire and Emergency Safety*, October 31–November 1, 2002, Sofia, Bulgaria.

Gordon, N.D., McMahon, T.A., and Finlayson, B.L., 1993, *Stream Hydrology, an Introduction for Ecologists*, John Wiley & Sons, New York.

Graf, F., Frei, M., and Böll, A., 2009, Effects of vegetation on the angle of internal friction of a moraine, *For. Snow Landsc. Res.*, 82(1), 61–77.

Gregory, K.J. and Gurnell, A.M., 1988, Vegetation and river channel form and process, *Biogeomorphology*, Viles, H.A., Ed., Blackwell, Oxford, U.K., pp. 11–42.

Gregory, S.V. et al., 1987, Influences of forest practices on aquatic production, *Streamside Management: Forestry and Fishery Interactions*, Salo, E.O. and Cundy, T.W., Eds., University of Washington Institute of Forest Resources Contributions, Seattle, WA, pp. 233–255.

Hall, J.D. and Baker, C.O., 1982, Influence of forest and rangeland management on anadromous fish habitat in Western North America: 12. Rehabilitating and enhancing stream habitat, 1. Review and evaluation, General Technical Report PNW-138, USDA Forest Service, Pacific NW Forest and Range Experiment Station, Portland, OR.

Harr, R.D. and Fredriksen, R.L., 1988, Water quality after logging small watersheds within the Bull Run Watershed, Oregon, *Water Resour. Bull.*, 24, 1103–1111.

Henley, W.F. et al., 2000, Effects of sedimentation and turbidity on lotic wood webs: A concise review for natural resource managers, *Rev. Fish. Sci.*, 8, 125–139.

Hession, W.C. et al., 2003, Influence of bank vegetation on channel morphology in rural and urban watersheds, *Geol. Soc. Am.*, 31(2), 147–150.

Hicks, B.J., Beschta, R.L., and Harr, R.D., 1991a, Long-term changes in streamflow following logging in western Oregon and associated fisheries implications, *Water Resour. Bull.*, 27, 217–226.

Hicks, B.J. et al., 1991b, Responses of salmonids to habitat changes, *Influences of Forest and Rangeland Management on Salmonid Fishes and Their Habitats*, Meehan, W.R., Ed., American Fisheries Society, Bethesda, MD, pp. 483–518.

Ice, G.G. et al., 2004, Effects of wildfire on soils and watershed processes, *J. For.*, 102(6), 16–20.

Illg, C. and Illg, G., 1997, Forest fire, water quality, and the incident at Buffalo Creek, *Am. For.*, 103(1), Winter/Spring, 33–35.

Karr, J.R. and Dudley, D.R., 1981, Ecological perspective on water quality goals, *Environ. Manage.*, 5, 55–68.

Karwan, D.L., Gravelle, J.A., and Hubbert, J.A., 2007, Effects of timber harvesting on suspended sediment loads in Mica Creek, Idaho, *For. Sci.*, 53(2), 181–188.

Kondolf, G.M., 2000, Assessing salmonid spawning gravel quality, *Trans. Am. Fish. Soc.*, 129, 263–281.

Lakel, W.A. III et al., 2010, Sediment trapping by streamside management zones of various widths after forest harvest and site Preparation, *For Sci.*, 56(6), 541–551.

Landsberg, J.D. and Tiedemann, A.R., 2000, Chapter 12: Fire management, *Drinking Water from Forests and Grasslands, a Synthesis of the Scientific Literature*, Dissmeyer, G.E., Ed., USFS, Southern Research Station, General Technical Report SRS-39, Blacksburg, VA, pp. 124–138.

Larsen, I.J., 2009, Causes of post-fire runoff and erosion: Water repellency, cover, or soil sealing? *Soil Sci. Soc. Am. J.*, 73(4), 1393–1407.

Layher, W.G. and Maughan, O.E., 1987, Spotted bass habitat suitability related to fish occurrence and biomass and measurement of physiochemical variables, *N. Am. J. Fish. Manage.*, 7, 238–251.

Lemly, A.D. and Hilderbrand, R.H., 2000, Influence of large woody debris on stream insect communities and benthic detritus, *Hydrobiologia*, 421, 179–185.

Lloyd, D.S., 1987, Turbidity as a water quality standard for salmonid habitats in Alaska, *N. Am. J. Fish. Manage.*, 7, 34–45.

Lynch, J.A. et al., 1975, Penn State experimental watersheds, *Municipal Watershed Management Symposium Proceedings*, USDA Forest Service General Technical Report NE-13, Portland, OR, pp. 32–46.

MacDonald, L.H., Smart, A.W., and Wissmar, R.C., 1991, Monitoring guidelines to evaluate effects of forestry activities on streams in the Pacific Northwest and Alaska, EPA 910/9-91-001, U.S. Environmental Protection Agency, Washington, DC.

Magilligan, F.J. et al., 2008, The geomorphic function and characteristics of large woody debris in low gradient rivers, coastal Maine, USA, *Geomorphology*, 97, 467–482.

Marsh, N., Rutherfurd, I., and Bunn, S., 2004, How does riparian revegetation affect suspended sediment in a Southeast Queensland Stream? Technical Report 04/13, Cooperative Research Centre for Catchment Hydrology, Canberra, Australia, 33 pp.

Matthews, K.R., 1996, Habitat selection and movement patterns of California golden trout in degraded and recovering streams selection in the Golden Trout Wilderness, California, *N. Am. J. Fish. Manage.*, 16, 579–590.

May, C.L. and Gresswell, R.E., 2003, Large wood recruitment and redistribution in headwater streams in the southern Oregon Coast Range, U.S.A., *Can. J. For. Res.*, 33, 1352–1362.

McCleary, R., Sherburne, C., and Bambrick, C., 2004, Long-term effects of riparian harvest on fish habitat in three Rocky Mountain foothills watersheds, *Forest Land-Fish Conference II, Ecosystem Stewardship through Collaboration, Proceedings*, Scrimgeour, G.J. et al., Eds., April 26–28, 2004, Canadian Forest Service, Edmonton, Alberta, Canada, pp. 189–198.

McClendon, D.D. and Rabeni, C.F., 1987, Physical and biological variables useful for predicting population characteristics of smallmouth bass and rock bass in an Ozark stream, *N. Am. J. Fish. Manage.*, 7, 46–56.

Micheli, E.R., Kirchner, J.W., and Larsen, E.W., 2004, Quantifying the effect of riparian forest versus agricultural vegetation on river meander migration rates, central Sacramento River, California, USA, *River Res. Appl.*, 20, 537–548.

Miranda, L.E. and Hunt, K.M., 2011, An index of reservoir habitat impairment, *Environ. Monit. Assess.*, 172(1–4), 225–234.

Montgomery, D.R., 1997, What's best on the banks? *Nature*, 388, 328–329.

Morisawa, M., 1985, *Rivers*, Longman, New York.

Moss, B., 1988, *Ecology of Fresh Waters, Man and Medium,* 2nd edn., Blackwell Scientific, Oxford, U.K.

Muñoz-Carpena, R. and Parsons, J.E., 2004, A design procedure for vegetative filter strips using VFSMOD-W, *Trans. ASAE*, 47(6), 1933–1941.

Murphy, M.L. and Hall, J.D., 1981, Varied effects of clear-cut logging on predators and their habitat in small streams of the Cascade Mountains, Oregon, *Can. J. Fish. Aquat. Sci.*, 38(2), 137–145.

Narver, D.W., 1971, Effects of logging debris on fish production, *Forest Land Uses and Stream Environment*, Oregon State University, Corvallis, OR, pp. 100–111.

Newbold, J.D., Erman, D.C., and Roby, K.B., 1980, Effects of logging on macro-invertebrates in streams with and without buffer strips, *Can. J. Fish Aquat. Sci.*, 37, 1076–1085.

Newcombe, C.P. and MacDonald, D.D., 1991, Effects of suspended sediments on aquatic ecosystems, *N. Am. J. Fish. Manage.*, 11, 72–82.

NIFC (National Interagency Fire Center), 2005, Wildland fire statistics: The total fires and acres 1960–2004, NIFC, Boise, ID, accessed at: http://www.nifc.gov/stats/wildlandfirestats.html

Norris, L.A. and Webb, W.L., 1989, Effects of fire retardant on water quality, *Proceedings of the Symposium on Fire and Watershed Management*, Berg, N.H., Ed., USFS, Pacific SW Forest Range and Experiment Station, General Technical Report PSW-109, Berkeley, CA, pp. 79–86.

Opperman, J.J. and Merenlender, A.M., 2004, The effectiveness of riparian restoration for improving instream fish habitat in four hardwood-dominated California streams, *N. Am. J. Fish. Manage.*, 24, 822–834.

Pirhalla, D.E., 2004, Evaluating fish–habitat relationships for refining regional indexes of biotic integrity: Development of a tolerance index of habitat degradation for Maryland stream fishes, *Trans. Am. Fish. Soc.*, 133, 144–159.

Plafkin, J.L. et al., 1989, Rapid bioassessment protocols for use in streams and rivers: Benthic macroinvertebrates and fish, EPA/444/4-89-001, U.S. Environmental Protection Agency, Washington, DC.

Rainwater, F.L., 2001, personal communication.

Raleigh, R.F., Zuckerman, L.D., and Nelson, P.C., 1984, Habitat suitability index models and instream flow suitability curves: Brown trout, U.S. Fish and Wildlife Service Biological Services Program FWS/OBS-82/10.71.

Ralph, S.C. et al., 1994, Stream channel condition and instream habitat in logged and unlogged basins of western Washington, *Can. J. Fish. Aquat. Sci.*, 51, 37–51.

Rankin, E.T., 1989, The qualitative habitat evaluation index (QHEI), rationale, methods, and application, Ohio Environmental Protection Agency, Division of Water Quality Planning and Assessment, Ecological Assessment Section, Columbus, OH.

Reeves, G.H. and Roelofs, T.D., 1982, Influence of forest and rangeland management on anadromous fish habitat in Western North America: 13. Rehabilitating and enhancing stream habitat, 2. Field application, General Technical Report PNW-140, USDA Forest Service, Pacific NW Forest and Range Experiment Station, Portland, OR.

Reinhart, K.G., Eschner, A.R., and Trimble, G.R., Jr., 1963, Effect on streamflow of four forest practices in the Mountains of West Virginia, Research Paper NE-1, U.S. Forest Service, Northeastern Forest Experiment Station, Upper Darby, PA.

Reynolds, J.B., Simmons, R.C., and Burkholder, A.R., 1989, Effects of placer mining discharge on health and food of Arctic grayling, *Water Resour. Bull.*, 25, 625–635.

Rinnie, J.N., 2002, Hydrology, geomorphology, and management: Implications for sustainability of native southwestern fishes, *Hydrol. Water Resour. SW*, 32, 45–50.

Rinnie, J.N., 2004, Forests, fish and fire: Relationships and management for fishes in the southwestern U.S.A., *Forest Land–Fish Conference II, Ecosystem Stewardship through Collaboration, Proceedings*, Scrimgeour, G.J. et al., Eds., April 26–28, 2004, Canadian Forest Service, Edmonton, Alberta, Canada, pp. 26–28.

Rosgen, D., 1996, *Applied River Morphology*, Wildland Hydrology, Pagosa Springs, CO.

Rosser, B. and O'Connor, M., 2007, Statistical analysis of streambed sediment grain size distributions: Implications for environmental management and regulatory policy, USDA Forest Service General Technical Report PSW-GTR-194, Portland, OR.

Rutherford, D.A., 2001, personal communication.

Sedell, J.R. et al., 1988, What we know about large trees that fall into streams and rivers, *From the Forest to the Sea: A Story of Fallen Trees*, USDA Forest Service General Technical Report PNW-GTR-229, Portland, OR, pp. 47–80.

Seehorn, M.E., 1992, *Stream Habitat Improvement Handbook*, U.S. Forest Service Technical Publication R8-TP 16, Portland, OR.

Seyedbagheri, K.A., 1996, Idaho forestry best management practices: Compilation of research on their effectiveness, USDA Forest Service General Technical Report INT-GTR-339, Portland, OR.

Scherer, R., 2004, Decomposition and longevity of instream woody debris: A review of literature from North America, *Forest Land–Fish Conference II, Ecosystem Stewardship through Collaboration, Proceedings*, Scrimgeour, G.J. et al., Eds., April 26–28, 2004, Canadian Forest Service, Edmonton, Alberta, Canada, pp. 127–133.

Shrimpton, J.M., Zydlewski, J.D., and Heath, J.W., 2007, Effect of daily oscillation in temperature and increased suspended sediment on growth and smolting in juvenile chinook salmon, *Oncorhynchus tshawytscha, Aquaculture*, 273, 269–276.

Simon, A. and Collison, A.J.C., 2002, Quantifying the mechanical and hydrologic effects of riparian vegetation on stream stability, *Earth Surf. Process. Landforms*, 27, 527–546.

Simon, A., Pollen, N., and Langendoen, E., 2006, Influence of two woody riparian species on critical conditions for streambank stability: Upper Truckee River, California, *J. Am. Water Resour. Assoc.*, 42, 99–113.

Sowa, S.P. and Rabeni, C.F., 1995, Regional evaluation of the relation of habitat to distribution and abundance of smallmouth bass and largemouth bass in Missouri streams, *Trans. Am. Fish. Soc.*, 124, 240–251.

Stednick, J.D., 2000, Timber management, *Drinking Water from Forests and Grasslands, A Synthesis of the Scientific Literature*, Dissmeyer, G.E., Ed., U.S. Forest Service, Southern Research Station, General Technical Report SRS-39, Portland, OR, pp. 103–119.

Stednick, J.D., Tripp, L.N., and McDonald, R.J., 1982, Slash burning effects on soil and water chemistry in southeastern Alaska, *J. Soil Water Conserv.*, 37, 126–128.

Steinblums, I.J., Froehlich, H.A., and Lyons, J.K., 1984, Designing stable buffer strips for stream protection, *J. For.*, 82, 49–52.

Swank, W.T., Vose, J.M., and Elliott, K.J., 2001, Long-term hydrologic and water quality responses following commercial clearcutting of mixed hardwoods on a southern Appalachian catchment, *For. Ecol. Manage.*, 143, 163–178.

Swanson, F.J., Lienkaemper, G.W., and Sedell, J.R., 1976, History, physical effects, and management implications of large organic debris in Western Oregon Streams, General Technical Report PNW-56, USDA Forest Service, Pacific NW Forest and Range Experiment Station, Portland, OR.

Swift, L.W., Jr., 1986, Filter strip widths for forest roads in the southern Appalachians, *South. J. Appl. For.*, 10, 27–34.

Thorne C.R., 1990, Effects of vegetation on riverbank erosion and stability, *Vegetation and Erosion: Processes and Environments*, Thornes, J.B., Ed., John Wiley & Sons, Chichester, U.K., pp. 125–144.

Tillma, J.S., Guy, C.S., and Mammoliti, C.S., 1998, Relations among habitat and population characteristics of spotted bass in Kansas streams, *N. Am. J. Fish. Manage.*, 18, 886–893.

Ursic, S.J., 1991, Hydrologic effects of two methods of harvesting mature southern pine, *Water Resour. Bull.*, 27, 303–315.

USDA, 1993, *Soil Survey Manual*, USDA-NRCS Agricultural Handbook #18, U.S. Government Printing Office, Washington, DC.

USEPA, 2000, The quality of our nation, EPA84 1-S-00-001, U.S. Environmental Protection Agency, Washington, DC.

USFWS, 2011, *Fire Management Handbook*, USDI, Fish and Wildlife Service, National Interagency Fire Center, Boise, ID, http://www.fws.gov/fire/handbook/index.shtml

Wang, L., Lyons, J., and Kanehl, P., 1998, Development and evaluation of a habitat rating system for low-gradient Wisconsin streams, *N. Am. J. Fish. Manage.*, 18, 775–785.

Waters, T.F., 1995, *Sediment in Streams: Sources, Biological Effects and Control*, American Fisheries Society, Bethesda, MD.

Wesche, T.A., Goertler, C.M., and Frye, C.B., 1987, Contribution of riparian vegetation to trout cover in small streams, *N. Am. J. Fish. Manage.*, 7, 151–153.

Westerling, A.L. et al., 2003, Climate and wildfire in the western United States, *Bull. Am. Meteorol. Soc.*, May, 595–604.

Westerling, A.L. et al., 2006, Warming and earlier spring increase western U.S. forest wildfire activity, *Science*, 313, 940–943.

14

Forests and Flooding

The forest–flood relationship is a classical issue in forest hydrology. Due to questions about flood frequency, severity, damage, and management, concerns about the relationship have increased greatly in recent years (Berz, 2000; Shankman and Liang, 2003; Bruijnzeel, 2004; FAO, 2005; Calder and Aylward, 2006; Eisenbies et al., 2007; Bathurst et al., 2010). These concerns have developed, for example, into legal actions of landowners against lumber companies (Mortimer and Visser, 2004) and the decision-making of government on forest harvest plans of lumber companies (The University of California, 1999). This chapter discusses floods and the impact of forest on flooding; the impact of flooding on tree survival and growth, forest regeneration, reproduction, and composition is not included in this text. (See Broadfoot and Williston, 1973; Teskey and Hinckley, 1977; Chang and Crowley, 1997; Tozlowski, 1997; Gergel et al., 2002; Burke et al., 2003; Marques et al., 2009 for more discussions on these issues.)

14.1 Folklore and Fallacies

The positive role of forests on flood mitigation was a popular perception in the nineteenth and early twentieth centuries. It was claimed that deforestation caused floods during wet seasons and made streams drier in summer and that the existence of forests prevented floods because root systems held soils in place and stored moisture in significant quantity in the soil. As a result, streamflows benefited in dry seasons from the slow release of stored water and from the increase of precipitation in forested areas. Thus, the disastrous flooding of the Ohio River in 1907 was attributed to logging activities in the Allegheny River and Monongahela River basins (Sartz, 1983).

This idea was derived from a combination of folklore, experience, speculation, inference, and conservatism. In France, the potential impact of forests on streamflow and floods was already claimed by naturists before the French revolution of 1789. As soon as the wars and political unrest were over, vigorous debates started on the forest–flood issue between foresters and engineers (Andréassian, 2004). In the United States, the book *Man and Nature*, written by George Perkins Marsh in 1864, discussed the consequences of deforestation on water, climate, ecology, and wildlife. It imposed a great impact on the conservation movement in U.S. history. However, the conservative role of forest impact on flooding has drawn many criticisms from hydrologists in the light of scientific studies and principles. This caused heated debates between Pinchot (1908, 1913), then chief of the U.S. Forest Service, and Chittenden (1909), of the U.S. Corps of Engineers, and Moore (1910), at the time the chief of the U.S. Weather Bureau (see Chapter 3, Section 3.1, Water in History—the United States). As a result of this conservation movement, the Organic Act (Forest Management Act) was passed in 1897 and the Weeks Act in 1911.

The heated disputes revealed a growing need to conduct field studies on the relationships between forests and streamflow. Consequently, the first experimental watershed study in the United States was established at Wagon Wheel Gap, headwaters of the Rio Grande River in Colorado, in 1911. The number of forested experimental watersheds in the conterminous United States grew to 150 by the 1960s (Sopper, 1970) and reached more than 400 at 51 different sites by the end of the twentieth century (see Chapter 16). These experimental watersheds were established to study the forest–water relationship under various vegetation, soil, climate, and management conditions. Similar studies have also been conducted in various other parts of the world. Results (Lull and Reinhart, 1972; Anderson et al., 1976; Hewlett and Bosch, 1984; Bruijnzeel, 1990; Calder and Aylward, 2006; Bathurst et al., 2011) show the following:

- The forest provides watersheds maximum opportunity for controlling runoff-producing storms.
- With reasonable care, forest cutting offers no flood threat to downstream areas and little detriment to site-protective capability.
- Forestry management activities such as road construction and site preparation that causes soil compaction are more likely to affect flood generation than is the removal of forests.
- Forests can reduce peak flows for storms of short duration and lower intensity, but cannot prevent the occurrence of floods that are produced from storms of high intensity and long duration over a large area.

The results are in general familiar only to hydrology communities, and the conservation perceptions are still widespread in the general public today (FAO, 2005). This issue often resurfaces whenever major floods occur, such as the floods in China's Yangtze River (Brown, 1998) and in the Himalayas (Myers, 1986). Marsh's (1864) traditional concepts about forests and water are still popular in modern society; the Harvard University Press reprinted his text in 1965, and the University of Washington printed it yet again in 2003. The flooding of large rivers in India, the Philippines, and other tropical areas (Hamilton and King, 1983) has been seen as the result of forest clearing in uplands. In China, the 1998 summer flood of the Yangtze River compelled the Chinese government to launch a new forest policy called the Natural Forest Protection Program (FNPP) to protect watersheds and ecosystems from flood damage (Yin et al., 2005). The FNPP classifies the upper reaches of the Yangtze and Yellow Rivers as ecosystem conservation areas in which logging activities are banned, restricted, or reduced. FNPP plans call for the ecosystems of clear sites to be rehabilitated and reforested.

In the United States, concerns about forest–flood relations have affected the decision-making of government and led to legal actions by private citizens in recent years. The magnitude and frequency of flooding in the Elk River and Freshwater Creek watersheds in Humboldt, California had increased during the 1990s. The California Department of Forestry then sent a letter to the Pacific Lumber Company and required the company to address potentially elevated peak flows and sedimentation rates in the two watersheds. They warned that the new timber harvest plans of the Pacific Lumber Company would not be approved until causality regarding current timber operations on flooding of homes was assessed (The University of California, 1999). In southern West Virginia, the July flood of 2001 drove thousands of landowners to file lawsuits in several counties against an array of timber, coal, and mining companies. They held those companies responsible for property and personal damages sustained after the flooding (Mortimer and Visser, 2004).

14.2 Flood Occurrences

14.2.1 Causes

Virtually all floods evolve from (1) an unusually large quantity of water delivered to a river channel, (2) slow movement of water in the channel that impedes the inflow of water from tributaries, thus exceeding its bank capacity, or (3) a combination of the two. Excessive quantities of water in watersheds are directly or indirectly related to storm systems and weather conditions, while slow streamflow drainage is largely caused by channel morphology and watershed conditions. In many cases, human activities can accelerate the movement of water to stream channels or slow drainage flow in stream channels. Flooding can be of hydrologic, as well as human, origins. The world's maximum observed floods are listed in Table 14.1.

14.2.1.1 Hydrologic Origins

14.2.1.1.1 Excessive Rainfall

Storms that deliver excessive amounts of rainfall to watersheds, resulting in the inundation of floodplains, can be from convective, tropical cyclonic, frontal, or monsoon-type activities. These storm types can be small- or large-scale anomalous atmospheric circulations, lower or upper atmosphere phenomena, and of high intensity and short duration or of moderate intensity and long duration. However, anomalous storm activities that result in excessive rainfall are not often caused by a single atmospheric condition; they are often a combination of a variety of meteorological elements on various scales. Severe material losses and damage in floods generated by such storm activities are, in many cases, impossible to avoid (Kundzewicz, 2002). The world records of the most intense rainfalls of various durations are given in Table 14.2 and Figure 14.1.

14.2.1.1.1.1 Thunderstorms   Thunderstorms are generated by the violent vertical uplifting of an air mass due to strong surface heating, orographic lift by mountains, the action of cold fronts, or temperature differences between the land and ocean. These severe storms occur mostly in spring and summer, can be local or cover wide areas, and are intense, with durations lasting only a few hours or less than an hour.

The June 9, 1972, flood at Rapid City, South Dakota was one of the most severe floods caused by thunderstorm activity in the twentieth century. Prior to the heavy rains of this storm, scattered showers had already occurred for several days throughout the Black Hills–Rapid City area. These showers saturated the soils and created potential conditions in the watersheds for flood generation.

On June 9, a group of nearly stationary thunderstorms formed over the eastern Black Hills along the Rapid Creek. These thunderstorms were formed from a strong, moist, low-level easterly airflow that was lifted orographically up by the elevations of the Black Hills. As the air rapidly elevated, it cooled, becoming very unstable and then releasing its moisture. The orographic process was further enhanced by light winds at higher atmospheric levels, enabling these storms to remain nearly stationary and continue to release heavy and abundant rains.

Precipitation totals for June 9 and 10 ranged from 102 to more than 305 mm in an area of about 155 km^2 around Rapid Creek. The heaviest rainfall was 381 mm recorded during a 6 h period at Nemo (30 km N of Rapid City), about four times the amount to be expected

TABLE 14.1

The World's Maximum Observed Floods

Site/River	State or Country	Basin Area (km²)	Discharge (m³/s)	K Value[a]	Year
San Rafael	California, United States	3.2	250	5.194	1973
L. San Gorgonio	California, United States	4.5	311	5.226	1969
Halawa R. ·	Hawaii, United States	12	762	5.494	1965
Wailua R.	Hawaii, United States	58	2,470	5.819	1963
Buey, San Miguel	Cuba	73	2,060	5.623	1963
Baisha, Guangdong	China	75.3	3,420	5.973	1894
San Bartolo	Mexico	81	3,000	5.859	1976
Suizhong, Lianing	China	171	7,000	6.263	1894
Cho Shui	Taiwan (China)	259	7,780	6.225	1979
Quaïème	New Caledonia, France[b]	330	10,400	6.389	1981
Yaté	New Caledonia, France[b]	435	5,700	5.810	1981
L. Nemaha, Syracuse	Nebraska, United States	549	6,370	5.826	1950
Shucheng, Anhui	China	1,110	12,100	6.130	1853
Mid. Fork. American	California, United States	1,360	8,780	5.770	1964
Cithuatlan	Mexico	1,370	13,500	6.156	1959
Pioneer, Pleystowe	Australia	1,490	9,840	5.840	1918
Hualien, Hualien Brdg.	Taiwan (China)	1,500	11,900	6.011	1973
Nyodo Ino	Japan	1,560	13,510	6.111	1963
Kiso, Imujama	Japan	1,680	11,150	5.910	1961
Nueces R., Bracketville	Texas, United States	1,800	15,600	6.156	1959
Taizhong	Taiwan (China)	1,980	18,300	6.306	1959
Tam Shui R. at Taipei	Taiwan (China)	2,110	16,700	6.199	1963
Shingu Oga	Japan	2,350	19,025	6.290	1959
Pedernales, Johnson	Texas, United States	2,450	12,500	5.873	1952
Qingtian, Zhejiang	China	3,225	19,200	6.174	1912
Yoshino at Iwazu	Japan	3,750	14,470	5.844	1974
Cagayan R., Isabella	Philippines	4,244	17,550	5.980	1959
Changjing/Guangdong	China	4,634	28,300	6.428	1887
Madhopur, Ravi	India	6,087	26,052	6.242	1988
Eel River at Scotia	California, United States	8,060	21,300	5.917	1964
Pecos R., Cornstock	Texas, United States	9,100	26,800	6.110	1954
Linyi, Shandong	China	10,315	30,000	6.180	1730
Toedong Gang	N. Korea	12,175	29,000	6.060	1967
Jhalawar, Chambal	India	22,584	37,000	6.073	1969
Jhelum at Mangla	Pakistan	29,000	31,100	5.739	1929
Hanjiang at Hankang	China	41,400	40,000	5.868	1583
Mangoky at Banyon	Madagascar	50,000	38,000	5.698	1933
Narmada Garudeshwar	India	88,000	69,400	6.210	1970
Amazonas R. at Obidos	Brazil	4,640,000	370,000	6.760	1953
Lena R. at Kusur	Russia	2,430,000	189,000	5.520	1967

Sources: Compiled from Rodier, J.A. and Roche, M.J., *World Catalogue of Maximum Observed Floods*, IAHS Publication No. 143, Institute of Hydrology, Oxfordshire, U.K., 1984; Costa, J.E., *J. Hydrol.*, 96, 101, 1987; Rakhecha, P.R., Highest floods in India, in *The Extremes of the Extremes: Extraordinary Floods*, Snorasson, A., Finnsdóttir, H.P., and Moss, M., Eds., IAHS Publication No. 271, Oxfordshire, U.K., pp. 167–172, 2002.

[a] See Equation 14.6.
[b] In the South Pacific Ocean, east of Australia.

TABLE 14.2

World Records of Extreme Rainfall for Various Durations

Duration	Total Rainfall (mm)	Total Rainfall (in.)	Location	Date (Year/Month/Day)
1 min	38.10	1.50	Barot, Guadeloupe	1970-11-26
5 min	61.72	2.43	Port Bells, Panama	1911-11-29
8 min	125.98	4.96	Fussen, Bavaria	1920-05-25
15 min	198.12	7.80	Plumb Point, Jamaica	1916-05-12
20 min	205.74	8.10	Curtea-de-Arges, Romania	1947-07-07
40 min	234.95	9.25	Guinea, Virginia, United States	1906-08-24
42 min	304.80	12.00	Holt, Missouri, United States	1947-06-22
60 min	401.00	15.79	Shangdi, Nei Monggol, China	1975-07-03
2 h 10 min	482.60	19.00	Rockport, West Virginia, United States	1889-07-18
2 h 45 min	558.80	22.00	D'Hanis, Texas, United States	1935-05-31
4 h	584.20	23.00	Bassetere, St. Kitts, W. Indies	1880-01-12
4 h 30 min	782.32	30.80	Smethport, Pennsylvania, United States	1942-07-18
9 h	1,086.87	42.79	Belouve, La Reunion	1964-02-28
10 h	1,400.00	55.12	Muduocaidang, China	1977-09-01
18 h 30 min	1,688.85	66.49	Belouve, La Reunion	1964-02-28 to 29
24 h	1,869.95	73.62	Cilaos, La Reunion	1952-03-15 to 16
48 h	2,499.87	98.42	Cilaos, La Reunion	1952-03-15 to 17
72 h	3,240.02	127.56	Cilaos, La Reunion	1952-03-15 to 18
96 h	3,721.10	146.50	Cherrapunji, India	1974-09-12 to 15
120 h	3,853.94	151.73	Cilaos, La Reunion	1952-03-13 to 18
144 h	4,055.11	159.65	Cilaos, La Reunion	1952-03-13 to 19
7 days	5,003.00	196.97	Commerson, La Reunion	1980-01-21 to 27
8 days	5,286.00	208.11	Commerson, La Reunion	1980-01-20 to 27
10 days	6,028.00	237.32	Commerson, La Reunion	1980-01-18 to 27
14 days	6,432.00	253.23	Commerson, La Reunion	1980-01-15 to 28
15 days	6,433.00	253.27	Commerson, La Reunion	1980-01-14 to 28
31 days	9,299.96	366.14	Cherrapunji, India	1861-07
61 days	12,766.80	502.63	Cherrapunji, India	1861-06 to 07
92 days	16,368.78	644.44	Cherrapunji, India	1861-05 to 07
122 days	18,737.58	737.70	Cherrapunji, India	1861-04 to 07
153 days	20,411.95	803.62	Cherrapunji, India	1861-04 to 08
183 days	22,454.36	884.03	Cherrapunji, India	1861-04 to 09
334 days	22,990.05	905.12	Cherrapunji, India	1861-01 to 11
365 days	26,461.21	1041.78	Cherrapunji, India	1860-08 to 1861-07
731 days	40,768.27	1605.05	Cherrapunji, India	1860 to 1861

Source: Data from van der Leeden, F.V. et al., *The Water Encyclopedia*, Lewis Publishers, Chelsea, MI, 1990; Courtesy of Australian Bureau of Meteorology, Melbourne, Victoria, Australia.

for a 100 year storm in the area. Flood waters suddenly reached Rapid City between 9:30 and 10:00 p.m. on June 9. Upstream, the Canyon Lake Dam failed at about 10:30 p.m., and a peak flow of 1415 m³/s passed through the city at around midnight on June 9—at a time when many people were sleeping and completely unaware of the impending flood. The water stage above Canyon Lake rose 3.96 m in 5 h. The water in Rapid Creek within Rapid City was back within its banks by 5 a.m. on June 10 (Schwartz et al., 1975).

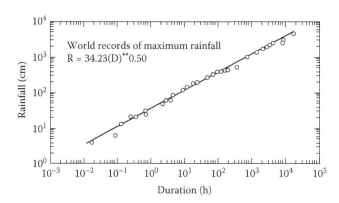

FIGURE 14.1
World records of maximum rainfall (R, in cm) as a function of rainfall duration (D, in h).

This flood caused the death of 238 people, injury to 3057 people, and the destruction of 1335 homes and 5000 automobiles. The total damage was estimated to be in excess of $160 million (not inflated, Schwartz et al., 1975). The peak flow at the river's gaging station near Rapid City was 1415 m³/s and was more than 4 times the previous maximum estimated during May to June, 1907, more than 15 times the previous maximum recorded on July 13, 1962, and about 16 times the maximum recorded since the 1972 flood (Carter et al., 2002). The flood magnitude was equivalent to a 500 year return period. This was the worst flash flood in the twentieth century in the United States.

14.2.1.1.1.2 Tropical Cyclones A tropical cyclone is the generic term to describe a low-pressure system of nonfrontal synoptic scale that develops over an ocean in tropical and subtropical regions. It often comes with thunderstorm activities and high-sustained cyclonic surface winds. It is called a *tropical depression* if the maximum sustained surface winds are less than 17 m/s, a *tropical storm* with an assigned name if winds are greater than 17 m/s, and a *hurricane*, a *typhoon*, a *severe tropical storm*, a *severe cyclonic storm*, or a *tropical cyclone* if winds are greater than 33 m/s. The term "hurricane" is used in regions covering the North Atlantic Ocean, the North Pacific Ocean east of the international dateline, and the South Pacific Ocean east of 160 E Longitude. All other terms for tropical cyclones are also used in specific regions.

Tropical low-pressure systems intensify their power and strength into storm and hurricane size when they remain stationary over the ocean but gradually lose energy as they move inland. The hazards of tropical storms and hurricanes may come in the forms of flooding, storm surge, high winds, and tornadoes. According to the U.S. National Weather Service, the tropical cyclone Denise generated 1144 mm of rain in 12 h over Foc-Foc, La Reunion Island (in the Indian Ocean, east of Madagascar), on January 7–8, 1966. Hurricane Camille smashed into the Mississippi Gulf Coast with a barometric pressure of 909 mb (millibars), winds up to 320 km/h, and 7.6 m tidal surges on August 17–19, 1969. Camille produced intense, torrential rainfall of more than 635 mm in 8 h in several areas and devastating floods along the James River system in Virginia. Total casualties from Camille were 256 deaths and $12.7 billion of property damage (inflated to 1998 level). However, the worst damage ever recorded in the United States was the storm surge of the September 8, 1900, tropical storm in Galveston, Texas, in which more than 8000 people drowned and property damage was estimated to be $30.9 million (not inflated). The storm surge (6–10 m in some regions) from Hurricane Katrina of August 23–31, 2005, breached levees around Lake Pontchartrain, Louisiana, inundated 80% of New Orleans, drowned more than 1300 people, and caused catastrophic damage of $75 billion along the Central Gulf Coast, the costliest hurricane in U.S. history.

Tropical Storm Allison of June 5–9, 2001, began as an eastward-moving tropical wave that slowly stalled over the Gulf of Mexico. It then developed into a small low-level, low-pressure system and then rapidly developed into a tropical storm. Allison came onshore near the west end of Galveston Island, Texas with winds of 80 km/h in the late afternoon of June 5 and then stalled again over the Houston metropolitan region, continuously drawing in moisture-laden air from the warm Gulf of Mexico for the next 5 days. Total precipitation in the Houston area ranged from 254 to 914 mm in 5 days, which caused extensive flooding as deep as 3 m in some parts of the city (RMS, 2001).

14.2.1.1.1.3 Frontal Storms A front is the boundary separating two air masses of different physical properties such as temperature, moisture content, and pressure. There are four types of fronts: cold, warm, occluded, and stationary (Figure 14.2). A cold front is cold air moving into an area occupied by warm air, and a warm front is warm air moving into an area occupied by cold air. A cold front often forces warm air to rise along a steep frontal boundary, producing cumulonimbus clouds and thunderstorms, while the slope of a warm front is gentle, producing steady gentle rains. Occluded fronts are rapidly moving cold air masses overtaking warm air masses, often associated with prolonged heavy showers. When the boundary between a warm and a cold front does not move appreciably, it is a stationary front.

Frontal systems often produce precipitation over a large area. The intensity and duration depend on the degree of moisture convergence along the front, the moisture content along and ahead of the front, the movement of the front, and the influence of upper-level winds. Slower moving fronts, with greater moisture convergence from surrounding high-pressure areas, tend to produce heavy and persistent rains.

The catastrophic Mississippi River flood of 1973 was a result of frequent and prolonged warm rains associated with extratropical cyclones and frontal activities over large areas

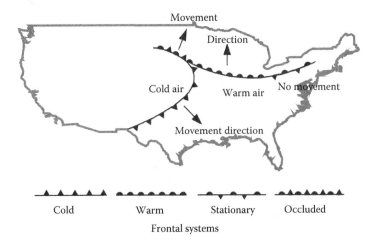

FIGURE 14.2
Frontal systems.

of the Mississippi River basin in March and April. It started with a mild and wet season prior to the flooding. Between October 1972 and February 1973, precipitation was above normal in every month except one. Water levels in many tributaries and reservoirs of the Mississippi were already well above normal levels prior to the heavy rains of that spring. In March and April 1973, atmospheric troughs repeatedly developed over the southeastern United States and had the effect of steering many fronts and extratropical cyclones across the basin (Hirschboeck, 1987). That March was extremely wet, with precipitation one to four times above normal for the entire Mississippi basin, except for the headwater region of the Missouri River. Flooding began along portions of the Upper Mississippi River in early March and, by April 3, the main stream was above flood stage along the entire course below Cairo, Illinois. The floodwaters remained above flood levels for 77 consecutive days at St. Louis, Missouri, 97 days at Chester, Illinois, and 88 days at Vicksburg, Missisippi, making this the longest flood duration in the river's history (Chin et al., 1975). The maximum flood stage at St. Louis was 13.177 m, higher than previous records and exceeded only by the great Mississippi flood of 1993 at 15.11 m (Perry et al., 2001). This 1973 flood inundated more than 30 million hectares (ha) of land, causing 28 deaths and $400 million of property damage.

14.2.1.1.1.4 Monsoon Rains   Like the land and sea breezes of daily circulation on a local scale, a monsoon is a seasonal thermal circulation on a continental scale and develops from differences in seasonal heating between continental and oceanic areas. In southern Asia and in a few other locations, substantial summer heating of the continent triggers an onshore inflow of moist air from the tropical oceans (called southwest monsoon) from May through September. As moist air moves onshore, it ascends orographically along the Himalayan Mountains, and adiabatic cooling causes condensation as monsoon rains. Cherrapunji is located at 1313 m above sea level in northeast India, about 400 km from the Bay of Bengal. Here a northward influx of warm, moist air and the orographic effect of the Khasi Hills during the southwest monsoon season make Cherrapunji the wettest spot on earth. It holds the world records of maximum rainfall for 11 different durations, including the maximum 365 day rainfall of 26,461.21 mm recorded during the period August 1860–July 1861 (Table 14.2).

   During the monsoon season, cyclonic depressions from the Bay of Bengal and the Arabian Sea steadily move across India, resulting in heavy monsoon rains of 400–800 mm/day. Most of the extreme floods that have occurred in India were in the northern and central basins, the areas most frequently impacted by the monsoon depressions from the Bay of Bengal. The flood of the Machhu River on August 10–12, 1979, generated a peak discharge of 16,307 m^3/s, a world record-breaking event. During the 3 day period (August 10–12), the Machhu River watershed (1930 km^2) received nearly four times the mean watershed rainfall for the entire month of August (Rakhecha, 2002).

14.2.1.1.2 Unusually Warm Winter and Spring

At middle and higher latitudes where winter precipitation is dominated by snow, an unexpected warm spell or rain shower can quickly melt much of the snowpack on the ground. In the wintertime, the ground is often saturated or frozen, evapotranspiration is at a minimum, and canopy interception may be limited. All of these factors generate much of the meltwater flowing over the land surface to nearby stream channels. If the warm spell is accompanied by prolonged and intense storms, flooding often results.

   The Christmas flood of December 24, 1964, in the Willamette Basin and the coastal streams of Oregon was one of the most severe wintertime floods. According to the

National Weather Service, the monthly rainfall recorded for that December at Salem, Oregon was 315 mm. This record was about 176% of the long-term December average (179 mm, 1928–2001) but was lower than the maximum 446 mm recorded in 1933. An unusually early and heavy snowpack in the lower elevations followed by warm weather and heavy rains crested the Willamette River above flood levels (Taylor and Hatton, 1999). Damage from this flood totaled over $157 million, and the flood caused the loss of 20 lives (State of Oregon, 2000).

14.2.1.1.3 Tidal Surges

When a hurricane moves swiftly across the ocean, winds on the right side of the storm track push surface water forward along with the storm, causing a rise of the sea surface referred to as tidal or storm surge. The surface of the sea near the center, or eye, of a hurricane (low air pressure area) is also much higher than the surrounding areas due to a difference in air pressure. The ocean rises 1.0 cm for every 1.0 mb of surface air pressure difference between the storm's center and its periphery. However, the rise of the surface level of the sea from the pressure difference in the hurricane center rarely exceeds 1.0 m, while tidal surges can exceed 10 m due to the dragging action of winds.

Tidal surges can also be generated by earthquake and volcanic activities on the ocean floor. When an earthquake occurs under the ocean, waves called tsunamis can be created. Such waves can travel in all directions for thousands of kilometers with sizes exceeding 30 m high at landfall and 1500 m long. A tsunami may travel from one side of the Pacific to the other in less than a day, or it may hit a coastal area in a few minutes. Tsunami waves are not very tall when they are traveling under the water but reach their maximum height when striking the land.

Tidal surges often cause backflow of river waters and inland flow of sea waters, resulting in the inundation of coastal areas, especially in areas with significant land subsidence problems due to excessive groundwater utilization. Recently, concern has been raised about the impact of global warming, which may induce changes in sea levels due to the thermal expansion of ocean water and changes in the quantities of water released from the melting of ice. These changes will impose significant inundation threats in coastal areas by the ocean.

The aforementioned hurricane that slammed into Galveston Island, Texas in 1900 had a maximum wind speed of 200 km/h and a storm surge of 6 m in height. Atmospheric pressure in the eye of the hurricane was 936 mb, one of the lowest on record. (In 1988, Hurricane Gilbert recorded the lowest atmospheric pressure ever of 888 mb or 666 mmHg in the United States.) Much of the city of Galveston was flooded with water from the storm or tidal surge as deep as 2.5–4.5 m. The entire eastern, southern, and western sections from two to five blocks inland were completely devastated.

14.2.1.1.4 Ice Jam

In areas where rivers are frozen in winter, unseasonably warm temperatures and heavy rain may cause rapid snow melting. The additional runoff from snowmelt and heavy rain increases the volume of flow in rivers, causing the frozen rivers to swell and ice to break apart. The broken chunks of ice float downstream and will often pile up and jam near narrow passages, obstructions, bridge pilings, or sharp curves. The ice jam may be thick enough to raise the water level behind the jam and result in flooding. Ice jams occur frequently in spring but may also occur in midwinter.

Some rivers, such as the Yellow River in China, have segments of middle channels flowing through regions of latitudes that are much higher than the upstream and downstream channels. When ice layers in the lower latitudes of upstream channels break and melt in the spring, the higher latitudes of the downstream channels can still be frozen. The ice jams and flooding potential of these rivers is very high.

The ice-jammed flood of the Susquehanna River Basin in Pennsylvania on January 19–21, 1996, was the third worst flood in the basin's history. This flood was the result of an unusual combination of heavy snowpack, high winds, unseasonably warm temperatures, heavy rainfall, and ice jams. The formation and breakup of ice jams caused a rise in the river level of 2.44 m in a single hour. The peak flow at Harrisburg, Pennsylvania during the flood was 16,131 m³/s, 20 times the normal flood level. This flood took 14 lives and caused $600 million in property damage.

14.2.1.1.5 Dam Failures

Dams can be of nature or man-built. Natural dams are formed by advancing glaciers that block local drainage patterns (glacier, ice dams), by debris accumulated as a result of glacier advance (moraine dams), or by dirt, gravels, and debris dropped from landslides or earthquakes (landslide dams). Artificial dams are structures built across a river channel to impound water for flood control, hydroelectric power generation, recreation, water supply, irrigation, fisheries, and other purposes. Both types of dams can fail due to a variety of reasons, naturally caused or humanly induced (Costa and Schuster, 1988; Walder and Costa, 1996). When a dam fails, the abrupt release of an enormous volume of water at hundreds or tens of feet high above the downstream channel level can not only inundate downstream areas, but also destroy everything standing in the way. The loss of property caused by a dam breach is certain, but the loss of livestock and life is greatly affected by the extent of the flooding area, the density of the population at risk, and the time available for warning (U.S. Department of Interior, 1998).

The major causes of dam failures in the United States, according to Costa (1985), are overtopping due to inadequate spillway capacity (34%), foundation defects (30%), and piping and seepage (28%). The U.S. Army Corps of Engineers, by the authorization of the National Dam Inspection Act of 1972 (P.L. 92-367), has compiled a list of about 77,000 dams under the National Inventory of Dams. All dams in the inventory are given a hazard rating in three hazard classes. A high hazard means dam failures would probably result in loss of life and major damage to property, a significant hazard could possibly cause some loss of life and property damage, while a low hazard is unlikely to cause loss of life and property damage.

Many floods have been caused by dam failures in the United States, but the collapse of the South Fork Dam that flooded the Johnstown area in Pennsylvania on May 31, 1889, and the breach of the tailings dam on Buffalo Creek, West Virginia on February 26, 1972, were the two most devastating in U.S. history.

14.2.1.1.5.1 Johnstown Flood, Pennsylvania, 1889

The South Fork Dam was an earthen dam located high in the Allegheny Mountains above Johnstown and near the community of South Fork, Pennsylvania on the Little Conemaugh River. The dam, measuring 21.95 m high, more than 274.32 m long, 6.10 m at the top, and 82.30 m at the base, was originally built by the State of Pennsylvania. At the time of collapse, the South Fork Fishing and Hunting Club owned the dam and its surrounding 65 ha; the club had owned them since 1879. There were known weaknesses in the structure, and the dam was under repair at the time of the breach (Floyd, 1990).

The rain was long and hard throughout the area on May 30, 1889. Water in Lake Conemaugh rose rapidly as runoff poured into the river from the surrounding mountains. Water overflowed the top of the dam shortly after 12:00 noon the next day. The center of the dam soon collapsed, and water in the lake emptied in about 45 min. The rushing wall of water was 12.19 m high at South Fork, with a flowing speed of 64 km/h. It rushed down through the towns of Mineral Point, East Conemaugh, and Woodvale, and arrived at Johnstown at 4:07 p.m. as a wall of water 11 m high. The raging water, along with mud, debris, and 91,000 kg of barbed wire picked up in East Conemaugh and swept away everything in its way, including buildings, mills, trains, railroads, animals, and people. Johnstown was covered under water as deep as 3.0 m in some places. The flood killed 2209 people in a matter of hours, second in casualties only to the hurricane-induced flood of Galveston, Texas of 1900.

14.2.1.1.5.2 Buffalo Creek Flood, West Virginia, 1972 Buffalo Creek is in southwestern West Virginia. It flows into the Guyandotte River at the edge of the town of Man, West Virginia, and then flows into the Ohio River near Huntington. There are about 16 communities along the Buffalo Creek valley. The slag-pile dam, 140 m long, 15.24–18.29 m high, and 97.54 m thick, was located at Buffalo Hollow near the mouth of Middle Fork at the headwaters of Buffalo Creek, about 64 km south of Charleston, West Virginia. It was owned and built by the Pittston Coal Co., in accordance with then-common mining practice, to reduce stream pollution by filtering wastewater from a local coal washing plant. The porous nature of the dam allowed the normal flow of the Middle Fork to filter through. Accordingly, an impoundment of water above the dam occurred only during periods of excessive runoff.

January/February 1972 was a very wet period in the region. The long-term average (1926–1971) of total January/February precipitation at Logan, West Virginia on the Guyandotte River was 174 mm, while the total precipitation for these 2 months in 1972 was 321 mm, the highest total ever recorded in 46 years. In southwestern West Virginia, precipitation increases with elevation at about 3.3% per 100 m. The elevation of the precipitation station at Logan was about 201 m, lower than the elevation of the dam at 533 m and lower than the highest ridge in the Middle Fork at 847 m. At the time, Middle Fork precipitation would be expected to have been around 356–388 mm or to exceed the measured precipitation at Logan by 11%–21%. A thorough analysis of the hydro-climatological conditions on, and prior to, the flood event can be found in a report by Chang et al. (1976).

In the early evening of February 25, 1972, a few thunderstorms developed in eastern and central Kentucky and moved eastward into West Virginia. As much as 142 mm of rain fell in Buffalo Creek beginning the evening of February 23 to 9:00 a.m. on February 26. Although no snowfall had been reported at Buffalo Hollow, local witnesses indicated an accumulation of about 152–305 mm of snow in the higher elevations prior to February 25 (National Weather Service, 1972). In addition, there was a rapid warming trend with all mean and maximum temperatures above freezing point beginning on February 21. A maximum air temperature of 20°C was recorded on the afternoon of February 25.

At about 8:00 a.m. on February 26, the massive coal waste dam breached and about 493,200 m^3 of impounded water along with 169,800 m^3 of coal waste exploded from the dam and discharged during the flood. A wall of floodwater, with coal and mud, that was about 10 m high raced through the Buffalo Creek valley. During a 3 h period, 125 lives were lost, 1000 homes were destroyed or damaged, and 4000 people became homeless. Total property damage exceeded $60 million (not inflated). The devastating flood was attributed to the combination of a number of factors, including

1. Extremely wet seasons prior to the flood
2. An excessive amount of rainfall immediately preceding the flood
3. Unusually warm temperatures that delivered additional snow-melted water to Buffalo Creek
4. The catastrophic dam collapse

It was the most devastating flood in West Virginia's history.

14.2.1.1.6 Watershed Conditions

A watershed with saturated soils, frozen ground, or poor drainage to stream channels always has great potential to turn thunderstorms into flooding. These conditions reduce the ability of a watershed to hold or absorb additional water from intense storms or snowmelt and thus increase the rise of water level in stream channels. Therefore, the flood-producing potential of intense storms is much greater in winter when soil and groundwater are at the highest levels, the ground surface is frozen, evaporation is negligible, and canopies of many hardwoods and some coniferous species are leafless.

Although the watershed conditions that cause watersheds to be more susceptible to flooding are mostly hydrologic, climatological, and geological in origin, some conditions have human origins. Humans engaging in various activities in watersheds can alter watershed conditions and flow regimes.

14.2.1.2 Human Origins

Watersheds provide and support many natural resources, including land, water, timber, vegetation, minerals, wildlife, fisheries, and recreation. Thus, they are important to humans. The use and management of watershed resources alters hydrologic budgets, lowers infiltration rates and soil water-holding capacity, and reduces channel capacities. As a result, watershed responses to storm rainfall are more dramatic, and flood potential is heightened.

There are many activities, such as deforestation, grazing, farming, urban development, mining, road and highway construction, forest fires, and stream channel alterations, which may alter watershed conditions important to runoff generation. The potential impact of these activities on flooding depends upon the severity of the disturbances and the extent of areal coverage. Impacts of these activities are more noticeable for small- to medium-sized storms with short duration. When the amount, intensity, and duration of precipitation increase, the relative influence of human activities on runoff volume diminishes. Increases in flow volume and peak caused by human activities are less evident downstream because of the cumulative net effect from other tributaries. This less significant effect downstream is because the relative size of a treated area tends to reduce as the overall size of the watershed area increases downstream, and the channel storage effect alternates the flood peak. Obviously, as discussed above, the impact of an activity is greater and the chance of flood occurrences is higher when these activities are conducted under wet antecedent conditions. Careful planning for the activities to be conducted during dry seasons can mitigate the impact.

The controversy over deforestation has become more intense and drawn more attention than the discussion of any other human activity and has involved the following considerations:

1. The effect of forests on streamflow regimes, being greatest among all the ecological systems
2. Forests covering about 30% of the Earth's total land surface, the largest type of land-use on Earth
3. Previous experiences with flooding and forest clear-cutting in upstream areas
4. Historical lessons on the abuse of forest resources in the Middle East
5. Conservation

Reports and research on flow response to human forest activities show little consensus. As discussed in Chapter 10, some reports indicate that deforestation causes increases in flow volumes and peak discharge, while other reports indicate little effect or even reductions in the flow of water.

14.2.2 Types of Floods

14.2.2.1 Flash Floods

Flash floods occur in mountain streams, upper tributaries, or canyons with relatively small watershed areas. Small watersheds have smaller water storage capacity than large watersheds do. The water storage capacity refers to the ability of a watershed to keep water or delay the movement of water within its drainage system (such as time of concentration) through vegetation, land surface, topography, soil profiles, and stream channels. Thus, smaller watershed storage means that water moves out of the watershed faster and in greater volume per unit area.

Most flash flooding is associated with slow-moving thunderstorms, intensive rains from hurricanes or tropical storms, a sudden release of water from snowpack, or a collapse of a dam or levee. Generally, thunderstorm activities are of high intensity, short duration, and small area. If the rate of an intense rain is greater than the infiltration rate of a watershed, the excessive rainwater will become overland flow and run to nearby stream channels. When thunderstorms occur in upper tributaries, which have small watershed storage capacity, overland flow from every portion of the watershed can arrive in stream channels in a short period of time. The enormous amount of water can easily exceed the channel capacity, resulting in flooding of its floodplain areas.

Flash floods in the western United States tend to be associated with storms 5–10 cm in size, 1–2 h in duration, and during afternoon and evening hours. Flash floods in the east, on the contrary, are nocturnal in nature and associated with storms of 10 cm or more and last for a few hours (Maddox and Chappell, 1979; Maddox et al., 1980). Flash floods can raise water to a height of 10 m or more in a few minutes or hours. These waters move at very fast speeds and can be accompanied by sediment, rock and stone, and debris. Floodwaters can travel downstream to areas where there are no storm activities at all. They rake trees, destroy buildings, and wipe out bridges. They occur with little or no warning and retreat rapidly.

14.2.2.2 River Floods

Unlike flash floods, river floods occur along major rivers and in lower tributaries of large watersheds. Since it takes more time for water to travel through large watersheds, the crest of a river flood may arrive at a downstream location days or even weeks after storms. Similarly, floodwaters may remain in floodplains for days or weeks before completely draining. The 1993 Mississippi River flood at St. Louis exceeded flood stage at 9.14 m on June 26 and briefly dropped below flood stage on September 13, a total duration above flood stage of 80 days (Koellner, 1996). That level broke the duration record of the 1973 Mississippi flood at St. Louis by 3 days (Chin et al., 1975).

River flooding is associated with torrential rains in spring, heavy thunderstorms, hurricanes, tropical storms, rain on frozen ground, and snow thaws. If waters generated from these storms in a watershed exceed the capacity of its channels, flooding results. However, as stated previously, a severe storm occurring in winter has a greater potential for flooding than if it occurs in summer because of saturated or frozen ground. Sometimes a failure of dams, or a jam of ice, may cause severe flooding. The failure is usually the result of negligence, poor management, inadequate design, or structural damage. Water from a collapsed dam can sweep everything downstream.

Most communities in the United States, in humid or arid areas, are subject to substantial risks of river flooding. Flooding is a natural phenomenon of rivers, and many components of riparian ecosystems depend on it. No rivers are exempted from periodic flooding. Since a river usually takes a great deal of time to build up to its flood stage, sufficient time is available for people to make necessary preparations. As a result, river floods are a greater threat to properties, while flash floods are more dangerous to human life.

14.2.2.3 Urban Floods

Urban communities are the major areas for human activity. About 50% of the world's population lives in urban areas and 75% of people in the United States live in 350 metropolitan areas with populations of 50,000 or more (The World Resources Institute et al., 1996). Unplanned and disruptive urbanization destroys vegetation cover on the ground and replaces the ground cover with concrete, asphalt, and roofing materials. These impervious areas prevent rainwater absorption into the soil profile, causing rapid runoff into drainage ditches and channels. When an intense storm lasts long enough to produce street and surface runoff greater than the capacity of drainage channels, flooding occurs (Figure 14.3).

In cities, runoff from streets is drained out through gutters, culverts, and conduits into drainage channels and creeks. Frequently, these conduits are too small to accommodate all rainwater at once if storms are too intense and long lasting. The flooding that results may rapidly cover streets, low-lying areas, or even bridges. Streets become fast-moving streams, and side ditches and creeks are covered with water, creating potential traps for vehicles and people. The suction of unseen underwater open sewer tops often swiftly pulls humans to a quick drowning.

Bridges can be a problem too. The existence of bridge piers in stream channels not only reduces channel drainage capacity but also blocks debris and forces water levels to rise. In that respect, cleaning out the debris that accumulates around bridge piers during the flood is an important task in flood mitigation programs. The current push by biologists and geomorphologists to increase the input and detention of large woody debris in upland channels as an important part of the stream morphology–habitat system often increases such need during the food seasons (Dr. Walt Megahan, personal communication; see Section 13.2.) Also, the danger of low-water bridge crossings should not be overlooked.

FIGURE 14.3
The urban flood caused by Tropical Storm Allison on June 5–9, 2001, inundated some areas in Houston to more than 3 m deep, with total damage of more than $4 billion. (Courtesy of Rebecca Chang.)

For those metropolitan areas in which excessive utilization of groundwater has caused serious land subsidence problems, zones that were above the 100 year floods may now lie in the floodplain areas. The Texas Water Development Board (1998) has discovered that the land surface elevation near the Galveston Bay–Houston area has been lowered as much as 3.5 m since 1906. This has consequently caused extensive flooding problems in urban subdivisions. Land subsidence may have caused the occurrence of floods after 5 cm storms that required intense storms of at least 10 cm or more in the past.

Urbanization generally causes watershed runoff about two to six times greater than that of natural terrain. Urban floods often occur shortly after storms or during storms. The damage and casualties caused by urban floods stem from the negligence of people overlooking the potential risk and unfamiliarity with the area. People usually are not aware of the risk that fast-moving water 15 cm deep can knock an adult off his feet, and 0.6 m of water can float a car.

14.2.2.4 Coastal Floods

Floods in coastal areas are caused either by tidal surges from the ocean or large lakes or rainfall runoff from rivers. Waters in the ocean can be driven inland through wave actions induced by winds in tropical storms and hurricanes, or as tsunamis created by earthquake and volcanic activities. When an earthquake or volcanic activity occurs in the ocean, the created waves, called tsunamis, can travel in all directions for thousands of kilometers. These waves, depending on how deep they are in the ocean, can travel at up to 800 km/h in the open ocean and have long wavelengths (100 km) and low amplitudes (less than 30 cm). However, upon reaching land, they can generate wave heights up to 30 m and can be as short as 1500 m in length. As a rule of thumb, deep-ocean tsunamis travel faster than the local tsunamis near the shore do, and the amplitude of a tsunami increases and wavelength decreases as it travels over the continental slope. A tsunami may travel from one side of the Pacific to the other in less than 1 day, or it may hit a coastal area in a few minutes.

Tsunamis are very destructive. They carry away people, animals, objects, and boats, and wipe out villages. A 9.0-magnitude earthquake occurred on the seafloor along the interface of the India and Burma plates near Aceh, Sumatra in northern Indonesia on December 26, 2004, a time when many worldwide tourists come to South Asia for Christmas vacation. It generated huge tsunami waves more than 6 m tall, hitting the coasts of Indonesia, Malaysia, Thailand, Burma, India, Sri Lanka, Bangladesh, and Maldives of South Asia and Somalia, Tanzania, and Kenya of east Africa without warning. Over 1 million people lost their homes and the dead/missing toll exceeded 300,000—the highest on record.

The Great East Japan Earthquake of March 11, 2011, an 8.9-magnitude, occurred at only 70 km east of the Oshika Peninsula and 400 km northeast of Tokyo. It triggered a massive tsunami with waves up to 38.9 m in height, travelling up to 10 km inland and rolling across the Pacific at 800 km/h to Hawaii and the U.S. West coast. Damages caused by the earthquake and tsunami in Japan included over 1,400 lives, 5,000 injuries, 9,000 people missing, 125,000 buildings destroyed or damaged, 1.5×10^6 households without electricity and water supplies, a dam ruptured, and a number of nuclear accidents. The total cost could exceed $300 billion.

14.2.2.5 Terminal-Lake Floods

Areas that have one or more rivers flowing into one or more lakes with no outlet are called *closed watersheds*, and the lakes in such a watershed are referred to as *terminal lakes*. In arid and semiarid areas where terminal lakes are not perennial and only contain water seasonally or only during rains, they are called *playas*. Some playas may hold water permanently if they are large and deep enough or their floors intersect the groundwater table; others may stay dry for years, but most are seasonal. Runoff from rainstorms is the main source of water that fills these terminal lakes, and water in the lakes is lost either to aquifers through percolation or to the air through evaporation.

Since there are no outlets, water levels in terminal lakes are dependent on and sensitive to storm runoff. During severe storms and in wet years, the water level may reach a stage that inundates surrounding areas, causing severe damage to personal property, farms, and public facilities. Floodwater in terminal lakes has duration much longer than that in rivers. For example, the USGS records show that the water level of the Great Salt Lake rose from 1277.52 m above sea level in 1963 to 1280.77 m in 1975 and again from 1280.16 m in 1983 to 1283.77 m in 1987. Steady increases in water levels over a number of consecutive years can cause flood damage to surrounding areas.

14.2.2.5.1 Playa Lakes

Playa lakes are common in arid and semiarid environments around the world. There are about 25,000–30,000 playa lakes in the playa lakes region of the southern High Plains, mostly in west Texas and a few in New Mexico, Oklahoma, Kansas, and Colorado. These playa lakes are circular, shallow (generally less than 1 m deep), small (averaging 6.3 ha in surface area; 87% are smaller than 12 ha), and flat-bottomed. The maximum size of a playa is limited by the volume of runoff into it, which is further determined by the size of the watershed. Watershed size in the playa lakes region ranges from 0.8 to 267 ha with an average of 55.5 ha (Haukos and Smith, 1992). Soils of the playa floor are predominantly clay, while the soils of the surrounding uplands are loams and sandy loams. These features provide an easy way to recognize playas from soil maps. These landforms are the result of a series of collective processes including wind, fluvial erosion and lacustrine deposition, pedogenesis, dissolution of soil carbonates, salt dissolution and subsidence, and animal activities (Gustavson et al., 1995). Playas are important wetlands in arid and semiarid areas, providing a vital habitat for wildlife (especially migrating birds), a vital water source for groundwater recharge, crop irrigation, and cattle grazing, and a collection basin for flood water and discharges of feedlot waste (Zartman et al., 2010). Flooding occurs during the rainy season every year with the water level depending on the amount of rainfall. In wet years with severe thunderstorm activity, playa flooding may reach a level that causes damage to agriculture and associated facilities.

14.2.2.5.2 The Great Salt Lake

Like playas, the Great Salt Lake is a terminal lake with a water level ranging from 1278 to 1284 m above sea level, depending upon precipitation, inflow of rivers, and surface runoff of the watershed. The main rivers entering the lake are the Bear River from the north, the Weber and Ogden Rivers from the east, and the Jordan River from the south, forming a drainage area of about 55,700 km^2. The accumulation of salts from runoff during historical times has made the salt content of the lake about 27% (depending on water level), eight times saltier than the ocean. It is the largest lake west of the Mississippi River and the 33rd largest lake in the world, measuring about 120 km long, 56 km wide, and 4 m deep.

Due to a period of above-average precipitation, the level of the Great Salt Lake began to rise in 1983. By 1986, the Lake rose nearly 3.66 m to reach a historical high of 1284 m. The high water level caused serious flooding, resulting in millions of dollars in property damage, disruption of major highway and railroad traffic, and inundation of farms, beach facilities, and waterfowl management areas. It forced the State of Utah to implement flood-control measures and the use of water pumps to lower the water level.

14.2.3 Flood Seasons

Floods can occur at any time and in any month or season of the year, but floods are more frequent in some seasons than others in a given region because of differences in weather patterns. Along the three states of the Pacific Coast where a marine climate dominates, summer and fall are dry and the cyclones and storms that bring moisture inland from the ocean occur mostly in the winter and early spring. Consequently, most of the floods in this coastal region occur during the winter and early spring. The pattern is different in the Rocky Mountain/North Plains region where a continental climate dominates. Here precipitation is dominated by snow, and 70%–80% of streamflow occurs during the snowmelt period in late spring and early summer. Thus, floods that occur during these seasons are more frequent than in other seasons.

In Texas, severe storms often occur in early spring due to frontal systems or in summer due to tropical storm activity from the Gulf of Mexico. The occurrence of floods in Texas, and extending to Oklahoma and Kansas, is relatively uniform in the six hot summer months of May to October. In fact, the moist air coming from the Gulf of Mexico in summer is orographically lifted along the Balcones Escarpment, the sharp gradation in terrain that marks the eastern limit of the Edwards Plateau in south-central Texas. This often produces some of the greatest rainfall intensities (Bomar, 1983) and the greatest flood concentrations (peak flows per unit area; O'Connor and Costa, 2004) in the United States. Combining the great rainfall intensity and sharp rises in elevation in the south-central region of Texas with the frequency of tropical storms and hurricanes in the summer makes the frequency and magnitude of floods in this region one of the greatest in the nation (Chang et al., 2004).

Weather conditions in the northern and eastern parts of the United States are often affected by factors including (1) air masses from the moist west, (2) cold and dry air masses from Canada, (3) the Great Lakes and Atlantic Ocean, and (4) warm, moist marine air masses from the Gulf of Mexico. Frontal storms, formed by cold and dry air masses from the north and warm, moist air masses from the south moving across the country, cause floods in the winter and spring more often than in other seasons. Also, water temperatures of the Great Lakes and Atlantic Ocean in the fall and winter are often higher than those of the overlying air. The moisture and heat from these water bodies tend to increase atmospheric instability, resulting in an increase in the amount of precipitation and the frequency of thunderstorm and hailstorm activities in the downwind areas during the winter.

The Southeast region lies under the influences of both the Gulf of Mexico and the Atlantic Ocean. Here annual precipitation is abundant, but rains in late summer and early fall are relatively less than in other seasons. Most of the summer rainfall is in the form of convectional showers, which often turn into flash floods. Also, tropical storms and hurricanes, which develop during summer months, often bring excessive rainfall inland, causing severe flooding in the region. Floods are most frequent in late summer and fall in Florida, but in early spring in other southeastern states (Estifanos, 2006).

The seasonal distributions of the 869 floods with return periods greater than 100 years in the conterminous United States used in Chang et al.'s, (2004) study are plotted by six geographical regions in Figure 14.4. In the three Pacific states, more than three quarters of the extreme floods (76%) occurred in the three winter months, December, January, and February. January alone accounted for 43% of the total. In the Rocky Mountains/North Plains, about two-thirds of the extreme floods (67%) occurred in late spring and early summer, i.e., May, June, and July, and 36% of the total occurred in June alone. However, the most frequent month for the extreme floods in the Southeast was April with 34% of the total, and March/April accounted for 52%.

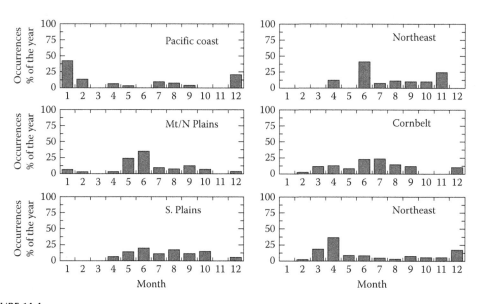

FIGURE 14.4
Monthly distributions of 869 floods with return periods greater than 100 years in 6 geographical regions of the conterminous United States.

14.3 Flood Measures

Floods are commonly measured in terms of stage height, discharge rate, duration, flooded area, flood frequency, and other parameters. In the United States, the stage height (flood elevation), discharge rate, and duration (time span above flood stage) of streamflow are available from around 9000 stream gaging stations operated by the U.S. Geological Survey. These data are then used, along with historical records and channel information, for analyses of flood frequency and other flood measurements to describe the characteristics and extremes of floods. Flood analyses for socioeconomic impact require data collection at the federal, state, and local levels. Such socioeconomic data may not be available immediately after the flood, and impact analyses of floods are therefore more difficult to assess.

14.3.1 Return Period

The size of floods, discharges, or stages is often assessed in terms of return period (recurrence interval) in years, a term related to the probability of occurrence for a given flood. A large flood has a small chance to occur in any given year, and the chances are that it will take more years to have a flood of that size occur again.

14.3.1.1 Definition

The term return period (T) is defined as the period in years, on the average, during which a given hydrological event (X) is expected to equal or exceed the observed value x. It is the reciprocal of the exceedance probability $P(X \geq x)$ of that event. In mathematical form

$$T = \frac{1}{P(X \geq x)} \tag{14.1}$$

and

$$P(X \geq x) = 1 - P(X \leq x) = \frac{1}{T} \tag{14.2}$$

Accordingly, the term "100-year flood" (X) means that there is a 1% probability (P = 0.01) that a flood of at least that observed size ($X \geq x$) is expected to occur in any given year (Figure 14.5). Values of P or T are obtained through the statistical analysis of past flood records at the stream gaging station in question. Note that floods in the analysis are assumed to be independent events. The flood that occurs in a particular year has nothing to do with the occurrence of flood in any other year. A 100 year flood does not mean it will occur in 100 year equal intervals. It implies that 10 events of 100 year size floods are expected to occur in a 1000 year period, and that two 100 year floods could occur in two consecutive years, although the probability is very small, only 0.01% ($P^2 = 0.01 \times 0.01$).

Methods of return period analysis can be found in texts given by Chow (1964a), Haan (1977), Bedient and Huber (1988), Stedinger et al. (1992), and many others.

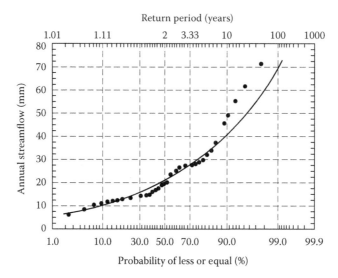

FIGURE 14.5
A typical frequency analysis by plotting the observed annual streamflow data versus probability (P ≤ x) in percent and return period in years, along with the best line fitting the observed values, on normal probability paper.

14.3.1.2 Frequency of Occurrences

With the probability of a flood event available, one may further be interested in knowing the probability of exactly m number of occurrences (successes) of that flood event in n-year trials. Since the occurrence of a flood is considered independent and the probability of a success (P) or failure (1 – P) outcome of that flood event in any given year is constant, such trials follow the Bernoulli process, or binomial distribution. The probability of m successes (B) is the product of the probability of any one success and the number of ways in which a sequence can occur, or

$$B(m \text{ floods in } n \text{ years}) = \frac{n!}{m!(n-m)!}P^m(1-P)^{n-m} \tag{14.3}$$

and the mean (E) and variance (Var) of m in the binomial distribution are

$$E(m) = nP \tag{14.4}$$

$$Var(m) = nP(1-P) \tag{14.5}$$

Example 14.1

What is the probability that a 25 year flood will occur exactly two times in a 50 year period?

Solution:

For n = 50, m = 2, probability of occurrence P = 1/25 = 0.04.

$$B(\text{2 floods in 50 years}) = \frac{50!}{2!(50-2)!}(0.04)^2(1-0.04)^{48} = 0.276, \text{ or } 27.6\%$$

14.3.2 Flood Extreme Indices

The return period discussed earlier is based on historical records at a given watershed, and the rarity of events refers only to the watershed itself. A flood discharge or flood stage may be equivalent to a 100 year flood in one watershed, while the same size of flood in another watershed may be equivalent to only a 50 year flood. Thus, for comparison purposes, the extremeness of floods needs to be defined, not only in terms of hydrological conditions, but also by the size of the watershed and socioeconomic consequences.

14.3.2.1 Hydrologic Indices

14.3.2.1.1 Discharge-Area Index

Hydrologic behaviors differ greatly between large watersheds and small watersheds. Generally, the streamflow in small watersheds tends to be very sensitive to intense storms of short duration and land-use conditions, while streamflow in large watersheds is greatly influenced by watershed and channel storage. This makes floods in large watersheds longer in duration and greater in total discharge as compared to those in small watersheds.

Differences in hydrographs produced from a 135.4 mm storm in two side-by-side watersheds, one small and one large, are illustrated in Figure 10.6. The peak average daily discharge was about $100\,\text{m}^3/\text{s}$ on December 22 on the small Garcitas Creek; it was about $600\,\text{m}^3/\text{s}$ on December 23 on the large Lavaca River, which is about six times greater in size than the small watershed. On the contrary, if discharge is expressed in per unit area, the ratio for the large watershed was $0.28\,\text{m}^3/\text{s}/\text{km}^2$, much less than $0.42\,\text{m}^3/\text{s}/\text{km}^2$ of the small watershed.

A simple index k to compare how abnormal the peak flow from a watershed of a given size is, in comparison to other watersheds of comparable size around the world, can be observed by the following equation (Framco and Rodier, 1967):

$$k = 10\left[1-\left(\frac{\log Q - 6}{\log A - 8}\right)\right] \tag{14.6}$$

where
Q is the largest flood (m^3/s)
A is the watershed area (km^2)

Floods with k-index values above six are extreme events on a global scale. Such floods have occurred mainly in tropical monsoon and equatorial watersheds. Under continental climates, extreme floods have a k-index between 2 and 4 (Papp, 2002). Rodier and Roche (1984) reviewed the maximum observed flood discharge at 1400 gaging stations around the world. They identified that the 41 largest floods on record occurred in watershed areas from 3.2 to $4,540,000\,\text{km}^2$, with k from 5.194 to 6.760. However, the envelope to all the largest floods of the world with watershed areas larger than $100\,\text{km}^2$ (34 watersheds) is a straight line corresponding approximately to $k=6$.

In a study on the spatial variation of maximum floods with return periods greater than 100 years in the conterminous United States, Chang et al. (2004) reported that the average k value for 869 maximum floods was 3.89. Fifty-two of them had k values greater than 5.0 and only two were greater than 6.0. Floods that occurred in the South Plains (Texas, Oklahoma, and Kansas), especially in Texas, had k values higher than those in other regions. There were 82 maximum floods in the South Plains, and the average k value was 4.53; 18 of those floods had values greater than 5.0, and the two floods with k values greater than 6.0 occurred in Texas. Floods in Texas are potentially larger than those in other regions.

14.3.2.1.2 Return Period Index

As discussed in Section 14.2.1, the extremeness of a flood at any given site along a stream channel is often expressed in terms of return period. A flood may be characterized as having a 30, 50, or 100 year return period. This allows a flood to be compared for its extremeness to other floods that occurred in the past at the same site. For a parameter to be comparable with other indices, however, the return period (T) in years is often converted into this simple index (r)

$$r = \log T \tag{14.7}$$

where T is the reciprocal of the annual probability of occurrence for a flood of equal or greater size.

The return period of a flood is based on streamflow records observed in the past. Among the earliest records of river stages preserved in the United States are those for the Mississippi River at Natchez, Mississippi from 1798. The first permanent streamflow gaging stations were installed in Eaton and Madison Brooks, Madison County, New York in 1835 (Boyer, 1964). Large-scale streamflow monitoring in the United States did not start until the U.S. Geological Survey was established in 1879. Most of our streamflow records are less than 100 years old, and there are very few records beyond 150 years. Thus, the estimates of flood may extend up to a 300 year period. In fact, a flood with the return period greater than a generation, or about 30 years, is customarily considered extreme, and the magnitude of r in rare events ranges from 1.5 to 3 (Papp, 2002).

14.3.2.2 Social Indices

14.3.2.2.1 Economic Risk

The extremeness of a flood should be also assessed in terms of its socioeconomic consequences. The impact of floods can be direct, indirect, and intangible. Direct impacts are those that can be evaluated in terms of cash value, while indirect impacts are those related to property depreciation and losses of a job or business. Effects on environmental aesthetics, environmental hazards, ecological habitat, species population, criminal activity, and disruption of routine activities that can be evaluated by cash are intangible impacts. In most cases, the economic damage assessment is difficult; it may require months or years to gather all the data required for the impact assessment. An economic risk parameter appears as (Papp, 2002)

$$e = P\left(\frac{L+M}{GDP}\right) \tag{14.8}$$

where
  e is the magnitude of economic risk
  P is the annual probability of the flood

L is the magnitude of loss

M is the cost of emergency measures

GDP is the annual gross domestic product of the economic unit (country)

14.3.2.2.2 Risk to Human Life

Information on human casualties caused by flooding is easier to obtain than information on economic losses and social impacts. Although the modern technology of flood forecasting has made the protection of human life much better than in the past, a single life is still far more precious than millions of dollars in property. Thus, evaluation of the extremeness of a flood should also include the risk to human life by the following index (Papp, 2002):

$$h = P\left(\frac{D}{N}\right) \tag{14.9}$$

where

h is the magnitude of risk to human life

P is the probability of occurrence of floods that cause loss of human life

D is the number of lives lost

N is the number of people in the flood-exposed area (floodplains)

14.3.2.2.3 Socioeconomic Risk

In order to provide a simple index approximately competitive to the hydrologic indices, Papp (2002) proposed the socioeconomic risk s, which combines the indices e and h together into

$$s = \log\left[1 + \left(\frac{e}{10^6}\right) + \left(\frac{h}{10^5}\right)\right] \tag{14.10}$$

A flood with $s \geq 2$ is considered extreme relative to the socioeconomic situation of the flooding unit (country).

The 1927 Mississippi River flood and the 1998 Yangtze River floods were two extremely large floods in the twentieth century. At the crest of the 1927 Mississippi River flood, the flow was $64,467\,m^3/s$ at 17.8 m at Vicksburg, Mississippi on June 4. This flood was about 2.13 times the size of the 1993 flood that occurred on the upper Mississippi–Missouri Rivers (Chin et al., 1975; Chang et al., 2004). During the 1927 flood, no fewer than 120 sections of levees on the lower Mississippi and its tributaries were breached during the flooding. The Caernarvon Levee was dynamited to divert floodwaters from getting into the city of New Orleans. Floodwaters spread over more than 41 million ha in seven states, up to 128 km in width and 5 m in depth in some places. The three states closest to the "Mighty Mississippi" river's mouth (Louisiana, Arkansas, and Mississippi) suffered the most. The flood drowned an estimated 250–500 people and drove 650,000 people out of their homes. Total losses to farmers included $102 million in crops, 1 million chickens, 9,000 work animals, 26,000 head of cattle, and 127,000 hogs.

The middle and lower Yangtze River flood in 1998 was the largest recorded flood in China since 1954. It lasted more than 60 days, drowned 1320 people, left 14 million people homeless, and caused a direct economic loss exceeding 177 billion Chinese yuan. Also, the floodwaters poured into the Yellow Sea, drifted to Cheju Island, South Korea, and reduced the salinity level of the sea from 3.1% to 2.5%, causing a significant reduction of the island's marine shellfish farming.

The s values for the 1927 Mississippi River flood and the 1998 Yangtze River flood were estimated to be 2.31 and 3.09, respectively (Papp, 2002).

14.3.3 Flood Forces

The damage of floods is due not only to the inundation of floodplains but also to the tremendous forces generated by the speed and mass of flowing water.

14.3.3.1 Shear Stress

The ability of a flow to erode particles is described as shear stress, tractive force, drag force, or shearing force. It has been discussed in Chapter 4. For your convenience, Equation 4.31 for calculating shear stress is shown here again.

$$\tau = \frac{F}{A} = \frac{w(\sin\alpha)}{A} = \frac{\rho g V(\sin\alpha)}{A} \tag{14.11}$$

$$\tau = \gamma RS$$

where
  τ is the shear stress (N/m^2)
  F is the force (N)
  w is the weight of flow = $\rho g V$ (N)
  ρ is the water density = $1000\,kg/m^3$ at 4°C
  V is the volume of flow (m^3)
  α is the channel slope, degrees (for small α, sin α = tan α = S)
  g is the acceleration due to gravity = $9.80\,m/s^2$
  A is the channel bed area (m^2)
  γ is the specific weight of water = ρg = $9800\,N/m^3$
  R is the hydraulic radius = channel cross-sectional area (m^2)/length of wetted perimeter (m)
  S is the energy slope, water-surface gradient, or channel gradient (dimensionless)

Note that water with heavy sediment loads may have a specific weight twice that of clear water (Baker and Costa, 1987). For river channels with a high width–depth ratio, flow depth (D) can be used to substitute hydraulic radius (R) with insignificant error, and then Equation 14.11 becomes

$$\tau = (\gamma)(D)(S) \tag{14.12}$$

The energy slope S for large watersheds is generally much smaller than that for small watersheds, and differences in S of more than 10-fold are not uncommon. Flash floods in small watersheds may have lower flow rates but are greater in shear stress and more destructive than river floods in large watersheds. For example, a large flash flood occurred on Ousel Creek, Montana (drainage area = $7.56\,km^2$) on June 8, 1964. The crest discharge = $118\,m^3/s$, mean depth = 1.8 m, mean velocity = 7.1 m/s, S = 0.2050, and τ = $2632\,N/m^2$. Also, an extremely large river flood of global scale occurred in the West Nueces River, Texas (drainage area = $1041\,km^2$) on June 14, 1935. The crest discharge = $16,426\,m^3/s$, mean depth = 10.8 m, mean velocity = 7.5 m/s, S = 0.00181, and τ = $189\,N/m^2$ (Baker and Costa, 1987). Although the

West Nueces flood had a crest discharge 139 times that of the Ousel Creek flood, the τ value was only 7.2%.

14.3.3.2 Flood Power

Another important index for the sediment transport capability of flows is flood power. Power is work done (or energy) per unit of time, and work is the product of force and the distance of displacement caused by the force. For mathematical expressions of work and power, see Chapter 4, Equations 4.32 and 4.33. Note that $1\,N$ of force $= 1\,kg\ m/s^2$, and work, kinetic energy, and potential energy all have the same unit, joule. $1\,J = 1\,kg\ m^2/s^2$, and $1\,J/s = 1\,W$. The power of floods is commonly expressed in per unit area or per unit length.

14.3.3.2.1 Flood Power as Per Unit of Streambed Area (ω_A, W/m^2)
It is calculated by

$$\omega_A = \left(\frac{F}{A}\right)V = (\tau)(V) \tag{14.13}$$

where
F is the force (N)
A is the streambed area (m^2)
V is the flood velocity (m/s)
τ is the shear stress (N/m^2)

The 1964 flood on Ousel Creek, Montana, mentioned earlier had a power per unit area (ω_A) equal to 18,582 W/m^2, 13 times greater than that of the 1935 extreme flood on West Nueces River, Texas (1421 W/m^2). Again, the forces of flash floods in small upstream basins can be much greater than those of river floods in large downstream basins. Evidence of the difference in forces along a stream channel is that the sizes of bedload are generally greater in upstreams than in downstreams.

14.3.3.2.2 Flood Power as Per Unit of Stream Length (ω_L, W/m)
It is calculated by

$$\omega_L = \frac{(F)(D)}{(t)(L)} = \frac{(kg\,m/s^2)(m)}{(s)(m)} = \frac{(kg\,m^3)(m/s^2)(m^3)(m)}{(s)(m)}$$

$$= (kg/m^3)(m/s^2)(m^3/s)(m/m)$$

$$\omega_L(\text{in } W/m) = (\rho)(g)(Q)(S) = (\gamma)(Q)(S) \tag{14.14}$$

where
ρ, g, S, and γ are defined in Equation 14.11
Q is flood (stream) discharge (m^3/s)
t is the time (s)
L is the length of stream (m)

Equation 14.14 is basically the rate of potential energy over a unit length of stream channel. It can be used in conjunction with bedload transport in streams (Bagnold, 1977). Using Equation 14.14 and the data given in the shear stress section, $\omega_L = 237,062\,\text{W/m}$ for the Ousel Creek, Montana flood in 1964, and $\omega_L = 291,364\,\text{W/m}$ for the West Nueces River, Texas flood in 1935.

14.3.3.3 Kinetic Energy

The kinetic energy of a flood is defined as its ability to do work by virtue of its motion. A flowing flood along the stream has kinetic energy carrying sediment and bedload. For a given mass of flood M (in kg) moving in a straight line with velocity V (in m/s), the kinetic energy E is

$$E = \frac{1}{2}(M)(V^2) \tag{14.15}$$

where E is in kg m²/s² (= joules). Thus, doubling flood velocity increases E by a factor of four. Dividing E in Equation 14.15 by time in seconds yields total power of the flood in watts.

Water weighs $1000\,\text{kg/m}^3$ ($62.4\,\text{lb/ft}^3$) and a typical flood flows downstream at 2.5–5.0 m/s. The kinetic energy (or work) generated by $1\,\text{m}^3$ of water at 3 m/s is equal to 4500 J, sufficient to move a mass of 4500 kg 1 m in distance with an acceleration of $1\,\text{m/s}^2$. A fast-moving floodwater of only 0.15 m in depth can knock an adult off his feet, and 0.60 m of water can carry away most automobiles easily. In a sluggish velocity of 1.0 m/s, a person would find it difficult to stand in a river 1.0 m deep. The tremendous amount of mass and velocity of flowing water makes flooding a very powerful, destructive, and dangerous force.

14.3.3.4 Hydropower

The kinetic energy of falling water can be converted into mechanical energy, and that mechanical energy can be further converted into electrical energy by the turbine of a generator. This is called hydropower, and it already provides about 10% of electricity in the United States and about 20% of all electricity worldwide. Although hydropower is not a concern in flooding, the capacity of a stream or a small dam in generating hydroelectricity may be of interest to many foresters and farmers in remote areas.

The hydropower of a dam is calculated by the following equation:

$$\omega_H = (\rho)(g)(Q)(H) \tag{14.16}$$

where
  ω_H is the hydropower (W)
  H is the distance of water fall (m)
  ρ, g, and Q are as defined in Equation 14.11

The ω_H calculated in Equation 14.16 can be termed the potential hydropower. It needs to be multiplied by an efficient coefficient to describe how well the turbine and generator can convert the power of falling water into electric power. The coefficient runs from 0.60 (older, poor maintained hydro plants) to 0.90 (well-operated plants).

Example 14.2

A dam of 5 m in height is used to produce electricity. The average amount of water flowing in the river above the dam is 10 m³/s. Assuming the generator runs at 0.80 efficiency, how much electricity (in kilowatts) can the dam produce?

Solution:

Apply the coefficient C = 0.8 to Equation 14.16.

$$\omega_H = (\rho)(g)(Q)(H)(C)$$

$$= (1000 \text{ kg}/\text{m}^3)(9.80 \text{ m}/\text{s}^2)(10 \text{ m}^3/\text{s})(5 \text{ m})(0.8)$$

$$= 392 \text{ kW}$$

14.4 Forest Impacts

14.4.1 Forests as a Nature Modifier

14.4.1.1 Forest Characteristics

Forest vegetation is environmentally distinguished in terms of large and thick canopies, tall and big stems, deep and spreading root systems, and rough and fluffy floors. These characteristics are used as a strong support for the impact of forests on flooding.

14.4.1.1.1 Canopies

The total surface area of leaves, or leaf area index (LAI), for forest stands can be 5–50 times greater than the ground area covered by the forest canopies. Since many of the biological processes and environmental functions of plants are conducted through leaves, values of LAI affect the degree of energy, water, and gas exchange between the ground and the air and between forest and surrounding areas. Impacts on transpiration, canopy interception, snow accumulation and snowmelt, light penetration, solar radiation, carbon sequestration, wind movement, ground shading, soil and water temperatures, raindrop impact energy, and horizontal precipitation (fog and cloud condensation) are highly related to LAI (Also see Chapter 7, Section 7.1.2.1 and Chapter 8, Section 8.2.1.1.3.).

14.4.1.1.2 Root Systems

Trees possess root systems that not only are deep and widespread but can also be massive. They usually penetrate more than 1 m below the ground surface, stretch more than 10 m wide, and spread out in an area more than five times greater than the ground area covered by the canopy. Based on 475 soil profiles located at 209 locations throughout the world, Schenk and Jackson (2002) found that over 90% of these profiles had 50% or more of all roots in the upper 0.3 m and 95% in the upper 2 m (Figure 14.6). The roots of *Boscia albitrunca* and *Acacia erioloba* have been found at depths of 68 and 60 m, respectively (Canadell et al., 1996). A 100 year old Scotch pine bore about 50,000 m of roots with 5 million root tips, while a 2 year old crested wheat grass plant possessed 500,000 m of roots in about 2.5 m³ of soil (Kramer and Boyer, 1995). Root area index (RAI), the ratio between total root surface area

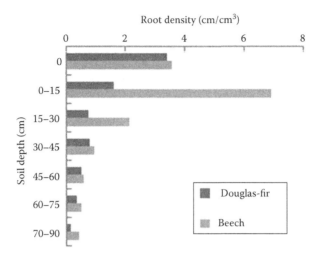

FIGURE 14.6
The distribution of fine root (<2.0 mm) densities in soils for pure Douglas-fir forest aged 60 and pure beech forest aged 64 stands in central Netherlands. (Data from Hendriks, C.M.A. and Bianchi, F.J.J.A., *Neth. J. Agric. Sci.*, 43, 321, 1995.)

and ground surface area in m^2/m^2, is at least comparable to LAI in all terrestrial systems (see Section 7.1.2).

Extended root systems enlarge a plant's moisture- and nutrient-absorbing sphere in the soil, increase soil infiltration rate and water-holding capacity, and reinforce soil aggregates and soil mass on hillside slopes and stream banks. As a result, transpiration is higher, soil moisture is lower, runoff is less, soil erosion is reduced, and stream banks are more stable in forested watersheds compared to nonforested watersheds.

14.4.1.1.3 Forest Floor

The forest floor accumulates vast quantities of litter along with animal remains that may temporarily retain their original shapes, gradually disintegrate above the soil surface, and eventually mix with the mineral soil through decomposition processes. The thickness of the organic horizon (forest floor) varies from little more than a thin litter layer to a very thick and well-developed layer. Thin forest floors most often occur in forests where fire is frequent and in areas where the climate is warm and humid. Tree leaves disappear completely in periods varying from 2 to 7 months in lowland tropical forests (Sydes and Grime, 1981). Up to 98.5% of leaf material in dipterocarp forests in tropical Malaysia is lost within 5 weeks (Gong, 1982). The decomposition and turnover rate of fresh litter can be from 1 to 3 years in temperate and cool climates (Pritchett, 1979). In coastal British Columbia, Pacific silver fir (*Abis amabilis*) retains foliage for about 6–9 years at the lower end of its altitudinal range (about 300 m), 13–19 years at midrange (1000 m), and 22–26 years at the upper altitudinal limit of continuous forest (1700 m) (Kimmins, 1987).

Thick forest floors are likely to occur in cool, moist forests where soils are acidic and decomposition is slow and incomplete (Currie et al., 2002). In the hardwood forests of the Great Lakes region, the forest floor can be up to 10 cm thick. The forest floor is as deep as 17 cm at old growth western hemlock and western red cedar stands in British Columbia, Canada (Plamondon et al., 1972). The decay rate of coniferous needle litter is slower than that of hardwood leaves. In the absence of recent disturbances, forest floor masses in humid regions tend to be greater under coniferous vegetation than under deciduous vegetation.

TABLE 14.3

Some Physical Characteristics of Soil Organic and Mineral Material

Parameters	Forest Floors	Mineral Material	
		Sand	Clay
Bulk density (g/cm^3)	0.10–0.90	1.30–1.80	1.10–1.50
Particle density (g/cm^3)	0.90–1.30	2.65–2.75	2.60–2.65
Porosity (cm^3/cm^3)	0.60–0.80	0.35–0.50	0.40–0.60
Maximum water content (Ratio by dry weight)	0.85–5.40	0.10–0.14	0.30–0.40
Available water storage (−0.33 to −15 bars,% cm^3)	0.15–0.20	0.04–0.08	0.16–0.25
Hydraulic conductivity, saturated (cm/s)	10^{-3}–10^0	10^{-3}–10^{-1}	10^{-9}–10^{-6}

Sources: Data compiled from Wang, Y. and Weng, J., The water and soil conservation functions of litter on forested land, in *Proceedings of the 12th International Soil Conservation Organization (ISCO) Conference*, Vol. II, Jiao, Y., Ed., Tsinghua University Press, Beijing, China, pp. 46–50, 2002; Soil Survey Staff, *Soil Taxonomy*, 2nd edn., USDA-NRCS Superintendent of Documents, U.S. Printing Office, Washington, DC, 1999; Laurén, A. and Heiskanen, K., *Can. J. Soil Sci.*, 77, 627, 1997; Heineman, J., Forest floor planting: A discussion of issues as they relate to various site-limiting factors, in *Forest Practices: Forest Site Management Section*, Silviculture Note 16, British Columbia Ministry of Forests, British Columbia, Canada, 1998; Potts, D.F., *Can. J. For. Res.*, 15, 464, 1984; Hanks, R.J. and Ashcroft, G.L., *Applied Soil Physics*, Springer-Verlag, New York, 1980; Fosberg, M.A., Heat and water transport properties in Conifer Duff and Humus, Research Paper RM-195, USDA Forest Service, Rocky Mountain Forest Range Experiment Station, Fort Collins, CO, 1977; Blow, F.E., *J. For.*, 53, 190, 1955; Lowdermilk, W.C., *J. For.*, 28, 474, 1930.

Within a forest, the forest floor is generally thinner on dry sites and on steep, upper, south-facing (N. Hemisphere), and windward slopes.

The physical properties of forest floors, depending upon the characteristics of the original litter material, degree of decomposition, and thickness, have a great impact on soil water retention, storage, and movement. Generally, the bulk density and particle density are low and the maximum water content, available water storage, and saturated hydraulic conductivity are high for the forest floor (Table 14.3). Also, the roughness of the forest floor is greater than that of mineral soils. A study conducted in Sanxia, China, cited by Wang and Weng (2002), showed that the roughness coefficient n is 0.2282 on mixed pine-oak stands, 0.1328 on shrub and grassland, 0.0723 on farmland, and 0.0544 on barren land. All these features allow forest soils to increase infiltration and percolation, keep more water in the watershed, and reduce overland flow volume and velocity. The available water capacity (AWC) of organic litter and mineral materials is given in Figure 14.7. The duration of runoff for an overland flow of 1 mm deep through a 1 m long and 20° slope with 3 cm deep litter could be prolonged 15.3 min compared to if the ground had no litter on its surface (Liu et al., 1991).

14.4.1.2 Impacts on Runoff Generation

The biological as well as environmental functions of forests are given in Table 7.1. The impact of forests on streamflow is outlined as follows:

- Canopy and litter interception reduce the amount of precipitation that reaches the mineral soil.
- Canopy transpiration and root systems make actual evapotranspiration in forested watersheds greater than that in nonforested watersheds.

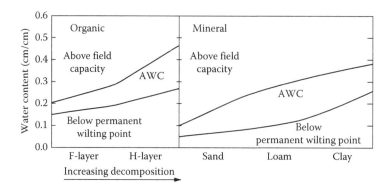

FIGURE 14.7
Available water capacity (AWC) of organic and mineral materials. (Modified from Heineman, J., Forest floor planting: A discussion of issues as they relate to various site-limiting factors, in *Forest Practices: Forest Site Management Section*, Silviculture Note 16, British Columbia Ministry of Forests, British Columbia, Canada, 1998.)

- Decayed roots, the macro- and micropore spaces created by organisms and small animals, low bulk density due to organic materials, and plant transpiration deplete soil-moisture content, which increases soil permeability, and reduce overland and total runoff.

- Forest litter increases surface roughness, slows down overland flow velocity, and increases water infiltration into the soil, resulting in reductions of overland flow and runoff energy for sediment-transport capacity.

- Canopy interception also reduces the impact energy of raindrops, which causes soil detachment. Coupling raindrop energy reduction with the reduction of overland flow and the increase of surface roughness results in a significant decrease in soil erosion and stream sediment.

- Roots increase the binding and cementing effect of soil aggregates and provide much of the shear strength for soil and stream bank stability. This keeps stream channels from being clogged by sediment and consequently reduces the flood potential of the stream.

- Canopy condensation in coastal areas and in fog-prone regions can add extra water to the ground by dripping, a process known as horizontal or occult precipitation, fog, or cloud drips.

Because they are the biggest in size and largest in distribution, the impact of forests in the hydrological cycle and on the physical environment should be the greatest among all natural cover types on Earth. However, the degree and significance of these various impacts, in many cases, may be overridden by other environmental factors such as climate, soils, topography, and geology. Climatic fluctuations, in many cases, may have caused a more significant impact on streamflow than forests. This requires a more detailed assessment.

14.4.2 Field Observations

14.4.2.1 Flood Occurrences

Timber harvesting and forest–agriculture conversion have been attributed to the occurrences of downstream flooding in many regions, such as Indonesia (van Noordwijk and

Agus, 2004), China (Sharp and Sharp, 1982), Brazil (Nordin and Meade, 1982), the central Himalayas (Haigh et al., 1990), Taiwan (Chen, 2003), and Vietnam (Tran et al., 2010). These forest activities remove forest canopies, destroy forest litter floors, compact mineral soils, and accelerate soil erosion, which causes a reduction in watershed and channel water-holding capacity and results in increase of flooding. This idea prevails among the general public, media, and governments. It affects forestry policies today, as it did in the past.

In assessing the impact of forests on flooding, all different types of flooding and their meteorological conditions need to be inclusively considered. Forests have little or no effect on the occurrences of floods that result from dam failures, ice jams, or a long duration of rainfall in saturated watersheds. However, coastal mangrove forests and coral reefs are able to deflect tsunami waves and mitigate flood damage in coastal areas. This is evident by the severe coastal flood, in historical magnitude, caused by tsunami waves in South Asia on December 26, 2004. Areas that were naturally protected by forests and reefs suffered less than did those that were more exposed. A village in India's Tamil Nadu state planted 80,244 saplings in 2002, thus entering the *Guinness World Records* book. BBC News reported that the people of this area, Naluvedapathy in Vedaranyam district, south of the Tamil Nadu's worst-affected areas of the 2004 South Asia tsunami, remained almost unscathed (Raman, 2005). Thus, it is with regard to flash floods and river floods that the impact of forest cover requires more studies and assessment. Frequently, local climatic conditions have more impact on the magnitude of floodwaters than forest coverage does, and hydrologic events caused by rain-on-snow in winter or by rain-on-snowmelt in spring could result in quiet different magnitudes of flooding.

14.4.2.1.1 Small versus Large Watersheds

Generally, flash floods occur in small, headwater watersheds, while river floods occur in large, downstream watersheds. The hydrologic behavior is significantly different between small and large watersheds. Streamflow in small watersheds is very sensitive to storm characteristics and land-use conditions, while streamflow of large watersheds is mainly dominated by watershed storage, where land-use conditions are less significant (Chow, 1964b). Also, small watersheds generally have steeper land slopes, higher percent of forest cover, and greater chances for the entire watersheds to be covered by severe storms than those of large watersheds. Consequently, deforestation in small watersheds can cause the frequency and magnitude of flooding to be greater.

The land use in large watersheds is more diversified, and forests may only cover a small portion of the entire watershed. It is doubtful that an increase of floodwaters in small upstream watersheds could have a significant impact downstream because of a small percentage of forest alteration, flood routing process, and the cumulative effect from other watersheds. When a watershed is saturated due to prolonged precipitation or in wet seasons, an excessive storm can result in severe flooding, no matter whether it is a forested, rangeland, or an agricultural watershed. The 1973 Mississippi River flood inundated more than 30 million hectares (ha) in 10 states (Chin et al., 1975), covering all land-use conditions. Lull and Reinhart (1972) listed many maximum floods in the eastern United States, which occurred in well-forested basins.

14.4.2.1.2 Rain-on-Snow versus Rain-on-Snowmelt

In snow-dominated areas, forest cutting reduces canopy–snow interception and increases direct solar radiation and turbulent heat transfer, resulting in increases of snow accumulation and rate of snowmelt. There, the elevation difference throughout the watershed and the degree of snowmelt at the time of storm occurrence play important roles in flooding in clear-cut watersheds. In watersheds of lower elevations, effects of forest cutting tend to be more pronounced because melt is occurring throughout the entire basin, while melt is occurring only in a portion of the basin at higher elevations.

The frequency and magnitude of storm flows may greatly increase on rain-on-snow events, especially in midwinter when snow accumulation is at a great depth. During the spring, if snow is already melted or in rain-dominated areas where there is no additional water melted from snowpack, the magnitude of floodwaters from rain-on-snowmelt events should not be as large as that from rain-on-snow events. Dr. Walter Megahan (personal communication) notes that if the rain occurs later in the snowmelt season when snow exists only in the uncut forest, effects of forest cutting could actually decrease flood peaks compared to the undisturbed conditions. Thus, the effects of timber harvest on the magnitude of floodwaters vary according to the cause of those floodwaters. Identifying the cause of floods is essential for sound forest management (MacDonale and Hoffman, 1995).

14.4.2.1.3 Stream Sedimentation

The impact of forest harvest and land-cover changes on soil erosion and stream sedimentation is well documented (see Chapter 12). An increase in erosion and stream sedimentation can cause streams to be wider in width or shallower in depth due to deposition of coarser sediments on the streambed, depending on the location along the stream. These affect channel water-holding capacity, resulting in a great flooding potential in the area. The control of soil erosion and stream sediment is recognized as a significant function of forests on the mitigation of flood damages.

A study was conducted to determine the effect of historical land-cover changes on flooding and sedimentation in the 122 km² North Fish Creek watershed in Wisconsin (Fitzpatrick et al., 1999). Before European settlement in the 1870s, the area was fully covered by balsam fir/white spruce forest. About 90% of the forest was cleared by 1890, followed by agricultural activities that peaked in the mid-1920s–1930s. The amount of forest has remained at about 60% since the 1930s. Using sediment core samples, field surveys, satellite imagery, and U.S. Corps of Engineers' HEC-1 rainfall/runoff model, the results indicate that modern flood peaks and sediment loads in the watershed may be double than that expected under presettlement forest cover. During the maximum agricultural period, flood peaks probably were about three times larger and sediment loads about five times larger than that expected under the presettlement forest cover. The upper reach of the streambed eroded down at least 3 m and the channel capacity at least doubled after European settlement, due to increased runoff and associated channel erosion. In the lower reach of the stream, however, the postsettlement sediment deposition on the floodplain and in the channel is four to six times the presettlement rate. About 0.5–1.4 m wide medium to very coarse sand has accumulated in the modern channel near the mouth of North Fish Creek to Lake Superior Chequamegon Bay. The large amount of sediment deposited in the lower reach of the creek added to the great magnitude of flooding in the postsettlement period.

In areas highly susceptible to road erosion, timber harvest above forest roads can cause a large portion of subsurface flow to be intercepted by the cut slope of the road and be transformed to surface runoff, resulting in increases in soil erosion, channel sedimentation, and peak flows (Megahan, 1983). A consideration of avoiding clear-cutting above forest roads or designing roads with special features to accommodate additional flows by incision of the subsurface flow zone may become necessary. On the other hand, runoff captured by the road ditch network may infiltrate below ditch relief culverts and cause changes in streamflow response to be minimum (La Marche and Lettenmaier, 2001). These conflicting results may reflect the complexity of road density, drainage area, storm intensity and duration, depth of groundwater table, and other site conditions. In any event, forest roads could be more important than vegetation removal in affecting storm flows because of their persistence and pervasiveness. Hydrologic modeling shows that road effects on flooding generally increase with flood return period, while vegetation effects decrease (La Marche and Lettenmaier, 2001).

14.4.2.2 Experimental Watersheds

Experimental watersheds provide a scientific approach to assess differences in peak flows between forested and nonforested watersheds relatively under the same environmental conditions. Such studies have shown that cutting forests may cause an increase, up to 880% in the United States (see peak flows in Section 10.3) and 1140% in Kenya (Bruijnzeel, 1990), in peak flows. However, decreases (Harr, 1982), insignificant effects (Harr and McCorison, 1979; Duncan, 1986), or disputable effects (Jones and Grant, 1996; Thomas and Megahan, 1998) on peak flows have also been reported in the literature. In many cases, climatic conditions such as rain-on-snow (effects to be exponential) or rain-on-snowmelt (effects to be linear) can superimpose forest cutting on hydrologic responses. This variation in results reveals the complexity of the problem, requiring detailed analyses and assessments of weather and site conditions, and perhaps the techniques of data analyses (Alila et al., 2009).

The location of forest alteration relative to the watershed, the degree and duration of the forest alteration, and the temporal and spatial distribution of precipitation often play important roles in the alteration outcome on peak flows. Generally, deforestation causes a significant increase in flood flows for storms that occur during the growing season, a time when the removal of forest canopies greatly decreases vaporization loss in a watershed. However, for storms that occur during the dormant season, a time when trees are leafless or transpiration rates are low and the difference in soil-moisture levels between forested and nonforested watersheds is insignificant, timber harvest causes little impact on peak flows, especially for large and infrequent flow events (Kochenderfer et al., 1997; Jackson et al., 2004). A 50% timber harvest or a partial cutting that is conducted on upper slopes causes less impact than a 100% harvest or partial cutting that is conducted on the lower slopes near the riparian zones (Black, 2004). Also, the size of peak flows can be exacerbated or ameliorated by altering the timing of runoff from different tributaries through harvesting different portions of the basin (Harr, 1981).

Generally, the impact of timber harvest on storm flows is most significant for small events (less than 2 year return period). As the size or duration of storms increases (about 5 year events or greater), watersheds gradually become saturated, regardless of forested or bare ground, and the impact of forest cover on storm flow decreases to insignificant levels. This has been reported in many experimental watershed studies throughout the United States, including those in Oregon (Thomas and Megahan, 1998; Jones, 2000), California (Wright et al., 1990), the central Rocky Mountains (Stednick and Troendle, 2004), and Georgia (Hewlett and Doss, 1984). However, most forest–storm flow studies have been conducted in small watersheds with sizes less than about 200 ha. Small watersheds are relatively homogeneous in environmental conditions. They are a small sub-hydrologic system within a large river basin in which the climatic patterns, geological formations, and vegetation are heterogeneous. Thus, it is difficult to transfer the results from small watershed studies to large river basins. The dilution and flood-routing effect may cause the increase in floodwater in small upstream watersheds to be insignificant when it routes to large downstream basins.

14.4.3 Spatial Analysis

14.4.3.1 What Is Spatial Analysis?

The experimental watershed studies mentioned earlier employ control watersheds as a reference for treatment assessments. Under the existing condition (with no control/treatment watersheds), the impact of forests along with other physical and climatic factors on floodwater can be assessed by watershed spatial analysis. The approach investigates hydrologic conditions and watershed characteristics (climatic, topographic, land use) at each gaging watershed. Once the data are collected for all watersheds in a given hydrologic region, the technique of regression analysis or other statistical methods are used to assess the impact of watershed characteristics on the variation of floodwaters. One of its outcomes is to develop a prediction equation for stream discharge based on a few selected watershed characteristics (see Section 16.3 in Chapter 16). Only those watershed characteristics that are significant in affecting the streamflow are retained in the prediction equation.

The spatial analysis, also called regional analysis or empirical modeling, usually uses much larger and as many gaged watersheds as possible. For our purpose, the developed equations can be simply expressed by the following general form:

$$Q_t = K[F^m]f(x's) \qquad (14.17)$$

where
  Q is the peak flow for a return period t (years)
  F is percent forest area
  $f(x's)$ is the general term for all other environmental parameters selected by the regression analysis in the prediction equation
  m and K are constants

Equation 14.17 describes the impact of F^m, along with the others, on the magnitude of peak flows Q_t. If the exponent m is positive, Q_t increases with increasing values of F and m. If m is negative, Q_t decreases with increasing F and decreasing m.

14.4.3.2 Analyzed Results

The spatial analysis has been employed by numerous investigators in the United States to study the variation of floodwaters. Some of these investigations, such as those in East Texas (Chang and Watters, 1984), South Dakota (Sando, 1998), Washington (Sumioka et al., 1998), Oklahoma (Lewis, 2010), Virginia (Bisese, 1995), and West Virginia (Wiley et al., 2002), have shown forests to have no significant effects on peak flows of longer return periods. Others have shown that forest has a significant effect on peak flows, with negative m values greater for longer return periods (Table 14.4). A greater negative m value for peak flows of longer return periods suggests that the impact of forest on floodwaters decreases as the size of the hydrologic events increases. These results of spatial analysis studies agree with the conclusion from experimental watershed studies that the impact of forest on storm flows is insignificant in large basins or decreases as the size of hydrologic events increases.

TABLE 14.4

The m Values of Equation 14.17 for a Few Floodwater Spatial Analysis Studies in the United States

Location	Station	Forest Percent (%)	2	5	10	25	50	100	500	References
E. Coastal Plain, Maryland	37	08–085	−0.464	−0.586	−0.667	−0.770	−0.847	−0.923	−1.100	Dillow (1996)
N. Panhandle, Idaho	21	24–098	−0.504	−0.885	−1.100	−1.360	−1.530	−1.670	−2.000	Berenbrock (2002)
Pennsylvania (Except the N.W. Region)	261	02–100			−0.971	−1.032	−1.082	−1.134	−1.267	Stuckey and Reed (2000)
Southwestern Wisconsin	39	00–057	−0.254	−0.260	−0.302	−0.308	−0.310	−0.312	−0.690	Walker and Krug (2003)
West Region, Montana	96	15–100	−0.508	−0.577	−0.605	−0.639	−0.652	−0.664		Parrett and Johnson (2004)

14.4.4 Watershed Modeling

14.4.4.1 Why Modeling?

Because of climatic variability, it is impossible to compare the occurrence of floods in a forested drainage basin with the occurrence of floods in the same drainage basin without forests. Although some statistical procedures can be used to account for some sources of the climatic variability, errors involved in these procedures are inevitable. No matter how sophisticated the design of an experimental watershed study using an untreated control watershed, it compares floodwaters of two completely different watersheds with respect to size, soil, geology, topography, vegetation, and climate, and not similar watersheds in different land-use conditions. Watershed simulation and modeling provides an approach other than the experimental watershed and spatial analysis methods to study flood events of a given watershed under various scenarios of land-use and climatic conditions. It uses physical principles and hydrologic mechanisms to study the response of a drainage system to various inputs and consequently make predictions of output in runoff production. In other words, it intends to answer questions like what would happen to streamflow if the watershed were under a specific land-use condition, different management scenarios, or climatic conditions? It provides useful information for management, development, and planning of natural resources. Hydrologic models can be physical (process-based) or empirical (observation data-based), deterministic or stochastic. They can focus on event or continuous modeling, individual or multiple processes, lumped (a watershed treated as a single unit) or distributed (a watershed treated as multiple units for spatial variation) simulation. For watershed modeling and its applications, see a state-of-the-art review by Daniel et al. (2011).

14.4.4.2 Simulated Results

Many watershed modeling studies have shown similar results to those produced by experimental watersheds and spatial analyses. They show that forest harvest causes an increase in floodwaters more significantly in small areas and for small storms. As the magnitude of storms, or drainage area, increases, the effect decreases. The effects of forest roads on peak flows may not be as significant as those of clear-cutting, but they are more persistent and pervasive than those related to vegetation removal.

The Distributed Hydrologic Soil Vegetation Model (DHSVM) is a physically based, spatially distributed hydrology model that simulates the interaction of topography, soils, vegetation, climate, and water movement in a watershed system (Wigmosta et al., 1994). Applying DHSVM to the Snoqualmie River at Carnation (1559 km^2) in Washington state, the result shows that the amount of forest harvesting that has actually occurred in the basin is not responsible for the observed peak flows in its 46 years of record (Storck et al., 1995). The effect of forest cutting on peak flows is more significant for smaller than for larger streamflow events. If the entire old-growth Douglas fir forests in the 1559 km^2 basin were cleared, it could, under various climatic conditions, cause a maximum increase in peak flows of about 300% and a maximum decrease of almost 100%. The largest increase in streamflow occurs during the early part of the spring melt season, while the largest decrease occurs during the later part of the snowmelt season. In reality, clear-cutting an area as large as 1559 km^2 is not likely to occur, not only because of the size but also because of the time required to complete the deforestation. Regrowth of vegetation in the clear-cut sites will cause the effect, if significant, to be reduced. Thus, with 10% (156 km^2) of the forested basin cleared, the simulation shows that the maximum increase in annual peak flows is about 5% (Storck et al., 1995).

The DHSVM, along with field data collection, also was used to simulate the effects of forest harvest and roads on flood flows in the extensively logged 149 km² basin of the Deschutes River in Washington state (La Marche and Lettenmaier, 2001). It shows that forest removal without introducing any forest roads would increase the mean annual flood in the Deschutes basin and its nine upstream sub-basins by 4%–18%, but the effect decreases for floods of greater return period. Forest roads, without forest harvest, would increase the mean annual flood by 3%–10%, but the effect increases with increasing flood magnitude. The effects of roads and harvest on flood flows are independent; therefore, the combined effects are additive.

Another modeling study on land-use change and peak flows was conducted for the Sand Creek watershed in Michigan. The hydrologic model was developed by the Hydrologic Unit of the Michigan Department of Environmental Quality using the Hydrologic Engineering Center's Hydrologic Modeling System (HEC-HMS) (Fongers, 2003). The watershed, 142 km², had 74% forest in 1800; the amount of forest declined, due to agricultural development, to 16% in 1998. The results showed that peak flows generated from 2 year 24 h, 10 year 24 h, and 25 year 24 h storms in 1998 would be 115%, 66%, and 50%, respectively, greater than those in 1800. A smaller 16.3 km² upstream watershed in the Sand Creek watershed had the same land-use pattern as the entire watershed, with 73% forested area in 1800 and 10% forest in 1998. The simulation showed that the change in land use in the small watershed (only 12% of the large watershed) had caused an increase in floodwaters for the three different sizes of storms to be 137%, 81%, and 61% greater than those in 1800, all greater than the changes in downstream basin.

14.4.5 Management Implications

The impacts of forest harvest on flood flows discussed earlier are summarized as follows for management considerations:

- Forests are more effective in flood mitigation than any other type of vegetation. However, major floods can occur in forested as well as nonforested watersheds as the result of large and extreme storm events. Thus, usually, forest cutting is not the cause of flooding, although it may increase floods in some situations.

- The occurrence of floods is affected by an array of environmental factors. Usually, the weather conditions are the major factors that cause flooding, regardless of land-use conditions.

- Studies conducted by experimental watersheds, spatial analysis, and watershed simulation have all supported the idea that the effect of forest harvest on flooding is most effective in small watersheds and for small storms. As the size of watersheds, or the size of storms, increases, the effect decreases to an insignificant level.

- Any impact of forest harvest on peak flows will be gradually mitigated due to the regrowth of vegetation on the harvested sites, especially in humid areas. This hydrologic recovery to its precut streamflow levels takes only about 4 years in humid East Texas and Pennsylvania and as long as 80 years in Colorado.

- The function of forests in erosion control, land stabilization, shoreline protection, and tsunami dissipation is much greater than the function of other types of vegetation. Increased land and channel erosion, due to forest harvest, can raise channel elevation, reduce channel capacity, and alter channel morphology, causing a greater flood potential and increased flood damage of the basin.

- Although maintaining forest coverage is a good measure and strategy for environmental protection and resource conservation, careful forest harvest practices can be used to maintain environmental protection and resource values. Application of Best Management Practices can reduce environmental damages to a minimum level. However, in environmentally fragile areas, such as some steep terrains, coastal shorelines, or landslide-prone slopes, forests are an economically attractive, functionally effective, and relatively maintenance-free means for controlling soil erosion, land stability, and stream sediment. In these cases, the potential impact of forests on erosion and downstream sedimentation stream can exceed the economic gains from timber harvest.

- In regions where precipitation is dominated by rainfall, deforestation causes a significant increase in floodwaters for storms that occur during the growing season. For storms that occur during the dormant season, a time when differences in soil-moisture levels between forested and nonforested watersheds are insignificant, timber harvest has little impact on peak flows, especially for large and infrequent storm events. In snowfall-dominated regions, forest harvest causes an increase in snow accumulation and snowmelt. Rainstorms that occur at the time when the ground is still covered by snow could cause a significant increase in peak flows.

- The timing of runoff in a watershed can be altered by conducting forest harvest at separate times, in different tributaries, and at upper slopes, consequently affecting the peak flows.

- Because of the limitations of forest on major floods generated from large and extreme storm activities, perhaps zoning or land-use planning should be considered an appropriate alternative for reducing flood damages.

References

Alila, Y. et al., 2009, Forests and floods: A new paradigm sheds light on age-old controversies, *Water Resour. Res.*, 45, W08416. DOI: 10.1029/2008WR007207.

Anderson, H.W., Hoover, M.D., and Reinhart, K.G., 1976, Forests and water: Effects of forest management on floods, sedimentation and water supply, USDA Forest Service, General Technical Report PSW-18, Washington, DC.

Andréassian, V., 2004, Waters and forests from historical controversy to scientific debate, *J. Hydrol.*, 291, 1–27.

Bagnold, R.A., 1977, Bed load transport by natural rivers, *Water Resour. Res.*, 13, 303–312.

Baker, V.R. and Costa, J.E., 1987, Flood power, in catastrophic flooding, *The Binghamton Symposia in Geomorphology*, Mayer, L. and Nash, D., Eds., International Series, No. 18, Allen & Unwin, Boston, MA, pp. 1–21.

Bathurst, J.C. et al., 2010, Forests and floods in Latin America: Science, management, policy and the EPIC FORCE project, *Water Int.*, 35(2), 114–131.

Bathurst, J.C. et al., 2011, Forest impact on floods due to extreme rainfall and snowmelt in four Latin American environments 1: Field data analysis, *J. Hydrol.*, 400(3–4), 281–291.

Bedient, P.B. and Huber, W.C., 1988, *Hydrology and Floodplain Analysis*, Addison-Wesley Publishing Co., Reading, MA.

Berenbrock, C., 2002, Estimating the magnitude of peak flows at selected recurrence intervals for streams in Idaho, USGS Water Resources Investigations Reports 02-4170, Denver, CO.

Berz, G., 2000, Flood disasters: Lessons from the past—Worries for the future, *Proc. Inst. Civil Eng., Water Marine Eng.*, 142(March), 3–8.

Bisese, J.A., 1995, Methods for estimating the magnitude and frequency of peak discharges of rural, unregulated streams in Virginia, USGS Water Resources Investigations Report 94-4148, Denver, CO.

Black, P.E., 2004, Forest and wildland watershed functions, *A Century of Forest and Wildland Watershed Lessons*, Ice, G.G. and Stednick, J.D., Eds., Society of American Foresters, Bethesda, MD, pp. 1–18.

Blow, F.E., 1955, Quantity and hydrologic characteristics of litter under upland oak forests in eastern Tennessee, *J. For.*, 53, 190–195.

Bomar, G.W., 1983, *Texas Weather*, University of Texas Press, Austin, TX.

Boyer, M.C., 1964, Streamflow measurement, *Handbook of Applied Hydrology*, Chow, V.T., Ed., McGraw-Hill, New York, pp. 15, 1–41.

Broadfoot, W.M. and Williston, H.L., 1973, Flooding effects on southern forests, *J. For.*, 71, 584–587.

Brown, L.R., 1998, *The Yangtze Flood: The Human Hand, Local and Global*, Worldwatch Institute, Washington, DC, http://www.worldwatch.org//alerts//pr98813.html

Bruijnzeel, L.A., 1990, Hydrology of moist tropical forests and effects of conservation: A state of knowledge review, UNESO International Hydrological Programme, Division of Water Sciences, Paris, France.

Bruijnzeel, L.A., 2004, Hydrological functions of tropical forests: Not seeing the soil for the trees? *Agric. Ecosyst. Environ.*, 104, 185–228.

Burke, M.K. et al., 2003, Vegetation, soil, and flooding relationships in a blackwater floodplain forest, *Wetlands*, 23, 988–1002.

Calder, I.R. and Aylward, B., 2006, Forest and floods: Moving to an evidence-based approach to watershed and integrated flood management, *Water Int.*, 31(1), 87–99.

Canadell, J. et al., 1996, Maximum rooting depth of vegetation types at the global scale, *Oecologia*, 108, 583–595.

Carter, J.M., Williams, J.E., and Teller, R.W., 2002, The 1972 Black Hills–Rapid City flood revisited, USGS Fact Sheet FS-037-02.

Chang, M. and Crowley, C.M., 1997, Downstream effects of a dammed reservoir on streamflow and vegetation in East Texas, *Sustainability of Water Resources under Increasing Uncertainty*, Rosbjerg, D. et al., Eds., IAHS Publication No. 240, Oxfordshire, U.K., pp. 267–275.

Chang, M., Lee, R., and Dickerson, W.H., 1976, Adequacy of hydrologic data for application in west Virginia, Water Research Institute Bulletin 7, West Virginia University, Morgantown, WV.

Chang, M., McDonald, D.L., and Mouton, A.G., 2004, The spatial variation for floods with return periods greater than 100-year in the conterminous United States, *Hydrology: Science and Practice for the 21st Century*, Webb, B. et al., Eds., Vol. I, British Hydrological Society, London, U.K., pp. 95–102.

Chang, M. and Watters, S.P., 1984, Forests and other factors associated with streamflows in East Texas, *Water Resour. Bull.*, 20, 713–719.

Chen, H.H., 2003, A study on how changes in land use impact water resources in Taiwan, *Water Int.*, 28, 422–425.

Chin, E.H., Skelton, J., and Guy, H.P., 1975, The 1973 Mississippi river basin flood: Compilation and analyses of meteorologic, streamflow, and sediment data, U.S. Geological Survey Professional Paper 937, Washington, DC.

Chittenden, H.M., 1909, Forests and reservoirs in their relation to streamflow with particular reference to navigable streams, *ASCE Trans.*, 62, 245–318.

Chow, V.T., 1964a, Statistical and probability analysis of hydrologic data: Part I, frequency analysis, *Handbook of Applied Hydrology*, Chow, V.T., Ed., McGraw-Hill, New York, pp. 8.1–8.42.

Chow, V.T., 1964b, Hydrology of flow control: Part I, flood characteristics and flow determination, *Handbook of Applied Hydrology*, Chow, V.T., Ed., McGraw-Hill, New York, pp. 25.1–25.33.

Costa, J.E., 1985, Floods from dam failures, USGS Open-file Report 85-560, Denver, CO.

Costa, J.E., 1987, A comparison of the largest rainfall–runoff floods in the United States with those of the People's Republic of China and the world, *J. Hydrol.*, 96, 101–115.

Costa, J.E. and Schuster, R.L., 1988, The formation and failure of natural dams, *Geol. Soc. Am. Bull.*, 100(7), 1054–1068.

Currie, W.S. et al., 2002, Processes affecting carbon storage in the forest floor and in downed woody debris, *The Potential of U.S. Forests to Sequester Carbon and Mitigate the Greenhouse Effect*, Kimble, J.M. et al., Eds., CRC Press, Boca Raton, FL.

Daniel, E.B. et al., 2011, Watershed modeling and its applications: A state-of-the-art review, *The Open Hydrol. J.*, 5, 26–50.

Dillow, J.J.A., 1996, Techniques for estimating magnitude and frequency of peakflows in Maryland, USGS Water Resources Investigations Report 95-4154, Denver, CO.

Duncan, S.H., 1986, Peak stream discharge during 30 years of sustained yield timber management in two fifth-order watersheds in Washington state, *Northwest Sci.*, 60, 258–264.

Eisenbies, M.H. et al., 2007, Forest operations, extreme flooding events, and considerations for hydrologic modeling in the Appalachians—A review, *For. Ecol. Manage.*, 242, 77–98.

Estifanos, M., 2006, The temporal variation of significant floods in the United States and impact of landuse on flooding in two watersheds in east Texas, unpublished MS thesis, Stephen F. Austin State University, Nacogdoches, TX.

FAO, 2005, *Forests and Floods, Drowning in Fiction or Thriving in Facts?* UN Food and Agriculture Organization, RAP Publication 2005/03, Forest perspectives 2, Bangkok, Thailand.

Fitzpatrick, F.A., Knox, J.C., and Whitman, H.E., 1999, Effects of historical land-cover changes on flooding and sedimentation, North Fish Creek, Wisconsin, USGS Water Resources Investigations Report 99-4083, Denver, CO.

Floyd, C., 1990, *America's Greatest Disaster*, Mallard Press, New York.

Fongers, D., 2003, A hydrologic study of the Sand Creek watershed, Geological and Land Management Division, Michigan Department of Environmental Quality, Lansing, MI.

Fosberg, M.A., 1977, Heat and water transport properties in Conifer Duff and Humus, Research Paper RM-195, USDA Forest Service, Rocky Mountain Forest Range Experiment Station, Fort Collins, CO.

Framcou, J. and Rodier, J., 1967, Essai de classification des maximales observées dans le monde (Report on the classification of the world's maximum observed floods), *Cahiers ORSTOM Série Hydrologie*, IV(3), 19–45, ORSTOM Bondy, France.

Gergel, S.E., Dixon, M.D., and Turner, M.G., 2002, Consequences of human-altered floods: Levees, floods, and floodplain forests along the Wisconsin River, *Ecol. Appl.*, 12, 1755–1770.

Gong, W.K., 1982, Leaf litter fall, decomposition, and nutrient element release in a lowland dipterocarp forest, *Malays. For.*, 45, 367–378.

Gustavson, T.C., Holliday, V.T., and Hovorka, S.D., 1995, Origin and development of playa basins, sources of recharge to the Ogallala aquifer, southern high plains, Texas and New Mexico, Bureau of Economic Geology, Report of Investigations No. 229, University of Texas, Austin, TX.

Haan, C.T., 1977, *Statistical Methods in Hydrology*, The Iowa State University Press, Ames, IA.

Haigh, M.J., Rawat, J.S., and Bight, H.S., 1990, Hydrological impact of deforestation in the central Himalaya, *Hydrology of Mountainous Areas*, IAHS Publication No. 190, Oxfordshire, U.K., pp. 419–433.

Hamilton, L.S. and King, P.N., 1983, *Tropical Forested Watersheds, Hydrologic and Soils Response to Major Uses or Conversions*, Westview Press, Boulder, CO.

Hanks, R.J. and Ashcroft, G.L., 1980, *Applied Soil Physics*, Springer-Verlag, New York.

Harr, R.D., 1981, Scheduling timber harvest to protect watershed value, *Proceedings, Interior West Watershed Management*, Spokane, WA, Baumgartner, D.M., Ed., Washington State University Cooperative Extension, Pullman, WA, pp. 269–282.

Harr, R.D., 1982, Fog drip in the Bull Run municipal watershed, Oregon, *Water Resour. Bull.*, 18, 785–788.

Harr, R.D. and McCorison, F.M., 1979, Initial effects of clearcut logging on size and timing of peak flows in a small watershed in western Oregon, *Water Resour. Res.*, 15, 90–94.

Haukos, D.A. and Smith, L.M., 1992, Ecology of playa lakes, *Waterfowl Management Handbook*, Fish and Wildlife Leaflet 13.3.7, p. 7, U.S. Fish and Wildlife Service, Washington, DC.

Heineman, J., 1998, Forest floor planting: A discussion of issues as they relate to various site-limiting factors, *Forest Practices: Forest Site Management Section*, Silviculture Note 16, British Columbia Ministry of Forests, British Columbia, Canada.

Hendriks, C.M.A. and Bianchi, F.J.J.A., 1995, Root density and root biomass in pure and mixed forest stands of Douglas-fir and beech, *Neth. J. Agric. Sci.*, 43, 321–331.

Hewlett, J.D. and Bosch, J.M., 1984, The dependence of storm flows on rainfall intensity and vegetal cover in South Africa, *J. Hydrol.*, 75, 365–381.

Hewlett, J.D., and Doss, R., 1984, Forests, floods, and erosion: A watershed experiment in the southeastern Piedmont, *For. Sci.*, 30, 424–434.

Hirschboeck, K.K., 1987, Catastrophic flooding and atmospheric circulation anomalies, *Catastrophic Flooding*, Mayer, L. and Nash, D., Eds., Allen & Unwin, Inc., Winchester, MA, pp. 23–56.

Jackson, C.R. et al., 2004, Fifty years of forest hydrology in the Southeast, *A Century of Forest and Wildland Watershed Lessons*, Ice, G.G. and Stednick, J.D., Eds., Society of American Foresters, Bethesda, MD, pp. 33–112.

Jones, J.A., 2000, Hydrologic processes and peak discharge response to forest removal, regrowth, and roads in 10 small experimental basins, western Cascades, Oregon, *Water Resour. Res.*, 36, 2621–2642.

Jones, J.A. and Grant, G.E., 1996, Peak flow responses to clear-cutting and roads in small and large basins, western Cascades, Oregon, *Water Resour. Res.*, 32, 959–974.

Kimmins, J.P., 1987, *Forest Ecology*, Macmillan Publishing Co., New York.

Kochenderfer, J.N., Edwards, P.J., and Wood, F., 1997, Hydrologic impacts of logging an Appalachian watershed using West Virginia's best management practices, *N. J. Appl. For.*, 14, 207–218.

Koellner, W.H., 1996, The flood's hydrology, *The Great Flood of 1993: Causes, Impacts, and Responses*, Changnon, S.A., Ed., Westview Press, Boulder, CO, pp. 68–100.

Kramer, P.J. and Boyer, J.S., 1995, *Water Relations of Plants and Soils*, Academic Press, New York.

Kundzewicz, Z.W., 2002, The flood of the floods—Poland, summer 1997, *The Extremes of the Extremes: Extraordinary Floods*, Snorasson, A., Finnsdóttir, H.G., and Moss, M., Eds., IAHS Publication 271, Oxfordshire, U.K., pp. 147–153.

La Marche, J.L. and Lettenmaier, D.P., 2001, Effects of forest roads on flood flows in the Deschutes River, Washington, *Earth Surf. Process. Landf.*, 26, 115–134.

Laurén, A. and Heiskanen, K., 1997, Physical properties of the mor layer in a Scots pine stand, I. Hydraulic conductivity, *Can. J. Soil Sci.*, 77, 627–634.

van der Leeden, F.V., Troise, F.L., and Todd, D.K., 1990, *The Water Encyclopedia*, Lewis Publishers, Chelsea, MI.

Lewis, J.M., 2010, Methods for estimating the magnitude and frequency of peak streamflows for ungaged streams in Oklahoma, USGS Scientific Investigations Report 2010-5137, Reston, VA, 42 pp.

Liu, X., Wu, Q., and Zhao, H., 1991, A study on hydro-ecological functions of litters of artificial Chinese pine forest on the loess plateau, *J. Soil Water Conserv.*, 5, 87–91.

Lowdermilk, W.C., 1930, Influence of forest litter on runoff, percolation, and erosion, *J. For.*, 28, 474–491.

Lull, H.W. and Reinhart, K.G., 1972, Forests and floods in the eastern United States, USDA Forest Service Research Paper NE-226, NE Forest Experiment Research Station, Upper Darby, PA.

MacDonale, L.H., and Hoffman, J.A., 1995, Causes of peak flows in northwestern Montana and northeastern Idaho, *Water Resour. Bull.*, 31, 79–95.

Maddox, R.A., Canova, F., and Hoxit, L.R., 1980. Meteorological characteristics of flash flood events over the western United States, *Monthly Weather Rev.*, 108, 1866–1877.

Maddox, R.A. and Chappell, C.F., 1979, Flash flood defenses, *Water Spectr.*, 11(2), 1–8.

Marques, M.M. et al., 2009, Dynamics and diversity of flooded and unflooded forests in a Brazilian Atlantic rain forest: A 16-year study, *Plant Ecol. Divers.*, 2(1), 57–64.

Marsh, G.P., 1864, *Man and Nature (Physical Geography as Modified by Human Action)*, Charles Scribner, New York.

Megahan, W.F., 1983, Hydrologic effects of clearcutting and wildfire on steep granitic slopes in Idaho, *Water Resour. Res.*, 19, 811–819.

Moore, W.L., 1910, A Report on the influence of forests on the climate and on floods, U.S. Weather Bureau, Washington, DC.

Mortimer, M.J. and Visser, R.J.M., 2004, Timber harvesting and flooding: Emerging legal risks and potential mitigations, *South. J. Appl. For.*, 28, 69–75.

Myers, N., 1986, Environmental repercussions of deforestation in the Himalaya, *J. World For. Resour. Management*, 2, 63–72.

National Weather Service, 1972, Report to the Administrator—Buffalo Creek Disaster on February 26, 1972, Buffalo Creek Disaster Hearings, Part I, 92nd Congress, U.S. Government Printing Office, Washington, DC, pp. 265–289.

van Noordwijk, M. and Agus, F., 2004, Dam-busting forest, water 'myth-understandings,' *Opinion, the Jakarta Post*, December 8, 2004, http://www.thejakartapost.com//detaileditorial.asp?field=20041208.F04&irec=3

Nordin, C.F. and Meade, R.H., 1982, Deforestation and increased flooding of the Upper Amazon, *Science*, 215, 426–427.

O'Connor, J.E. and Costa, J.E., 2004, Spatial distribution of the largest rainfall–runoff floods from basins between 2.6 and 26,000 km^2 in the United States and Puerto Rico, *Water Resour. Res.*, 40, W01107.

Papp, F., 2002, Extremeness of extreme floods, *The Extremes of the Extremes: Extraordinary Floods*, Snorasson, A., Finnsdóttir, H.P., and Moss, M., Eds., IAHS Publication No. 271, Oxfordshire, U.K., pp. 373–378.

Parrett, C. and Johnson, D.R., 2004, Methods for estimating flood frequency in Montana based on data through water year 1998, USGS Water Resources Investigations Report 03-4308, Denver, CO, 109 pp.

Perry, C.A., Aldridge, B.N., and Ross, H.C., 2001, Summary of significant floods in the United States, Puerto Rico, and the Virgin Islands, 1970 through 1989, USGS Water-Supply Paper 2502, Washington, DC.

Pinchot, G., 1908, The relation of forests to stream control, American Academy of Political and Social Science, *Annals*, 31, 219–227.

Pinchot, G., 1913. Forests as life-savers, *Outlook*, May 17, pp. 103–104.

Plamondon, P.A., Black, T.A., and Goddell, B.C., 1972, The role of hydrologic properties of the forest floor in watershed hydrology, *Watersheds in Transition*, American Water Resources Association, Proceedings Series No. 14, Bethesda, MD, pp. 341–348.

Potts, D.F., 1984, Water potential of forest duff and its possible relationship to regeneration success in the northern Rocky Mountains, *Can. J. For. Res.*, 15, 464–468.

Pritchett, W.L., 1979, *Properties and Management of Forest Soils*, John Wiley & Sons, New York.

Rakhecha, P.R., 2002, Highest floods in India, *The Extremes of the Extremes: Extraordinary Floods*, Snorasson, A., Finnsdóttir, H.P., and Moss, M., Eds., IAHS Publication No. 271, Oxfordshire, U.K., pp. 167–172.

Raman, S., 2005, Tsunami villagers give thanks to trees, *BBC News*, February 16, 2005, available at: http://news.bbc.co.uk/2/hi/south_asia/4269847.stm, retrieved on July 18, 2012.

RMS, 2001, Tropical Storm Allison, June 2001, RMS event report, Risk Management Solutions, Inc., Newark, CA.

Rodier, J.A. and Roche, M.J., 1984, *World Catalogue of Maximum Observed Floods*, IAHS Publication No. 143, Institute of Hydrology, Oxfordshire, U.K.

Sando, S.K., 1998, Techniques for estimating peak-flow magnitude and frequency relations for South Dakota streams, USGS Water Resources Investigations Report 98-4055, Denver, CO.

Sartz, R.S., 1983, Watershed management, *Encyclopedia of American Forest and Conservation History*, Vol. II, Davis, R.C., Ed., MacMillan Publishing Co., New York, pp. 680–684.

Schenk, H.J. and Jackson, R.B., 2002, The global biogeography of roots, *Ecol. Monogr.*, 72, 311–328.

Schwartz, F.K. et al., 1975, The Black Hills–Rapid City flood of June 9–10, 1972—A description of the storm and flood, USGS Professional Paper No. 877, Washington, DC.

Shankman, D. and Liang, Q., 2003, Landscape changes and increasing flood frequency in China's Poyang Lake region, *The Prof. Geogr.*, 55, 434–455.

Sharp, D. and Sharp, T., 1982, The desertification of Asia, *ASIA 2000*, 1, 40–42.

Soil Survey Staff, 1999, *Soil Taxonomy*, 2nd edn., USDA-NRCS Superintendent of Documents, U.S. Printing Office, Washington, DC.

Sopper, W.E., 1970, Education in forest hydrology in the United States, *Am. Geophys. Union Trans.*, 51, 494–500.

State of Oregon, 2000, Natural hazards mitigation plan: Flood chapter—State flood damage reduction plan, Emergency Management Plan, Salem, OR.

Stedinger, J.R., Vogel, R.M., and Foufoila-Georgiou, E., 1992, Frequency analysis of extreme events, *Handbook of Hydrology*, Maidment, D.R., Ed., McGraw-Hill, New York, pp. 18.1–18.66.

Stednick, J.D. and Troendle, C.A., 2004, Water yield and timber harvesting practices in the subalpine forests of the central Rocky Mountains, *A Century of Forest and Wildland Watershed Lessons*, Ice, G.G. and Stednick, J.D., Eds., Society of American Foresters, Bethesda, MD, pp. 169–186.

Storck, P. et al., 1995, Implications of forest practices on downstream flooding, Phase II Final Report, Timber Fish Wildlife Service Report, TFW-SH20-96-001, Washington Department of Natural Resources, Olympia, WA.

Stuckey, M.H. and Reed, L.A., 2000, Techniques for estimating magnitude and frequency of peakflows for Pennsylvania streams, USGS Water Resources Investigations Report 00-4189, U.S. Geological Survey, Washington, DC.

Sumioka, S.S., Kresch, D.L., and Kasnick, K.D., 1998, Magnitude and frequency of floods in Washington, USGS Water Resources Investigations Report 97-4277, Tacoma, WA, 91 pp.

Sydes, C. and Grime, J.P., 1981, Effects of tree leaf litter on herbaceous vegetation in deciduous woodland, I. Field investigations, *J. Ecol.*, 69, 237–248.

Taylor, G.H. and Hatton, R.R., 1999, *The Oregon Weather Book: A State of Extremes*, Oregon State University, Corvallis, OR.

Teskey, R.O. and Hinckley, T.M., 1977, Impact of water level changes on woody riparian and wetland communities, Vol. I, II, III. FWS/OBS-77/58, 77/59. 77/60, U.S. Fish and Wildlife Service, Columbia, MO.

Texas Water Development Board, 1998, Groundwater Management Plan, Harris-Galveston Coast Subsidence District, Austin, TX.

The University of California, 1999, Review of an analysis of flooding in Elk river and fresh creek watersheds, Humboldt County, California, by the Pacific Lumber Company, March 1999, Committee on the Scientific Basis for the Analysis and Prediction of Cumulative Watershed Effects, UC Center for Forestry, College of Natural Resources, University of California, Berkeley, CA.

The World Resources Institute et al., 1996, *World Resources, 1996–97*, Oxford University Press, Oxford, U.K.

Thomas, R.B. and Megahan, W.F., 1998, Peak flow responses to clear-cutting and roads in small and large basins, western Cascades, Oregon: A second opinion, *Water Resour. Res.*, 34, 3393–3403.

Tozlowski, T.T., 1997, Responses of woody plants to flooding and salinity, *Tree Physiol. Monogr.*, 1, 1–29.

Tran, P., Marincioni, F., and Shaw, R., 2010, Catastrophic flood and forest cover change in the Huong river basin, central Viet Nam: A gap between common perceptions and fact, *J. Environ. Manage.*, 91(11), 2186–2200.

U.S. Department of Interior, 1998, Prediction of embankment dam breach parameters: A literature review and needs assessment, Water Resources Research Laboratory, Dam Safety Office, DSO-98-004, Washington, DC.

Walder, J.S., and Costa, J.E., 1996, Outburst floods from glacier-dammed lakes: The effect of mode drainage of flood magnitude, *Earth Surf. Process. Landf.*, 21, 701–723.

Walker, J.F. and Krug, W.R., 2003, Flood-frequency characteristics of Wisconsin streams, USGA Water Resources Investigations Report 03-4250, Denver, CO.

Wang, Y. and Weng, J., 2002, The water and soil conservation functions of litter on forested land, *Proceedings of the 12th International Soil Conservation Organization (ISCO) Conference*, Vol. II, Jiao, Y., Ed., Tsinghua University Press, Beijing, China, pp. 46–50.

Wigmosta, M.S., Vail, L.W., and Lettenmaier, D.P., 1994, A distributed hydrology–vegetation model for complex terrain, *Water Resour. Res.*, 30, 1665–1679.

Wiley, J.B., Atkins, J.T., Jr., and Newell, D.A., 2002, Estimating the magnitude of annual peak discharges with recurrence intervals between 1.1 and 3.0 years for rural, Unregulated Streams in West Virginia, USGA Water Resources Investigations Report 02-4164, Denver, CO.

Wright, K.A. et al., 1990, Logging effects on streamflow: Storm runoff at Caspar Creek in northwest California, *Water Resour. Res.*, 26, 1657–1667.

Yin, R., Xu, J., and Liu, C., 2005, China's ecological rehabilitation: The unprecedented efforts and dramatic impacts of reforestation and slope protection in western China, *China Environmental Series*, 7, 17–32.

Zartman, R., Villarreal, C., and Hudnall, W., 2010, Cropping management system influences on playa sediments in US southern high plains, *Symposium 3.1.2 Farm System and Environmental Impacts, the 19th World Congress of Soil Science, Soil Solutions for a Changing World*, International Union of Soil Sciences, August 1–6, 2010, Brisbane, Queensland, Australia, pp. 37–40.

Zon, R., 1927, Forests and water in the light of scientific investigation, U.S. Forest Service, Lake States Forest Experiment Station, U.S. Government Printing Office, Washington, DC.

15

Watershed Management Planning
and Implementation

Forest and water issues are tightly linked. In the United States, more than 60% of drinking water originates from forested watersheds. Many forestry activities in upstream watersheds will inevitably affect water quantity, quality, timing, and aquatic life in downstream. Therefore, forested watershed protection and the management of natural resources are matters of vital interest to the Nation, societies, and individuals.

The forest and water relations discussed in the previous chapters provide the base and principles for managing forested watersheds. They are the foundation for developing watershed management measures relevant to water quantity and quality, flood mitigation, and stream habitats. Watershed management projects seeking technical and financial assistance as authorized by, among others, the Watershed Protection and Flood Prevent Act of 1954 (PL83-566), the Flood Control Act of 1944 (PL78-543), and the Clean Water Act of 1977 (PL 95-217) can go to the standard procedures and guidelines provided by the Natural Resources Conservation Service (NRCS, 2009). Also, EPA (2008) provides a thorough handbook for developing watershed plans to restore and protect the Nation's waters. This chapter only deals with the key elements and general activities involved in the watershed management planning process as illustrated in Figure 15.1.

15.1 Watershed Programs/Projects

A watershed program is to obtain certain overall goals and benefits for the watershed and people through a number of watershed projects, each addressing a different aspect of watershed issues and problems. These watershed projects collectively and coordinately manage the watershed to achieve those goals and benefits not attainable by each project alone. Individuals, organizations, or whoever has interests and concerns regarding the prevention, restoration, and enhancement of natural resources in watersheds can initiate watershed management programs/projects. However, managing watersheds is often too complex and too expensive for one person or one organization to tackle alone because (a) there are multidisciplinary and multiple natural resources involved, (b) activities may affect downstream watersheds, (c) severe weather events are unpredictable and may cause devastating impacts to the watersheds, (d) interests and objectives on watershed management are different among landowners, (e) watersheds may be covered by multiple political jurisdictions, (f) it may require compliance with a variety of regulations enacted by federal, state, and local governments, and (g) current environmental regulations and standards may be modified and new ones enacted during the planning and implementation process. It is therefore essential for the project sponsor to bring in new ideas, inputs, assistance, and support from different sectors and disciplines. This process can strengthen

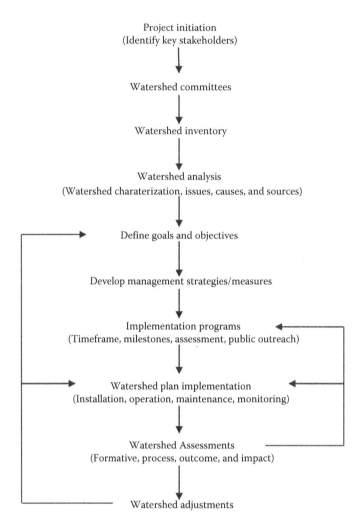

FIGURE 15.1
General steps in watershed planning and implementation process.

the collaborative levels of the project and increase public understanding, interests, and commitment toward solutions of watershed problems.

15.1.1 Stakeholders

Watershed management plans often bring together people and organizations that have interests, concerns, and knowledge on watershed activities to join the planning, implementation, and assessment processes. They are called stakeholders. Potential stakeholders may include landowners, representatives of local, state and federal agencies, people from environmental groups, business and industries, experts/scholars, and community and religious leaders, etc. The stakeholders who are identified and invited to join the watershed group are affected by the project and will be responsible for the watershed planning and implementation. They are knowledgeable about existing programs and able to provide the information and technical and financial supports.

15.1.2 Watershed Committees

Once stakeholders are invited, they need to be organized into groups so that their participations in the watershed planning and implementation processes can make the project more effective and successful. The makeup of these groups depends on the stakeholders' willingness and contribution levels, and the time and financial resources available of the project. The following three groups are common in watershed planning:

- Executive committee: Decision-making, governing board of the project. Committee members require representation from a broad array of public agencies and private entities.
- Advisory committee: Provides advice to the executive committee on direction and management of the project but does not have authority to vote on project matters.
- Technical committee: Responsible for works on project data collections, watershed analyses, defining watershed problems, developing management strategies, formulating management options, implementation, and assessments.

15.2 Watershed Inventory

A successful watershed management project depends a great deal on how much we know about the watershed and its problems, how these problems are generated, and what impacts they have on watershed resources and environmental conditions. Once a watershed management project is initiated and working committees are formed, the first task is to conduct watershed data collections so that the watershed conditions and problems can be understood.

15.2.1 Watershed Data

Data typically solicited to characterize watershed conditions, problems, and resources in watershed management are given in Table 15.1. How detailed and extensive the collections of these data are dependent on watershed size and its resource, the severity of watershed impairment, potential treatments, and affected population. The availability, reliability, and representativeness of these data can greatly affect the success of the planning and implementation of a watershed program/project. An insufficient, unreliable, or unrepresentative set of data can lead to incorrect planning, improper designs, unsuitable measures, unnecessary operation delay, and increased project costs. On the other hand, a collection of data more than necessary is costly and wasting time.

15.2.2 Existing Data

Once the types of data required for watershed characterization and watershed planning are determined, many of these data may have already been collected, reported, and archived by governments at various levels, industries, special interest groups, and individuals. The use of existing data is a must in most watershed management projects. They save time and costs in the data collection programs.

TABLE 15.1

Data Typically Solicited in Watershed Inventory

Data Type	Coverage
Watershed settings	Watershed geographic locations, boundaries, areas, river systems, topographic map, transportation, political jurisdiction, etc.
Topography/soil	Elevations, slopes, aspects, other topographic features; geologic formations, soil types, soil water storage, soil erosion, soil stability.
Hydrology	Streamflow records, gaging stations, daily, monthly, and annual streamflows, flow durations, water bodies, flood stages, water tables, etc.
Climate	Annual and monthly precipitation, rainfall intensities, snowfall, extreme rainfalls, annual and monthly temperatures, maximum and minimum temperatures, prevailing wind directions and speeds, severe weather events.
Land use	Land-use types, areas and history, forest species composition and ages, site index, logging roads, existing management practices, production, land ownerships.
Stream habitats	Riparian shading, aquatic life, stream segment alterations, impairments, bed materials, sedimentation, in-stream flow alteration.
Wildlife	Endangered species.
Water pollution	Protected water bodies and associated water quality standards, potential pollution sources; impaired stream segments, causes and sources (303d list); existing TMDL reports.
Social community	Cities, population statistics, industries, air quality, highway.

15.2.2.1 At Federal Level

Many federal government agencies are responsible for monitoring environmental conditions of the Nation. Some have to do so in conjunction with their assigned jurisdictions. Major sources of environmental data needed for developing watershed management programs/projects are given as follows:

- *Climatic data*: The National Weather Service (NWS), a branch of the National Oceanic and Atmosphere Administration (NOAA), is responsible for climatological observations and flood forecasting of the Nation. NWS maintains a total of around 10,700 precipitation stations and 7,000 air temperature stations throughout the Nation. Other observations include snow depth, wind movements, pan evaporation, solar radiation, and soil temperature. Two NWS publications, by state, are of importance to watershed managers, "Monthly Climatological Data" and "Hourly Precipitation." Readers can go to the website of the National Climatic Center (under NWS, NOAA) at Ashville, TN for daily and monthly records of basic meteorological variables for stations across the United States.

  In addition to NWS, the National Water and Climate Center (NWCC) of the Natural Resources Conservation Service (NRCS) also maintains a network of about 600 climate stations in the western United States called SNOTEL (SNOwpackTELemetry). Data on daily and hourly precipitation, along with pan evaporation, are also available at dams and reservoirs operated by the U.S. Army Corps of Engineers (eastern United States) and Bureau of Reclamation (western United States).

- *Streamflow data*: Chartered in 1879 by Congress, the U.S. Geological Survey (USGS) is the Nation's leading earth science information agency. The USGS operates and maintains more than 85% of the Nation's stream gaging stations, or 7292 stations dispersed throughout the Nation and 4200 of which are equipped with earth satellite radios that provide real-time communications. Records of discharge and stages of streams, lakes, and reservoirs measured by the USGS are published in a series of papers entitled, "Surface Water Supply of the United States." Before 1960, these water supply papers were published in an annual series; later records are in a 5-year series. The USGS streamflow records have been published as "Water Resources Data," on a statewide basis, since 1964. Today, long-term stream and river discharge data at many gage stations can be directly downloaded from the USGS National Water Information System (NWIS) at http://waterdata.usgs.gov/nwis.

Water quality data are also available at many USGS gaging stations. Historical records from those stations can be downloaded from the NWIS site. USGS also provides a new Water Quality Watch website (http://waterwatch.usgs.gov/wqwatch/) on real-time water quality measurements, including water temperature, pH, specific conductance, turbidity, and nitrate, at more than 1300 stations.

- *Soil information*: Soils in each county are mapped by the NRCS and data relevant to soil types, locations, physical properties, uses, and use limitations are published as "Soil Survey" reports on a county-wide basis. The reports can be obtained free of charge from an NRCS local office. Since the 1990s, the soil maps have been digitalized and the survey data have been entered into three geo-referenced soil databases: STATSGO (State Soil Geographic Data Base), SSURGO (Soil Survey Geographic Data Base), and National Soil Geographic Data Base (NATSGO). While STATSGO and SSURGO are two most used soil datasets, the latter provides much detailed spatial information and has been found to be more accurate in watershed-scale estimation of soil organic matter (SOM) (Zhong and Xu, 2011).

- *Topographic maps*: The topographic maps needed for land-use planning and watershed topography characterization can be obtained from the USGS. The best known USGS topographic maps are the 1:24,000 scale, commonly known as 7.5 min quadrangles. Officially completed in 1992, the 7.5 s series cover the entire 48 conterminous states with more than 55,000 maps in considerable detail.

  Moreover, the USGS has recently made a new set of topographic information, *The National Map*, available for the Nation. *The National Map* can be used from recreation to scientific analysis and to emergency response. It is easily accessible for display on the web, as products and services, and as downloadable data. The geographic information available from *The National Map* includes aerial photographs, elevation, geographic names, hydrography, boundaries, transportation, structures, and land cover:

- *Wetlands information*: Section 404 of the Clean Water Act (CWA) regulates the discharge of dredged or fill material into waters of the United States. This program is jointly implemented by EPA and the U.S. Army Corps of Engineers. The U.S. Fish and Wildlife Service has continuously maintained wetland inventory and has developed a series of geospatial maps to show wetlands and deepwater habitats in the United States. The digital data can be viewed and downloaded by the public from the agency's web page at http://www.fws.gov/wetlands/ for use in management, research, policy development, education, and planning activities.

In addition, the National Marine Fisheries Service, NRCS, and state resource agencies have important advisory roles. These agencies can help identify what management programs exist or what data are available in your studied watersheds.

15.2.2.2 At State Level

There are many state agencies that routinely collect biological, hydrological, climatological, aquatic, land use, wildlife, and agricultural information in the state. These agencies are typically responsible for environmental quality, agriculture and forestry, wildlife and fisheries, and transportation. Some federal agencies also have offices or divisions in states. They can be of assistance with environmental data, management projects in progress, and environmental regulations.

15.2.2.3 At Local Level

Offices of local water and soil conservation districts, watershed management clubs, flood control boards, and other nongovernmental and environmental groups may be able to provide information and data useful to watershed management. Some industries and environmental consulting firms may have reports on TMDLs (Total Maximum Daily Loads, Section 303 of the Clean Water Act) and nonpoint pollution planning (Section 208, CWA). Local universities' research programs on water, soil, forest, wildlife, and aquatic environment often require short- and long-term monitoring on environmental conditions. They may have data and information useful to the projected watersheds.

Other informal information, especially on extreme values, may be obtained from local newspapers, diaries of pioneers, files of local and other historical societies, flood marks on trees and buildings, private collections of memories of seniors, local agricultural extension agency, or soil conservation personnel are often able to provide information on where and from whom to obtain environmental data.

15.2.3 Field Data

Since the existing data are important and fundamental, they need to be reviewed, examined, and assessed for (a) data quality, (b) adequacy on the delineation of spatial and temporal variations of the watershed conditions, (c) ability to identify and quantify the water quantity and quality problems of the watershed, and (d) linkage on watershed problems to specific sources in the watershed. If data are found insufficient to meet the needs for watershed characterization and watershed planning, then a field sampling plan to fill data gaps needs to be set up and operated.

Data collection is of fundamental importance not only for immediate needs in watershed characterization and planning, but also for long-term tracking needs on watershed management progress and assessments. Thus, a sampling plan should, at least, consider (a) project goals and objectives, (b) measurement errors, (c) data representativeness, (d) duration and frequency of the data monitoring programs, and (e) quality control and the standard operation procedures in laboratory analyses. Detailed information on sampling designs is given by EPA (1997) for monitoring the effectiveness of nonpoint source control, by EPA (2002) for environmental data collection, by Rex and Carmichael (2002) for monitoring fine sediment deposition in streams, by Argonne National Laboratory (2004) for habitat restoration monitoring, and by Kilgour et al. (2007) for monitoring aquatic environmental impacts.

15.3 Watershed Analysis

The data collected from the studied watersheds are records of environmental conditions over time and space. These data need to be analyzed for their confined environmental significance and the revelation of watershed conditions and problems in a time–space context. Watershed analysis is to make the collected data meaningful and useful, a necessary step in watershed management.

15.3.1 Data Deficiency

The collected data, both existing and monitoring, should be reviewed and examined for data deficiency on two aspects, data quality and data quantity.

15.3.1.1 Data Quality

Error and abnormal records can usually be identified by visual reviews and graphical examinations (Helsel and Hirsch, 2002). Missing data may be estimated by surrounding stations (Chang and Wong, 1983), if available, or by graphical/statistical techniques (McCuen, 1998). Data consistency is another concern on long-term precipitation/streamflow records and can be tested and adjusted by the classic double-mass analysis (Merriam, 1937; Chang and Lee, 1974).

15.3.1.2 Data Quantity

There are two concerns regarding this issue: (a) number of monitoring stations required in a watershed and (b) number of observations required at each station. The collected data need to be representative for a population with a sufficient variation in time and space.

Collecting the entire population is impossible and population mean is always unknown. Thus, we can only collect samples with certain confidence that the collected sample mean is deviated from the true mean by an acceptable error limit. Based on statistical theory of sampling error, sample size for a continuous data population can be determined by

$$N = \frac{t^2 C^2}{E^2} \qquad (15.1)$$

where
  N is the number of observations required to ensure a probability that the sample mean will be within a desired% of the true mean (% error limit)
  t is the Student's t value with $N - 1$ degrees of freedom at the preselected probability level
  C is the coefficient of variation of the sample (= sample standard deviation/sample mean)
  E is the desired limit of accuracy, a ratio of error limit to the sample mean

Equation 15.1 shows that sample size N increases with increasing t (higher% of confidence) and C (data more dispersed from the sample mean), but with decreasing E (smaller error limit). Values of t and E are preselected by the investigator, but C is calculated based on the observed data. Usually, Equation 15.1 is used to test and make adjustments for the present observations to ensure with the preselected % confidence that the current network is sufficient to estimate sample mean within E% error.

For estimating a concerned or occurred proportion in the population, the sample size N required to ensure a percent of confidence that the investigation will be within a desired error E of the true proportion can be estimated by

$$N = \frac{(z^2)(1-p)p}{E^2} \tag{15.2}$$

where
  p is estimated % affected or occurred value
  E is % error
  z is standard error associated with the chosen level of confidence

Values of p and E are expressed in fraction and z is 1.96 for a 95% confidence.

Example 15.1

The mean and standard deviation of annual precipitation observed from a network of 25 stations in a watershed are 1025 and 154 mm, respectively. If one is interested in estimating mean annual precipitation of the watershed, is the present network adequate to ensure that for 19 samples out of 20 (95%), the mean will be within 5% of the true mean?

Solution:

Using Equation 15.1, where C = 154/1025 = 0.15, E = 0.05, t @ 95%, d.f. = 25−1 = 24 = 2.064 (obtained from the distribution of t table available in most textbooks in statistics), then the number of stations required to ensure the degree of accuracy is

$$N = \frac{(2.064^2)(0.15^2)}{0.05^2} = 38.34 \cong 38 \text{ (stations)}$$

Therefore, the current network of 25 stations is not adequate. It requires 38−25 = 13 additional stations to ensure a 95% confidence that the sample mean will be within 5% from the true mean.

Example 15.2

A public survey is designed to estimate the proportion of people having interest in supporting a plan for watershed management. How many persons should be selected if the estimate is expected to be less than 5% from the true proportion with 95% confidence?

Solution:

From the information provided, we get E = 0.05, Z = 1.96 (corresponding to the 95% confidence interval), and p = 0.50 (% of people having interest is unknown). Using Equation 15.2, the number of people required is estimated at

$$N = \frac{(0.5)(1-0.5)(1.96^2)}{0.05^2} = 384.16 \cong 384 \text{ (people)}$$

15.3.2 Watershed Topography

15.3.2.1 Topographic Parameters and Significance

The topography of a watershed can be physically characterized by a number of parameters. A few such parameters and their environmental significance are listed as follows:

- *Watershed area*: The size of watersheds is a very popular parameter used in hydrologic modeling and predictions. Large watersheds have greater watershed storage and thus a greater runoff volume and a greater peak flow than small watersheds do. However, stream sediment loss, flow volume, and flood damage, if all expressed in per unit area, decrease with increasing watershed size.

  In small, upstream watersheds, the average slope is greater, the average rainfall intensity is higher, the soil depth is shallower, the waterway is shorter and steeper, and the climate conditions are more variable than in large watersheds. Thus, the hydrologic behaviors of small watersheds are very sensitive to rainfall intensity and change in land use, consequently causing floods to be flash, duration to be short, and peak flows per unit area to be high.

- *Slope*: Slope topography can be characterized by slope steepness (gradient, in percent or degrees), aspect, length, position, shape, and roughness. These parameters generally affect the downward movements of water, soil, and elements and the isolation of ground surfaces. At 50°N latitude, the annual potential solar radiation irradiated at a 20% south-facing slope is greater than that received at a horizontal surface 10° latitude farther south. Steep slopes are generally associated with shallow soils, low soil moisture contents, and low infiltration rates, which can lead to greater runoff, higher peak flows, and more soil and nutrient losses in watersheds (Chang et al., 1989). Also, OM in soil is usually low on steep slopes (Zhong and Xu, 2009).

- *Elevation*: Precipitation usually increases with elevation and air temperature decreases, due to adiabatic rates, 0.6°C (saturated air) to 1.0°C (dry air) for every 100 m increase in elevation.

- *Drainage density*: It is defined as the ratio of total length of stream channels within a watershed to the watershed area. Drainage density is normally higher in areas with high precipitation, poor permeability, and rugged terrains. Watersheds with a higher drainage density often have a hydrograph with steep falling limb and short flow duration, and thus a great potential risk for flash flood. Drainage density can also reflect the quantity and distribution of SOM in a watershed (Zhong and Xu, 2009), consequently affecting soil nutrient and stream water quality conditions.

- *Stream frequency*: It is a measure of the total number of stream channels (segments) per unit watershed area. Stream frequency is affected by the soil erodibility and permeability, bed rocks, and climatic conditions, a parameter relevant to surface runoff processes. For watersheds with low values of stream frequency, surface runoff is likely to have slow movements, making it susceptible to flooding, gully erosion, and landslides (Eze and Efiong, 2010).

- *Watershed shape*: The shape of watersheds has something to do with watershed storages. There are a number of quantitative terms that have been proposed to describe watershed forms and shapes. For example, the *coefficient of compaction* (CFC) is defined as the ratio between the length of watershed perimeter (L_p, km) and perimeter of a circle whose area (A, km^2) is equal to that of the watershed, or CFC $= 0.282 \times L_p/A^{0.5}$. *Basin form factor* (BFF) is the ratio between watershed area (A, km^2) and the square value of maximum length (L$_m$, km) of the watershed, or BFF $= A/(L_m)^2$. For watersheds with a perfect circle in shape, CFC $= 1$ and BFF $= 0.7854$. It causes the time of concentration on overland flow as well as the time to peak flow short, the rising limb of flow hydrographs steep, and watershed storages small. As a result, peak flows are high and low baseflows are low.

- *Channel gradient*: The USGS usually determines main channel slope from elevations at points 10% and 85% of the distance along the channel from the gaging station to the divide. The difference in elevations between these two points divided by their horizontal distance is the channel gradient. Channel gradient affects flow movement in streams and has been used as an important parameter in hydrology modeling, predictions, and calculations.

- *Relief ratio*: Differences in elevations between two reference points is relief. Watershed total relief (maximum and minimum elevations) divided by the horizontal distance on which it is measured is termed "relief ratio." It relates to erosion potential, infiltration, and runoff processes.

15.3.2.2 Methods of Characterization

15.3.2.2.1 Topo Map Analyses

Traditionally, watershed topographic parameters can be obtained by interpreting information from watershed topographic maps. First, topo maps covering the entire watershed should be solicited and the watershed boundaries above a point of interest (such as stream gaging station, reservoir, etc.) should be delineated on the maps or on tracing papers overlaid on top of the topo maps. Watershed boundaries are the topographic divides (lines of the highest elevations) above the stream gaging station and the enclosed area above the station is the drainage area of the watershed (Figure 15.2).

Once the watershed boundaries are defined, many topographic data and information can be read, interpreted, or measured from the maps by use of a planimeter (measuring watershed area), a map measurer (measuring length of contour lines), a straight edge (linear distances), and a scientific calculator. These data and information are then used to calculate or derive watershed topographic parameters.

15.3.2.2.2 Digital Elevation Models

As part of the national mapping program, the USGS has converted information on terrain elevations into digital data files called the "digital elevation models" (DEMs). A DEM is a raster (grid) representation of terrain elevations for ground position at regularly spaced intervals. It typically consists of thousands of grid cells that represent the topography of an area. These data files are available at various resolutions (10, 30, 90 m, and others) for the entire United States from the USGS Seamless Data Distribution System.

The grid-based DEMs are often used with Geographic Information Systems (GIS). They can be displayed as an elevation layer in GIS and watershed topographic analyses performed, such as watershed delineation, areas, channel and watershed slope, etc., by use of developed hydrologic models. Model tools available for watershed topographic analyses include Basinsoft (Harvey and Eash, 2005), HEC-GepRAS (Ackerman, 2009), Arc Hydro (Maidment, 2002), and many others.

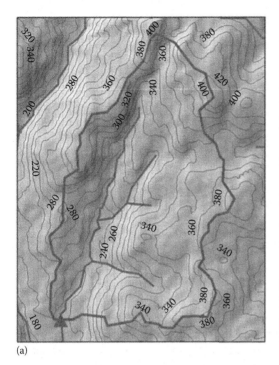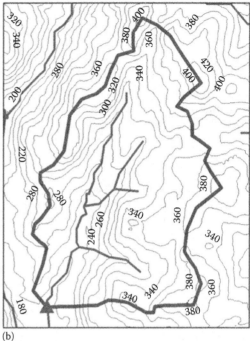

(a)   (b)

FIGURE 15.2
Delineation of a watershed boundary by (a) the USGS "Digital Elevation Models" and by (b) manual topo map analyses. (Courtesy of Dr. Yanli Zhang.)

15.3.3 Data Analyses

All climatic and hydrologic data, and others if applicable, collected should be statistically analyzed and/or graphically examined so that (a) the characteristics of collected data can be revealed, (b) the environmental and managerial significance of collected data can be identified, (c) the relationship between watershed issues and environmental conditions can be established, (d) the watershed impairments and problem areas can be located, (e) the sources of watershed impairments can be determined, and (f) the assessments on watershed conditions before and after treatments or comparisons with other watersheds can be made.

Statistical analyses cover one or all of the following tasks:

- Descriptive statistics, including mean, median, standard deviation, coefficient of variation, maximum, minimum, range, coefficient of skewness, and the central 50% observations (the difference between 25th and 75th percentiles)
- Temporal variations, including daily, monthly, seasonal, and interannual variations, and long-term trends in climatic, hydrologic and other time-series environmental data
- Spatial variations, including geospatial analysis with GIS technology to assess characteristics in climatic, hydrologic, and other spatially referenced data
- Interrelationships, including correlation and multivariate analyses to identify watershed impairments in relation to environmental conditions, land use versus stream sediment and water quality, climate change versus streamflow, watershed characteristics versus impairments

- Estimates, including streamflow, soil erosion, stream sediment, and nutrient loads
- Other analyses that help find locations and causes of watershed impairments

What data analyses and what level of detail is required to conduct depends on the potential problems in the watershed. For example, concern on watershed erosion and stream sedimentation problems requires analyses on rainfall intensity, soils, topography, and vegetation. Frequently, the concerns, interests, and goals of watershed stakeholders may have a great role on the determination of what kinds of analyses are needed.

15.3.4 Watershed Issues

Based on data analyses, field observations, local residents or stakeholders' inputs, and governmental reports, a set of watershed issues/impairments could be produced along with the causes, sources, and environmental impacts of these issues. These issues can be characterized by the following four categories:

- *Natural origins*: Problems derived from severe weather conditions and activities, climate change, earthquakes, fires, events beyond human control, land instability, steep slopes, bad/discrete lands, etc.
- *Anthropogenic origins*: Problems derived from forestry activities, shifting land use, poor road construction and maintenance, mining activities, drainage of wetlands, over-grazing, etc.
- *Socioeconomic origins*: Problems associated with land tenure, poverty, education, low acceptance of innovations, seasonal shortages of labor, and water rights (could be major obstacles in management processes and implementation), etc.
- *Stakeholder origins*: Problems identified in accordance with the goals and interests of stakeholders.

These problems of different origins should be prioritized for considerations in formulating watershed management plans.

15.3.5 Developing Management Options

Once watershed problems and sources are characterized and quantified, the next step is to develop management options for considerations in formulating a watershed management plan. The developed management options should cover management (1) goals, (2) objectives, and (3) measures.

15.3.5.1 Goals

A watershed management project cannot succeed without a clear declaration of project goals. Goals are the broad, long-term, and ultimate targets of a management project. They can be derived from the personal insight of watershed professionals, surfaced in the process of receiving public and stakeholder's inputs, or driven by compliance with the mandates of environmental regulations. Thus, goals are more or less a qualitative description of desirable watershed conditions on forest, land, water, wildlife, and aquatic resources in the future. They may require multiple cycles of implementation to achieve or approach.

TABLE 15.2

Goal #1 and Its Relevant Objectives, Black River Watershed Management Plan, Michigan

Goal #1	Objectives
Improve water quality and habitat for fish and indigenous aquatic life and wildlife in the watershed by reducing the amount of nutrients, sediment, and chemical pollutants entering the system.	(a) Stabilize priority stream bank erosion sites through the installation of corrective measures.
	(b) Establish a road/stream crossing improvement program to correct identified problems.
	(c) Assist drain commissioners in identifying areas to improve (and limit erosion).
	(d) Work to limit or control direct livestock access to the river and tributaries.
	(e) Install corrective measures to reduce runoff at agricultural sites of concern.
	(f) Encourage farmers to participate in the Michigan Agriculture Environmental Assurance Program.
	(g) Reestablish greenbelts/conservation buffers at sites in critical areas.
	(h) Work with communities to reduce polluted storm water entering local waterways.
	(i) Identify and improve failing septic systems.
	(j) Encourage the creation of local sanitary sewer systems on densely populated inland lakes.

Source: Data extracted from Fuller, E., Black River watershed management plan, Van Buren Conservation District, Paw Paw, MI, Tracking Code 2002-0067, Michigan Department of Environmental Quality, MI, 2005.

15.3.5.2 Objectives

They are the short-term, measurable, and specific targets set to achieve the long-term goals. One long-term goal may require works from a number of short-term objectives (Table 15.2). Thus, each objective contributes a share to complete the designated goal. For these objectives, the targeted areas, maintenance needs, and timetable for implementation need to be clearly and specifically stated. Measurable indicators and both short-term and long-term milestones of assessment also need to be included in the statement, with the priority of objectives identified. This provides the watershed group with a foundation for formulating a watershed management plan.

15.3.5.3 Measures

For each objective, various measures need to be developed as to how the proposed objective can be accomplished. Measures can be preventive, restorative or enhancive, depending on the nature of the objective. They must be technologically sound as not to cause adverse effects to the watershed. Listed as follows are a few principles that can be considered while developing management measures:

- Forests are the best type of vegetation, inexpensive, effective, and relatively permanent for erosion control and water quality protection. Riparian vegetation is an effective measure to protect sediment and nutrient losses, stream temperature, and aquatic habitats.
- The occurrence of floods are due to a large quantity of water delivered to stream channels; however, the stream channel either does not have sufficient capacity

to accommodate the inflow of water nor the ability to drain the excess water fast enough downstream. Thus, flood control or mitigation measures could be designed to use all techniques to retain as much storm water and surface runoff in the watershed (soils or reservoirs) as possible. When the watershed water holding capacity reaches its capacity, the excess water should be drained out of the watershed as quick as possible.

• Soil erosion is due to the kinetic energy of raindrops and overland flow (also streamflow on bank erosion) greater than the energy that holds soil particles together. Thus, any measures that can reduce the energy levels of rainfall and overland flow, reduce the overland flow sediment carrying capacity (volume and speed), or increase the bonding ability of soil particles can effectively reduce soil erosion.

• Stream habitat management covers all three aspects: protection, restoration, and enhancement. Aquatic organisms are evolved and adapted in stream habitats created by the geomorphic (the interaction of water, sediment, and wood debris that creates channel and shoreline structure) and biological (species interactions, nutrient cycling, and riparian vegetation dynamics) processes. Any measures that make and protect the continuation of these processes (i.e., free flow of water, natural stream bank erosion, species and habitat diversity, and dynamics of riparian vegetation, etc.) provide a long-term survival habitat for organisms.

• In areas where water is in short supply and reducing forest transpiration is one of the management options, activities can include adopting a short-rotation forestry, planting shallow-rooted species, carrying a heavy thinning in timber stands on well-watered, north-facing upland sites, and cutting streamside vegetation (except the buffer zone). When flooding or damaging runoff is a major concern of land management, measures may include maintaining a dense timber stand with deep-rooted species, using single-tree selection cutting, and adopting long-rotation forestry, especially on shallow soils with poor infiltration capacities.

• In areas where downward shear forces on a slope land are greater than the upward resisting strength, mass movements (landsliding) are a venerable watershed problem. It is important that disturbances be avoided on the fragile areas by road constructions, forest cutting, grazing, and other human activities such as adding excessive weight above the slope, digging at mid-slope or at the foot of the slope. Any measures that can reduce the downward shear forces (such as reducing water content of the soil by proper land drainage and water diversion) or increase the shearing resistance (such as lower groundwater table, higher soil internal cohesion and the angle of internal friction of the soil) can make slopes more stable. Note that the internal friction angle is also known as the *angle of repose*, the maximum angle formed when particles are piled on the ground, and is related to the density, surface area and shapes of the soil particles, and the coefficient of friction of the material.

15.3.6 Cost/Benefit Analysis

All proposed management projects/options are required to be accompanied with a cost-and-benefit analysis (CBA) for objective and rational assessments on their desirability. CBA is a common economic tool for decision making. It evaluates all relevant costs and benefits, including goods, services, and attributes, tangible and intangible, incurred by proposed projects, the total impact on society as a whole. A project is desirable if expected benefits are greater than expected costs.

Primary elements of a CBA are given as follows:

- All benefits and costs of a project should be measured in terms of their equivalent monetary values and be of a particular time.
- All future values (V_f) of costs and benefits should be discounted with a proper interest rate (r, in fraction) to the present values (V_p) of that particular time, or $V_p = V_f (1 + r)^{-t}$, where t is time in years of the discounted period.
- The valuation of benefits and costs that do not have a clear monetary value should represent peoples' behavior or choices that have been made. For example, the indirect prices and nonmarket values for benefits provided by forests can be estimated using surrogate prices, replacement or avoided costs, and surveys for willingness to pay for a given event or improvement (Cavatassi, 2004).
- The analysis of a project should include a "with" versus "without" comparison.
- The analysis should not concentrate solely on the financial implication; all other tangible and intangible externalities and benefits in the future must also be assessed.
- Project costs include all expenditures during the planning phase (R&D, consultation, training, etc.), implementation/operation phase, and maintenance/management phase.

The CBA described previously requires the assignment of monetary values to measure management outcomes (benefits). Sometimes, it is not easy on one hand and frequently subjective on the other hand. An alternative approach is the "cost-effectiveness analysis" (CEA). It is a technique that analyzes the desirability among a number of management strategies by calculating and comparing each respective cost-effectiveness ratio (CER), or CER = cost of a strategy/effect (outcome) of the strategy. In other words, the ratio is the price per unit of strategy outcome. If the ratio is low, the strategy is considered "cost-effective." However, a strategy being cost-effective is not necessary money-saving, and a strategy being money-saving is not necessary cost-effective. The final selection may not be the most cost-effective one; there are other subjective considerations too, such as time frame, policies, and preference, etc.

For example, a simulation study using the Water Erosion Prediction Project model and 5 years of field measurement data on a 6.4 ha agriculture watershed in eastern Iowa, United States (Zhou et al., 2009). The results showed that, without supplemental conservation measures, the predicted sediment yield for the field under chisel plow management was 22.5 ton/ha/year. An adoption of conservation practices for grassed waterways, filter strips, and terraces required a net cost, respectively, of $22.8, $119.8, and $184.7 ha/year, but reduced sediment production by 8.8, 12.1, and 14.3 ton/ha/year. Thus, the grassed waterway is considered the most cost-effective with a CER of $2.6/t, compared to $9.9/t for filter strips and $12.9/t for terraces.

15.4 Watershed Management Strategies

Discussed as follows are some issues and strategies worthy of considerations in developing watershed management plans. These strategies, if adopted, will also affect how a watershed should be managed, what resource should be taken care of first, and whose interests should be respected most.

15.4.1 Single versus Multiple Uses

Many forests are used for a single purpose, such as a national or state park for recreation opportunities and a lumber company or private plantation for timber productions. However, forests possess many environmental functions (Chapter 7, Section 7.2) and are associated with many natural resources (Chapter 7, Section 7.1). As such, it is wasteful to manage forests for a single use and it may affect other resources due to its activities.

In the United States, the multiple-use concept of forest was legislatively recognized in the Organic Act of 1897, the Weeks Act of 1911, the Multiple Use and Sustained Yield of 1960, and the National Forest Management Act of 1976. Under the Organic Act, national forests shall be established for the purpose of "securing favorable conditions of water flows, and to furnish a continuous supply of timber" for U.S. citizens. The Weeks Act authorized the U.S. Department of Agriculture to purchase lands within the watersheds of navigable streams, in order to regulate the flow of navigable streams or to produce timber, and add to national forests. The concept of multiple-use, which was more explicit in 1960, stated that national forests be established and administered for outdoor recreation, range, timber, watershed, wildlife, and fish purposes, and it further strengthened in 1976.

Although the concept and implementation of multiple-use management is carried on in national forests, they have also been widely adopted in recent decades to private forests in developed as well as developing countries (Garcia-Fernandez et al., 2008), especially dealing with land and forest. The soil type, land topography, climate variation, and landowner's interests are complex in a watershed, and consequently multiple-use management must be a logical practice at the watershed scale.

15.4.1.1 Objectives

The concept of multiple-use can be applied both to land and natural resources. A land can be managed for growing crops, fruits, grazing; water in streams can be used for drinking, irrigation, hydropower, recreation, and industrial processes; and a forest can be grown for timber, pulpwood, fuel, and wildlife habitats. In reality, the multiple-use concept often applies to managing both land and natural resources in watersheds. For instance, a watershed in the U.S. Coweeta Experimental Forests was successfully managed for the multiple uses of water and timber production, hunting, fishing, and hiking (Swank, 1998).

Many envision multiple-use as a preferable alternative to the traditional practice of timber-dominated management. The main objectives of multiple-use management are

- Maximizing the yield of multiple outputs and benefits to landowners
- Fully utilizing land and natural resources
- Keeping sustained yields of outdoor recreation, timber, range, watershed, wildlife, and fish values for current and future generations
- Satisfying the demands and interests from multiple stakeholders
- Balancing the conflicts between conservation and development on lands and natural resources

15.4.1.2 Practices

- All management measures taken on a given area should not impair the productive capacity of the site nor should the measures on one natural resource impair other resources and their objectives.

- The whole watershed can be divided into natural units by landscapes. Multiple-use can then be prescribed at the landscape level in accordance with land capability; single use is allowed on particular sites to which they are not suited for other usages.

- Thinning and clear-cutting improve deer habitat in hardwood forests by producing abundant browse. However, most of the browse has grown out of the reach of deer after about age 10 in the clear-cut sites. A rotation schedule for thinning and clear-cutting around forests can provide sustainable browse to deer population.

- Environmental stability should be incorporated into multiple-use management activities.

- In those critical areas where vegetation is rare, soil erosion is severe, bed rocks are exposed, and land is discrete, the establishment of vegetation cover is the first priority on land management. Multiple-use can then be applied once ground vegetation cover is established.

- The involvement of multilevels and multi-stakeholders is a must.

- Projects should target policy and institutional support, including the development of incentive mechanisms for multiple-use adoption and income generation at the local level.

15.4.2 Multiple-Use versus Ecosystem Management

Since multiple-use management intends to maximize the outputs of natural resources on a sustainable basis, it usually results in an overemphasis on plant and animal species with great economic values, or paying more attention to those natural resources with monetary market values. These exercises are at the expenses of environmental variability and issues of equitability in sharing the benefits supplied by various resources. Ecosystem management brings in ecological values, environmental integrity, and biodiversity as general guidelines in management processes. It adds health of the total ecosystem as a new ingredient to the multiple-use strategies. Ecosystem management is currently the core concept of operations to strengthen multiple-use objectives for the U.S. national forests.

15.4.2.1 Objectives

Ecosystem management is a new approach to land and forest management and is a response to today's deepening biodiversity crisis (Grumbine, 1994). It takes the entire ecosystem under consideration and places greater concern on the ecological values and biodiversity of the areas. At the same time, it also recognizes the many demands and expectations that the society places on forests. Thus, the managed ecosystem must not only be ecologically sound, but it must be economically feasible and socially acceptable as well.

The primary objectives of ecosystem management involve (a) sustaining the long-term productivity and health of the entire ecosystem, (b) enhancing overall biological diversity, variety, abundance, and distribution of biological entities, (c) recognizing the human entity and its needs in the ecosystem, and (d) maintaining evolutionary and ecological processes across the area.

These objectives concern both human welfare as well as the welfare of natural systems and the needs of currents and future generations.

15.4.2.2 Principles

The objectives of ecosystem management described earlier can be achieved through (a) integrated land evaluation on inventory, classification, and analysis, (b) optimal land-use planning, (c) the creation of landscape structure and process, and (d) considerations of society's expectations and the constraints of the land's ecology (Jensen and Everett, 1994). By these approaches, sociological, ecological, technological, and economic information must be integrated in planning, management, and implementation processes. It is also important that a balance is needed between consumption and conservation of resources and between the welfares of current and future generations.

The following six principles are proposed by Overbay (1992) to describe the initial components of an ecosystem management plan:

- Multiple-use, sustained-yield management of lands and resources depends on sustaining the diversity and productivity of ecosystems at many geographic scales.

- The natural dynamics and complexity of ecosystems means that conditions are not perfectly predictable and that any ecosystem offers many options for uses, values, products, and services, which can change over time.

- Descriptions of desired conditions for ecosystems at various geographic scales should integrate ecological, economic, and social considerations into practical statements that can guide management activities.

- Ecosystem connections at various scales and across ownerships make coordination of goals and plans for certain resources essential to success.

- Ecological classifications, inventories, data management, and analysis tools should be integrated to support integrated management of lands and resources.

- Monitoring and research should be integrated with management to continually improve the scientific basis of ecosystem management.

Besides these components, the human factor is another one that is important to the processes of land evaluation and planning. Human needs must be respected, perceptions must be considered, and participations must be encouraged. However, human needs must be compatible with the ecosystem potentials, a great challenge to resources and watershed managers.

15.4.3 Upstream versus Downstream

Identifying differences in upstream and downstream of a drainage basin is an important issue in watershed management. Frequently, upstream land-use activities are perceived as producing heavy economic costs, such as reduction of economic lives of reservoirs, irrigation system, power plants, and flood control measures, for downstream residents. Although deforestation and soil erosion are recognized as significant problems, the welfare of people in upstream watersheds is rarely given adequate priority, especially in developing countries (Gibbs, 1986). Also, there exists a conflict regarding the degree of impact of upland watershed management activities on downstream water quantity, quality, and aquatic habitats. This is partly because of difficulties in predicting with reasonable accuracy the results of these activities. Until the magnitudes of natural and human-induced erosion and subsequent sedimentation can be reliably quantified in a watershed, the issue is likely to remain a major concern in watershed management.

Thus, two questions regarding upstream and downstream watersheds often pop up in the watershed planning process: (a) whether inhabitants at upstream should receive the most benefits from the management program or those at downstream? and (b) which section of the watershed should be more emphasized in watershed treatments? Watershed conditions, patterns of natural resource usage, property values, and stakeholders' interests between upstream and downstream are significantly different, making the balance between upstream and downstream requires special considerations.

15.4.3.1 Characteristics

On average, upstream watersheds usually have steep land slopes, shallow soils, great channel gradients, and high rainfall intensity, subjecting it to greater potential for soil erosion than downstream watersheds. Streamflow in upstream watersheds is typically smaller in volume, faster in velocity, shallower in depth, cooler in temperature, and better in water quality. The substrate is dominated by coarse particles, pebbles, gravels, and stones in upstream reaches, but by finer particles such as silts and clays in downstream reaches. This may cause diversity of temperate stream fish communities to be highest in higher-order streams rather than lower-order, headwater streams (Moyle and Cech, 1988).

Land use on upstream watersheds is generally predominated by forests, with the rest being agriculture and range. Development and urbanization on upstream watersheds are small. This makes those who live at the upstream watershed more interested in land resources. Their land activities may have a great impact on the downstream water quantity and quality. On the other hand, downstream watersheds are more developed and urbanized. Consequently, those who live and work at downstream watersheds are more interested in watershed resources. They depend on the good conduct of those who use the upper watersheds to ensure adequate quantity and quality of water flowing from upstream to downstream reaches (Table 15.3).

15.4.3.2 Strategies

From the integrated watershed management perspective point of view, there are no upstream and downstream problems, there are watershed problems. Watershed management needs to take into concern the welfares of the entire watershed under consideration and respect the interests of all people, no matter upstream or downstream. To be site specific and community oriented are two major themes in developing watershed treatment measures. The downstream section of a watershed implies higher property values, greater population density, and numerous community infrastructures. It requires dams, levees, or other measures to protect property and people from flood threats, and requires water supplies in clean and sufficient quantity to meet population and community demands. On the other hand, land production is in general the major activity upstream. It requires vegetation treatments and soil conservation measures to enhance land productivity upstream and at the same time protecting water supplies and reservoirs downstream.

For a management project to be successful, both upstream and downstream measures need to be implemented and mutual benefits are considered (Sheng, 2001). Those upstream

TABLE 15.3

Characteristics of Upstream and Downstream Watersheds

Item	Upstream Watersheds	Downstream Watersheds
Watershed	Low-order streams, small in area, less watershed storage	High-order streams, consisting of many small watersheds, large in area, greater watershed storage
Topography	Steep slopes, shallow soils with low water holding capacity, more mass wasting	Gentle slopes, deep soils with high water holding capacity, more bank erosion
Climate	Cold air temperature, great weather variations, high average rainfall intensity	Warm air temperature, less weather variation, low average rainfall intensity
Channels	High ingredient, substrata with gravels/course particles, greater amount of large woody debris	Low ingredient, substrata with silt/fine particles, length and diameter of LWD increased
Streamflow	Shallow and fast moving flow, low flow volume, cold water temperature, high dissolved oxygen, low concentration of nutrients, high accuracy in flow measurements, high in peak flow/area, impacts of land management more significant	Deep and slow moving flow, high flow volume, warm water temperature, low dissolved oxygen, high concentration of nutrients, accuracy in flow measurements declined, relative low in peak flow/area, impacts of land management insignificant or difficult to detect
Fish community	Low diversity of temperate community; great abundance of minnows, sunfish, and darters, 10%–20% of the fish are stonerollers	High diversity of temperate community; algae-grazed stonerollers most abundant, 20%–50% of the community
Land use	Predominately forest, followed by pasture and agriculture	Predominately pasture, followed by agriculture, pasture, and urban
Resources use	Predominately land resources	Predominately water resources

who participate in watershed programs should receive some incentives as a compensation for their inputs on conservation programs. Since conservation practices are costly and downstream residents enjoy some of the generated benefits, the use of incentives could induce the enthusiasms of people at upstream participating watershed projects. The social, economic, and institutional linkages between upstream land users and downstream water users need to be developed. Kiersch (2000) identifies six approaches that have been applied to develop such linkages in watershed projects:

- Regulatory instruments (command and control measures)
- Economic instruments (i.e., subsides, taxes, land-use rights on sensitive areas)
- Educational and awareness-building measures
- Mechanisms to increase market access (upstream farmers to downstream markets)
- Building organizational structures
- Participatory approaches

Many instruments of upstream–downstream linkages require a stable institutional and legal framework for implementation. For example, the New York City Watershed Agreement allows the city to purchase land in sensitive areas in upland watersheds to protect its water supply. These lands, still belonging to the original owners, are set aside as "conservation easement" and can only be used for certain recreational activities. Also, institutions are required to enforce the regulations established under the regulatory instruments.

15.4.4 Vegetative versus Structural

Erosion control on land with steep slopes and rough terrains often employs vegetative or structural measures. Vegetation measures, such as forests, agro-forests, grass waterways, cover cropping, and grass strips, etc., are inexpensive and naturally appealing, but may not be easy to establish on rough terrains, may require favorable weather conditions to establish, and are not effective before established. Structural measures, such as bench terraces, check dams, hillside ditches, and concrete or ballasted waterways, etc., are effective, but are expensive, have environmental appeal problems, and may interfere with mechanical farming. Conservation practices with vegetative or structural measures could be controversial in decision-making (Sheng, 2001).

15.4.4.1 Erosion Control Treatments

The quantity, quality, and timing of streamflow are an integrated indicator of watershed management conditions. Silted and rapid flow in stream channels, especially after storms, implies the watershed having serious soil erosion and poor vegetation management problems.

The control of watershed erosion starts with field reconnaissance to identify eroded areas, the severity of erosion, and causes. Once erosion causes are identified, control measures can be locally developed through dissipating erosion forces, reducing the carrying capacity of overland flow, or increasing soil ability to resist erosion.

The prescription of watershed erosion control measures can be, according to the degrees of severity, divided into the following six levels:

Level I: No treatments required. The watershed is stable, forest and ground cover is good, no signs of accelerated erosion are noticeable.

Level II: Only regulations and control management such as limiting the land carrying capacity, visitor control, or using crop rotation systems, etc., are required. The watershed soil erosion is light.

Level III: Vegetation treatments. Soil erosion is light to moderate; it requires increases in ground cover and revegetation to curtail the erosion problems. Once established, vegetation is the most effective in erosion control.

Level IV: Soil treatments. Soil erosion is moderate to heavy and conservation practices, such as terracing, contour cropping, and strip cropping, etc., are required to conserve soil and water.

Level V: Bioengineering treatments. This is an advanced technique combining vegetation and structural measures to control erosion and stabilizing slopes/banks, such as live staking, vegetated rock gabions, live cribwalls, and revetment with slope face planting, woven geotextiles, etc.

Level VI: Structural treatments required. The erosion is so severe that it cannot be effectively controlled by vegetation and conservation practices.

Treatments at Levels II–IV are collectively referred to as the so-called *Best Management Practices.*

15.4.4.2 Strategies

The selection of these different treatments listed earlier is based on site conditions, resources, and causes of erosion, and not by personal preferences. For those who prefer vegetative to structural measures because of fewer costs, Sheng (2001) states that "structures are not necessarily more expensive than vegetative measures" and "terraces are more economical in the long run than cover cropping." The bioengineering techniques listed under Level VI employ living plant materials to provide some engineering function on erosion control and slope stabilization. Some use vegetation and structures, others use live plants and logs only (Gray and Sotir, 1996). They are effective, but not all expensive (Petrone and Preti, 2010).

15.4.5 Preventive versus Restorative

Prevention and restoration are two branches of watershed management. Prevention involves protecting sensitive areas and resources from degradation and damages on the one hand and enhancing the quality of the ecosystem and the value of resources on the other hand. Restoration is to apply remedial measures to damaged or deteriorated areas/ resources that have already occurred in the watershed. Watershed prevention (protection and enhancement), no matter how costly, will almost always be cheaper and more cost-effective than the restoration of a degraded watershed or of a social system that is undermined in the course of degradation.

15.4.5.1 Watershed Degradation

Watersheds may be degraded due to natural events, such as torrential rains, floods, hurricanes, wildfires, and other natural disasters, as well as anthropogenic reasons, such as abusive forest cutting, overgrazing, malpractice of agriculture, and urban development. Damages caused by natural disasters are acute and could be huge, but in time nature will gradually adjust, adopt, and remedy most of the damage to form a new balance with no human interferences. Damages caused by human activities are chronic and accelerate with time. As a result, the degradation will get worse, the watershed will have to be abandoned from production, and it will impose a great threat to water supplies and community safety downstream.

15.4.5.2 Strategies

For those watersheds already having severe erosion problems, the implementation of restorative measures is the logical and feasible strategy. The purposes of these restorative measures are to control soil erosion, to upgrade water quality, to improve stream habitats, and to bring back land productivity. Natural disasters may cause severe damage to watersheds, including soil erosion, gullies, mass movement, roads, stream channels, and wildlife and aquatic habitats. Restorative measures to these damages are necessary.

In practice, both preventive and restorative measures are required in managing watershed land and water resources. Preventive care can be applied to watersheds with little disturbances, while restorative measures are applied to those that have already been damaged or deteriorated. Perhaps a better preventive measure is to use land in accordance with its capability as defined in land-use classification systems (Sheng, 2001). Landowners are encouraged to do cultivation with conservation practices on classified arable lands and set aside or do forestation and protective measures on nonarable lands such as steep slopes with high potential of severe erosion and mass movement.

15.4.6 Short Term versus Long Term

Besides goals and objectives, management strategies also need to be conducted at two levels, a short-term and a long-term. Short-term level refers to those practices that will be applied to selected sites that were identified during the watershed field inventories. It is a local-wide strategy. Long-term level refers to those practices that cover the entire watershed or something related to the future and basic. It is a watershed-wide strategy.

Local-wide or site-specific strategies are the initial actions to be completed within the first 1 or 2 years of implementation. They provide examples and demonstrations for the types of projects that could be implemented at similar sites throughout the watershed. Local-wide practices can have both short- and long-term benefits.

Long-term strategies are basic practices that address issues throughout the entire watershed and those continued practices of any additional projects that are necessary to meet watershed management objectives. Programs such as progress assessment, regulations, public education and outreach, and environmental monitoring activities are included in this category. Such watershed-wide strategies are intended to be completed during a 5–10 years time frame and beyond. Their benefits are long term and cumulative.

15.5 Watershed Management Plans

Once watershed data are collected and analyzed, watershed problems and causes are identified, management goals and objectives are defined, and options for management strategies and measures are developed, a plan to achieve the targeted goals and objectives can be finalized.

15.5.1 Selection of Final Strategies

The final management strategies selected for implementation in a management plan must be:

- Able to meet the management objectives
- Cost-effective
- Feasible for implementation
- Without adverse side effects
- With the consensus of stakeholders

Since there are quite a few management strategies options resulting from *watershed analyses*, an assessment process is required to eliminate those that do not meet the decision criteria listed earlier. Watershed and hydrologic models are often chosen to evaluate the potential outcomes of management practices. EPA (2008) has summarized the strengths and limitations for a number of current watershed models that have been frequently employed for evaluations of management practices. The selection of these methods depends on (a) types and locations of the practices, (b) indicators used to measure the performance, (c) watershed scale, complexity, and expensiveness of the management project, and (d) data required in the modeling. Note that quantitative watershed models are developed in watersheds with physical environments not quite comparable to the applied watersheds, so too do the types of management practices and intensities. Thus, the result of evaluation needs to be precautious in interpretation, especially since it can only be verified after management practices are implemented.

In cases where appropriate models are not available, the data required in modeling are absent, or there is a lack of modeling techniques and expertise in the management committees, a common alternative is conducting literature reviews on existing research to obtain expected percentage or range values of effectiveness typically associated with each type of management practice and pollutant. Note that many published studies may be based on data collected from a few months of the year or from a few storms; they do not represent the yearly values or the annual averages.

15.5.2 Implementation Programs

Planning and implementation are two sequential linkages of a watershed program. Once work on watershed planning is done, it needs to be implemented. It is the implementation that converts planning into actions and makes watershed management goals and objectives a reality.

15.5.2.1 Watershed Implementation

A well-designed implementation program intends to provide the conditions and momentum necessary for successful watershed management. It specifies who will do what works, and when accomplishment is expected. These tasks fall into one of these five categories: (a) public outreach, (b) installation, (c) operation, (d) maintenance, and (e) assessment. Typical tasks of implementation programs may involve items such as

- Organizations responsible for each management activity
- Time frames required for accomplishing each activity, objective, and goal
- A monitoring scheme for environmental indicators
- A periodic check-out schedule for evaluating the progress of implementation
- Contingency plans in case organizations are unable to do the designated works
- New or modified ordinances, codes, guidelines, or rules
- Formation of a new group on implementation tasks
- Training/education programs, public outreach

Some of these activities may be effectively implemented by each organization acting alone and using its own internal procedures. Others may require considerable coordination involving all of the implementation organizations together. Also since implementation activities are a long-term effort, organizational continuity and commitment are the keys to success. All of these make it necessary to organize an implementation team to carry out implementation tasks and oversee activities. The team may need to bring in some new, skilled, and enthusiastic personnel to work with the current key partners. New faces can come with new ideas and energy, which may help to move the ball forward. For stability and efficiency, the formed implementation team may need to be institutionalized.

15.5.2.2 Establish a Schedule

The implementation schedule is to define time scales of management projects. It should cover two components, a target time frame in which each watershed practice will be implemented and accomplished, and an interim progress assessment program in which the progress along the course of implementation can be periodically reviewed and the plan adjusted, if necessary.

Implementation tasks can be designated as short-term (1–2 years), mid-term (3–5 years), and long-term (6–10 years or longer). Once management activities are being implemented, evaluation programs should be activated through a predetermined strategy to (a) track if implementation activities are meeting the designed standards, (b) determine if the plan is meeting its goals and objectives, (c) check if the current measures are cost-effective, and (d) review the coordination and efficiency between the administration and partnership. The evaluation strategy includes (a) developing a series of milestones to measure project progress, (b) identifying indicators, both environmental and administrative, that will be used to track the implementation, and (c) having a clear assessment schedule. For such purposes, monitoring schemes are often installed and operated to collect field data for long-term, objective assessments on the effectiveness of the original strategies. They also provide evidence and justification for modification of the original plans, if necessary.

15.5.3 Participations of Stakeholders

Stakeholders are those people who are affected by the watershed plan, or have direct interest or knowledge in the watershed. Implementation of many watershed management plans commonly rests with members of the local community. If they are involved in the beginning of the planning process and their concerns addressed, they will be more likely to participate in developing management options and supporting plan implementation. The increase in public understanding and public commitment to the watershed solutions is of utmost importance to the success of a watershed project.

To encourage the participation of stakeholders, one can begin by appointing a few key stakeholders to the watershed planning committees. This provides opportunity for concerns to be heard, interests addressed, and benefits balanced. Once a watershed plan is formulated, it should be open to the public for reviews, comments, and suggestions. Developing a reader-friendly summary version of the watershed plan, a short executive summary, or a list of frequently asked questions are easy ways to reach out to various audiences. The distribution can be done through public posting with computer-accessible files, local newspaper announcements, handouts at community events, press releases, and town meetings.

15.5.4 Costs and Financial Sources

Funding is a critical factor affecting management plans and implementation. There are a number of activities that require financial budgets to get the plan rolling, such as administration, personnel, watershed inventory and analysis, public outreach, the implementation of management measures, instruments and equipments, supplies, watershed monitoring, and laboratory data analysis, etc. The costs should be estimated at the total and annual bases for each category. Note that some of the costs of implementing watershed plans can be defrayed by leveraging existing efforts and seeking in-kind services, such as using existing data and studies, technical supports from partners, and service contributions from volunteers.

There are hundreds of funding sources at the federal, state, local, and private levels that can help fund the implementation of watershed projects. The *Guidebook of Financial*

Tools: Paying for Sustainable Environmental Systems, available for download at http://www. epa.gov/efinpage/guidbkpdf.htm, is developed by EPA to provide tools, databases, and information about sources of funding to watershed practitioners in the public and private sectors. More information on funding sources for watershed programs is posted at EPA's Grants and Funding website through http://www.water.epa.gov/aboutow/owow/funding.cfm. Note that many such organizations do not provide full funding to an entire watershed project. They may fund watershed projects at planning or implementation stages, or the implementation stage during different phases (i.e., short-term, mid-term, and long-term). It is then quite often that watershed projects are funded through different sources in their lifetime.

15.6 Watershed Assessment

Watershed management is a continual process of information gathering, monitoring of implementation outcomes, and revision of management activities. Thus, watershed assessment is a key element in watershed management plans. It is the mechanism used to check if the planned strategies work properly, if the implementation proceeds with the designed standards and schedules, if the progress of implementation can meet the stated goals and objectives, and if there is a need to modify the current management plan.

15.6.1 Types of Assessment

Watershed assessment is not necessary to be limited to the progress of implementation activities. It can be conducted any time and at multiple times in the management program. Different approaches, data collection, and analytical techniques can be developed for special purposes in each phase of the management process.

Ideally, a watershed management program is evaluated, at least four times, at four different phases. The first evaluation starts when the watershed plan is just being developed, followed by the implementation of project activities, then again after the project activities have been completed, and finally the long-term impact evaluation. These four types of assessment have been described as *formative, process, outcome,* and *impact*, respectively (Davenport, 2003).

15.6.1.1 First Assessment (Formative, Prior)

The first stage of assessment is launched once the watershed plan has already been developed but has not begun implementation yet. It intends to test approaches, materials, and ideas, to reallocate resources for priority needs, and to determine if local residents are ready, supportive, and committed to the upcoming changes and activities in the watershed. It includes the administrative and technical procedures used to secure agreements with landowners, develop specifications, engage contractors, and others. In other words, this is a preparatory step to get the plan ready for implementation as well as to ready local residents. If local residents are not ready, then it may be necessary to launch public outreach programs, such as workshops, trainings, field demonstration plots, in addition to other public information and communication approaches.

15.6.1.2 Second Assessment (Process, During)

The second stage of assessment is conducted during the implementation of project activities. The intention is to provide feedback on activities so that changes can be made if there is a need for increasing cost-effectiveness. Thus, if changes are made, then periodic reassessment of the new implementation strategies may be necessary to assure cost-effectiveness. The ultimate purpose of this assessment is to improve the ongoing implementation effort so that the overall goals can be achieved.

The assessment needs to develop a series of milestones to measure a project's progress and identify indicators that will be used to track the effectiveness of implementation activities (Table 15.4). The milestones are like subtasks, a schedule of what needs to be accomplished over time so that the practice or management measure can be fully implemented. Frequently, the use of multiple indicators may be needed to account for the complexity of watershed processes. The assessment involves actual measurements and field data collections.

15.6.1.3 Third Assessment (Outcome, Afterward)

The third stage of assessment is after the project activities have been completed. Its purpose is to provide some measures of project effectiveness. Basically, this is a review process to check (a) if the project followed the targeted timeframe; (b) if budgets spent were appropriate and effective; (c) if stakeholders are happy with the results; (d) if the projects go beyond the desired effects, (e) if there are any changes in awareness, attitudes, skills, or behavior toward the watershed; and (f) what needs to be done differently if it were started over again. These assessments usually rely on readily available

TABLE 15.4

Example of Indicators to Measure Progress in Reducing Pollutants

Issue	Suite of Indicators
Eutrophication	Phosphorus load
	Number of nuisance algae blooms
	Transparency of water body or Secchi depth
	Hypolimnetic dissolved oxygen in a lake or reservoir
	Soil test phosphorus in agricultural fields
	Frequency of taste and odor problems in water supply
Pathogens (related to recreational use)	Bacteria counts
	Compliance with water quality standards (single sample or geometric mean)
	Number and duration of beach closings
	Number of shellfish bed re-openings
	Incidence of illness reported during recreational season
Sediment	Total suspended solids concentration and load
	Raw water quality at drinking water intake
	Frequency and degree of dredging of agricultural ditches, impoundments, and water supply intake structures

Source: Adapted from EPA, *Handbook for Developing Watershed Plans to Restore and Protect Waters*, EPA 841-B-08-002, U.S. Environmental Protection Agency, Washington, DC, 2008, Chap. 12, Table 12.1, available at: http://www.epa.gov/owow/nps/watershed_handbook

information, or public survey, and rarely measure real progress in environmental indicators (Davenport, 2003).

15.6.1.4 Fourth Assessment (Impact, Long Term)

The last stage of assessment is to observe the long-term impacts of a management effort after the project has been completed for years or decades. This is the most difficult aspect of the evaluation to complete because of lack of long-term funding. However, the chance to get this done is much greater if a partnership of watershed management is built into a sustaining organization to maintain continuity and stability through the years. The assessment is of importance for promoting watershed management in the future. If the project fails to achieve the targeted goals and objectives, it may be due to the designed management strategies, the implementation of these strategies, both, or other factors. They should be identified.

15.6.2 Communication

Continuous communication with watershed stakeholders and the public can (a) build up the credibility of the watershed planning process, (b) increase awareness of the management plan, (c) gain support for implementation activities, (d) stimulate more stakeholders to get involved in the effort by offering new ideas or suggestions, and (e) help to strengthen accountability among watershed partners by keeping them actively engaged. Lack of communication can hinder the participation of stakeholders as well as reduce the likelihood of successful implementation.

Communication is required during all phases of the management process—before, during, after, and post project. Regular reports, implementation plans, interim progress, assessment results, changes in environmental regulations and ordinances, and project status are all of concern and interest to stakeholders. These progress and implementation results can be shared through various channels such as local newspapers, TV or radio announcements, press release, community meetings, computer files, and flyers, fact sheets, brochures, community releases and calendars. It is also important to open a channel for stakeholders to make comments and suggestions, ask questions, and contribute new ideas. When people see progress, they will continue to commit to the project and work together to making the plan a success.

15.6.3 Plan Adjustments

If watershed assessments have shown that the management practices do not meet the implementation milestones or do not make progress toward interim targets (e.g., sediment reductions, water quality improvements, watershed conditions, and other goals), it is first necessary to examine the causes of these deficiencies before making any adjustments and modifications. A number of factors may contribute to the shortfalls and deficiencies, such as

- The treatment results may need a longer time to be detectable.
- The installed management practices are not being activated or properly maintained.
- It may be weather related, causing implementation delays or installation damages.
- There was a shortage in technical assistance or anticipated funding.
- The milestones and goals are not realistic, time frames are misjudged.

- There was lack of participation and support from stakeholders.
- Management measures are not implemented correctly.
- Inappropriate environmental indicators or parameters were used to measure the progress.

Some of the shortages listed earlier require no actions, some require minor adjustments or additional efforts, and some require revisiting the original assumptions that led to the designed measures and targets.

References

Ackerman, C.T., 2009, *HEC-GeoRAS-GIS Tools for Support of HEC-RAS using Arc GIS, User's Manual, Version 4.2*, Hydrologic Engineering Center, U.S. Army Corps of Engineers, Davis, CA, CPD-83, available at: http://www.hec.usace.army.mil/software/hec-ras/documents/HEC-GeoRAS42_UsersManual.pdf

Argonne National Laboratory, 2004, *Guidance for Habitat Restoration Monitoring: Framework for Monitoring Plan Development and Implementation*, Naval Facilities Engineering Command, Washington, DC.

Cavatassi, R., 2004, Valuation methods for environmental benefits in forestry and watershed investment projects, ESA Working Paper No. 04-01, FAO, the United Nations, 52 pp., available at: http://www.fao.org/es/esa

Chang, M. and Lee, R., 1974, Objective double-mass analysis, *Water Resour. Res.*, 10(6), 1123–1126.

Chang, M., Watters, S.P., and Sayok, A.K., 1989, A comparison of methods of estimating mean watershed slope, *Water Resour. Bull.*, 25(2), 327–333.

Chang, M. and Wong, K.L., 1983, Effects of land use and watershed topography on sediment delivery ratio in East Texas, *Beitr. Hydrol.*, 9, 55–69.

Davenport, T.E., 2003, *The Watershed Project Management Guide*, Lewis Publishing, New York.

EPA, 1997, *Monitoring Guidance for Determining the Effectiveness of Nonpoint Source Controls*, EPA 841-B-96-004, U.S. Environmental Protection Agency, Washington, DC.

EPA, 2002, *Guidance on Choosing a Sampling Design for Environmental Data Collection for Use in Developing a Quality Assurance Project Plan*, EPA QA/R-5S, EPA/240/R-02/005, U.S. Environmental Protection Agency, Washington, DC, 166 pp.

EPA, 2008, *Handbook for Developing Watershed Plans to Restore and Protect Waters*, EPA 841-B-08-002, U.S. Environmental Protection Agency, Washington, DC, available at: http://www.epa.gov/owow/nps/watershed_handbook

Eze, E.B. and Efiong, J., 2010, Morphometric parameters of the Calabar River basin: Implication for hydrologic processes, *J. Geogr. Geol.*, 2(1), 18–26.

Garcia-Fernández, C., Ruiz-Pérez, M., and Wunder, S., 2008, Is multiple-use forest management widely implementable in the tropics? *For. Ecol. Mange.*, 256, 1468–1476.

Gibbs, C.J.N., 1986, Institutional and organizational concerns in upper watershed management, Chapter 7, *Watershed Resources Management, An Integrated Framework with Studies from Asia and the Pacific*, Easter, K.W., Dixon, J.A., and Hufschmidt, M., Eds., Westview Press, Boulder, CO, pp. 91–102.

Gray, D.H. and Sotir, R.B., 1996, *Biotechnical and Soil Bioengineering Slope Stabilization*, John Wiley & Sons, New York.

Grumbine, R.D., 1994, What is ecosystem management? *Conserv. Biol.*, 8(1), 27–38.

Harvey, C.A. and Eash, D.A., 2005, Description of basin soft, a computer program to quantify drainage-basin characteristics, U.S. Geological Survey, accessed via web at: http://gis.esri.com/library/userconf/proc96/to100/pap072/p72.htm

Helsel, D.R. and Hirsch, R.M., 2002, *Statistical Methods in Water Resources Techniques of Water Resources Investigations*, Book 4, Chapter A3, U.S. Geological Survey, 522 pp., available at: http://water.usgs.gov/pubs/twri/twri4a3

Jensen, M.E. and Everett, R., 1994, An overview of ecosystem management principles, *Vol. II: Ecosystem Management: Principles and Applications*, Jensen, M.E. and Bourgeron, P.S., Eds., U.S. Forest Service Pacific NW Research Station, General Technical Report PNW-GTR-318, Portland, OR, pp. 6–15.

Kiersch, B., 2000, Instruments and mechanisms for upstream-downstream linkages: A literature review, discussion paper 2, *Land-Water Linkages in Rural Watersheds, Electronic Work Shop*, September 18–October 27, 2000, FAO, Rome, Italy, available at: http://www.fao.org/ag/agl/watershed/watershed/papers/paperbck/papbcken/kiersch2.pdf

Kilgour, B.W. et al., 2007, Aquatic environmental effects monitoring guidance for environmental assessment practitioners, *Environ. Monit. Assess.*, 130, 423–436.

Maidment, D.R., 2002, *ArcHydro: GIS for Water Resources*, ESRI Press, Redlands, CA.

McCuen, R.H., 1998, *Hydrologic Analysis and Design*, 2nd edn., Prentice Hall, Upper Saddle River, NJ.

Merriam, C.F., 1937, A comprehensive study of the rainfall on the Susquehanna Valley, *Am. Geophys. Union Trans.*, 18(2), 471–476.

Moyle, P.B. and Cech, J.R., 1988, *Fishes: An Introduction to Ichthyology*, Prentice Hall, Englewood Cliffs, NJ.

NRCS, 2009, *National Watershed Program Manual*, 3rd edn., M-390, the Natural Resources Conservation Service, Washington, DC, available at: http://www.nrcs,usda.gov/programs/watershed.PDFs/NWPM_PDF_3-22-10.pdf

Overbay, J.C., 1992, Ecosystem management, *Proceedings of the National Workshop: Taking an Ecological Approach to Management*, April 27–30, 1992, Salt Lake City, UT; WO-WSA-3, U.S. Department of Agriculture, Forest Service, Watershed and Air Management Washington, DC, pp. 3–15.

Petrone, A. and Preti, F., 2010, Soil bioengineering for risk mitigation and environmental restoration in a humid tropical area, *Hydrol. Earth Syst. Sci.*, 14, 239–250.

Rex, J.F. and Carmichael, N.B., 2002, *Guidelines for Monitoring Fine Sediment Deposition in Streams*, Version 1.3, Field test edition, Resources Information Standards Committee, the Province of British Columbia, Canada.

Sheng, T.C., 2001, Important and controversial watershed management issues in developing countries, *Sustaining the Global Farm*, Stott, D.E., Mohtar, R.H., and Steinhardt, G.G., Eds., Purdue University and USDA-ARS National Soil Erosion Research Laboratory, West Lafayette, IN, pp. 49–52.

Swank, W.T., 1998, Multiple use forest management in a catchment context, *Proceedings of an International Conference*, September 11–13, 1996, Aberdeen Research Consortium, Land Management and Environmental Sciences Research Centre, The Macaulay Land Use Research Institute, Aberdeen, Scotland, U.K., pp. 27–37.

Zhong, B. and Xu, Y.J., 2009, Topographic effects on soil organic carbon in Louisiana watersheds, *Environ. Manage.*, 43, 662–672.

Zhong, B., and Xu, Y.J., 2011, Scale effects of geographical soil datasets on soil carbon estimation in Louisiana, USA—A comparison of STATSGO and SSURGO, *Pedosphere*, 21, 491–501.

Zhou, X. et al., 2009, Cost-effectiveness and cost-benefit analysis of conservation management practices for sediment reduction in an Iowa agricultural watershed, *J. Soil Water Conserv.*, 64(5), 314–323.

16

Research in Forest Hydrology

Research is systematic investigation to confirm or refute hypotheses and theories. Two of the earliest research studies in forest hydrology were an 1862 precipitation–interception study in Germany (Friedrich, 1967) and an 1867 experimental-watershed study in Czechoslovakia (Němec et al., 1967). Studies on fog interception by vegetation were conducted on Table Mountain in South Africa between 1901 and 1904 (Marloth, 1907, cited by Olivier, 2002). An investigation on the water budget of young trees was conducted with small-scale lysimeters at Eberswalde, Germany in 1907 (Müller and Bolte, 2009). The first U.S. watershed study—a joint effort by the Forest Service and the Weather Bureau—began in 1911 at Wagon Wheel Gap in Colorado. The importance of understanding forest hydrology is demonstrated by the scale of research and number of experimental watersheds dedicated to this issue. Figure 16.1 plots over 400 major experimental watersheds located at 51 different sites throughout the United States that are engaged in studying the interactions between water and forests under a variety of environmental and forest conditions. These watersheds are mainly operated by the U.S. Forest Service, universities, Tennessee River Authority, Oak Ridge National Laboratory, and lumber companies. The Agricultural Research Service (ARS) of the U.S. Department of Agriculture (USDA) also maintains a network of 140 active experimental watersheds at 17 different locations in the conterminous United States (Weltz and Bucks, 2003). These watersheds are mainly in agricultural and rangeland environments. Lists of watershed studies and history of watershed research can also be found in reports by the Forest Service (1977), Callaham (1990), Binkley (2001), Ice and Stednick (2004), and Adams et al. (2004).

16.1 Research Issues

Forest hydrology is a pure as well as an applied science and therefore consists of both basic and applied types of research. An example of basic research might be a study of how channel roughness influences average streamflow velocities. An example of applied research might be a study of how large-wood removal from channels during harvesting affects peak flows. Yet, the results of the basic research and a hypothesis derived from them can be used to explain the expected peakflow response found in the applied research. Similarly, the results from the applied research can be used to confirm the patterns predicted by the basic understanding of watershed processes. Forest hydrology research continues to refine our understanding of water in forests.

FIGURE 16.1
Major experimental forested watersheds in the conterminous United States.

1. Beer Creek, Muscle Shoals. 2 WS
2. Alum Creek, etc., Fayetterville. 14 WS
3. Ten Mle Creek, Monticello. 6 WS
4. Beaver Creek, etc., Flagstaff. 34 WS
5. Three Bar, etc., Tempe. 17 WS
6. San Dimas, Glendora. 17 WS
7. Slerra National Forest, Fresno. 2 WS
8. Shasta-Trinity Forest, Redding. 4 WS
9. Putah Creek, Ukiah. 5 WS
10. UC Berkeley. 4 WS
11. UC Davis. 1 WS
12. Fraser Exp. Forest, Fort Collins. 4 WS
13. Grant Forest, U. of Georgia. 2 WS
14. Bolse. 12 WS
15. Mc Call. 1 WS
16. Grangerville. 15 WS
17. Pine River. Cadillac. 12 WS
18. Marcell, Grand Rapids. 6 WS
19. Oxford. 12 WS
20. Coffeeville. 5 WS
21. Goodwin Creek. 14 WS
22. U. of MO. 4 WS
23. Hubbard Brook. 8 WS
24. San Luis Basins, Albuquerque. 3 WS
25. Coweeta. 23 WS
26. Alsea, U. of Oregon. 6 WS
27. Coyote Creek. 4 WS
28. H. J. Andrews. 10 WS
29. Fox Creek. 3 WS
30. HI-15 Basins, Portland. 3 WS
31. Penn State U. 5 WS
32. Upper Beer Creek, etc., (TVA). 14 WS
33. Walker Branch, Oak Ridge. 3 WS
34. Davis County Exp. WS, Logan. 8 WS
35. Femow, Parsons. 9 WS
36. Elkins. 56 WS
37. Entiat Exp. Forest, Entiat. 3 WS
38. Laramie. 2 WS
39. Allen Creek, Rhinelander. 5 WS
40. Reno, U. of Nevada10 WS
41. Coshocton. 1 WS
42. Alto. 13 WS
43. Robinson Forest, U. of Kentucky. 3 WS
44. Bluewater Creek, Grants. 13 WS
45. Mc Cellan Creek, Helena. 4 WS
46. Sante Exp. Forest, Huger. 3 WS
47. New Bem, Weyerhaeuser. 6 WS
48. Bradford Forest WS., Starke. 3 WS
49. Shenandoah Park. 4 WS
50. Upper Big Run. 1 WS
51. Oklahoma S. U., 5 WS

Issues in forest hydrology studies can be discussed in terms of research objectives or research subjects. Research objectives can be separated into three categories:

1. Cause and effect of plant–water–watershed relationships
2. Predictions of hydrologic consequences caused by changes in climatic conditions, in vegetation communities, or in human activities
3. Management of watershed natural resources and hydrological conditions to serve the best interests and needs of people

Subject issues can also be separated into three categories:

1. Water functions
2. Forest functions
3. Watershed functions

16.1.1 Objectives

16.1.1.1 Cause and Effect

Cause-and-effect studies are the fundamental studies in forest hydrology. The objectives of cause-and-effect studies are to comprehend the hydrological cycle in forested watersheds, to explain variations in hydrological phenomena among watersheds, and to find physical laws that govern forest–water–climate–soil–topography relationships (Riekerk, 1989; Keppeler and Ziemer, 1990; Ursic, 1991a; Landers et al., 2007; Groom et al., 2011). Studies for these purposes may involve evaluations of watershed water-balances, energy budgets, runoff processes, rainfall–runoff relationships, roles of forests in the hydrological cycle, topographic effects, soil-water movements, etc. They require long-term watershed observations over a wide range of environmental conditions. Such requirements are intended to explain the effects of temporal and spatial variations and to avoid biased interpretations that are due to unrepresentative data.

16.1.1.2 Prediction

Prediction is a major task of all sciences. Based on knowledge learned in cause-and-effect studies, many studies are conducted to develop methodology and models for predicting consequences induced under various management activities and environmental conditions. In the forest–water interface, those predictions include evaporation from soil and water (Penman, 1956; Idso et al., 1979; Shuttleworth, 1993; Latha et al., 2011), forest transpiration (Stewart, 1984), evapotranspiration (Abtew, 1996; Oishi et al., 2008), interception loss (Massman, 1983; Van Dijk and Bruijnzeel, 2001), effects of land use on runoff generation (Douglas, 1983; Bren et al., 2006), climate changes and water resources (McNulty et al., 1997; Dettinger et al., 2004), snow accumulation and snowmelt (Stegman, 1996; Jost et al., 2007), sediment movement and yields (Clayton and Megahan, 1997; Kim et al., 2009), nutrient transport (Bakke and Pyles, 1997; Bhat et al., 2007), and other events. The developed models must be calibrated, tested, and validated using data observed in experimental watersheds (Lee, 1970; Galbraith, 1975).

Predictions provide a basis for management, control, and adjustment. Besides on-site and downstream effects, forest and land activities can have impacts upslope as well (Chang, 1997). Up-site, on-site, and down-site impacts should all be considered in prediction studies.

16.1.1.3 Management

Management studies are designed to seek or develop methods that can be used to augment water supply, reduce watershed deterioration, stabilize watershed conditions, improve timing of streamflow, enhance multiple use of watersheds, restore stream habitats, or increase economic benefits to land owners (Lakel et al., 2010; Newbold et al., 2010). Management must be based on either cause-and-effect or prediction capability and must be evaluated in terms of economic feasibility and environmental integrity (Gary, 1975). Frequently, conventional management approaches need to be revised and reassessed to meet standards set up by new environmental regulations or executive orders.

16.1.2 Subjects

16.1.2.1 Water Functions

The water molecule's polarity gives water many unique properties, and it performs many functions important to the environment and human society. Water functions are discussed in Chapter 2 and water properties in Chapter 4. Important research questions about water functions include the following:

- How are water properties affected in different environments?
- What is the spatial and temporal variation of water properties?
- What are the interrelations between water properties and the functions of water?
- What governs the distribution and occurrence of water?
- How do we establish and confirm water-quality standards for different uses?
- How are water pollutants generated and routed in watersheds and how do they interact with one another and the environment?
- What are the impacts of climate changes and wet deposition on water resources?

16.1.2.2 Forest Functions

Forest functions and functional forests are discussed in Chapter 7. Forest environmental functions are directly affected by the morphological and physiological characteristics of individual species, species composition, and forest density. Thus, studies on plant and forest characteristics will lead to a better understanding of forest performance in the hydrological environment. Topics such as measurements and predictions of leaf area index, vapor diffusion on the stoma–ambient air interface, the distribution of raindrop diameter and rainfall intensity under forest canopies, evapotranspiration variations among species, the climatic impact on forest growth and distribution, problems on forest fragmentation, and forest adaption to the environment are some areas important to modeling and management. For example, with the increasing concern over global warming, carbon sequestration in forests is of great interest to foresters and environmentalists (Davin and de Noblet-Ducoudré, 2010). The impacts of deforestation and elective cutting on forest fragmentation and edge effects need to be quantified (Broadbent et al., 2008).

The root system is an area that has had little attention in forest function study. The horizontal and vertical distribution of root systems plays an important role in soil erosion (especially on mass movement, bank erosion, and channel morphology). Root patterns may also be important in subsurface runoff modeling and peakflow assessments. However, our knowledge of the growth and decay of root systems in different soils; the distribution, density, root area index, and shear strength among species; and the effect on soil-water movement and uptake are so limited that the reliability of hydrological modeling and the efficiency of watershed management have been hindered.

16.1.2.3 Watershed Functions

The functions of a watershed include collection and storage of precipitation; discharge of surface and groundwater; support of water chemistry and habitat niches; attenuation of nutrients and pollutants (Black, 1997); storage of water, nutrients and sediments; and dissipation and transfer of solar radiation. However, the ultimate function of a watershed is to serve as a system to convey water, sediment, and nutrients to lower areas. Such functions result from the integrated effects of climate, soil, topography, and vegetation acting upon a watershed system. Studies of the integrated effects of environmental factors on watershed functions, the variation among watersheds, cumulative effects, riparian and aquatic habitats, human impacts, ecosystem management, watershed restoration and remedies, nutrient budgets, sediment losses, and water quality are some subjects that have drawn much attention. Studies are needed on the application of geographical information systems (GIS) for archiving silvicultural activities, manipulating watershed information, and modeling hydrologic processes (Swanson, 1998).

Improvement in fundamental measurements in the dynamic processes of storm rainfall and subsurface flow pathway across diverse landscapes is increasingly needed in watershed modeling (Beschta, 1998). A 10% error in precipitation input can cause a 20% error in streamflow output, depending on precipitation intensity, duration, and distribution. Also, studies on how to extrapolate results of small-watershed studies to larger watersheds, to downstream reaches, and to watersheds with insufficient hydrologic data are particularly valuable to watershed managers. "Watershed cumulative effects" is a topic of concern in watershed management and decision making (Litschert, 2009).

16.2 Principles of Field Studies

Forest hydrology studies often involve field-data collection, land and vegetation treatments, and data analyses and interpretations. They are conducted to test new hypotheses, to verify other study results, or to discover new relationships among parameters. All field studies are subject to a certain degree of error due to

1. The heterogeneous nature of the study material
2. Climatic variations
3. The accuracy problem of instruments
4. Human factors
5. No watersheds of absolute alike for comparisons

Thus, studies conducted in the field often use the three principles of experimental design to reduce study errors: control, randomization, and replication.

16.2.1 Control

All studies require a control or reference standard for assessment of treatment effects. The control can be set up by theoretical considerations, existing standards, specific targets, or levels defined by the National Bureau of Standards. Since it is impossible to have two watersheds with identical vegetation and hydrological conditions over time, an experimental watershed study requires not only a control watershed but also a calibration period. The calibration period is used to establish a reference relationship on streamflow between treatment and control watersheds for later assessment of treatment effect. However, if the size of the study area is reduced from a watershed scale to a small-plot scale relatively uniform in soils, topography, and vegetation conditions, then the calibration period can be omitted without considerable errors, provided the assumption of environmental similarity is valid.

16.2.2 Randomness

Randomization improves a study's accuracy by reducing subjectivity and personal preference in conducting the study. It is the basis for obtaining a valid estimate of treatment effects and error variation. Absolute randomness is difficult to achieve because not all watersheds are suitable for hydrological studies for a specific purpose. Researchers usually establish certain criteria for site selection, group the study sites into blocks in accordance with certain characteristics, and then randomly arrange the treatments in each block.

16.2.3 Replication

Replication means the repetition of the basic watershed experiment at more than one study site. It provides for a more accurate estimate of experimental error, and it enables researchers to obtain a more precise estimate of the mean effect of the study (Ostle and Malone, 1988). Usually three or four replicas are used to cover sample variations; these replicas are randomized to avoid any subjectivity. Replication also makes the collected data more representative. The entire study may need to be repeated by other researchers to validate the results, and to be performed in different climatic conditions to assess its applicability.

16.3 Research Methods

Numerous studies have been conducted to investigate the forest–water yield relationship. The methods employed can be grouped into six categories:

1. Experimental watersheds
2. Upstream–downstream approach
3. Experimental plots
4. Regional analysis
5. Watershed simulation
6. GIS/remote sensing

Experimental watersheds, including the paired-, single-, and replicated-watershed approaches, are referred to as the hydrometric method, while experimental plots are referred to as the gravimetric method. Regional analysis and watershed simulation are known as the statistical (empirical) and physical methods, respectively. GIS/remote sensing is a distance data collection and analysis approach that can be applied to all of the other five study methods.

16.3.1 Experimental Watersheds

In this method, one or multiple representative watersheds are selected to study hydrological processes utilizing streamflow-gaging stations and other hydroclimatic instruments. This method has been employed as a principal means of monitoring baseline data on long-term hydroclimatological phenomena, evaluating the hydrological impact of various land uses and management activities, and providing data that can serve as a database for calibrations and validations of hydrological or ecological models. There are three approaches: paired watershed, single watershed, and replicated watershed.

16.3.1.1 Paired-Watershed Approach

The paired-watershed approach chooses two or more watersheds to be gaged for study. One of them is designated as the control watershed and the other(s) as the treatment watershed(s). The whole study is divided into two periods, a calibration period and a treatment period. The purpose of the calibration period is to establish a reliable hydrological relationship (calibration equations) between the control and treatment watersheds through statistical analysis. It is used as a reference base to assess treatment effects on streamflow in the treatment watersheds during the treatment period. After each pair of watersheds is satisfactorily calibrated, then the treatment watershed is treated as designated. The treatment effects are calculated between the predicted (from the calibration equation) and observed values and are evaluated for significance through statistical tests. Details are discussed in the following sections.

16.3.1.1.1 Watershed Selection

All watersheds selected for the study must be representative of the region in question so that the results are transferable to ungaged watersheds. Also, the paired watersheds must be close to each other to ensure similarity with respect to vegetation cover, soil types, topographic characteristics, and climatic conditions. Since environmental variability is much greater in watersheds of large size and the hydrological behaviors of large watersheds are much different from those of small watersheds, it is important that the control and treatment watersheds be comparable in size. Watersheds with potential water leakage due to geological structures should not be used in the study. Valid results are based upon the assumption that all runoff (surface and subsurface) merges at the gaging stations.

The experimental watersheds used for water-augmentation studies in the United States are mostly first-order or other low-order streams. They range in size from 1 ha (Ursic and Popham, 1967) to more than 500 ha (Schneider and Ayer, 1961; TVA, 1961; Troendle and King, 1985), but are seldom larger than 1000 ha. Watersheds of less than 100 ha are the most common size used in hydrologic studies. If the experimental watershed is too small, the environmental conditions tend to be oversimplified, making the transferability of results to other areas questionable. Furthermore, smaller watersheds frequently have watershed-leakage problems, causing serious errors in water measurements and water-balance computations. On the other hand, it is more difficult to estimate areal precipitation, to measure streamflow accurately, and to control treatments on large watersheds (Bosch and Hewlett, 1982).

16.3.1.1.2 Calibration Methods

Once experimental watersheds are selected, streamflow gaging stations (Figure 16.2) and a network of climatic instruments are then installed to monitor hydroclimatic conditions

FIGURE 16.2
Stream-gaging station consisting of a 3 ft H-flume, a 2 ft Coshocton wheel, a stilling well equipped with stage-recording potentiometer, an ISCO runoff sampler, and a data logger installed at the Alto Experimental Watershed, Texas.

in each watershed for a sufficient period of years (the sufficient period will be discussed later). Inventories and analyses of forests, land use, soil, and topography are conducted to provide information necessary for applying treatments and analyzing and interpreting results. The hydroclimatic data are used to calibrate a reliable relationship between the treatment and the control watersheds. This calibrated relationship serves as a reference standard for evaluation of treatment effects.

The most common calibrated relationship between the paired watersheds is based on annual streamflow in a simple linear-regression analysis. It appears as follows:

$$y_x = a + b(x) \tag{16.1}$$

where
y_x is estimated annual streamflow for the treatment watershed
x is observed streamflow for the control watershed
a and b are regression constants

Once a satisfactory relationship is achieved between the paired watersheds, the vegetation (or other considerations) on the treatment watershed is then treated to its designated degree of intensity. This launches the beginning of the "treatment period," and all the hydroclimatic data collection proceeds as in the previous period.

Although linear regressions are most commonly used in the calibration, curvilinear regressions may fit the streamflow better if it is under the influence of extreme climatic conditions or if the physiographic conditions of the paired watersheds are not similar. This led Rich and Gottfried (1976) to use the second-order polynomial regression to calibrate the Workman Creek experimental watersheds in central Arizona. They fit the following equation:

$$y_x = a + b_1(x) + b_2(x)^2 \tag{16.2}$$

Another version of the analysis is to use streamflow of the control watershed, and precipitation differences $(P_t - P_c)$ between treatment and control watersheds, to calibrate streamflow of the treatment watershed (Reinhart, 1967). Its mathematical expression appears as follows:

$$y_x = a + b_1(x) + b_2(P_t - P_c) \tag{16.3}$$

The term $(P_t - P_c)$ should be relatively small or approaching zero if the two watersheds lie side by side (Hewlett, 1970).

Using Equation 16.1 in the treatment assessment has been criticized for not considering precipitation variation between the two watersheds. Muhamad and Chang (1989) proposed a calibration of runoff coefficient (runoff–rainfall ratio, Q/P) instead of annual streamflow in the simple regression analysis, or

$$\left(\frac{Q}{P}\right)_t = a + b\left(\frac{Q}{P}\right)_c \tag{16.4}$$

where subscripts t and c refer to treatment and control watersheds, respectively. By using runoff coefficient (Q/P), the impact of climatic variations on watershed calibration could be minimized. It provides a more reliable result if the two watersheds are having considerable climate variations or are located with a great distance in between. The treatment effects on streamflow are calculated as the product of the estimated runoff coefficients, Q/P, and precipitation.

All calibration methods mentioned here are based on annual streamflow. The same methods can be applied to calibrate experimental watersheds based on seasonal and monthly streamflows, peak flows, lowflows, or even stream sediments and other water quality data. Breaking down annual streamflow into seasonal or monthly streamflows increases sample points in the regression analysis, but because of auto-correlation problems, reliance may not be placed upon analysis of data by months (Reinhart, 1958), due to antecedent flow effects.

16.3.1.1.3 Treatment Methods

The most common methods used for vegetation treatments are mechanical (Pond, 1961), prescribed burning (Hibbert et al., 1982), and chemical (Kochenderfer and Wendel, 1983; Davis, 1993). In order to minimize the adverse effects on stream sediment and water quality, precautionary measures need to be applied in these treatments. Such measures include using a mosaic treatment pattern (Davis, 1993), leaving a buffer strip along stream channels, using a ridgeline treatment (Hibbert et al., 1986) and partial cutting (Troendle and King, 1987), or using progressive or alternative strips (Hornbeck et al., 1987).

16.3.1.1.4 Duration of Calibration and Treatment Periods

The calibration period generally lasts about 7 or more years, depending on weather conditions during the study period, data variability, and the level of precision desired. Studies have taken as long as 23 year in the Holly Springs National Forest in northern Mississippi (Ursic, 1991a) and 15 year in Workman Creek Experimental Watershed in central Arizona (Rich and Gottfried, 1976), and as short as 5 year in Coweeta Experimental Watershed (Hoover, 1944). If too short a period is used, the experiment may not cover sufficient climatic variations, and the results may be questionable. Besides, there may not be enough

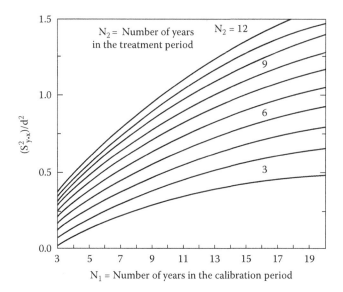

FIGURE 16.3
Graphic solution for the Wilm (1949) equation to determine the minimum length of calibration and treatment periods.

data points to define a clear relationship. On the other hand, if the calibration period is longer than necessary, it increases experimental costs, delays the results, and makes the watersheds subject to alterations by natural hazards.

Wilm (1949) has presented an analytical method, based on standard error of estimates and an assigned value of smallest treatment effect (difference expected as a result of treatment), to determine the minimum length of calibration and treatment periods. The graphical solution of Wilm's equation, prepared by Kovner and Evans (1954), is given in Figure 16.3. It is a family of curves for values of N_2 (length of treatment period) with F-value at the 0.95 probability level. The vertical axis is the squared ratio–standard errors of estimate ($S_{y \cdot x}$) divided by the smallest worthwhile difference (d, treatment effect). Values of standard errors of estimate are calculated using the data already collected in the calibration period, while values of d must be assigned based on experience. Usually, 5% or 10% of the mean yield for the calibration period of the watershed is chosen as the smallest difference expected as a result of treatment. In other words, if the d value is not reached, the investigator is willing to forego determination of its significance.

For example, using Figure 16.3, if the error of variance, $S_{y \cdot x}^2$, for an 8 year calibration period is 5.00 cm, the mean yield is 50.00 cm, and the smallest worthwhile difference d is 5% of 50.00 cm, or 2.50 cm, then $S_{y \cdot x}^2/d^2 = 5.0/2.5^2 = 0.80$. With $N_1 = 8$ and $S^2/d^2 = 0.8$, the point intersected in Figure 16.3 shows that 11 year would be required in the treatment period (N_2).

When Figure 16.3 is used for the determination of calibration period, it is necessary to use Wilm's (1949) successive approximations by taking $N_1 = N_2$. The value $S_{y \cdot x}^2/d^2$, as calculated previously, is entered on the vertical scale and moved horizontally along this ordinate until the curve is intersected at $N_1 = N_2$.

Bethlahmy (1963) suggested a storm-hydrograph method for calibrating watersheds with a maximum period of 2–3 year. This method computes a simple regression line on maximum stage rise and time to peak obtained from storm hydrographs between the two watersheds for the pretreatment year and a regression line for the posttreatment years. Pretreatment and posttreatment regression lines are then compared for statistical significance of the treatment. In this method, only storms that cover both watersheds may be used in the analysis.

Calibration and analysis based on individual runoff events is particularly useful in watersheds where changes in overland flow are of primary concern. Ursic and Popham (1967) showed that the accumulation of 30 consecutive runoff events is sufficient to define a reliable relationship. In the southern Coastal Plain, storms occurring in a given year may not be fully representative, but 2 or 3 year should be sufficient.

16.3.1.1.5 Treatment Effects

The differences in streamflow between the observed and the predicted (for the conditions as if the watershed had not been treated) values calculated by calibration equations such as Equation 16.1 are the treatment effects, $\delta q = y - y_x$. To determine statistical significance of the treatment effects, a posttreatment regression on streamflow between the control and treated watersheds can be developed. The posttreatment regression is then compared to the pretreatment (calibration) regression on their slopes and adjusted means (elevations) by covariance analysis for statistical significance at a preselected alpha level (Li, 1964; Troendle and King, 1985; Bent, 1994). A numerical example for the statistical tests is given in Section 16.4. It is advisable, however, that the periods before and after treatment be as close to the same length as possible to allow an equal opportunity for exposure to variations in climate and other environmental factors (Wilm, 1949). Also, a difference between the observed and estimated streamflow can be considered statistically significant if the deviation exceeds a preselected (usually 95%) confidence interval about the regression line (Harr et al., 1979; Hornbeck et al., 1987).

As was explained earlier, deforestation is generally expected to cause an increase in streamflow. The magnitudes of increase, although highly variable, seem to be positively affected by annual precipitation and are gradually reduced to levels at the pretreatment period due to regrowth of vegetation. Thus, a two-step analysis was used to study the effect of forest understory cutting and regrowth on streamflow at Coweeta Hydrologic Laboratory, North Carolina (Johnson and Kovner, 1954). First, the increase in annual streamflow was estimated using a simple linear-regression model (similar to Equation 16.1) developed for the calibration period. Then the estimated treatment effect in annual water yield (δq) was regressed with annual precipitation (P) and a time factor in the following fashion:

$$\delta q_{(t)} = a + b_1(P) + b_2(t) \tag{16.5}$$

where
    t is the time since treatment in years
    P is annual precipitation
    a, b_1, and b_2 are regression coefficients

The time factor (t) and streamflow from the control watershed (x) were also used together in the following multiregression model to evaluate treatment effects (Schneider and Ayer, 1961; Baker, 1986):

$$\delta q_{(t)} = a + b_1(x) + b_2(t)(x) \tag{16.6}$$

The technique of double-mass analysis (Merriam, 1937; Chang and Lee, 1974) has also been applied to evaluate the significance of treatment effects (Anderson, 1955; Chang and Crowley, 1997). This method plots the accumulated annual streamflow of the control

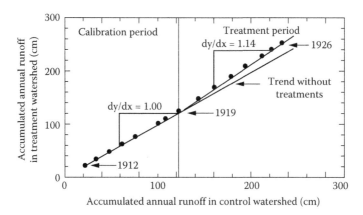

FIGURE 16.4
Plot of double-mass analysis for the watershed study at Wagon Wheel Gap, Colorado. (Data from Bates, C.G. and Henry, A.J., Forest and streamflow experiments at wagon wheel gap, Colorado, U.S. Weather Bureau No. 30, Washington, DC, 1928.)

watershed against the accumulated annual streamflow of the treatment watershed. The plot appears as a straight line with minor variations if there was no treatment effect or the treatment effect was insignificant. A definite break in slopes on the plotted trend line indicates the effects of treatment, and the magnitude of treatment effect is evaluated by the proportionality between these two slopes (Figure 16.4).

16.3.1.1.6 Criticisms of the Approach

Though the paired-watershed approach is the most reliable form of watershed investigation on forest–water relationships, it is still subject to a number of criticisms (Reigner, 1964; Muller, 1966; Hewlett et al., 1969; Hewlett, 1971):

- The time required to produce the results is excessively long.
- It is difficult to locate paired watersheds ideal for the study; they are often unrepresentative.
- It is expensive, especially because of the additional costs for maintaining the control.
- A fire or insect infestation can alter the watershed character.
- The vegetation on the control watershed can change after the calibration.
- The procedure does not take into account the effects of climatic variation between the control and treatment watersheds, especially the temporal variations in weather between watersheds.
- It is difficult to transfer results obtained from experimental watersheds to other areas.
- The integrated results obtained from experimental watersheds conceal hydrological processes.
- The approach can be confounded in regions where small convective storms cause events in one watershed and cloudburst events in an adjacent watershed. As watersheds get larger, it becomes more difficult to characterize weather patterns and match watersheds as a pair.

16.3.1.2 Single-Watershed Approach

The single-watershed approach also involves two study periods: a calibration period and a treatment period. It uses only one watershed throughout the study; no control watershed is required. During the calibration period, streamflow is related statistically to climatic variables through multiple-regression analysis. The developed hydroclimatic model is then used to estimate streamflow for the treatment period. The treatment effects on streamflow are calculated by the differences between observed and estimated values.

By way of example, the following equation was developed to estimate streamflow for the 1964–1972 period of the forested La Nana Creek watershed in east Texas (Chang and Sayok, 1990):

$$Q = -396.16 + 194.23 \frac{P}{T^2} \tag{16.7}$$

where
Q and P were annual streamflow and precipitation, respectively (mm)
T was annual air temperature (°C)

Equation 16.7 was then employed to estimate streamflow for the 1973–1984 period, in which the lower portion of the watershed went through a rapid and intensive urbanization. Comparing the observed streamflow to that estimated by Equation 16.7, urbanization in the lower section of the 80 km² watershed increased streamflow by 85 mm/year.

As previously indicated, the treatment effect can also be evaluated by plotting the accumulated values of precipitation against accumulated runoff in a double-mass analysis (Potts, 1984). A break in slopes on the double-mass curve indicates the effect of land-use treatments and is calculated by the difference in slopes between these two periods (Bernt and Swank, 1970; Betson, 1979). However, the use of double-mass curves is rather crude and provides no measure of random error (Reigner, 1964). In northern California, Pitt et al. (1978) used runoff/precipitation ratio as a criterion to evaluate the effects of converting woody vegetation into brush and weather patterns on streamflow among three study periods on a single watershed. They found that grassy watershed released 39% more total runoff than did woody vegetation.

Thornthwaite's water-balance model has also been employed to evaluate treatment effects in four single watersheds in New York (Muller, 1966). The model is based on the concept of potential evapotranspiration (PE) computed by mean air temperature and by a temperature-based heat index. By using monthly PE and actual evapotranspiration and precipitation, the monthly distribution of water deficit, soil-water utilization, soil moisture recharge, and surplus water can be characterized for the area (Thornthwaite and Mather, 1955, 1957). The method takes into account much of the monthly and seasonal variation of the hydroclimatological elements in the analysis. However, the evaporation equation used in this method fails to explain the difference in the plant cover on the watersheds (Hewlett et al., 1969).

The single-watershed approach requires a calibration period longer than the paired-watershed approach, and hence is subject to additional costs and greater environmental risks. This may be due to two reasons. First, it is more difficult to estimate streamflow from climatic variables than from streamflow of a nearby watershed because of data variability. Second, because of budget constraints and the consideration of site accessibility, the climatic data used in the watershed study are usually collected from a single site near the stream-gaging station or even out of the watershed. In this case, the climatic data are most likely not representative of the climatic conditions of the entire watershed.

In the paired-watershed approach, if forests in one watershed are destroyed due to natural or man-made disasters, study of the other watershed can proceed using the single-watershed approach.

16.3.1.3 Replicated-Watershed Approach

The results obtained from the paired- and single-watershed approaches refer to the specific site conditions. Measurements made at a single site may not be representative of the entire area, and the experimental errors are difficult to assess. This leads to the use of the replicated-watershed approach (Blackburn et al., 1986; Ursic, 1991b). Replication, randomization, and control are the three basic principles of experimental design. However, this approach is not common in watershed studies partly due to the variation of watershed environmental conditions. Ideal watersheds for replication are very difficult to find. Replication is general practice in small-plot studies (Shahlaee et al., 1991).

Careful design of a replicated watershed study does not preclude a calibration period as was discussed in the paired-watershed approaches. Replication allows researchers to evaluate whether or not the treatments have caused a significant effect. It does not reveal the magnitude of the treatment effects. Without such information, watershed-modeling studies cannot be validated.

16.3.2 Upstream–Downstream Approach

The upstream–downstream approach is often used in studying water quality, habitat quality, and aquatic organisms as affected by existing land-use conditions, land treatments, or natural disasters. It selects two sampling sites along a reach for study, one above and one below the project area. If feasible, use three sampling positions, one above, one within, and one or more than one below the treatment (Groom et al., 2011). The upstream site is a control for the detection of changes due to downstream treatments. Differences in concentrations or other quality parameters between downstream (or the treatment stream) and upstream are attributable to the conditions of the project (treatment) area. Key elements of this approach are the similarity and simultaneous monitoring of the sampling sites along a reach. Thus, the reach with upstream and downstream sites selected for the study should be straight in morphology and alike in riparian vegetation, bank materials, substrates, and flow patterns. They should be located as close to the project area as practicable so that the confounding differences between the two sites can be minimized (MacDonald et al., 1991). However, the upstream control site should be far enough from the downstream treatment effect. The method generally does not use a calibration period; measurements from an untreated reach are adequate for comparisons if the site is properly located (Gluns and Toews, 1989; EPA, 1997). The approach is often criticized for streamflow autocorrelation between the upstream and downstream sites. If practicable, three to four replicates should be established in the same area for statistical inference.

16.3.3 Experimental Plots

Experimental plots are blocks of undisturbed soil surrounded by bricks, concrete blocks, or even plywood installed in the field. The lower end of each plot is equipped with instruments or containers to monitor surface and subsurface runoff and to catch sediment

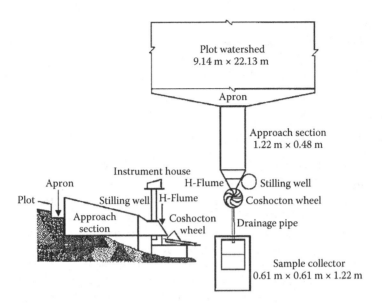

FIGURE 16.5
Layout of a plot-watershed. (Adapted from Chang, M. and Ting, J.C., Applications of the universal soil loss equation to various forest conditions, in *Monitoring to Detect Changes in Water Quality Series (Proceedings of the Budapest Symposium)*, July 1986, Budapest, Hungary, IAHS, Oxfordshire, U.K., pp. 165–174, 1986.)

under various forest conditions (Figure 16.5). Since these plot watersheds are relatively small—0.01–0.1 ha in size—it is possible to install them in relatively homogeneous conditions with sufficient replicates with respect to soils, topography, vegetation, and climatic conditions. Such homogeneous plots, with sufficient replicates, allow the effects of vegetation manipulation on runoff to be monitored directly from these plot watersheds without need for a calibration period. Compared to the paired- and single-watershed approaches, the costs and time required in plot studies are greatly reduced.

Lysimeters are another version of the plot-study approach. A lysimeter is a whole block of soil (bottom and sides) completely sealed from its surroundings. Usually a scale-balance and drainage pipes are installed below the block; evapotranspiration (ET) of the block can then be calculated by the loss in weight, and soil-water percolation can be completely collected, between two consecutive periods. A design like this is referred to as a weighing lysimeter (Harrold and Dreibelbis, 1958; Kirkham et al., 1984). Sometimes the whole soil block is in an inner tank that floats on an outer tank filled with liquid and installed in the soil. ET of the soil tank is evaluated by the fluctuation of water level in the water tank through Archimedes' principle. This type of tank is called a floating lysimeter (Chang et al., 1997). Since the hydrologic components can be measured and calculated precisely, lysimeters are a common tool in ET and soil-percolation studies. It is important that the soil in a lysimeter be undisturbed, that it be installed to simulate natural conditions, and that the sensitivity of the weighing device be adequate to detect a change in weight of water equivalent to 0.5 mm or less.

If small plots (i.e., 0.1–10.0 ha in size) are used to measure snow accumulation and snowmelt (Gary and Troendle, 1982) or to measure water use by various plants through soil moisture changes (Veihmeyer, 1953), no soil boundary is required. Since small plots are homogeneous and can be isolated under relatively controlled conditions, they can be used to obtain basic understanding of certain hydrological processes. Plot studies are often used as a first step in testing a hypothesis relating to watershed management, or to complement watershed studies (Ward, 1971). Because of the difficulty of relating study plots to watershed conditions, a follow-up study using experimental watersheds is often recommended if results from plot studies are promising.

16.3.4 Regional Analysis

The regional-analysis approach develops models for a region based on streamflow records from existing stream-gaging stations and information on vegetation, topography, and climatic characteristics in each watershed. It is applicable to a region where streamflow records and climatic observations are available over a wide area of coverage. This creates a set of data with streamflow as the dependent variable and watershed topographic, climatic, and land-use characteristics as the independent variables for statistical analyses. The statistical techniques employed include correlation coefficients; simple, multiple, and curvilinear regressions (Chang and Boyer, 1977; Kennedy and Neville, 1986); multivariate analysis (DeCoursey and Deal, 1971); and covariance analysis (Muda et al., 1989). Once quantitative effects of vegetation and physioclimatic parameters on streamflow are established through statistical analyses, management application can be made to watersheds in the region.

By using this method, Lull and Sopper (1967) analyzed the streamflow of 137 watersheds with areas of less than 259 km^2 in the northeastern United States. Of the 14 environmental parameters analyzed, annual streamflow was found to be positively affected by annual precipitation, percent of forest cover, and elevation and was negatively affected by mean maximum temperature for July. About 61% of annual streamflows were accounted for by the four parameters. The positive effect of percent forest area on annual streamflow, which seems to be contradictory to the central concept discussed earlier, is due to the forest environment of the region. In the northeast, forest cover generally increases with elevation, which in turn is associated with greater precipitation, steeper slopes, shallower soils, cooler temperatures, and lower evapotranspiration. These environmental factors may result in greater streamflow.

In forested east Texas, Chang and Waters (1984) developed the following equation with a standard error of estimate 11.8% of the observed mean to estimate annual streamflow for 19 unregulated, natural streams:

$$Q = -2018 + 4.16(P_s) + 50.91(T) - 1.98(FO) \qquad (16.8)$$

where
  Q and P_s are annual streamflow and spring precipitation (March + April + May), respectively (mm)
  T is annual air temperature (°C)
  FO is percent forest area

The results show that a reduction of watershed forested area by 20%, other things being equal, would increase water yield about 40 mm/year, and the difference in annual streamflow between full forest cover and bare watersheds could be as much as 200 mm. This difference was much less than that reported in western Oregon (Harr, 1983) and Coweeta, North Carolina (Hewlett and Hibbert, 1961), about the same as that in West Virginia (Kochenderfer and Aubertin, 1975), and greater than that in Arizona and New Mexico (Hibbert, 1981). Equation 16.8 also shows the forest impact on stream analogous to the forest impact on evapotranspiration in the entire southeastern United States (see Equation 9.42 in Chapter 9).

16.3.5 Watershed Simulation

Experimental watershed studies are costly, time consuming, weather dependent, and difficult to replicate. Hydrologic simulation provides an alternative approach for rapidly assessing impacts of forest practices and other land management in the hydrologic cycle. It is a means of predicting watershed response to various weather conditions and management strategies. The simulation models can be for estimating streamflow, sediment, water quality, and other hydrologic components of interest through physically based (processes) or empirical approaches. The time frame of the estimates can be for (1) single rainfall–runoff event and routing programs, (2) continuous simulation programs, or (3) long-term averages. The spatial scale of watershed models can be lumped (a watershed is treated as a single unit with watershed parameters averaged for the computation) or distributed (a watershed is divided into subunits in accordance with the environmental variation).

By way of example, Figure 16.6 is the conceptualization of a watershed monthly hydrologic model developed by Allred and Haan (1996). The model, "SWMHMS—small watershed monthly hydrologic modeling system," simulates monthly runoff from a small nonurban watershed. The rainfall input in the model is partitioned into three components: (1) surface and vegetation interception loss, (2) soil infiltration and soil-water reservoir, and (3) surface runoff. Soil water is first subject to ET loss and then to soil percolation as interflow and groundwater if the excess soil water is above field capacity. Base flow from the interflow and groundwater reservoir is then added to surface runoff to become the total runoff. All components of the model are calculated on a daily basis and are summed to obtain monthly values.

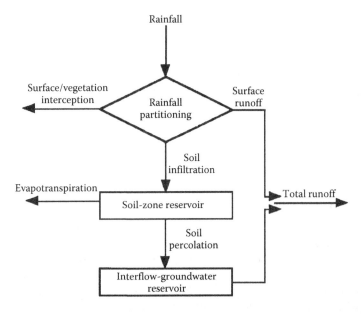

FIGURE 16.6
Hydrologic conceptualization of SWMHMS. Note that rectangle means calculations and diamond means decision making. (From Allred, B. and Haan, C.T., *Water Resour. Bull.*, 32, 465, 1996. Reproduced by permission of AWRA.)

The first major hydrologic model to simulate the whole phase of the hydrologic cycle of a watershed was the Stanford Watershed Model developed in the 1950s (Crawford and Linsley, 1966). Since then, at least 69 computer models have been developed to estimate watershed response to rainfall events and changes in land use (Singh and Woolhiser, 2002; Eisenbies et al., 2007). Some include:

- AGPNPS, Agricultural Nonpoint Source (Young et al., 1987; Vennix and Northcott, 2004)
- ANSWERS, Areal Nonpoint Source Watershed Environmental Response Simulation (Beasley et al., 1980)
- BASINS, Better Assessment Science Integrating Point and Nonpoint Sources (Whittemore and Beebe, 2000)
- CREAM, Chemicals, Runoff, and Erosion from Agricultural Management Systems (Knisel, 1980; Connolly et al., 1997)
- DHSVM, Distributed Hydrologic Soil Vegetation Model (Wigmosta et al., 1994, 2002; Cuo et al., 2008)
- EPIC, Erosion/Productivity Impact Calculator (Williams et al., 1985; Jain and Dolezal, 2000)
- ERHYM-II, Ekalaka Rangeland Hydrology and Yield Model (Wight, 1987)
- GLEAMS, Groundwater Loading Effects on Agricultural Management Systems (Leonard et al., 1987)
- HSPF, Hydrological Simulation Program—Fortran (NCASI, 2001)
- PRMS, Precipitation/Runoff Modeling System (Leavesley et al., 1983)
- SPUR, Simulation of Production and Utilization of Rangelands (Wight and Skiles, 1987)
- SWRRB, Simulator for Water Resources in Rural Basins (Williams et al., 1984)
- SWAT, Soil Water Assessment Tool (Neitsch et al., 2002; Gassman et al., 2007)
- WEEP, Water Erosion Prediction Project (Foster and Lane, 1987), forest applications (Elliot and Hall, 1997; Dun et al., 2009)

Detailed summaries on the characteristics, features, strengths, weaknesses, and applications of many models listed here, and some that are unlisted, are given by Borah and Bera (2003) and Daniel et al. (2011).

16.3.6 Remote Sensing/GIS

Remote sensing is the science and art of obtaining information about an object, a region, or an event through analyses of data acquired by a sensor not in direct contact with the target of investigation. (*Note*: Geophysical methods that measure electrical, magnetic, and gravity force fields are excluded from remote sensing.) A GIS is a computer-based system for capturing, storing, manipulating, checking, analyzing, and displaying spatial information on Earth, or is a computer-based tool for mapping and analyzing the variation of events, conditions, and processes in a geographical region. The technology can scan maps and photos into digital files in a computer to be integrated with other sources of data for analyses and modeling. In addition, the field-referenced data from GPS receivers, remote sensing sensors, and devices such as lasers, multispectral scanners, and radars can be directly loaded into GIS and manipulated for further analysis.

One GIS and spatial analysis technique with wide application in hydrology is geostatistics (Myers, 1996; Goovaerts, 1997; Siska, 2004) and there are powerful versions of new GIS software with sophisticated modules and algorithms, such as kriging, cokriging, and Gaussian simulations, which can be readily used for interpolation of spatially correlated data and stochastic modeling. In other words, these GIS techniques have the ability to:

1. Collect data about spatial and temporal variations from spectral information on Earth's surface
2. Transfer the collected information into digital data
3. Perform qualitative and quantitative analyses, statistical tests, modeling, and assessment
4. Provide graphical illustrations

The data collection and analyses can be conducted in a manner that covers a large area, is in great detail, and can be completed in a short period of time—far beyond the reach of traditional methods. All these capabilities are enhanced due to the advancement of computer engineering, software development, and satellite technology.

Accordingly, remote sensing/GIS has become a new tool with wide applications in hydrology, such as:

1. Assessment of the hydrologic state, processes, or responses of a basin or region
2. Characterization of spatial and temporal variations of environmental information
3. Estimates of hydrologic events

Discussions on the principles and uses of remote sensing/GIS in hydrology and water resource management are beyond the scope of this text; readers can refer to those works published by Grant (1988), Gurnell and Montgomery (2000), Kovar and Nachtnebel (1996), Lyon (2000), Vieux (2005), and others.

16.4 Wagon Wheel Gap Study: Example of Calibration and Analyses

Experimental-watershed studies require watershed calibration and assessment of treatment effects. The analysis methods depend on whether it is a paired-watershed, single-watershed, or replicated-watershed study. This section uses the data collected from Wagon Wheel Gap Experimental Watersheds in Colorado (Bates and Henry, 1928), the first paired-watershed study in the United States, as an example to illustrate the procedures in watershed calibration and treatment assessment. The basic features of this study were as follows:

Objective: Testing the effect of forest clear-cutting on annual runoff

Location: Rio Grande basin, Mineral County, southern Colorado, the United States

Methodology: A simple paired-watershed study consisting of a control and a treatment watershed, a calibration period, and a treatment period

Watershed: Control watershed = 90 ha; treatment watershed = 81 ha, located side by side

Soils: Clay loam texture, well drained, quite permeable, derived from augite-quarta-latite

TABLE 16.1

Annual Precipitation and Runoff (All in cm) Data at the Wagon
Wheel Gap Experimental Watersheds in Colorado, the
United States, 1912–1924

| | Calibration Period | | | | | Treatment Period | | | |
| | Control Watershed | | Treatment Watershed | | | Control Watershed | | Treatment Watershed | |
Year	Q	P	Q	P	Year	Q	P	Q	P
1912	21.3	54.2	21.3	54.6	1919	15.4	53.7	15.2	53.7
1913	12.1	47.4	13.2	49.9	1920	20.0	57.1	21.7	55.3
1914	14.3	47.7	14.1	48.6	1921	17.5	57.6	21.1	57.1
1915	13.6	50.7	13.7	50.4	1922	17.3	54.5	22.3	52.1
1916	14.2	57.7	14.1	58.8	1923	15.5	61.8	18.2	60.5
1917	24.5	58.1	25.0	57.9	1924	18.0	43.2	20.4	42.6
1918	08.1	48.0	09.0	47.9	1925	10.8	55.6	12.6	56.9
					1926	11.1	46.4	12.8	45.9

Elevation: 2700–3350 m Vegetation: about 49% and 61% occupied by aspen in the control and treatment watersheds, respectively. Other species include Douglas fir and bristlecone pine

Duration: Calibration: 1912–1918; treatment: clear-cutting, July 1919–1920

16.4.1 Hydroclimatic Data

The annual precipitation (P) and runoff (Q) data for the two experimental watersheds during the entire study period are given in Table 16.1.

16.4.2 Calibration Equation

During the calibration period, a calibration equation should be developed that mathematically relates runoff in the treatment watershed to runoff and other hydroclimatic parameters in the control watershed. In this example, we shall calibrate the two watersheds based on annual runoff in a simple linear relationship as depicted in Equation 16.1, the most popular approach. The simple linear equation is written as follows:

$$y_x = \bar{y} + b(x - \bar{x}) = (\bar{y} - b\bar{x}) + bx = a + bx \tag{16.9}$$

where
y_x is the estimated runoff for the treatment watershed
$\bar{y}$ is the runoff mean for the treatment watershed
b is the regression coefficient (slope)
x and $\bar{x}$ are the runoff and runoff mean for the control watershed, respectively

The derived equation appears as follows:

$$y_x = 15.77 + 0.97\,(x - 15.44)$$

or

$$y_x = 0.79 + 0.97(x) \qquad (16.10)$$

All calculations for developing the equation and the statistic for testing the significance of the regression coefficient b are summarized in Table 16.2. The F-test shows that the regression slope 0.97 is significantly different from zero at a probability greater than 99%. This means that the annual runoff in the treatment watershed can be satisfactorily estimated by runoff from the control watershed if there are no alterations in either watershed. Now the watershed is ready for treatments.

TABLE 16.2

Summary of Computations and Developing a Calibration Equation for the Wagon Wheel Gap Experimental Watersheds, Colorado, the United States

Control Watershed (x)		Treatment Watershed (y)		Control (x) × Treatment (y)	
Statistic	Value	Statistic	Value	Statistic	Value
Calibration Period					
n =	7	n	7	n	7
Σx =	108.10	Σy	110.40		
$(\Sigma x)^2$ =	11,685.61	$(\Sigma y)^2$	12,188.16	$(\Sigma x)(\Sigma y)$	11,934.24
$(\Sigma x)^2/n$ =	1,669.37	$(\Sigma y)^2/n$	1,741.17	$(\Sigma x)(\Sigma y)/n$	1,704.89
Σx^2 =	1,857.05	Σy^2	1,919.24	$\Sigma(xy)$	1,886.98
$SS_x = \Sigma(x-\bar{x})^2$ =	187.68	$SS_y = \Sigma(y-\bar{y})^2$ =	178.07	$SP = \Sigma[(x-\bar{x})(y-\bar{y})]$ =	182.09

Derivations for a Calibration Equation

1. Regression coefficient $b = SP/SS_x = 182.09/187.68 = 0.97$
2. Mean of the treatment watershed $\bar{y} = \Sigma y/n = 110.4/7 = 15.77$
3. Mean of the control watershed $\bar{x} = \Sigma x/n = 108.1/7 = 15.44$
4. Correlation coefficient $r = SP/\sqrt{(SS_x)(SS_y)} = 182.09/(187.68 \times 178.02)^{0.5} = 0.996$
5. Regression SS (sum of squares) = $\Sigma(y_x - \bar{y})^2 = (SP)^2/SS_x = 182.09/187.68 = 176.67$
6. Residual SS = $\Sigma(y - y_x)^2 = SS_y -$ Regression SS = $178.07 - 176.67 = 1.40$
7. Estimated variance $S^2 = $ (Residual SS)$/(n-2) = 1.40/(7-2) = 0.28$
8. Test of hypothesis $\beta = 0$
   F = Regression SS$/S^2 = 176.67/0.28 = 630.96$ with 1 and 5 degrees of freedom (d.f.),
   Critical F @ 1 and 5 d.f.; 1% probability (one-tail) = 16.258,
   Since F > critical F, the hypothesis is rejected and $\beta \neq 0$.
9. The derived calibration $y_x = \bar{y} + b(x - \bar{x}) = 15.77 + 0.97(x - 15.44)$

Treatment Period					
n =	8	n	8	n	8
Σx =	125.60	Σy	144.30		
$(\Sigma x)^2$ =	15,775.36	$(\Sigma y)^2$	20,822.49	$(\Sigma x)(\Sigma y)$	18,124.08
$(\Sigma x)^2/n$ =	1,971.92	$(\Sigma y)^2/n$	2,602.81	$(\Sigma x)(\Sigma y)/n$	2,265.51
Σx^2 =	2,046.80	Σy^2	2,714.43	$\Sigma(xy)$	2,350.58
$SS_x = \Sigma(x-\bar{x})^2$ =	74.88	$SS_y = \Sigma(y-\bar{y})^2$ =	111.62	$SP = \Sigma[(x-\bar{x})(y-\bar{y})]$ =	85.07

Note: $\Sigma(x-\bar{x})^2 = \Sigma x^2 - (\Sigma x)^2/n$; $\Sigma[(x-\bar{x})(y-\bar{y})] = \Sigma(xy) - (\Sigma x)(\Sigma y)/n$; $y_x =$ estimated y. All values are in centimeters except N in years.

16.4.3 Assessment of Treatment Effects

16.4.3.1 How Much Effect?

During and after treatments, the collection of hydrologic data in the two watersheds continued as usual. Posttreatment monitoring lasted for 8 year in the Wagon Wheel Gap study. The treatment effect is first assessed here by a preliminary comparison based on simple ratios of hydrologic data between the treatment and control watersheds during the calibration and treatment periods, then by the standard procedure based on the calibration equation and the double-mass analysis.

16.4.3.1.1 Preliminary Assessment

First, the means of annual runoff, precipitation, and runoff/precipitation ratios are calculated for each of the two watersheds during the calibration period and the treatment period. Also, the three hydrologic variables in each study period were expressed as ratios between treatment and control watersheds (T/C ratio). All these values are given in Table 16.3. The effect of forest clear-cutting on streamflow can be examined preliminarily by comparing the differences in means of these ratios between the two study periods.

Precipitation ratios between the treatment and control watersheds (T/C ratio) were virtually the same during the two study periods, 1.01 versus 0.99. Thus, any increase in runoff in the control watershed during the treatment period could not be attributed to the difference in precipitation. The runoff ratio between the treatment and control watersheds (T/C ratio) was 1.03 in the calibration period and 1.15 in the treatment period. When runoff is expressed as a fraction of precipitation (runoff coefficient), the ratio between treatment and control watersheds was 1.17 in the treatment period and 1.02 in the calibration period. The higher values of runoff and runoff as a fraction of precipitation during the posttreatment period, by 12% and 15%, respectively, than those during the calibration period, must be attributable to the watershed treatment.

16.4.3.1.2 Based on the Calibration Equation

Equation 16.10 can be used to predict annual runoff in the treatment watershed for the conditions as if the watershed had not been treated. Therefore, differences between the observed and predicted annual runoff in the treatment watershed must be the treatment effects (δq), assuming precipitation patterns between the two watersheds stay the same as in the past, or

$$\delta q = y - y_x = y - (0.79 - 0.97x) \tag{16.11}$$

TABLE 16.3

Comparisons of Hydroclimatic Records between the Calibration Period and Treatment Period

Variable	Calibration Period			Treatment Period		
	Control (C)	Treatment (T)	T/C Ratio	Control (C)	Treatment (T)	T/C Ratio
Runoff	15.44	15.77	1.03	15.70	18.04	1.15
Precipitation	51.97	52.59	1.01	53.74	53.01	0.99
Runoff/Precipitation	0.29	0.30	1.02	0.29	0.34	1.17

TABLE 16.4

Computations, Based on the Calibration Equation, for the Treatment Effect of Annual Runoff (cm)

Watersheds	1919	1920	1921	1922	1923	1924	1925	1926	Mean
Control (x)	15.4	20.0	17.5	17.3	15.5	18.0	10.8	11.1	15.7
Treatment									
y (Obs.)	15.2	21.7	21.1	22.3	18.2	20.4	12.6	12.8	18.0
y_x (Pred.)	15.7	20.2	17.8	17.6	15.8	18.3	11.3	11.6	16.0
δq	−0.5	1.5	3.3	4.7	2.4	2.1	1.3	1.2	2.0
% Change	−3.2	7.4	18.5	26.7	15.2	11.5	11.5	10.3	12.2

where y is the observed annual runoff for the treatment watershed during the treatment period. The results showed that forest clear-cutting caused a negative effect (−0.5 cm) on annual runoff in the treatment year, 1919, and a positive effect thereafter. Streamflow was increased from 1.5 cm in the second year (1920) to a peak 4.7 cm in the fourth year (1922); it then gradually decreased to 1.2 cm in the eighth year (1926), the last year of the study. The treatment started in the middle of 1919; its impact on runoff was expected to be minimal or insignificant. Overall, the average increase (1919–1926) in annual streamflow due to forest clearcutting was 2.0 cm/year, or about 12.2% of the predicted mean (Table 16.4).

16.4.3.1.3 Based on the Double-Mass Analysis

The technique of double-mass analysis was originally developed for testing data consistency and consequently making adjustments. It is often applied to test the significance of environmental changes within a watershed or between two watersheds over a period of years. In the current application, the accumulated annual runoff of the entire record of the treatment watershed is plotted against the accumulated annual runoff of the control watershed. The plotted graph shows a definite break on slopes and trend at year 1919, the beginning of the treatment period (Figure 16.4). It is therefore interpreted that the clear-cutting caused significant effects on streamflow during the posttreatment period. Interpreting the plotting shows that the slope for the calibration period is about $dy/dx = 1.00$, while the slope for the treatment period is about $dy/dx = 1.14$. Thus, the forest clear-cutting caused an average 14% increase in runoff during the posttreatment period. Since the average runoff in the treatment watershed was 18.04 cm/year during the posttreatment period, the increase in runoff was $18.04(1 − 1/1.14) = 2.22$ (cm/year).

The double-mass analysis is a graphical technique; it is rather subjective because the significance of the slope break is based on visual judgment. To overcome the weakness, the significance of the slope break can be tested by the analysis of covariance (Searcy and Hardison, 1960). Furthermore, using a computerized stepwise technique and making a subsequent data adjustment can help reduce the subjectivity (Chang and Lee, 1974).

The results of the three methods are very consistent: 2.16 cm/year (12%) by runoff ratios, 2.00 cm/year (12.2%) by the calibration equation, and 2.22 cm/year (12.3%) by the double-mass analysis.

16.4.3.2 Is the Effect Significant?

Two common techniques are employed here to test whether or not the treatment effect is significant.

16.4.3.2.1 Covariance Analysis

The covariance analysis tests the homogeneity of the two regression lines between the calibration period and treatment period. First, it tests the significance of the two regression coefficients. If the hypothesis that the two regression coefficients are equal is rejected at a selected probability level, then the treatment effect is significant. An acceptance of the hypothesis means that the two regression lines are parallel to each other. The analysis needs to proceed with a test for the elevations of the two regression lines. If the test shows that the two elevations also have no significant difference, then the two lines are virtually identical and there are no significant effects of the treatment. Otherwise, the runoff in one watershed is always higher or lower than runoff in the other watershed by a value equal to the difference in elevation between the two regression lines. This may be due to treatments, watershed precipitation, soils, topography, or other reasons. There is no need to test the regression elevations if two regression coefficients are significantly different.

Test for Two Regression Coefficients—The two regression lines, one for the calibration period and one for the treatment period, are plotted in Figure 16.7. The statistic used in testing the hypothesis that two populations (k = 2) regression coefficients are equal is

$$F = \frac{\text{Among regression coefficient SS}/(k-1)}{\text{Pooled residual SS}/(\Sigma n - 2k)} \tag{16.12}$$

with $(k-1)$ and $(\Sigma n - 2k)$ degrees of freedom (DF). All calculations required for solving Equation 16.12 are given in Table 16.5. Thus

$$F = \frac{1.48/1}{21.37/(15-4)} = 0.76$$

with 1 and 11 DF in a one-tailed test. The critical F for 1 and 11 DF and 95% probability is 4.84 (from an F-distribution table in any statistical textbook). Since the calculated F is smaller than the critical F, the hypothesis that the two coefficients are equal is accepted.

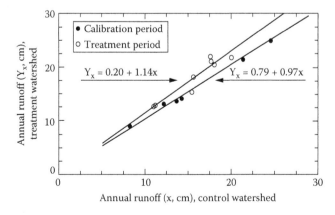

FIGURE 16.7
Simple-regression lines on annual runoff between the treatment and control watersheds.

TABLE 16.5

Test of Homogeneity of Regression Coefficients

Sample	n	$\bar{y}$	SS_x	SP	SS_y	b	Regression SS	Regression DF	Residual SS	Residual DF
Calibration	7	15.77	187.68	182.09	178.07	0.97	176.67	1	1.40	5
Treatment	8	18.04	74.88	85.07	111.62	1.14	96.65	1	19.97	6
Sum	15						273.32	2	21.37	11
Pooled		16.98	262.56	267.16	289.69	1.02	271.84	1		
Difference (SS among b)							1.48	1		

Notes: Regression $SS = SP^2/SS_x$.
Residual $SS = SSy -$ regression SS.
Pooled $\bar{y} = \Sigma(n\bar{y})/\Sigma n$.
Pooled $b = \bar{b} = \Sigma[(SS_x)(b)]/\Sigma SS_y$.
SS among regression coefficient $b = \Sigma[(SS_x)(b-\bar{b})^2] = (SP_{x1})^2/(SS_{x1}) + (SP_{x2})^2/(SS_{x2}) - \Sigma(SP)^2/\Sigma SS_x$.
SS among sample mean $= \Sigma[n(\bar{y} - \bar{\bar{y}})^2]$.

Test for the Two Regression Elevations—The acceptance for the test for the two regression coefficients leads to the test for regression elevations. The statistic used is

$$F = \frac{\text{Residual SS among sample}/(k-1)}{\text{Residual SS within sample}/(\Sigma n - 2k)} \qquad (16.13)$$

with $(k - 1)$ and $(\Sigma n - 2k)$ degrees of freedom. All computations required to perform the test for the Wagon Wheel Gap are given in Table 16.6, and the F value of Equation 16.13 is

$$F = \frac{14.99}{1.62} = 9.25$$

with 1 and 11 degrees of freedom in a one-tailed test. Since the calculated F is larger than the critical $F = 4.84$ at the 95% probability level, the hypothesis is rejected and the elevation of the regression line in the treatment period is significantly higher than that in the calibration period. Thus, the average treatment effect (δq) of forest clear-cutting on streamflow can be calculated by

$$\delta q = (\bar{y}_2 - \bar{y}_1) - b_p(x_2 - \bar{x}_1)$$

$$= (18.04 - 15.77) - 1.02(15.70 - 15.44) = 2.00 \text{ cm} \qquad (16.14)$$

where
$\bar{y}$ and $\bar{x}$ are the mean streamflow for the treatment watershed and control watershed, respectively
subscripts 1 or 2 refer to the calibration period or treatment period
b_p is the pooled (common) regression coefficient (Table 16.5)

TABLE 16.6

Test for Elevation on the Two Regression Lines

Source of Variation	SS$_x$	SP	SS$_y$	DF	Residual SS	Residual DF	Residual MS	F
Between sample	0.24	2.17	19.17	1	14.99	1	14.99	9.25
Within sample	262.56	267.16	289.69	12	17.85	11	1.62	
Total	262.80	269.33	308.86	13	32.84	12		

Notes: SS$_x$ between sample $= [\Sigma(x_1^2)/n_1 + (x_2^2)/n_2] - (\Sigma n)^2/n$
$\qquad\qquad\qquad\qquad = [1669.37 + 1971.92] - 233.702/15 = 0.240.$
SSx within sample $= \Sigma(x)^2 - [\Sigma(x_1^2)/n_1 + \Sigma(x_2^2)/n_2]$
$\qquad\qquad\qquad\qquad = 3903.85 - [1669.37 + 1971.92] = 262.56.$
SP between sample $= [(\Sigma x_1 \Sigma y_1/n_1) + (\Sigma x_2 \Sigma y_2/n_2)] - \Sigma x \Sigma y/n$
$\qquad\qquad\qquad\qquad = [(108.10)(110.40)/7 + (125.60)(144.32)/8] - (233.7)(254.7)/15 = 2.17.$
SP within sample $= \Sigma xy - \Sigma x \Sigma y/n = 4237.56 - (233.7)(254.7)/15 = 267.16.$
Residual SS (total) $= SS_y - SP^2/SS_x = 308.86 - 269.332/262.80 = 32.84.$
Res. SS (between) $=$ Total residual SS $-$ within-sample residual SS
$\qquad\qquad\qquad\qquad = 32.84 - 17.85 = 14.99.$
Residual MS $=$ Residual SS/DF.
Subscripts 1 and 2 $=$ Calibration period and treatment period, respectively.
x, y, n $=$ the data or the number of years for the entire study period
(i.e., n $= n_1 + n_2$).

The average 2.00 cm/year increase in streamflow from the covariance analysis is identical to the result calculated by the calibration equation (Table 16.4).

16.4.3.2.2 *Paired t-Test*

Assuming the samples of annual runoff in the study are drawn independently from a normal population, then the observed versus predicted annual runoff can be tested for the difference in population means being equal to zero, or the treatment does not increase the annual runoff, by the paired t-test. Using the predicted and observed annual runoff as a pair, the statistic used in the test is

$$t = \frac{\overline{D}}{\sqrt{(S^2/n)}} \qquad\qquad (16.15)$$

where

the quantity t follows Student's t-distribution (a sample distribution described in most statistical textbooks) with $(n-1)$ degrees of freedom
n is the number of pairs (years) in the treatment period
$\overline{D}$ and S^2 are the mean of the differences of all pairs (observed – predicted values, D_i) and the variance of the population of differences, respectively

or

$$\overline{D} = \frac{\sum(\text{observed} - \text{predicted})}{n} = \frac{\sum D_i}{n} \qquad\qquad (16.16)$$

$$S^2 = \frac{\sum (D_i - \bar{D})^2}{n-1} \tag{16.17}$$

The denominator $[S^2/n]^{0.5}$ is referred to as the standard error of the average difference. Data required to solve Equation 16.15 for the Wagon Wheel Gap experimental watersheds are $D_i = \delta q$ in Table 16.4, $\bar{D} = \Sigma D_i/n = 2.00$, $S^2 = 2.40$, $N = 8$, and $t = 3.65$.

The calculated $t = 3.65$ is larger than the critical $t = 3.00$ at the 99% probability level with 8 DF (use a one-tailed test). The hypothesis is rejected and the average increase in annual runoff 2.00 cm/year due to forest clear-cutting in the treatment watershed is significant at the 99% probability level.

References

Abtew, W., 1996, Evapotranspiration measurements and modeling for three wetland systems in South Florida, *Water Resour. Bull.*, 32, 465–473.

Adams, M.B., Loughry, L.H., and Plaugher, L.L., 2004, Experimental forests and ranges of the USDA forest service, General Technical Report NE-321, U.S. Forest Service, Northeastern Research Station, Parsons, WV.

Allred, B. and Haan, C.T., 1996, SWMHMS—Small watershed monthly hydrologic modeling system, *Water Resour. Bull.*, 32, 541–552.

Anderson, H.W., 1955, Detecting hydrologic effects of changes in watershed conditions by double-mass analysis, *Trans. Am. Geophys. Union*, 36(1), 119–125.

Baker, M.B., Jr., 1986, Effects of Ponderosa pine treatments on water yield in Arizona, *Water Resour. Res.*, 22, 67–73.

Bakke, P.D. and Pyles, M.R., 1997, Predictive model for nitrate load in the Bull Run Watershed, Oregon, *J. Am. Water Resour. Assoc.*, 33, 897–906.

Bates, C.G. and Henry, A.J., 1928, Forest and streamflow experiments at wagon wheel gap, Colorado, U.S. Weather Bureau No. 30, Washington, DC.

Beasley, D.B., Huggins, L.F., and Monke, E.J., 1980, ANSWERS: A model for watershed planning, *Trans. ASAE*, 23, 938–944.

Bent, G.C., 1994, Effects of timber cutting on runoff to Quabbin reservoir, central Massachusetts, *Effects of Human-Induced Changes on Hydrologic Systems, Proc., AWRA 1994 Annual Summer Symposium*, June 26–29, 1994, Jackson Hole, WY, pp. 187–196.

Bernt, H.W. and Swank, G.W., 1970, Forest land use and streamflow in central Oregon, Research paper PNW-93, Pacific Northwest Forest and Range Experiment Station, Portland, OR.

Beschta, R.L., 1998, Forest hydrology in the Pacific Northwest: Additional research needs, *J. Am. Water Resour. Assoc.*, 34, 729–741.

Bethlahmy, N., 1963, Rapid calibration of watersheds for hydrologic studies, *Bull. Int. Assoc. Sci. Hydrol.*, 8, 38–42.

Betson, R.P., 1979, The effects of clear cutting practices on Upper Bear Creek, Alabama Watersheds, Division of Water Resources Report No.WR28–1–550–101, Tennessee Valley Authority, Knoxville, TN.

Bhat, S. et al., 2007, Surface runoff contribution of nitrogen during storm events in a forested watershed, *Biogeochemistry*, 85, 253–262.

Binkley, D., 2001, Patterns and processes of variation in nitrogen and phosphorus concentrations in forested streams, NCASI (National Council for Air and Stream Improvement) Technical Bulletin No. 836, Research Triangle Park, NC.

Black, P.E., 1997, Watershed functions, *J. Am. Water Resour. Assoc.*, 34, 1–11.

Blackburn, W.H., Wood, J.C., and Dehaven, M.G., 1986, Storm flow and sediment losses from site-prepared forestland in east Texas, *Water Resour. Res.*, 22, 776–784.

Borah, D.K. and Bera, M., 2003, Watershed-scale hydrologic and nonpoint source pollution models: Review of mathematical bases. *Trans. ASAE*, 46(6), 1553–1566.

Bosch, J.M. and Hewlett, J.D., 1982, A review of catchment experiments to determine the effect of vegetation changes on water yield and evapotranspiration, *J. Hydrol.*, 55, 3–23.

Bren, L., Lane, P., and McGuire, D., 2006, An empirical, comparative model of changes in annual water yield associated with pine plantations in southern Australia, *Australian For.*, 69(4), 275–284.

Broadbent, E.N. et al., 2008, Forest fragmentation and edge effects from deforestation and selective logging in the Brazilian Amazon, *Biol. Conserv.*, 141, 1745–1757.

Callaham, R.Z., Ed., 1990, Case studies and catalog of watershed projects in western provinces and states, Report No. 22, Wildland Resources Center, University of California, Berkeley, CA.

Chang, M., 1997, Improper soil management causing severe water erosion problems, *Proceedings of Conference 28, the International Erosion Control Association*, February 25–28, 1997, Nashville, TN, pp. 495–501.

Chang, M. and Boyer, D.G., 1977, Estimates of low flows using watershed and climatic parameters, *Water Resour. Res.*, 13, 997–1001.

Chang, M. and Crowley, C.M., 1997, Downstream effects of a dammed reservoir on streamflow and vegetation in east Texas, *Sustainability of Water Resources under Increasing Uncertainty*, Rosbjerg, D. et al., Eds., IAHS Publication No. 240, Oxfordshire, U.K., pp. 267–275.

Chang, M. and Lee, R., 1974, Objective double-mass analysis, *Water Resour. Res.*, 10, 1123–1126.

Chang, M. and Sayok, A.K., 1990, Hydrological responses to urbanization in forested La Nana Creek watershed, Nacogdoches, Texas, *Tropical Hydrology and Caribbean Water Resources, Proceedings of the American Water Resources Association*, San Juan, Puerto Rico, pp. 131–140.

Chang, M. and Ting, J.C., 1986, Applications of the universal soil loss equation to various forest conditions, *Monitoring to Detect Changes in Water Quality Series (Proceedings of the Budapest Symposium)*, July 1986, Budapest, Hungary, IAHS, Oxfordshire, U.K., pp. 165–174.

Chang, M. and Waters, S.P., 1984, Forests and other factors associated with streamflows in east Texas, *Water Resour. Bull.*, 20, 713–719.

Chang, M. et al., 1997, Evapotranspiration of herbaceous mimosa (*Mimosa strigillosa*), a new drought-resistant species in the southeastern United States, *Resour. Conserv. Recycl.*, 21, 175–184.

Clayton, J.L. and Megahan, W.F., 1997, Natural erosion rates and their prediction in the Idaho Batholith, *J. Am. Water Resour. Assoc.*, 33, 689–703.

Connolly, R.D., Silburn, D.M., and Ciesiolka, C.A.A., 1997, Distributed parameter hydrology model (ANSWERS) applied to a range of catchment scales using rainfall simulator data, III. Application to a spatially complex catchment, *J. Hydrol.*, 193(1–4), 183–203.

Crawford, N.H. and Linsley, R.K., 1966, Digital simulation hydrology: Stanford water-shed model 4, Technical Report 39, Stanford University, Department of Civil Engineering, Stanford, CA.

Cuo, L. et al., 2008, Hydrologic prediction for urban watersheds with the Distributed Hydrology–Soil–Vegetation Model, *Hydrol. Process.*, 22(21), 4205-4213.

Daniel, E.B. et al., 2011, Watershed modeling and its applications: A state-of-the-art review, *The Open Hydrol. J.*, 5, 26–50.

Davin, E.L. and de Noblet-Ducoudré, N., 2010, Climatic impact of global-scale deforestation: Radiative versus nonradiative processes. *J. Climate*, 23, 97–112.

Davis, E.A., 1993, Chaparral control in mosaic pattern increased streamflow and mitigated nitrate loss in Arizona, *Water Resour. Bull.*, 29, 391–399.

DeCoursey, D.G. and Deal, R.B., 1971, General aspects of multivariate analysis with applications to some problems in hydrology, *Proceedings by Symposium on Statistical Hydrology*, Tucson, Arizona, USDA, Agricultural Research Service, Miscellaneous Publication No. 1275, Washington, DC, pp. 47–68.

Dettinger, M. et al., 2004, Simulated hydrologic responses to climate variations and change in the Merced, Carson, and American River Basins, Sierra Nevada, California, 1900–2099. *Climate Change*, 6, 283–317.

Douglas, J.E., 1983, The potential for water yield augmentation from forest management in the eastern United States, *Water Resour. Bull.*, 19, 351–358.

Dun, S. et al., 2009, Adapting the water erosion prediction project (WEPP) model for forest application, *J. Hydrol.*, 366, 46–54.

Eisenbies, M.H. et al., 2007, Forest operations, extreme flooding events, and considerations for hydrologic modeling in the Appalachians—A review, *For. Ecol. Manage.*, 242, 77–98.

Elliot, W.J. and Hall, D.E., 1997, Water erosion prediction project (WEEP) forest applications, USDA Forest Service, Intermountain Research Station, General Technical Reports Draft, Ogden, UT, p. 15.

EPA, 1997, Monitoring guidance for determining the effectiveness of nonpoint source controls, EPA 841-B-96-004, U.S. Environmental Protection Agency, Washington, DC.

Forest Service, 1977, Nonpoint water quality modeling in wildland management: A state-of-the-art assessment, Vol. II, EPA-600/3-77-078, Washington, DC.

Foster, G.R. and Lane, L.J., 1987, User requirements USDA water erosion prediction project (WEEP), NSERL Report No. 1, National Soil Erosion Research Laboratory, West Lafayette, IN.

Friedrich, W., 1967, Forest hydrology research in Germany F.R, *International Symposium on Forest Hydrology*, University Park, PA, Sopper, W.E. and Lull, H.W., Eds., Pergamon Press, New York.

Galbraith, A.F., 1975, Methods for predicting increases in water yield related to timber harvesting and site conditions, *Watershed Management, Proceedings*, Logan, UT, ASCE, New York, pp. 169–184.

Gary, H.L., 1975, Watershed management problems and opportunities for the Colorado front range Ponderosa pine zone: The status of our knowledge, Research Paper RM-139, USDA Forest Service, Rocky Mountain Forest and Range Experiment Station, Fort Collins, CO.

Gary, H.L. and Troendle, C.A., 1982, Snow accumulation and melt under various stand density in Lodgepole pine in Wyoming and Colorado, Research Note RM-417, U.S. Forest Service, Rocky Mountain Forest and Range Experiment Station, Fort Collins, CO.

Gassman, P.W. et al., 2007, The soil and water assessment tool: Historical development, applications and future research directions, *Trans. Am. Soc. Agric. Biol. Eng.*, 50(4), 1211–1250.

Gluns, D.R. and Toews, D.A.A., 1989, Effect of a major wildlife on water quality in southeastern British Columbia, *Symposium Proceedings on Headwaters Hydrology*, Missoula, MT, Woessner, W.W. and Potts, D.F., Eds., American Water Resources Association, Bethesda, MD, pp. 487–499.

Goovaerts, P., 1997, *Geostatistics for Natural Resources Evaluation*, Oxford University Press, New York.

Grant, G., 1988, The RAPID technique: A new method for evaluating downstream effects of forest practices on Riparian zones, U.S. Forest Service Pacific NW Research Station, General Technical Report PNW-GTR-220, Portland, OR.

Groom, J.D., Dent, L., and Madsen, L.J., 2011, Stream temperature change detection for state and private forests in the Oregon Coast Range, *Water Resour. Res.*, 47, W01501. DOI: 10.1029/2009WR009061.

Gurnell, A.M. and Montgomery, D.R. (Eds.), 2000, *Hydrological Applications of GIS*, John Wiley & Sons, New York.

Harr, R.D., 1983, Potential for augmenting water yield through forest practices in western Washington and western Oregon, *Water Resour. Bull.*, 19, 383–394.

Harr, R.D., Fredriksen, R.L., and Rothacher, J., 1979, Changes in streamflow following timber harvest in southwest Oregon, Research Paper PNW-249, Pacific Northwest Forest and Range Experiment Station, Forest Service, Portland, OR.

Harrold, L.L. and Dreibelbis, F.R., 1958, Evaluation of agricultural hydrology by monolith lysimeters, U.S. Department of Agriculture Technical Bulletin #1179, U.S. Government Printing Office, Washington, DC.

Hewlett, J.D., 1970, Review of the catchment experiment to determine water yield, *Proceedings of the Joint U.N. Food Agriculture Organization—USSR, International Symposium on Forest Influences and Watershed Management*, August–September, 1970, Moscow, Russia, pp. 145–155.

Hewlett, J.D., 1971, Comments on the catchment to determine vegetal effects on water yield, *Water Resour. Bull.*, 7, 376–377.

Hewlett, J.D. and Hibbert, A.R., 1961, Increases in water yield after several types of forest cutting, *Int. Assoc. Sci. Hydrol. Bull.*, 6, 5–17.

Hewlett, J.D., Lull, H.W., and Reinhart, K.G., 1969, In defense of experimental watersheds, *Water Resour. Res.*, 5, 306–316.

Hibbert, A.R., 1981, Opportunity to increase water yield in the southwest by vegetation management, *Interior West Watershed Management, Symposium Proceedings*, Spokane, WA, Washington State University, Pullman, WA, pp. 223–230.

Hibbert, A.R., Davis, E.A., and Knipe, O.D. 1982, Water yield changes resulting from treatment of Arizona chaparral, *Proceedings of Symposium on Dynamic and Management of Mediterranean-Type Ecosystems*, June 22–26, 1981, San Diego, CA, General Technical Report PSW-58, U.S. Forest Service, Pacific SW Forest and Range Experiment Station, Berkeley, CA, pp. 382–389.

Hibbert, A.R., Knipe, O.D., and Davis, E.A., 1986, Streamflow response to control of chaparral shrubs along channels and upper slopes, *Proceedings, Chaparral Ecosystems Research Conference*, May 16–17, 1985, Santa Barbara, CA, Report No. 62, University of California, Davis, CA, pp. 95–103.

Hoover, M.D., 1944, Effect of removal of forest vegetation upon water yields, *Trans. Am. Geophys. Union*, 26, 969–975.

Hornbeck, J.W. et al., 1987, The northern hardwood forest ecosystem: Ten years of recovery from clearcutting, USDA, Forest Service, Northeastern Forest Experiment Station, NE-RP-596, Broomall, PA.

Ice, G.G. and Stednick, J.D., 2004, Forest watershed research in the United States, *For. History Today*, spring/fall, 16–26.

Idso, S.B., Reginato, R.J., and Jackson, R.D., 1979, Calculation of evaporation during three stages of soil drying, *Water Resour. Res.*, 15, 487–488.

Jain, S.E. and Dolezal, F., 2000, Modeling soil erosion using EPIC supported by GIS, Bohemia, Czech Republic, *J. Environ. Hydrol.*, 8(2), 11.

Johnson, E.A. and Kovner, J.L., 1954, Increasing water yield by cutting forest vegetation, *Georgia Miner. Newslett.*, 7, 145–148.

Jost, G. et al., 2007, The influence of forest and topography on snow accumulation and melt at the watershed-scale, *J. Hydrol.*, 347, 101–115.

Kennedy, J.B. and Neville, A.M., 1986, *Basic Statistical Methods for Engineers and Scientists*, 3rd edn., Harper & Row, New York.

Keppeler, E.T. and Ziemer, R.R., 1990, Logging effects on streamflow: Water yield and summer low flows at Caspar Creek in northwestern California, *Water Resour. Res.*, 26, 1669–1679.

Kim, J.G. et al., 2009, Development of a SWAT patch for better estimation of sediment yield in steep sloping watersheds, *J. Am. Water Resour. Assoc.*, 45(4), 963–972.

Kirkham, R.R., Gee, G.W., and Jones, T.L., 1984, Weighing lysimeters for long-term water balance investigations at remote sites, *Soil Sci. Soc. Am. J.*, 48, 1203–1205.

Knisel, W.G., 1980, CREAMS: A field-scale model for chemicals, runoff, and erosion from agricultural management systems, Conservation Research Report No. 26, USDA Science and Education Administration, Washington, DC.

Kochenderfer, J.M. and Aubertin, G.M., 1975, Effects of management practices on water quality and quantity: Fernow experimental forest, west Virginia, *Municipal Watershed Management, Symposium Proceedings*, U.S. Forest Service General Technical Report NE-13, Northeastern For. Exp. Sta., Upper Darby, PA, pp. 14–24.

Kochenderfer, J.N. and Wendel, G.W., 1983, Plant succession and hydrologic recovery on a deforested and herbicided watershed, *For. Sci.*, 29, 545–558.

Kovar, K. and Nachtnebel, H.P., Eds., 1996, *Application of Geographic Information Systems in Hydrology and Water Resources Management*, Proceedings of the HydroGIS'96 Conference, April 16–19, 1996, Vienna, Austria, IAHS Publication No. 235, Intern'l Assoc. Hydrol. Sci., Oxfordshire, U.K.

Kovner, J.L. and Evans, T.C., 1954, A method for determining the minimum duration of watershed experiments, *Trans. Am. Geophys. Union*, 35, 608–612.

Lakel, W.A. III et al., 2010, Sediment trapping by streamside management zones of various widths after forest harvest and site preparation, *For. Sci.*, 56(6), 541–551.

Landers, M.N., Ankcorn, P.D., and McFadden, K.W., 2007, Watershed effects on streamflow quantity and quality in six watersheds of Gwinnett County, Georgia, Scientific Investigations Report, 2007-5132, U.S. Geological Survey, Reston, VA, p. 62.

Latha, C.J., Saravanan, S., and Palanichamy, K., 2011, Estimation of spatially distributed monthly evapotranspiration, *Int. J. Eng. Sci. Technol.*, 3(2), 877–883.

Leavesley, G.H. et al., 1983, *Precipitation Runoff Modeling Systems: User's Manual*, USGS Water Resources Investigations Report 83-4238, Denver, CO.

Lee, R., 1970, Theoretical estimates vs. forest water yield, *Water Resour. Res.*, 6, 1327–1334.

Leonard, R.A., Knisel, W.G., and Still, D.A., 1987, GLEAMS: Groundwater loading effects of agricultural management systems, *Trans. ASAE*, 30, 1403–1418.

Li, J.C.R., 1964, *Statistical Inference*, Edwards Brothers, Ann Arbor, MI.

Litschert, S.C., 2009, Predicting cumulative watershed effects in small forested watersheds, Unpublished PhD dissertation, Colorado State University, Fort Collins, CO.

Lull, H.W. and Sopper, W.E., 1967, Prediction of average annual and seasonal streamflow of physiographic units in the northeast, *International Symposium on Forest Hydrology*, University Park, PA, Sopper, W.E. and Lull, H.W., Eds., Pergamon Press, New York, pp. 507–521.

Lyon, J.G., Ed., 2000, *GIS for Water Resources and Watershed Management*, Taylor & Francis Group, New York.

MacDonald, L.H., Smart, A.W., and Wissmar, R.C., 1991, Monitoring guidelines to evaluate effects of forestry activities on streamflows in the Pacific Northwest and Alaska, EPA 910/9–91–001, Washington, DC.

Marloth, R., 1907, Results of further experiments on Table Mountain for ascertaining the amount of moisture deposited from the SE clouds, *Trans. S. Afr. Philos. Soc.*, 16, 97–105.

Massman, W.J., 1983, The derivation and validation of a new model for interception of rainfall by forests, *Agric. Meteorol.*, 28, 261–286.

McNulty, S.G., Vose, J.M., and Swank, W.T., 1997, Regional hydrologic response of loblolly pine to air temperature and precipitation changes, *J. Am. Water Resour. Assoc.*, 33, 1011–1022.

Merriam, C.F., 1937, A comprehensive study of rainfall in the Susquehanna valley, *EOS, Trans. AGU*, 18, 471–476.

Muda, A.B., Chang, M., and Watterston, K.G., 1989, Effects of six forest-site conditions on nutrient losses in East Texas, *Headwaters Hydrology*, Woessner, W.W. and Potts, D.F., Eds., American Water Resources Association, Bethesda, MD, pp. 55–64.

Muhamad, A.I. and Chang, M., 1989, On the calibration of forested watershed: Single- vs. paired-watershed approaches, *Regional Seminar on Tropical Forest Hydrology*, Forest Research Institute, Malaysia, Kuala Lumpur.

Muller, R.A., 1966, *The Effects of Reforestation on Water Yield—A Case Study Using Energy and Water Balance Models of the Allgheny Plateau*, New York, Publications in Climatology Laboratory of Climatology, Centerton, NJ, Vol. 19, pp. 251–304.

Müller, J. and Bolte, A., 2009, The use of lysimeters in forest hydrology research in north-east Germany, *Agric. For. Res.*, 59(1), 1–10.

Myers, D.E., 1996, Geostatistics applied to groundwater pollution and flow: Problems and tools, *Statistics for the Environment 3: Aspects of Pollution*, Barnett, V. and Turkman, K.F., Eds., John Wiley & Sons, Chichester, U.K., pp. 225–239.

NCASI, 2001, Hydrological Simulation Program—FORTRAN (HSPF) calibration for Mica Creek, Idaho, Special Report No. 01–01 National Council for Air and Stream Improvement, Research Triangle Park, NC.

Neitsch, S.L. et al., 2002, Soil water assessment tool user's manual, Version 2000, Texas Water Research Institute, TWRI Report, TR-192, College Station, TX.

Němec, J., Pasák, V., and Zelen'y, V., 1967, Forest hydrology research in Czechoslovakia, *International Symposium on Forest Hydrology*, University Park, PA, Sopper, W.E. and Lull, H.W., Eds., Pergamon Press, New York, pp. 31–33.

Newbold, J. D. et al., 2010, Water quality functions of a 15-year-old riparian forest Buffer system. *J. Am. Water Resour. Assoc.*, 46(2), 299–310.

Oishi, A.C., Ram Oren, R., and Stoy, P.C., 2008, Estimating components of forest evapotranspiration: A footprint approach for scaling sap flux measurements, *Agric. For. Meteorol.*, 148, 1719–1732.

Olivier, J., 2002, Fog-water harvesting along the West Coast of South Africa: A feasibility study, *Water SA*, 28(4), 349–260.

Ostle, B. and Malone, L.C., 1988, *Statistics in Research*, Iowa State University Press, Ames, IA.

Penman, H.L., 1956, Estimating evaporation, *Trans. Am. Geophys. Union*, 37, 43–50.

Pitt, M.D., Burgy, R.H., and Heady, H.F., 1978, Influences of brush conversion and weather patterns on runoff from a northern California watershed, *J. Range Manage.*, 31, 23–27.

Pond, F.W., 1961, Mechanical control of Arizona chaparral and some results from brush clearing, *Proceedings of the Fifth Annual Arizona Watershed Symposium, Modern Techniques in Water Management*, Phoenix, AZ, pp. 39–41.

Potts, D.F., 1984, Hydrologic impacts of large-scale mountain pine beetle (*Dendroctonus ponderosa* Hopkins) epidemic, *Water Resour. Bull.*, 20, 373–377.

Reigner, I.C., 1964, Calibration of a watershed using climatic sata, Research Paper NE-15, NE Forest Experiment Station, Upper Darby, PA.

Reinhart, K.G., 1958, Calibration of five small forested watersheds, *Trans. Am. Geophys. Union*, 38, 933–936.

Reinhart, K.G., 1967, Watershed calibration methods, *International Symposium on Forest Hydrology*, University Park, PA, Sopper, W.E. and Lull, H.W., Eds., Pergamon Press, New York, pp. 715–723.

Rich, L.R. and Gottfried, G.J., 1976, Water yields resulting from treatments on the Workman Creek experimental watersheds in Central Arizona, *Water Resour. Res.*, 12, 1053–1060.

Riekerk, H., 1989, Influences of silvicultural practices on the hydrology of pine flatwoods in Florida, *Water Resour. Res.*, 25, 713–719.

Schneider, J. and Ayer, G.R., 1961, Effect of reforestation on streamflow in central New York, USGS Water Supply Paper No. 1602, Washington, DC.

Searcy, C.F. and Hardison, C.H., 1960, Double-mass curves, Water Supply Paper 1541-B, U.S. Geological Survey, Washington, DC, pp. 27–66.

Shahlaee, A.K. et al., 1991, Runoff and sediment production from burned forested sites in the Georgia Piedmont, *Water Resour. Bull.*, 27, 485–493.

Shuttleworth, W.J., 1993, Evaporation, *Handbook of Hydrology*, Maidment, D.R., Ed., McGraw-Hill, New York, pp. 4.1–4.53.

Singh, V.P. and Woolhiser, M., 2002, Mathematical modeling of watershed hydrology, *J. Hydrol. Eng.*, 7, 270–292.

Siska, P., 2004, A multivariate spatial model for determining urban development sites in flood prone coastal areas, *Papers of the Applied Geography Conferences*, Montz, E.B. and Tobin, G.A., Eds., Saint Louis, MO, October 20–24, 2004, University of Southern Illinois, Edwardsville, IL, Vol. 27, pp. 75–83.

Stegman, S.V., 1996, Snowpack changes resulting from timber harvest: Interception, redistribution, and evaporation, *Water Resour. Bull.*, 32, 1353–1360.

Stewart, J.B., 1984, Measurement and prediction of evaporation from forested and agricultural watersheds, *Agric. Water Manage.*, 8, 1–28.

Swanson, R.H., 1998, Forest hydrology issues for the 21st century: A consultant's viewpoint, *J. Am. Water Resour. Assoc.*, 34, 755–763.

Thornthwaite, C.W. and Mather, J.R., 1955, The water balance, *Publications in Climatology*, Vol. 8, No. 1, Laboratory of Climatology, Drexel Institute of Technology, Centerton, NJ.

Thornthwaite, C.W. and Mather, J.R., 1957, *Instruction and Tables for Computing Potential Evapotranspiration and the Water Balance*, Laboratory of Climatology XIX (3), Centerton, NJ, pp. 185–311.

Troendle, C.A. and King, R.M., 1985, The effect of timber harvest on the Fool Creek watershed, 30 years later, *Water Resour. Res.*, 21, 1915–1922.

Troendle, C.A. and King, R.M., 1987, The effect of partial and clearcutting on streamflow at Dead-Horse Creek, Colorado, *J. Hydrol.*, 90, 145–157.

TVA, 1961, Forest cover improvement influences upon hydrologic characteristics of white hollow watershed 1935–58, Report No. 0–5163A, Division of Water Control Planning, Tennessee Valley Authority, Cookeville, TN.

Ursic, S.J., 1991a, Hydrologic effects of two methods of harvesting mature southern pine, *Water Resour. Bull.*, 27, 303–315.

Ursic, S.J., 1991b, Hydrologic effects of clearcutting and stripcutting loblolly pine in the Coastal Plain, *Water Resour. Bull.*, 27, 925–938.

Ursic, S.J. and Popham, T.W., 1967, Using runoff events to calibrate small forested catchments, *Proceedings of the International Union of Forestry Research Organizations Congress*, Munich, Germany, pp. 319–324.

Van Dijk, A.I.J.M. and Bruijnzeel, L.A., 2001, Modeling rainfall interception by vegetation of variable density using an adapted analytical model, I. Model description, *J. Hydrol.*, 247, 230–238.

Veihmeyer, F.J., 1953, Use of water by native vegetation vs. grasses and forbs on watersheds, *Trans. Am. Geophys. Union*, 34, 201–212.

Vennix, S. and Northcott, W., 2004, Prioritizing vegetative buffer strip placement in an agricultural watershed, *J. Spatial Hydrol.*, 4(1), 1–19.

Vieux, B.E., 2005, *Distributed Hydrologic Modeling Using GIS*, Water Science and Technology Library, Vol. 38. Kluwer Academic Publishers, Boston, MA.

Ward, R.C., 1971, Small watershed experiments—An appraisal of concepts and research developments, Occasional Papers in Geography No. 19, University of Hull, England, U.K.

Weltz, M.A. and Bucks, D.A., 2003, The USDA Agricultural Research Service watershed research program, *First Interagency Conference on Research in the Watersheds*, Benson, AZ, USDA Agricultural Research Service, Washington, DC, pp. 2–9.

Whittemore, R.C. and Beebe, J., 2000, EPA's BASINS model: Good science or serendipitous modeling? *J. Am. Water Resour. Assoc.*, 36, 493–499.

Wight, J.R., 1987, ERHYM-II: Model description and user guide for the basic version, USDA, Agricultural Research Service, ARS 59, Washington, DC.

Wight, J.R. and Skiles, J.W., Eds., 1987, SPUR: Simulation of production and utilization of rangelands. Documentation and User Guide, USDA Agricultural Research Service, ARS 63, Washington, DC.

Wigmosta, M.S., Nijssen, B., and Storck, P., 2002, The distributed hydrology soil vegetation model, *Mathematical Models of Small Watershed Hydrology and Applications*, Singh, V.P. and Frevert, D.K., Eds., Water Resources Publications, Littleton, CO, pp. 7–42.

Wigmosta, M.S., Vail, L.W., and Lettenmaier, D.P, 1994, A distributed hydrology–soil–vegetation model for complex terrain, *Water Resour. Res.*, 30, 1665–1679.

Williams, J.R., Jones, C.A., and Dyke, P.T., 1984, A modeling approach to determining the relationship between erosion and soil productivity, *Trans. ASAE*, 27(1), 129–144.

Williams, J.R., Nicks, A.D., and Arnold, J.G., 1985, Simulator for water resources in rural basins, *J. Hydraulic Eng.*, 111, 970–986.

Wilm, H.G., 1949, How long should experimental watersheds be calibrated? *Trans. Am. Geophys. Union*, 30, 272–278.

Young, R.A. et al., 1987, AGNPS, Agricultural nonpoint sources pollution model: A large watershed analysis tool, Conservation Research Report No. 35, USDA Agriculture Research Service, Washington, DC.

Appendix A: Precipitation Measurements

Measurements of precipitation are conventionally made by precipitation gages installed at specific sites (*point measurements*), and measurements from single gages are then employed to estimate average precipitation for the watershed. Thus, the reliability of watershed-precipitation estimates depends on the accuracy of point-precipitation measurements and the areal representativeness of each measurement. Precipitation gages provide the prime data used in watershed hydrologic studies.

Other modern techniques such as Doppler radar observations (Creutin et al., 1988; Xiao et al., 2005; Nikolopoulos et al., 2008), satellite thermal infrared and passive microwave sensors (Dai et al., 2007; Schuster et al., 2011), optical disdrometers (Krajewski et al., 2006; Thurai et al., 2011), and underwater ambient noises (Nystuen, 1986; Murugan et al., 2010) are remote sensing measurements that do not involve in catching raindrops. The WMO conducted a field intercomparison of 26 catching and non-catching rainfall intensity gages against a set of reference catching type rain gauges positioned in a standard pit from October 2006 to April 2009 in Italy. The results showed "suitably post-processed weighing gauges and tipping-bucket rain gauges had acceptable performance, while none of the non-catching rain gauges agreed well with the reference" (Lanza et al., 2010). Thus, precipitation measurements at the ground level are indispensable in hydroclimatological studies, despite the development of remote-sensing techniques. Those techniques are more useful for weather and flood forecasts and for regions where the conventional precipitation networks at the ground level are deficient, inaccessible, or poorly distributed. This appendix discusses only the technique and accuracy of point-precipitation measurements using the conventional gages.

A.1 Precipitation Gages

The first known measurement of rainfall in history is believed to have been in India, where bowls as wide as 18 in. (45.72 cm) were set in the open to catch rainfall as a basis for land taxation as early as 400 BC (Kurtyka, 1953). In China, local county officers were asked to report the amount of rainfall measured between the vernal equinox (March 21) and the autumnal equinox (September 23) to the monarchal government during the East Hann dynasty, about AD 25. In the South Sung dynasty (1127–1279), a network of rain gages was set up at cities in all provinces. Several designs of rain gages were reported in 1241. One of the rain gages measured 2–3 chi (old Chinese length scale, close to a foot) for the orifice diameter, 1.2 chi for the bottom diameter, and 1.8 chi in depth (Liu, 1983). The Chinese rain gages were earlier than the first rain gage designed by B. Castelli in 1639 in Europe (Kurtyka, 1953) by as much as 400 years.

Conventional precipitation gages in use today are of three types: nonrecording, recording, and storage gages.

A.1.1 Nonrecording Gages

The basic design of nonrecording precipitation gages is an open cylinder with a sealed bottom installed in the open. Precipitation falling into the cylinder is collected and manually measured in terms of depth. However, the diameter of opening (orifice), the depth of cylinder (collector), the gage height above the ground, and the way that precipitation depth is measured are different among countries. Sevruk and Klemn (1989) listed 50 different types of nonrecording precipitation gages in use around the world today. A few gages are given in Table A.1. Nonrecording gages are generally accepted as the standard in precipitation measurements.

The standard nonrecording precipitation gage used in the United States consists of five parts (Figure A.1):

1. An 8 in. (20.32 cm) diameter orifice as the sensor
2. A funnel as the transmitter to direct rainfall catch from the orifice to the measuring tube
3. A measuring tube (inner cylinder) as the amplifier
4. An overflow can (outer cylinder) for storage of rainwater overflowing from the measuring tube
5. A measuring stick as the recording device

The cross-sectional area of the measuring tube is exactly one-tenth that of the orifice, which is 2.53 in. (6.43 cm) in diameter. Thus, a 0.1 in. (0.254 cm) rainfall will fill the tube to a depth

TABLE A.1

Few Standard Nonrecording Rain Gages in Use around the World

Country	Gage Name	Material	Orifice Area (cm²)	Collector Depth (cm)	Gage Height (cm)
Australia	Manual 1508	Galvanized iron	324	7.0	30.0
Brazil	Ville Paris	Stainless steel	400	13.0	63.0
Canada	Nipher	Copper	127	50.0	52.0
China	China	Galvanized iron	314	22.0	59.0
France	SPIEA MN	Fiberglass	400	14.0	44.0
Germany	Hellman	Galvanized iron	200	27.0	43.0
Israel	Small	Hardened PVC	7	4.5	12.0
Israel	Amir	Hardened PVC	100	8.0	50.0
Malaysia	S 203 DRG	Brass	324	14.0	38.0
Mexico	Mexican	Galvanized iron	400	15.0	36.0
New Zealand	Nylex 1000	Plastic	81	10.0	35.0
United Kingdom	Mark 2	Copper	127	15.0	30.0
United States	WB-8-inch	Copper and steel	324	20.0	68.0
Russia	Tretjakov	Galvanized iron	200	24.0	40.0
Sweden	SMHI	Aluminum	200	15.0	35.0
Switzerland	Tognini	Aluminum	200	27.0	43.0

Source: After Sevruk, B. and Klemn, S., Types of standard precipitation gages, in *Precipitation Measurements*, Sevruk, B., Ed., *WMO/IAHS/ETH Workshop*, St. Moritz, Switzerland, pp. 227–232, 1989.

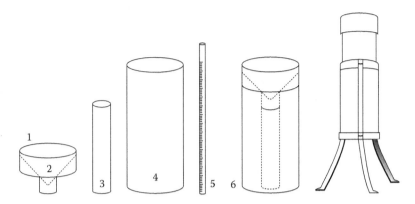

FIGURE A.1
The U.S. National Weather Service stand nonrecording 8 in. rain gage (6) including an orifice (1), funnel (2), measuring tube (3), outer container (4), and measuring stick (5).

of 1 in. (2.54 cm), making the smallest measurable rainfall 0.01 in. (0.025 cm). The measuring stick is graduated to directly read water in the measuring tube in rainfall depth.

Rainwater collected in the measuring tube is read by inserting the measuring stick into the tube. The measuring tube has a capacity of 2 in. (5.08 cm) of rainfall. Water will overflow from the measuring tube into the outer cylinder if rainfall exceeds 2 in. The overflow rainwater needs to be poured into the measuring tube for measurement. Total storm rainfall is then obtained by summing up all measurements from the measuring tube. When freezing temperatures or snow are likely to occur, the precipitation-gage receiver (the orifice and funnel, 1 and 3 in Figure A.1) and measuring tube should be removed, leaving the overflow can (4) for catching snowfall.

A.1.2 Recording Gages

In many cases, total depth of rainfall for a given storm obtained from nonrecording gages is not sufficient in hydroclimatological applications, so recording gages are used to provide additional information on time of occurrence, duration, and intensity. They have a mechanical, electrical, or electronic device to convert weight or volume of the collected rainfall into depth over time.

There are two main types of recording rain gages in the United States: weighing-type and tipping-bucket. The weighing-type recording gage consists of an orifice of 8 in. (20.32 cm) diameter through which precipitation is funneled into a bucket (collector) mounted on a weighing mechanism (Figure A.2a). The weight of the catch is recorded on a clock-driven chart with a capacity of 12 in. (30.48 cm) in 192 h. There is another type of weighing gage called Fisher & Porter Punched Tape Precipitation Gage, in which the weight of collected rainfall is converted to a code disk position. This code disk position is recorded on punched paper tape in a standard binary-decimal code at selected intervals.

The tipping-bucket rain gage has two small buckets under the orifice (Figure A.2b). Each bucket has a fixed capacity equivalent to 0.01 in. (0.254 mm) or 1 mm (0.04 in.) rain. A full volume of water will tip down to empty the bucket and push the other adjoining bucket up to catch rain. Each tip closes a circuit, and the pulse is recorded on a counter.

For data consistency, recording gages should have an orifice diameter, material, and gage height above the ground identical to those of the standard nonrecording gage.

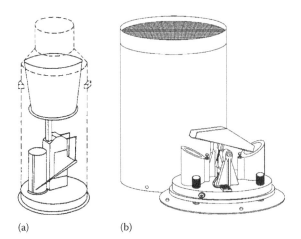

(a)   (b)

FIGURE A.2
Weighing-type recording rain gage (a) and tipping-bucket rain gage (b).

A.1.3 Storage Gages

In mountainous areas or remote sites where accessibility and frequent visits are a major problem, precipitation gages are designed to have a collector large enough to last for a month, a season, or even a half-year. For example, the Octapent gage used in Britain has a capacity of 1270 mm (50 in.). The Fisher & Porter Precipitation Gage-Recorder used in the United States can accommodate 25 in. (635 mm) of precipitation, but the recording apparatus is limited to 0–19.5 in. (495.3 mm). Storage gages require antifreeze such as a mixture of 24 oz of ethylene glycol and 8 oz of oil to prevent damage from freezing and to retard evaporation (National Weather Service, 1972). However, recent developments in technology have given automatic rainfall monitoring systems the capability to digitize temporal rainfall distribution, power the system through solar energy, perform data management and acquisition, and have data accessed from a central station. This system may gradually replace traditional storage gages for monitoring rainfall in mountainous areas.

A.2 Accuracy

A.2.1 Sources of Error

Precipitation gages are intended to measure the "true" amount of precipitation at the gage sites as if the gages were not there. However, the presence of the gage can cause errors due to gage geometry, site exposure, and wind effect. A classic study reviewing more than 1000 references dating back to the 1600s showed that measurement error caused by evaporation, adhesion, color, inclination, and splash are usually ±1% each, while error caused by gage exposure alone, which is mainly wind effect, ranges from −5% to −80% (Kurtyka, 1953). More recent studies showed that wind-induced losses of precipitation can be up to 35% (Kreuels and Breuer, 1986) and errors caused by wetting to be 2% (Sokollek, 1986). Differences in precipitation catch between south (windward) and north (leeward) slopes can be as much as 100% in West Virginia.

Errors in precipitation measurements are either random or systematic in nature. Random errors are caused by precipitation sampling, as reflected by differences in gage exposure and site representativeness. They can include the variation due to topography, surrounding vegetation, aspect, altitude, and human sources. Systematic errors are associated with factors such as wind effects, wetting, evaporation, splash, gage inclination, and other geometry of the gage. Random errors can be kept to a minimum through the use of standard installation practices and careful site selection. Among systematic errors, wind effect is the most critical and difficult issue to address.

A.2.2 Wind Effects

Wind affects precipitation catch in two ways. First, precipitation gages are designed to catch raindrops falling in a vertical trajectory. However, the sky in stormy weather is generally not calm, and raindrops are blown away by wind from vertical trajectories into an angle of inclination. The inclined raindrops make the effective gage-orifice smaller than if the drops had come in vertically. This results in catch deficiency.

Let V_h = horizontal wind speed in centimeters per second and V_t = terminal velocity of raindrop in centimeters per second, then the angle of raindrop inclination α due to wind effect can be calculated by

$$\alpha = \tan^{-1}\left(\frac{V_h}{V_t}\right) \tag{A.1}$$

and catch deficiency d in percent is

$$d\ (\%) = (1 - \cos\alpha)100 \tag{A.2}$$

Thus, the observed precipitation P_o can be adjusted to correct the wind effect by

$$P_a = \frac{P_o}{\cos\alpha} \tag{A.3}$$

where P_a is the adjusted precipitation due to wind effect.

The terminal velocity of raindrops (V_t) is a function of raindrop diameter, and raindrop diameter is related to rainfall intensity by Equation 12.2. Once the information on raindrop diameter is available, V_t (cm/s) can be obtained using Figure 8.2 or estimated by the following empirical equation:

$$V_t = -20.30 + 484.43(D_{50}) - 84.55(D_{50})^2 + 4.98(D_{50})^3 \tag{A.4}$$

where D_{50} is the median raindrop diameter in millimeters. Equation A.1 shows that α increases with wind speed and decreases with raindrop size. Consequently, the wind effect is greater for solid precipitation than for liquid precipitation and smaller for precipitation of high intensity than of low intensity.

TABLE A.2

Typical Values of P (Equation A.6) for Varying Terrain Conditions

Terrain	p Value
Smooth, hard ground, lake, or ocean	0.10
Short grass on untilled ground	0.14
Level country with foot-height grass, occasional trees	0.16
Tall row crops, hedges, a few trees	0.22–0.24
Woody country, small towns, and shrubs	0.28–0.30
Urban areas with tall buildings	0.40

Data source: Bechraki, D.A. and Sparis, P.D., *Wind Eng.*, 24(2), 127, 2000.

The horizontal wind speed V_h in Equation A.1 refers to wind speeds at the orifice level. If not available, it can be estimated from measurements at a nearby station by the widely used power wind-profile equation:

$$\frac{V_2}{V_1} = \left(\frac{Z_2}{Z_1} \right)^p \tag{A.5}$$

where
  V_2 is the wind speed at height Z_2
  V_1 is the measured (referenced) wind speed at height Z_1 of the nearby station

The exponent p, depending on the stability of the atmosphere and surface roughness, can be calibrated by actual measurements through this logarithmic arrangement:

$$p = \frac{\ln(V_2 / V_1)}{\ln(Z_2 / Z_1)} \tag{A.6}$$

For neutral stability conditions, $p = 1/7$, or 0.143. Typical p values for a number of terrain conditions are given in Table A.2.

Second, the presence of precipitation gages in the field interferes with the general wind pattern and movement. It causes eddies and turbulence around the gage orifice. These eddies, updrafts, and downdrafts within and around the gage orifice carry away precipitation particles that would otherwise dump into the gage, causing a deficiency in catch. Coupling the turbulence created by the rain gage with raindrop-inclination effects of wind makes point-precipitation measurements always deficient.

A.2.3 Slope Terrain

In hilly regions, rain gages are installed vertically, but the land surface is composed of many facets with different slopes and aspects. Combining the topographic features with wind effects makes the accuracy of precipitation measurements more difficult to attain. The precipitation caught by rain gages is most likely to be less than the precipitation intercepted by the slope. Deficiencies are affected by slope steepness (α) and aspect (Z_α) along with the incoming storm direction relative to the zenith (β) and orientation (Z_β). When precipitation falls normal to the surface (i.e., $\alpha = \beta$) with an orientation parallel to the slope aspect, the deficiency is highest. In equation form, the incident precipitation on a slope P_i can be expressed by (Sharon, 1980)

$$P_i = P_o \, (1 + \tan \alpha \times \tan \beta \times \cos (Z_\alpha - Z_\beta)) \tag{A.7}$$

where
  P_o is the observed precipitation by gage on slope
  α, β, Z_α, and Z_β are all in degrees

Applications of the equation in Israel showed results in good agreement with tilted (directional) gages, receiving precipitation with the P_i/P_o ratios up to 1.40 (or 40% more incident rain) at some sites (Sharon et al., 2000). However, Blocken et al. (2006) noted that the method is suitable for a small-scale topography with rectilinear raindrop trajectories.

A.2.4 "True" Precipitation

Because of wind effects, "true" precipitation at a station is the amount of catch by a gage as if the gage were not there, or wind is calm and other error sources had no effect on precipitation measurements. This is the amount of precipitation actually reaching the ground surface. With special gage designs and under certain environmental conditions, "true" precipitation for point measurements may be attainable. However, "true" precipitation for a watershed is an immeasurable quantity because of precipitation distribution, intrinsic variation of storm systems, topography, vegetation, and gage density. The average density of nonrecording rain-gage networks in the conterminous United States is about 1.39 gages/1000 km² (Chang, 1981); the ratio of represented area to sampled area is $2.12(10)^{10}$ to 1. Apparently, many convective storms with diameters less than 30 km are never recorded. Thus, point accuracy and areal representativeness are two critical factors affecting watershed precipitation estimates.

A.3 Measurements

A.3.1 Gage Installation

When installing a gage, one should bear in mind that wind is the most critical factor in precipitation measurements, and wind speed increases with height above the ground and is lower on the leeward of a belt of trees or shrubs. Thus, wind-prone sites such as hilltops or wind-blown alleys should be avoided. Proper protection from surrounding vegetation can reduce wind effects on precipitation measurements. On the other hand, a gage located too close to surrounding objects or in a deep valley can cause a catch deficiency due to precipitation interception and rain-shadow effect.

It is advantageous to keep the gage orifice as low as possible to reduce wind effects but high enough to reduce raindrop splash into the gage, snow cover, and animal intrusion. As a general rule in rural areas, a gage should be located at a distance equal to about twice the height (2H) of the surrounding objects above the gage. The requirement of 2H diameter is impractical in forests, and Wilson (1954) suggested that a diameter equal to the height of a tree (1H) in a forest clearing would be sufficient. Each country has a standardized installation procedure including gage height above the ground and proper exposure from the surroundings as a reflection of its physical environment, experience, or tradition. Those procedures should be followed closely to make the collected data consistent with others.

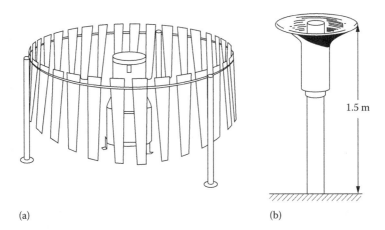

(a)   (b)

FIGURE A.3
Rain gage equipped with (a) Alter shield and (b) Nipher shield.

Also, it is important that the gage be upright and the orifice be level. An orifice inclined 20° from the vertical would cause a deficiency in catch of about 6% compared to a level gage.

A.3.2 Improvements

A.3.2.1 Shielded Gages

A set of windshields is often attached to the gage at the orifice height to reduce the effects of eddies, drafts, and turbulence on precipitation catch. However, windshields generally are not effective for wind speed greater than 32 km/h (Larson and Peck, 1974). Two types of windshields are in general use today: Nipher (rigid) and Alter (flexible) (Figure A.3). The Nipher shield is less effective for snowfall measurement because snow tends to build up on the flat top of the shield. A gage equipped with windshields increases rainfall catch less than 5% in general (Chang and Flannery, 1998) and has been reported to cause no significant differences at all (Larson, 1971).

A.3.2.2 Pit Gages

Since wind speed decreases to a minimum level as it approaches the ground surface, a gage placed in a pit with its orifice at ground level is expected to have the least amount of wind effect (Figure A.4). Precipitation measured by a pit gage is considered the most accurate, and it has been employed as a reference standard in precipitation studies (De Bruin, 1986). However, pit gages require special effort on installation, and they can be clogged with leaves and other debris or disturbed by animals. The redistribution of snow on the ground surface makes them questionable for snowfall measurements.

A.3.2.3 Tilted Gages

Installing 19 gages perpendicular to the slope caught an average precipitation 15% greater than 19 vertical gages did in the San Dimas Experimental Forest, southern California (Hamilton, 1954). Some gages have stereo orifices with funnels of rectangular or elliptical cross section cut off parallel to the slope (Sevruk, 1972). Tilted gages have been suggested for use in mountainous areas (Ambroise and Adjizian-Gerard, 1989). Wind changes speed

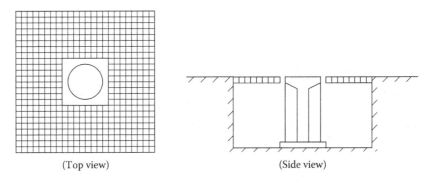

(Top view) (Side view)

FIGURE A.4
Pit gage.

and direction with time and space. Tilted gages are effective only when facing toward the incoming storm direction and the angle of inclination. The hillside slope and direction need not be perpendicular to the incoming storm direction. During the period when downslope winds dominate, stereo orifices can result in a reduction of precipitation catch due to the shadow effect of the higher side of the receiver (Hibbert, 1977).

A.3.2.4 Lysimeters

The lysimeter is a ground-level soil tank with devices to measure changes in weight of the whole tank. It can be employed to study plant transpiration and soil percolation under natural conditions as well as to detect the gain in weight due to precipitation. The catch in precipitation by a lysimeter represents the actual amount of precipitation at the ground surface without the effect of eddies and turbulence created by a gage. Studies showed lysimeter catch to exceed rain gage catch by 7%–9% (Kurtyka, 1953; McGuinness, 1966). Unfortunately, lysimeters are difficult to install, expensive, mechanically fallible, high-maintenance, and inadequate for snow measurements. All these disadvantages cause lysimeters to be limited in widespread application.

A.3.2.5 Wind-Speed Adjustment

Wind causes raindrops to fall at an angle of inclination, making the effective diameter of the gage orifice smaller than the raindrops that would fall vertically into the gage. The result is a reduction in precipitation catch. If information on rainfall intensity and wind speed during the storm is available, then the wind effect on precipitation catch by a standard gage can be corrected by the procedures given in Equations A.1 through A.4. Applying this procedure to 59 storms in east Texas, Chang and Flannery (1998) showed that the catch deficiency of the standard gage was reduced from 11% to 6% of the referenced pit gage.

Although oversimplified, the approach is easy to apply. The effectiveness of this approach depends on the accuracy of wind-speed measurements and the angle-of-inclination estimates. The most effective wind speed is at the gage height and around the gage orifice. If wind speed at the gage site is not available, then the wind speed at a nearby station can be used by applying Equations A.5 and A.6. Also, since the effect of wind speed on catch deficiency is not linear, adjustments based on an average angle of inclination for the entire storm probably are not as effective as those based on several inclination angles by the distribution of rainfall intensity of the storm. One may use Equation A.7 if gages are located on a slope.

A.3.2.6 Wind-Driven Rain Gages

Wind-driven rain is rain that falls with inclined angles due to wind effect. Many types of rain gages have been designed in an attempt to solve the wind effect on rainfall measurements. Some are vectopluviometers, which use a vertical as well as two to four horizontal openings to catch rainfall. Others attach a vane to the gage to direct the opening so it is always facing the wind (Kurtyka, 1953; Blocken and Carmeliet, 2004). Another design in recent years is spherical rain gages.

Because wind causes the effective orifice of the rain gage to be smaller than its actual orifice, a gage orifice must always face the incoming direction of raindrops to keep the effective orifice the same as its original. This can be done if the orifice of a rain gage is spherical in shape. A spherical orifice always has the same effective diameter regardless of wind speed and direction (Figure A.5).

Chang and Flannery (2001) developed two spherical and two semispherical orifices to be mounted on top of the standard gage and others in use today (Figure A.6). The semispherical orifices, installed side by side with a standard gage, correct 50% of catch deficiencies made by the standard gage (Figure A.5). Based on 115 storms, the four gages recorded rainfall with average differences ranging from −3% to 4% from a pit gage and 8%–16% from the standard gage. The spherical orifice with cylinders (Model A) recorded about 10% more rainfall than the standard gage and about 1% less than the pit gage and is the most effective and accurate among the four devices. The catch of spherical gages also was not significantly affected by three gage heights at 0.91, 1.83, and 2.74 m above the ground (Chang and Harrison, 2005).

These gages seem to be very promising for rainfall measurements. Their adoption does not abandon the current gages, alter rain gages in use today, or change the way rainfall depth is recorded. They are simple, inexpensive, easy to operate, and suitable for large-scale

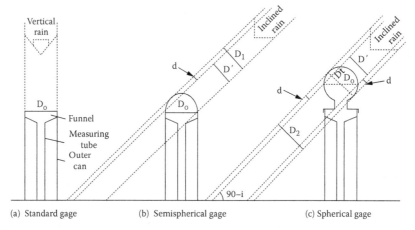

FIGURE A.5
Wind effects on rainfall catch on (a) the standard gage, (b) semispherical gage, and (c) spherical gage. D_o = orifice diameter of the standard gage; D' = effective diameter of the standard gage when rains come at an angle; D_t = the "true" rainfall; D_1 = catch by the semispherical gage; D_2 = catch by the spherical gage; d = catch deficiency; when rain is vertical, $D_t = D_o = D_1 = D_2$; when rain is inclined, $D_1 = D' + d = D_t - d$; $D_2 = D' + 2d = D_t$. (Adapted from Chang, M. and Flannery, L.A., *Hydrol. Processes*, 15, 643, 2001.)

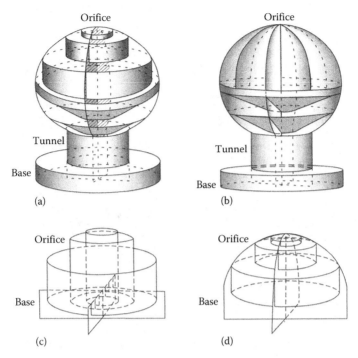

FIGURE A.6
Side views of (a) spherical orifice with cylinder, (b) spherical orifice with vanes, (c) semispherical orifice with cylinders, and (d) semispherical orifice with vanes and cylinders. (Adapted from Chang, M. and Flannery, L.A., *Hydrol. Processes*, 15, 643, 2001.)

applications. However, these gages have been tested only in humid areas; additional tests need to be conducted in different environments such as in arid climates, cold and frigid regions, and areas with high wind speed, high rainfall intensity, or misty weather. Also, owing to greater surface wetting and evaporation loss, the spherical gages may underestimate rainfall catch by a standard gage for small storms (<5.0 mm). Using polyethylene or other synthesized materials to construct the spherical orifice may improve the catch for small storms.

A.3.2.7 Globoscopic Pictures

Chang and Lee (1975) employed an optographic technique to evaluate the gage exposure and consequently make adjustments for wind effect on point precipitation. The technique involves (1) taking globoscopic pictures above the orifice of a rain gage, (2) obtaining wind speed in eight compass directions at the study site, and (3) developing equations to estimate the effect of shelterbelt density on wind speeds and catch error by wind speed in eight directions. The globoscopic pictures (Figure A.7) are stereographic projections of the hemisphere above the orifice of a rain gage. They are used to determine shelter belt density (three categories) around the gage site and distance from the gage to shelter belt, in all eight directions. These data obtained from the globoscopic picture were used in conjunction with the developed equations to estimate catch error caused by gage exposure and wind effect.

The design of the globoscope was described in Lee's (1978) work. It is a photographic device consisting of a paraboloidal mirror at the base of the instrument and a camera mounted above the mirror to record a polar stereographic image of the gage site, sky, and

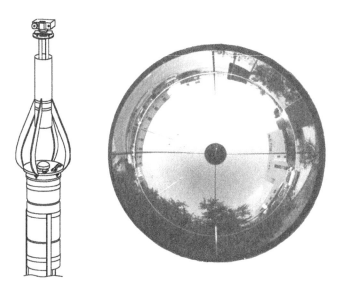

FIGURE A.7
Globoscope and a globoscopic picture.

surroundings (Figure A.7). The linear distance, D, of any point on the picture from the center (gage site) is given by

$$D = r \tan Z \qquad (A.8)$$

where
  r is the radius of the projection
  Z is the zenith distance in degrees, read from a grid system of concentric circles

The distance of a rain gage to surrounding obstructions has been suggested to be 1H (Leonard and Reinhart, 1963), 2H (National Weather Service, 1972), and 4H (Brooks, 1938), where H is the height (H) of the obstruction. In fact, the proper distance between a gage and the surrounding obstruction is affected by wind speeds and the density of surrounding obstructions, which can vary with seasons. This variation can cause a fixed distance appropriate in one season to be inappropriate in other seasons. This seasonal exposure problem can be improved by the optographic technique because of its objectivity and its ability to make corrections due to various exposures (distances). However, the procedure is tedious and complicated, and its general application is therefore doubtful.

References

Ambroise, B. and Adjizian-Gerard, J., 1989, Test of a trigonometrical method of slope rainfall in the small Ringelbach catchment (High Vosges, France), *Precipitation Measurement*, Sevruk, B., Ed., *WMO/IAHS/ETH Workshop*, St. Moritz, Switzerland, pp. 81–86.

Bechraki, D.A. and Sparis, P.D., 2000, Simulation of the wind speeds at different heights using artificial neutral networks, *Wind Eng.*, 24(2), 127–136.

Blocken, B. and Carmeliet, J., 2004, A review of wind-driven rain research in building science, *J. Wind Eng. Ind. Aerodyn.*, 92, 1079–1130.

Blocken, B., Poesen, J., and Carmeliet, J., 2006, Impact of wind on the spatial distribution of rain over micro-scale topography—Numerical modelling and experimental verification, *Hydrol. Process.*, 20(2), 345–368.

Brooks, C.F., 1938, Need for universal standards for measuring precipitation, snowfall, and snow cover, *IUGG Bull.*, 23, 7–58.

Chang, M., 1981, A survey of the U.S. national precipitation network, *Water Resour. Bull.*, 17, 241–243.

Chang, M. and Flannery, L.A., 1998, Evaluating the accuracy of rainfall catch by three different gages, *J. Am. Water Resour. Assoc.*, 34, 559–564.

Chang, M. and Flannery, L.A., 2001, Spherical gages for improving the accuracy of rainfall measurements, *Hydrol. Processes*, 15, 643–654.

Chang, M. and Harrison, L., 2005, Field assessments on the accuracy of spherical gauges in rainfall measurements, *Hydrol. Processes*, 19, 403–412.

Chang, M. and Lee, R., 1975, Representativeness of watershed precipitation samples, Bulletin 4, Water Resources Institute, West Virginia University, Morgantown, WV.

Creutin, J., Delrieu, G., and Lebel, T., 1988, Rainfall measurement by raingauge–radar combination: A geostatistical approach, *J. Atmos. Oceanic Technol.*, 5, 102–115.

Dai, A., Lin, X., and Hsu, K.-L., 2007, The frequency, intensity, and diurnal cycle of precipitation in surface and satellite observations over low- and mid-latitudes, *Clim. Dyn.*, 29, 727–744.

De Bruin, H.A.R., 1986, Results of the international comparison of rain gauges with a reference pit gauge, part A: Basic stations, *Correction of Precipitation Measurements*, Sevruk, B., Ed., *ETH/IAHS/WMO Workshop*, Zurich, Switzerland, pp. 97–100.

Hamilton, E.L., 1954, Rainfall sampling on rugged terrain, USDA Technical Bulletin No. 1096.

Hibbert, A.R., 1977, Distribution of precipitation on rugged terrain in Central Arizona, *Hydrol. Water Resour. Arizona Southwest*, 7, 163–173.

Krajewski, W.F. et al., 2006, DEVEX-disdrometer evaluation experiment: Basic results and implications for hydrologic studies, *Adv. Water Resour.*, 29, 311–325.

Kreuels, R.K. and Breuer, L.J., 1986, Wind influenced rain gage errors in heavy rain, *Correction of Precipitation Measurements*, Sevruk, B., Ed., *ETH/IAHS/WMO Workshop*, Zurich, Switzerland, pp. 105–107.

Kurtyka, J.C., 1953, Precipitation measurements study, Annual Report of Investigation No. 20, State Water Survey Division, Urbana, IL.

Lanza, L.G., Vuerich, E., and Gnecco, I., 2010, Analysis of highly accurate rain intensity measurements from a field test site, *Adv. Geosci.*, 25, 37–44.

Larson, L.W., 1971, Shielding precipitation gages from adverse wind effects with snow fences, Water Resources Series 25, Water Resources Research Institute, University of Wyoming, Laramie, WY, pp. 1–161.

Larson, L.W. and Peck, E.L., 1974, Accuracy of precipitation measurements for hydrologic modeling, *Water Resour. Res.*, 10, 857–863.

Lee, R., 1978, *Forest Microclimatology*, Columbia University Press, New York.

Leonard, R.E. and Reinhart, K.G., 1963, Some observations on precipitation measurement on forested experimental watersheds, Research Note NE-6, U.S. Forest Service, Northeastern Forest Experiment Station, Upper Darby, PA.

Liu, T.M., 1983, The inventories of meteorological instruments and observational equipment in ancient China, *The History of Chinese Technology*, Vol. II, Wu, T.L., Ed., The Natural Sciences and Cultural Enterprises, Inc., Taipei, Taiwan, pp. 149–161 (in Chinese).

McGuinness, J.L., 1966, A comparison of lysimeter catch and rain gage catch, USDA Agricultural Research Service ARS 41–124.

Murugan, S.S., Natarajan, V., and Joseph, L., 2010, Interpretation of underwater noise spectra at different rainfall and wind speed conditions based on time series hydrophone voltage signal processing, *Int. J. Adv. Eng. Technol.*, 1(II), 184–194.

National Weather Service, 1972, Substation observations, supersedes circular B, U.S. Department of Commerce, NOAA, Washington, DC.

Nikolopoulos, E.I. et al., 2008, Comparative rainfall data analysis from two vertically pointing radars, an optical disdrometer, and a rain gauge, *Nonlinear Processes Geophys.*, 15, 987–997.

Nystuen, J.A., 1986, Rainfall measurements using underwater ambient noise, *J. Acoust. Soc. Am.*, 79, 972–982.

Schuster, G. et al., 2011, Application of satellite precipitation data to analyse and model arbovirus activity in the Tropics, *Int. J. Health Geographics*, 10, 8.

Sevruk, B., 1972, Precipitation measurements by means of storage gauges with stereo and horizontal orifices in the Baye De Montreux Watershed, *Distribution of Precipitation in Mountainous Areas*, Vol. II, WMO/OMM No. 326, Geneva, Switzerland, pp. 86–95.

Sevruk, B. and Klemn, S., 1989, Types of standard precipitation gages, *Precipitation Measurements*, Sevruk, B., Ed., *WMO/IAHS/ETH Workshop*, St. Moritz, Switzerland, pp. 227–232.

Sharon, D., 1980, The distribution of effective rainfall incident on sloping ground, *J. Hydrol.*, 46, 165–188.

Sharon, D., Margalit, A., and Arazi, A., 2000, The study of rainfall distribution in small watersheds in Israel: From early observations to model simulations, *The Hydrology—Geomorphology Interface: Rainfall, Floods, Sedimentation, Land Use (Proceedings of the Jerusalem Conference)*, May 1999, IAHS Publication No. 261, Oxfordshire, U.K., pp. 13–28.

Sokollek, V., 1986, Problems of precipitation measurement for water budget studies in the Highlands of Hessen, *Correction of Precipitation Measurements*, Sevruk, B., Ed., *ETH/IAHS/WMO Workshop*, Zurich, Switzerland, pp. 89–94.

Thurai, M. et al., 2011, Drop size distribution comparisons between Parsivel and 2-D video disdrometers, *Adv. Geosci.*, 30, 3–9.

Wilson, W.T., 1954, Analysis of winter precipitation observations in cooperative snow investigation, *Mon. Weather Rev.*, 82, 183–195.

Xiao, Q. et al., 2005, Assimilation of Doppler radar observations with a regional 3DVAR system: Impact of Doppler velocities on forecasts of a heavy rainfall case, *J. Appl. Meteorol.*, 44, 768–788.

Appendix B: Streamflow Measurements

The measurement of streamflow basically consists of:

1. River stage
2. Flow velocity
3. Water discharge
4. The stage–discharge relationship

These measurements are then extended to various streamflow characteristics by incorporating time factors. This appendix covers only the basic measurements.

B.1 River Stage

River stage refers to the elevation of water surface at a given site relative to a reference datum, usually mean sea level. The data are required for navigation, river-sport activities, floodplain management, flood forecasting, calibration of stage–discharge relationships, river-fluctuation studies, sediment-transport studies, and many others. River stages can be measured by manual or automatic instruments, depending on the type of data required, degree of accuracy, and the frequency of measurements.

B.1.1 Manual

River stages can be measured directly with a rod while the investigator wades in the stream, or with a wire-weight gage and chain gage lowered from a bridge or a boat to the stream. When these options are impractical, a bucket can be suspended along a cable installed across the stream; the investigator lowers a graduated tape with weight from the bucket to the water surface for measurements. A staff gage is commonly installed on a streambank or painted on a bridge pier or wall for periodic observations at a permanent site. To cover the full range of floodwaters, it may be necessary to install sectional staff-gages, i.e., more than one staff gage at various elevations.

The manual readings on staff gages often miss data at higher stages, especially for small streams in times of severe storms and flash floods. Also, if on-site visits are required for several sites in a watershed study, the lag time in visiting each site can cause the data collected not to be comparable. In such cases, using automatic methods may become necessary. For information on crest stages, a water-sensitive substance, such as fluorescein ($C_{20}H_{12}O_5$, orange-red compounds) and potassium permanganate ($KMnO_4$, black compounds), can be painted on a permanent structure. The paint will change color (from orange to green and from black to deep purple, respectively) when it gets wet, and the highest crest is so identified. In many cases, examining watermarks, soil stain, or debris on rocks, walls, tree trunks, or banks can help identify the maximum river stage.

B.1.2 Automatic

The automatic method, either mechanical or electrical, allows the continuous monitoring of river stages. The mechanical method uses a float as the sensor sitting on the water surface, and the float is attached by chain to mechanisms including a stage ratio gear, a time clock, and a recording device (Figure B.1). The vertical movement of the float, triggered by the water-level fluctuation and a counterweight attached to the other end of the chain, is continuously recorded by an ink trace on a chart. The float is housed in a stilling well to prevent erratic measurements from wave action and turbulence of the flow movement. The water in the stilling well is connected to the stream by pipes (Figure B.2).

Some instruments digitally record electrical signals triggered by a small change of water level in the stilling well. The electrical signals are stored and converted into depths in a data-logger and can be accessed or downloaded by a laptop computer for data management and analyses. Others use pressure (bubbler) or ultrasonic sensors instead of floats and stilling well. The bubbler uses a small compressor to pump air into a reservoir. The air is released slowly into a small flexible tube submerged in the stream. As the water level in the stream increases, the amount of air pressure required to force the bubble from the tube also increases. A pressure transducer along with electronic circuitry and computer software converts the pressure into streamflow depth above the air tube.

Flowmeters that use an ultrasonic sensor to detect water level do not need to contact the stream. The sensor is installed above the stream, and the transmitted sound pulse will be reflected when it reaches the water surface. The elapsed time between sending a pulse and receiving an echo determines the water level of the stream.

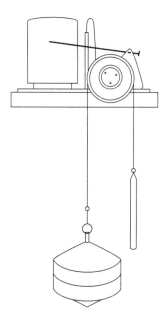

FIGURE B.1
The mechanical FW-1 water-level recorder.

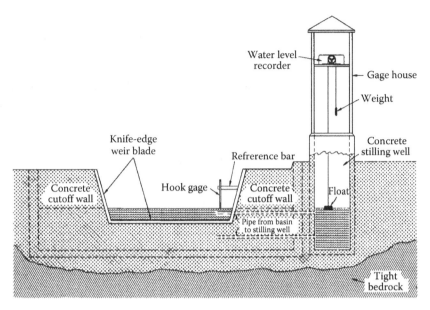

FIGURE B.2
A typical streamgaging station used on the Coweeta Hydrological Laboratory near Otto, North Carolina. (From U.S. Forest Service, Coweeta Hydrological Laboratory: A guide to research program, SE Forest Experiment Station, Asheville, NC, 1984.)

B.2 Flow Velocity

B.2.1 Velocity Variation

Affected by the character of the streambank, stream bed, and channel configuration, flow velocity varies with time, along a stream channel, across a cross-section, and throughout a vertical profile. In general, velocity tends to increase as channel slope increases, and decreases as resistance to water movement increases. Thus, the flow velocity in a stream with rocky beds is lower than in one with earth beds. In a cross-section, velocities are lower near the banks and the bottom. The maximum velocity tends to occur around the center of the stream and below the water surface (Figure B.3a).

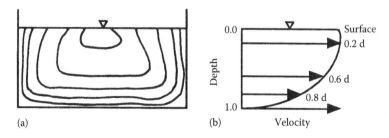

FIGURE B.3
Velocity distributions as depicted by lines of equal velocity in a channel's cross-section (a) and over a vertical profile (b).

The mean flow velocity for a stream must be obtained by estimating mean velocities for different segments across the stream. In practice, the mean flow velocity in a vertical stream column is often estimated at 0.6 (60%) depth from the surface for streams less than 1 m deep and by the average reading at 0.2 and 0.8 depth for streams 1–3 m deep (Figure B.3b). More than two measurements may be required to estimate mean velocity for streams deeper than 3 m.

B.2.2 Site Selection

An ideal site for determining flow velocity in mountainous streams includes the following:

1. Straight reach of about 20 m above and below the measuring site
2. Uniform channel in terms of width, bank materials, vegetation, and profile
3. Stable with respect to bank erosion and cross-section profile
4. Free from back flow caused by downstream tributaries or the damming effect of rocks, logs, debris, and other materials
5. Far below the mixing effect of upper tributaries

For a permanent gaging station, a site with all flows contained within the banks will make measurements more reliable, although the maximum stage of severe floods often exceeds the channel capacity.

B.2.3 Methods

B.2.3.1 Floats

Flow velocities can be determined approximately by timing the travel of a float over a known distance. It is the simplest method, but it is not recommended unless other methods are not feasible or practical. Two types of floats, surface floats submerged one-fourth or less of the flow depth and rod floats submerged more than one-fourth the depth but not touching the bottom, can be used. The measurement should be conducted on windless days and the cross-section should be divided into at least three, preferably five, sections of equal width. Several measurements should be made in each section and all measurements summed to obtain the average value. However, the float method gives only surface velocity and must be multiplied by a correction factor of 0.7 for streams of 1 m depth and 0.8 for streams of 6 m or deeper to estimate the average velocity for the column (Bureau of Reclamation, 1981).

B.2.3.2 Velocity Head

The total energy level on a cross-section of stream can be expressed in terms of the sum of elevation head, pressure head, and velocity head. The pressure head is the energy due to water depth, while velocity head (h, m) is a function of flow velocity (V, m/s).

$$h = \frac{V^2}{2g} \tag{B.1}$$

where g is the acceleration due to gravity (9.80 m/s^2). Accordingly, if the velocity head in streams is measured, then the flow velocity can be obtained by

$$V = (2gh)^{0.5} \tag{B.2}$$

or

$$V = 4.427(h)^{0.5} \tag{B.3}$$

Velocity heads can be measured using a *velocity rod* or a *pitot tube* (Figure B.4). The rod is used to measure flow velocity in streams where wading is possible. It is a regularly graduated rod with a sharp edge and a flat edge on its cross-section. When the sharp edge faces upstream, it measures the depth of the stream. Facing the flat edge upstream, it measures the depth plus velocity head (or *hydraulic jump*) of the stream. The difference in readings between flat edge and sharp edge is the velocity head. Using a velocity rod, the calculated V from Equation B.3 needs to be multiplied by a correction factor 0.85 as an estimate for the mean velocity for the column of water (Wilm and Storey, 1944).

The simplest form of a pitot tube is a transparent tube with a right-angle bend. The tube is submerged in the stream to a desired depth with the bend pointing upstream, and the rise of water in the tube is the velocity head due to flow velocity. However, velocity heads in the simple form of a pitot tube are difficult to read when flow velocities in the open channel are low. A pitot-static tube that consists of an inner tube with an opening in the longitudinal direction and an outer tube with holes in the vertical direction greatly improves the measurement (Figure B.4). When pointing upstream, the inner tube receives the impact pressure from the flow, while the outer tube is exposed only to the local static pressure. The difference in pressures between the tubes is due to velocity head and is directly converted into flow velocity by connecting to a pressure transducer and a manometer or other devices for electronic data processing. Pitot tubes give satisfactory measurements at drops, chutes, or other stations with fairly large velocity heads and are common in pipe-flow measurements (Bureau of Reclamation, 1981; Bradner and Emerson, 1988; Vogel, 1994). The use of pitot tubes in forestry is rare.

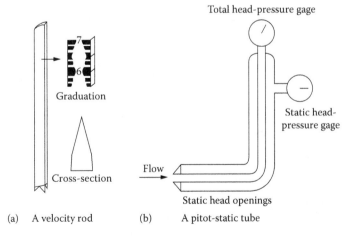

(a) A velocity rod   (b)   A pitot-static tube

FIGURE B.4
A velocity rod (a) and a pitot-static tube (b).

B.2.3.3 Current Meters

Properly calibrated, current meters provide reliable measurements sensitive to velocity changes. A current meter may use four to six cups rotating around a vertical axis, or a propeller spinning on a horizontal axis, as the sensor. In practice, the meter is submerged in the stream to a desired depth. The number of revolutions of the cups is signaled by a battery-operated buzzer, which is detected by the observer through an earphone or is directly recorded by a digital counter. Flow velocities (V, m/s) are calculated based on the number of revolutions per second (N) by

$$V = a + b(N) \tag{B.4}$$

where a and b are constants specifically calibrated for each instrument.

Each current meter has a velocity-detection range for reliable measurements. In small, forested streams, the Pygmy meters, which measure flows with velocities up to 1.3 m/s and are the smallest commercial instruments available, are often used. The meter is fastened to a graduated wading rod, and the flow velocity at various depths in the stream can be measured by adjusting meter position on the rod (Figure B.5). For streams with high flow velocities or greater depth where wading is not possible, measurements can be made from a boat, bridge, or cableway. In such streams, a sounding weight, ranging from 7 to 136 kg in size, is suspended below the current meter to keep it stationary in the water.

B.2.3.4 Others

The three methods mentioned earlier are the most common techniques used in watershed studies. Other methods include the following:

B.2.3.4.1 Tracers

In shallow, braided, or rocky streams, or streams with high flow velocities, turbulence, and poor accessibility in which a current meter is impossible or difficult to use, various tracers can be employed to measure flow velocities or discharge directly. Common tracers include

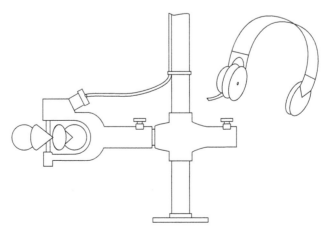

FIGURE B.5
A Pygmy current meter on wading rod.

salts (Spence and McPhie, 1997), dyes, and radioactive materials, although handling problems have limited widespread use of radioactive tracers in flow measurements (Kilpatrick and Cobb, 1985).

The two approaches to the use of tracers in flow measurements are trace-velocity and trace-dilution. The *salt-velocity method* is based on the addition of salts causing an increase in the electrical conductivity of water. Thus, flow velocity is the time required for salts (usually sodium chloride) to travel between two sites along a channel as detected by electrodes.

In the *salt-dilution method*, a concentration (C_1) of salt solution is added to a stream channel at a given rate (q). The stream discharge (Q) can be calculated by measuring the natural salt concentration (C_o) of the stream and the new concentration (C_2) at a downstream site where the added salt is completely mixed with the stream:

$$Q(C_o) + q(C_1) = (Q + q)(C_2)$$

$$Q = \frac{q(C_1 - C_2)}{C_2 - C_o} \tag{B.5}$$

Mean velocity of the stream is then equal to the stream's discharge divided by the stream's cross-section area (A), or $V = Q/A$.

As in the float method, a colored dye such as fluorescein (a red powder that becomes a green color when dissolved in slightly alkaline water) can be poured into the stream. The time between the instant the dye is added and the instant the center of the mass of the colored water passes the downstream station is determined. The average time required for the dye to pass the course is the mean velocity of the stream. Since detecting the position of the center of the mass of the colored water is not easy, especially in high-velocity flows, the observed velocity may be just the surface velocity rather than the mean velocity.

B.2.3.4.2 Hot-Wire Anemometers

In a hot-wire anemometer, constant AC (alternating current) voltage heats a thermopile. When the thermopile is submerged in a flowing fluid, its temperature is cooled and resistance changed. The degree of cooling, as indicated by the thermopile voltage output, is related to the flow velocity. Hot-wire anemometers are widely applied in airflow measurements but have some difficulties in streamflow measurements because of hysteresis effects and frailty of the thin wire (Gordon et al., 1993).

B.2.3.4.3 Electromagnetic Method

Water is an electrical conductor. Based on the Faraday law of electromagnetic induction, a conductor will induce a voltage when it flows across a magnetic field. The magnitude of this induced voltage is proportional to the strength of the magnetic field and the average velocity of the flow (WMO, 1980).

A typical electromagnetic flowmeter (Figure B.6) uses an electromagnetic coil that generates the magnetic field as a sensor and a pair of electrodes as a detector to measure the voltage produced by the movement of the flow conductor. When water flows across the sensor, the direction of the flow, the magnetic field, and the induced voltage are mutually perpendicular to one another. The output voltage, detected by the electrodes, is processed by the electronics to give a direct reading of flow velocity in meters per second or feet per second.

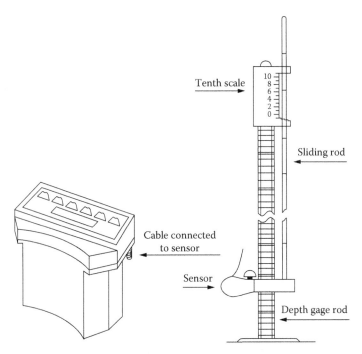

FIGURE B.6
A Marsh–McBirney electromagnetic flowmeter mounted on a depth gage rod (top-setting rod).

B.2.3.4.4 Acoustic Flowmeters

Acoustic flowmeters operate on the principle that the difference in arrival time of two simultaneously created sound pulses traveling in opposite directions through the water can be related to flow velocity. Two special transducers that transmit and receive sound pulses are installed in or on the channel side-slopes, one in an upstream direction and the other across the river in a downstream direction (Figure B.7). In the downstream direction from X to Y, the effective speed of an acoustic pulse is the flow path velocity V_p plus the speed of sound C, or $V_p + C$, and the time (T_1) for a pulse to travel within this distance L is (WMO, 1980; Laenen, 1985):

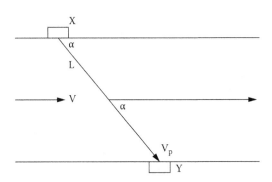

FIGURE B.7
Ultrasonic method for velocity measurements.

$$T_1 = \frac{L}{V_p + C} \tag{B.6}$$

On the other hand, the effective speed of an acoustic pulse in the upstream direction (from Y to X) is $C - V_p$, and the time (T_2) for a pulse to travel within this distance L is

$$T_2 = \frac{L}{C - V_p} \tag{B.7}$$

Thus, the difference in transit time (δT) of the two simultaneously created pulses is

$$\delta T = T_1 - T_2 = \frac{L}{C + V_p} - \frac{L}{C - V_p} = 2L \frac{V_p}{C_2 - V_p^2} \tag{B.8}$$

Since C^2 is much greater than V_p^2, Equation B.8 can be simplified as

$$\delta T = 2L \frac{V_p}{C^2} \tag{B.9}$$

or

$$V_p = \delta T \frac{C^2}{2L} \tag{B.10}$$

The average flow velocity (V) at a given depth can be corrected by the inclined angle of the path line (α)

$$V = \frac{V_p}{\cos \alpha} \tag{B.11}$$

B.2.3.4.5 *Acoustic Doppler Current Profiler*

The acoustic Doppler current profiler (ADCP) system is a modern technique developed by the U.S. Geological Survey (USGS) to monitor horizontal and vertical water velocities at 1-m vertical intervals from a moving vessel. The ADCP transmits acoustic pulses, in the range 300–3000 kHz, into the water column. Part of the transmitted acoustic energy is reflected back toward the transducers by particulate matter in the water that is assumed to move with the water. The Doppler effect causes the frequency of these reflected signals to shift, and the magnitude of the frequency shift is a function of the speed of the particulate matter along the acoustic beam. The frequency shifts are then converted into water speeds by the computer software and hardware of the ADCP system. The system is designed to overcome the difficulty, time consumption, and dangers associated with discharge measurements in large rivers and estuaries (Simpson and Oltmann, 1993). Measurements using a mechanical current meter require a minimum of 20 individual readings across the river, and could take as long as several hours to complete. The ADCP measurement is dramatically faster, made in a matter of minutes rather than hours without losing accuracy (Hirsch and Costa, 2004).

Note that the Doppler effect, named after the Austrian physicist Christian J. Doppler (1803–1853), describes our experience of the relative motion between the source and the observer.

When the source and observer are moving toward each other, the frequency heard by the observer is higher than the frequency of the source. On the other hand, when they are moving away from each other, the frequency heard by the observer is lower than the frequency of the source. Some of the common applications of the Doppler effect include measurements of rainfall intensity, ocean waves and currents, flow velocity, and automobile speed.

B.3 Discharge

B.3.1 Volumetric and Gravimetric Methods

In volumetric and gravimetric methods, the entire flow in a stream or a study plot is collected in a container over a given time period. Sometimes it is necessary to build an artificial control so that all the flows can be diverted into a calibrated container of known volume or weight. Discharge (Q, m³ or L/s) is calculated by the time (T, s) required to fill a known volume (ϖ, m³, or L) of flow read directly from the container or calculated indirectly by placing the container on a weighing scale

$$\varpi = \frac{W_2 - W_1}{W} \tag{B.12}$$

$$Q = \frac{\varpi}{T} \tag{B.13}$$

where
   W_2 is the weight of water and container (kg)
   W_1 is the weight of empty container (kg)
   W is the weight of water (1000 kg/m³)

The container can be graduated so that the depth of water can note volume increments. For continuous measurements, a tipping bucket gage operated with a datalogger is often employed for overland runoff studies. The two buckets in the gage are used to sense runoff and send an electrical signal to the datalogger. The size of the two buckets needs to be modified to meter a volume of runoff equivalent to a known depth generated from the study plot.

Possible applications of these methods are at a spring, V-notch weir, overfall, pipe, overland runoff collector, or drainage ditch, or at the deep pool behind a broad-crested weir where water on the crest is too shallow and water in the pool is too slow to be measured by a current meter. These methods are considered the most accurate methods for measurements of small flows (Gordon et al., 1993).

B.3.2 Velocity–Area Method

The velocity–area method estimates discharge (Q, m³/s) as the product of flow velocity (V, m/s) and cross-sectional area (A, m²), or

$$Q = (V)(A) \tag{B.14}$$

Since flow velocity in a cross-sectional area varies with location and depth, the method in practice breaks the cross-sectional area into several segments (15–20 are common; 25–30 are adequate even for extremely large channels) with no more than 10% of the total flow in each segment. However, the ideal would be that no partial section contains more than 5% of the total flow (USGS, 1980). Few sections may be used for channels with a smooth cross-section and good velocity distribution. Measurements are then made on the flow velocities, width, and depth of each segment. These data are then used to calculate the discharge in each segment by Equation B.14, and the total discharge (Q) for the cross-sectional area is the sum of discharges for all segments (Σq_n), or

$$Q = \Sigma q_n = (v_1)(a_1) + (v_2)(a_2) + (v_3)(a_3) + \cdots + (v_n)(a_n) \tag{B.15}$$

where v and a refer, respectively, to the mean velocity and area of segments 1 through n.

The discharge calculation for each segment can be made in several ways. The mean-section method and mid-section method are the most common. These are illustrated in Figure B.8, and an example of the calculation of these two methods is given in Table B.1.

B.3.3 Slope–Area Method

The slope–area method provides an "after-the-fact" estimate for peak discharges or an indirect estimate for flood flows in which direct discharge measurements are impossible. It uses physical characteristics of the stream channel to estimate mean flow velocities and multiply the estimated flow velocity by the cross-sectional area to yield flood discharges. The most popular equation to estimate flood velocities in the United States is the Manning equation. Use of the Manning equation for estimating flood discharges is given in Equation 10.33 and its subsequent discussion.

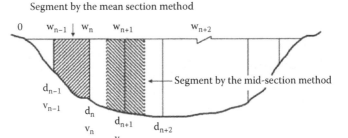

Segment by the mean section method

Segment by the mid-section method

Discharge calculations for each segment

Mean section method:

$$q = \left(\frac{v_{n-1} + v_n}{2}\right)\left(\frac{d_{n-1} + d_n}{2}\right)\left(w_n - w_{n-1}\right)$$

Mid-section method:

$$q = \left(v_{n+1}\right)\left(d_{n+1}\right)\left(\frac{w_{n+2} - w_n}{2}\right)$$

FIGURE B.8
Streamflow measurements and discharge calculations using two different methods.

TABLE B.1

An Example for Velocity–Area Discharge Calculations Using the Two Methods Illustrated in Figure B.8

Width[a] (w_n, m)	Depth (d_n, m)	Velocity[b] (v_n, m/s)	Mean-Section Method				Mid-Section Method	
			$(v_{n-1} + v_n)/2$	$(d_{n-1} + d_n)/2$	$(w_n - w_{n-1})/2$	q_n	$(w_{n+1} - w_{n-1})/2$	q_n
0.0	0.0	0.0					0.0	0.0
			0.076	0.378	2.0	0.057		
2.0	0.756	0.152					2.50	0.287
			0.219	0.890	3.0	0.585		
5.0	1.024	0.286					3.00	0.879
			0.346	1.173	3.0	1.218		
8.0	1.321	0.405					3.00	1.605
			0.484	1.445	3.0	2.098		
11.0	1.568	0.563					3.00	2.648
			0.538	1.495	3.0	2.413		
14.0	1.422	0.512					3.00	2.184
↓								
29.0	1.025	0.206						
			0.191	0.920	3.0	0.527		
32.0	0.814	0.175					3.00	0.427
			0.140	0.711	3.0	0.299		
35.0	0.607	0.104					2.75	0.174
			0.052	0.304	2.5	0.019		
37.5	0.0	0.0					0.0	0.0
Total discharge (m³/s)							15.805	16.054

[a] Having the same width is the easiest approach, but in the field widths of segments would vary based on the depth and velocity of water in a channel (Dr. Don Turton, personal communication).

[b] Mean flow velocity of each segment may be based on one measurement at 0.6 depth or two measurements at 0.2 depth and 0.8 depth from the surface.

B.4 Stage–Discharge Relation

The streamflow measurements described earlier are tedious and time-consuming. It is extremely difficult, if not impossible, to apply these procedures to measure streamflow at both low and high flow stages in a drainage system. Since streamflow volume and stage height are affected by the hydraulic geometry of the stream channel, the measurement procedure can be simplified by relating discharge volume to stage height alone through a calibration procedure conducted in the field or laboratory.

B.4.1 Channel Hydraulic Geometry

The measurement of streamflow (Q) is based on:

$$Q = AV = (WD)(V) \tag{B.16}$$

where the cross-sectional area A is calculated by the channel width W and the average depth D. Accordingly, at a particular station along a stream, flow velocity V tends to change abruptly in accordance with changes in the cross-sectional width W and stream depth D. The hydraulic geometry of streams can be expressed by a series of empirical equations (Leopold and Maddock, 1953):

$$W = aQ^b \tag{B.17}$$

$$D = cQ^f \tag{B.18}$$

$$V = kQ^m \tag{B.19}$$

$$WDV = Q = ack(Q)^{(b+f+m)} \tag{B.20}$$

where a, c, k, b, f, and m are empirical constants, and they must have the following relationships:

$$ack = 1 \tag{B.21}$$

$$b + f + m = 1 \tag{B.22}$$

to validate Equation B.20.

Average values of the exponents b, f, and m derived for some streams in the conterminous United States are given in Table B.2. In general, depth seems to change more rapidly than width at a station, but width changes more rapidly than depth in a downstream direction. Velocity generally increases with depth as depicted by the Manning equation. But, for a given Q, an increase in channel slope or a reduction in channel width will cause a consequent increase in velocity. This is the general principle used to induce a critical or supercritical state of flow in the design of flumes for discharge measurements. The occurrence of critical or supercritical flow in the flume allows the calibration of a definitive head–discharge relationship, and measurements can be based on one head reading.

TABLE B.2

Averages of Hydraulic Geometry Exponents for Some Streams in the Conterminous United States

Location	Condition	b (Width)	f (Depth)	m (Velocity)	Source
Midwest	At-a-station	0.26	0.40	0.34	Leopold and Maddock (1953)
	Downstream	0.50	0.40	0.10	
Frazer Exp. Forest, CO	At-a-station	0.16	0.35	0.49	David et al. (2010)
Niobrara River, NE	At-a-station	0.14	0.40	0.45	Alexander et al. (2009)
	Downstream	0.54	0.25	0.21	
Nonerodible Δ-cross-section	At-a-station	0.375	0.375	0.250	Lane and Foster (1980)
	Downstream	0.4954	0.4594	0.0813	
674 stations in 15 states, USA	Downstream	0.557	0.341	0.1035	Allen et al. (1994)

Note: Hydraulic geometry components b, f, and m are given in Equations B.17, B.18, and B.19, respectively.

Note that the state of flow is generally described by three conditions: subcritical, critical, and supercritical. The conditions are separated by a dimensionless parameter, the *Froude number* (F), defined as the ratio of inertial force to the force of gravity

$$F = \frac{V}{(gD)^{0.5}} \tag{B.23}$$

where
  V is the mean flow velocity (m/s)
  g is the acceleration due to gravity (9.80 m/s²)
  D is the hydraulic depth (m)

The inertial forces describe the compulsive ability of water, while the gravitational forces pull water downhill:

If F < 1, the flow is in subcritical condition—slow, tranquil and taming—where the flow is predominated by the gravity.

If F = 1, the flow is in critical condition, where the specific energy is minimum for a given discharge.

If F > 1, the flow is in supercritical condition—high in velocity, turbulent, and torrential—where the flow is dominated by the inertial forces.

B.4.2 The Rating Equation

The rating equation makes stream discharge Q a function of channel depth D (an inverse relationship of Equation B.18), or

$$Q = mD^n \tag{B.24}$$

where m and n are constants specifically calibrated for the site in question. They are derived by a least-squares fit using the measured Q versus mean stages D. For stations where Q is not zero when D = 0, a correction stage d should be included in the calibration.

$$Q = m(D \pm d)^n \tag{B.25}$$

where d is the stage reading at zero flow. Once Equation B.24 or Equation B.25 is calibrated, discharges can be obtained by measuring only stream stages.

Traditionally, streamflow discharges are plotted against the corresponding mean stream stages on graph paper, and a curve is fitted visually through the plotted points. The fitted curve, approximately parabolic, is called the *rating curve*, relating discharges Q to stages D through graphical solutions. If log–log paper is used, then the fitted curve appears as a straight line, making interpretations of the line easier.

It is important that the plotting procedure include some measurements at high flood stages to depict most flow conditions. The result may produce more than one curve as a reflection of the physical effects of channel and floodplain on flow movements. Two rating curves, a curve for low to intermediate flows and a curve for intermediate to high flows, are common for most gaging stations (Figure B.9). Two rating equations may be required to provide more reliable estimates at higher stages.

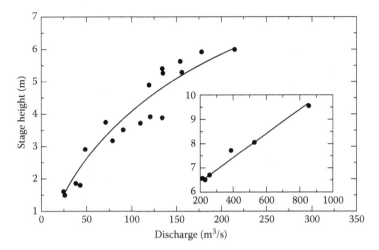

FIGURE B.9
Stage–discharge rating curves.

B.4.3 Precalibrated Devices

The rating equation and the rating curve described earlier are valid only if the channel cross-section is relatively stable over time. Should the shape of the channel cross-section be altered due to flood flows or forest harvesting, then the calibrated stage–discharge relationship is no longer applicable to the new cross-section. This leads to an attempt to improve the reliability of the *discharge–stage relationship* by artificial control of the stream-flow. If flows in the stream channel are controlled by a rigid, permanent structure of special shape and characteristics, then there is no need to worry about the alteration of the channel cross-section, and the stage–discharge relation can be standardized and tested in the laboratory.

There are many hydraulic devices of special shape and characteristics whose discharge–stage relations are precalibrated for measurements in relatively small creeks, shallow rivers, and open channels. These hydraulic devices may be categorized into three types: weirs, flumes, and orifices. Weirs and flumes are used in open channels, while orifices are associated with submerged sluices, regulated or unregulated gates, and pipe flows. The use of weirs and flumes depends on stream size, flow range, channel gradient, and sediment load. When orifices are not submerged, they may act as weirs.

B.4.3.1 Weirs

B.4.3.1.1 Features

A weir is a dam built across an open channel for flow measurements. Weirs are barriers on the bottom of stream channels that cause the overflow of water to accelerate. Components of a weir include

1. A crest of specific shape, usually high above the bottom of the stream channel, for water-head determination
2. A bulkhead to form a water pool behind the weir
3. An approach section long enough to allow water approaching the weir with a minimum velocity free from turbulence

4. Two side walls to ensure that all overland and subsurface runoff passes through the weir

5. A stilling well, equipped with an electrical water-level recorder, installed a few feet behind the weir

6. An opening beneath the crest of the weir bulkhead to sluice accumulated sediment

The discharge rate of the channel is determined by measuring the water head above the weir crest. The head can then be converted into discharge rate by a standard equation specifically calibrated for the shape and size of the weir. Weirs are not suitable for installations in channels with flat slopes or carrying excessive solid materials and sediment (Grant, 1992). Accordingly, openings beneath the crest are a mechanism to clean the accumulation of sediment, and the crest must be kept free from fibers, stringy materials, and other articles. If installed properly, weirs are the most reliable structures for measuring water in canals and ditches.

B.4.3.1.2 Types of Weirs

Weirs can be classified in several ways. By the shapes of crest, they can be rectangular, triangular (V-notch), or trapezoidal (Cipoletti). By the contact of flows to the crest, they can be sharp-crested (thin-plate weirs) or broad-crested weirs. A sharp-crested weir has a vertical flat plate with a sharpened upper edge installed along the crest so that water does not contact any part of the weir structure downstream. Broad-crested weirs have a solid broad section of concrete or other material such that water flows over the entire surface of the crest. Broad-crested weirs are not as popular as sharp-crested weirs, which can be rectangular, triangular, or trapezoidal, and are often existing structures such as dams or levees. Discharges are obtained by a calibration procedure in the field.

Classified by the crest length of the rectangular weir with respect to the channel cross-section, there are weirs without end contractions (suppressed weirs) and weirs with end contractions (contracted weirs). A suppressed weir extends the weir's crest length to the full span of the channel, while a weir with its crest length less than the width of the channel is said to have end contractions.

Classified by the downstream water level below the weir's crest, weirs are referred to as free or critical and submerged or subcritical. The sheet of water passing the weir crest is called the *nappe*. When the water level downstream from the weir is low enough to allow free air movement beneath the nappe, the underside of the nappe or water jet is exposed to the atmospheric pressure, and the flow is referred to as free or critical. This means that the weir is under free discharge conditions, which is the best condition for weir operations.

If the downstream water level rises to a height at which air does not flow freely beneath the nappe, and the nappe is not ventilated, the discharge can increase due to the low pressure beneath the nappe. When the water level downstream is above the crest, the flow is considered submerged or subcritical. This can affect the discharge rate and is not desirable for standard operations. The determination of discharge rate under submerged conditions requires measurements of both upstream and downstream heads and use of submerged weir tables.

B.4.3.1.3 Flow Rates

Weirs with sharp crests and free discharge conditions are the most popular devices for streamflow measurements. The dimensions of four sharp-crested weirs, as suggested by Grant (1992), are given in Figure B.10; their discharge equations are summarized next for illustration. Detailed derivations and discussions of these equations along with flow rates for other types of weirs can be obtained from standard texts and specialist reference books.

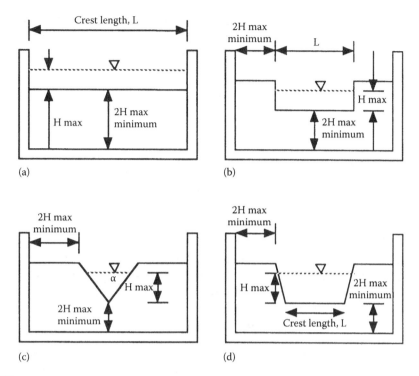

FIGURE B.10
The shapes and dimensions of weirs. (a) Suppressed rectangular weir (without end contractions), (b) contracted rectangular weir (with end contractions), (c) V-noich sharp-crested weir (triangular, α in degrees), and (d) cipolletti sharp-crested weir (trapezoidal).

For free-discharge, sharp-crest, rectangular weirs with end contractions,

$$Q = K(L - 0.2H)(H^{1.5}) \tag{B.26}$$

where
  Q is the flow rate
  L is the crest length of the weir
  H is the head of the weir
  K is a constant depending on units of measurements

In SI units, L and H are in meters, K is 1.838, to give Q in m^3/s. In British units, L and H are in feet, $K = 3.330$, and Q is in ft^3/s.
  For free-discharge, sharp-crested, rectangular weirs without end contractions,

$$Q = K(L)(H^{1.5}) \tag{B.27}$$

As in Equation B.26, $K = 1.838$ in SI units and 3.330 in British units.
  For free-discharge, sharp-crested triangular (V-notch) weirs,

$$Q = K(H^{2.5}) \tag{B.28}$$

where K is a constant, depending on the angle of notch (α) and the unit of measurement. In SI units, K = 0.3730, 0.7962, and 1.3794 for weirs with α = 30°, 60°, and 90°, respectively. In British units, K = 0.6760, 1.443, and 2.50, in that order. For free discharge, sharp-crested trapezoidal weirs,

$$Q = KL(H^{1.5}) \tag{B.29}$$

where
   Q is the flow rate (m³/s, or ft³/s)
   L is the crest length of the weir (m or ft)
   H is the head on the weir (m or ft)
   K is 1.8589 in SI units and 3.367 in British units

B.4.3.2 Flumes

Unlike weirs, which cause a considerable loss of head by damming water across a stream channel, flumes are specifically shaped devices of flow section that are installed along a stream channel for discharge measurements. The head loss of flumes is smaller than that of weirs, an advantage for channels where the available head is limited. Thus, flumes are used in channels where the use of weirs is not feasible.

B.4.3.2.1 Parshall Flumes

The Parshall flume is particularly suitable for flow measurement in irrigation canals, sewers, and industrial discharges. It consists of three sections: (1) a level, converging section upstream, (2) a constricted, downward-sloping throat section in the middle, and (3) an upward diverging section downstream (Figure B.11). The narrow, upstream, converging

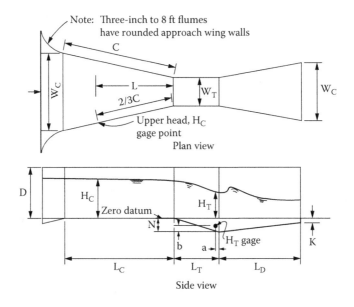

FIGURE B.11
The configuration of Parshall flumes. (Reproduced from Kilpatrick, F.A. and Schneider, V.R., Use of flumes in measuring discharge, Chapter A14 in *Techniques of Water-Resources Investigations*, U.S. Geological Survey, Book 3, 1983.)

TABLE B.3

Dimensions (m) and Capacities (m³/s) of Standard Parshall Flumes

Widths (m)			Axial Lengths (m)			Vertical Distance Below Crest (m)	
W_T	W_C	W_D	L_C	L_T	L_D	N	K
0.152	0.396	0.393	0.610	0.305	0.457	0.114	0.076
0.305	0.844	0.610	1.344	0.610	0.914	0.229	0.076
1.524	2.301	1.829	1.945	0.610	0.914	0.229	0.076
3.048	4.755	3.658	4.267	0.914	1.829	0.341	0.152
6.096	9.144	7.315	7.620	1.829	3.658	0.686	0.305

W_T	D	C	Gage Points (m)			Free-Flow Capacity (m³/s)	
			H_C	H_T, a	H_T, b	Minimum	Maximum
0.152	0.610	0.719	0.415	0.051	0.076	0.001	0.110
0.305	0.914	1.372	0.914	0.051	0.076	0.003	0.456
1.524	0.914	1.981	1.320	0.051	0.076	0.045	2.422
3.048	1.219	2.743	1.829	—	—	0.170	8.490
6.096	2.134	4.267	2.844	—	—	0.283	37.922

Source: Kilpatrick, F.A. and Schneider, V.R., Use of flumes in measuring discharge, Chapter A14 in *Techniques of Water-Resources Investigations*, U.S. Geological Survey, Book 3, 1983.

Note: H located 2/3 C distance from crest for all sizes; distance is wall length, not axial. Flumes size 0.076–2.438 m have approach aprons rising at 25% slope and the following entrance roundings: 0.152 m, radius = 0.405 m; 0.305 m, radius = 0.509 m; 1.524 m, radius = 0.610 m. W_T = throat width, W_C = width of upstream end, W_D = width of downstream end, L_C = axial length of converging section, L_T = axial length of throat section, L_D = axial length of diverging section, N = vertical distance below crest, dip at throat, K = vertical distance below crest at lower end of flume, D = wall depth in converging section, C = converging wall length, H_C = gage point located at wall length upstream of crest, H_T = head at the throat section.

section of the flume accelerates the entering water, eliminating the effect of sediment deposition on measurement accuracy. The throat section induces greater flow velocity and a differential head that can be related to discharge. A downward-sloping floor in the throat gives the flume its ability to withstand relatively high degrees of submergence without affecting the flow rate. The diverging section below the throat assures lower water level in the downstream section than in the converging section.

The size of Parshall flumes is designated by the throat width (W_T), and standard dimensions are given by the USGS for sizes ranging from 2.54 cm (1 in.) to 15.24 m (50 ft) (Kilpatrick and Schneider, 1983). A few sizes with their dimensions corresponding to the letters in Figure B.11 are given in Table B.3. The determination of discharge rate depends on flow conditions. For free-flow conditions in which there is insufficient backwater depth downstream to reduce the discharge rate, only one head reading at the upstream gaging location (H_C) is needed to measure and use this head reading to determine discharge rate from standard tables or equations. Hydraulic jumps or standing waves occurring downstream from the flume are an indicator of the free-flow condition. The standard equations for the three throat sizes are given in Table B.4, in both British and SI units.

TABLE B.4

Discharge Q as a Function of Throat Width W_T and Flow Head H_C for Parshall Flumes under Free-Flow Conditions

Throat Width	SI Units (m, m³/s)	British Units (ft, ft³/s)
0.15 m (6 in.)	$Q = 0.381(H_C)^{1.580}$	$Q = 2.060\ (H_C)^{1.580}$
0.30–2.44 m (1–8 ft)	$Q = 4[0.3048]^{(2-1.57X)}\left[W_T H_C^{1.57X}\right]$	$Q = 4(W_T)(H_C)^{1.52X}$
3.05–15.24 m (10–50 ft)	$Q = [2.2927W_T + 0.4738]H_C^{1.6}$	$Q = [3.6875W_T + 2.5]H_C^{1.6}$

Note: $X = W_T^{0.026}$.

For submerged flows in which the water surface downstream from the flume is high enough to cause a backflow and a reduction in discharge rate, the determination of flow rate requires both depth measurements at upstream gaging site (H_C) and in the throat (H_T). The ratio of the downstream depth to the upstream depth (H_T/H_C) is used to determine the two flow conditions and to make a correction from standard nomographs and equations (USDA, 1979; Bureau of Reclamation, 2001).

B.4.3.2.2 HS, H, and HL Flumes

These flumes were developed by the U.S. Soil Conservation Service in the 1930s to measure runoff from small watersheds and experimental plots. They have a flat bottom; the vertical sidewalls are converged toward the control opening, and the converging sidewalls slant upward from the lip of the outlet to form a trapezoidal shape for the opening and the sidewalls (Figure B.12). The flat bottom allows a better passage of sediment than in weirs, the converging sidewalls accelerate flow velocity, and the trapezoidal opening (narrow at the bottom and wide and pitched toward upstream at the top) provides a capacity to measure the full range of a flow event. Thus, they combine the sensitivity and accuracy of sharp-crested weirs with the self-cleaning features of flumes.

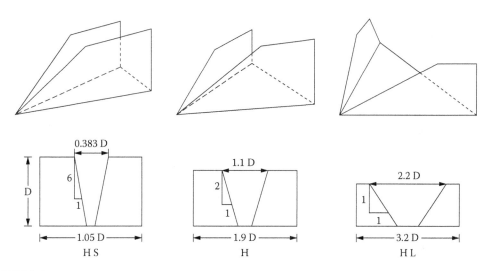

FIGURE B.12
The HS, H, and HL flumes (side views and front views).

The three H-type flumes are designed to measure different flow capacities. The HS flumes are for relatively small flows (maximum flow rates less than $0.0232\,\mathrm{m^3/s}$), the H flumes for medium flows (maximum flow rates ranging from 0.0099 to $0.8773\,\mathrm{m^3/s}$), and the HL flumes for larger flows (maximum flow rates ranging from 0.5858 to $3.3111\,\mathrm{m^3/s}$). Each of the flumes is described for its maximum depth (head) attainable in the flume. A 3 ft (0.9144 m) H flume has a maximum head of 3 ft, and its maximum flow rate is $30.70\,\mathrm{ft^3/s}$ ($0.8688\,\mathrm{m^3/s}$). The standard dimensions for each type of flume and the rating tables relating water heads in the flume to discharge rates are given in USDA (1979).

On installation, each flume is attached with a stilling well for water-level measurements. The approach section (channel) should have the same depth and width as the flume and a length three to five times the depth of the flume. The water entering the flume must be subcritical with no turbulence, and the water leaving the flume should be free in falling downstream.

B.4.3.3 Orifices

An orifice used for discharge measurements is a sharp-edged opening in a wall, bulkhead, or barrier in a stream through which flow occurs under pressure; it can be a hole in a detention pond as outlet controls (Figure B.13). Orifices can be used to measure flow rate when the size and shape of the openings and the heads acting upon them are known. The discharge of water from the orifice can be either free into the air or submerged under the water. The submerged orifice causes no head loss in installation and is therefore used where there is insufficient fall for a weir, the cost is unjustifiable for a Parshall flume, or there is a need for small or controlled discharges.

If a small opening is made under a water tank 1 m below the water surface, water will spout from the opening with a velocity of 4.427 m/s, equal to the terminal velocity of a free-falling rock from 1 m in the air. It would be 6.261 m/s if depth were 2 m below the surface. The velocity (V, m/s) increases with depth H by this relationship.

$$H = \frac{V^2}{2g} \tag{B.30}$$

where
  g is the acceleration due to gravity ($9.80\,\mathrm{m/s^2}$)
  H is measured from water surface to the center line of an orifice (m)

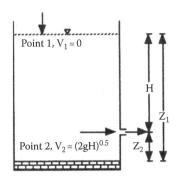

Based on Bernoulli's equation
at points 1 and 2:

$$P_1 + V_1^2/(2g) + Z_1 = P_2 + V_2^2/(2g) + Z_2$$

The pressure head $P_1 = P_2$
    = The atmospheric pressure
The velocity head at point $1 = V_1^2/(2g) = 0$
  The $Z_1 = V_1^2/(2g) + Z_2$
      $V_2^2/(2g) = Z_1 - Z_2 = H$
      $V_2 = (2gH)^{0.5}$

FIGURE B.13
The velocity for a submerged orifice.

With flow velocity V (m/s) and orifice area A (m^2), the discharge Q (m^3/s) for a submerged orifice under approach-velocity-negligible condition can be calculated by

$$Q = AV = C_dA \, (2gH)^{0.5} \qquad (B.31)$$

where C_d is the discharge coefficient due to water viscosity and orifice characteristics such as shape, size, and edge. Values of C_d range from 0.5 to 1.0, and a value of 0.60 is used for most applications in the United States. If the velocity of approach is significant (V_1 in Figure B.13), the head due to velocity of approach (h) should be included in the calculation, or

$$Q' = C_d \, A \, [2g(H + h)]^{0.5} \qquad (B.32)$$

The value h is obtained first by calculating velocity of approach V (V = Q/A, where Q is from Equation B.31 and A is cross-sectional area of approach pipe in m^2), then using V to calculate h (h = V^2/(2 g)). The value h is then used in Equation B.32 to solve for Q'.

B.5 USGS Streamflow Measurements

The USGS is the federal agency responsible for collecting and assembling streamflow data (quantity and quality) over the entire nation for flood prediction, water management and allocation, engineering design, operation of locks and dams, and recreational safety and enjoyment. USGS started its first streamgage in 1889 on the Rio Grande River in New Mexico and now operates over 8000 streamgages nationwide (Figure B.14). The streamflow data collected are available to the general public; they are very valuable to forest hydrology research and watershed management planning.

FIGURE B.14
Location map of 8168 stream gaging stations operated by the USGS. (Courtesy of Estifanos, Medhin.)

B.5.1 USGS Streamgages

There are five types of gages in use today as follows:

- *Staff gages*: The height of water in a stream is manually recorded in a definite or indefinite period of time. They also can be used as stream depth awareness to the public.
- *Crest stage gages*: They provide information on the highest flow since the gages were last visited.
- *ALERT (automated local evaluation in real-time) streamgages*: They are designed to send a warning when the water level reaches a predetermined level or changes rapidly.
- *Partial record streamgages*: They are those operated and quality assured only for given flow regimes such as high flow, peak flow, low flow, etc. The period (or date) of the quality assured flow value(s) may be variable from one year to another, or only during a particular season of the year.
- *Real-time streamgages*: They are designed to measure river stages to 0.01 in. (0.254 mm) at 15–60 min intervals, stored onsite in an electric data recorder, and then transmitted to USGS offices every 1–4 h, depending on the data relay technique used. Recording and transmission times may be more frequent during critical events, such as 5 min. Data from real-time sites are relayed to USGS offices via satellite, telephone, and/or radio telemetry. They are available for viewing within minutes of arrival. The majority of the USGS streamgaging stations is operated with the real-time devices (Figure B.15).

B.5.2 Streamflow Data

The USGS streamflow data are public accessible through the National Water Information System (NWIS) (http://waterdata.usgs.gov/nwis/sw). They include

FIGURE B.15
The USGS real-time stream gaging station at La Nana Creek, Nacogdoches, TX.

- *Real-time data*: They are time-series data from automated equipment and represent the most current hydrologic conditions. Measurements are commonly recorded at 15–60 min intervals and transmitted to the NWIS database every 1–4 h. Real-time data are available online for 120 days.

- *Daily data*: They are summarized from real-time data for each day and for the period of record, presented as the daily mean, median, maximum, minimum, and/or other derived values.

- *Statistics*: They are computed from approved daily mean time-series data at each site. These links provide summaries of approved historical daily values for daily, monthly, and annual (water year or calendar year) time periods.

- *Peakflow data*: They are the annual maximum instantaneous peak streamflow and gage height at each site.

References

Alexander, J.S., Zelt, R.B., and Schaepe, N.J., 2009, Geomorphic segmentation, hydraulic geometry, and hydraulic microhabitats of the Niobrara River, Nebraska—Methods and initial results, U.S. Geological Survey Scientific Investigations Report 2009-5008, Reston, VA.

Allen, P.M., Arnold, J.G., and Byars, B.W., 1994, Downstream channel geometry for use in planning-level models, *Water Resour. Bull.*, American Water Resources Association, 30(4), 663–671.

Bradner, M. and Emerson, L., 1988, Flow measurement, *Fluid Mechanics Source Book*, Parker, S.P., Ed., McGraw-Hill, New York, pp. 203–221.

Bureau of Reclamation, 2001, *Water Measurement Manual*, 3rd edn, U.S. Department of Interior, Denver, CO.

David, C.C.L. et al., 2010, Controls on at-a-station hydraulic geometry in a steep headwater streams, Colorado, USA, *Earth Surf. Process. Landforms.*, 35(15), 1820–1837.

Gordon, N.D., McMahon, T.A., and Finlayson, B.L., 1993, *Stream Hydrology*, John Wiley & Sons, New York.

Grant, D.M., 1992, *ISCO Open Channel Flow Measurement Handbook*, 3rd edn., Isco, Inc., Lincoln, NE.

Hirsch, R.M. and Costa, J.E., 2004, Stream flow measurement and data dissemination improve, *EOS*, 85(20), 197, 203.

Kilpatrick, F.A. and Cobb, E.D., 1985, Measurement of discharge using tracers, Chapter A16 in *Techniques of Water-Resources Investigations*, U.S. Geological Survey, Reston, VA, Book 3.

Kilpatrick, F.A. and Schneider, V.R., 1983, Use of flumes in measuring discharge, Chapter A14 in *Techniques of Water-Resources Investigations*, U.S. Geological Survey, Reston, VA, Book 3.

Laenen, A., 1985, Acoustic velocity meter systems, Chapter A17 in *Techniques of Water-Resources Investigations*, U.S. Geological Survey, Reston, VA, Book 3.

Lane, L.J. and Foster, G.R., 1980, Modeling channel processes with changing land use, *Symposium on Watershed Management 1980*, Boise, ID, ASCE, Reston, VA, pp. 200–214.

Leopold, L.B. and Maddock, T.J., 1953, Hydraulic geometry of stream channels and some physiographic implications, U.S. Geological Survey Professional Paper 252.

Simpson, M.R. and Oltmann, R.N., 1993, Discharge-measurement system using an acoustic Doppler current profiler with applications to large rivers and estuaries, U.S. Geological Survey Water-Supply Paper 2395.

Spence, C. and McPhie, M., 1997, Streamflow measurement using salt dilution in tundra streams, Northwest Territories, Canada, *J. Am. Water Resour. Assoc.*, 33, 285–291.

USDA, 1979, *Field Manual for Research in Agricultural Hydrology*, Agriculture Handbook No. 224, U.S. Department of Agriculture, Washington, DC.

U.S. Forest Service, 1984, Coweeta Hydrologic Laboratory: A guide to research program, SE Forest Experiment Station, Asheville, NC.

USGS, 1980, *National Handbook of Recommended Methods for Water-Data Acquisition*, U.S. Department of Interior, Reston, VA, pp. 1–130.

Vogel, S., 1994, *Life in Moving Fluids, the Physical Biology of Flow*, 2nd edn., Princeton University Press, Princeton, NJ.

Wilm, H.G. and Storey, H.C., 1944, Velocity-head rod calibrated for measuring streamflow, *Civil Eng.*, 14, 475–476.

WMO, 1980, Manual on stream gauging, Operational Hydrology Report No. 13, WMO-No. 519, World Meteorological Organization, Geneva, Switzerland.

Appendix C: Measurements of Stream Sediment

Stream sediment, or fluvial sediment, refers to the fragmental material and particles that are retained and transported in stream channels. These materials are derived primarily from the physical and chemical disintegration of rocks and delivery to stream channels via various erosion phases including raindrop splash, interrill, rill, ephemeral gullies, gullies, channel, and mass movement. Sediment information is required for

1. Evaluation of land-use management
2. The design of hydraulic structures
3. Water-quality management
4. Water utilization
5. Aquatic-life management

Sediments in streams appear as washload, suspended load, and bedload. Washload and suspended load are those fine sediments that are transported in streams with no direct contact with the streambed for an appreciable length of time, while bedload consists of those particles rolling and bouncing along the streambed. In measurement, sediments in suspended load and washload are separated from bedload.

C.1 Suspended Sediment

Suspended sediments are supported by the fluid and carried along above the layer of laminar flow. They have settling velocities less than the buoyant velocity of the flow and stay in suspension in the water for an appreciable length of time. Suspended sediments in the stream are measured in terms of sediment concentration in milligrams per liter (mass/volume of streamflow) or in terms of turbidity units such as nephelometric turbidity unit (NTU) or formazin turbidity unit (FTU). Sediment concentration is based on water samples collected in the field and gravimetrically determined in the laboratory. The sediment concentration is then weighted by the flow rates measured simultaneously to obtain sediment yield (or discharge) in kilograms per day (or other units in mass/time). Finally, the sediment yield in mass/time is divided by the watershed area to give the result in mass/time/area for comparisons with other studies. Turbidity is an optical unit measuring clarity or the passage of light through the stream.

C.1.1 Sediment Variation

The concentration of suspended sediment varies with depth and location on a stream cross section, site of a given stream, and time. It is affected by flow velocity, turbulence, particle size and shape, channel geometry, physical and chemical properties of water, watershed soil and topographic characteristics, and climatic conditions. The variation is greater in small watersheds than in large watersheds because of small watersheds' sensitivity to storm characteristics. For a given cross section, the greatest concentration is near the bed, and the concentration decreases to a minimum at the surface. Concentration is greater in the segments where velocity is high and depth is low. During high flows, total suspended sediment can be greater than that in low flows by a factor of more than 1000 (Bonta, 2000). Consequently, the majority of annual sediment yield can be attributable to only a few major storms (McBroom et al., 2003; Beschta and Jackson, 2008). A 200 km² watershed in Wisconsin, cited by Garde and Ranga Raju (1985), generated 90% of the total sediment load of a 15 month period in 10 d or 2.2% of the time. Likewise, a 7 year study on the River Creedy in Devon, United Kingdom, showed that 80% of the sediment load was carried in 3% of the time (Walling and Webb, 1981).

C.1.2 Sampling

C.1.2.1 Site Selection

Sampling sites are selected with consideration of the data needs and the nature of the flow and channel conditions. Because streamflow discharge is also required in most sediment measurements, existing stream-gaging stations are logical sites for sampling suspended sediments. At such stations, samples should be taken upstream of still ponds (Gordon et al., 1992). If no suitable gaging station already exists, then the sampling site should be located in the middle section of a straight reach of length at least four times the width of the channel, but no less than 150 m (Singhal et al., 1981). In mountainous regions where straight reaches are difficult to find, a length of 50 m should be acceptable.

Flow conditions at sites such as those downstream or upstream from the confluence of two streams can be disturbed by unusual water-surface slopes, lateral flows, back flows, and mixing effects. These conditions will likewise affect the normal movement of sediment and may require additional sediment measurements. Also, sites should not be adjacent to hydraulic structures. A measuring section at a bend should be avoided because of uneven distribution of stream energy. Likewise, sites with unstable cross sections, such as those with active bank erosion, silting, and scouring, should not be used for measurements. Because more samples are required during periods of high flows, it is imperative that sites be accessible during times of flooding.

If samples are required for the assessment of land-treatment effects, then the sampling sites are obvious. They should be close to the treatments, both upstream and downstream locations. Locating too far from the treatment sites can cause the dilution effect of streamflow to be significant. Changes in hydraulic and channel conditions between sites increase with distance.

C.1.2.2 Number of Verticals

The number of verticals required in sampling suspended sediment along a stream's cross section depends on the type of information needed, the degree of accuracy required, and the physical aspects of the river. It may be determined based on statistical approaches or field experience. The number has ranged from a single vertical for small streams to more than 15 verticals for streams with great sediment variations. If only the smaller fractions such as silts and clays are of concern, then one grab sample taken near the water surface in the center of the stream is considered sufficient (Gordon et al., 1992). Sampling at a single vertical is also adequate if a stream's cross section is stable and its lateral suspended-sediment distribution is rather uniform. If the distribution of sediment concentration or particle size across the stream is important, then samples are needed at several verticals.

For sand-bed streams, the bed configuration can vary significantly across and along the stream, as can the sediment concentration. Such streams may require more sampling verticals to estimate the sediment concentration of the cross section. Samplings at 6–12 verticals are usually sufficient to produce a reliable estimate of discharge-weighted suspended sediment for most small streams (USDA, 1979).

C.1.2.3 Sampling Frequency

Suspended-sediment concentration (SSC) in a given stream is highly variable with time. It is often much higher during floods, on the increasing discharge during a flood (the rising limb of the hydrograph), and for storms proceeded by a long, dry period (MacDonald et al., 1991), but is lower for runoff generated from snowstorms than from rainstorms. Suspended-sediment sampling is therefore required more frequently during these high-concentration periods and for streams with greater sediment variation. Because of the great temporal variation of suspended sediments in streams, an adequate sampling program may need three different schemes:

1. A routine scheme for sampling baseflows
2. A special scheme for sampling storm runoff
3. A stratification scheme to adjust the other two schemes for meeting special conditions and needs

The routine sampling scheme is commonly on a weekly basis, but may be as short as a day or other longer intervals. This scheme is set up in compliance with the general monitoring program of the region, flow conditions, and the available resources. However, the routine sampling is a biased estimate of stream sediment and may cause underestimates of up to 50% (Ferguson, 1986). In France, Coynel et al. (2004) showed that the sampling frequency for suspended matter must be at least one in 3 days for a large watershed and every 7 h for a small watershed for an error of less than 20%. Thus, a special, intensified scheme must be taken to sample sediments during floods.

The sampling frequency should be short enough to take samples along the rising limb, at or near the crest, and in the recession limb of the flood hydrograph. In general, more samples should be obtained on and near the crest, and the time intervals should be shorter on the rising stage than on the recession stage and in small streams than in large streams. If automatic water samplers are used for the sampling, the sampler can be programmed to pump water samples based on increments of flood stage, discharge rate, or time intervals.

The stratification scheme of sampling improves the accuracy of the sediment estimates by adjusting the frequencies used in the routine and flood-sampling schemes for special environmental conditions. For example, in regions where snowstorms dominate for an extended period of time, or hurricane season often brings substantial amounts of intense rainfall, it becomes necessary to set up sampling programs to reflect different sampling frequencies or initial baseflow stages between seasons.

C.1.2.4 Types of Samples

There are five types of samples taken from streams for sediment-concentration analyses: (1) depth-integrated samples, (2) point-integrated samples, (3) grab samples, (4) discrete samples, and (5) composite samples. Depth-integrated samples, most common in routine suspended-sediment measurements, are the samples integrated from the water surface to about 10 cm above the streambed by use of a depth-integrating sampler. The sample covers the total suspended sediment over the entire vertical, except the 10 cm layer near the bottom that cannot be reached due to the height of the sampler. Point-integrated samples are water-sediment samples that are collected by a point-integrating sampler at a specific depth in the stream. Usually, a number of point-integrated samples are collected to define the sediment-concentration distribution in a vertical. If the velocity distribution in the vertical is measured simultaneously, then the sediment concentration versus flow velocity profile can be defined, and sediment load can be calculated. The point-integrating method is slow and is therefore used in large streams where water is too deep or too swift for use of the depth-integrating sampler.

In small headwater streams where sediment is well mixed and dominated by clays and silts, water samples can be grabbed by lowering an open container into the water. The grab sample represents an instantaneous sample at one point in the stream. If sand is in suspension, the grab sampler must lower to near the streambed (UNESCO, 1982). A *discrete* water sample is the sample collected at a specific time during a storm runoff event. The combination of a few discrete samples collected during a storm runoff event, in the same quantity, into a single sample creates a *composite* water sample. Also, depth-integrating samples from the same stream cross section can be combined into a single sample for laboratory analyses if they are collected at the same transit rate at verticals of equal width.

C.1.2.5 Procedures for Depth-Integrated Samplers

In this method, the sampler is lowered from the water surface to the streambed and raised to the surface at a uniform speed. This makes the sampling rate at any point in the vertical proportional to the stream velocity and produces a sample whose concentration is velocity weighted throughout the entire depth. The location of the sampler in the cross section depends on the number of verticals. If only one vertical is used in the measurement, it is located either at midstream or at the point of greatest depth. If multiple verticals are used, then the location of each vertical is determined either by the equal-discharge-increment method (EDI) or by the equal-transit-rate method (ETR).

C.1.2.5.1 EDI Method

The EDI method first determines the number of verticals required in the measurement. The entire cross section is then divided into verticals of equal discharges, and the boundaries (locations) of each vertical are identified. Water samples are obtained at the centroid (50% discharge rate) of each vertical. Thus, some knowledge of the streamflow discharge distribution is required in this method, and the channel cross section is preferably uniform and stable.

If prior knowledge of stream discharge is not available for the use of the EDI method, sequential discharge measurements across the stream need to be made. The water discharges are then accumulated from the initial point of measurement to the other side of the stream. Next, the accumulated discharges of each sequential measurement are expressed as a percentage of the total discharge and plotted against distance from the initial point on a graph. The plotting can be used to determine the locations of each equal discharge vertical along the stream's cross section (Guy and Norman, 1970).

Based on data given in Table B.1, Figure C.1 is a plot of the cumulative discharges expressed as percentages of the total against distances from the initial point. If four verticals are used in the measurements, each vertical should have 25% of the total discharge,

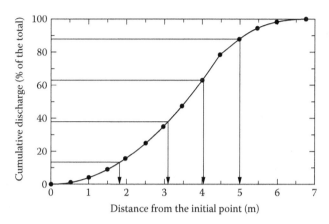

FIGURE C.1
Determination of verticals' locations and boundaries in the EDI method.

and the three boundaries separating these four verticals are located at approximately 2.5, 3.6, and 4.4 m from the initial point. Water samples are taken in the median discharge of each vertical, or at 12.5%, 37.5%, 62.5%, and 87.5% of the total discharge located at 1.8, 3.1, 4.0, and 5.0 m from the initial point, respectively.

C.1.2.5.2 ETR Method

The ETR method, more widely used than the EDI method, obtains samples at equally spaced verticals with a volume proportional to the amount of discharge at each vertical. The transit rate of the sampler should be at a uniform speed within a vertical and among all verticals. This gives a discharge-weighted sample for the flow cross section, and all samples from a cross section can be composited to form a single sample for laboratory analyses.

C.1.2.6 Procedures for Point-Integrating Samplers

Point-integrating samplers are equipped with a valve that can open and close to trap a sample at a specific depth in the stream. The sampling process also requires a measurement of the stream velocity at the sampling point so that the sediment discharge can be calculated. This method is generally used in streams that are too deep (>5 m) or too turbulent for use of a depth-integrating sampler.

Both the EDI and ETR methods can be used for point-integrating samplers to determine the location of verticals on the cross section. The number of sampling points required in each vertical can be determined by one-point, two-point, three-point, and five-point methods. In the one-point method, the sample is taken at the surface or at 0.6 times the depth below the surface (0.6 d). One-point sampling may be sufficient in small streams or necessary during high flood stages. Singhal et al. (1981) stated that the average SSC (ppm) in India lies at 0.5 d in rivers and 0.54 d in canals, and is about 2.353 times the sediment concentration at the surface.

When streams are shallow, samples are taken at two points in the vertical, that is, 0.2 and 0.8 d from the surface. For deep streams, three points (at surface, bottom, and mid-depth) or five points (at surface, 2, 6, and 8 d, and bottom) may be required in routine measurements. It is a common practice to determine the number of points required in the sampling through a trial procedure. A sufficient number of points are sampled so that the vertical distribution of sediment concentration at various flow stages can be defined. Once the vertical-distribution profiles of the suspended sediment are available, the most practical ways to estimate the mean sediment concentration in the vertical can be determined.

C.1.3 Measuring Devices

All devices for sampling the SSC can be separated as either hand operated or automatic. The techniques of remote sensing used to estimate suspended sediment are beyond the scope of this book.

C.1.3.1 Hand-Operated Samplers

Hand-operated samplers are of two types: depth-integrating and point-integrating. These samplers are operated by hand either for wading measurements in streams or for measurements by a hand line suspended from a bridge or from a reel on a cable car.

The most common depth-integrating samplers used in the United States are the U.S. DH-48 (Figure C.2), the U.S. D-49, and the U.S. DH-59, while the point-integrating samplers are the U.S. P-61 (Figure C.3) and the U.S. P-63. They were all developed by the Federal Inter-Agency Sedimentation Project, St. Anthony Falls Hydraulic Laboratory in Minnesota (USDA, 1979). The depth-integrating samplers have an open nozzle pointing upstream and a one-pint milk bottle for sample collection. Air is exhausted from the sample-collection bottle to admit the water–sediment mixture at stream velocity. The sampler is lowered from the water surface to the streambed and raised back to the surface at a uniform rate, and the collected sample is velocity weighted throughout the entire vertical. The point-integrating samplers are similar to the depth-integrating samplers in design except that the nozzle is equipped with an electric remote-controlled valve. Thus, the samplers can collect point-integrated samples at a designated depth in the stream by controlling the valve; they can collect depth-integrated samples if the nozzle is left open continuously.

Hand-operated samplers can provide a satisfactory estimate of the SSC in a cross section, if sampled properly. But it is very difficult for on-site visits to cover the wide range of

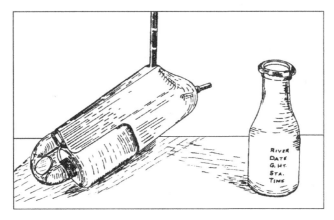

FIGURE C.2
U.S. DH-48 depth-integrated suspended-sediment sampler.

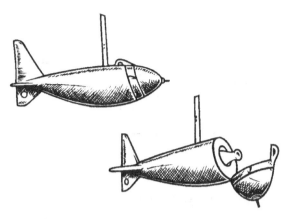

FIGURE C.3
U.S. P-61 point-integrated, suspended-sediment sampler.

flow conditions, and these samplers cause significant underestimates of sediment yields. Although the timing problem can be overcome by automatic samplers, they too have weaknesses (Thomas, 1985).

C.1.3.2 Automatic Samplers

C.1.3.2.1 Mechanical Samplers

Two major types of automatic samplers use mechanical principles to collect water samples: Coshocton-wheel runoff samplers and rising-stage samplers. The Coshocton-wheel runoff samplers were developed by the former Soil Conservation Service in the 1950s to sample runoff from experimental plots and small watersheds. They are installed, slightly inclined, below an H flume. The discharge jet from the H flume falls on the slightly tilted wheel, causing the sampler to rotate automatically. As the wheel passes the flow jet, the elevated slot mounted on the wheel extracts a small fraction of runoff, and the aliquot sample is routed through the base of the wheel to a storage tank (Figure C.4).

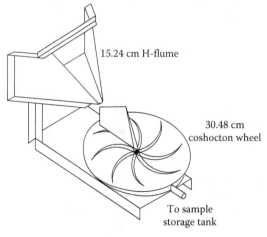

15.24 cm H-flume

30.48 cm coshocton wheel

To sample storage tank

FIGURE C.4
Coshocton N-1 runoff sampler.

TABLE C.1

Size Schedule for Coshocton-Wheel Runoff Samplers

Model	H flume (m)	Wheel Diameter (m)	Capacity to Sample Peak Runoff (m³/s)	Headroom Requirement (m)	Aliquot (%)
N-1	0.152	0.305	0.0094	0.457	1.0
N-2	0.305	0.610	0.0566	0.762	0.5
N-3	0.457	0.914	0.1557	1.143	0.3

Source: Parsons, D.A., Coshocton-type runoff samplers, laboratory investigations, USDA, Soil Conservation Service, SCS-TP-124, 1954.

The sampler aliquot is the ratio between the slot width (w) and the circumference ($2\pi r$) of the wheel at any radius (r), or

$$k = \frac{w}{2\pi r} \tag{C.1}$$

where k is the aliquot, independent of the average turning rate of the wheel. Thus, the slot width w can be calculated for a given k value designated for different models. Three basic models, N-1, N-2, and N-3, have been designed for use with 0.5, 1.0, and 1.5 ft (15.2, 30.5, and 45.7 cm) H flumes. The diameter, maximum peak flow capacity, and aliquot of the three models are given in Table C.1. Model N-1 has been used extensively in plot studies, while model N-2 is popular for studies in small creeks.

The single, rising-stage suspended-sediment sampler, developed by the Federal Inter-Agency Sedimentation Project, operates on the siphon principle to collect water samples automatically from small streams and at remote sites. The sampler consists of a 1 pt glass bottle to sample at only one depth level, or several bottles mounted vertically on a suitable support to collect samples at several stages as the water rises. Each bottle is fitted with a two-hole stopper—one hole for intake tubing and the other for air-outlet tubing. The intake tubing is connected to a board and pointed into the flow at a designated elevation, which must be higher than the sampler. When water rises to the designated elevation, the siphon effect causes water to flow into the sampler.

Since samples are taken at the water surface, the sampler is suitable for sampling suspended sediment finer than 52 μm (silts and clays). It is inadequate for sampling larger particles because they are more concentrated at the lower levels, and the intake velocities are not the same as the flow velocities. The intake tubing is also susceptible to plugging by debris near the surface.

C.1.3.2.2 *Electrical Pumping Samplers*

For flash floods in small streams, peak flows in large streams, and streams in remote areas where on-site visits are not readily available, pumping-type samplers are often used to obtain water samples from one fixed point in the stream cross section. The major components of these samplers include an intake nozzle or strainer located in the stream, a power supply, a pumping and purging system, a sample handling system, and a control unit. The sampler is usually activated by a float-controlled switch or by a water-pressure sensor when the water stage reaches a predetermined elevation (or volume). Samples are pumped into containers (usually 473 [1 pt milk bottle], 500, or 1000 mL) at regular, predetermined time intervals. The sampling stops when the water level (or volume) recedes or the capacity of the sample handling system is reached, usually 28, 24, or 12

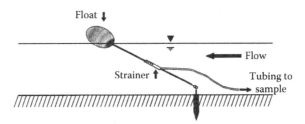

FIGURE C.5
Sample setup to keep the strainer always at 0.5 d of the vertical.

bottles. Before a sample is pumped in, the intake system is purged a few flushes to clear the line of any debris and sediment that may have accumulated in the system.

Since the sample strainer (intake nozzle) is often located at one depth (point) near a stream bank, the correlation between the point measurement and the mean concentration of the cross section must be calibrated by conventional measurements at different flow stages. To maintain the sampling point at 0.5 d as the mean concentration of the vertical, a free-swinging rod with length equivalent to the full depth of the stream can be secured to the streambed on one end and attached to a float on the other end. The strainer is fastened firmly in the middle of the rod. As the water stage rises, so does the rod with float. This enables the strainer to keep at the 0.5 d (Figure C.5). To relate measurements from one vertical to the cross section, a calibration procedure must be taken so that a more appropriate location can be identified or a correction factor can be applied.

Several electrical samplers have been developed by the Federal Inter-Agency Sedimentation Project. They are discussed in the *Agricultural Handbook No. 224* (USDA, 1979). Some private industries also have developed models that combine a pumping sampler and flow meter into a compact, single unit with computer-processing capability. Those models can be programmed, based on water stages, flow volume, and time intervals, to monitor discharge rates and pump water samples simultaneously following the progress of a storm runoff hydrograph.

C.1.3.2.3 Optical Turbidity Systems

The measurement of SSCs requires lengthy field work and laboratory analysis. Also, the temporal variation of suspended sediments is not easy to monitor in small mountainous watersheds where SSC and streamflow can change rapidly. It is then logical to use turbidity as a surrogate for SSC measurements. Turbidity is relatively easy to measure in the field or laboratory, more cost effective, and less time consuming (Spear et al., 2008).

Measurements—In optical turbidity systems, a light is scattered, absorbed, and transmitted as it passes through water. The scatter and absorption of light in water, due to the presence of fine suspended particles and other materials, such as organic matter, plankton, and microorganisms, cause the water to be cloudy and turbid. Other factors, such as the shape and color of particles and air bubbles entrained in the water, also significantly affect light absorption and scattering. Thus, *turbidity* refers to the relative amount of light that is scattered or absorbed by a fluid. It is an optical property of the fluid and can be measured by turbidity meters on water samples or monitored continuously by a turbidity system.

Units—Turbidity is usually measured in NTU or FTU, depending on the method or instrument used. The NTU is measured with an electronic nephelometer. Water sample is placed in a standard container. A light beam passes through the sample and strikes a sensor on the other side of the container and a second sensor is mounted at right angles to the beam. These two sensors measure light scattered by particles in the sample, and the ratio, with reference to a standard, between the light intensities at the two sensors is used to determine the turbidity in NTU.

The measurement in FTU uses absorptiometric methods with a spectrophotometer. To calibrate the devices, a primary standard solution is needed, formazin polymer or sty-rene divinylbenzene polymer for NTU and formazin polymer for FTU. The turbidity for forested streams in mountainous areas is about 1 NTU, and it is around 10 NTU for the Mississippi River under dry-weather conditions. These two units, NTU and FTU, are roughly equivalent. The Jackson turbidity unit is no longer in use today.

Instruments—A continuous turbidity-monitoring system usually consists of a pumping and recirculating system for water samples from the stream, sedimentation chamber, and detector and recording devices (Grobler and Weaver, 1981; WMO, 1981). Some turbidimeter systems are able to link a number of sensors into a high-level, two-way communications network that provides real-time monitoring from a central station (HACH, 1998). Also, multiparameter instruments with submersible sondes that can accommodate a turbidity sensor, commonly referred to as a water-quality probe or turbidity probe, can be used in conjunction with a data logger for in situ measurements. Most multiparameter instruments are microprocessor-based, with the calibration parameters stored in instrument memory. The standards for turbidity are user-selectable in some probes, but are internally established and cannot be changed in some others. Detailed information on turbidity measurements is given by the USGS (Wilde and Radtke, 1998).

Turbidity: SSC Relations—Studies on turbidity and SSC have shown that meaningful relationships can be established in some areas (Kunkle and Comer, 1971; Marquis, 2005), but not in all (Weigel, 1984; Williamson and Crawford, 2011). Such relationships, if established, vary by site, within storms, and between storms (Beschta, 1980; Gao et al., 2008). In the Pacific Northwest, about 80% of the variability of the SSC is attributable to turbidity (MacDonald et al., 1991). There, turbidity was found to be a better surrogate than discharge for estimating SSC in Oregon (Uhrich and Bragg, 2003), and the root mean square errors were 1.9%–7.7% for SSC–turbidity and 8.8%–23.2% for SSC–discharge relationships in the Caspar Creek Experimental Watersheds in northern California (Lewis, 1996). Regulatory agencies, researchers, and land owners in California have adopted turbidity monitoring to determine impacts of forest activities on suspended-sediment loads and water quality at various scales (Harris et al., 2007).

The variation in sediment–turbidity relationships is probably due to the nature of turbidity and its contributing sources. Other factors such as particle shape, organic compounds, algae, color, and phytoplankton also affect the scattering and absorption of light in water. Organic substances in streams vary in seasons, between the rising and recession stages in runoff events, within a watershed, and among watersheds. Prediction equations should therefore be developed on a site-by-site basis.

C.1.3.2.4 *Acoustic Doppler Current Profiler*

With the acoustic Doppler current profiler (ADCP), sound bursts are transmitted into the water column; echoes are produced when the sound bursts hit suspended particles carried by water currents. The ADCP senses simultaneously in four orthogonal directions in which particles moving toward the instrument record greater frequencies than those moving away from the instrument. Lag times and frequencies of these echoes due to the Doppler shift are detected by the monitoring system to estimate suspended-solids concentration. ADCPs have the ability to measure flow velocities and suspended-solids concentrations throughout the water column and across the stream channel.

C.1.3.2.5 Radioactive Turbiometric Gages

In streams with sediment concentrations in excess of $100\,g/L$, nuclear gages are the most suitable means for in situ measurements. These gages can monitor a wide range of sediment concentrations and resist bedload transport consisting of boulder-size material. They are fully automatic, are independent of the color and diameter of the suspended particles, and practically do not disturb the fluid (Tazioli, 1981).

Nuclear suspended-sediment gages measure the attenuation of radiation intensity (gamma rays) either backscattered or absorbed by the water–sediment mixture. An artificial source of radiation (^{137}Cs, ^{60}Co, ^{109}Cd, or ^{241}Am) is emitted to the water, and its attenuation, due to the sediment concentration, is detected by a counter installed in the downstream side of the gage or in a separate rod. Counts are converted directly into sediment concentrations by comparison to reference samples in a precalibration procedure. The volume of measurements, varying from a few liters to several cubic meters, depends on the source–detector distance, the radiation energy, and the time interval in counting. Gages developed in China have shown excellent agreement with the conventional method of sampling for sediment concentrations from 15 to $500\,kg/m^3$ in the Yellow River (Lu et al., 1981).

C.1.3.2.6 Real-Time Densimetric Method

An automatic, real-time densimetric technique, described by Lewis and Rasmussen (1999) and Calhoun (2000), detects changes in fluid densities between two fixed depths using a precision differential pressure transducer. It allows for corrections of the small temporal changes in fluid densities caused by temperature, dissolved solids, and gasses, making the remaining differences in density to be due to primarily suspended sediments. The technique meets both the temporal and depth-integration needs of sediment sampling. Tests in a laboratory showed good linear relations between SSC and the output current readings from the densimeter, especially at high SSC levels (Hsu and Cai, 2010).

C.1.4 Calculations

The collection of suspended-sediment water samples and the measurement of flow discharge are often conducted simultaneously in a few verticals along a stream cross section. The collected water samples are brought to the laboratory for sediment-concentration determinations and are expressed in mass/volume (mg/L). The SSC and discharge rate are then used to calculate the suspended load.

C.1.4.1 Suspended-Sediment Load for a Cross Section

The SSC (c_s) in each vertical is multiplied by its mean discharge rate (q, m^3/s) to obtain suspended-sediment load, or yield (q_{SL}) in mass/time for that vertical. In common practice, c_s is in mg/L, q in m^3/s, and q_{SL} in t/day calculated by

$$q_{SL} = 0.0864(q)(c_s) \qquad (C.2)$$

The constant 0.0864 converts the units from mg/s to t/day, assuming q and c_s stay the same for the day. Summing up all values of q_{SL} in each vertical yields the total suspended-sediment load for the cross section, Q_{SL}, or

$$Q_{SL} = (q_{SL})_1 + (q_{SL})_2 + (q_{SL})_3 + \cdots + (qq_{SL})_n \qquad (C.3)$$

C.1.4.2 Mean Suspended-Sediment Concentration

For the mean SSC (C_s) of the entire cross section, all values of c_s need to be weighted by each corresponding discharge rate as a percent of the total discharge rate for the entire cross section, or

$$C_s = \frac{(c_s)_1(q)_1 + (c_s)_2(q)_2 + (c_s)_3(q)_3 + \cdots}{q_1 + q_2 + q_3 + \cdots} \tag{C.4}$$

where the numerical subscripts 1, 2, 3, etc., refer to the different sampling verticals. They can also refer to different time intervals if mean concentrations for a day or other periods of time are needed.

C.1.4.3 Suspended-Sediment Load for a Runoff Event

Often, more water samples are collected during a major runoff event. In a runoff event, water samples are usually not collected at equal time intervals along the rise and fall of the discharge hydrograph, and the sediment load sampled at time 1 needs to be averaged with the sediment load sampled at time 2 for that time interval. Thus, averages of sediment load are calculated for all time intervals for the entire runoff event. The summation of the averages of the suspended-sediment load for all time intervals is the total suspended-sediment load for the runoff event. In equation form,

$$Q_{SL} = \sum \left[\frac{\left[(c_s)_1(q)_1 + (c_s)_2(q)_2 \right](t_{1-2})}{2} + \frac{\left[(c_s)_2(q)_2 + (c_s)_3(q)_3 \right](t_{2-3})}{2} + \cdots \right] \tag{C.5}$$

where the numerical subscripts 1, 2, 3, etc., refer to the sampling period at time 1, 2, 3, etc. Thus, t_{1-2} is the time interval between periods 1 and 2.

C.2 Bedload

Bedload is the sediment transported downstream by sliding, saltation, and rolling along the streambed. The transported material remains in contact with the streambed. For practical purposes, particles with a diameter greater than 1.0 mm are transported as bedload, while those with diameters less than 0.1 mm are transported as suspended load, and those with diameters between 1.0 and 0.1 mm can be transported either as bedload or suspended load, depending on the hydraulic conditions (MacDonald et al., 1991). Particles that are less than 0.0625 mm in diameter will be carried in suspension by the stream as washload and may never settle out.

Bedload transport can be further differentiated into sand transport (1–2 mm) and coarse bedload transport (2 mm to boulders). Sand transport is more controlled by hydraulic conditions and can be initiated in ordinary flows, while coarse bedload is usually limited to high flows that occur less frequently (Bunte, 1996). The amount and size of bedloads are important factors affecting channel morphology, streambed configuration, reservoir storage capacity, and aquatic habitat. Generally, fine material tends to fill up the interstitial spaces between large particles, causing less habitat space to be available for juvenile fish and macroinvertebrates and lower dissolved-oxygen levels for benthic communities and salmon spawning (Rosser and O'Conner, 2007).

Compared to suspended sediment, the measurement of bedloads is more difficult, time and labor intensive, and expensive. Thus, many sediment load studies concentrated only on the suspended mode, the bedload is either ignored or taken as a fraction of the total. The fractions are often cited as 10%–20% in general and 20%–40% for mountain streams (Turowski et al., 2010).

C.2.1 Bedload Variation

Bedload transport depends greatly on stream discharge, but not in a predictable way. It moves in irregular waves and sheets due to the streambed configuration and the variation in shear stress of the flow. Thus, the transport varies considerably with time and across the section. In fact, bedload generally occurs in part of the stream cross section only, except during floods. It tends to be higher during the recession stage than at the rising stage of a runoff event, and the majority is transported during the very few largest flows in a given year. There is no consistent relationship between bedload and discharge or between bedload and suspended load (Williams, 1989).

As an example of the temporal variation, Figure C.6 is a plot of bedload transport rate sampled consecutively with a Halley–Smith sampler in a 3 min interval between samples on a sand-bed stream in western Tennessee (Carey, 1985). These samples were collected during a period of essentially constant peak discharge. The data show that bedload transport changed from 1.15 to 0.02 lb/s/ft (1.72–0.03 kg/s/m) between two consecutive samples, a drop of more than 57 times. Individual rates compared to the overall mean

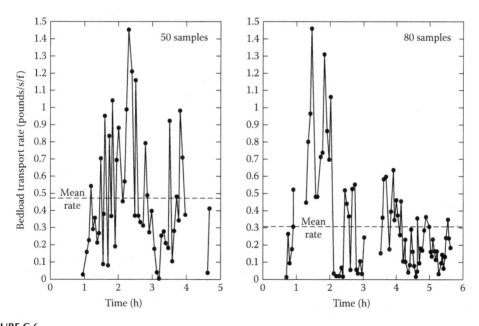

FIGURE C.6
Temporal variability of bedload transport rates consecutively collected by a Helley–Smith sampler from the Fork Obion River in western Tennessee. (From Carey, W.P., *Water Resour. Bull.*, 21, 39, 1985. Reproduced by permission of AWRA.)

varied from about zero to four times, and 60% of the sampled rates were equal to or less than the mean. Since the bedload transport is expected to be in equilibrium with the flow conditions, the extreme fluctuation of bedload transport must be indicative of the inherent variability in the transport process. The large variation requires many observations to establish an accurate estimate of the mean rate at a given location.

C.2.2 Sampling

Bedload is measured by the amount of sediment trapped in a sampler, or other devices, located at one or more points on the streambed. The sampled bedload is dried and weighed in the laboratory. The dried-sample weight is then divided by the time for which the sampler was left on the streambed and by the width of the sampler to obtain the bedload discharge per unit stream width per unit time at the gaging station. Finally, the average value is converted into the total bedload discharge for the entire cross section.

The measurement of bedload transport is more difficult than that of suspended sediment. First, unless continuous monitoring devices with high temporal resolutions are used (Bunte, 1996), the great temporal and spatial variation in bedload transport makes it very difficult for the sampling schedule to provide a representative spectrum with conventional samplers. The majority of bedload in most streams often occurs during the two to three largest flows in a given year (MacDonald et al., 1991). Sampling bedload in a stream with great depths and swift velocities is difficult, if not impossible. UNESCO (1970) suggests that samples be taken in all velocity verticals of the gaging station no fewer than 10 times a year, including at flood periods. Perhaps one should start with a biweekly schedule with additional measurements for severe storms. The sample frequency is then properly reduced if the data variation is permissible. The USGS (1978) suggests that the initial number of positions in a cross section be no fewer than 20, but WMO (1981) states that 3–10 sampling points should be sufficient. All these suggestions reflect the complexity of sampling problems.

Second, placing a sampler on the bed is problematic. As a sampler is lowered into the water, it is dragged downstream by the flow, but the drag decreases as the velocity decreases downward. The sampler will then move in the upstream direction by the weight of the sampler and the elasticity of the suspended cable. As a result, some bed material will be scooped up unless the upstream motion is stopped properly. Upon the presence of the sampler on the streambed, the general pattern of water and sediment movements is disturbed, and the transport rate and distribution of sediment upstream from the sampler are altered. This causes the sampler to be less efficient than if the sampler were not there. The geometry of the sampler, the hydraulic conditions of flow, and the size of particles are the major factors affecting the efficiency of a sampler. No samplers are efficient enough to sample bedload in full spectrum (size range), and the selection of an appropriate sampler becomes important. Generally, the Dutch Arnhem sampler (Figure C.7a) is used for sand-bed rivers, while the Helley–Smith sampler (Figure C.8) and the VUV sampler (Figure C.7b) are for sizes between coarse sand and coarse gravel (WMO, 1981; Vericat et al., 2006). Their performances are generally affected by bedload transport rates, bedload particle-size fractions, and bed-material characteristics of the study streams (Bunte and Abt, 2009).

Third, many ripples, dunes, and bars are formed on the streambed by the moving sediment. This causes the bed configuration to be irregular, sediment transport to be variable, and the sampler's position to be inconsistent and unstable. In other words, the

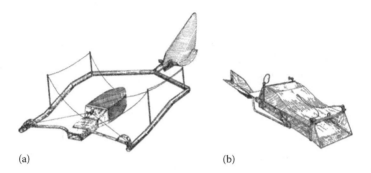

(a)   (b)

FIGURE C.7
Arnhem sampler (a) and VUV sampler (b). (Reproduced from Hubbell, D.W., Apparatus and techniques for measuring bedload, U.S. Geological Survey Water-Supply Paper 1748, 1964.)

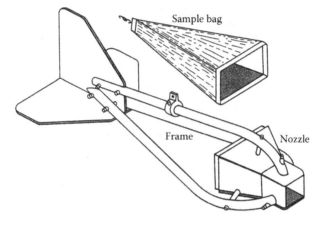

FIGURE C.8
Helley–Smith sampler. (Reproduced from Emmett, A field calibration of the sediment-trapping characteristics of the Helley–Smith bedload sampler, U.S. Geological Survey Professional Paper 1139, 1980.)

sampler's position may be upward, downward, or level in respect to whether the front edge of the sampler is on the crest or in the trough of a dune. Also, the turbulence of flow and drag force can cause the sampler to move back and forth and scoop up bed material. The scooping is most likely if the sampler is lowered into a trough in the bed (Hubbell, 1964). Thus, a single measurement of bedload sample is unrepresentative. The experience of USGS shows that systematic traverses of the riverbed have to be repeated several times and averaged. The integrated result is close to that obtained by the direct trapping across the entire river (Bagnold, 1977).

The sampling time is usually 5–10 min (WMO, 1981), but can be as short as 15 (Ferguson et al., 1989) or 30 s (Carey, 1985), depending on the size of the sampler and the transport rate. It should last long enough to collect at least 50 g (UNESCO, 1970), but not to fill the sampler in excess of 30% (Garde and Ranga Raju, 1985). The efficiency of samplers decreases rapidly after the sampler is 30% full.

Since bedload samplers do not provide an estimate to the true value, it is necessary to know the sampler efficiency so that the true value of the transport rate can be calculated. The sampler efficiency can be calibrated in the laboratory (Hubbell et al., 1987) or in the field (Emmett, 1980).

C.2.3 Measuring Devices

Many devices have been developed for estimating bedload transport in streams; none is completely satisfactory for sampling large particles and fine particles with the same efficiency. Devices that are briefly introduced here include bedload samplers, sediment traps, tracers, and ultrasonic measurement.

C.2.3.1 Bedload Samplers

Direct bedload samplers are of three main types: box or basket, pan or tray, and pressure difference.

C.2.3.1.1 Basket-Type Samplers

This type of sampler, made of wire mesh material in a rectangular shape attached to a rigid steel frame, is suspended on a cable for operation. The mesh has small openings sufficient for suspended particles, but not bedload material, to pass through. The front end is completely open, which allows bedload material to enter into the sampler. They are the oldest types of samplers, suitable for measuring coarse particles because of their large capacity. The average efficiency is about 45%, varying between 20% and 90% depending on the particle size and discharge of the bedload.

C.2.3.1.2 Tray Samplers

Tray samplers are wedge-shaped along the longitudinal direction with a slot or slots open to the top and located at the downstream end of the sampler. The bedload is trapped into the slot when it rolls and slides from the tip of the wedge to the end of the sampler. They are best suited for sand material and have average efficiencies of 38%–75%.

C.2.3.1.3 Pressure-Difference Samplers

In the basket and tray samplers, the flow velocity at the entrance is reduced due to the flow resistance of the sampler, causing some diversion or deposition of sediment at the entrance. The pressure-difference sampler is designed to maintain the flow velocity and the entrance velocity at the same level by constructing the sampler walls diverging toward the rear and producing a pressure drop at the exit of the sampler.

Introduced in 1971 by modifying the then-popular Arnhem sampler, the Helley–Smith sampler has become a standard for direct bedload measurement. The sampler has a stainless steel nozzle leading to a mesh bag (0.25 mm mesh, 46 cm long) and three steel-plate tail fins held together by solid-steel frames to maintain a 30 kg total weight (Figure C.8). The sampler's nozzle has several models, such as 7.62×7.62, 15.24×15.24, and 30.48×30.48 cm (Hubbell et al., 1987). Smaller ones are operated by hand, while larger ones can be operated by a backhoe from a bridge. A field calibration shows that the 7.62 cm square-orifice Halley–Smith sampler has a near-perfect sediment-trapping efficiency for particle sizes between 0.50 and 16.00 mm (Emmett, 1980).

C.2.3.2 Bedload Traps

In this type of device, samplers are installed under the streambed with orifices on the top of the sampler to trap bedload sediment moving in contact with the bed. Because they do not affect the flow, the results are therefore more accurate than results of other bedload samplers. Bedload traps can be a few individual pit samplers (Georgiev, 1990), a long trough installed across the stream, a local weighing device, or a portable bedload trap (Bunte et al., 2004). Pit samplers have been used to determine the size distribution of bedload or the bedload discharge by measuring the time required for a pit of known volume to fill.

The conveyor-belt bedload sampler installed in 1973 across the East Fork River in the Wind River Range of Wyoming (Klingeman and Emmett, 1982) is a unique example of a trough-trap sampler. It is constructed with a concrete trough across the stream, normal to the flow and 0.4 m wide by 0.6 m deep with its lip at the streambed surface. The trough has a slot 0.25 m wide and 14.6 m long with eight individually hydraulic control gages each 1.83 m long. Bedload material falling into the open slot is removed by an endless rubber belt installed along the bottom of the trough to a sump constructed in the riverbank. The system can accommodate loads of as much as 150 kg/min and has been used to calibrate other samplers in the field (Emmett, 1980).

Using the principle of pressure lysimeters, the Birkbeck bedload sampler is a continuous device that records both flood hydrographs and the combined pressure heads due to trapped bedload material and water (Reid et al., 1980). The sampler consists of an inner box and a concrete liner installed in a pit opening on the streambed. The inner box is placed on top of a rubber pressure pillow so that it is free to move up and down in response to the changing mass of both the trapped sediment and the water column. The fluid pressure of the pillow is transmitted to a transducer and to a mechanical recording device. Also, a synchronous record of stream-water stage provides an independent measure of the pressure changes due to stage rises and falls alone. Differences in pressure head between the trapper and the stage level are related proportionally to the mass of trapped sediment. The sampler is designed mainly for small streams. It is sensitive and robust but requires periodic clearing out of the accumulated sediment and the prevention of particles from settling onto the pressure pillow.

Trough-trap samplers provide the most direct method for bedload measurements; their efficiency is virtually 100%, but the cost is high, the installation is difficult, and the maintenance is cumbersome. Except for sampler calibration and sediment research, the installation of this sampler for general use is limited.

C.2.3.3 Tracers

In this technique, sediment grains are marked or injected by luminescent (Shteinman et al., 1998), magnetic (Ferguson et al., 1996; Froehlich, 2010), iron, radio (Schmidt and Ergenzinger, 1992), or radioactive (Shteinman et al., 2000) substances and are placed onto the streambed. The movement of the treated tracers is repeatedly sampled or detected for subsequent analysis. In Italy, for example, permanent magnets were cemented into holes drilled in cobbles and were introduced into the river. When the artificially magnetic cobbles pass over a detector consisting of iron-cored coils of wire, a measurable electrical current is generated and transmitted to an amplifier, filter, and recorder. The count rate can be converted to a mass-transport rate by relating the mass, size, and magnetic field intensity of the particles in the sample to signals produced from the detector through a precalibration procedure (Ergenzinger and Conrady, 1982). This technique can also obtain information such as the movement of tracer particles in relation to various hydraulic conditions. There are many streams in andesitic and basaltic terrain or with geologic formations around the world that have natural sources of magnetic material. Ergenzinger and Custer

(1983) showed that the technique also could be applied to estimate bedload transport by using naturally magnetic tracers in Montana. More recent developments in the magnetic tracer technique are given by Spieker and Ergenzinger (1990).

C.2.3.4 Acoustic and Ultrasonic Measurement

In sand-bed streams where the movement of bedload characteristically forms many well-defined dunes, the ultrasonic technique can be used to monitor changes in vertical depth along a selected cross section of the stream. Thus, data on the profile of the dunes and the velocity of their movement can be obtained. The sand-dune profile can be used to compute the sand-dune area, and multiplying the area by the mean velocity of the sand-dune movement gives the bedload discharge per unit width of the stream channel (WMO, 1981). An application of this technique is to map vertical bathymetric changes across a stream channel through a 250 kHz Geoswath echo sounder mounted in a RTK GPS positioned boat. The change in elevations between two profiles at the same location taken at two different times is an indication of scour or deposition; consequently bedload transport can be calculated (Abraham et al., 2011).

In coarse-grained streams, the moving and colliding particles cause clicking noises when they hit an acoustic (hydrophonic) detector consisting of a metal sheet installed on the streambed. A microphone is built into the detector to amplify the noises, which are transmitted to a headphone or a magnetic tape recorder. The frequency of noises gives some qualitative information on the intensity of bedload movement (Banziger and Burch, 1990; Møen et al., 2010) and can help to select bedload sampling points along a cross section. The passive acoustic technique allows for continuous monitoring coarse particles during flood events (Froehlich, 2010).

C.2.4 Calculations

Measurements for bedload discharge by a sampler are made at several verticals of the stream cross section. Bedload discharge at each measurement point is calculated by

$$q_{BL} = \left(\frac{M}{wt}\right)(\varepsilon) \tag{C.6}$$

where
q_{BL} is the bedload discharge in mass per unit width per unit time
M is the mass of the bedload caught by the sampler
w is the orifice width of the sampler
t is the sampling duration at the measuring point
ε is the efficiency of the sampler

Total bedload discharge (Q_{BL}, in mass/time) for the cross section is calculated by

$$Q_{BL} = \sum \left[\frac{[(q_{BL})_1 + (q_{BL})_2](d_{1-2})}{2} + \frac{[(q_{BL})_2 + (q_{BL})_3](d_{2-3})}{2} + \cdots \right] \tag{C.7}$$

where
$(q_{BL})_1$ is the bedload discharge at measuring point 1 calculated by Equation C.6
d_{1-2} is the distance between measuring points 1 and 2

References

Abraham, D., Kuhnle, R.A, and Odgaard, A.A., 2011, Validation of bed-load transport measurements with time-sequenced bathymetric data, *J. Hydrol. Eng.*, 137(7), 723–728.

Bagnold, R.A., 1977, Bed load transport by natural rivers, *Water Resour. Res.*, 13, 303–312.

Banziger, R. and Burch, H., 1990, Acoustic sensors (hydrophones) as indicators for bed load transport in a mountain torrent, *Hydrology in Mountainous Regions*, IAHS, Wallingford, U.K., pp. 207–214.

Beschta, R.L., 1980, Turbidity and suspended sediment relationships, *Symposium on Watershed Management*, Vol. I, ASCE, New York, pp. 271–282.

Beschta, R.L. and Jackson, W.L., 2008, Forest practices and sediment production in the Alsea Watershed Study, *Hydrological and Biological Responses to Forest Practices: The Alsea Watershed Study*, Stednick, J.D., Ed., Springer Science+Business Media, New York, *Ecol. Stud.*, 199, 55–66.

Bonta, J.V., 2000, Impact of coal surface mining and reclamation on suspended sediment in three Ohio watersheds, *J. Am. Water Resour. Assoc.*, 36, 869–887.

Bunte, K., 1996, Analyses of the temporal variation of coarse bedload transport and its grain size distribution, General Technical Report RM-GTR-288, USDA Forest Service, Rocky Mountain Forest and Range Experiment Station, Fort Collins, CO.

Bunte, K. and Abt, S.R., 2009, Transport relationships between bedload traps and a 3-inch Helley–Smith sampler in coarse gravel-bed streams and development of adjustment functions, Completion Report 218, Colorado Water Institute, Colorado State University, Fort Collins, CO.

Bunte, K. et al., 2004, Measurement of coarse gravel and cobble transport using portable bedload traps, *J. Hydraul. Eng.*, 130(9), 879–893.

Calhoun, D.L., 2000, In situ monitoring of suspended sediments: Development of a densimetric instrument, Unpublished MS thesis, University of Georgia, Athens, GA.

Carey, W.P., 1985, Variability in measured bedload-transport rates, *Water Resour. Bull.*, 21, 39–48.

Coynel, A. et al., 2004, Sampling frequency and accuracy of SPM flux estimates in two contrasted drainage basins, *Sci. Total Environ.*, 330, 233–247.

Emmett, 1980, A field calibration of the sediment-trapping characteristics of the Helley–Smith bedload sampler, U.S. Geological Survey Professional Paper 1139.

Ergenzinger, P.J. and Conrady, J., 1982, A new tracer technique for measuring bedload in natural channels, *Vatena*, 9, 77–90.

Ergenzinger, P.J. and Custer, S.G., 1983, Determination of bedload transport using naturally magnetic tracers: First experiences at Squaw Creek, Gallatin County, Montana, *Water Resour. Res.*, 19, 187–193.

Ferguson, R.I., 1986, River loads underestimated by rating curves, *Water Resour. Res.*, 22, 74–76.

Ferguson, R.I., Prestegaard, K.L., and Ashworth, P.J., 1989, Influence of sand on hydraulics and gravel transport in a braided gravel bed river, *Water Resour. Res.*, 25, 635–643.

Ferguson, R. et al., 1996, Field evidence for rapid downstream running of river gravels through selective transport, *Geology*, 24, 179–182.

Froehlich, W., 2010, Monitoring of bed load transport within a small drainage basin in the Polish Flysch carpathians, U.S. Geological Survey Scientific Investigations Report 2010-5091, pp. 185–194.

Gao, P. et al., 2008, Estimating suspended sediment concentration using turbidity in an irrigation-dominated southeastern California watershed, *J. Irrig. Drain. Eng.*, 134(2), 250–259.

Garde, R.J. and Ranga Raju, K.G., 1985, *Mechanics of Sediment Transportation and Alluvial Stream Problems*, John Wiley & Sons, New York.

Georgiev, B.V., 1990, Reliability of bed load measurements in mountain rivers, *Hydrology in Mountainous Regions*, IAHS, Wallingford, U.K., pp. 263–270.

Gordon, N.D., McMahon, T.A., and Finlayson, B.L., 1992, *Stream Hydrology, an Introduction for Ecologists*, John Wiley & Sons, New York.

Grobler, D.C. and Weaver, A. van B., 1981, Continuous measurement of suspended sediment in rivers by means of a double beam turbidity meter, *Erosion and Sediment Transport Measurement*, *Proceedings of the Florence Symposium*, IAHS, Wallingford, U.K., pp. 97–104.

Guy, H.P. and Norman, V.W., 1970, Field methods for measurement of fluvial sediment, *Techniques of Water-Resources Investigations*, U.S. Geological Survey, Chapter C2, Book 3, Applications of Hydraulics, U.S. Government Printing Office, Washington, DC.

HACH, 1998, *Products and Analysis*, HACH Company, Loveland, CO.

Harris, R.R. et al., 2007, Applications of turbidity monitoring to forest management in California, *Environ. Manage.*, 40(3), 531–543.

Hsu, Y.-S. and Cai, J.-F., 2010, Densimetric monitoring technique for suspended-sediment concentrations, *J. Hydraul. Eng.*, 136(1), 67–73.

Hubbell, D.W., 1964, Apparatus and techniques for measuring bedload, U.S. Geological Survey Water-Supply Paper 1748.

Hubbell, D.W. et al., 1987, Laboratory data on coarse-sediment transport for bedload-sampler calibrations, U.S. Geological Survey Water-Supply Paper 2299.

Klingeman, P.C. and Emmett, W.W., 1982, Gravel bedload transport processes, *Gravel Bed River. Fluvial Processes, Engineering and Management*, Hey, R.D., Bathurst, J.C., and Thorne, C.R., Eds., John Wiley & Sons, New York, pp. 141–179.

Kunkle, S.H. and Comer, G.H., 1971, Estimating suspended sediment concentration in streams by turbidity measurements, *J. Soil Water Conserv.*, 26, 18–20.

Lewis, A.J., 1996, Turbidity-control suspended sediment sampling for runoff-event load estimation, *Water Resour. Res.*, 32, 2299–2310.

Lewis, A.J. and Rasmussen, T.C., 1999, Determination of suspended sediment concentrations and particle size distributions using pressure measurements, *J. Environ. Qual.*, 28, 1490–1496.

Lu, Z. et al., 1981, The development of nuclear sediment concentration gauges for use on the Yellow River, *Erosion and Sediment Transport Measurement*, Proceedings of the Florence Symposium, IAHS, Wallingford, U.K., pp. 83–90.

MacDonald, L.H., Smart, A.W., and Wissmar, R.C., 1991, Monitoring guidelines to evaluate effects of forestry activities on streams in the Pacific Northwest and Alaska, EPA/910/9-91-001.

Marquis, P., 2005, Turbidity and suspended sediment as measures of water quality, *Streamline Watershed Manage. Bull.*, 9(1), 21–23.

McBroom, M. et al., 2003, Runoff and sediment losses from annual and unusual storm events from the Alto Experimental Watersheds, Texas: 23 years after silvicultural treatments, *Proceedings of the First Interagency Conference on Research in the Watersheds*, Benson, AZ, http://www.tucson.ars.ag.gov/icrw/proceedings.htm.

Møen, K.M. et al., 2010, Bedload measurement in rivers using passive acoustic sensors, USGS Science Investigations Report 2010-5091, pp. 336–351.

Parsons, D.A., 1954, Coshocton-type runoff samplers, laboratory investigations, USDA, Soil Conservation Service, Washington, DC, SCS-TP-124.

Reid, I., Layman, J.T., and Frostick, L.E., 1980, The continuous measurement of bedload discharge, *J. Hydraul. Res.*, 18, 243–249.

Rosser, B. and O'Conner, M., 2007, Statistical analysis of streambed sediment grain size distribution: Implications for environmental management and regulatory policy, USDA Forest Service General Technical Report PSW-GTR-194.

Schmidt, K.H. and Ergenzinger, P., 1992, Bedload entrainment, travel lengths, step lengths, set periods, studied with passive (iron, magnetic) and active (radio) tracer techniques, *Earth Surf. Process. Landforms*, 17, 147–165.

Shteinman, B., Wynne, D., and Kamenir, Y., 2000, Study of sediment dynamics in the Jordan River–Lake Kinneret contact zone using tracer methods, *The Hydrology–Geomorphology Interface: Rainfall, Floods, Sedimentation, Land Use*, IAHS, Wallingford, U.K., pp. 275–284.

Shteinman, B.S. et al., 1998, A modified fluorescent tracer approach for studies of sediment dynamics, *Isr. J. Earth Sci.*, 46, 18–31.

Singhal, H.S.S., Goshi, G.C., and Verma, R.S., 1981, Sediment sampling in rivers and canals, *Erosion and Sediment Transport Measurement*, Proceedings of the Florence Symposium, IAHS, Wallingford, U.K., pp. 169–175.

Spear, B. et al., 2008, Turbidity as a surrogate measure for suspended sediment concentration in Elkhorn Slough, CA, Publ. No. WI-2008-04, Watershed Institute, California State University Monterey Bay, Seaside, CA.

Spieker, R. and Ergenzinger, P.J., 1990, New developments in measuring bedload by the magnetic tracer technique, *Erosion, Transport, and Deposition Processes*, IAHS, Wallingford, U.K., pp. 171–180.

Tazioli, G.S., 1981, Nuclear techniques for measuring sediment transport in natural streams—Examples from instrumented basins, *Erosion and Sediment Transport Measurement, Proceedings of the Florence Symposium*, IAHS, Wallingford, U.K., pp. 63–81.

Thomas, R.B., 1985, Measuring suspended sediment in small mountain streams, General Technical Report PSW-83, U.S. Forest Service, Pacific Southwest Forest and Range Experiment Station, Berkeley, CA.

Turowski, J.M., Rickenmann, D., and Dadson, S.J., 2010, The partitioning of the total sediment load of a river into suspended load and bedload: A review of empirical data, *Sedimentology*, 57, 1126–1146.

Uhrich, M.A. and Bragg, H.M., 2003, Monitoring instream turbidity to estimate continuous suspended-sediment loads and yields and clay-water volumes in the Upper North Santiam River Basin, Oregon, 1998–2000, USGS Water-Resources Investigation Reports 03-4098.

UNESCO, 1970, *Representative and Experimental Basin, an International Guide for Research and Practice*, Toebes, C. and Ouryvaev, V., Eds., UNESCO, Paris, France.

UNESCO, 1982, *Sedimentation Problems in River Basins*, Project 5.3 of the International Hydrological Programme, UNESCO, Paris, France.

USDA, 1979, Field manual for research in agricultural hydrology, *Agriculture Handbook No. 224*, U.S. Department of Agriculture, Washington, D.C.

USGS, 1978, *National Handbook of Recommended Methods for Water-Data Acquisition*, USGS, Reston, VA, pp. 1–100.

Vericat, D., Church, M., and Betalla, R.J., 2006, Bed load bias: Comparison of measurements obtained using two (76 and 152 mm) Helley–Smith samplers in a gravel bed river, *Water Resour. Res.*, 42, W01402, 13. DOI: 10.1029/2005WR004025

Walling, D.E. and Webb, B.W., 1981, The reliability of suspended sediment load data, *Erosion and Sediment Transport Measurement, Proceedings of the Florence Symposium*, IAHS, Wallingford, U.K., pp.177–194.

Weigel, J.F., 1984, Turbidity and suspended sediment in the Jordan River, Salt Lake County, Utah, USGS Water-Resources Investigation Report 84-4019.

Wilde, F.D. and Radtke, D.B., Eds., 1998, *National Field Manual for the Collection of Water-Quality Data*, U.S. Geological Survey, Reston, VA.

Williams, G.P., 1989, Proportion of bedload to total sediment load in rivers, *Eos*, 70, 1106.

Williamson, T.N. and Crawford, C.G., 2011, Estimation of suspended-sediment concentration from total suspended solids and turbidity data for Kentucky, 1978–1995, *J. Am. Water Resour. Assoc.*, 47(4), 739–749.

WMO, 1981, Measurement of river sediments, World Meteorological Organization of Operational Hydrology Report No. 16, WMO-No. 561, Geneva, Switzerland.

Appendix D: Measurements of Forest Interception

H. Krutsch of Germany conducted an investigation on forest rainfall interception in 1863 (Molchanov, 1963; Friedrich, 1967), which is believed to be the earliest scientific study of forest hydrology. It compared rainfall catch in the open and under forest canopies using rainfall and throughfall gages. Over an observational period of 16 months, the results showed that throughfall was only 9% in rainfall of less than 0.5 mm, 18%–57% in rainfall of 1–7 mm, and 80%–90% during strong showers. Krutsch's approach of comparing rainfall to throughfall measurements has been widely used in forest interception studies, and much of our knowledge on forest interception among species was obtained by this method. In recent decades, newer methods, including gamma-ray attenuation (Calder and Wright, 1986), lysimeters (Calder, 1976; Hall, 1985), a microwave transmission method (Bouten et al., 1991), and physically based analytical models (Gash et al., 1995; van Dijk and Bruijnzeel, 2001; Limousin et al., 2008; Ahrends and Penne, 2010), have all been used. Lundberg (1993) listed methods for measuring evaporation of intercepted snow; many of them are also applicable for intercepted rain. This appendix discusses only the technique of interception measurements; studies on interception simulation are not included.

D.1 Precipitation Interception

On event basis, precipitation intercepted by a forest, or forest interception (I_F), is the sum of canopy interception (I_C) and litter interception (I_L), or

$$I_F = I_C + I_L \tag{D.1}$$

Canopy interception is estimated by the difference between gross precipitation in the open (P_G) and canopy throughfall (P_{TH}), with a correction for stemflow (P_S), or

$$I_C = P_G - (P_{TH} + P_S) \tag{D.2}$$

Combining Equations D.1 and D.2 yields

$$I_F = P_G - (P_{TH} + P_S) + I_L \tag{D.3}$$

Thus, the estimate of total loss due to forest interception involves measurements of four components—P_G, P_{TH}, P_S, and I_L. However, this conventional method is liable to large errors for storms at high wind speeds and in low intensities. The canopy variation and the variation of ground coverage require a very dense network of throughfall gages for reliable estimates. Equations D.2 and D.3 give estimates on total interception per storm basis.

They do not describe interception processes that may be required in many hydrologic modeling. Various measuring techniques are introduced in the following sections.

D.1.1 Gage Comparisons

D.1.1.1 Interception Components

D.1.1.1.1 Gross Precipitation

Precipitation in the open should be measured at sites close to, in, or among the study plots. For proper gage exposure to account for wind effects, the U.S. National Weather Service (1972) requires that the height of surrounding objects above the gage not exceed half their distance to the gage. An opening with this distance could be larger than 1 ha in size and is difficult to find in forests, unless one is willing to clear the trees. Thus, a subtending angle of 45° from the gage orifice to the top of surrounding trees in all directions is used in most practices. Sites overprotected by surrounding trees may create a rain shadow that reduces rainfall catch. Some researchers monitor gross precipitation at the top of forest canopies (Xiao et al., 2000). Since wind speed increases with height, precipitation catch at canopy level is expected to be much less than that at ground level, depending on canopy height (Chang and Harrison, 2005). Thus, canopy precipitation does not seem to represent true precipitation for ground-level studies.

Throughfall plots are usually less than 1 ha in size and most likely covered by entire storms. Thus, one or two rain gages should be sufficient to sample gross precipitation for the study plots. If two gages are used, they should be located along the prevailing wind direction, one on the windward side and one on the leeward of the plot. On hilly terrain, one should be upslope and the other downslope. If one gage is used, it should be at a central location. For throughfall that is monitored for water budget studies on a watershed scale, more gages should be used, with considerations for both horizontal as well as vertical distribution within the watershed, especially in mountainous regions. In this case, gross precipitation P_G for the watershed should be estimated by the area-weighted Thiessen or isohyetal methods (Shaw, 1983).

D.1.1.1.2 Canopy Throughfall

Throughfall is the portion of precipitation that passes through or drips from canopies to the ground. It is sampled by placing standard 8 in. rain gages or trough gages of certain lengths under the canopy. However, substantial variations within forest canopies make throughfall more difficult to sample and require many more gages than would be needed to measure gross precipitation. To reduce the cost of the large number of gages required in throughfall studies, rain gages made of 2 in. PVC (polyvinylchloride) tubes have been used as a substitute for the standard rain gages (Roth and Chang, 1981). Artificial catchments consisting of two panels with sloping sides have been used; they are installed under each tree to collect canopy throughfall (Xiao et al., 2000).

The number of gages required in throughfall studies is largely dependent on the variability of vegetation canopies. Thus, the requirement may be different for conifers versus hardwoods, summer versus winter, young versus mature, and mixed forests versus plantations. If the canopy variation is not uniform throughout the entire watershed, one may need to designate sampling plots on map or air photos representing the various forest conditions and determine the number of gages for each plot. Throughfall gages should be randomly or systematically located along transit lines in each plot. After the measurement of each storm event, these gages need to be relocated to a new site, through a predetermined scheme, to sample more throughfall variation in the forest.

Determining the number of throughfall gages required is a two-stage process, involving a trial stage and an adjustment stage. In the beginning of a throughfall study, the number of gages is determined by empirical judgments. It is subjective and is considered as a trial. Once the initial network is installed in the forest, the collected throughfall data can be used to calculate the number of gages required to sample throughfall at a predetermined accuracy and confidence level. The initial gage networks are then adjusted according to the calculated result. The estimation of an optimum sample size is discussed in the next section.

D.1.1.1.3 Stemflow

A portion of canopy-intercepted precipitation will flow to the ground through branches and boles as stemflow. Stemflow can be measured using circular gutters secured on the bark. The collected stemflow from these gutters is then led to containers for volume measurements. The total volume from all sampled trees is further converted into depth for the whole plot for comparisons to gross precipitation and throughfall. Note that the gutter should be as close to the ground as possible but high enough to accommodate the height of collectors. Frequently, overflow may occur, and leaves, insects, and dead animals may clog the outlet tubes of the gutters. Thus, the gutters often require frequent care and management, covering nets, or special designs to offset interfering problems.

D.1.1.1.4 Litter Interception

Litter interception is often estimated by small gravimetric plots 30 cm × 30 cm that are randomly or systematically located throughout the forest interception plots to assure data representativeness. The difference in weights between wet and dry litters needs to be converted into water depth for comparisons with gross rainfall and throughfall. Since litter interception is only a fraction of gross precipitation (Helvey, 1971), the accuracy level is not as critical as that required in throughfall measurements.

D.1.1.2 Number of Gages Required

For estimating the number of throughfall gages (N) required to ensure a certain confidence level that the estimated throughfall will be within a given degree of accuracy (E), Equation 15.1 can be used:

$$N = \frac{(t^2)(C_v^2)}{(E^2)} \tag{D.4}$$

where

C_v is the coefficient of variation of the sample (sample standard deviation divided by sample mean)

E is the desired limit of accuracy (a ratio of error limit to the mean)

t is the Student's t value with $N - 1$ degrees of freedom

The value t is greater for higher probabilities (more confidence) and smaller for greater degrees of freedom (more observations used to calculate C_v). Equation D.4 indicates that N values are higher for data of greater variation, more confidence, and higher level of

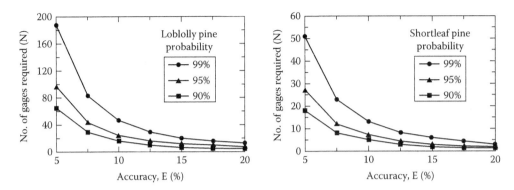

FIGURE D.1
Based on an initial network of 15 gages in a throughfall study in east Texas, an optimum number of through-fall gages can be determined through a series of Ns at various accuracy and probability levels calculated by Equation D.4.

accuracy. This method can be utilized for other sampling protocols of the study as well (i.e., litter interception).

Since the coefficient of variation required in Equation D.4 is based on data collected from an initial network, the calculated N is a test on the adequacy of the current network for meeting a desired level of accuracy. For a given C_v (as determined by the observational data), N is affected by the accuracy level E and the confidence level described by t. Thus, one may estimate N for various values of E and probability and plot them on graph paper. From the plotted curves, an optimum network of N suitable for the study could be determined in regions where slopes change. For example, the numbers of throughfall gages required for various accuracy levels on a loblolly pine and shortleaf pine plantation are plotted in Figure D.1. A critical change in slopes occurs at around E = 10% (or 7.5%), where a further reduction of accuracy levels does not significantly reduce N values, or the sacrifice of accuracy does not gain a sufficient return in gage reduction. An optimum gage network can thus be determined by examining changes in slopes for all probability levels.

D.1.2 Gamma-Ray Attenuation

As gamma rays are emitted from a source and pass through a substance, their intensity is attenuated due to the chemical composition and density of the substance. Generally, attenuation is greater for materials of higher density and for measurements of smaller volume. Thus, the densities of wet and dry forest canopies can be determined through the count rate of gamma-ray attenuation and a precalibration procedure to relate the count rate to the density of the substance in the measurement. The difference in densities between wet and dry canopies is the density of intercepted water in mass/area and is expressed in depth of water.

The gamma-ray measurement system includes:

1. A nuclear source, usually Cs^{137}, emitting gamma rays

2. A detector

3. A counter or scale

4. A radiation shield and housing

The isotope Cs[137] emits low-energy gamma rays (660 keV) and is more suitable for water measurement than are isotopes with high-energy rays such as Ra–Ba isotopes. In practice, the source and detector can be suspended from towers, which are separately installed in the forest at a distance long enough to cover the canopy variation for interception studies. These units should be movable up and down the towers to allow the beam to scan different levels in the canopy. The mass per unit area (M) of material in the beam of attenuation can be calculated by the following equation (Calder and Wright, 1986):

$$M\ (\mathrm{kg/m^2}) = \frac{1}{\mu} \log\left[\frac{n - nb}{na - nb}\right] \tag{D.5}$$

where
   n is the dead-time-corrected count rate of the gamma-ray beam attenuated by the forest
      canopy, sec^{-1}
   na is the dead-time-corrected count rate of pulses in air, sec^{-1}
   nb is the dead-time-corrected background count rate of pulses, sec^{-1}
   μ is the attenuation coefficient for water, or 0.008564 m^2/kg

With a computer control of data acquisition system incorporated into the gamma-ray attenuation and detection units, the total mass values of wet canopy per unit ground area, on subtraction of the dry canopy mass per unit ground area, are direct measurements of (a) the canopy storage of water, (b) the canopy storage capacity, (c) the evaporation rate of the intercepted water, or (d) the total loss of canopy interception. Apparently, the determination of evaporation rates of intercepted water (during the drying phase of an event) is a major advantage of the gamma-ray method that is not attainable by the gage comparison method. The approach may introduce 1%–3% errors. It is not suitable for unattended automated monitoring because of safety standards.

D.1.3 Lysimeters

Lysimeters are usually designed to detect evapotranspiration from and percolation into an isolated block of soil with or without growing vegetation in the field through gravimetrical, volumetric, or hydrologic budget approaches. Since the movement of water in the lysimeter is largely isolated, the same device can be used to measure rainfall, and differences in the hydrologic budgets between bare-soil and vegetated lysimeters can be used to detect canopy storage and interception loss. Hall (1985) used a wet-surface weighing lysimeter system to estimate four canopy parameters required in interception modeling. For the estimates to be reliable, lysimeters should:

1. Be large enough to cover the variation of the plant community and to minimize boundary effects
2. Have soil structure close to that of the surrounding area
3. Keep the vegetation gap along the boundary walls as small as possible

However, because of the costs, lysimeter size is a more difficult parameter to comply with than the other two requirements. It is suitable only for small vegetation such as seedlings, bushes, and grasses.

D.1.4 Microwave Transmission

Microwaves are electromagnetic radiation with wavelengths ranging from 0.1 to 100 cm, frequencies from 300 to 0.3 GHz, and photon energy from 10^{-3} to 10^{-6} electron volts (eV) (Strangeways, 2000). Note that the frequency and energy level of radiation are inversely related to wavelength, and the dielectric properties of a medium affect its radiation emissivity and reflectivity. Since the dielectric constant of water is relatively higher than that of forest canopies or other porous media such as soils, the attenuation of microwaves (at dry and wet canopies), which is due to radiation absorption and scattering by water inside and on trees, can be used to detect canopy interception.

Bouten et al. (1991) developed a technique of using microwave attenuation to determine canopy wetness in a 34 year Douglas fir plantation in the Netherlands. Two towers, 12.5 m apart, were equipped with electric winches to carry instrumentation at different heights in the plantation. One winch carried an 8.9 GHz microwave generator with a built-in direction isolator, a variable attenuator, and transmitting antennas. The other carried receiving antennas, a low-noise amplifier, an envelope detector, an indicator, and a data logger. The data logger sampled the analogue output of the receiver amplifier per second and averaged it for every 30 s period. These averages were used to calculate attenuation by use of a calibration curve. The results showed that the vertically integrated microwave attenuation profiles of the stand increased upon wetting and decreased due to dripping and evaporation. Thus, a canopy wetness period could be established from the measurements, and the attenuation could be converted to canopy storage amounts by an established calibration. The calibration showed relations between microwave attenuation and canopy water storage through high-resolution rainfall and throughfall measurements. Once the calibration is available, canopy water storage can be automatically scanned and recorded.

Microwave attenuation has been used for remotely sensed soil moisture monitoring since the 1970s (Schmugge et al., 1980; Rasmussen and Campbell, 1987).

D.2 Fog Interception

The term "fog interception" refers to the deposition of fog, cloud, and other water vapor on forest canopies that may drip to the ground in liquid state or vaporize to the atmosphere. It is also termed canopy condensation, horizontal precipitation, occult precipitation, negative precipitation, fog precipitation, fog drips, and cloud drips in literature. Because of its ecological and hydrological significance, interest in studies of fog interception has increased greatly in recent decades, especially in tropical montane forests (González, 2000; Bruijnzeel, 2001; Holder, 2004). Although fog interception may add more than 100% of rainfall input to water budgets in some areas (see Table 8.3), quantification and comparisons of forest impact are difficult because of a lack of standardized monitoring instrumentation and a lack of control on cloud deposition in natural environments.

One of the earliest studies on fog deposition by vegetation was conducted on Table Mountain in South Africa by Marloth between 1901 and 1903. Marloth (1904, cited by Olivier, 2002) used two rain gages—one was left open in the usual manner, while a bunch of reeds was suspended above the other. Between November 1901 and February 1903, the open gage collected 126 mm of water while the reed gage collected 2028 mm.

Four approaches are generally employed in monitoring fog and cloud interception:

1. Comparisons of canopy drip in a stand to rainfall in the open by throughfall gages
2. Fog collectors
3. Gravimetric measurements
4. Direct measurements

They are discussed in the following section.

D.2.1 Gage Comparisons

D.2.1.1 Estimates

Like canopy interception, canopy deposition (fog interception) can be indirectly estimated by comparing net rainfall (P_N) in the forest and that in a nearby open site (Holder, 2003). Throughfall (P_{TH}) and stemflow (P_S) gages are randomly positioned in the forest. The sum of P_{TH} and P_S (or P_N) is used to compare the gross rainfall (P_G) measured at a nearby open site for estimating canopy condensation (C_C) or canopy interception (I_C), depending on the relative magnitudes between P_G and P_N.

During rainy periods, P_G is greater than P_N (or $P_{TH} + P_S$), and the difference is canopy interception I_C, as shown in Equation D.2. During foggy periods with no rainfall, $P_G = 0$ and the measured $P_{TH} + P_S$ is the fog interception dripped from canopies. In this case, $P_G - P_N$ is negative, a reverse phenomenon of canopy interception loss. The sum of all negative quantities of canopy interception is the estimate of total fog or cloud precipitation that drip to the ground during the study period. To account for canopy interception of fog that accompanies with rains, automatic recording devices are needed in the monitoring network, or measurements should be conducted as frequently as possible, for example, daily.

If canopy condensation is the only purpose of the study, then there is no need for stemflow measurements because P_S is usually negligible and can be offset by the sample size. The frequency of data collections should not be too long to cause significant loss of evaporation. A long observational interval may mix canopy throughfall in rainfall events with drips of canopy condensation. It may underestimate fog interception when all the catches in the forest are summed to compare with gross rainfall in the open. Table D.1 shows that canopy condensation is 7.8 mm by daily measurements and zero by weekly measurements.

D.2.1.2 Gage Variations

Gages used for collecting canopy deposition drips and canopy throughfall vary among investigators in terms of type, size, material, and height above ground. Some of them have included 5 gal totalized gages with straight-sided funnels (Oberlander, 1956), the standard 8 in. (20.32 cm) nonrecording rain gages (Juvik and Ekern, 1978), weighing-type rain gages, digitizing recording rain gages, tipping-bucket recording rain gages (Azevedo and Morgan, 1974), 24.4 m long collector troughs (Harr, 1982), plastic funnels with a diameter of 200 mm, and 20 L plastic containers (Holder, 2004). It is important that gages used for throughfall and gross precipitation measurements should be comparable in terms of orifice size, material, and height above the ground. In order to account for the canopy

TABLE D.1

Difference in Canopy Depositions due to Daily or
Weekly Observations (All Units are in mm)

Date	Gross Rainfall (P_G)	Net Rainfall (P_N)	$P_G - P_N$ Interception (+)	Deposition (−)
Daily				
1	15.0	11.0	4.0	
2	0.0	2.0		2.0
3	0.0	1.5		1.5
4	21.0	16.0	5.0	
5	0.0	1.8		1.8
6	10.0	7.0	3.0	
7	0.0	2.5		2.5
Sum	46.0	41.8	12.0	<u>7.8</u>
Weekly	46.0	41.8	4.2	<u>0.0</u>

variation, the number of throughfall gages and a periodic relocation of these gages should be conducted in accordance with the outlines described in Section D.1.1.2.

D.2.2 Fog Gages

D.2.2.1 Basic Designs

There are two basic models for collecting deposition of fog and cloud water: (a) the wire mesh cylinder type, and (b) the wire harp type. The cylinder or the wire harp, acting as a sensor, is mounted on top of a collector. The condensed water of fog and cloud from the sensor is drained into a collector. Another collector (rain gage) without the fog and cloud sensor monitors rainfall at a nearby site. Volume differences between fog and rain collectors are the amount of fog deposition detected by the sensor.

The cylinder gage originally used by Grünow (1952, cited by González, 2000) consisted of a small metallic, gauze cylinder of 20 cm height, 200 cm² of cross-sectional area, and 1.55 mm of mesh size mounted on top of a conventional rain gage. However, the Grünow gage has been criticized because the sensor is too small, the surface area for interception may be changed when the mesh holes fill up with water, and the gauze can intercept wind-driven rainfall, leading to overestimation of fog interception (Cavelier et al., 1997; González, 2000).

The harp gage designed by Goodman (1985) used 0.8 mm diameter monofilament nylon as the impaction mechanism. It consisted of two flat harps, each 1 m², mounted on top of a trough collector 1 m above ground level. Each harp was composed of 500 vertically strung wires at 1.2 mm intervals, and the catch of fog water was piped to a jar. The harp sensors should be fixed with an orientation normal to the prevailing wind direction. This harp gage collected a total of 252.15 L as compared to 511.5 L for a fog gaged with three concentric cylinders 1 m tall during a period of 3 summer months in California (Goodman, 1985). The author stated that in areas with clear prevailing wind direction, harp gages would appear to be most efficient, while in those locations with wide variations in wind direction, a single-layer cylinder is a preferable sensor for fog collection. Harp sensors can be kept oriented normal to the prevailing wind direction by a wind vane (Ekern, 1964).

D.2.2.2 Gage Variations

There are many modifications on fog gages in terms of material, mesh dimension, cylinder size, collector, and height above ground. On the cylinder gage, for example, Juvik and Ekern (1978) used a louvered-screen cylinder of 41.9 cm in height and 12.7 cm in diameter mounted within a stainless steel drainage funnel of 15.2 cm in diameter. A flexible plastic hose drained the fog gage funnel into a (covered) standard 20.3 cm rain gage. The center of the cylinder was set at 3 m above the ground. The cross-sectional area of the fog gage drainage funnel is 179.5 cm^2 and the surface areas of the screen is 1672 cm^2, which are 56% and 500% of the cross-sectional area of the standard rain gage (324.3 cm^2), respectively. Thus, the actual fog deposition (F) is calculated by $F = (f - 0.56R)/5$, where f is the fog catch in depth read from the standard rain gage connected to the cylinder and R is the rainfall depth measured from a companion standard rain gage.

Other cylinder gages include (a) a 40 mesh/cm^2, 8 cm diameter and 8 cm tall plastic window screen cylinder by Cavelier and Goldstein (1989) in Colombia and Venezuela; (b) a single open-ended cylinder sensor of 18 mesh/cm^2 stainless screen, 80 cm tall and 10 cm diameter by Azevedo and Morgan (1974) in California; (c) a 40 mesh/cm^2 plastic screen cylinder of 8.4 cm diameter and 8 cm height mounted on top of plastic funnels 56 cm^2 in a cross-sectional area 1.5 m above the ground by Cavelier et al. (1996) in Panama; (d) a wire cylinder consisting of 48 nylon wires, 100 cm long and 1 mm wide, hanging vertically from a circular 100 cm perimeter metallic ring, drained into a circular tubing through 5 mm holes, and mounted 2 m above the floor by González (2000) in Colombia; and (e) a wire sensor consisting of three 100 cm tall concentric cylinders—a diameter of 63.6 cm with 1000 wires for the outer cylinder, a diameter of 53.8 cm with 844 wires for the middle cylinder, and a diameter of 44.1 cm with 694 wires for the inner cylinder mounted 100 cm above ground level by Goodman (1985) in California.

The collection of cloud water usually uses sensors much higher above the ground than the collection of fog water does. The cloud water collectors used by Falconer and Falconer (1980) and Clark et al. (1998) were 30 cm tall, 10 cm in diameter, mounted above a polypropylene funnel (16.3 cm in diameter) and a 4 L bottle. Each collector (sensor) consisted of an external ring of 100 Teflon monofilament lines (0.05 cm diameter) and an internal ring of eight acrylic plastic rods (1.0 cm diameter). These collectors were located in a plastic basket at 17 m height on a 27 m meteorological tower located in a gap in the forest.

A standard fog collector was proposed by Schemenauer and Cereceda (1994); it consists of a 1 m × 1 m frame and is covered with a double layer of polypropylene mesh (60% mesh coverage) at 2 m above the ground. Immediately below the frame is a collection trough 10 cm deep × 15 cm wide × 1.04 m long for the fog water. The trough should have a slight slope to drain the water to one end and pipe to a container. This gage may be inefficient in areas with great variations in wind direction. It cannot partition rainfall from cloud deposition (González, 2000).

D.2.2.3 Estimates

The inconsistency in terms of sensor dimension and height of sensor base above the ground among various fog gages makes the measurements of fog condensation incomparable. Even if fog gages and installation procedures are standardized, the collected fog water can only be referred to as the fog water intercepted by the sampled air, expressed in volume of fog water/area of the sensor, not canopy condensation. In other words, the collected fog deposition cannot be related to forest fog interception, especially judging the unique differences on canopy configurations among forest communities. Volume of fog water/area of the sensor is then converted into depth in millimeters, to compare with rainfall measurements.

Fog gages are designed to collect fog interception as well as rainfall, while rain gages collect only rainfall. If the air is calm during a rainfall event, all raindrops fall vertically and fog gages presumably collect rainfall comparative to that collected by rain gages. Then the difference in measurements between a fog gage and a rain gage is an estimate of fog condensation. If the air is not calm, as in most stormy weather conditions, wind causes an inclination of raindrops, which will be intercepted by the vertical sensor and dripped to the collector. In other words, the rainfall collected by fog gages will include the raindrops falling directly into the collector and the raindrops intercepted by the vertical sensor and dripped into the collector, while rain gages collect only the raindrops falling directly into the orifice. Accordingly, fog gages and rain gages are likely to collect unequal amounts of rainfall, making the calculation of fog condensation erroneous. One solution to this problem, as mentioned earlier, is to exclude rainfall events from fog measurements, as determined by an automatic recording device.

D.2.3 Gravimetric Methods

D.2.3.1 Whole Plant Measurements

Weighing lysimeters with growing plants have been used to study fog deposition by solving the following water balance equation (Fowler et al., 1990; Fahey et al., 1996):

$$(F + E) = \Delta S - (P + Q) \tag{D.6}$$

where the sum of fog deposition F and evaporation E is estimated by the difference between the change in lysimeter storage ΔS and the sum of precipitation P and drainage outflow Q. When $(F+E) > 0$, the inputs to the lysimeter exceed the measured precipitation, fog deposition is expected. The approach assumes evaporation negligible during the foggy period.

Of the three variables in the right-hand side of Equation D.6, P is the most difficult to measure accurately. Standard rain gages usually catch fewer amounts than the true rainfall that would actually reach the lysimeter, especially under windy conditions. Also, precipitation in the form of snow is not recorded until it melts. Both situations can cause fog deposition to be overestimated. Thus, when precipitation occurs as snow, data with $(F+E)>0$ cannot be included in calculating total fog deposition. Also, using two lysimeters, one with plants and one without plants, can solve the rainfall and snowfall measurement issue. The difference in weight between the planted and unplanted lysimeters can be used as an estimate of fog deposition.

D.2.3.2 Individual Leaf Measurements

This method measures the increase in weight of plant leaves exposed to foggy conditions as an estimate of fog deposition. The fog deposition rate of individual leaves is then upscaled to the entire forest by multiplying its biomass. In practice, a few leaves (15–20 g) of each plant species are collected, air dried, hung over a rack, and mounted on an electric balance. Under foggy conditions, the weight of these leaves is continuously recorded in a portable computer or a data logger. After the fog is over, the dry weight (65°C oven-dried) of the leaves is measured and the difference in weight between wet and dry conditions is the estimated fog deposition rate of that leaf. Multiplying the average fog deposition rate of individual leaves with the estimated biomass of the plant in t/ha yields the deposition rate of the forest in mm/h (Chang et al., 2006).

In such experiments, sampled leaves need to be placed at various canopy heights and on different locations inside the forest. The edge effect of forest on fog deposition is significant in both horizontal and vertical dimensions. In a California redwood forest, fog deposition decreases along a windward edge-to-interior gradient (Ewing et al., 2009). Parameters such as the wind speed, the liquid water content (LWC) of fog droplets, and the estimated biomass of forest can affect the outcomes of the study.

D.2.4 Direct Measurements

A new technique that quantifies fog deposition by observing fog LWC through optical principles has become available in recent decades (Beswick et al., 1991; Burkard et al. 2002). This technique, known as the "eddy covariance method," employs a cloud particle spectrometer (model FM-100, Droplet Measurement Technologies, Inc., Boulder, CO), a 3D ultrasonic anemometer, and a laptop computer with software for data processing and computations. The advanced technique, which can directly measure fog flux, has been used in many fog deposition studies worldwide (Bruijnzeel, 2006; Klemm and Wrzesinsky, 2007; Katata et al., 2009; Westbeld et al., 2009), including testing collection efficiency for other passive fog gages (Holwerda et al., 2006; Frumau et al., 2011).

The water droplet spectrometer is based on the property that the amount of light scattered by a fog droplet is proportional to its size, composition, and shape. The spectrometer is able to count and characterize fog droplets with a diameter roughly ranging from 2 to $50\,\mu m$ in up to 40 user-selectable size classes when air is forced through a laser beam at a constant speed, around $13\,m/s$. In practice, the spectrometer and ultrasonic anemometer are mounted on a turntable to align the instruments with the wind direction every 10–30 min to minimize flow distortion. The LWC for each size class is obtained by multiplying the mean droplet volume with the counted number of droplets. Summing the LWCs of all classes yields total LWC.

The vertical flux of fog droplets between the atmosphere and a forest canopy consists of two main components, a turbulent diffusion component (D_{turb}) of smaller droplets (typically $<10\,\mu m$) suspended in the atmosphere and a nonturbulent flux component (D_{sed}) of larger droplets settled gravitationally to canopy and ground. The total flux of fog water deposition is the sum of total turbulent flux and total gravitational flux (Westbeld et al., 2009), or

$$D_{total} = D_{sed} + D_{turb} \tag{D.7}$$

The gravitational settling of fog droplets was calculated from Stokes' law to determine the settling velocity for droplets of known diameter:

$$V_s = \frac{gd^2(\rho_{water} - \rho_{air})}{18\mu_{air}} \tag{D.8}$$

where
  V_s is the settling velocity (m/s)
  g is the acceleration due to gravity (m/s²)
  d is the droplet diameter (m)
  ρ is density (kg/m³)
  μ is the dynamic viscosity (kg/m/s)

Multiplying the settling velocity (m/s) with LWC (mg/m³) yields the gravitational flux (mg/m²/s) and summing the gravitational fluxes of all size classes (d > 10 µm) gives the total gravitational contribution to the fog water deposition. In equation form, it can be expressed as

$$D_{sed} \ (mg/m^2/s) = \sum [(V_s)_i \ (LWC)_i] \qquad (D.9)$$

where the subscript i refers to the various droplet size classes. The turbulent flux (F_t) was calculated as the covariance of vertical wind speed (w) and LWC, or

$$F_t = \overline{(w)'(LWC)'} \qquad (D.10)$$

where the overbar denotes the time average (30 min in general) and primes denote the instantaneous or turbulent deviation from that average, thus $(w)' = w - \overline{w}$ and $(LWC)' = LWC - \overline{(LWC)'}$ for every record within an averaging interval. Summing all values of F_t is the total D_{turb}.

D.3 Grass Interception

Measurements of interception by grass or other groundcovers can be conducted in the field as well as under controlled conditions.

D.3.1 Field Measurements

D.3.1.1 Small Surface Drainage Basin

Like forest interception measurements, grass interception also can be estimated by the comparison approach. However, canopy interception, stemflow, and litter interception are separate items in forest interception studies, and these components are a combined item in grass interception measurements.

For grass interception measurements, a 20.32 cm (8 in.) diameter metal collar is driven into the soil to act as a plot boundary. Each collar extends about 2.54 cm above the soil surface. The ground surface in each collar plot is sealed by material such as a neoprene latex emulsion to prevent rainwater from entering into the ground (Crouse et al., 1966; Corbett and Crouse, 1968). The emulsion should be applied on top of a thin layer of moist sand in each collar basin so that the seal is flexible, durable, nontoxic, and watertight, with no effect on normal plant development. Throughfall and stemflow from the collar surface basin are channeled from a drain at the base of the basin to a collector located below ground level. The volume of throughfall and stemflow is then converted into depth to compare with gross precipitation for calculating grass interception.

In order to remove undesirable species and make the ground plot easier to seal, the study area may need to be burned or tilled and seeded with desirable species. The collar plots are installed after the seedlings are established. Pit gages (at ground level) should be used for gross precipitation estimates for a more accurate comparison.

D.3.1.2 Weighing Devices

A measuring device consisting of a permeable upper basin filled with forest floor and a watertight lower basin equipped with a drain valve was used for litter interception studies by Gerrits et al. (2006). The two aluminum basins are mounted above each other and are weighed continuously with two sets of three strain gage sensors and a data logger. Water in the upper basin can percolate into the lower basin through its geotextile bottom and can be collected from the valve. The change in weight between precipitation and the sum of water storage in the two basins is a measure of litter interception. The upper basin can also grow grasses for interception studies.

Weighing lysimeters discussed in the fog deposition section are also suitable for grass interception studies.

D.3.2 Laboratory Measurements

For studying interception under controlled rainfall conditions, grasses can be clipped at the soil surface or excavated and placed into wire mesh containers to simulate natural conditions. If grasses are clipped, they are transported to the laboratory, weighed, and subjected to simulated rainfall at various intensities and durations. After 30 min of simulated rainfall, the clippings are weighed again, and the difference in weight before and after the exposure is the estimated rainfall interception and a rainfall versus interception as % of rainfall curve can be developed. If grasses are excavated to sample containers, then the soil surface needs to be sealed to prevent water from entering the soil.

The developed curve should show that interception increases with increasing amount of rainfall. The increase in interception will eventually level off and that leveling off point is the maximum canopy storage capacity. With the interception information from the developed curve, grass interception (I_G) can be calculated by

$$I_G = (P_G)(C_S)(A_G) \tag{D.11}$$

where
    P_G is gross precipitation
    C_S is % of canopy storage (interception) obtained from the developed curve
    A_G is % areal canopy coverage of the grass species

Values of A_G can be estimated by the line-intersect method along 200 m transects (West and Gifford, 1976).

Another method involving whole grass excavation adopts the following procedures (Thurow et al., 1987):

1. Excavate and place grasses into a wire mesh container in such a way that the natural configuration of the grasses can be maintained.

2. Clip representative grass samples from the field for moisture determination in the laboratory by weight before and after drying at 60°C for 48 h, where the moisture content of the clipped grass samples is an estimate of the moisture content of the excavated grasses.

3. Expose the excavated grasses to simulated rainfall within minutes of collection.

4. At the end of rainfall simulation, place the excavated grasses and containers in a freezer (−46°C) that will freeze the intercepted water in 5 min.

5. Clip the frozen grasses while in the freezer and weigh.

6. Determine the moisture content of the frozen grasses at 60°C, 48 h.

7. Subtract the moisture content of the field sample obtained under Step 2 from the contents of the frozen samples under Step 6 to yield grass interception.

References

Ahrends, B. and Penne, C., 2010, Modeling the impact of canopy structure on the spatial variability of net forest precipitation and interception loss in Scots pine stands, *Open Geogr. J.*, 2010(3), 115–124.

Azevedo, J. and Morgan, D.L., 1974, Fog precipitation in coastal California forests, *Ecology*, 55, 1135–1141.

Beswick, K. M. et al., 1991, Size-resolved measurements of cloud droplet deposition velocity to a forest canopy using an eddy-correlation technique, *Q. J. R. Meteorol. Soc.*, 117, 623–645.

Bouten, W., Swart, P.J.F., and De Water, E., 1991, Microwave transmission, a new tool in forest hydrological research, *J. Hydrol.*, 124, 119–130.

Bruijnzeel, L.A., 2001, Hydrology of tropical montane cloud forests: A reassessment, *Land Use Water Resour. Res.*, 1, 1.1–1.18.

Bruijnzeel, L.A., 2006, Hydrological impacts of converting tropical montane cloud forest to pasture, with initial reference to Northern Costa Rica, Final Technical Report DFID-FRP Project No. R7991, Vrije Universiteit, Amsterdam, the Netherlands, 51 pp.

Burkard, R. et al., 2002. Vertical divergence of fogwater fluxes above a spruce forest, *Atmos. Res.*, 64(1–4), 133–145.

Calder, I.R., 1976, The measurement of water losses from a forested area using a "natural lysimeter," *J. Hydrol.*, 30, 311–325.

Calder, I.R. and Wright, I.R., 1986, Gamma ray attenuation studies of interception from sitka spruce: Some evidence for additional transport mechanism, *Water Resour. Res.*, 22, 409–417.

Cavelier, J. and Goldstein, G., 1989, Mist and fog interception in elfin cloud forests in Colombia and Venezuela, *J. Trop. Ecol.*, 5, 309–322.

Cavelier, J., Solis, D., and Jaramillo, M.A., 1996, Fog interception in montane forests across the Central Cordillera of Panamá, *J. Trop. Ecol.*, 12, 357–369.

Cavelier, J. et al., 1997, Water balance and nutrient inputs in bulk precipitation in tropical montane cloud forest in Panama, *J. Hydrol.*, 193, 83–96.

Chang, M. and Harrison, L., 2005, Field assessments on the accuracy of spherical gages in rainfall measurements, *Hydrol. Process.*, 19, 403–412.

Chang, S.-C. et al., 2006, Quantifying fog water deposition by in situ exposure experiments in a mountainous coniferous forest in Taiwan, *For. Ecol. Manage.*, 224, 11–18.

Clark, K.L. et al., 1998, Atmospheric deposition and net retention of ions by the canopy in a tropical montane forest, Monteverde, Costa Rica, *J. Trop. Ecol.*, 14, 27–45.

Corbett, E.S. and Crouse, R.P., 1968, Rainfall interception by annual grass and chaparral... Losses compared, Pacific U.S. Forest Service, SW Forest and Range Experiment Station, Research Paper PSW-48, Berkeley, CA, 12 pp.

Crouse, R.P., Corbett, E.D., and Seegrist, D.W., 1966, Methods of measuring and analyzing rainfall interception by grass, *Bull. IAHS*, XI, 110–120.

Ekern, P.C., 1964, Direct interception of cloud water on Lanaihale, Hawaii, *Soil Sci. Soc. Am. Proc.*, 28, 419–421.

Ewing, H.A. et al., 2009, Fog water and ecosystem function: Heterogeneity in a California redwood forest, *Ecosystems*, 12, 417–433.

Fahey, B.D., Murray, G.L., and Jackson, R.M., 1996, Detecting fog deposition to tussock by lysimeter at Swampy Summit near Dunedin, New Zealand, *J. Hydrol. (NZ)*, 35(1), 85–102.

Falconer, R.E. and Falconer, P.D., 1980, Determination of cloud water acidity at a mountain observatory in the Adirondack Mountains of New York State, *J. Geophys. Res.*, 85, 943–969.

Fowler, D. et al., 1990, Measurements of cloud water deposition on vegetation using a lysimeter and a flux gradient technique, *Tellus*, 42(b), 285–293.

Friedrich, W., 1967, Forest hydrology research in Germany F.R, *International Symposium on Forest Hydrology*, Sopper, W.E. and Lull, H.W., Eds., Pergamon Press, Oxford, U.K., pp. 45–47.

Frumau, K.F.A. et al., 2011, Fog gauge performance under fog and wind-driven rain conditions, *Tropical Montane Cloud Forests: Science for Conservation and Management*, Bruijnzeel, L.A., Scatena, F.N., and Hamilton, L.S., Eds., Cambridge University Press, London, U.K., pp. 293–301.

Gash, J.H.L., Lloyd, C.R., and Lachaud, G., 1995, Estimating sparse forest rainfall interception with an analytical model, *J. Hydrol.*, 170, 79–86.

Gerrits, A.M.J. et al., 2006, Measuring forest floor interception in a beech forest in Luxembourg, *Hydrol. Earth Syst. Sci. Discuss.*, 3, 2323–2341.

Goodman, J., 1985, The collection of for drip, *Water Resour. Res.*, 21, 392–394.

González, J., 2000, Monitoring cloud interception in a tropical montane cloud forest of the southwestern Colombian Andes, *Adv. Environ. Monit. Model.*, 1, 97–117.

Grünow, J., 1952, Nebelniederschlag: Bedeutung und Erfassung einer Zustzkomponente des Niederschlags, *Ber. Dtsch. Wetterdienstes (US-Zone)*, 7(42), 30–34.

Hall, R.L., 1985, Further interception studies of heather using a wet-surface weighing lysimeter system, *J. Hydrol.*, 81, 193–210.

Harr, R.D., 1982, Fog drip in the Bull Run municipal watershed, Oregon, *Water Resour. Bull.*, 18, 785–789.

Helvey, J.D., 1971, A summary of rainfall interception by certain conifers of North America, *Biological Effects in the Hydrological Cycle, Proceedings of the 3rd International Seminar for Hydrology Professors*, Purdue University, West Lafayette, IN, pp. 103–113.

Holder, C.D., 2003, Fog precipitation in the Sierra de lass Minas Biosphere Reserve, Guatemala, *Hydrol. Process.*, 17, 2001–2010.

Holder, C.D., 2004, Rainfall interception and fog precipitation in a tropical montane cloud forest of Guatemala, *For. Ecol. Manage.*, 190, 373–384.

Holwerda, F. et al., 2006, Estimating fog deposition at a Puerto Rican elfin cloud forest site: Comparison of the water budget and eddy covariance methods, *Hydrol. Process.*, 20, 2669–2692.

Juvik, J.O. and Ekern, P.C., 1978, A climatology of Mountain Fog on Mauna Loa Hawai'i Island, Technical Report No. 118, Water Resources Research Center, University of Hawaii at Manoa, HI.

Katata, G. et al., 2009, Application of a land surface model that includes fog deposition over a tree heath-laurel forest in Garajonay National Park (La Gomera, Spain), *Estudios en la Zona no Saturada del Suelo*, Vol. IX, Silva, O. et al., Eds., Barcelona, Spain, November, 18–20, 2009.

Klemm, O. and Wrzesinsky, T., 2007, Fog deposition fluxes of water and ions to a mountainous site in Central Europe, *Tellus B*, 59, 705–714.

Limousin, J.-M. et al., 2008, Modeling rainfall interception in a Mediterranean *Quercus ilex* ecosystem: Lesson from a throughfall exclusion experiment, *J. Hydrol.*, 357, 57–66.

Lundberg, A., 1993, Evaporation of intercepted snow—Review of existing and new measurement methods, *J. Hydrol.*, 151, 267–290.

Marloth, R., 1904, Results of experiments on Table Mountain for ascertaining the amount of moisture deposited from the SE clouds, *Trans. SA Philos. Soc.*, 14, 403–408.

Molchanov, A.A., 1963, *The Hydrological Role of Forests*, Translated from Russian, USDA OTS 63-11089, Washington, DC.

Oberlander, G.T., 1956, Summer fog precipitation on the San Francisco Peninsula, *Ecology*, 37, 851–852.

Olivier, J., 2002, Water harvesting along the west coast of South Africa: A feasible study, *Water SA*, 28(4), 349–360.

Rasmussen, V.P. and Campbell, R.H., 1987, A simple microwave method for the measurement of soil moisture, *International Conference on the Measurement of Soil and Plant Water Status*, Vol. I, Utah State University, Logan, UT, pp. 275–277.

Roth, F.A., II and Chang, M., 1981, Throughfall in planted stands of four southern pine species in East Texas, *Water Resour. Bull.*, 17, 880–885.

Schemenauer, R.S. and Cereceda, P., 1994, A proposed standard fog collector for used in high elevation regions, *J. Appl. Meteorol.*, 33, 1313–1322.

Schmugge, T.J., Jackson, T.J., and McKim, H.L., 1980, Survey of methods for soil moisture determination, *Water Resour. Res.*, 16, 961–979.

Shaw, E.M., 1983, *Hydrology in Practice*, Van Nostrand Reinhold, New York.

Strangeways, I., 2000, *Measuring the Natural Environment*, Cambridge University Press, Cambridge, U.K.

Thurow, T.L. et al., 1987, Rainfall interception by midgrass, shortgrass, and live oak mottes, *J. Range Manage.*, 49, 455–460.

U.S. National Weather Service, 1972, National weather service observing handbook no. 2: Substation observations, supersedes circular B, U.S. Department of Commerce, Silver Spring, MD.

Van Dijk, A.I.J.M. and Bruijnzeel, L.A., 2001, Modelling rainfall interception by vegetation of variable density using an adapted analytical model. Part 1. Model description, *J. Hydrol.*, 247, 230–238.

West, N.E. and Gifford, G.F., 1976, Rainfall interception by cool-desert shrubs, *J. Range Manage.*, 29, 171–172.

Westbeld, A. et al., 2009, Fog deposition to a Tillandsia carpet in the Atacama Desert, *Ann. Geophys.*, 27, 3571–3576.

Xiao, Q. et al., 2000, Winter rainfall interception by two mature open-grown trees in Davis, California, *Hydrol. Process.*, 14, 763–784.

Index

Printed in the United States
by Baker & Taylor Publisher Services